Lecture Notes in Computer Science 11653

Ding-Zhu Du · Zhenhua Duan ·
Cong Tian (Eds.)

Computing
and Combinatorics

25th International Conference, COCOON 2019
Xi'an, China, July 29–31, 2019
Proceedings

 Springer

Editors
Ding-Zhu Du
The University of Texas at Dallas
Richardson, TX, USA

Zhenhua Duan
Xidian University
Xi'an, China

Cong Tian
Xidian University
Xi'an, China

ISSN 0302-9743 ISSN 1611-3349 (electronic)
Lecture Notes in Computer Science
ISBN 978-3-030-26175-7 ISBN 978-3-030-26176-4 (eBook)
https://doi.org/10.1007/978-3-030-26176-4

LNCS Sublibrary: SL1 – Theoretical Computer Science and General Issues

This Springer imprint is published by the registered company Springer Nature Switzerland AG
The registered company address is: Gewerbestrasse 11, 6330 Cham, Switzerland

Preface

This volume contains the papers presented at the 25th International Computing and Combinatorics Conference (COCOON 2019), held during July 29–31, 2019, in Xi'an, China. COCOON 2019 provided a forum for researchers working in the areas of algorithms, theory of computation, computational complexity, and combinatorics related to computing.

The technical program of the conference included 55 contributed papers selected by the Program Committee from 124 full submissions received in response to the call for papers. All the papers were peer reviewed by at least three Program Committee members or additional reviewers. The papers cover various topics, including algorithm design, approximation algorithm, graph theory, complexity theory, problem solving, optimization, computational biology, computational learning, communication network, logic, and game theory. Some of the papers were selected for publication in special issues of *Algorithmica, Theoretical Computer Science* (TCS), and *Journal of Combinatorial Optimization* (JOCO), with the journal version of the papers being in a more complete form.

We would like to thank all the authors for contributing high-quality research papers to the conference. We express our sincere thanks to the Program Committee members and the additional reviewers for reviewing the papers. We thank Springer for publishing the proceedings in the *Lecture Notes in Computer Science* series. We thank Xidian University for hosting COCOON 2019. We are also grateful to all members of the Organizing Committee and to their supporting staff.

The conference-management system EasyChair was used to handle the submissions, to conduct the electronic Program Committee meetings, and to assist with the assembly of the proceedings.

June 2019

Ding-Zhu Du
Zhenhua Duan
Cong Tian

Organization

Program Committee Chairs

Ding-Zhu Du   University of Texas at Dallas, USA
Zhenhua Duan   Xidian University, China
Cong Tian   Xidian University, China

Program Committee

Pavan Aduri   Iowa State University, USA
Chee Yap   New York University, USA
Zhizhong Chen   Tokyo Denki University, Japan
Shuaicheng Li   City University of Hong Kong, SAR China
Boting Yang   University of Regina, Canada
Peng Zhang   Shandong University, China
Youming Qiao   University of Technology Sydney, Australia
Ryuhei Uehara   Japan Advanced Institute of Science and Technology,
                       Japan
Jarek Byrka   University of Wroclaw, Poland
Jianxin Wang   Central South University, China
Xiuzhen Huang   Arkansas State University, USA
Hans-Joachim   ETH Zurich, Switzerland
  Boeckenhauer
Dennis Komm   ETH Zurich, Switzerland
Donghyun Kim   Kennesaw State University, USA
Marc Uetz   University of Twente, The Netherlands
Bhaskar Dasgupta   University of Illinois at Chicago, USA
Kazuo Iwama   Kyoto University, Japan
Iyad Kanj   DePaul University, Illinois, USA
Lenwood Heath   Virginia Tech University, USA
Ming-Yang Kao   Northwestern University, USA
Joong-Lyul Lee   University of North Carolina at Pembroke, USA
Viet Hung Nguyen   Sorbonne Université, France
Xianyue Li   Lanzhou University, China
Xianmin Liu   Harbin Institute of Technology, China
Meghana Nasre   Indian Institute of Technology, India
Pavel Skums   Georgia State University, USA
Wei Wang   Xi'an Jiaotong University, China
Gerhard Woeginger   Eindhoven University of Technology, The Netherlands
Zhao Zhang   Zhejiang Normal University, China
Martin Ziegler   Korea Advanced Institute of Science and Technology,
                       South Korea

Yong Chen   Hangzhou University of Electronic Science
                            and Technology, China
Ovidiu Daescu   University of Texas at Dallas, USA
Thomas Erlebach   University of Leicester, UK
Neng Fan   University of Arizona, USA
Stanley Fung   University of Leicester, UK
Meng Han   Kennesaw State University, USA
Micheal Khachay   Ural Federal University, Russia
Guangmo Tong   University of Texas at Dallas, USA

Contents

Fully Dynamic Arboricity Maintenance

Niranka Banerjee[(✉)], Venkatesh Raman, and Saket Saurabh

The Institute of Mathematical Sciences, HBNI,
CIT Campus, Taramani, Chennai 600 113, India
{nirankab,vraman,saket}@imsc.res.in

Abstract. Given an undirected graph, its arboricity is the minimum number of edge disjoint forests, its edge set can be partitioned into. We develop the first fully dynamic algorithms to determine the arboricity of a graph under edge insertions and deletions. While our insertion algorithm is based on known static algorithms to determine the arboricity, our deletion algorithm is, to the best of our knowledge, new.

Our algorithms take $\tilde{O}(m)$ time ($\tilde{O}$ notation ignores logarithmic factors.) to insert or delete an edge where m is the number of edges in the graph while the best static algorithm to compute arboricity takes $O(m^{3/2}\log(n^2/m))$ time [7].

We complement our upper bound with a lower bound result of amortized $\Omega(\log n)$ for any algorithm that maintains a forest decomposition of size arboricity of the graph under edge insertions and deletions.

1 Introduction

The arboricity of a graph is defined as the minimum number of edge disjoint forests that the edges of the graph can be partitioned into [13]. A decomposition of a graph of arboricity **a** has **a** such spanning forests and we call it an arboricity **a** decomposition. Intuitively, it is a measure of sparseness of a graph. An equivalent definition [13] for an arboricity **a** of a graph $G = (V, E)$ is

$$\max_{\{J \subseteq V, |J| \geq 2\}} \lceil |E(J)|/|J| - 1 \rceil \tag{1}$$

where $E(J)$ is the set of edges in the subgraph induced by the vertex set $J \subseteq V$.

For a given graph, a 2-approximation of the arboricity can be easily calculated in $O(m+n)$ time [5], whereas more complex algorithms exist for calculating the exact arboricity in $O(m^{3/2}\log(n^2/m))$ time [7]. Faster algorithms for computing the arboricity are known for planar graphs. Grossi and Lodi [8] gave an algorithm to show how to decompose a planar graph on n vertices into three edge disjoint spanning forests in $O(n \log n)$ time. But they do not compute the exact arboricity. In particular the algorithm fails to differentiate between a planar graph of arboricity two or three.

In the world of dynamic graph algorithms, edges are inserted or deleted over time, and the goal is to maintain some property of the graph under these modifications in time faster than the static algorithm that starts from scratch.

© Springer Nature Switzerland AG 2019
D.-Z. Du et al. (Eds.): COCOON 2019, LNCS 11653, pp. 1–12, 2019.
https://doi.org/10.1007/978-3-030-26176-4_1

A dynamic algorithm is fully dynamic if it is able to support both insertion and deletion of edges; if it only supports insertions, it is called an incremental algorithm and if it supports only deletions, it is called a decremental algorithm.

For dynamic graphs, the study of arboricity in the literature has been mainly through works in orienting undirected graphs. An **a**-bounded orientation of a graph G is an orientation of its edges where all out-degrees of vertices have degree at most **a**. If a graph has arboricity **a**, by rooting each tree in the forest decomposition arbitrarily and orienting each edge towards the root, we can get an **a**-bounded orientation. It can also be shown that any graph with an **a**-bounded orientation can be decomposed into $2\mathbf{a} - 1$ forests and hence this gives a 2-approximation of arboricity [5].

Study on maintaining Δ-bounded orientations of a graph under edge insertions and deletions was initiated by Brodal and Fagerberg in [2]. On insertion or deletion of an edge, the goal was to reorient the edges of the graph efficiently so that a Δ-bounded orientation or an approximation of it was maintained. There has been subsequently a number of other works in this topic [1,9–11] but none of them maintain the exact arboricity. So it was a natural question to look at computing the exact arboricity of the graph in the dynamic setting. Erickson [6] asks "whether there are efficient algorithms to maintain graph arboricity and/or minimal forest decompositions in dynamic graphs".

This paper is an attempt towards answering this and in the process we give a new perspective on existing static algorithmic techniques by showing that, if carefully implemented, they give a fully dynamic algorithm.

1.1 Our Results

We develop a fully dynamic algorithm for computing the exact arboricity of a graph i.e. on every edge insertion or deletion of the graph we compute the exact arboricity **a** of the graph by maintaining a decomposition into **a** edge disjoint spanning forests.

While our insertion algorithm is based on a known static algorithm to determine the arboricity due to Roskind and Tarjan [15], we prove certain invariants about the algorithm that helps to maintain the arboricity under deletion as well. Both insertion and deletion take $\tilde{O}(m)$ time in the worst case where m is the number of edges in the graph at that time. These bounds are an improvement over the best static bounds of $O(m^{3/2} \log(n^2/m))$ time [7] for computing arboricity.

We complement our upper bounds with a lower bound result. We give an amortized bound of $\Omega(\log n)$ for the cost of answering the arboricity query under edge insertions and deletions.

1.2 Previous Work, Methodology and Organization of the Paper

Let G be a graph of arboricity **a** given with the arboricity **a** decomposition of its edges, to which an edge (u, v) is to be inserted. If u and v are in different trees of a forest in the decomposition, then we have nothing more to do than adding that edge to that forest. Otherwise, first we try adding this to a forest of the decomposition and remove an edge that is part of the cycle formed and try inserting that edge into another forest. This process is repeated until the endpoints of a new edge belong to two trees of a forest or somehow we can conclude that the existing decomposition is not good enough to take this new edge and that the arboricity of the graph increases. This process has the flavour of finding an augmenting path in matchings. This idea of an updating algorithm that uses augmenting sequences dates back to the classical static algorithm to find arboricity due to Nash-Williams [12,13] and Tutte [17].

Our algorithm for dynamically maintaining arboricity can be put into larger perspective. In the static setting, this is an instantiation of the matroid partitioning problem. In this problem, the goal is to partition the elements of a matroid into as few independent sets as possible. In our setting, edges of the given graph forms the universe of the matroid and the forests of the graph form the independent sets. This matroid is usually known as graphic matroid. The known algorithms for the matroid partitioning problem, given an independence oracle for the matroid, also uses the idea of augmenting sequences in an appropriate auxiliary graph [3,4]. Some of the steps of our dynamic algorithm for maintaining arboricity do extend to the dynamic variant of the matroid partitioning problem, but we restrict ourselves to graphic matroids, as it allows us to use some additional structural properties.

In Sect. 2.3, we give a brief description of the insertion algorithm by casting this algorithm as an algorithm to find a specific path in an auxiliary graph whose vertex set is the edge set of the given graph. We use this auxiliary graph terminology for the insertion and deletion algorithm we describe later. A naive implementation as in Sect. 2.3 results in an $O(m^2)$ insertion algorithm where m is the number of edges in the original graph. In Sect. 3, we use ideas from the static algorithm of Roskind and Tarjan [15] to improve the insertion time to $\tilde{O}(m)$. Deletion of an edge involves checking after deleting the said edge, whether the arboricity of the graph decreases. In Sect. 4, we give a fully dynamic algorithm that supports insertion and deletion in $\tilde{O}(m)$ time, by first modifying the insertion algorithm of Sect. 3 slightly. Finally in Sect. 5, we give a lower bound of $\Omega(\log n)$ for the cost of insertion into a graph with arboricity **a**.

2   Preliminaries

The following lemma follows easily from the alternate definition of arboricity (expression (1) in the introduction), and we use it extensively in some of our algorithms.

Lemma 1 *[1] [15]. *Let* **a** *be the arboricity of the given graph* $G(V, E)$*. Given a decomposition of the edges of* G *into* **a** *edge disjoint forests, the arboricity of* $G \cup (u, v)$ *for some* $(u, v) \notin E$ *increases if there is a subset* $S \subseteq V$ *containing both* u *and* v *such that* S *forms a subtree in every forest of the decomposition.*

This set S is referred to as a *clump* [15] which will be used later in the paper.

---

[1] Proofs of results marked ⋆ are deferred to the full version.

2.1 Data Structure

Throughout our upper bound results we will be using Link-Cut trees [16] as our data structure to maintain the forest decomposition. In this data structure, one can, in $O(\log n)$ time, for a graph on n vertices,

- find whether a pair of vertices are in the same tree in a forest,
- add an edge that joins two trees of a forest or delete an edge from a tree in a forest or
- find the least common ancestor (lca) of any pair of vertices in a tree of the forest,
- make a vertex u the root of a tree it belongs to, in a forest.

We begin the development of our algorithm by revising the classical static algorithm due to Nash-Williams [12,13] and Tutte [17] to compute arboricity.

2.2 The Auxiliary Graph

Suppose for a graph $G = (V, E)$, an edge disjoint forest structure is given and its arboricity is **a**. On insertion of an edge (u, v) into the graph, we need to determine whether the arboricity of the graph stays the same or increases, and obtain the resulting optimal decomposition.

The insertion algorithm of Nash-Williams [12,13] and Tutte [17] essentially looks for a certain kind of an "augmenting path"(defined below) and we find it convenient to explain it using an auxiliary graph $G'_{ed} = (V'_{ed}, E'_{ed})$ with respect to a decomposition of G into edge disjoint forests. A vertex of the directed graph $G'_{ed} = (V'_{ed}, E'_{ed})$ represents an edge (u, v) $(u, v \in V)$ of the original graph. We color the vertices of the graph G'_{ed} with two different colors. Specifically,

- $C((u, v)) = $ red, if u and v are in the same component in every forest in the decomposition of G,
- $C((u, v)) = $ green, otherwise.

In G'_{ed}, $((a, b), (c, d))$ is a directed edge (i.e. in E'_{ed}), if and only if (c, d) is on the path from a to b in some tree of the forest decomposition. This means that when (a, b) is added to a tree, (c, d) is an edge in the cycle formed. Note that in this directed graph G'_{ed}, if there is a directed path from (u, v) to a vertex (edge in G) marked green, then the path gives a way to insert (u, v) into the existing decomposition. The path from (u, v) to this green vertex path is defined as our "augmenting" path. It turns out that when there is no such path the arboricity increases.

Theorem 1. * *On insertion of an edge (u, v) in G, its arboricity increases if and only if in the directed graph $G'_{ed} = (V'_{ed}, E'_{ed})$ there is no directed path from (u, v) to any vertex marked green. If no such path exists, then the end points of all the vertices (edges of G) reachable from (u, v) form a clump S.*

2.3   An Incremental Algorithm

The following pseudocode summarises our insertion algorithm and forms the basis of the improved algorithms we develop later. In the improved algorithms, we sometimes change the auxiliary graph, or find an augmenting path more cleverly.

Algorithm 1.
1 Insert edge (u, v)
2 Check whether the auxiliary graph has a path from (u, v) to a green vertex
3 **if** *Yes* **then**
4   Arboricity remains same
5   Find augmenting path and fix decomposition
6 **else**
7   Arboricity increases
8   Insert (u, v) in a new forest.

As the auxiliary graph G'_{ed} can have up to m^2 edges, and doing a BFS (breadth-first-search) naively on G'_{ed} takes $O(m^2)$ time, we have the following theorem. Note that the neighbors of a vertex $x = (u, v)$ in G'_{ed} can be obtained by finding, in each forest, the least common ancestor of u and v and following the parent pointer from u and v to their least common ancestor and listing the edges accessed in these paths.

Theorem 2 [4]. *Given a graph $G = (V, E)$ on n vertices and m edges with arboricity $\mathbf{a}$, along with an arboricity decomposition and an edge (u, v) to be inserted, we can compute the arboricity of the modified graph $G = (V, E \cup (u, v))$ in $\tilde{O}(m^2)$ time.*

In Sects. 3 and 4 we will define auxiliary graphs G'_{rt1} and G'_{rt2} which will further reduce the runtime of the algorithm.

3   Insertion in $\tilde{O}(m)$ Time

Roskind and Tarjan [15] gave a faster static algorithm which translated in the language of our auxiliary graph meant using a modification of G'_{ed} where the neighbors of a vertex (u, v) are the edges which form a cycle with (u, v) but just in the next forest, i.e. if $(u, v) \in F_i$, the ith forest(in some arbitrary ordering of the forests), then there is a directed edge from (u, v) to (x, y) if $(x, y) \in F_{i+1}$ and is part of the cycle formed if (u, v) is inserted in F_{i+1}. If $(u, v) \in F_a$ which is the last forest or $(u, v) \notin G$, then the neighbors of (u, v) in G'_{ed} come from F_1. We refer to this modified G'_{ed} graph as G'_{rt1}. We also modify the coloring of vertices in G'_{rt1}: for an edge $(u, v) \in G, C((u, v)) =$ green if $(u, v) \in$ forest F_i and u and v are in different components in F_{i+1} and is red otherwise. For an edge $(u, v) \notin G, C((u, v)) =$ green if u and v are in different components in F_1 and is red otherwise.

In what follows, we state a series of lemmas, leading to the correctness of Algorithm 1 using our auxiliary graph terminology.

Lemma 2. * *Let G be a graph with an optimal arboricity decomposition of forests $F_1, F_2, \ldots F_a$, and let (u, v) (which may or may not be an edge in G) be such that there is no path from (u, v) to a vertex marked green in G'_{rt1}. Then*

1. *vertices u and v are in the same component in all the forests F_1 to F_a.*
2. *Every edge in every u-v path in F_i, for all i from 1 to a, is reachable from (u, v) in G'_{rt1}.*
3. *The end points of all edges in the u-v path in F_i are in the same component as u and v in $F_{(i+1 \mod a)}$ for $i = 1$ to $\mathbf{a}$.*

From Lemma 2, we have that if there is no path from (u, v) to a green vertex in G'_{rt1}, then u and v are in the same component in all the forests (first part of the lemma) and hence (u, v) is red in G'_{ed}. This leads to the following corollary,

Corollary 1. *If (u, v) is a green vertex in G'_{ed} then there exists a path from (u, v) to a green vertex in G'_{rt1}.*

Next we will show that if (u, v) is a red vertex but has a path to a green vertex in G'_{ed}, then there is such a path in G'_{rt1}. We prove this by showing the following lemma strengthening Lemma 2.

Lemma 3. * *Let G be a graph with an optimal arboricity decomposition of forests $F_1, F_2, \ldots, F_a$, and let (u, v) be such that there is no path from (u, v) to a vertex marked green in G'_{rt1}. Let $R((u, v))$ be the set of all vertices in G'_{rt1} reachable from (u, v) (note that every vertex in $R(u, v)$ is marked red). Let S denote the union of all the end points of vertices (i.e. edges of G) in G'_{rt1}. Then*

– *all vertices of S are in the same component in all the forests F_i ($i = 1$ to a).*
– *between any pair (a, b) of vertices of S, every path between a and b in every forest contains vertices only from S. I.e. $G[S]$, the induced subgraph on S, forms a subtree in all the forests F_i ($i = 1$ to a).*

Lemma 3 is crucially used in the following theorem which shows the correctness of Algorithm 1.

Theorem 3. *There exists a path from (u, v) to a green vertex in G'_{ed} if and only if there exists a path from (u, v) to a green vertex in G'_{rt1}. Furthermore if there is no such path, then the set of all end points of all vertices (i.e. edges of G) reachable from (u, v) in G'_{rt1} including u and v form a clump S.*

Proof. As G'_{rt1} is a subgraph of G'_{ed} and any *green* vertex in G'_{rt1} is also a green vertex in G'_{ed}, any path in G'_{rt_1} from (u, v) to a *green* vertex also exists in G'_{ed} as a path from (u, v) to a *green* vertex. This shows one direction of the theorem that if there exists a path from (u, v) to a *green* vertex in G'_{rt1} then there exists a path from (u, v) to a *green* vertex in G'_{ed}.

The proof of the converse part of the theorem follows from Lemma 3 because if there is no path from (u, v) to a green vertex in G'_{rt1}, then there exists a clump S (as in Lemma 3)and hence by Lemma 1 and Theorem 1, there can not be a path from (u, v) to a green vertex in G'_{ed} as well.   □

The outdegree of any vertex in G'_{rt1} is at most $n-1$ as a vertex's neighbors are only in one forest. This immediately implies that the number of edges in the auxiliary graph is $O(mn)$ and so our BFS algorithm to find the augmenting path will take $\tilde{O}(mn)$ time improving from $\tilde{O}(m^2)$.

The idea to improve the runtime to $\tilde{O}(m)$ involves traversing through the unvisited vertices of G'_{rt1} only once in the BFS traversal. We show this in the following lemma adapted from [15].

Lemma 4 [15]. *In the BFS traversal starting from (u, v) in G'_{rt1}, a vertex $\in$ G'_{rt1} is seen only once.*

Proof. Whenever a new edge (u, v) is being added to the forest decomposition, we first make u the root of the trees they belong to in each of the forest decompositions F_1 to F_a.

If u and v are in different components in F_1 then the BFS stops. We start a BFS traversal in G'_{rt1} from vertex (u, v). Its neighbors are precisely those edges in the u-v path in F_1. We can return all these neighbors by following parent pointers from v to u. We add these edges in the BFS queue in the order from u to v. Let this path be $u = x_0, x_1, x_2, ..., x_k, v = x_{k+1}$. To further explore BFS, we need to find the neighbors of $(u, x_1), (x_1, x_2), ... (x_k, v)$ in G'_{rt1} from F_2. The BFS traversal starts from (u, x_1) and ends with (x_k, v). Either u and x_1 are in different components in F_2 in which case we stop or else we find the neighbors of (u, x_1) in F_2. The algorithm now exploits the fact(follows from Theorem 1 that if we are exploring the neighbors in $F_{(i+1 \mod a)}$ of the edge (x_i, x_{i+1}) in F_i on the u-v path, then the neighbors of all edges $(u, x_0), (x_0, x_1)$ up to (x_{i-1}, x_i) form a subtree in $F_{(i+1 \mod a)}$. This ensures that to visit the new neighbors of (x_i, x_{i+1}), we simply start in F_2 from x_{i+1} and traverse the path to u via parent pointers. We stop when we hit an already visited edge. This ensures that everytime we explore neighbors of an edge, we explore only unvisited vertices of G'_{rt1}. □

Running Time: To make u the root of each tree takes $\tilde{O}(a)$ total time using Link-Cut trees. Each vertex is encountered exactly twice by the algorithm, once on adding to the BFS queue and then on removing them from the queue. Using Link-Cut trees adding vertices take $O(1)$ time, while removing them and checking whether they belong to different components take $O(\log n)$ time. Using Lemma 4 the runtime to find the augmenting path is proportional to the number of vertices in G'_{rt1} which is $\tilde{O}(m)$.

Theorem 4. *Given a graph $G = (V, E)$, on n vertices and m edges with arboricity a and an edge (u, v) to be inserted, we can compute the arboricity of the modified graph $G = (V, E \cup (u, v))$ in $\tilde{O}(m)$ time.*

4   Fully Dynamic Algorithm

To perform the deletion efficiently, we modify the insertion algorithm slightly so that a specific invariant (described below) is maintained.

We modify the auxiliary graph G'_{rt1}, calling it G'_{rt2} and pay attention to the order in which we visit the neighbors of a vertex in the BFS of G'_{rt2} to implement our new insertion algorithm.

Note that the way we defined G'_{rt1} in Sect. 3, the edges in the 'last' forest has neighbors in the first forest. So, as edges are inserted, the 'last' forest itself can change. For example, if F_i was the last forest, its edges had neighbors in F_1, but when some new edges create the forest F_{i+1}, the neighbors of edges in F_i are in F_{i+1} and not in F_1, the way we have defined G'_{rt1}. But note that we never store G'_{rt1}, but only use it during BFS, and hence there is no need to update the neighborhood.

However Roskind and Tarjan, in their (other) algorithm that computes arboricity, found it convenient to keep the neighbors of edges in F_i in F_1 even after a new forest is created and the edges of F_i have neighbors in F_{i+1}. In short G'_{rt1} is augmented with additional edges as defined below. An edge in this directed graph is directed from a vertex (a, b) in forest F_i, for some i, to a vertex (c, d) if and only if (c, d) is on the path from a to b in some tree of F_{i+1} or F_1. This graph is referred to as G'_{rt2}.

We also modify the coloring of vertices as below in G'_{rt2}: for an edge $(u, v) \in G, C((u, v)) = green$ if $(u, v) \in$ forest F_i and u and v are in different components in F_{i+1} or F_1 and is red otherwise. For an edge $(u, v) \notin G$, $C((u, v)) = green$ if u and v are in different components in F_1 and is red otherwise.

Note that G'_{rt2} is a supergraph of G'_{rt1}, and hence a theorem analogous to Theorem 3 holds true even with this new auxiliary graph proving the correctness of the insertion algorithm.

Theorem 5. * *There exists a path from (u, v) to a green vertex in G'_{rt1} if and only if there exists a path from (u, v) to a green vertex in G'_{rt2}. If no such path exists, then the union of all vertices of G which are part of the reachability vertex set from (u, v) in G'_{rt1} and G'_{rt2} inclusive of u and v form a clump S.*

We modify the BFS so that when an edge is inserted in forest F_i in the process, we first check whether the edges of F_i that form a cycle can be inserted in F_1 (i.e. explore neighbors in F_1 first). This process ensures that any new edge is inserted in the first few forests if possible before inserting into the next forest. In G'_{rt2} we have increased the degree of every vertex by at most n, but Lemma 4 still holds true for the graph G'_{rt2}. The running time of the BFS algorithm remains $\tilde{O}(m)$.

It turns out this insertion algorithm satisfies the following stronger invariant that is helpful for deletion. Towards that, we first define the notion of $clump_i$. Let $F_1, F_2, \ldots F_j$ be the collection of edge disjoint forests at any point of time in the algorithm (after inserting a subset of edges). A set S of vertices is said to form a $clump_i$ $i \leq j$, if the induced subgraph $G[S]$ is connected in each of the forests $F_1, F_2, \ldots F_i$.

Lemma 5. *An edge $e = (u, v)$ of a graph is present in forest F_i if and only if before insertion of e in forest F_i, u and v are together in $clump_{i-1}$, but not in any $clump_i$.*

Proof. Suppose $e = (u, v)$ is in F_i. The algorithm maintains a decomposition of forests by maintaining that the first i forests in the decomposition contains as many edges as possible. Let us look at the insertion of an edge (x, y) which pushes (u, v) to F_i. When (x, y) is inserted into the decomposition it first tries to find a path from (x, y) to a *green* vertex with respect to the forest F_1. If not, then it tries to find a *green* vertex in F_1 and F_2. In general if there is no (x, y) to *green* path in forests $F_1, ...F_{i-1}$, then it tries to find a *green* path in forests $F_1, F_2, \ldots, F_i$ if it exists.

The algorithm (x, y) did not find any *green* path in $F_1, \ldots F_{i-1}$ for it to push (u, v) into any of the forests F_1 to F_{i-1}. Hence, by Theorem 5, (x, y) is part of a $clump_{i-1}$. As (u, v) was pushed into F_i, (u, v) was reachable from (x, y) in G'_{rt2}. Therefore, in forests $F_1, \ldots, F_{i-1}$, u,v,x and y are all part of the same clump S. The number of edges of S became at least $(i - 1)(|S| - 1) + 1$ (after insertion of (u, v)). So (u, v) from S had to be inserted into F_i.

For (u, v), the same S now forms a $clump_{i-1}$. To show the second condition, say an edge is present in F_i. Suppose before insertion of the edge in F_i the endpoints were in some $clump_i$. But by Lemma 1 the edge can not be present in forest F_i leading to a contradiction.

We now prove the converse. If the endpoints of an edge before insertion in F_i is part of some $clump_{i-1}$ then it can not be present in any forest $\leq i - 1$. Again if the endpoints of the edge do not belong to any $clump_i$, then there is an "augmenting" sequence (from Lemma 1 and Theorem 4) in $F_1, \ldots, F_i$. So the edge will get inserted in any forest from $F_1, \ldots, F_i$. Combining both we get that the edge will get inserted in forest F_i.   □

Before going into the algorithm we prove the following lemma, which prevents us from exploring useless edges.

Lemma 6. *Suppose an edge (u, v) gets deleted from forest F_i. If there are no edges in forest F_{i+1} that can merge the components containing u and v, then there are no such edges in any forests from F_{i+2} to F_a.*

Proof. Suppose that there exists an edge, say (w, x) in F_{i+2} such that w belongs to the component of u and x the component of v in F_i. As $(w, x) \in F_{i+2}$, they are part of the same $clump_{i+1}$ by Lemma 5 and hence a part of some $clump_i$. Also as clumps form a subtree, the vertices w, x, u and v are part of the same $clump_i$ as well. As w and x are part of the same $clump_{i+1}$ all vertices in the path from w to x in F_i are also part of the same $clump_{i+1}$. So there is at least one $clump_{i+1}$ containing the endpoints of such an edge along with u and v. As the decomposition is edge disjoint, the path between u and v in F_{i+1} is separated by at least one vertex w. Thus there is at least one edge in this path that goes across the two new components in F_i. This contradicts that no edges of F_{i+1} can merge the two components containing u and v.   □

Now we give our algorithm to delete an edge (u, v),

– Find the forest F_i, the edge (u, v) belong to. Delete (u, v) from forest F_i which breaks the tree into two parts, one containing u and the other containing v.

- For $k \in [i + 1, ..., \mathbf{a}]$ do the following,
    - In F_k check if there is an edge going across the two newly formed components of F_{k-1}. If not, stop and return the same decomposition.
    - If there is such an edge (w, x), remove it from F_k and attach it to F_{k-1} to attach the components containing u and v.
- If $F_\mathbf{a}$ is empty then report that arboricity decreases.

Theorem 6. *Given a graph $G = (V, E)$ with arboricity* **a** *and an edge (u, v) to be deleted the above algorithm computes the arboricity of the modified graph $G = (V, E \setminus (u, v))$ correctly in $O(m \log n)$ time.*

Proof. **Correctness:** On deletion of an edge (u, v), to check whether the arboricity of the graph is correctly maintained it is sufficient to check if $F_\mathbf{a}$ is non empty after the deletion algorithm, then arboricity does not decrease. This is shown by maintaining the invariant of Lemma 5 for each of the edges after deletion. Then it follows that the end points of edges of F_a form a $clump_{a-1}$ which implies that the arboricity has to be at least **a**.

Now we show how the algorithm maintains the invariant of Lemma 5 for all edges of the graph and the correctness follows from there. The edges from $F_1, ..., F_i$ are not touched by the algorithm and thus by default they maintain the invariant. Before deletion of (u, v) we know by Lemma 5 that the endpoints of all edges in F_k, $k \in [i + 1, ..., \mathbf{a}]$ are part of some $clump_{k-1}$.

We know that u and v are part of the same $clump_{i-1}$. If all edges of forest F_{i+1} are not in the same $clump_{k-1}$, then by Lemma 6, the decomposition remains as is and there is no further work to be done. All such edges remain in forest F_{i+1} and the invariant of Lemma 5 is maintained. Otherwise, there is at least one edge in F_{i+1} such that the endpoints are in the same $clump_i$ as both u and v. This can be divided into two cases:

- There is no edge in F_{i+1} that goes across the trees containing u and v after deletion of (u, v). In this case again by Lemma 6 the decomposition remains as is. The invariant of Lemma 5 is maintained.
- Otherwise, there is an edge in F_{i+1} that goes across the two newly split trees in F_i. The endpoints of such an edge has to be a part of the same $clump_i$ containing u and v (as by definition, clumps form a subtree). The algorithm attaches this edge to F_i connecting the two trees containing u and v. For this edge, the invariant of Lemma 5 is maintained as it can not still settle in any level $< i$. At this point, the invariant continues to hold for edges in F_{i+1} as each of the end points are in a $clump_i$ containing u and v (as we have deleted an edge and added an edge to the clump). This in turn splits F_{i+1} into two components which is again merged similarly above, treating this like a new edge deletion, and in a similar way we can show that the invariant is maintained for every edge in subsequent forests where edges are deleted to merge two components in the previous forest.

Implementation and Running Time: We look at data structures that help us maintain each forest in the decomposition. Each of the forests, F_k, $k \in [1, ...\mathbf{a}]$ is represented by dynamic link-cut tree data structure [16].

For the algorithm, we need to update the link-cut tree data structure of forest F_i after (u, v) is deleted. For each edge starting from F_{i+1} we need to check if the endpoints are in the same or different trees in F_i and possibly do an insertion. For each such check we spend $O(\log n)$ time. The algorithm makes such a check for each edge from F_{i+1} to F_a at most once. Thus the total runtime is $O(m \log n)$.   □

Combining both results of Sects. 3 and 4 we get,

Theorem 7. *Given a graph $G = (V, E)$ with arboricity $\mathbf{a}$ and an edge (u, v) to be either inserted or deleted we can compute the arboricity of the modified graph $G = (V, E \cup (u, v))$ or $G = (V, E \setminus (u, v))$ correctly in $\tilde{O}(m)$ time.*

5   An $\Omega(\log n)$ Lower Bound

Theorem 8 [14]. *Consider any dynamic data structure that performs a sequence of n edge insertions and deletions that maintain the forest structure starting from an edgeless graph. Suppose the structure also supports queries of the form whether a pair of vertices are in the same connected component. Then such a structure requires $\Omega(n \log n)$ expected time to support a sequence of n query and update operations in the cell probe model of word size $\log n$.*

We use this observation to give a reduction to the problem of detecting whether the arboricity of a graph is 1 or 2 to prove a similar lower bound.

Theorem 9. *Any dynamic structure that maintains the arboricity of a graph under edge insertions and deletions requires $\Omega(\log n)$ amortized time per update in the cell probe model of word size $\log n$.*

Proof. Given an instance I of the fully dynamic connectivity problem on forests with n vertices, we create a graph I' on the same n vertices. Whenever an edge $\{u, v\}$ is added to I, we call the addition of edge $\{u, v\}$ to I'. Whenever an edge $\{u, v\}$ is deleted from I, we delete the same edge from I'. The forest maintenance property of the instance I ensures that these addition or deletion of edges always ensures a forest is maintained in I' as well.

When a query between a pair of vertices u and v comes, we simply add the edge $\{u, v\}$ and ask whether the resulting graph has arboricity 1 or 2. If it is 1, then we declare that u and v are in different components of the forest, and otherwise if arboricity is 2 then they are in the same component. We then delete the edge $\{u, v\}$ from the graph. This proves the correctness of the reduction. □

6   Conclusion

We have initiated a study to compute arboricity in fully dynamic graphs. We have recasted two of the existing algorithms in the language of an auxiliary graph of the given graph and augmenting paths. It is an interesting open question as to whether one can bridge the gap between an $\tilde{O}(m)$ upper bound and an $\Omega(\log n)$ lower bound. Another interesting question would be to design faster dynamic algorithms for special classes of graphs like for planar graphs under edge updates.

References

1. Berglin, E., Brodal, G.S.: A simple greedy algorithm for dynamic graph orientation. In: 28th International Symposium on Algorithms and Computation, ISAAC 2017, Phuket, Thailand, 9–12 December 2017, pp. 12:1–12:12 (2017)
2. Brodal, G.S., Fagerberg, R.: Dynamic representation of sparse graphs. In: 6th International Workshop on Algorithms and Data Structures, WADS 1999, Vancouver, British Columbia, Canada, 11–14 August 1999, Proceedings, pp. 342–351 (1999)
3. Edmonds, J.: Lehman's switching game and a theorem of Tutte and Nash-Williams. Natl Bur. Stan. **69B**, 73–77 (1965)
4. Edmonds, J.: Minimum partition of a matroid into independent subsets. Natl Bur. Stan. **69B**, 67–72 (1965)
5. Eppstein, D.: Arboricity and bipartite subgraph listing algorithms. Inf. Process. Lett. **51**(4), 207–211 (1994)
6. Erickson, J.: http://jeffe.cs.illinois.edu/teaching/datastructures/2006/problems/bill-arboricity.pdf
7. Gabow, H.N.: Algorithms for graphic polymatroids and parametriscs-sets. J. Algorithms **26**(1), 48–86 (1998)
8. Grossi, R., Lodi, E.: Simple planar graph partition into three forests. Discrete Appl. Math. **84**(1–3), 121–132 (1998)
9. He, M., Tang, G., Zeh, N.: Orienting dynamic graphs, with applications to maximal matchings and adjacency queries. In: Ahn, H.-K., Shin, C.-S. (eds.) ISAAC 2014. LNCS, vol. 8889, pp. 128–140. Springer, Cham (2014). https://doi.org/10.1007/978-3-319-13075-0_11
10. Kopelowitz, T., Krauthgamer, R., Porat, E., Solomon, S.: Orienting fully dynamic graphs with worst-case time bounds. In: Esparza, J., Fraigniaud, P., Husfeldt, T., Koutsoupias, E. (eds.) ICALP 2014. LNCS, vol. 8573, pp. 532–543. Springer, Heidelberg (2014). https://doi.org/10.1007/978-3-662-43951-7_45
11. Kowalik, L.: Adjacency queries in dynamic sparse graphs. Inf. Process. Lett. **102**(5), 191–195 (2007)
12. Nash-Williams, C.S.J.A.: Edge-disjoint spanning trees of finite graphs. J. Lond. Math. Soc. **36**, 445–450 (1961)
13. Nash-Williams, C.S.J.A.: Decomposition of finite graphs into forests. J. Lond. Math. Soc. **39**(1), 12 (1964)
14. Patrascu, M., Demaine, E.D.: Logarithmic lower bounds in the cell-probe model. SIAM J. Comput. **35**(4), 932–963 (2006)
15. Roskind, J., Tarjan, R.E.: A note on finding minimum-cost edge-disjoint spanning trees. Math. Oper. Res. **10**(4), 701–708 (1985)
16. Sleator, D.D., Tarjan, R.E.: A data structure for dynamic trees. J. Comput. Syst. Sci. **26**(3), 362–391 (1983)
17. Tutte, W.T.: On the problem of decomposing a graph into n connected factors. J. Lond. Math. Soc. **36**, 221–230 (1961)

A Lower Bound on the Growth Constant of Polyaboloes on the Tetrakis Lattice

Gill Barequet[1(✉)] and Minati De[2]

[1] Department of Computer Science, Technion—Israel Institute of Technology, 3200003 Haifa, Israel
barequet@cs.technion.ac.il
[2] Department of Mathematics, Indian Institute of Technology Delhi, New Delhi, India
minati@maths.iitd.ac.in

Abstract. A "lattice animal" is an edge-connected set of cells on a lattice. In this paper we consider the *Tetrakis* lattice, and provide the first lower bound on λ_τ, the growth constant of polyaboloes (animals on this lattice), proving that $\lambda_\tau \geq 2.4345$. The proof of the bound is based on a concatenation argument and on calculus manipulations. If we also rely on an unproven assumption, which is, however, supported by empirical data, we obtain the conditional slightly-better lower bound 2.4635.

Keywords: Lattice animals · Polyaboloes · Concatenation · Growth constant

1 Introduction

Lattice animals are edge-connected sets of cells on a lattice. (One can regard the lattice as a graph. In the dual graph, that is, in the cell-adjacency graph of the original lattice, cells (faces) become sites (vertices). Hence, "site animals" are also a common name for lattice animals made of connected cells.)

The study of lattice animals began in the mid 1950s in the community of statistical physics. For example, Temperley [15] investigated the mechanics of macro-molecules, and Broadbent and Hammersley [6] studied percolation processes. Mathematicians began to show interest in lattice problems at about the same time. Harary [10] composed a list of unsolved problems in the enumeration of graphs, and Eden [8] analyzed cell growth problems.

Fixed animals are considered distinct if they have different *shapes* or *orientations*. In this paper we consider only fixed animals, hence, we omit hereafter the adjective "fixed" when we refer to lattice animals.

Most attention was given in the literature to the cubical lattices in two dimensions (see Fig. 1(a)) and higher dimensions. Less attention was given to other lattices, such as the triangular and hexagonal lattices in two dimensions.

Work on this paper by the first author has been supported in part by ISF Grant 575/15 and by BSF Grant 2017684. Work by the second author has been supported in part by Grant DST-IFA-14-ENG-75 and new faculty Seed Grant NPN5R.

© Springer Nature Switzerland AG 2019
D.-Z. Du et al. (Eds.): COCOON 2019, LNCS 11653, pp. 13–24, 2019.
https://doi.org/10.1007/978-3-030-26176-4_2

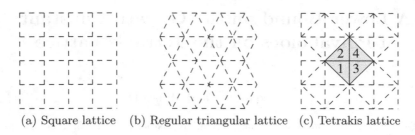

(a) Square lattice   (b) Regular triangular lattice   (c) Tetrakis lattice

Fig. 1. Two-dimensional lattices

The size of an animal is the number of cells it contains. Let (a_n) be an integer sequence. The limit $\lim_{n\to\infty} a_{n+1}/a_n$, if it exists, is called the *growth constant* of (a_n). In the context of sequences that enumerate lattice animals by size, the existence of a growth constant was discussed for the first time for the square lattice, on which animals are called *polyominoes* and the number of polyominoes of size n is denoted by $A(n)$. Klarner [11] showed that $\lambda := \lim_{n\to\infty} \sqrt[n]{A(n)}$ exists. Only three decades later, Madras [13] proved that $\lim_{n\to\infty} A(n+1)/A(n)$ exists, and, hence, is equal to λ. A main open problem in this area is to determine the exact value of this elusive constant. The currently best-known lower and upper bounds on λ are 4.0025 [5] and 4.6496 [12], respectively.

Similarly, a *polyiamond* of size n is an edge-connected set of n cells on the regular two-dimensional triangular lattice (Fig. 1(b)), in which every cell is an equilateral triangle. Let $T(n)$ denote the number of polyiamonds of size n. The existential results of Klarner [11] and Madras [13] extend to all periodic lattices, and, in particular, to the triangular lattice. Hence, the sequence $(T(n))$ has a growth constant, $\lambda_T := \lim_{n\to\infty} T(n+1)/T(n)$. The currently best-known lower and upper bounds on λ_T are 2.8424 [4] and 3.6108 [3], respectively.

In this paper, we consider animals on the Tetrakis lattice (see Fig. 1(c)), also known as *kisquadrille* [7, §21]. Throughout the paper, we refer to these animals as "polyaboloes." It is important to note that we restrict ourselves to polyaboloes that can be embedded in the Tetrakis lattice, forbidding triangle neighborhoods like the one between the two dark-gray triangles in Fig. 4(c). Let $\tau(n)$ denote the number of polyaboloes of size n. Figure 2 shows polyaboloes of size up to 3. The first twenty elements of $(\tau(n))$ appear as sequence A197467 in the On-Line Encyclopedia of Integer Sequences [1], referring to these animals as "poly-[4.8^2]-tiles"; See also Grünbaum and Shephard [9, §§2.7,6.2,9.4]. More values of this sequence are provided in the Appendix of this paper. By Klarner [11] and Madras [13], we know that the sequence $(\tau(n))$ has a growth constant, that is, the limit $\lambda_\tau := \lim_{n\to\infty} \tau(n+1)/\tau(n) = \lim_{n\to\infty} \sqrt[n]{\tau(n)}$ exists. In this paper, we set a lower bound on λ_τ, showing that $\lambda_\tau \geq 2.4345$. Under some unproven assumption, which is, however, supported by empirical data, we obtain a conditional improved lower bound, $\lambda_\tau \geq 2.4635$.

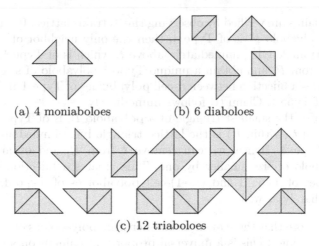

(a) 4 moniaboloes   (b) 6 diaboloes

(c) 12 triaboloes

Fig. 2. Polyaboloes of size up to 3

2   Preliminaries

As is shown in Fig. 1(c), the Tetrakis lattice contains cells of four distinct types, which we denote as cells of Type 1 through Type 4. We also define a lexicographic order on the cells of the lattice as follows: A triangle t_1 is *smaller* than triangle $t_2 \neq t_1$ if either

1. The column of t_1 is to the left of the column of t_2; or
2. Both t_1 and t_2 are in the same column, and t_1 lies below t_2.

We say that a polyabolo is of Type i (for $1 \leq i \leq 4$) if its lexicographically-*smallest* triangle is of Type i, and denote by $\tau_i(n)$ the number of such polyaboloes of size n. Furthermore, we split every Type-i polyaboloes into four complementing Subtypes (i,j) polyaboloes $(1 \leq j \leq 4)$, in which the lexicographically-smallest triangle is of Type i (as before), and the lexicographically-*largest* triangle is of Type j. We also denote by $\tau_{i,j}(n)$ (for $1 \leq i,j \leq 4$) the number of polyaboloes of Type (i,j) and size n.

A concatenation of two polyaboloes P_1, P_2 is the union of the cell set of P_1 and the cell set of a translated copy of P_2, such that the *largest* cell of P_1 is attached to the *smallest* cell of P_2, and all cells of P_1 are smaller than the translates of cells of P_2. It is easy to verify that the concatenation of a given pair of polyaboloes is either impossible or unique; see Fig. 4 and Table 1.

3   Properties

First, we observe a close relationship between three of the types of polyaboloes.

Lemma 1. $\tau_3(n+2) \stackrel{(ii)}{=} \tau_4(n+1) \stackrel{(i)}{=} \tau_1(n)$ *(for $n \geq 1$).*

Proof. Both claims are verified by observing the Tetrakis lattice. If t, the smallest triangle of a polyabolo P, is of Type 4, then the only neighbor of t (within P) is the lattice triangle lying immediately above t, which is of Type 1. Hence, we can remove t from P, and obtain a unique Type-1 polyabolo of size smaller by one. This creates a bijection between n-cell polyaboloes of Type 4 and $(n-1)$-cell polyaboloes of Type 1. Claim (i) follows immediately.

Similarly, if t, the smallest triangle of a polyabolo P, is of Type 3, then the only neighbor of t (within P) is the lattice triangle lying immediately above t, which is of Type 4. Hence, we can remove t from P, and obtain a unique Type-4 polyabolo of size smaller by one. This creates a bijection between n-cell polyaboloes of Type 3 and $(n-1)$-cell polyaboloes of Type 4. Claim (ii) follows immediately as well.   □

Second, we note that the sequences enumerating polyaboloes of each type are monotone increasing. (This is a universal property of animals on all lattices.)

Observation 1. $\tau_i(n) \leq \tau_i(n+1)$, *for* $1 \leq i \leq 4$ *and all* $n \in \mathbb{N}$.

Indeed, consider n-cell polyaboloes of Type i. By stacking one more triangle on top of the lexicographically-largest triangle of each such polyabolo, we create polyaboloes of size $n+1$ of the same type and *without* repetitions. In fact, for each Type i of polyaboloes, there exists a nominal size $n_0 = n_0(i)$, such that for all $n \geq n_0$, there exist polyaboloes of size $n+1$ that cannot be built this way: All such polyaboloes whose largest triangle cannot be removed without breaking the polyabolo into two pieces. Hence, $\tau_i(n) < \tau_i(n+1)$, for $1 \leq i \leq 4$ and all $n \geq n_0(i)$.

Third, we find relations between the numbers of polyaboloes of different types.

Corollary 1. $\tau_3(n) \leq \tau_4(n) \leq \tau_1(n)$ *(for every* $n \in \mathbb{N}$*).*

Proof. The claim follows from Lemma 1 and Observation 1.   □

Finally, we observe a simple equality of numbers of some polyaboloes of different subtypes.

Observation 2. $\tau_{i,j}(n) = \tau_{5-j,5-i}(n)$, *for* $1 \leq i, j \leq 4$ *and all* $n \in \mathbb{N}$.

Observation 2 is easily justified by rotating the plane by $180°$.

4   The Bound

We are now ready to prove our main result, setting a lower bound on λ_τ.

Theorem 1. $\lambda_\tau \geq 2.4345$.

Proof. We proceed with a concatenation argument tailored to the specific lattice under consideration. Note that the only valid concatenation options are those specified in Table 1. Hence, the only families of polyaboloes which are closed under concatenation are those of Types (2,1), (3,1), (3,2), (4,2), (4,3), and (1,4). Consider one such family of Type (i,j) (the exact values of i, j will be determined later). By the same arguments used by Klarner [11] and Madras [13], we know that the sequence $(\tau_{i,j}(n))$ has a growth constant. In addition, we obviously have that $\tau_{i,j}(n) \leq \tau(n)$. Using elementary facts from calculus, we conclude that the growth constant of $(\tau_{i,j}(n))$, which we denote by $\lambda_{\tau_{i,j}}$, is at most λ_τ. Therefore, any lower bound we set on $\lambda_{\tau_{i,j}}$ will also be a lower bound on λ_τ.

Table 1. Number of valid concatenation options for all cases of triangle j with triangle i

$j \setminus i$	1	2	3	4
1	0	1	1	0
2	0	0	1	1
3	0	0	0	1
4	1	0	0	0

Let P_1, P_2 be two polyaboloes of Type (i,j) and size n, and let Q be the polyabolo of size $2n$ that is the result of concatenating P_1 and P_2. It is crucial to observe that Q *cannot* be the result of concatenating any other pair of polyaboloes, both of size n (but it may be represented as the concatenation of polyaboloes of different sizes). However, there exist polyaboloes of size $2n$, which cannot be represented as the concatenation of any pair of polyaboloes of size n, because their lexicographically-smallest (or largest) n triangles do not form a connected set of triangles; see, for example, Fig. 3. (In our setting, there is another possible reason for this: It may be the case that the lexicographically-smallest (or largest) n triangles of a polyabolo of Type (i,j) and size $2n$ is not a polyabolo of the same type.) Hence,

$$\tau_{i,j}^2(n) \leq \tau_{i,j}(2n).$$

By simple manipulations of this relation, we obtain that

$$\left(\tau_{i,j}(n)\right)^{1/n} \leq \left(\tau_{i,j}(2n)\right)^{1/(2n)}.$$

Thus, $\left(\tau_{i,j}(k)\right)^{1/k}, \left(\tau_{i,j}(2k)\right)^{1/(2k)}, \left(\tau_{i,j}(4k)\right)^{1/(4k)}, \ldots$ is monotone increasing for any value of k, and, as a subsequence of $\left(\left(\tau_{i,j}(n)\right)^{1/n}\right)$, it converges to $\lambda_{\tau_{i,j}}$ as well. Therefore, any term of the form $\left(\tau_{i,j}(n)\right)^{1/n}$ is a lower bound on $\lambda_{\tau_{i,j}}$. (See the Appendix for some of the available values of the sequences $(\tau_{i,j}(n))$.) Now, we simply choose the values of (i,j) that provide the best lower bound out of the six options. It turns out that $(1,4)$ is the best choice. In particular, $\lambda_{\tau_{1,4}} \geq (\tau_{1,4}(36))^{1/36} = 81530581108477^{1/36} \geq 2.4345$, and the claim follows. $\square$

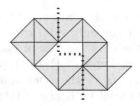

Fig. 3. A sample 16-triangle polyabolo which cannot be represented as the concatenation of two 8-triangle polyaboloes

5   A Conditional Better Bound

If we refine the proof in the previous section to consider simultaneously all possible concatenation options, we may achieve a better (higher) lower bound on λ_τ. However, the refined proof requires a reasonable assumption which we were unable to prove so far.

Assumption 1. $\tau_1(n) \geq \tau_2(n)$ for all $n \in \mathbb{N}$.

Theorem 2. *Under Assumption 1, we have that $\lambda_\tau \geq 2.4635$.*

Proof. As in the proof of Theorem 1, we proceed with a concatenation argument. Since not all pairs of polyaboloes of size n can be concatenated (as can be seen in Table 1), let us count systematically those pairs of polyaboloes that can be concatenated. It can easily be verified that pairs of polyaboloes can be concatenated in at most one way. Here are two examples.

- A polyabolo whose largest triangle is of Type 2 cannot be concatenated at all with a polyabolo whose smallest triangle is of Type 1 (see Fig. 4(a)).
- A polyabolo whose largest triangle is of Type 3 can be concatenated in one way with a polyabolo whose smallest triangle is of Type 4 (see Fig. 4(b)).

Note that some "plausible" concatenations are in fact not allowed on our lattice. For example, there is no way to concatenate a polyabolo whose largest triangle is of Type 1 with a polyabolo whose smallest triangle is of Type 4. Indeed, the former triangle can be attached to the latter triangle either to the left of it or below it, but neither composition is valid (see Fig. 4(c)).

By rotational symmetry, the number of polyaboloes of size n, whose *largest* (top-right) triangle is of Type j, is $\tau_{5-j}(n)$. Indeed, such a rotation converts triangles of Type 1 to triangles of Type 4 (and vice versa), and triangles of Type 2 to triangles of Type 3 (and vice versa). Therefore, the total number of concatenations of two polyaboloes of size n is

$$\tau_4(n)\tau_2(n) + \tau_4(n)\tau_3(n) + \tau_3^2(n) + \tau_3(n)\tau_4(n) + \tau_2(n)\tau_4(n) + \tau_1^2(n)$$
$$= \tau_1^2(n) + \tau_3^2(n) + 2\tau_4(n)(\tau_2(n) + \tau_3(n)).$$

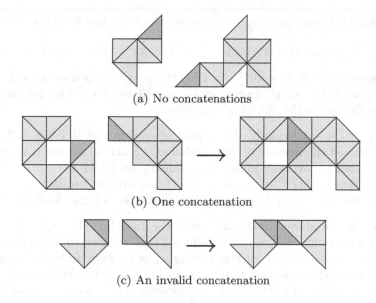

(a) No concatenations

(b) One concatenation

(c) An invalid concatenation

Fig. 4. Valid and invalid concatenations

Let P_1, P_2 be two polyaboloes of size n, which can be concatenated in some way ξ (one of the options listed in Table 1), and yield a polyabolo Q of size $2n$. As observed in the proof of Theorem 1, the polyabolo Q can be represented as the concatenation of two polyaboloes of size n only by the triple (P_1, P_2, ξ). However, there exist polyaboloes of size $2n$, which cannot be represented as the concatenation of two polyaboloes of size n, because their lexicographically-smallest (or largest) n triangles do not form a connected set of triangles. Hence, we conclude that

$$\tau_1^2(n) + \tau_3^2(n) + 2\tau_4(n)(\tau_2(n) + \tau_3(n)) \leq \tau(2n). \tag{1}$$

Let us now find a good lower bound on the number of concatenations. Denote by $x_i(n)$ (for $1 \leq i \leq 4$) the fraction of polyaboloes of Type i out of all polyaboloes of size n, that is, $x_i(n) = \tau_i(n)/\tau(n)$. The left-hand side of Eq. (1) can be rewritten as

$$\left(x_1^2(n) + x_3^2(n) + 2x_4(n)(x_2(n) + x_3(n))\right)\tau^2(n). \tag{2}$$

Obviously, we have that

$$x_4(n) = 1 - x_1(n) - x_2(n) - x_3(n). \tag{3}$$

Substituting Eq. (3) into the count of concatenations (Eq. (2)), we find that the left-hand side of Relation (1) is equal to

$$\left(x_1^2(n) + x_3^2(n) + 2(1 - x_1(n) - x_2(n) - x_3(n))(x_2(n) + x_3(n))\right)\tau^2(n).$$

Our next goal is to find a lower bound on the trivariate function

$$f(x_1, x_2, x_3) := x_1^2 + x_3^2 + 2(1 - x_1 - x_2 - x_3)(x_2 + x_3)$$

in the range $[0, 1] \times [0, 1] \times [0, 1]$. (In fact, the range is open on all sides since there are polyaboloes of all four types for all values of n.) This minimization problem is subject to the following constraints:

1. $x_1 + x_2 + x_3 \leq 1$: Obviously, the number of polyaboloes of Types 1, 2, and 3 cannot exceed the number of all polyaboloes. (Since all x_is are nonnegative, this constraint is actually weaker than constraints 2 and 3.)
2. $x_3 \leq x_4$, that is, $2x_3 \leq 1 - x_1 - x_2$: This follows from Corollary 1.
3. $x_4 \leq x_1$, that is, $1 - x_2 - x_3 \leq 2x_1$: This also follows from Corollary 1.

Subject to the above three constraints, the function $f(x_1, x_2, x_3)$ assumes at $(0, 1, 0)$ a minimum of 0, which is useless. Thus, we need to add a constraint which keeps x_2 away from 1, or keeps either x_1 or x_3 away from 0. Empirically (see the Appendix), we see that the sequences $(x_1(n))$ and $(x_2(n))$ are monotone increasing (and the limits of both are visually around 0.4), while $x_4(n)$ and $x_3(n)$ are monotone decreasing (and their limits are visually around 0.15 and 0.05, respectively), and the existing data suggest that $x_1(n) > x_2(n) > x_4(n) > x_3(n)$ for $n \geq 3$. If we rely on Assumption 1 and add to the above the additional constraint $x_1 \geq x_2$, the function $f(x_1, x_2, x_3)$ now assumes a minimum of $1/4$ at two points: (0.5,0.5,0) and (0.5,0,0). (The second minimum point seems to be superfluous.) Hence,

$$\frac{1}{4}\tau^2(n) \leq \tau(2n).$$

By simple manipulations of this relation, we obtain that

$$\left(\frac{1}{4}\tau(n)\right)^{1/n} \leq \left(\frac{1}{4}\tau(2n)\right)^{1/(2n)}.$$

This implies that the sequence $\left(\frac{1}{4}\tau(k)\right)^{1/k}, \left(\frac{1}{4}\tau(2k)\right)^{1/(2k)}, \left(\frac{1}{4}\tau(4k)\right)^{1/(4k)}, \ldots$ is monotone increasing for any value of k, and, as a subsequence of $\left(\left(\frac{1}{4}\tau(n)\right)^{1/n}\right)$, it converges to λ_τ as well. Therefore, any term of the form $\left(\frac{1}{4}\tau(n)\right)^{1/n}$ is a lower bound on λ_τ. (See the Appendix for available values of the sequence $(\tau(n))$.) In particular, $\lambda_\tau \geq (\frac{1}{4}\tau(36))^{1/36} = (499003797597583/4)^{1/36} \geq 2.4635$, and the claim follows. $\qquad \square$

6   Conclusion

In this paper, we prove a lower bound on the growth constant of polyaboloes on the Tetrakis lattice, namely, that $\lambda_\tau \geq 2.4345$. Under some empirical assumption, we obtain an improved conditional lower bound of 2.4635. Future work includes improving the lower bound and finding a good upper bound on λ_τ.

Appendix: Computing Elements of $(\tau(n))$

We have implemented Redelmeier's algorithm [14] for counting polyominoes, and adapted it to the Tetrakis lattice.[1] The algorithm was implemented in C on a server with four 2.20 GHz Intel Xeon processors and 512 GB of RAM. The software consisted of about 200 lines of code.

Table 2. Split of $\tau(36)$ into the 16 subtypes $\tau_{i,j}(36)$, $1 \leq i, j \leq 4$

$i \setminus j$	1	2	3	4	Total
1	30,137,895, 510,778	11,148,998, 271,068	78,876,705, 335,638	81,530,581, 108,477	201,694,180, 225,961
2	29,162,541, 569,420	10,790,377, 951,489	76,306,266, 133,179	78,876,705, 335,638	195,135,890, 989,726
3	4,127,703, 521,100	1,529,507, 765,873	10,790,377, 951,489	11,148,998, 271,068	27,596,587, 509,530
4	11,148,998, 271,068	4,127,703, 521,100	29,162,541, 569,420	30,137,895, 510,778	74,577,138, 872,366
Grand total					**499,003,797,597,583**

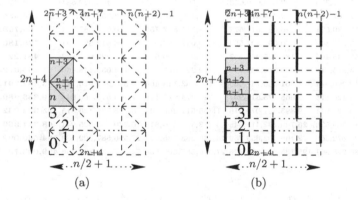

Fig. 5. Cell IDs for Redelmeier's Algorithm

Assume, for simplicity, that we wanted to count polyaboloes up to size n, where n is divisible by 4. Then, we created the graph, dual of the portion of

---

[1] Originally, the algorithm was proposed for counting polyominoes (site animals on the square lattice). However, as was already noted elsewhere (see, e.g., [2]), this algorithm can be adapted to any lattice, once it is formulated as an algorithm for counting connected subgraphs of a given graph, that contain one marked vertex in the graph. The reader is referred to the reference cited above for further details.

Table 3. Counts of polyaboloes (values of $\tau(21)$–$\tau(36)$ (in bold) are new)

n	$\tau_1(n)$	$\tau_2(n)$	$\tau_3(n)$	$\tau_4(n)$	$\tau(n)$
1	1	1	1	1	4
2	2	2	1	1	6
3	5	4	1	2	12
4	10	8	2	5	25
5	22	19	5	10	56
6	52	48	10	22	132
7	125	121	22	52	320
8	311	304	52	125	792
9	781	759	125	311	1,976
10	1,965	1,905	311	781	4,962
11	4,986	4,844	781	1,965	12,576
12	12,765	12,424	1,965	4,986	32,140
13	32,904	32,049	4,986	12,765	82,704
14	85,303	83,072	12,765	32,904	214,044
15	222,145	216,224	32,904	85,303	556,576
16	580,700	565,062	85,303	222,145	1,453,210
17	1,523,496	1,482,251	222,145	580,700	3,808,592
18	4,010,346	3,900,592	580,700	1,523,496	10,015,134
19	10,587,019	10,292,607	1,523,496	4,010,346	26,413,468
20	28,019,133	27,227,765	4,010,346	10,587,019	69,844,263
21	74,323,315	72,197,057	10,587,019	28,019,133	**185,126,524**
22	197,565,811	191,849,795	28,019,133	74,323,315	**491,758,054**
23	526,189,451	510,796,099	74,323,315	197,565,811	**1,308,874,676**
24	1,403,920,681	1,362,392,571	197,565,811	526,189,451	**3,490,068,514**
25	3,751,867,755	3,639,699,653	526,189,451	1,403,920,681	**9,321,677,540**
26	10,041,587,514	9,738,372,232	1,403,920,681	3,751,867,755	**24,935,748,182**
27	26,912,890,591	26,092,611,572	3,751,867,755	10,041,587,514	**66,798,957,432**
28	72,223,625,842	70,002,807,553	10,041,587,514	26,912,890,591	**179,180,911,500**
29	194,053,148,466	188,035,944,757	26,912,890,591	72,223,625,842	**481,225,609,656**
30	521,974,915,118	505,660,330,038	72,223,625,842	194,053,148,466	**1,293,912,019,464**
31	1,405,512,260,944	1,361,250,747,068	194,053,148,466	521,974,915,118	**3,482,791,071,596**
32	3,788,326,126,815	3,668,175,811,997	521,974,915,118	1,405,512,260,944	**9,383,989,114,874**
33	10,220,263,525,941	9,893,931,070,016	1,405,512,260,944	3,788,326,126,815	**25,308,032,983,716**
34	27,596,587,509,530	26,709,792,413,846	3,788,326,126,815	10,220,263,525,941	**68,314,969,576,132**
35	74,577,138,872,366	72,166,102,653,759	10,220,263,525,941	27,596,587,509,530	**184,560,092,561,596**
36	201,694,180,225,961	195,135,890,989,726	27,596,587,509,530	74,577,138,872,366	**499,003,797,597,583**

the Tetrakis lattice, that consists of $n/2$ columns, each of height $2n + 4$. Cells (triangles) of this portion of our lattice were numbered as is shown in Fig. 5(a). In fact, the cell-adjacency graph was identical to the one shown in Fig. 5(b), where a thick edge means that the two cells sharing this edge were not considered as neighbors. In order to count polyaboloes of Types 1, 2, 3, or 4, we fixed their smallest triangle at cell number $n + 1$, $n + 2$, $n + 3$, or n, respectively. This ensured that animals of size n would never spill over the allocated portion of the Tetrakis lattice.

In fact, we needed to count only polyaboloes of two out of the four types, as is implied by Lemma 1. Thus, we ran our program for counting polyaboloes of Types 1 and 2, and computed counts of polyaboloes of Types 3 and 4 by applying the lemma. Then, we summed up the results to finally obtain $\tau(n) = \sum_{i=1}^{4} \tau_i(n)$. The running times of our program were 27.5 and 26.25 days, for computing $\tau_1(n)$ and $\tau_2(n)$ for $1 \leq n \leq 36$, respectively, for a total of 53.75 days for computing $\tau(n)$ for this range of n.

Table 2 provides the split of $\tau(36)$ into all 16 subtypes. Table 3 shows the total counts of polyaboloes, produced by our program, extending significantly the previously-published counts [1, Sequence A197467]. Figure 6 plots the known values of $\sqrt[n]{\tau(n)}$ and $\tau(n)/\tau(n-1)$ for $2 \leq n \leq 36$, demonstrating the convergence of the two sequences. Figure 7 plots the 36 known values of the sequences $(x_i(n))$ $(i = 1, \ldots, 4)$, showing the tendencies of these sequences.

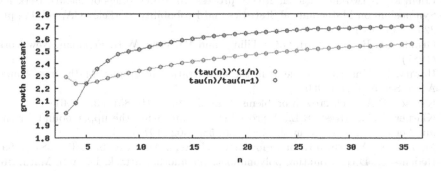

Fig. 6. Convergence of $\sqrt[n]{\tau(n)}$ and $\tau(n)/\tau(n-1)$

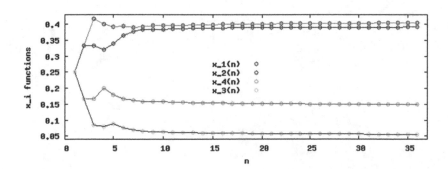

Fig. 7. Empirical monotonicity and convergence of $(x_i(n))$ (for $1 \leq i \leq 4$)

References

1. The On-Line Encyclopedia of Integer Sequences. http://oeis.org
2. Aleksandrowicz, G., Barequet, G.: Counting d-dimensional polycubes and nonrectangular planar polyominoes. Int. J. Comput. Geom. Appl. **19**, 215–229 (2009)
3. Barequet, G., Rote, G., Shalah, M.: An improved upper bound on the growth constant of polyiamonds (2019)
4. Barequet, G., Shalah, M., Zheng, Y.: An improved lower bound on the growth constant of polyiamonds. J. Comb. Optim. **37**, 424–438 (2019)
5. Barequet, G., Rote, G., Shalah, M.: $\lambda > 4$: an improved lower bound on the growth constant of polyominoes. Commun. ACM **59**, 88–95 (2016)
6. Broadbent, S.R., Hammersley, J.M.: Percolation processes: I. Crystals and mazes. Proc. Cambridge Philos. Soc. **53**, 629–641 (1957)
7. Conway, J.H., Burgiel, H., Goodman-Strass, C.: The Symmetries of Things. A.K. Peters/CRC Press, New York (2008)
8. Eden, M.: A two-dimensional growth process. In: Proceedings of the 4th Berkeley Symposium on Mathematical Statistics and Probability, Berkeley CA, vol. IV, pp. 223–239 (1961)
9. Grünbaum, B., Shephard, G.C.: Tilings and Patterns. W.H. Freeman, New York (1987)
10. Harary, F.: Unsolved problems in the enumeration of graphs. Math. Inst. Hung. Acad. Sci. **5**, 1–20 (1960)
11. Klarner, D.A.: Cell growth problems. Can. J. Math. **19**, 851–863 (1967)
12. Klarner, D.A., Rivest, R.L.: A procedure for improving the upper bound for the number of n-ominoes. Can. J. Math. **25**, 585–602 (1973)
13. Madras, N.: A pattern theorem for lattice clusters. Ann. Comb. **3**, 357–384 (1999)
14. Redelmeier, D.H.: Counting polyominoes: yet another attack. Discrete Math. **36**, 191–203 (1981)
15. Temperley, H.N.V.: Combinatorial problems suggested by the statistical mechanics of domains and of rubber-like molecules. Phys. Rev. **2**(103), 1–16 (1956)

An FPTAS for a General Class
of Parametric Optimization Problems

Cristina Bazgan[1], Arne Herzel[2]([✉]), Stefan Ruzika[2], Clemens Thielen[2],
and Daniel Vanderpooten[1]

[1] Université Paris-Dauphine, PSL Research University, CNRS, LAMSADE,
75016 Paris, France
{bazgan,vanderpooten}@lamsade.dauphine.fr
[2] Department of Mathematics, University of Kaiserslautern, Paul-Ehrlich-Str. 14,
67663 Kaiserslautern, Germany
{herzel,ruzika,thielen}@mathematik.uni-kl.de

Abstract. In a (linear) parametric optimization problem, the objective
value of each feasible solution is an affine function of a real-valued parameter and one is interested in computing a solution for each possible value
of the parameter. For many important parametric optimization problems including the parametric versions of the shortest path problem, the
assignment problem, and the minimum cost flow problem, however, the
piecewise linear function mapping the parameter to the optimal objective value of the corresponding non-parametric instance (the *optimal
value function*) can have super-polynomially many breakpoints (points
of slope change). This implies that any optimal algorithm for such a
problem must output a super-polynomial number of solutions.

We provide a (parametric) fully-polynomial time approximation
scheme for a general class of parametric optimization problems for which
(i) the parameter varies on the nonnegative real line, (ii) the non-
parametric problem is solvable in polynomial time, and (iii) the slopes
and intercepts of the value functions of the feasible solutions are nonnegative, integer values below a polynomial-time computable upper bound.
In particular, under mild assumptions, we obtain the first parametric
FPTAS for each of the specific problems mentioned above.

Keywords: Parametric optimization · Approximation scheme ·
Parametric assignment problem ·
Parametric minimum-cost flow problem ·
Parametric shortest path problem

This work was supported by the bilateral cooperation project "Approximation methods
for multiobjective optimization problems" funded by the German Academic Exchange
Service (DAAD, Project-ID 57388848) and by Campus France, PHC PROCOPE 2018
(Project no. 40407WF) as well as by the DFG grants TH 1852/4-1 and RU 1524/6-1.

D.-Z. Du et al. (Eds.): COCOON 2019, LNCS 11653, pp. 25–37, 2019.
https://doi.org/10.1007/978-3-030-26176-4_3

1   Introduction

In a linear parametric optimization problem, the objective function value of a feasible solution does not only depend on the solution itself but also on a parameter $\lambda \in \mathbb{R}$, where this dependence is given by an affine linear function of λ. The goal is to find an optimal solution for each possible parameter value, where, under some assumptions (e.g., if the set of feasible solutions is finite), an *optimal* solution can be given by a finite collection of intervals $(-\infty, \lambda_1], [\lambda_1, \lambda_2], \ldots, [\lambda_{K-1}, \lambda_K], [\lambda_K, +\infty)$ together with one feasible solution for each interval that is optimal for all values of λ within the corresponding interval.

The function mapping each parameter value $\lambda \in \mathbb{R}$ to the optimal objective value of the non-parametric problem induced by λ is called the *optimal value function* (or the *optimal cost curve*). The above structure of optimal solutions implies that the optimal value function is piecewise linear and concave in the case of a minimization problem (convex in case of a maximization problem) and its *breakpoints* (points of slope change) are exactly the points $\lambda_1, \ldots, \lambda_K$ (assuming that K was chosen as small as possible).

There is a large body of literature that considers linear parametric optimization problems in which the objective values of feasible solutions are affine-linear functions of a real-valued parameter. Prominent examples include the parametric shortest path problem [4,13,17,23], the parametric assignment problem [8], and the parametric minimum cost flow problem [3]. Moreover, parametric versions of general linear programs, mixed integer programs, and nonlinear programs (where the most general cases consider also non-affine dependence on the parameter as well as constraints depending on the parameter) are widely studied – see [16] for an extensive literature review.

The number of breakpoints is a natural measure for the complexity of a parametric optimization problem since it determines the number of different solutions that are needed in order to solve the parametric problem to optimality. Moreover, any instance of a parametric optimization problem with K breakpoints in the optimal value function can be solved by using a general method of Eisner and Severance [5], which requires to solve $\mathcal{O}(K)$ non-parametric problems for fixed values of the parameter.

Carstensen [3] shows that the number of breakpoints in the optimal value function of any parametric binary integer program becomes linear in the number of variables when the slopes and/or intercepts of the affine-linear functions are integers in $\{-M, \ldots, M\}$ for some constant $M \in \mathbb{N}$. In most parametric problems, however, the number of possible slopes and intercepts is exponential and/or the variables are not binary. While there exist some parametric optimization problems such as the parametric minimum spanning tree problem [6] or several special cases of the parametric maximum flow problem [2,7,15,22] for which the number of breakpoints is polynomial in the input size even without any additional assumptions, the optimal value function of most parametric optimization problems can have super-polynomially many breakpoints in the worst case – see, e.g., [4,19] for the parametric shortest path problem, [8] for the parametric assignment problem, and [20] for the parametric minimum cost flow problem. This, in particular, implies that there cannot exist any polynomial-time algorithm for these problems even if P = NP, which provides a strong motivation for the design of approximation algorithms.

So far, only very few approximation algorithms are known for parametric optimization problems. Approximation schemes for the parametric version of the 0-1 knapsack problem have been presented in [9,12], and an approximation for the variant of the 0-1 knapsack problem in which the weights of the items (instead of the profits) depend on the parameter has recently been provided in [10]. Moreover, an approximation scheme for a class of parametric discrete optimization problems on graphs, whose technique could potentially be generalized to further problems, has been proposed in [11].

Our Contribution. We provide a (parametric) fully-polynomial time approximation scheme (FPTAS) for a general class of parametric optimization problems. This means that, for any problem from this class and any given $\varepsilon > 0$, there exists an algorithm with running time polynomial in the input size and $1/\varepsilon$ that computes a partition of the set $\mathbb{R}_{\geq 0}$ of possible parameter values into polynomially many intervals together with a solution for each interval that $(1 + \varepsilon)$-approximates all feasible solutions for all values of λ within the interval.

Our FPTAS can be viewed as an approximate version of the well-known Eisner-Severance method for parametric optimization problems [5]. It applies to all parametric optimization problems for which the parameter varies on the nonnegative real line, the non-parametric problem is solvable in polynomial time, and the slopes and intercepts of the value functions of the feasible solutions are nonnegative, integer values below a polynomial-time computable upper bound. In particular, under mild assumptions, we obtain the first parametric FPTAS for the parametric versions of the shortest path problem, the assignment problem, and a general class of mixed integer linear programming problems over integral polytopes, which includes the minimum cost flow problem as a special case. As we discuss when presenting the applications of our method in Sect. 4, the number of breakpoints can be super-polynomial for each of these parametric problems even under our assumptions, which implies that the problems do not admit any polynomial-time exact algorithms.

2 Preliminaries

In the following, we consider a general parametric minimization or maximization problem Π of the following form:

$$\min / \max f_\lambda(x) := a(x) + \lambda \cdot b(x)$$
$$\text{s. t. } x \in X \tag{1}$$

We assume that the functions $a, b : X \to \mathbb{N}_0$ defining the intercepts and slopes of the value functions, respectively, take only nonnegative integer values and are polynomial-time computable. Moreover, we assume that we can compute a rational upper bound UB such that $a(x), b(x) \leq$ UB for all $x \in X$ in polynomial time. In particular, this implies that UB is of polynomial encoding length.[1]

For any fixed value $\lambda \geq 0$, we let Π_λ denote the non-parametric problem obtained from Π by fixing the parameter value to λ. We assume that, for each $\lambda \geq 0$, this non-parametric problem Π_λ can be solved exactly in polynomial time by an algorithm ALG. The running time of ALG will be denoted by T_{ALG}, where we assume that this running time is at least as large as the time needed in order to compute the objective value $f_\lambda(x) = a(x) + \lambda \cdot b(x)$ of any feasible solution x of Π_λ.[2]

The above assumptions directly imply that the optimal value function mapping $\lambda \in \mathbb{R}_{\geq 0}$ to the optimal objective value of the non-parametric problem Π_λ is piecewise linear, increasing, and concave (for minimization problems) or convex (for maximization problems): Since there are at most UB $+ 1$ possible integer values for each of $a(x)$ and $b(x)$, the function mapping λ to the objective value $f_\lambda(x) = a(x) + \lambda \cdot b(x)$ of a given feasible solution $x \in X$ is one of at most $(\mathrm{UB} + 1)^2$ many different affine functions. Consequently, the optimal value function given as $\lambda \mapsto \min / \max\{f_\lambda(x) : x \in X\}$ is the minimum/maximum of finitely many affine functions. The finitely many values of λ at which the slope of the optimal value function (or, equivalently, the set of optimal solutions of Π_λ) changes are called *breakpoints* of the optimal value function.

Even though all our results apply to minimization as well as maximization problems, we focus on minimization problems in the rest of the paper in order to simplify the exposition. All our arguments can be straightforwardly adapted to maximization problems.

Definition 1. *For $\alpha \geq 1$, an α-approximation $(I^1, x^1), \ldots, (I^k, x^k)$ for an instance of a parametric optimization problem Π consists of a cover of $\mathbb{R}_{\geq 0}$ by finitely many intervals $I^1, \ldots, I^k$ together with corresponding feasible solutions $x^1, \ldots, x^k$ such that, for each $j \in \{1, \ldots, k\}$, the solution x^j is an α-approximation for the corresponding instance of the non-parametric problem Π_λ for all values $\lambda \in I^j$, i.e.,*

$$f_\lambda(x^j) \leq \alpha \cdot f_\lambda(x) \text{ for all } x \in X \text{ and all } \lambda \in I^j.$$

An algorithm A that computes an α-approximation in polynomial time for every instance of Π is called an α-approximation algorithm for Π.

A polynomial-time approximation scheme (PTAS) for Π is a family $(A_\varepsilon)_{\varepsilon > 0}$ of algorithms such that, for every $\varepsilon > 0$, algorithm A_ε is a $(1 + \varepsilon)$-approximation

---

[1] Note that, the *numerical value* of UB can still be *exponential* in the input size of the problem, so there can still exist exponentially many different slopes $b(x)$ and intercepts $a(x)$.

[2] This technical assumption – which is satisfied for most relevant algorithms – is made in order to be able to express the running time of our algorithm in terms of T_{ALG}. If the assumption is removed, our results still hold when replacing T_{ALG} in the running time of our algorithm by the maximum of T_{ALG} and the time needed for computing any value $f_\lambda(x)$.

algorithm for Π. A PTAS $(A_\varepsilon)_{\varepsilon>0}$ for Π is called a fully polynomial-time approximation scheme (FPTAS) *if the running time of A_ε is additionally polynomial in $1/\varepsilon$.*

3 An FPTAS

We now present our FPTAS for the general parametric optimization problem (1). To this end, we first describe the general functioning of the algorithm before formally stating and proving several auxiliary results needed for proving its correctness and analyzing its running time.

The algorithm, which is formally stated in Algorithm 1, starts by computing an upper bound UB on the values of $a(\cdot)$ and $b(\cdot)$, which is possible in polynomial time by our assumptions on the problem. It then starts with the initial parameter interval $[\lambda_{\min}, \lambda_{\max}]$, where $\lambda_{\min} := \frac{1}{\text{UB}+1}$ is chosen such that an optimal solution $x^{\min}$ of the non-parametric problem $\Pi_{\lambda_{\min}}$ is optimal for Π_λ for all $\lambda \in [0, \lambda_{\min}]$ and $\lambda_{\max} := \text{UB} + 1$ is chosen such that an optimal solution $x^{\max}$ of the non-parametric problem $\Pi_{\lambda_{\max}}$ is optimal for Π_λ for all $\lambda \in [\lambda_{\max}, +\infty)$ (see Lemma 3).

The algorithm maintains a queue Q whose elements $([\lambda_\ell, \lambda_r], x^\ell, x^r)$ consist of a subinterval $[\lambda_\ell, \lambda_r]$ of the interval $[\lambda_{\min}, \lambda_{\max}]$ and optimal solutions x^ℓ, x^r of the respective non-parametric problems $\Pi_{\lambda_\ell}, \Pi_{\lambda_r}$ at the interval boundaries. The queue is initialized as $Q = \{([\lambda_{\min}, \lambda_{\max}], x^{\min}, x^{\max})\}$, where $x^{\min}, x^{\max}$ are optimal for $\Pi_{\lambda_{\min}}, \Pi_{\lambda_{\max}}$, respectively.

Afterwards, in each iteration, the algorithm extracts an element $([\lambda_\ell, \lambda_r], x^\ell, x^r)$ from the queue and checks whether one of the two solutions x^ℓ, x^r obtained at the boundaries of the parameter interval $[\lambda_\ell, \lambda_r]$ is a $(1 + \varepsilon)$-approximate solution also at the other boundary of the interval. In this case, Lemma 1 below implies that this boundary solution is a $(1 + \varepsilon)$-approximate solution within the whole interval $[\lambda_\ell, \lambda_r]$ and the pair consisting of the interval $[\lambda_\ell, \lambda_r]$ and this $(1 + \varepsilon)$-approximate solution is added to the solution set S. Otherwise, $[\lambda_\ell, \lambda_r]$ is bisected into the two subintervals $[\lambda_\ell, \lambda_m]$ and $[\lambda_m, \lambda_r]$, where $\lambda_m := \sqrt{\lambda_\ell \cdot \lambda_r}$ is the geometric mean of λ_ℓ and λ_r. This means that an optimal solution x^m of Π_{λ_m} is computed and the two triples $([\lambda_\ell, \lambda_m], x^\ell, x^m)$ and $([\lambda_m, \lambda_r], x^m, x^r)$ are added to the queue in order to be explored.

This iterative subdivision of the initial parameter interval $[\lambda_{\min}, \lambda_{\max}]$ can be viewed as creating a binary tree: the root corresponds to the initialization, in which the two non-parametric problems $\Pi_{\lambda_{\min}}$ and $\Pi_{\lambda_{\max}}$ are solved. Each other internal node of the tree corresponds to an interval $[\lambda_\ell, \lambda_r]$ that is further subdivided into $[\lambda_\ell, \lambda_m]$ and $[\lambda_m, \lambda_r]$, which requires the solution of one non-parametric problem Π_{λ_m}. In order to bound the total number of non-parametric problems solved within the algorithm, it is, thus, sufficient to upper bound the number of nodes in this binary tree.

We now prove the auxiliary results mentioned in the algorithm description above. The first lemma shows that a solution that is optimal at one boundary of a parameter interval and simultaneously α-approximate at the other interval boundary must be α-approximate within the whole interval.

Algorithm 1. An FPTAS for parametric optimization problems

---

input : an instance of a parametric optimization problem Π as in (1), $\varepsilon > 0$, an exact
algorithm ALG for the non-parametric version of Π
output: a $(1 + \varepsilon)$-approximation for Π

1 Compute an upper bound UB such that $a(x), b(x) \leq \text{UB}$ for all $x \in X$
2 $\lambda_{\min} \leftarrow \frac{1}{\text{UB}+1}$; $\lambda_{\max} \leftarrow \text{UB} + 1$
3 $x^{\min} \leftarrow \text{ALG}(\lambda_{\min})$; $x^{\max} \leftarrow \text{ALG}(\lambda_{\max})$
4 $Q \leftarrow \{([\lambda_{\min}, \lambda_{\max}], x^{\min}, x^{\max})\}$ /* queue of intervals still to be considered */
5 $S \leftarrow \{([0, \lambda_{\min}), x^{\min}), [\lambda_{\max}, +\infty), x^{\max})\}$ /* solution set   */
6 **while** $Q \neq \emptyset$ **do**
7 $\quad$ Extract some element $([\lambda_\ell, \lambda_r], x^\ell, x^r)$ from Q
8 $\quad$ **if** $f_{\lambda_\ell}(x^r) \leq (1 + \varepsilon) \cdot f_{\lambda_\ell}(x^\ell)$ **then**
9 $\quad\quad \mid \quad S \leftarrow S \cup \{([\lambda_\ell, \lambda_r], x^r)\}$
10 $\quad$ **else if** $f_{\lambda_r}(x^\ell) \leq (1 + \varepsilon) \cdot f_{\lambda_r}(x^r)$ **then**
11 $\quad\quad \mid \quad S \leftarrow S \cup \{([\lambda_\ell, \lambda_r], x^\ell)\}$
12 $\quad$ **else**
13 $\quad\quad \mid \quad \lambda_m \leftarrow \sqrt{\lambda_\ell \cdot \lambda_r}$
14 $\quad\quad \mid \quad x^m \leftarrow \text{ALG}(\lambda_m)$
15 $\quad\quad \mid \quad Q \leftarrow Q \cup \{([\lambda_\ell, \lambda_m], x^\ell, x^m), ([\lambda_m, \lambda_r], x^m, x^r)\}$

16 **return** S

Lemma 1. *Let $[\underline{\lambda}, \bar{\lambda}] \subset \mathbb{R}_{\geq 0}$ be an interval, and let $\underline{x}, \bar{x} \in X$ be optimal solutions of $\Pi_{\underline{\lambda}}$ and $\Pi_{\bar{\lambda}}$, respectively. Then, for any $\alpha \geq 1$:*

(1) If $f_{\bar{\lambda}}(\underline{x}) \leq \alpha \cdot f_{\bar{\lambda}}(\bar{x})$, then $f_\lambda(\underline{x}) \leq \alpha \cdot f_\lambda(x)$ for all $x \in X$ and all $\lambda \in [\underline{\lambda}, \bar{\lambda}]$.
(2) If $f_{\underline{\lambda}}(\bar{x}) \leq \alpha \cdot f_{\underline{\lambda}}(\underline{x})$, then $f_\lambda(\bar{x}) \leq \alpha \cdot f_\lambda(x)$ for all $x \in X$ and all $\lambda \in [\underline{\lambda}, \bar{\lambda}]$.

Proof. We only prove (1) – the proof of (2) is analogous.

Let $f_{\bar{\lambda}}(\underline{x}) \leq \alpha \cdot f_{\bar{\lambda}}(\bar{x})$, i.e., $a(\underline{x}) + \bar{\lambda} b(\underline{x}) \leq \alpha \cdot (a(\bar{x}) + \bar{\lambda} b(\bar{x}))$. Fix some $\lambda \in [\underline{\lambda}, \bar{\lambda}]$. Then $\lambda = \gamma \underline{\lambda} + (1 - \gamma) \bar{\lambda}$ for some $\gamma \in [0, 1]$ and, for each $x \in X$, we have:

$$
\begin{aligned}
f_\lambda(\underline{x}) &= a(\underline{x}) + \lambda b(\underline{x}) \\
&= a(\underline{x}) + (\gamma \underline{\lambda} + (1 - \gamma) \bar{\lambda}) \cdot b(\underline{x}) \\
&= \gamma \cdot [a(\underline{x}) + \underline{\lambda} b(\underline{x})] + (1 - \gamma) \cdot [a(\underline{x}) + \bar{\lambda} b(\underline{x})] \\
&\leq \gamma \cdot [a(\underline{x}) + \underline{\lambda} b(\underline{x})] + (1 - \gamma) \cdot \alpha \cdot [a(\bar{x}) + \bar{\lambda} b(\bar{x})] \\
&\leq \gamma \cdot [a(x) + \underline{\lambda} b(x)] + (1 - \gamma) \cdot \alpha \cdot [a(x) + \bar{\lambda} b(x)] \\
&\leq \alpha \cdot [a(x) + (\gamma \underline{\lambda} + (1 - \gamma) \bar{\lambda}) b(x)] \\
&= \alpha \cdot f_\lambda(x)
\end{aligned}
$$

Here, the first inequality follows by the assumption of (1), and the second inequality follows since $\underline{x}, \bar{x}$ are optimal solutions of $\Pi_{\underline{\lambda}}$ and $\Pi_{\bar{\lambda}}$, respectively. $\square$

The following lemma shows that a solution that is optimal for some value λ^* of the parameter is always $(1 + \varepsilon)$-approximate in a neighborhood of λ^*.

Lemma 2. For $\lambda^* \in [0, \infty)$ and $\varepsilon > 0$, let $\underline{\lambda} = \frac{1}{1+\varepsilon} \cdot \lambda^*$ and $\bar{\lambda} = (1+\varepsilon) \cdot \lambda^*$. Also, let x^* be an optimal solution for Π_{λ^*}. Then $f_\lambda(x^*) \leq (1+\varepsilon) \cdot f_\lambda(x)$ for all $x \in X$ and all $\lambda \in [\underline{\lambda}, \bar{\lambda}]$.

Proof. First, note that x^* is $(1 + \varepsilon)$-approximate for $\Pi_{\bar{\lambda}}$: Let $\bar{x}$ be an optimal solution for $\Pi_{\bar{\lambda}}$. Then we have

$$
\begin{aligned}
f_{\bar{\lambda}}(x^*) &= a(x^*) + \bar{\lambda} \cdot b(x^*) \\
&= a(x^*) + (1+\varepsilon) \cdot \lambda^* \cdot b(x^*) \\
&\leq (1+\varepsilon) \cdot (a(x^*) + \lambda^* \cdot b(x^*)) \\
&= (1+\varepsilon) \cdot f_{\lambda^*}(x^*) \\
&\leq (1+\varepsilon) \cdot f_{\bar{\lambda}}(\bar{x}),
\end{aligned}
$$

where the last inequality is due to the monotonicity of the optimal cost curve. Moreover, x^* is $(1 + \varepsilon)$-approximate for $\Pi_{\underline{\lambda}}$: Let $\underline{x}$ be an optimal solution for $\Pi_{\underline{\lambda}}$ and let x^0 be an optimal solution for Π_0. Then, since $\underline{\lambda} = \frac{1}{1+\varepsilon} \cdot \lambda^* = \frac{1}{1+\varepsilon} \cdot \lambda^* + \frac{\varepsilon}{1+\varepsilon} \cdot 0$, we have

$$
\begin{aligned}
f_{\underline{\lambda}}(x^*) &\leq f_{\lambda^*}(x^*) \\
&\leq f_{\lambda^*}(x^*) + \varepsilon \cdot f_0(x^0) \\
&= (1+\varepsilon) \cdot \left(\frac{1}{1+\varepsilon} \cdot f_{\lambda^*}(x^*) + \frac{\varepsilon}{1+\varepsilon} \cdot f_0(x^0) \right) \\
&\leq (1+\varepsilon) \cdot f_{\underline{\lambda}}(\underline{x}),
\end{aligned}
$$

where the first inequality is due to the monotonicity of $\lambda \mapsto f_\lambda(x^*)$ and the last inequality is due to the concavity of the optimal cost curve. Now, the claim follows from applying Lemma 1. $\square$

The next result justifies our choice of the initial parameter interval $[\lambda_{\min}, \lambda_{\max}]$.

Lemma 3. Let $\lambda_{\min} := \frac{1}{\mathrm{UB}+1}$ and $\lambda_{\max} := \mathrm{UB} + 1$ as in Algorithm 1. Then the solution $x^{\min} = \mathrm{ALG}(\lambda_{\min})$ is optimal for Π_λ for all $\lambda \in [0, \lambda_{\min}]$ and the solution $x^{\max} = \mathrm{ALG}(\lambda_{\max})$ is optimal for Π_λ for all $\lambda \in [\lambda_{\max}, +\infty)$.

Proof. Let $\lambda \in [\lambda_{\max}, +\infty)$ and $x \in X$ be an arbitrary solution. Since $x^{\max}$ is optimal for $\Pi_{\lambda_{\max}}$, we have $f_{\lambda_{\max}}(x^{\max}) \leq f_{\lambda_{\max}}(x)$, i.e., $a(x^{\max}) + (\mathrm{UB} + 1) \cdot b(x^{\max}) \leq a(x) + (\mathrm{UB} + 1) \cdot b(x)$. Reordering terms, and using that $0 \leq a(x), a(x^{\max}) \leq \mathrm{UB}$, we obtain that

$$
b(x^{\max}) - b(x) \leq \frac{a(x) - a(x^{\max})}{\mathrm{UB} + 1} \leq \frac{\mathrm{UB}}{\mathrm{UB} + 1} < 1.
$$

Since $b(x^{\max}) - b(x) \in \mathbb{Z}$ by integrality of the values of b, this implies that $b(x^{\max}) - b(x) \leq 0$, i.e., $b(x^{\max}) \leq b(x)$. Consequently, using that $x^{\max}$ is optimal for $\Pi_{\lambda_{\max}}$, we obtain

$$f_\lambda(x^{\max}) = \underbrace{f_{\lambda_{\max}}(x^{\max})}_{\leq f_{\lambda_{\max}}(x)} + \underbrace{(\lambda - \lambda_{\max})}_{\geq 0} \cdot \underbrace{b(x^{\max})}_{\leq b(x)}$$
$$\leq f_{\lambda_{\max}}(x) + (\lambda - \lambda_{\max}) \cdot b(x)$$
$$= f_\lambda(x).$$

Since $x \in X$ was arbitrary, this proves the optimality of $x^{\max}$ for Π_λ.

Now, consider the interval $[0, \lambda_{\min}]$. We know that $x^{\min}$ is optimal for $\Pi_{\lambda_{\min}}$. Analogously to the above arguments, we can show that $a(x^{\min}) \leq a(x)$, i.e., $f_0(x^{\min}) \leq f_0(x)$, for all $x \in X$. Thus, for any $\lambda \in [0, \lambda_{\min}]$, we obtain $f_\lambda(x^{\min}) \leq f_\lambda(x)$ by applying Lemma 1.   □

We are now ready to show that Algorithm 1 yields an FPTAS for the parametric optimization problem (1):

Theorem 1. *Algorithm 1 returns a $(1 + \varepsilon)$-approximation of the parametric problem in time*

$$\mathcal{O}\left(T_{\mathrm{UB}} + T_{\mathrm{ALG}} \cdot \frac{1}{\varepsilon} \cdot \log \mathrm{UB}\right),$$

where T_{UB} denotes the time needed for computing the upper bound UB and T_{ALG} denotes the running time of ALG.

Proof. In order to prove the approximation guarantee, we first note that, at the beginning of each iteration of the while loop starting in line 6, the intervals corresponding to the first components of the elements of $Q \cup S$ form a cover of $\mathbb{R}_{\geq 0}$. Since $Q = \emptyset$ at termination, the approximation guarantee follows if we show that, for each element $(\hat{I}, \hat{x})$ in the final solution set S returned by the algorithm, we have $f_\lambda(\hat{x}) \leq (1 + \varepsilon) \cdot f_\lambda(x)$ for all $x \in X$ and all $\lambda \in \hat{I}$.

Consider an arbitrary element $(\hat{I}, \hat{x}) \in S$. If $(\hat{I}, \hat{x}) = ([0, \lambda_{\min}], x^{\min})$ or $(\hat{I}, \hat{x}) = ([\lambda_{\max}, +\infty), x^{\max})$, then the solution $\hat{x}$ is optimal for Π_λ for all $\lambda \in \hat{I}$ by Lemma 3.

Otherwise, $\hat{I} = [\hat{\lambda}_\ell, \hat{\lambda}_r]$ and $(\hat{I}, \hat{x})$ was added to S within the while loop starting in line 6, i.e., either in line 9 or in line 11. In this case, the solution $\hat{x}$ must be optimal for $\Pi_{\hat{\lambda}_\ell}$ or for $\Pi_{\hat{\lambda}_r}$: Whenever an element $([\lambda_\ell, \lambda_r], x^\ell, x^r)$ is added to the queue Q, the solution x^ℓ is optimal for Π_{λ_ℓ} and the solution x^r is optimal for Π_{λ_r}. Consequently, whenever $([\lambda_\ell, \lambda_r], x^\ell)$ or $([\lambda_\ell, \lambda_r], x^r)$ is added to S, the contained solution is optimal for Π_{λ_ℓ} or for Π_{λ_r}, respectively. Thus, if $(\hat{I}, \hat{x}) = ([\hat{\lambda}_\ell, \hat{\lambda}_r], \hat{x})$ was added to S in line 9 or 11, the solution $\hat{x}$ satisfies $f_\lambda(\hat{x}) \leq (1 + \varepsilon) \cdot f_\lambda(x)$ for all $x \in X$ and all $\lambda \in \hat{I}$ by the if-statement in the previous line of the algorithm and Lemma 1.

We prove now the bound on the running time. Starting with the initial interval $[0, \lambda_{\max}]$, the algorithm iteratively extracts an element $([\lambda_\ell, \lambda_r], x^\ell, x^r)$ from the queue Q and checks whether the interval $[\lambda_\ell, \lambda_r]$ has to be further bisected into two subintervals $[\lambda_\ell, \lambda_m]$ and $[\lambda_m, \lambda_r]$ that need to be further explored. This process of bisecting can be viewed as creating a full binary tree: the root corresponds to the initial element $([\lambda_{\min}, \lambda_{\max}], x^{\min}, x^{\max})$, for which the two non-parametric problems $\Pi_{\lambda_{\min}}$ and $\Pi_{\lambda_{\max}}$ are solved. Each other node corresponds to an element $([\lambda_\ell, \lambda_r], x^\ell, x^r)$ that is extracted from Q in some iteration of the while loop, where x^ℓ and x^r are optimal for Π_{λ_ℓ} and Π_{λ_r}, respectively. If the interval $[\lambda_\ell, \lambda_r]$ is not bisected into $[\lambda_\ell, \lambda_m]$ and $[\lambda_m, \lambda_r]$, the corresponding node is a leaf of the tree, for which no non-parametric problem is solved. Otherwise, the node corresponding to $([\lambda_\ell, \lambda_r], x^\ell, x^r)$ is an internal node, for which the optimal solution x^m of the non-parametric problem Π_{λ_m} with $\lambda_m = \sqrt{\lambda_\ell \cdot \lambda_r}$ is computed, and whose two children correspond to $([\lambda_\ell, \lambda_m], x^\ell, x^m)$ and $([\lambda_m, \lambda_r], x^m, x^r)$. In order to bound the total number of non-parametric problems solved within the algorithm, it is, thus, sufficient to bound the number of (internal) nodes in this full binary tree. This is done by bounding the height of the tree.

In order to bound the height of the tree, we observe that due to Lemma 2, the algorithm never bisects an interval $[\lambda_\ell, \lambda_r]$, for which $\lambda_\ell \leq (1 + \varepsilon) \cdot \lambda_r$, i.e., for which the ratio between λ_r and λ_ℓ satisfies $\frac{\lambda_r}{\lambda_\ell} \leq 1 + \varepsilon$. Also note that $\lambda_m = \sqrt{\lambda_\ell \cdot \lambda_r}$ and, thus, for any subdivision $\{[\lambda_\ell, \lambda_m], [\lambda_m, \lambda_r]\}$ of $[\lambda_\ell, \lambda_r]$ in the algorithm, we have

$$\frac{\lambda_m}{\lambda_\ell} = \frac{\lambda_r}{\lambda_m} = \sqrt{\frac{\lambda_r}{\lambda_\ell}} = \left(\frac{\lambda_r}{\lambda_\ell}\right)^{\frac{1}{2}}.$$

Hence, for both intervals $[\lambda_\ell, \lambda_m]$ and $[\lambda_m, \lambda_r]$ in the subdivision, the ratio between the upper and lower boundary equals the square root of the corresponding ratio of the previous interval $[\lambda_\ell, \lambda_r]$.

This implies that an interval $[\lambda_{\ell,k}, \lambda_{r,k}]$ resulting from k consecutive subdivisions of an interval $[\lambda_\ell, \lambda_r]$ satisfies

$$\frac{\lambda_{\ell,k}}{\lambda_{r,k}} = \left(\frac{\lambda_\ell}{\lambda_r}\right)^{\frac{1}{2^k}}.$$

Thus, starting from the initial interval $[\lambda_{\min}, \lambda_{\max}] = [\frac{1}{UB+1}, UB + 1]$, for which the ratio is $\frac{\lambda_{\max}}{\lambda_{\min}} = (UB + 1)^2$, there can be at most $\left\lceil \log_2\left(\frac{\log((UB+1)^2)}{\log(1+\varepsilon)}\right)\right\rceil$ consecutive subdivisions until the ratio between the interval boundaries becomes less or equal to $1 + \varepsilon$, which upper bounds the height of the tree by $\left\lceil \log_2\left(\frac{\log((UB+1)^2)}{\log(1+\varepsilon)}\right)\right\rceil + 1$.

Since any binary tree of height h has at most $2^{h-1} - 1$ internal nodes, we obtain an upper bound of

$$2^{\left\lceil \log_2\left(\frac{\log((UB+1)^2)}{\log(1+\varepsilon)}\right)\right\rceil} - 1 \in \mathcal{O}\left(\frac{\log UB}{\varepsilon}\right)$$

for the number of internal nodes of the tree generated by the algorithm.

Adding the time T_{UB} needed for computing the upper bound UB at the beginning of the algorithm, this proves the claimed bound on the running time.

□

4   Applications

In this section, we show that our general result applies to the parametric versions of many well-known, classical optimization problems including the parametric shortest path problem, the parametric assignment problem, and a general class of parametric mixed integer linear programs that includes the parametric minimum cost flow problem. As will be discussed below, for each of these parametric problems, the number of breakpoints in the optimal value function can be super-polynomial, which implies that solving the parametric problem exactly requires the generation of a super-polynomial number of solutions.

Parametric Shortest Path Problem. In the single-pair version of the parametric shortest path problem, we are given a directed graph $G = (V, R)$ together with a source node $s \in V$ and a destination node $t \in V$, where $s \neq t$. Each arc $r \in R$ has a parametric length of the form $a_r + \lambda \cdot b_r$, where $a_r, b_r \in \mathbb{N}_0$ are nonnegative integers. The goal is to compute an s-t-path P_λ of minimum total length $\sum_{r \in P_\lambda} (a_r + \lambda \cdot b_r)$ for each $\lambda \geq 0$.

Since the arc lengths $a_r + \lambda \cdot b_r$ are nonnegative for each $\lambda \geq 0$, one can restrict to *simple* s-t-paths as feasible solutions, and an upper bound UB as required in Algorithm 1 is given by summing up the $n - 1$ largest values a_r and summing up the $n - 1$ largest values b_r and taking the maximum of these two sums, which can easily be computed in polynomial time. The non-parametric problem Π_λ can be solved in polynomial time $\mathcal{O}(m + n \log n)$ for any fixed $\lambda \geq 0$ by Dijkstra's algorithm, where $n = |V|$ and $m = |R|$ (see, e.g., [21]). Hence, Theorem 1 yields an FPTAS with running time $\mathcal{O}\left(\frac{1}{\varepsilon} \cdot (m + n \log n) \cdot \log nC\right)$, where C denotes the maximum among all values a_r, b_r.

On the other hand, the number of breakpoints in the optimal value function is at least $n^{\Omega(\log n)}$ in the worst case even under our assumptions of nonnegative, integer values a_r, b_r and for $\lambda \in \mathbb{R}_{\geq 0}$ [4,19].

Parametric Assignment Problem. In the parametric assignment problem, we are given a bipartite, undirected graph $G = (U, V, E)$ with $|U| = |V| = n$ and $|E| = m$. Each edge $e \in E$ has a parametric weight of the form $a_e + \lambda \cdot b_e$, where $a_e, b_e \in \mathbb{N}_0$ are nonnegative integers. The goal is to compute an assignment A_λ of minimum total weight $\sum_{e \in A_\lambda} (a_e + \lambda \cdot b_e)$ for each $\lambda \geq 0$.

Similar to the parametric shortest path problem, an upper bound UB as required in Algorithm 1 is given by summing up the n largest values a_r and summing up the n largest values b_r and taking the maximum of these two sums. The non-parametric problem Π_λ can be solved in polynomial time $\mathcal{O}(n^3)$ for any fixed value $\lambda \geq 0$ (see, e.g., [21]). Hence, Theorem 1 yields an FPTAS with running time $\mathcal{O}\left(\frac{1}{\varepsilon} \cdot n^3 \cdot \log nC\right)$, where C denotes the maximum among all values a_e, b_e.

On the other hand, applying the well-known transformation from the shortest s-t-path problem to the assignment problem (see, e.g., [14]) to the instances of the shortest s-t-path problem with super-polynomially many breakpoints presented in [4,19] shows that the number of breakpoints in the parametric assignment problem can be super-polynomial as well (see also [8]).

Parametric MIPs over Integral Polytopes. A very general class of problems our results can be applied to are parametric mixed integer linear programs (parametric MIPs) with nonnegative, integer objective function coefficients whose feasible set is of the form $P \cap (\mathbb{Z}^p \times \mathbb{R}^{n-p})$, where $P \subseteq \mathbb{R}^n_{\geq 0}$ is an integral polytope. More formally, consider a parametric MIP of the form

$$\min / \max \ (a + \lambda b)^T x$$
$$\text{s.t.} \quad Ax = d$$
$$Bx \leq e$$
$$x \geq 0$$
$$x \in \mathbb{Z}^p \times \mathbb{R}^{n-p}$$

where A, B are rational matrices with n rows, d, e are rational vectors of the appropriate length, and $a, b \in \mathbb{N}^n_0$ are nonnegative, integer vectors. We assume that the polyhedron $P := \{x \in \mathbb{R}^n : Ax = d, \ Bx \leq e, \ x \geq 0\} \subseteq \mathbb{R}^n$ is an integral polytope, i.e., it is bounded and each of its (finitely many) extreme points is an integral point.

Since, for each $\lambda \geq 0$, there exists an extreme point of P that is optimal for the non-parametric problem Π_λ, one can restrict to the extreme points when solving the problem. Since $\bar{x} \in \mathbb{N}^n_0$ for each extreme point $\bar{x}$ of P and since $a, b \in \mathbb{N}^n_0$, the values $a(\bar{x}) = a^T \bar{x}$ and $b(\bar{x}) = b^T \bar{x}$ are nonnegative integers. In order to solve the non-parametric problem Π_λ for any fixed value $\lambda \geq 0$, we can simply solve the linear programming relaxation $\min / \max\{(a + \lambda b)^T x : x \in P\}$ in polynomial time. This yields an optimal extreme point of P, which is integral by our assumptions. Similarly, an upper bound UB as required in Algorithm 1 can be computed in polynomial time by solving the two linear programs $\max\{a^T x : x \in P\}$ and $\max\{b^T x : x \in P\}$ and taking the maximum of the two resulting (integral) optimal objective values.

While Theorem 1 yields an FPTAS for any parametric MIP as above, it is well known that the number of breakpoints in the optimal value function can be exponential in the number n of variables [3,18].

An important parametric optimization problem that can be viewed as a special case of a parametric MIP as above is the parametric minimum cost flow problem, in which we are given a directed graph $G = (V, R)$ together with a source node $s \in V$ and a destination node $t \in V$, where $s \neq t$, and an integral desired flow value $F \in \mathbb{N}_0$. Each arc $r \in R$ has an integral capacity $u_r \in \mathbb{N}_0$ and a parametric cost of the form $a_r + \lambda \cdot b_r$, where $a_r, b_r \in \mathbb{N}_0$ are nonnegative integers. The goal is to compute a feasible s-t-flow x with flow value F of minimum total cost $\sum_{r \in R}(a_r + \lambda \cdot b_r) \cdot x_r$ for each $\lambda \geq 0$. Here, a large variety of (strongly) polynomial algorithms exist for the non-parametric problem, see, e.g., [1]. An upper bound UB can either be obtained by solving two linear programs as above, or by taking the maximum of $\sum_{r \in R} a_r \cdot u_r$ and $\sum_{r \in R} b_r \cdot u_r$. Using the latter and applying the enhanced capacity scaling algorithm to solve the non-parametric problem, which runs in $\mathcal{O}((m \cdot \log n)(m + n \cdot \log n))$ time on a graph with n nodes and m arcs [1], Theorem 1 yields an FPTAS with running time $\mathcal{O}\left(\frac{1}{\varepsilon} \cdot (m \cdot \log n)(m + n \cdot \log n) \cdot \log mCU\right)$, where C denotes the maximum among all values a_r, b_r, and $U := \max_{r \in R} u_r$.

On the other hand, the optimal value function can have $\Omega(2^n)$ breakpoints even under our assumptions of nonnegative, integer values a_r, b_r and for $\lambda \in \mathbb{R}_{\geq 0}$ [20].

Acknowledgments. We thank Sven O. Krumke for pointing out a possible improvement of the running time of a first version of our algorithm.

References

1. Ahuja, R.K., Magnanti, T.L., Orlin, J.B.: Network Flows: Theory, Algorithms and Applications. Prentice Hall, Englewood Cliffs (1993)
2. Arai, T., Ueno, S., Kajitani, Y.: Generalization of a theorem on the parametric maximum flow problem. Discrete Appl. Math. **41**(1), 69–74 (1993)
3. Carstensen, P.J.: Complexity of some parametric integer and network programming problems. Math. Program. **26**(1), 64–75 (1983)
4. Carstensen, P.J.: The complexity of some problems in parametric, linear, and combinatorial programming. Ph.D. thesis, University of Michigan (1983)
5. Eisner, M.J., Severance, D.G.: Mathematical techniques for efficient record segmentation in large shared databases. J. ACM **23**(4), 619–635 (1976)
6. Fernández-Baca, D., Slutzki, G., Eppstein, D.: Using sparsification for parametric minimum spanning tree problems. In: Karlsson, R., Lingas, A. (eds.) SWAT 1996. LNCS, vol. 1097, pp. 149–160. Springer, Heidelberg (1996). https://doi.org/10.1007/3-540-61422-2_128
7. Gallo, G., Grigoriadis, M.D., Tarjan, R.E.: A fast parametric maximum flow algorithm and applications. SIAM J. Comput. **18**(1), 30–55 (1989)
8. Gassner, E., Klinz, B.: A fast parametric assignment algorithm with applications in max-algebra. Networks **55**(2), 61–77 (2010)
9. Giudici, A., Halffmann, P., Ruzika, S., Thielen, C.: Approximation schemes for the parametric knapsack problem. Inf. Process. Lett. **120**, 11–15 (2017)
10. Halman, N., Holzhauser, M., Krumke, S.O.: An FPTAS for the knapsack problem with parametric weights. Oper. Res. Lett. **46**(5), 487–491 (2018)
11. Hertrich, C.: A parametric view on robust graph problems. Bachelor's thesis, University of Kaiserslautern, August 2016
12. Holzhauser, M., Krumke, S.O.: An FPTAS for the parametric knapsack problem. Inf. Process. Lett. **126**, 43–47 (2017)
13. Karp, R.M., Orlin, J.B.: Parametric shortest path algorithms with an application to cyclic staffing. Discrete Appl. Math. **3**(1), 37–45 (1981)
14. Lawler, E.: Combinatorial Optimization: Networks and Matroids. Dover Publications, New York (2001)
15. McCormick, S.T.: Fast algorithms for parametric scheduling come from extensions to parametric maximum flow. Oper. Res. **47**(5), 744–756 (1999)

16. Mitsos, A., Barton, P.I.: Parametric mixed-integer 0–1 linear programming: the general case for a single parameter. Eur. J. Oper. Res. **194**(3), 663–686 (2009)
17. Mulmuley, K., Shah, P.: A lower bound for the shortest path problem. J. Comput. Syst. Sci. **63**(2), 253–267 (2001)
18. Murty, K.: Computational complexity of parametric linear programming. Math. Program. **19**(1), 213–219 (1980)
19. Nikolova, E., Kelner, J.A., Brand, M., Mitzenmacher, M.: Stochastic shortest paths via quasi-convex maximization. In: Azar, Y., Erlebach, T. (eds.) ESA 2006. LNCS, vol. 4168, pp. 552–563. Springer, Heidelberg (2006). https://doi.org/10.1007/11841036_50
20. Ruhe, G.: Complexity results for multicriterial and parametric network flows using a pathological graph of Zadeh. Zeitschrift für Oper. Res. **32**(1), 9–27 (1988)
21. Schrijver, A.: Combinatorial Optimization: Polyhedra and Efficiency. Algorithms and Combinatorics, vol. 24, 1st edn. Springer, Heidelberg (2003)
22. Scutellà, M.G.: A note on the parametric maximum flow problem and some related reoptimization issues. Ann. Oper. Res. **150**(1), 231–244 (2007)
23. Young, N.E., Tarjan, R.E., Orlin, J.B.: Faster parametric shortest path and minimum-balance algorithms. Networks **21**(2), 205–221 (2006)

Geodesic Fault-Tolerant Additive Weighted Spanners

Sukanya Bhattacharjee and R. Inkulu[✉]

Department of Computer Science and Engineering, IIT Guwahati, Guwahati, India
{bsukanya,rinkulu}@iitg.ac.in

Abstract. Let S be a set of n points and let w be a function that assigns non-negative weights to points in S. The additive weighted distance $d_w(p, q)$ between two points $p, q \in S$ is defined as $w(p) + d(p, q) + w(q)$ if $p \neq q$ and it is zero if $p = q$. Here, $d(p, q)$ is the geodesic Euclidean distance between p and q. For a real number $t > 1$, a graph $G(S, E)$ is called a *t-spanner* for the weighted set S of points if for any two points p and q in S the distance between p and q in graph G is at most $t.d_w(p, q)$ for a real number $t > 1$. For some integer $k \geq 1$, a t-spanner G for the set S is a (k, t)-*vertex fault-tolerant additive weighted spanner*, denoted with (k, t)-VFTAWS, if for any set $S' \subset S$ with cardinality at most k, the graph $G \setminus S'$ is a t-spanner for the points in $S \setminus S'$. For any given real number $\epsilon > 0$, we present algorithms to compute a $(k, 4 + \epsilon)$-VFTAWS for the metric space (S, d_w) resulting from the following: (i) points in S are in the free space of the polygonal domain, (ii) points in S lying on a terrain.

Keywords: Computational geometry · Geometric spanners · Approximation algorithms

1 Introduction

When designing geometric networks on a given set of points in a metric space, it is desirable for the network to have short paths between any pair of nodes while being sparse with respect to the number of edges. Let $G(S, E)$ be an edge-weighted geometric graph on a set S of n points in $\mathbb{R}^d$. The weight of any edge $(p, q) \in E$ is the Euclidean distance $|pq|$ between p and q. The distance in G between any two nodes p and q, denoted by $d_G(p, q)$, is defined as the length of a shortest (that is, minimum-weighted) path between p and q in G. The graph G is called a *t-spanner* for some $t \geq 1$ if for any two points $p, q \in S$ we have $d_G(p, q) \leq t.|pq|$. The smallest t for which G is a t-spanner is called the *stretch factor* of G, and the number of edges of G is called its size.

Peleg and Schäffer [40] introduced spanners in the context of distributed computing and by Chew [27] in a geometric context. Althöfer et al. [9] first

This research is supported in part by SERB MATRICS grant MTR/2017/000474.

D.-Z. Du et al. (Eds.): COCOON 2019, LNCS 11653, pp. 38–51, 2019.
https://doi.org/10.1007/978-3-030-26176-4_4

attempted to study sparse spanners on edge-weighted graphs that have the triangle-inequality property. The text by Narasimhan and Smid [39], handbook chapter [31], and Gudmundsson and Knauer [32] detail various results on Euclidean spanners, including a $(1 + \epsilon)$-spanner for the set S of n points in $\mathbb{R}^d$ that has $O(\frac{n}{\epsilon^{d-1}})$ edges for any $\epsilon > 0$.

Many variations of sparse spanners have been studied, including spanners of low degree [11,19,25,42], spanners of low weight [18,30,33], spanners of low diameter [13,14], planar spanners [10,27,29,36], spanners of low chromatic number [17], fault-tolerant spanners [3,16,28,35,37,38,43], low power spanners [7,41,45], kinetic spanners [2,5], angle-constrained spanners [26], and combinations of these [12,15,21–24]. When the doubling dimension of a metric space is bounded, results applicable to the Euclidean settings are given in [44].

As mentioned in Abam et al., [4], the cost of traversing a path in a network is not only determined by the lengths of the edges on the path but also by the delays occurring at the nodes on the path. The result in [4] models these delays with the additive weighted metric. Let S be a set of n points in $\mathbb{R}^d$. For every $p \in S$, let $w(p)$ be the non-negative weight associated to p. The following additive weighted distance function d_w on S defining the metric space (S, d_w) is considered in [4] and by Bhattacharjee and Inkulu in [16]: for any $p, q \in S$, $d_w(p, q)$ equals to 0 if $p = q$; otherwise, it is equal to $w(p) + |pq| + w(q)$.

Recently, Abam et al. [6] showed that there exists a $(2 + \epsilon)$-spanner with a linear number of edges for the metric space (S, d_w) that has bounded doubling dimension in [34]. And, [4] gives a lower bound on the stretch factor, showing that $(2 + \epsilon)$ stretch is nearly optimal. Bose et al. [20] studied the problem of computing spanners for a weighted set of points. They considered the points that lie on the plane to have positive weights associated to them; and defined the distance d_w between any two distinct points $p, q \in S$ as $d(p, q) - w(p) - w(q)$. Under the assumption that the distance between any pair of points is non-negative, they showed the existence of a $(1 + \epsilon)$-spanner with $O(\frac{n}{\epsilon})$ edges.

A simple polygon $P_\mathcal{D}$ containing $h \geq 0$ number of disjoint simple polygonal holes within it is termed the *polygonal domain* $\mathcal{D}$. (When h equals to 0, the polygonal domain $\mathcal{D}$ is a simple polygon.) The free space $\mathcal{F}(\mathcal{D})$ of the given polygonal domain $\mathcal{D}$ is defined as the closure of $P_\mathcal{D}$ excluding the union of the interior of polygons contained in $P_\mathcal{D}$. Essentially, a path between any two given points in $\mathcal{F}(\mathcal{D})$ needs to be in the free space $\mathcal{F}(\mathcal{D})$ of $\mathcal{D}$. Given a set S of n points in the free space $\mathcal{F}(\mathcal{D})$ defined by the polygonal domain $\mathcal{D}$, computing geodesic spanners in $\mathcal{F}(\mathcal{D})$ is considered in Abam et al. [1]. For any two distinct points $p, q \in S$, $d_\pi(p, q)$ is defined as the geodesic Euclidean distance along a shortest path $\pi(p, q)$ between p and q in $\mathcal{F}(\mathcal{D})$. [1] showed that for the metric space (S, π), for any constant $\epsilon > 0$, there exists a $(5 + \epsilon)$-spanner of size $O(\sqrt{h}n(\lg n)^2)$. Further, for any constant $\epsilon > 0$, [1] gave a $(\sqrt{10} + \epsilon)$-spanner with $O(n(\lg n)^2)$ edges when $h = 0$ i.e., the polygonal domain is a simple polygon with no holes. Given a set S of n points on a polyhedral terrain $\mathcal{T}$, the geodesic Euclidean distance between any two points $p, q \in S$ is the distance along any shortest path between p and q on $\mathcal{T}$. [6] showed that for a set of unweighted points on a polyhedral terrain, for any constant $\epsilon > 0$, there exists a $(2+\epsilon)$-geodesic spanner with $O(n \lg n)$ edges.

For a network to be vertex fault-tolerant, i.e., when a subset of nodes is removed, the induced network on the remaining nodes requires to be connected. Formally, a graph $G(S, E)$ is a k-vertex fault-tolerant t-spanner, denoted by (k, t)-VFTS, for a set S of n points in $\mathbb{R}^d$ if for any subset S' of S with size at most k, the graph $G \setminus S'$ is a t-spanner for the points in $S \setminus S'$. Algorithms in Levcopoulos et al. [37], Lukovszki [38], and Czumaj and Zhao [28] compute a (k, t)-VFTS for the set S of points in $\mathbb{R}^d$. These algorithms are also presented in [39]. [37] devised an algorithm to compute a (k, t)-VFTS of size $O(\frac{n}{(t-1)^{(2d-1)(k+1)}})$ in $O(\frac{n \lg n}{(t-1)^{4d-1}} + \frac{n}{(t-1)^{(2d-1)(k+1)}})$ time and another algorithm to compute a (k, t)-VFTS with $O(k^2 n)$ edges in $O(\frac{kn \lg n}{(t-1)^d})$ time. [38] gives an algorithm to compute a (k, t)-VFTS of size $O(\frac{kn}{(t-1)^{d-1}})$ in $O(\frac{1}{(t-1)^d}(n \lg^{d-1} n \lg k + kn \lg \lg n))$ time. The algorithm in [28] computes a (k, t)-VFTS having $O(\frac{kn}{(t-1)^{d-1}})$ edges in $O(\frac{1}{(t-1)^{d-1}}(kn \lg^d n + nk^2 \lg k))$ time with total weight of edges upper bounded by a $O(\frac{k^2 \lg n}{(t-1)^d})$ multiplicative factor of the weight of MST of the given set of points.

For a real number $t > 1$, a graph $G(S, E)$ is called a t-spanner for the weighted set S of points if for any two points p and q in S the distance between p and q in graph G is at most $t.d_w(p, q)$ for a real number $t > 1$. For some integer $k \geq 1$, a t-spanner G for the set S is a (k, t)-vertex fault-tolerant additive weighted spanner, denoted with (k, t)-VFTAWS, if for any set $S' \subset S$ with cardinality at most k, the graph $G \setminus S'$ is a t-spanner for the points in $S \setminus S'$. In [16], Bhattacharjee and Inkulu devised the following algorithms: one for computing a $(k, 4 + 5\epsilon)$-VFTAWS when the input points are located in $\mathbb{R}^d$, and the other for computing a $(k, 4 + 14\epsilon)$-VFTAWS when the given points are located in a simple polygon. In this paper, we extend the results in [16] for points in polygonal domains and terrains.

Our Results. The spanners computed in this paper are first of their kind as we combine vertex fault-tolerance with the additive weighted set of points in the context of points are located either in a polygonal domain or on a terrain. Here, we generalize the results obtained in [16] for points located in $\mathbb{R}^d$ and points residing in a simple polygon. In specific, we devise the following algorithms for computing (k, t)-VFTAWS, for any $\epsilon > 0$ and $k \geq 1$:

* Given a set S comprising of n points in the polygonal domain $\mathcal{P}$ that has h simple polygonal holes while each point in S is associated with a non-negative weight, we compute a $(k, 4 + \epsilon)$-VFTAWS having $O(\frac{k\sqrt{h}n}{\epsilon^2} \lg n)$ edges.
* Given a set S of n points located on a terrain with a positive weight associated to each point, we compute a $(k, 4 + \epsilon)$-VFTAWS having $O(\frac{kn}{\epsilon^2} \lg n)$ edges.

The Euclidean distance between two points p and q is denoted by $|pq|$. The distance between two points p, q in the metric space X is denoted by $d_X(p, q)$.

The length of the shortest path between p and q in a graph G is denoted by $d_G(p, q)$.

Section 2 details algorithm for $(k, 4 + \epsilon)$-VFTAWS for points located in the polygonal domain. Section 3 presents an algorithm to compute a $(k, 4 + \epsilon)$-VFTAWS when the input points are located on a terrain.

2   Vertex Fault-Tolerant Additive Weighted Spanner for Points in a Polygonal Domain

We devise an algorithm to compute a geodesic $(k, (4 + \epsilon))$-vertex fault-tolerant spanner for a set S of n points lying in the free space $\mathcal{D}$ of the given polygonal domain $\mathcal{P}$ while each input point is associated with a non-negative weight. The polygonal domain $\mathcal{P}$ consists of a simple polygon and h simple polygonal holes located interior to to that polygon. Our algorithm depends on the algorithm given in [1] to compute a $(5 + \epsilon)$-spanner for a set of unweighted points lying in $\mathcal{D}$. We decompose the free space $\mathcal{D}$ into simple polygons using $O(h)$ splitting segments such that no splitting segment crosses any of the holes of $\mathcal{D}$ and each of the resultant simple polygons has at most three splitting segments bounding it. As part of this decomposition, two vertical line segments are drawn, one upwards and the other downwards, respectively from the leftmost and rightmost extreme (along the x-axis) vertices of each hole. If any of the resulting simple polygons has more than three splitting segments on its boundary, then that simple polygon is further decomposed. To achieve efficiency, a splitting segment is chosen such that it has around half of its bounding splitting segments on either of its sides. This algorithm results in partitioning $\mathcal{D}$ into $O(h)$ simple polygons. Further, a graph $\mathcal{G}_d$ is constructed where each vertex of $\mathcal{G}_d$ corresponds to a simple polygon of this decomposition. Each vertex v of $\mathcal{G}_d$ is associated with a weight equal to the number of points that lie inside the simple polygon corresponding to v. Two vertices are connected by an edge in $\mathcal{G}_d$ whenever their corresponding simple polygons are adjacent to each other in the decomposition. We note that $\mathcal{G}_d$ is a planar graph. Hence, we use the following theorem from [8] to compute a $O(\sqrt{h})$-separator R of $\mathcal{G}_d$.

Theorem 1. *Suppose $G = (V, E)$ is a planar vertex-weighted graph with $|V| = m$. Then, an $O(\sqrt{m})$-separator for G can be computed in $O(m)$ time. That is, V can be partitioned into sets P, Q and R such that $|R| = O(\sqrt{m})$, there is no edge between P and Q, and $w(P), w(Q) \leq \frac{2}{3}w(V)$. Here, $w(X)$ is the sum of weights of all vertices in X.*

We compute a $O(\sqrt{h})$-separator R for the graph $\mathcal{G}_d$ using Theorem 1. Let P, Q, and R be the sets into which the vertices of $\mathcal{G}_d$ is partitioned. For each vertex $r \in R$, we collect the bounding splitting segments of the simple polygon corresponding to r into H i.e., $O(\sqrt{h})$ splitting segments are collected into a set H. For each splitting segment l in H, we proceed as follows. For each point p that lies in the given simple polygon, we find the projection p_l of p on l; we assign the weight $w(p) + d_\pi(p, p_l)$ to point p_l and include p_l into the set S_l corresponding to points projected on to line l. We compute the $(k, 4 + \epsilon)$-VFTAWS $\mathcal{G}_l$ for the set S_l of points using the algorithm from [16] for additive weighted points located in $\mathbb{R}^d$. For every edge (r, s) in $\mathcal{G}_l$, we introduce an edge (p, q) in G, where r (resp. s) is the projection of p (resp. q) on l. Recursively, we compute vertex-fault tolerant additive weighted spanner for points lying in simple polygon corresponding to P (resp. Q). The recursion is continued till P (resp. Q) contains exactly one vertex. We first prove that this algorithm computes a $(k, (12 + \epsilon))$-vertex fault-tolerant spanner. Further, we modify this algorithm to compute a $(k, (4 + \epsilon))$-vertex fault-tolerant spanner.

Lemma 1. *The spanner G is a geodesic $(k, (12+15\epsilon))$-vertex fault-tolerant additive weighted spanner for points in $\mathcal{D}$.*

Proof. Using induction on the number of points, we show that there exists a $(12 + 15\epsilon)$-spanner path between p and q in $\mathcal{G} \setminus S'$. The induction hypothesis assumes that for the number of points $k' < |S|$, there exists a $(12 + 15\epsilon)$-spanner path between any two points belonging to $G\ S$. Consider a set $S' \subset S$ such that $|S'| \leq k$ and two arbitrary points p and q from the set $S \setminus S'$. Here, as described above, P, Q, and R correspond to vertices of a planar graph $\mathcal{G}_d$. The union of simple polygons that correspond to vertices of P (resp. Q, R) is denoted with $poly(P)$ (resp. $poly(Q), poly(R)$). Also, the set H is as described in the algorithm. Based on the location of p and q, the following cases arise: (i) both p and q are lying in $P' \in \{poly(P), poly(Q),$ and $poly(R)\}$ and the geodesic Euclidean shortest path between p and q does not intersect any splitting segment from the set H, and (ii) p is lying in $P' \in \{poly(P), poly(Q), poly(R)\}$ and q is lying in $P'' \in \{poly(P), poly(Q), poly(R)\}$ where $P' \neq P''$. In case (i), if P' is a simple polygon, then we can apply algorithm for simple polygons from [16] and obtain a $(4 + 14\epsilon)$-path between p and q. Otherwise, from the induction hypothesis, there exists a $(12 + 15\epsilon)$-path between p and q. In case (ii), a shortest path from p and q intersects at least one of the $O(\sqrt{h})$ splitting segments in H, say l. Let $\pi(p, q)$ be a shortest path between p and q that intersects l at some point. Let r be this point of intersection. Since $\mathcal{G}_l$ is a $(k, (4 + 5\epsilon))$-VFTAWS, there exists a path P' between p_l and q_l in $\mathcal{G}_l$ with length at most $(4 + 5\epsilon)d_{l,w}(p_l, q_l)$. By replacing each vertex x_l of P' by $x \in S$ such that x_l is the projection of x on l, gives a path between p and q in $\mathcal{G} \setminus S'$. Thus, the length of the path $d_{\mathcal{G} \setminus S'}(p, q)$ is less than or equal to the length of the path P' in $\mathcal{G}_l$. For every $x, y \in S$,

$$
\begin{aligned}
d_{\pi,w}(x, y) &= w(x) + d_\pi(x, y) + w(y) \\
&\leq w(x) + d_\pi(x, x_l) + d_\pi(x_l, y_l) + d_\pi(y_l, y) + w(y) \\
&\quad \text{[by the triangle inequality]} \\
&= w(x_l) + d_\pi(x_l, y_l) + w(y_l) \\
&\quad \text{[since the weight associated with projection } z_l \text{ of every point } z \text{ is} \\
&\quad w(z) + d_\pi(z, z_l)] \\
&= d_{l,w}(x_l, y_l) \tag{1}
\end{aligned}
$$

This implies,

$$d_{\mathcal{G} \setminus S'}(p, q) = \sum_{x_l, y_l \in P} d_{\pi, w}(x, y)$$

$$\leq \sum_{x_l, y_l \in P} d_{l,w}(x_l, y_l)$$

[from (1)]

$$\leq (4 + 5\epsilon).d_{l,w}(p_l, q_l) \tag{2}$$

[since $\mathcal{G}_l$ is a geodesic $(k, (4 + 5\epsilon))$-VFTAWS]

$$= (4 + 5\epsilon) \cdot [w(p_l) + d_l(p_l, q_l) + w(q_l)]$$

$$= (4 + 5\epsilon) \cdot [w(p_l) + d_\pi(p_l, q_l) + w(q_l)]$$

[since P contains l, shortest path between p_l and q_l along l
is same as the geodesic shortest path between p_l and q_l]

$$= (4 + 5\epsilon) \cdot [w(p) + d_\pi(p, p_l) + d_\pi(p_l, q_l) + d_\pi(q_l, q) + w(q)] \tag{3}$$

[since the weight associated with projection z_l of every point z is
$w(z) + d_\pi(z, z_l)$].

Since r is a point belonging to both l as well as to $\pi(p, q)$,

$$d_\pi(p, p_l) \leq d_\pi(p, r) \text{ and } d_\pi(q, q_l) \leq d_\pi(q, r). \tag{4}$$

Substituting (4) into (3),

$$d_{\mathcal{G} \setminus S'}(p, q) \leq (4 + 5\epsilon) \cdot [w(p) + d_\pi(p, r) + d_\pi(p_l, q_l) + d_\pi(r, q) + w(q)]$$

$$\leq (4 + 5\epsilon) \cdot [w(p) + d_\pi(p, r) + w(r) + d_\pi(p_l, q_l) + w(r) + d_\pi(r, q) + w(q)]$$

[since the weight associated with every point is non-negative]

$$= (4 + 5\epsilon) \cdot [d_{\pi,w}(p, r) + d_\pi(p_l, q_l) + d_{\pi,w}(r, q)]$$

$$= (4 + 5\epsilon) \cdot [d_{\pi,w}(p, q) + d_\pi(p_l, q_l)]$$

[since $\pi(p, q)$ intersects l at r, by optimal substructure property of shortest
paths, $\pi(p, q) = \pi(p, r) + \pi(r, q)$]. $\tag{5}$

Consider

$$d_\pi(p_l, q_l) \leq d_\pi(p_l, p) + d_\pi(p, q) + d_\pi(q, q_l)$$

[since π follows triangle inequality]

$$\leq d_\pi(r, p) + d_\pi(p, q) + d_\pi(q, r)$$

[using (4)]

$$\leq w(r) + d_\pi(r, p) + w(p) + w(p) + d_\pi(p, q) + w(q) + w(q) + d_\pi(q, r) + w(r)$$

[since weight associated with every point is non-negative]

$$= d_{\pi,w}(p, r) + d_{\pi,w}(p, q) + d_{\pi,w}(r, q)$$

$$= d_{\pi,w}(p, q) + d_{\pi,w}(p, q)$$

[since $\pi(p, q)$ intersects l at r, by optimal substructure property of shortest paths, $\pi(p, q) = \pi(p, r) + \pi(r, q)$]

$$= 2 d_{\pi,w}(p, q). \tag{6}$$

Substituting (6) into (5), $d_{G \setminus S'}(p, q) \leq 3(4 + 5\epsilon) \cdot d_{\pi,w}(p, q)$. □

We further improve the stretch factor of $\mathcal{G}$ by applying the refinement given in [6] to the above-described algorithm. In doing this, for each point $p \in S$, we compute the geodesic projection p_γ of p on the splitting line γ and we construct a set $S(p, \gamma)$ as defined herewith. Let $\gamma(p) \subseteq \gamma$ be $\{x \in \gamma : d_{\gamma,w}(p_\gamma, x) \leq (1 + 2\epsilon) \cdot d_\pi(p, p_\gamma)\}$. Here, for any $p, q \in S$, $d_{\gamma,w}(p, q)$ is equal to 0 if $p = q$; otherwise, it is equal to $w(p) + d_\gamma(p, q) + w(q)$. We divide $\gamma(p)$ into c pieces with $c \in O(1/\epsilon^2)$: each piece is denoted by $\gamma_j(p)$ for $1 \leq j \leq c$. For every piece j, we compute the point $p_\gamma^{(j)}$ nearest to p in $\gamma_j(p)$. The set $S(p, \gamma)$ is defined as $\{p_\gamma^{(j)} : p_\gamma^{(j)} \in \gamma_j(p)$ and $1 \leq j \leq c\}$. For every $r \in S(p, \gamma)$, the non-negative weight $w(r)$ of r is set to $w(p) + d_\pi(p, r)$. Let S_γ be $\cup_{p \in S} S(p, \gamma)$. We replace the set S_l in computing G with the set S_γ and compute a $(k, (4 + 5\epsilon))$-VFTAWS $\mathcal{G}_l$ for the set S_l using the algorithm for points in $\mathbb{R}^d$ from [16]. Further, for every edge (r, s) in $\mathcal{G}_l$, we add the edge (p, q) to G such that $r \in S(p, l)$ and $s \in S(q, l)$. The rest of the algorithm remains the same.

Theorem 2. *Let S be a set of n points in a polygonal domain $\mathcal{D}$ with non-negative weights associated to points via weight function w. For any fixed constant $\epsilon > 0$, there exists a $(k, (4 + \epsilon))$-vertex fault-tolerant additive weighted geodesic spanner with $O(\frac{kn\sqrt{h}}{\epsilon^2} \lg n)$ edges for the metric space $(S, d_{\pi,w})$.*

Proof. Let $S(n)$ be the size of $\mathcal{G}$. Our algorithm adds $O(\frac{kn\sqrt{h}}{\epsilon^2})$ edges at each recursive level except for the last level. At every leaf node l of the recurrence tree, we add $O(\frac{kn_x}{\epsilon^2} \lg n_x)$ edges, where n_x is the number of points inside the simple polygon corresponding to l. Hence, the number of edges of G is $O(\frac{kn\sqrt{h}}{\epsilon^2} \lg n)$.

Next, we prove the stretch factor of the spanner. Consider any set $S' \subset S$ such that $|S'| \leq k$ and two arbitrary points p and q from the set $S \setminus S'$. We show that there exists a $(4 + 14\epsilon)$-spanner path between p and q in $\mathcal{G} \setminus S'$.

If $r \notin l(p)$, then we set p_l' (resp. q_l') equal to p_l (resp. q_l). Otherwise, p_l' (resp. q_l') is set as the point from $S(p, l)$ (resp. $S(q, l)$) that is nearest to p (resp. q). (The r is defined before the theorem statement.)

$$d_{l,w}(p_l', q_l') = w(p_l') + d_l(p_l', q_l') + w(q_l')$$
$$\leq w(p_l') + d_l(p_l', r) + d_l(r, q_l') + w(q_l')$$
[by the triangle inequality]
$$\leq w(p_l') + d_l(p_l', r) + w(r) + w(r) + d_l(r, q_l') + w(q_l')$$
[since the weight associated with each point is non-negative]
$$= w(p) + d_\pi(p, p_l') + d_l(p_l', r) + w(r) + w(r) + d_l(r, q_l')$$
$$+ d_\pi(q_l', q) + w(q)$$
[due to the assignment of the weight to the projection of any point]
$$\tag{7}$$

From the triangle inequality, we know the following:

$$d_\pi(p, p_l') + d_l(p_l', r) \leq d_\pi(p, r), \text{ and} \tag{8}$$
$$d_l(r, q_l') + d_\pi(q_l', q) \leq d_\pi(r, q). \tag{9}$$

Substituting (8) and (9) in (7),

$$d_{l,w}(p_l', q_l') \leq w(p) + d_\pi(p, r) + w(r) + w(r) + d_\pi(r, q) + w(q)$$
$$= d_{\pi,w}(p, r) + d_{\pi,w}(r, q)$$
$$= d_{\pi,w}(p, q)$$
[since $r \in l \cap \pi(p, q)$, by the optimal substructure property of shortest paths, $\pi(p, q) = \pi(p, r) + \pi(r, q)$]. $\tag{10}$

Replacing p_l (resp. q_l) by p_l' (resp. q_l') in inequality (2),

$$d_{\mathcal{G} \setminus S'}(p, q) \leq (4 + 5\epsilon).d_{l,w}(p_l', q_l')$$
$$\leq (4 + 5\epsilon)d_{\pi,w}(p, q) \quad \text{[from (10)]}.$$

Thus, G is a geodesic $(k, (4 + \epsilon))$-VFTAWS for S. $\qquad \square$

3 Vertex Fault-Tolerant Additive Weighted Spanner for Points on a Terrain

In this section, we present an algorithm to compute a geodesic $(k, (4 + \epsilon))$-VFTAWS with $O(\frac{kn \lg n}{\epsilon^2})$ edges for any given set S of n non-negative weighted points lying on a polyhedral terrain $\mathcal{T}$. We denote the boundary of $\mathcal{T}$ with $\partial \mathcal{T}$. The following distance function $d_{\mathcal{T},w} : S \times S \to \mathbb{R} \cup \{0\}$ is used to compute the geodesic distance on $\mathcal{T}$ between any two points $p, q \in S$: $d_{\mathcal{T},w}(p, q) = w(p) + d_{\mathcal{T}}(p, q) + w(q)$. Here, $w(p)$ (resp. $w(q)$) is the non-negative weight of $p \in S$ (resp.

$q \in S$). We denote a geodesic Euclidean shortest path between any two points a and b on $\mathcal{T}$ with $\pi(a,b)$. For any two points $x, y \in \partial\mathcal{T}$, we denote the closed region lying to the right (resp. left) of $\pi(x,y)$ when going from x to y, including the points lying on the shortest path $\pi(x,y)$ with $\pi^+(x,y)$ (resp. $\pi^-(x,y)$). The *projection* p_π of a point p on the shortest path π between two points lying on the polyhedral terrain $\mathcal{T}$ is defined as a point on π that is at the minimum geodesic distance from p among all the points located on π. For three points $u, v, w \in \mathcal{T}$, the closed region bounded by shortest paths $\pi(u,v)$, $\pi(v,w)$, and $\pi(w,u)$ is termed *sp-triangle*, denoted with $\Delta(u,v,w)$. If the points $u, v, w \in \mathcal{T}$ are clear from the context, we denote the sp-triangle with Δ. In the following, we restate a Theorem from [6], which is useful for our analysis.

Theorem 3. *For any set P of n points on a polyhedral terrain $\mathcal{T}$, there exists a balanced sp-separator: a shortest path $\pi(u,v)$ connecting two points $u, v \in \partial\mathcal{T}$ such that $\frac{2n}{9} \leq |\pi^+(u,v) \cap P| \leq \frac{2n}{3}$, or a sp-triangle Δ such that $\frac{2n}{9} \leq |\Delta \cap P| \leq \frac{2n}{3}$.*

Thus, an sp-separator is either bounded by a shortest path (in the former case) or by three shortest paths (in the latter case). Let γ be a shortest path that belongs to an sp-separator. First, a balanced sp-separator as given in Theorem 3 is computed. The sets S_{in} and S_{out} comprising of points are defined as follows: if the sp-separator is a shortest path then define S_{in} to be $\gamma^+(u,v) \cap S$; otherwise, S_{in} is $\Delta \cap S$; points in S that do not belong to S_{in} are in S_{out}. For each $p \in S$, we compute the projection p_γ of p on every shortest path γ of sp-separator, and associate a weight $d_\mathcal{T}(p, p_\gamma)$ with p_γ. Let S_γ be a set defined as $\cup_{p \in S} p_\gamma$. Our algorithm computes a $(2 + \epsilon)$-spanner $\mathcal{G}_\gamma$ for the weighted points in S_γ. Further, for each edge (p_γ, q_γ) in $\mathcal{G}_\gamma$, an edge (p,q) is added to $\mathcal{G}$, where p_γ (resp. q_γ) is the projection of p (resp. q) on γ. The spanners for the sets S_{in} and S_{out} are computed recursively, and the edges from these spanners are added to $\mathcal{G}$. In the base case, if $|S| \leq 3$ then a complete graph on the set S is constructed. We first obtain a $(k, (12 + 15\epsilon))$-vertex fault-tolerant additive weighted spanner for the set S of points lying on the terrain $\mathcal{T}$. (This construction is later modified to compute a $(k, (4 + \epsilon))$-VFTAWS.) In specific, with every projected point p_γ, instead of associating $d_\mathcal{T}(p, p_\gamma)$ as the weight of p_γ, we associate $w(p) + d_\mathcal{T}(p, p_\gamma)$ as the weight of p_γ. The rest of the algorithm in constructing $\mathcal{G}$ remains the same as in [6].

To prove the graph $\mathcal{G}$ is a geodesic $(k, (12 + 15\epsilon))$-VFTAWS for the points in S, we use induction on the number of points. Consider any set $S' \subset S$ such that $|S'| \leq k$ and two arbitrary points p and q from the set $S \setminus S'$. We show that there exists a path between p and q in $\mathcal{G} \setminus S'$ such that $d_\mathcal{G}(p,q)$ is at most $(12 + 15\epsilon)d_{\pi,w}(p,q)$. The induction hypothesis assumes that for the number of points $k' < |S|$ in a region of $\mathcal{T}$, there exists a $(12 + 15\epsilon)$-spanner path between any two points belonging to the given region in $\mathcal{G} \setminus S'$. As part of the inductive step, we extend it to n points. For the case of both p and q are on the same side of a bounding shortest path γ of the balanced separator, i.e., both are in S_{in} or S_{out}, by induction hypothesis (as the number of points in S_{in} or S_{out} is less than $|S|$), there exists a $(12 + 15\epsilon)$-spanner path between p and q in $\mathcal{G} \setminus S'$. The only case remains to be proved is when p lies on one side of γ and q lies on the other side of γ, i.e., $p \in S_{in}$ and $q \in S_{out}$ or, $q \in S_{in}$ and $p \in S_{out}$. W.l.o.g., we assume that the former holds. Let r be a point on γ at which the geodesic shortest path $\pi(p, q)$ between p and q intersects γ. Since $\mathcal{G}_\gamma$ is a $(k, (4+5\epsilon))$-VFTS, there exists a path P between p_γ and q_γ in $\mathcal{G}_\gamma$ of length at most $(4 + 5\epsilon).d_{\gamma,w}(p_\gamma, q_\gamma)$. Let P' be the path obtained by replacing each vertex x_γ of P by $x \in S$ such that x_γ is the projection of x on γ. Note that the path P' is between nodes p and q in $\mathcal{G} \setminus S'$. The length $d_{\mathcal{G} \setminus S'}(p, q)$ of path P' is less than or equal to the length of the path P in $\mathcal{G}_\gamma$. In the following, we show that $d_{\mathcal{G} \setminus S'}(p, q) \leq (12 + 15\epsilon)d_{T,w}(p, q)$.

For every $x, y \in S$,

$$
\begin{aligned}
d_{T,w}(x, y) &= w(x) + d_T(x, y) + w(y) \\
&\leq w(x) + d_T(x, x_\gamma) + d_T(x_\gamma, y_\gamma) + d_T(y_\gamma, y) + w(y)
\end{aligned}
$$

[by the triangle inequality]

$$
= w(x_\gamma) + d_T(x_\gamma, y_\gamma) + w(y_\gamma)
$$

[since the weight associated with projection z_γ of every point z is $w(z) + d_T(z, z_\gamma)$]

$$
= d_{\gamma,w}(x_\gamma, y_\gamma). \tag{11}
$$

This implies,

$$
d_{\mathcal{G} \setminus S'}(p, q) = \sum_{x_\gamma, y_\gamma \in P} d_{T,w}(x, y)
$$

$$
\leq \sum_{x_\gamma, y_\gamma \in P} d_{\gamma,w}(x_\gamma, y_\gamma)
$$

[from (11)]

$$
\leq (4 + 5\epsilon).d_{\gamma,w}(p_\gamma, q_\gamma) \tag{12}
$$

[since $\mathcal{G}_\gamma$ is a $(k, (4 + 5\epsilon))$-vertex fault tolerant geodesic spanner]

$$
= (4 + 5\epsilon).[w(p_\gamma) + d_T(p_\gamma, q_\gamma) + w(q_\gamma)]
$$

[since γ is a shortest path on T, shortest path between any two points on γ is a geodesic shortest path on T]

$$
= (4 + 5\epsilon).[w(p) + d_T(p, p_\gamma) + d_T(p_\gamma, q_\gamma) + d_T(q_\gamma, q) + w(q)] \tag{13}
$$

[since the weight associated with projection z_γ is $w(z) + d_T(z, z_\gamma)$].

By the definition of projection of any point on γ, we know that

$$
d_T(p, p_\gamma) \leq d_T(p, r) \text{ and } d_T(q, q_\gamma) \leq d_T(q, r). \tag{14}
$$

Substituting (14) into (13),

$$d_{\mathcal{G}\setminus S'}(p,q) \leq (4+5\epsilon).[w(p) + d_{\mathcal{T}}(p,r) + d_{\mathcal{T}}(p_\gamma, q_\gamma) + d_{\mathcal{T}}(r,q) + w(q)]$$

$$\leq (4+5\epsilon).[w(p) + d_{\mathcal{T}}(p,r) + w(r) + d_{\mathcal{T}}(p_\gamma, q_\gamma) + w(r) + d_{\mathcal{T}}(r,q) + w(q)]$$

[since the weight of every point is non-negative]

$$= (4+5\epsilon).[d_{\mathcal{T},w}(p,r) + d_{\mathcal{T}}(p_\gamma, q_\gamma) + d_{\mathcal{T},w}(r,q)]$$

$$= (4+5\epsilon).[d_{\mathcal{T},w}(p,q) + d_{\mathcal{T}}(p_\gamma, q_\gamma)] \tag{15}$$

[since $\pi(p,q)$ intersects γ at r]

paths $\pi(p,q) = \pi(p,r) + \pi(r,q)$].

Further,

$$d_{\mathcal{T}}(p_\gamma, q_\gamma) \leq d_{\mathcal{T}}(p_\gamma, p) + d_{\mathcal{T}}(p,q) + d_{\mathcal{T}}(q, q_\gamma)$$

[by the triangle inequality]

$$\leq d_{\mathcal{T}}(r,p) + d_{\mathcal{T}}(p,q) + d_{\mathcal{T}}(q,r)$$

[using (14)]

$$\leq w(r) + d_{\mathcal{T}}(r,p) + w(p) + w(p) + d_{\mathcal{T}}(p,q) + w(q) + w(q)$$
$$+ d_{\mathcal{T}}(q,r) + w(r)$$

[since the weight of every point is non-negative]

$$= d_{\mathcal{T},w}(p,r) + d_{\mathcal{T},w}(p,q) + d_{\mathcal{T},w}(r,q)$$

$$= d_{\mathcal{T},w}(p,q) + d_{\mathcal{T},w}(p,q)$$

[since $\pi(p,q)$ intersects γ at r]

$$= 2d_{\mathcal{T},w}(p,q). \tag{16}$$

Substituting (16) into (15) yields, $d_{\mathcal{G}\setminus S'}(p,q) \leq 3(4+5\epsilon).d_{\mathcal{T},w}(p,q)$.

We improve the stretch factor of $\mathcal{G}$ by applying the same refinement as the one used in the algorithm in Sect. 2. Again, we denote the graph resulted after applying that refinement with $\mathcal{G}$.

Theorem 4. *Let S be a set of n weighted points on a polyhedral terrain $\mathcal{T}$ with non-negative weights associated to points via weight function w. For any fixed constant $\epsilon > 0$, there exists a $(k, (4+\epsilon))$-vertex fault-tolerant additive weighted geodesic spanner with $O(\frac{kn}{\epsilon^2} \lg n)$ edges.*

Proof. The argument for the number of edges is same as in the proof of Theorem 2. To prove that the graph $\mathcal{G}$ is a geodesic $(k, (4+14\epsilon))$-VFTAWS for the points in S, we use induction on $|S|$. W.l.o.g., we assume that $p \in S_{in}$ and $q \in S_{out}$. Let r be the point at which the geodesic shortest path $\pi(p,q)$ between p and q intersects r. Since $\mathcal{G}_\gamma$ is a $(k, (4+5\epsilon))$-VFTS, there exists a path R between p_γ and q_γ in $\mathcal{G}_\gamma$ of length at most $(4+5\epsilon).d_{\gamma,w}(p_\gamma, q_\gamma)$. By replacing each vertex x_γ of R by $x \in S$ such that x_γ is the projection of x on γ, yields a path R' between p and q in $\mathcal{G}\setminus S'$. The length $d_{\mathcal{G}\setminus S'}(p,q)$ of path R' is less than or equal to the length of the path R

in $\mathcal{G}_\gamma$. If $r \notin \gamma(p)$, point p'_γ (resp. q'_γ) is set as p_γ (resp. q_γ). Otherwise, p'_γ (resp. q'_γ) is set as the point in $S(p, \gamma)$ (resp. $S(q, \gamma)$) that is nearest to p (resp. q).

$$d_{\gamma,w}(p'_\gamma, q'_\gamma) = w(p'_\gamma) + d_\gamma(p'_\gamma, q'_\gamma) + w(q'_\gamma)$$
$$\leq w(p'_\gamma) + d_\gamma(p'_\gamma, r) + d_\gamma(r, q'_\gamma) + w(q'_\gamma)$$
[by the triangle inequality]
$$\leq w(p'_\gamma) + d_\gamma(p'_\gamma, r) + w(r) + w(r) + d_\gamma(r, q'_\gamma) + w(q'_\gamma)$$
[since the weight of each point is non-negative]
$$= w(p) + d_T(p, p'_\gamma) + d_\gamma(p'_\gamma, r) + w(r) + w(r) + d_\gamma(r, q'_\gamma)$$
$$+ d_T(q'_\gamma, q) + w(q) \tag{17}$$
[due to the association of weight to the projections of points].

From the triangle inequality, we know that $d_T(p, p'_\gamma) + d_\gamma(p'_\gamma, r) \leq d_T(p, r)$, and $d_\gamma(r, q'_\gamma) + d_T(q'_\gamma, q) \leq d_T(r, q)$. Hence (17) is written as

$$d_{T,w}(p'_\gamma, q'_\gamma) \leq w(p) + d_T(p, r) + w(r) + w(r) + d_T(r, q) + w(q)$$
$$= d_{T,w}(p, r) + d_{T,w}(r, q)$$
$$= d_{T,w}(p, q) \tag{18}$$
[since $\pi(p, q)$ intersects γ at r].

Replacing p_γ (resp. q_γ) by p_γ' (resp. q_γ') in inequality (12),

$$d_{\mathcal{G} \setminus S'}(p, q) \leq (4 + 5\epsilon).d_{\gamma,w}(p'_\gamma, q'_\gamma)$$
$$\leq (4 + 5\epsilon)d_{T,w}(p, q) \quad \text{[from (18).]}$$

Thus $\mathcal{G}$ is a geodesic $(k, (4 + \epsilon))$-VFTAWS for S.   □

References

1. Abam, M.A., Adeli, M., Homapour, H., Asadollahpoor, P.Z.: Geometric spanners for points inside a polygonal domain. In: Proceedings of Symposium on Computational Geometry, pp. 186–197 (2015)
2. Abam, M.A., de Berg, M.: Kinetic spanners in $\mathbb{R}^d$. Discrete Comput. Geom. **45**(4), 723–736 (2011)
3. Abam, M.A., de Berg, M., Farshi, M., Gudmundsson, J.: Region-fault tolerant geometric spanners. Discrete Comput. Geom. **41**(4), 556–582 (2009)
4. Abam, M.A., de Berg, M., Farshi, M., Gudmundsson, J., Smid, M.H.M.: Geometric spanners for weighted point sets. Algorithmica **61**(1), 207–225 (2011)
5. Abam, M.A., de Berg, M., Gudmundsson, J.: A simple and efficient kinetic spanner. Comput. Geom. **43**(3), 251–256 (2010)
6. Abam, M.A., de Berg, M., Seraji, M.J.R.: Geodesic spanners for points on a polyhedral terrain. In: Proceedings of Symposium on Discrete Algorithms, pp. 2434–2442 (2017)

7. Abu-Affash, A.K., Aschner, R., Carmi, P., Katz, M.J.: Minimum power energy spanners in wireless ad hoc networks. Wirel. Netw. **17**(5), 1251–1258 (2011)
8. Alon, N., Seymour, P.D., Thomas, R.: Planar separators. SIAM J. Discrete Math. **7**(2), 184–193 (1994)
9. Althöfer, I., Das, G., Dobkins, D., Joseph, D., Soares, J.: On sparse spanners of weighted graphs. Discrete Comput. Geom. **9**(1), 81–100 (1993)
10. Arikati, S., Chen, D.Z., Chew, L.P., Das, G., Smid, M., Zaroliagis, C.D.: Planar spanners and approximate shortest path queries among obstacles in the plane. In: Diaz, J., Serna, M. (eds.) ESA 1996. LNCS, vol. 1136, pp. 514–528. Springer, Heidelberg (1996). https://doi.org/10.1007/3-540-61680-2_79
11. Aronov, B., et al.: Sparse geometric graphs with small dilation. Comput. Geom. **40**(3), 207–219 (2008)
12. Arya, S., Das, G., Mount, D.M., Salowe, J.S., Smid, M.H.M.: Euclidean spanners: short, thin, and lanky. In: Proceedings of Annual ACM Symposium on Theory of Computing, pp. 489–498 (1995)
13. Arya, S., Mount, D.M., Smid, M.: Dynamic algorithms for geometric spanners of small diameter: randomized solutions. Comput. Geom. **13**(2), 91–107 (1999)
14. Arya, S., Mount, D.M., Smid, M.H.M.: Randomized and deterministic algorithms for geometric spanners of small diameter. In: Proceedings of Annual Symposium on Foundations of Computer Science, pp. 703–712 (1994)
15. Arya, S., Smid, M.H.M.: Efficient construction of a bounded-degree spanner with low weight. Algorithmica **17**(1), 33–54 (1997)
16. Bhattacharjee, S., Inkulu, R.: Fault-tolerant additive weighted geometric spanners. In: Pal, S.P., Vijayakumar, A. (eds.) CALDAM 2019. LNCS, vol. 11394, pp. 29–41. Springer, Cham (2019). https://doi.org/10.1007/978-3-030-11509-8_3
17. Bose, P., Carmi, P., Couture, M., Maheshwari, A., Smid, M., Zeh, N.: Geometric spanners with small chromatic number. Comput. Geom. **42**(2), 134–146 (2009)
18. Bose, P., Carmi, P., Farshi, M., Maheshwari, A., Smid, M.: Computing the greedy spanner in near-quadratic time. Algorithmica **58**(3), 711–729 (2010)
19. Bose, P., Carmi, P., Chaitman, L., Collette, S., Katz, M.J., Langerman, S.: Stable roommates spanner. Comput. Geom. **46**(2), 120–130 (2013)
20. Bose, P., Carmi, P., Couture, M.: Spanners of additively weighted point sets. In: Gudmundsson, J. (ed.) SWAT 2008. LNCS, vol. 5124, pp. 367–377. Springer, Heidelberg (2008). https://doi.org/10.1007/978-3-540-69903-3_33
21. Bose, P., Fagerberg, R., van Renssen, A., Verdonschot, S.: On plane constrained bounded-degree spanners. Algorithmica **81**, 1392 (2018)
22. Bose, P., Gudmundsson, J., Smid, M.: Constructing plane spanners of bounded degree and low weight. Algorithmica **42**(3), 249–264 (2005)
23. Bose, P., Smid, M., Xu, D.: Delaunay and diamond triangulations contain spanners of bounded degree. Int. J. Comput. Geom. Appl. **19**(02), 119–140 (2009)
24. Carmi, P., Chaitman, L.: Bounded degree planar geometric spanners. CoRR, arXiv:1003.4963 (2010)
25. Carmi, P., Chaitman, L.: Stable roommates and geometric spanners. In: Proceedings of the 22nd Annual Canadian Conference on Computational Geometry, pp. 31–34 (2010)
26. Carmi, P., Smid, M.H.M.: An optimal algorithm for computing angle-constrained spanners. J. Comput. Geom. **3**(1), 196–221 (2012)
27. Chew, L.P.: There are planar graphs almost as good as the complete graph. J. Comput. Syst. Sci. **39**(2), 205–219 (1989)
28. Czumaj, A., Zhao, H.: Fault-tolerant geometric spanners. Discrete Comput. Geom. **32**(2), 207–230 (2004)

29. Das, G., Joseph, D.: Which triangulations approximate the complete graph? In: Djidjev, H. (ed.) Optimal Algorithms. LNCS, vol. 401, pp. 168–192. Springer, Heidelberg (1989). https://doi.org/10.1007/3-540-51859-2_15

30. Das, G., Narasimhan, G.: A fast algorithm for constructing sparse Euclidean spanners. Int. J. Comput. Geom. Appl. **7**(4), 297–315 (1997)

31. Eppstein, D.: Spanning trees and spanners. In: Sack, J.-R., Urrutia, J. (eds.) Handbook of Computational Geometry, pp. 425–461. Elsevier, Amsterdam (1999)

32. Gudmundsson, J., Knauer, C.: Dilation and detour in geometric networks. In: Gonzalez, T. (ed.) Handbook of Approximation Algorithms and Metaheuristics. Chapman & Hall, Boca Raton (2007)

33. Gudmundsson, J., Levcopoulos, C., Narasimhan, G.: Fast greedy algorithms for constructing sparse geometric spanners. SIAM J. Comput. **31**(5), 1479–1500 (2002)

34. Har-Peled, S., Mendel, M.: Fast construction of nets in low-dimensional metrics and their applications. SIAM J. Comput. **35**(5), 1148–1184 (2006)

35. Kapoor, S., Li, X.-Y.: Efficient construction of spanners in d-dimensions. CoRR, arXiv:1303.7217 (2013)

36. Keil, J.M., Gutwin, C.A.: The Delaunay triangulation closely approximates the complete Euclidean graph. In: Dehne, F., Sack, J.-R., Santoro, N. (eds.) WADS 1989. LNCS, vol. 382, pp. 47–56. Springer, Heidelberg (1989). https://doi.org/10.1007/3-540-51542-9_6

37. Levcopoulos, C., Narasimhan, G., Smid, M.H.M.: Improved algorithms for constructing fault-tolerant spanners. Algorithmica **32**(1), 144–156 (2002)

38. Lukovszki, T.: New results of fault tolerant geometric spanners. In: Proceedings of Workshop on Algorithms and Data Structures, pp. 193–204 (1999)

39. Narasimhan, G., Smid, M.H.M.: Geometric Spanner Networks. Cambridge University Press, Cambridge (2007)

40. Peleg, D., Schäffer, A.: Graph spanners. J. Graph Theory **13**(1), 99–116 (1989)

41. Segal, M., Shpungin, H.: Improved multi-criteria spanners for ad-hoc networks under energy and distance metrics. In: IEEE INFOCOM, pp. 6–10 (2010)

42. Smid, M.: Geometric spanners with few edges and degree five. In: Gudmundsson, J., Jay, C.B. (eds.) CATS 2006, pp. 7–9. Australian Computer Society (2006)

43. Solomon, S.: From hierarchical partitions to hierarchical covers: optimal fault-tolerant spanners for doubling metrics. In: Proceedings of Symposium on Theory of Computing, pp. 363–372 (2014)

44. Talwar, K.: Bypassing the embedding: algorithms for low dimensional metrics. In: Proceedings of ACM Symposium on Theory of Computing, pp. 281–290 (2004)

45. Wang, Y., Li, X.-Y.: Minimum power assignment in wireless ad hoc networks with spanner property. J. Comb. Optim. **11**(1), 99–112 (2006)

Diameter of Colorings Under Kempe Changes

Marthe Bonamy[1], Marc Heinrich[2], Takehiro Ito[3(✉)], Yusuke Kobayashi[4], Haruka Mizuta[3], Moritz Mühlenthaler[5], Akira Suzuki[3], and Kunihiro Wasa[6]

[1] CNRS, LaBRI, Université de Bordeaux, Talence, France
marthe.bonamy@u-bordeaux.fr
[2] Université Lyon 1, LIRIS, UMR5205, Lyon, France
marc.heinrich@univ-lyon1.fr
[3] Graduate School of Information Sciences, Tohoku University, Sendai, Japan
{takehiro,a.suzuki}@ecei.tohoku.ac.jp, haruka.mizuta.s4@dc.tohoku.ac.jp
[4] Research Institute for Mathematical Sciences, Kyoto University, Kyoto, Japan
yusuke@kurims.kyoto-u.ac.jp
[5] Fakultät für Mathematik, TU Dortmund University, Dortmund, Germany
moritz.muehlenthaler@math.tu-dortmund.de
[6] National Institute of Informatics, Tokyo, Japan
wasa@nii.ac.jp

Abstract. Given a k-coloring of a graph G, a Kempe-change for two colors a and b produces another k-coloring of G, as follows: first choose a connected component in the subgraph of G induced by the two color classes of a and b, and then swap the colors a and b in the component. Two k-colorings are called Kempe-equivalent if one can be transformed into the other by a sequence of Kempe-changes. We consider two problems, defined as follows: First, given two k-colorings of a graph G, KEMPE REACHABILITY asks whether they are Kempe-equivalent; and second, given a graph G and a positive integer k, KEMPE CONNECTIVITY asks whether any two k-colorings of G are Kempe-equivalent. We analyze the complexity of these problems from the viewpoint of graph classes. We prove that KEMPE REACHABILITY is PSPACE-complete for any fixed $k \geq 3$, and that it remains PSPACE-complete even when restricted to three colors and planar graphs of maximum degree six. Furthermore, we show that both problems admit polynomial-time algorithms on chordal graphs, bipartite graphs, and cographs. For each of these graph classes, we give a non-trivial upper bound on the number of Kempe-changes needed in order to certify that two k-colorings are Kempe-equivalent.

Supported partially by JSPS and MAEDI under the Japan-France Integrated Action Program (SAKURA). Also supported partially by the ANR Project GrR (ANR-18-CE40-0032) operated by the French National Research Agency (ANR), by JSPS KAKENHI Grant Numbers JP16H03118, JP16K16010, JP17K12636, JP17K19960, JP18H04091, JP18H05291, JP19J10042, JP19K11814, and JP19K20350, Japan, and by JST CREST Grant Numbers JPMJCR1401, JPMJCR1402, and JPMJCR18K3, Japan.

© Springer Nature Switzerland AG 2019
D.-Z. Du et al. (Eds.): COCOON 2019, LNCS 11653, pp. 52–64, 2019.
https://doi.org/10.1007/978-3-030-26176-4_5

1   Introduction

The technique of "Kempe-changes" has been introduced in 1879 by A.B. Kempe in his attempt to prove the four color theorem [8]. Even though his proof turned out to be incomplete, the Kempe-change technique is known to be powerful and useful in, for example, the proof of the five color theorem and a short proof of Brooks' Theorem [10]. In addition, applications of the Kempe-change technique can be found in theoretical physics [13,14], in the study of Markov chains [19], and in timetables [15].

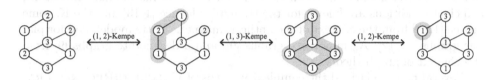

Fig. 1. A sequence of Kempe-changes for 3-colorings, where "(a, b)-Kempe" indicates a Kempe-change for two colors a and b, and the vertices in the gray area are recolored from the immediate left 3-coloring.

Recall that for a positive integer k, a *k-coloring* of a graph G assigns a color in $\{1, 2, \ldots, k\}$ to each vertex of G such that no two adjacent vertices receive the same color. For a k-coloring of G, a *Kempe-change* for two colors $a, b \in \{1, 2, \ldots, k\}$ first chooses a connected component C in the subgraph of G induced by two color classes of a and b, and then swaps the colors a and b in C. (See Fig. 1.) Two k-colorings of G are called *Kempe-equivalent* if one can be transformed into the other by a sequence of Kempe-changes. If any two k-colorings of a graph are Kempe-equivalent, then we say that the graph is *k-Kempe-mixing*. The question whether certain graphs are k-Kempe-mixing has been subject to much scrutiny. However, its complexity remains open, as pointed out by van den Heuvel [7]:

> *Very little is known about the complexity of determining if a graph is k-Kempe-mixing. The same holds for the "path"-version of the problem (determining if two given k-colourings can be transformed into one another by a sequence of Kempe chain recolourings).*

We will refer to the former problem as KEMPE CONNECTIVITY: Given a graph and a positive integer k, the problem asks whether the graph is k-Kempe-mixing. For example, the graph drawn in Fig. 1 is 3-Kempe-mixing, and hence all the 3-colorings (not only four 3-colorings in the figure) are Kempe-equivalent. On the other hand, we will refer to the "path"-version as KEMPE REACHABILITY: Given a graph and two k-colorings of the graph, the problem asks whether they are Kempe-equivalent. Note that if the answer to an instance of KEMPE CONNECTIVITY is yes for a graph G and an integer k, then any instance of KEMPE REACHABILITY given by G and two k-colorings of G is a yes-instance.

1.1 Related Work

Most known results give sufficient conditions on the number k of colors for which graphs are k-Kempe-mixing (i.e., the answer to KEMPE CONNECTIVITY is yes). One of the most prominent motivations for studying the Kempe-equivalence of k-colorings is the Wang-Swendsen-Kotecký (WSK) algorithm [1,13,14] from statistical mechanics, which is a Markov-chain Monte Carlo algorithm that simulates the antiferromagnetic Potts model at zero-temperature. For the WSK algorithm to be valid, the Markov-chains need to be ergodic, which translates to the requirement that certain graphs are k-Kempe-mixing. Hence the importance to give such sufficient conditions. Although we do not explain the details, sufficient conditions are known for the triangular lattice [1,13] and the Kagomé lattice [1,14] from the context of the antiferromagnetic Potts model; and for planar graphs [11,12], K_5-minor-free graphs [9], and d-degenerate graphs [9] from the context of graph theory.

Several papers clarified the complexity status of COLORING RECONFIGURATION [2–5], which asks if two given k-colorings of a graph can be transformed into each other by recoloring a single vertex at a time, while keeping k-colorings during the transformation. This problem can be seen as a restricted variant of KEMPE REACHABILITY, where the restriction is that we can apply a Kempe-change only to a connected component consisting of a single vertex; we call such a Kempe-change an *elementary recoloring*. For example, the rightmost Kempe-change in Fig. 1 recolors only a single vertex, so it is an elementary recoloring; while the other two Kempe-changes are not. Indeed, under elementary recolorings, there is no transformation between the leftmost and rightmost 3-colorings in the figure, because we cannot recolor any vertex in the triangle (using only three colors). Conversely, any transformation under elementary recolorings can be seen as a transformation under (unrestricted) Kempe-changes.

1.2 Our Contribution

The main purpose of this paper is to determine the polynomial-time solvability of KEMPE CONNECTIVITY and KEMPE REACHABILITY from the viewpoint of graph classes. Furthermore, we give upper bounds on the Kempe-distance between any two Kempe-equivalent k-colorings of a graph, where the *Kempe-distance* between two k-colorings is the minimum number of Kempe-changes needed to transform one into the other.

We prove that KEMPE REACHABILITY is PSPACE-complete for any fixed $k \geq 3$. Note that any instance of KEMPE CONNECTIVITY (and hence KEMPE REACHABILITY) with $k \leq 2$ is a yes-instance [12]. Therefore, under standard complexity assumptions, our hardness result gives a sharp threshold for the polynomial-time solvability with respect to the number k of colors. In addition, this result implies that there are pairs of Kempe-equivalent k-colorings of super-polynomial Kempe-distances assuming PSPACE $\neq$ NP. By a more sophisticated reduction, we show that KEMPE REACHABILITY is PSPACE-complete even when restricted to 3-colorings of planar graphs of maximum degree six.

We provide positive results for three subclasses of perfect graphs. We prove that any instance of KEMPE CONNECTIVITY (and hence KEMPE REACHABIL-ITY) on chordal graphs, bipartite graphs, or cographs is a yes-instance. We also remark that there are no-instances for perfect graphs in general. Our proofs are constructive, and give polynomial upper bounds on Kempe-distances. Note that bipartite graphs are known to be k-Kempe-mixing for any $k \geq 2$ [12], but no explicit bound was given.

Due to the page limitation, we omit proofs for the claims marked with ($*$).

2   Preliminaries

Let G be a simple graph with vertex set $V(G)$ and edge set $E(G)$. We denote by $\chi(G)$ the chromatic number of G, that is, the minimum number k such that G admits a k-coloring.

Let k be a positive integer. For a k-coloring of a graph G and two colors $a, b \in \{1, 2, \ldots, k\}$, consider the subgraph $G_{a,b}$ of G induced by two color classes of a and b. Then, each connected component in $G_{a,b}$ is called an (a, b)-Kempe-chain. A Kempe-change for two colors a and b (sometimes referred as an (a, b)-Kempe-change) swaps two colors a and b assigned to the vertices in an (a, b)-Kempe-chain. For two k-colorings f and f' of G, a sequence $\langle f_0, f_1, \ldots, f_q \rangle$ of k-colorings of G is called a transformation from f to f' if $f_0 = f$, $f_q = f'$, and f_i can be obtained from f_{i-1} by a single Kempe-change for some pair of colors, for every $i \in \{1, 2, \ldots, q\}$. The length of this transformation is q, that is, the number of Kempe-changes. The k-Kempe-diameter of a graph G is defined to be the maximum length of a shortest transformation between any two k-colorings of G. If G is not k-Kempe-mixing, then we define the k-Kempe-diameter of G as $+\infty$.

3   PSPACE-Completeness

In this section, we first prove that KEMPE REACHABILITY is PSPACE-complete for $k = 3$ and planar graphs, and then prove the PSPACE-completeness for any fixed $k \geq 3$ and general graphs. Recall that any instance of KEMPE REACHA-BILITY with $k \leq 2$ is a yes-instance [12]. Therefore, our theorems give a sharp complexity analysis with respect to the number k of colors.

Observe that KEMPE REACHABILITY can be solved in (most conveniently, nondeterministic [18]) polynomial space, and hence it is in PSPACE. Therefore, we will prove the PSPACE-hardness of the problem.

3.1   PSPACE-Hardness for $k = 3$ and Planar Graphs

We will give the following theorem in this subsection.

Theorem 1. KEMPE REACHABILITY is PSPACE-complete even for three colors and planar graphs of maximum degree 6.

As a proof of the theorem, we will give a polynomial-time reduction from NONDETERMINISTIC CONSTRAINT LOGIC [6,20]. Our proof consists of two stages. In the first stage, we introduce the "list" variant of KEMPE REACH-ABILITY, and prove its PSPACE-hardness for three colors and planar graphs of bounded bandwidth and maximum degree 4. The list constraint restricts the colors that can be assigned to each vertex, and makes the construction and analysis of gadgets easier. In the second stage, we remove the list constraint. To do so, the bandwidth becomes unbounded, and the maximum degree increases to six; while we can keep the planarity and the number of colors.

Definition of Nondeterministic Constraint Logic

We now define NONDETERMINISTIC CONSTRAINT LOGIC (NCL for short) [6,20], as follows. An NCL "machine" M is specified by an undirected graph together with an assignment of weights from $\{1,2\}$ to each edge of the graph. A *configuration* of M is an orientation (direction) of the edges such that the sum of weights of incoming arcs at each vertex is at least 2. We sometimes call a configuration of M a *valid* orientation (or a *valid* configuration) of M to emphasize the constraint of incoming weights. Figure 2(a) illustrates a configuration of an NCL machine, where each weight-2 edge is depicted by a thick (blue) line and each weight-1 edge by a thin (orange) line. Given an NCL machine M and its two configurations C_s and C_t, it is known to be PSPACE-complete to determine whether there exists a sequence of configurations of M which transforms C_s into C_t, where every intermediate configuration is obtained from the previous one by changing the direction of a single edge [6,20].

In fact, NCL remains PSPACE-complete even for planar graphs of bounded bandwidth consisting only of two types of vertices, called "AND nodes" and "OR nodes" [20]. A vertex of degree three is called an *AND node* if its three incident edges have weights 1, 1 and 2. (See Fig. 2(b).) An AND node v behaves as a logical AND, in the following sense: the weight-2 edge can be directed outward for v if and only if both two weight-1 edges are directed inward for v. Note that, however, the weight-2 edge is not necessarily directed outward even when both weight-1 edges are directed inward. A vertex of degree three is called an *OR node* if its three incident edges have weights 2, 2 and 2. (See Fig. 2(c).) An OR

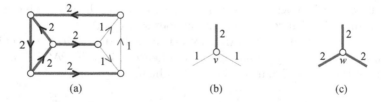

(a)   (b)   (c)

Fig. 2. (a) A configuration of an NCL machine, (b) an AND node v, and (c) an OR node w. (Color figure online)

node w behaves as a logical OR: one of the three edges can be directed outward for w if and only if at least one of the other two edges is directed inward for w. For example, the NCL machine in Fig. 2(a) consists of only AND/OR nodes. From now on, we assume that such a graph is given as an input. We sometimes call an edge in an NCL machine an *NCL edge*.

The First Stage: List Variant
We now introduce the list variant of the problem. Assume that each vertex v of a graph G has a set $L(v) \subseteq \{1, 2, \ldots, k\}$ of colors, called the *list* of v. Then, a k-coloring f of G is called a *list coloring* of G if $f(v) \in L(v)$ holds for every vertex $v \in V(G)$. In the *list variant* of KEMPE REACHABILITY, an (a, b)-Kempe-change for two colors a and b can be applied to an (a, b)-Kempe-chain C if and only if all the vertices in C have both colors a and b in their lists.

Theorem 2. *The list variant of* KEMPE REACHABILITY *is* PSPACE-*complete, even restricted to planar graphs of bounded bandwidth and maximum degree 4, and only two kinds of lists chosen from three colors.*

Construction of Our Reduction for Theorem 2
Suppose that we are given an instance of NCL, that is, an NCL machine M and two configurations C_s and C_t of M. We build a corresponding graph G, by replacing each of NCL edges and AND/OR nodes with its corresponding gadget; if an NCL edge e is incident to a node v, then we glue the corresponding gadgets for e and v together by identifying a pair of vertices, called *connectors* between v and e (or sometimes called (v, e)-*connectors*), as illustrated in Fig. 3. Thus, each edge gadget has two pairs of connectors, and each AND/OR gadget has three pairs of connectors. Our gadgets are all edge-disjoint, and share only connectors.

Figure 4 shows our three types of gadgets which correspond to NCL edges and AND/OR nodes. Notice that our three gadgets are planar and of maximum degree at most 4. Since M is assumed to be planar, the constructed graph G is also planar and of maximum degree 4. In addition, the bandwidth of G is bounded by a fixed constant, because the input NCL machine M is of bounded bandwidth and each gadget consists of a constant number of edges.

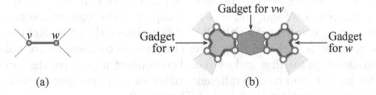

(a)   (b)

Fig. 3. (a) An NCL edge vw, and (b) its corresponding gadgets, where the connectors are depicted by (red) double-lined circles. (Color figure online)

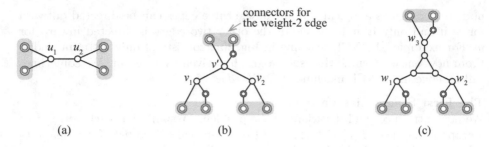

Fig. 4. (a) Edge gadget, (b) AND gadget, and (c) OR gadget, where each gray area represents a pair of connectors. In the AND gadget, the top pair of connectors will be identified with that of the weight-2 edge.

In Fig. 4, all vertices depicted by (both red and black) double-lined circles have the list $\{1, 2\}$, while the other vertices have the list $\{1, 2, 3\}$. Thus, there are only two kinds of lists chosen from three colors 1, 2 and 3. We make two list colorings f_s and f_t of G which correspond to two given configurations C_s and C_t of the NCL machine M. In our reduction, we construct the correspondence between configurations of M and list colorings of G, as follows. We regard that the direction of an NCL edge $e = vw$ is inward for v if the two (v, e)-connectors receive different colors. On the other hand, we regard that the direction of $e = vw$ is outward for w if the two (w, e)-connectors receive the same color. Note that there are (in general, exponentially) many list colorings of G which correspond to the same configuration of M. In contrast, the correspondence implies that no two distinct configurations of M correspond to the same list coloring of G. We arbitrarily choose two list colorings f_s and f_t of G which correspond to C_s and C_t, respectively.

This completes the construction of our corresponding instance. The construction can be performed in polynomial time.

Correctness of Our Reduction

We first roughly explain the key behavior of gadgets. In Fig. 4, eight vertices u_1, u_2, v', v_1, v_2, w_1, w_2 and w_3 are called *gate* vertices, that play an important role. By coloring some gate vertices with the color 3, we can ensure that a $(1, 2)$-Kempe-change remains local, and does not propagate throughout the graph. The gate vertex can be colored with either 1 or 2 (and hence the $(1, 2)$-Kempe-change can propagate further) if and only if its closest two connectors are colored as the inward direction for that gadget (i.e., two connectors receive the same color for the edge gadget, and receive different colors for node gadgets). Formally, the following lemma completes the proof of Theorem 2.

Lemma 1 (∗). *There exists a desired sequence of configurations of M between C_s and C_t if and only if two list colorings f_s and f_t of G are Kempe-equivalent.*

The Second Stage: Removing the List Constraint

We now describe how to remove the list constraint. The key point for the removal is to notice that it is *not* important to forbid the color 3 to be assigned to the (double-lined circle) vertices having the list of size 2, but the *same* color $c \in \{1, 2, 3\}$ is forbidden at *all* such vertices in the whole graph G. Then, at the gate vertices, the color c will control the propagations of Kempe-changes swapping the other two colors. To accomplish this property, the easiest way is to add a new vertex to G, and join it with all the (double-lined circle) vertices having the list of size 2. However, this does not preserve the planarity, and produces a high-degree vertex. Therefore, a more elaborate way is required.

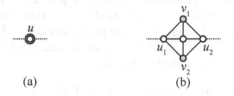

(a)   (b)

Fig. 5. (a) A vertex u in V_2, and (b) its replacement without list.

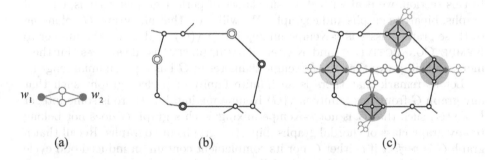

(a)   (b)   (c)

Fig. 6. (a) A diamond, (b) a face of G whose boundary has four vertices from V_2, and (c) its replacement without the list constraint by a chain of diamonds.

Let V_2 be the set of vertices in G that have the list of size 2. Notice that, in our reduction, every vertex in V_2 is of degree 2. We first replace each vertex in V_2 by the gadget as in Fig. 5. Observe that in any 3-coloring, the vertices u_1 and u_2 in Fig. 5(b) always receive the same color. Thus, u_1 is recolored by a Kempe-change if and only if u_2 is. Similarly, both gray vertices v_1 and v_2 always receive the same color in any 3-coloring. Therefore, if we can force that all such gray vertices in the whole graph receive the same color c at any time during a transformation, then it corresponds to forbidding the color c to be assigned to any vertex in V_2.

For this purpose, we will use a diamond, as illustrated in Fig. 6(a). Note that two vertices w_1 and w_2 always receive the same color in any 3-coloring. Consider a plane embedding of G. For each face F of G, we link the gray vertices on the boundary of F together using a chain of diamonds, as shown in Fig. 6(b) and (c) for example. This can be done while preserving the planarity of the graph. Additionally, the maximum degree is 6, which occurs when three diamond meet at a single vertex. This completes the proof of Theorem 1. □

3.2 PSPACE-Hardness for Any Fixed $k \geq 3$

In this subsection, we prove the PSPACE-hardness of KEMPE REACHABILITY for any fixed integer $k \geq 3$, although we lose the planarity. We note that losing the planarity is reasonable: It is known that any instance of the problem with k colors is a yes-instance if a given planar graph admits a $(k-1)$-coloring [12]. Therefore, any instance with $k \geq 5$ is a yes-instance for planar graphs. For general graphs, we give the following theorem.

Theorem 3 (∗). KEMPE REACHABILITY *is* PSPACE-*complete for k colors, where $k \geq 3$ is an arbitrary fixed integer.*

4 Classes of Perfect Graphs

In this section, we deal with three subclasses of perfect graphs, that is, chordal graphs, bipartite graphs and cographs. We will show that any graph G belonging to these graph classes is k-Kempe-mixing if $k \geq \chi(G)$, and hence the answer to KEMPE CONNECTIVITY (and KEMPE REACHABILITY) is always yes. Furthermore, we will prove that the k-Kempe-diameter of G has a polynomial length.

Let us remark that there is an infinite family of perfect graphs such that any graph G from this family is $\chi(G)$-Kempe-mixing, but there is some integer $k > \chi(G)$ such that G is not k-Kempe-mixing; such a graph G does not belong to any graph class of chordal graphs, bipartite graphs or cographs. Recall that a graph G is *perfect* if neither G nor its complement contain an induced odd cycle of length at least five. We have the following proposition.

Proposition 1 (∗). *There is a perfect graph G that is $\chi(G)$-Kempe-mixing but not $(\chi(G) + 1)$-Kempe-mixing.*

4.1 Chordal Graphs

A graph is *chordal* if it contains no induced cycle of length at least four. We give the following theorem, which appeared in the PhD thesis of one of the authors [16].

Theorem 4. *Let G be a chordal graph with n vertices, and let $k \geq \chi(G)$ be any integer. Then, G is k-Kempe-mixing, and its k-Kempe-diameter is at most n.*

We give a constructive proof of Theorem 4. Our main tool is the well-known characterization of chordal graphs in terms of perfect elimination orderings [17]. Let G be a chordal graph with n vertices. A linear ordering $v_1 \prec v_2 \prec \ldots \prec v_n$ of the vertices of G is called a *perfect elimination ordering* (PEO), if for every i, $1 \le i \le n$, the neighbors of v_i in G that appear after v_i in the ordering induce a clique. A graph is chordal if and only if it admits a perfect elimination ordering. Furthermore, such an ordering can be computed in linear time [17].

It suffices to prove the theorem for a connected chordal graph G. Let f and g be any two k-colorings of G. We give an actual transformation from f to g of length at most n. Let $v_1 \prec v_2 \prec \ldots \prec v_n$ be a PEO of the vertices of G. For each i from n downto 1, if the current color of v_i is different from its target color $g(v_i)$, we exchange the current color with $g(v_i)$ via a single Kempe-change. The procedure executes at most n Kempe-changes, and the following claim implies that this sequence of Kempe-changes gives a transformation from f to g.

Lemma 2 (∗). *The Kempe-change that assigns to v_i its target color $g(v_i)$ does not alter the color of any vertex v_j, $j > i$.*

4.2 Bipartite Graphs

It is already known that any bipartite graph G is k-Kempe-mixing for every $k \ge \chi(G)$ [12]. However, its k-Kempe-diameter was not analyzed. We thus give the following theorem, whose proof is different from [12]. We note that our constructive proof blow produces an actual transformation in polynomial time.

Theorem 5. *Let G be a bipartite graph with n vertices, and let $k \ge \chi(G)$ be any integer. Then, G is k-Kempe-mixing and its k-Kempe-diameter is at most $3n/2$.*

Proof. It suffices to prove the theorem for a connected bipartite graph G. Let f and g be any two k-colorings of G. To prove the theorem, we will construct a transformation from f to g of length at most $3n/2$.

Let $A, B \subseteq V(G)$ be the bipartition of G; we assume without loss of generality that $|A| \le |B|$, and hence $|A| \le n/2$. Choose an arbitrary color c. We first transform f into f' via at most $|A|$ Kempe-changes, where f' is a k-coloring of G such that all vertices in A have the same color c. If all vertices $v \in A$ satisfy $f(v) = c$, then f is already a desired k-coloring. We thus consider the case where there is at least one vertex in A which has a color other than c. In this case, we perform a single Kempe-change that changes the color of v to c. We note that this Kempe-change does not recolor any vertex in A which already has the color c, because the distance from v to such a vertex is even. We apply such a Kempe-change to each vertex $v \in A$ having a color other than c, and hence we can obtain a desired k-coloring f' of G via at most $|A|$ Kempe-changes.

By the similar way, we can transform g into g' via at most $|A|$ Kempe-changes, where g' is a k-coloring of G such that all vertices in A have the same color c. Note that we here chose the same color c for g' as for f'. Since the transformation is reversible, this gives a transformation from g' to g via at most $|A|$ Kempe-changes.

Finally, we construct a transformation from f' to g' of length at most $|B|$. Assume that $f' \neq g'$ holds, otherwise we are done. Then, there is at least one vertex in B which has different colors in f' and g'. We perform a single Kempe-change that changes the color of v from $f'(v)$ to $g'(v)$. Notice that this Kempe-change recolors only the single vertex v, because $f'(v) \neq c$, $g'(v) \neq c$, and all neighbors (in A) of v are colored with c. We apply such a Kempe-change to each vertex $v \in B$ having a color other than $g'(v)$, and hence we can obtain the k-coloring g' via at most $|B|$ Kempe-changes.

The total length of the transformation above is at most $2|A| + |B| \leq 3n/2$, since $|A| + |B| = n$ and $|A| \leq n/2$. □

We conclude this subsection by showing that forests (which are bipartite and chordal) admit a better upper bound on the Kempe-diameter. For two k-colorings f and g of a graph G, we denote by $\Delta(f, g)$ the number of vertices whose color differs with respect to f and g, that is, $\Delta(f, g) :=$ $|\{v \in V(G) : f(v) \neq g(v)\}|$. Since $\Delta(f, g) \leq |V(G)|$, the following upper bound is always at least as good as the ones from Theorems 4 and 5.

Theorem 6 (∗). *Let G be a forest, and let $k \geq \chi(G)$ be any integer. For any two k-colorings f and g of G, there is a transformation from f to g of length at most $\Delta(f, g)$.*

4.3 Cographs

The class of cographs is also known as P_4-free graphs. In this subsection, we give the following theorem.

Theorem 7 (∗). *Let G be a cograph with n vertices, and let $k \geq \chi(G)$ be any integer. Then, G is k-Kempe-mixing, and its k-Kempe-diameter is at most $2n \log n$.*

Our proof of the theorem can be sketched as follows. Let G be any cograph on n vertices. Using the recursive characterization of cographs, we prove the following property by induction on n: For any k-coloring f and any $\chi(G)$-coloring g of G, there is a transformation from f to g of length at most $\max(1, n \log n)$ with only using colors appeared in f or g. Note that Theorem 7 follows from this stronger property, simply by choosing some intermediate $\chi(G)$-coloring.

5 Conclusion

We investigated the complexity of KEMPE REACHABILITY and KEMPE CONNECTIVITY on several classes of graphs. For the former problem, we gave a sharp threshold on the polynomial-time solvability with respect to the number k of colors, and also proved the PSPACE-completeness for 3-colorings on planar graphs of maximum degree six. For the latter problem, we proved that any instance on chordal graphs, bipartite graphs, or cographs is a yes-instance; thus any instance of KEMPE REACHABILITY on these graphs is a yes-instance, too. We also gave polynomial upper bounds on k-Kempe-diameters on these graphs.

Combining Theorem 1 with the results from [12] leaves open the complexity status of KEMPE REACHABILITY for 4-colorings of planar graphs that admit no 3-coloring. Another interesting question is whether KEMPE REACHABILITY on perfect graphs admits a polynomial-time algorithm.

References

1. Bonamy, M., Bousquet, N., Feghali, C., Johnson, M.: On a conjecture of Mohar concerning Kempe equivalence of regular graphs. J. Comb. Theory Ser. B **135**, 179–199 (2019)
2. Bonsma, P., Cereceda, L.: Finding paths between graph colourings: PSPACE-completeness and superpolynomial distances. Theor. Comput. Sci. **410**, 5215–5226 (2009)
3. Bonsma, P., Paulusma, D.: Using contracted solution graphs for solving reconfiguration problems. In: Proceedings of MFCS 2016, LIPIcs, vol. 58, pp. 20:1–20:15 (2016)
4. Cereceda, L., van den Heuvel, J., Johnson, M.: Finding paths between 3-colorings. J. Graph Theory **67**(1), 69–82 (2011)
5. Hatanaka, T., Ito, T., Zhou, X.: The coloring reconfiguration problem on specific graph classes. IEICE Trans. Inf. Syst. **E102–D**(3), 423–429 (2019)
6. Hearn, R., Demaine, E.: PSPACE-completeness of sliding-block puzzles and other problems through the nondeterministic constraint logic model of computation. Theor. Comput. Sci. **343**(1–2), 72–96 (2005)
7. van den Heuvel, J.: The complexity of change. Surv. Comb. **2013**, 127–160 (2013)
8. Kempe, A.B.: On the geographical problem of the four colours. Am. J. Math. **2**(3), 193–200 (1879)
9. Las Vergnas, M., Meyniel, H.: Kempe classes and the Hadwiger conjecture. J. Comb. Theory Ser. B **31**(1), 95–104 (1981)
10. Melnikov, L.S., Vizing, V.G.: New proof of Brooks' theorem. J. Comb. Theory **7**(4), 289–290 (1969)
11. Meyniel, H.: Les 5-colorations d'un graphe planaire forment une classe de commutation unique. J. Comb. Theory Ser. B **24**(3), 251–257 (1978)
12. Mohar, B.: Kempe equivalence of colorings. In: Graph Theory in Paris. Trends in Mathematics, pp. 287–297 (2007). Proc. of a Conference in Memory of Claude Berge
13. Mohar, B., Salas, J.: A new Kempe invariant and the (non)-ergodicity of the Wang-Swendsen-Kotecký algorithm. J. Phys. A Math. Theor. **42**(22), 225204 (2009)
14. Mohar, B., Salas, J.: On the non-ergodicity of the Swendsen-Wang-Kotecký algorithm on the Kagomé lattice. J. Stat. Mech. Theory Exp. **2010**(05), P05016 (2010)
15. Mühlenthaler, M., Wanka, R.: The connectedness of clash-free timetables. In: Proceedings of PATAT 2014, pp. 330–346 (2014)
16. Mühlenthaler, M.: Fairness in Academic Course Timetabling. Lecture Notes in Economics and Mathematical Systems, vol. 678. Springer, Cham (2015). https://doi.org/10.1007/978-3-319-12799-6
17. Rose, D.J., Tarjan, R.E., Lueker, G.S.: Algorithmic aspects of vertex elimination on graphs. SIAM J. Comput. **5**(2), 266–283 (1976)

18. Savitch, W.J.: Relationships between nondeterministic and deterministic tape complexities. J. Comput. Syst. Sci. **4**, 177–192 (1970)
19. Vigoda, E.: Improved bounds for sampling colorings. J. Math. Phys. **41**(3), 1555–1569 (2000)
20. van der Zanden, T.C.: Parameterized complexity of graph constraint logic. In: Proceedings of IPEC 2015, LIPIcs, vol. 43, pp. 282–293 (2015)

Dominating Set on Overlap Graphs
of Rectangles Intersecting a Line

Dibyayan Chakraborty$^{(\boxtimes)}$, Sandip Das, and Joydeep Mukherjee

Indian Statistical Institute, Kolkata, India
dibyayancg@gmail.com

Abstract. A graph $G = (V, E)$ is called a *rectangle overlap* graph if there is a bijection between V and a set $\mathcal{R}$ of axis-parallel rectangles such that two vertices in V are adjacent if and only if the corresponding rectangles in $\mathcal{R}$ *overlap* i.e. their boundaries intersect.

In this article, assuming the *Unique Games Conjecture* to be true we show that it is not possible to approximate the MINIMUM DOMINATING SET (MDS) problem on rectangle overlap graphs with a factor $(2 - \epsilon)$ for any $\epsilon > 0$. Previously only APX hardness was known for this problem due to Erlebach and Van Leeuwen (LATIN 2008) and Damian and Pemmaraju (Inf. Process. Lett. 2006). We give an $O(n^5)$-time 768-approximation algorithm for the MDS problem on *stabbed rectangle overlap graphs* i.e. overlap graphs of rectangles intersecting a common straight line. Here n denotes the number of vertices of the input graph.

Our second result is the first constant factor approximation for MDS problem on *stabbed rectangle overlap graphs* which is a strict generalisation of a graphclass considered by Bandyapadhyay et al. (MFCS 2018).

1 Introduction and Results

An *overlap representation* $\mathcal{R}$ of a graph $G = (V, E)$ is a family of sets $\{R_u\}_{u \in V}$ such that $uv \in E$ if and only if $R_u \cap R_v \neq \emptyset$, $R_u \nsubseteq R_v$ and $R_v \nsubseteq R_u$. When $\mathcal{R}$ is a collection of geometric objects, it is said to be a *geometric overlap representation* of G. Particular cases of relevance to this paper are when $\mathcal{R}$ is a collection of *real intervals*, in which case it is called an *interval overlap representation* and when $\mathcal{R}$ is a collection of axis-parallel rectangles, in which case, it is called a *rectangle overlap representation*. A graph G is an *interval overlap graph* (respectively a *rectangle overlap graph*) if G has an interval overlap representation (respectively a rectangle overlap representation). Similarly, a *geometric intersection representation* $\mathcal{R}$ of a graph $G = (V, E)$ is a family of objects $\{R_u\}_{u \in V}$ such that $uv \in E$ if and only if $R_u \cap R_v \neq \emptyset$. The notions of *rectangle intersection representation* and *rectangle intersection graphs* are defined accordingly.

A *dominating set* of a graph $G = (V, E)$ is a subset D of vertices V such that each vertex in $V \setminus D$ is adjacent to some vertex in D. The MINIMUM DOMINATING SET (MDS) problem is to find a minimum cardinality dominating set of a graph. It is not possible to approximate the MDS problem on general graphs with n vertices within $(1 - \alpha) \ln n$ unless $NP \subseteq DTIME(n^{O(\log \log n)})$ [6].

© Springer Nature Switzerland AG 2019
D.-Z. Du et al. (Eds.): COCOON 2019, LNCS 11653, pp. 65–77, 2019.
https://doi.org/10.1007/978-3-030-26176-4_6

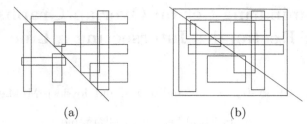

Fig. 1. (a) A set of diagonally anchored rectangles (b) A set of stabbed rectangles.

Erlebach and Van Leeuwen [10] proved that the MDS problem is APX-hard on rectangle intersection graphs. Their reduction procedure can be modified to show that the MDS problem remains APX-hard on rectangle overlap graphs. The same result also follows from Damian and Pemmaraju [8]. In this paper, we prove the following theorem.

Theorem 1. *Assuming the* Unique Games Conjecture *to be true, it is not possible to have a polynmial time* $(2 - \epsilon)$-*approximation algorithm for the* MDS *problem on rectangle overlap graphs, even if a rectangle overlap representation is given as input.*

It is a challenging open question to determine if there is a constant factor approximation algorithm for the MDS problem on rectangle intersection graphs and rectangle overlap graphs. Erlebach et al. [10] gave an $O(c^3)$-approximation for the MDS problem on intersection graphs of rectangles with aspect-ratio at most c. The MDS problem admits PTAS on the intersection graphs of *non-piercing*[1] rectangles [11]. Damian-Iordache and Pemmaraju [9] gave a $(2 + \epsilon)$-approximation for the MDS problem on interval overlap graphs. Bousquet et al. [3] studied the parametrized complexity of the MDS problems on interval overlap graphs.

A set $\mathcal{R}$ of rectangles is a set of *diagonally anchored* rectangles if there is a straight line l with slope -1 such that intersection of any $R \in \mathcal{R}$ with l is exactly one corner of R. See Fig. 1(a) for an example. Surprisingly, the MDS problem remains NP-Hard on intersection graphs of diagonally anchored rectangles [16]. Bandyapadhyay et al. [1] gave a $(2 + \epsilon)$-approximation algorithm for the same.

In this paper, we consider the following generalisation of interval overlap graphs and intersection graphs of diagonally anchored rectangles. A set $\mathcal{R}$ of rectangles is *stabbed* if there is a straight line that intersects all rectangles in $\mathcal{R}$. See Fig. 1(b) for an example. A graph G is a *stabbed rectangle overlap graph* if G has a stabbed rectangle overlap representation. We prove the following.

Theorem 2. *Given a stabbed rectangle overlap representation of a graph* $G = (V, E)$, *there is an* $O(|V|^5)$-*time 768-approximation algorithm for the* MDS *problem on* G.

---

[1] Two rectangles R and R' are non-piercing if both $R \setminus R'$ and $R' \setminus R$ are connected.

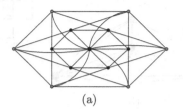 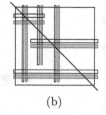

(a)   (b)

Fig. 2. A graph which is a stabbed rectangle overlap graphs but neither an interval overlap graph nor an intersection graph of diagonally anchored rectangles.

We note that interval overlap graphs and intersection graphs of diagonally anchored rectangles are both strict subclasses of stabbed rectangle overlap graphs. Figure 2(a) is an example of a graph which is a stabbed rectangle overlap graph (representation shown in Fig. 2(b)) but neither an interval overlap graph nor an intersection graph of diagonally anchored rectangles [7]. We also note that approximation algorithms for optimization problems like MAXIMUM INDEPENDENT SET and MINIMUM HITTING SET on "stabbed" geometric objects have been studied [4,5,7,13,15].

In Sect. 2, we prove Theorem 1 and in Sect. 3, we prove some lemmas required to prove Theorem 2. Then in Sect. 4, we prove Theorem 2.

Throughout this paper, ILP stands for Integer Linear Program and LP stands for Linear Program. Moreover, $OPT(Q)$ and $OPT(Q_l)$ denote the cost of the optimum solution of an ILP Q and LP Q_l respectively.

2 Proof of Theorem 1

A *vertex cover* of a graph $G = (V, E)$ is a subset C of V such that each edge in E has an endvertex which lies in C. The MINIMUM VERTEX COVER problem is to find a minimum cardinality vertex cover of a graph. Assuming *Unique Games Conjecture* to be true, the MINIMUM VERTEX COVER has no polynomial time $(2-\epsilon)$-approximation algorithm for any $\epsilon > 0$ [14]. We shall reduce the MINIMUM VERTEX COVER problem to the MDS problem on rectangle overlap graphs.

Given a graph $G = (V, E)$, construct another graph $G' = (V', E')$ as follows. Define $V' = V \cup E$. Define $E' = \{uv: u, v \in V\} \cup \{ue: u \in V, e \in E$ and u is an endvertex of e in $G\}$. Observe that, G has a vertex cover of size k if and only if G' has a dominating set of size k.

Therefore, we will be done by showing that G' is a rectangle overlap graph. Let $V = \{v_1, v_2, \ldots, v_n\}$ and for each $v_i \in V$ define $R_{v_i} = [i, n+1] \times [-i, 0]$ (See Fig. 3(c) for illustration). Observe that, for $i < j$, the bottom boundary of R_{v_i} intersects the left boundary of R_{v_j} at a point.

Notice that, each vertex $u \in V' \setminus V$, has degree two and is adjacent to exactly two vertices of V. For each vertex $u \in V' \setminus V$, introduce a rectangle R_u which overlaps only with R_{v_i} and R_{v_j} where $\{v_i, v_j\}$ is the set of vertices adjacent to u with $i < j$. This is possible as R_u can be kept around the unique

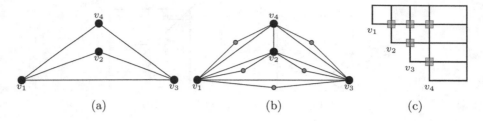

Fig. 3. Reduction procedure for Theorem 1. (a) Input graph G, (b) The graph G' and (c) rectangle overlap representation of G'.

intersection point of the bottom boundary of R_{v_i} and the left boundary of R_{v_j} (see Fig. 3(c) for illustration). Formally, for each $u \in V' \setminus V$, define $R_u = [p - \epsilon, p + \epsilon] \times [q - \epsilon, q + \epsilon]$ where $\epsilon = \frac{1}{|V|}$ and (p, q) is the intersection point of the bottom boundary of R_{v_i} and the left boundary of R_{v_j}. Observe that the set of rectangles $\mathcal{R}' = \{R_{v_i} : v_i \in V\} \cup \{R_u : u \in V' \setminus V\}$ is a rectangle overlap representation of G'. This completes the proof.

3 Necessary Lemmas to Prove Theorem 2

To prove Theorem 2, we need to study two related problems described below and prove two lemmas.

The *Local Vertical Segment Covering* (LVSC) Problem: In this problem, the inputs are a set H of disjoint horizontal segments intersecting a common straight line and a set V of disjoint vertical segments. The objective is to select a minimum number of horizontal segments that intersect all vertical segments. Throughout this article, we let $\mathcal{LVSC}(V, H)$ denote an LVSC instance.

The *Local Horizontal Segment Covering* (LHSC) Problem: In this problem, the inputs are a set H of disjoint horizontal segments all intersecting a common straight line and a set V of disjoint vertical segments. The objective is to select a minimum number of vertical segments that intersect all horizontal segments. Throughout this article, we let $\mathcal{LHSC}(V, H)$ denote an LHSC instance.

Lemma 1. *Let $\mathcal{C}$ be an ILP formulation of an $\mathcal{LVSC}(V, H)$ instance. There is an $O(n^5)$ time algorithm to compute a set $D \subseteq H$ which gives a feasible solution of $\mathcal{C}$ and $|D| \leq 8 \cdot OPT(\mathcal{C}_l)$ where $n = |V \cup H|$ and $\mathcal{C}_l$ is the relaxed LP formulation of $\mathcal{C}$.*

Lemma 2. *Let $\mathcal{C}$ be an ILP formulation of an $\mathcal{LHSC}(V, H)$ instance. There is an $O(n^5)$ time algorithm to compute a set $D \subseteq V$ which gives a feasible solution of $\mathcal{C}$ and $|D| \leq 8 \cdot OPT(\mathcal{C}_l)$ where $n = |V \cup H|$ and $\mathcal{C}_l$ is the relaxed LP formulation of $\mathcal{C}$.*

To prove Theorem 2 (the main result of this paper), the above lemmas are enough. However, to prove Lemma 1 we need to study the *stabbing segment with rays* (SSR) problem introduced by Katz et al. [12].

The *Stabbing Segment with Rays* (SSR) Problem. In this problem, the inputs are a set R of disjoint leftward-directed rays and a set V of disjoint vertical segments. The objective is to select a minimum number of leftward-directed rays that intersect all vertical segments. Throughout this article, we let $SSR(R, V)$ denote an SSR instance. We shall prove the following lemma in Sect. 3.1.

Lemma 3. *Let C be an ILP formulation of an $SSR(R, V)$ instance. There is an $O((n + m) \log(n + m))$ time algorithm to compute a set $D \subseteq R$ which gives a feasible solution of C and $|D| \leq 2 \cdot OPT(C_l)$ where $n + m = |V \cup H|$ and C_l is the relaxed LP formulation of C.*

We shall prove Lemma 1 in Sect. 3.2 using Lemma 3. To prove Lemma 2, we need to study the *stabbing rays with segments* (SRS) problem also introduced by Katz et al. [12].

The *Stabbing Rays with Segments* (SRS) problem: In this problem, the inputs are a set R of disjoint leftward-directed rays and a set V of disjoint vertical segments. The objective is to select a minimum number of vertical segments that intersect all leftward-directed rays. Throughout this article, we let $SRS(R, V)$ denote an SRS instance. We shall prove the following lemma.

Lemma 4. *Let C be an ILP formulation of an $SRS(R, V)$ instance. There is an $O(n \log n)$ time algorithm to compute a set $D \subseteq V$ which gives a feasible solution of C and $|D| \leq 2 \cdot OPT(C_l)$ where $n = |V|$ and C_l is the relaxed LP formulation of C.*

Due to space constraints, the proofs of Lemmas 4 and 2 are omitted.

3.1   Proof of Lemma 3

In this section, we represent a *leftward-directed ray* by simply a *ray* and a *vertical segment* by a *segment* in short. Let R be a set of disjoint rays and V be a set of disjoint vertical segments. We assume each segment intersects at least one ray in R and no two segments in V has the same x-coordinate.

To prove Lemma 3, first we present an iterative algorithm consisting of three main steps. The first step is to include all rays $r \in R$ in heuristic solution S whenever some segments in V intersect precisely a single ray r in that iterative step. In the next step, delete all segments intersecting any ray in S from V. In the final step, find a ray in $R \setminus S$ whose x-coordinate of the right endpoint is the smallest among all rays in $R \setminus S$ and delete it from R (when there are multiple such rays, choose anyone arbitrarily). We repeat the above three steps until V is empty. The above algorithm takes $O((|R| + |V|) \log(|R| + |V|))$ time (using segment trees [2]) and outputs a set S of rays such that all segments in V intersect at least one ray in S.

We describe the above algorithm formally in Algorithm 1. Below we introduce some notations used to describe the algorithm. We assign *token* $T_r = \{r\}$ for each $r \in R$ initially. For $i \geq 1$, let R_i, V_i, S_i be the set of rays, the set of segments and the heuristic solution constructed by this Algorithm 1, respectively at the *end* of ith iteration. A ray $r \in R_i$ is *critical* if there is a segment $v \in V_i$ such that r is the only ray in R_i that intersects v. We describe a *discharging technique* below.

Let D be a subset of R. A ray $r \in D$ lies *between* two rays $r', r'' \in D$ if the y-coordinate of r lies between those of r', r''. A ray $r \in D$ lies *just above* (resp. *just below*) a ray $r' \in D$ if y-coordinate of r is greater (resp. smaller) than that of r' and no other ray lies between r, r' in D. Two rays $r, r' \in D$ are *neighbours* of each other if r lies just above or below r'.

Discharging Method: Let $r \in R_{i-1} \backslash S_i$ be a ray whose x-coordinate of the right endpoint is the smallest. The phrase "r discharges the token to its neighbours" in the ith iteration means the following operations in the given order.

(i) Let r' lie just above r and r'' lie just below r in $R_{i-1} \setminus S_i$. For all $x \in T_r$ (x and r not necessarily distinct) do the following. If there is a segment in V_i that intersects x, r' and r then assign $T_{r'} = T_{r'} \cup \{x\}$ and if there is a segment in V_i that intersects x, r'' and r then $T_{r''} = T_{r''} \cup \{x\}$.
(ii) Make $T_r = \emptyset$ after performing the above step.

Algorithm 1. SSR-Algorithm

Input: A set R of leftward-directed rays and a set V of vertical segments.
Output: A subset of R that intersects all segments in V.
1: $T_r = \{r\}$ for each $r \in R$ and $i \leftarrow 1, V_0 \leftarrow V, R_0 \leftarrow R, S \leftarrow \emptyset, S_0 \leftarrow \emptyset$ ▷
   Initialisation.
2: **while** $V_{i-1} \neq \emptyset$ **do**
3:    $S \leftarrow S \cup \{r : r \in R_{i-1}, r$ is critical after $(i-1)^{th}$ iteration$\}$ and $S_i \leftarrow S$.
                                                                ▷ Critical ray collection.
4:    $V_i \leftarrow$ the set obtained by deleting all segments from V_{i-1} that intersect a ray
      in S_i.
5:    Find a $r \in R_{i-1} \setminus S_i$ whose x-coordinate of the right endpoint is the smallest.
6:    r discharges the token to its neighbours.
7:    $R_i \leftarrow$ The set obtained by deleting $\{r\} \cup S_i$ from R_{i-1}.
                                                                ▷ Discharging token step.
8:    $i \leftarrow i + 1$;
9: **end while**
10: **return** S

We have the following lemma whose proof will appear in the extended version of the paper.

Lemma 5. *Let S is the set returned by SSR-algorithm with rays R and segments V as input. Then (a) for a ray r, there are at most two tokens containing r, and (b) $|S| \leq 2|OPT|$, where OPT is an optimum solution of $SSR(R, V)$.*

The proof of Lemma 3 shall follow directly from the proof of Lemma 5. The proof of Lemma 1 in the next section shall use Lemma 3.

3.2   Proof of Lemma 1

Let l be the straight line that intersects all horizontal segment in H. Notice that if l is a horizontal line then any vertical line segment intersects at most one horizontal line segment in H. This is because horizontal lines in H are disjoint. But, in this case, there is nothing to prove.

Therefore, without loss of generality, we assume that l passes through the origin. at an angle in $[\frac{\pi}{2}, \pi)$. For a vertical segment $v \in V$, let $N(v)$ denote the set of horizontal segments intersecting v, $A(v)$ be the set of horizontal segments that intersect v *above* l and $B(v) = N(v) \setminus A(v)$. Observe that for a vertical segment v and a horizontal segment $h \in B(v)$, h intersects v *on* or *below* l.

Based on these consider the following ILP formulation, Q, of the $\mathcal{LVSC}(V, H)$ instance. For each horizontal segment $h \in H$ let $x_h \in \{0, 1\}$ denote the variable corresponding to h. Objective is to minimize $\sum_{h \in H} x_h$ with constraints

$$\sum_{h \in A(v)} x_h + \sum_{h \in B(v)} x_h \geq 1, \ \forall v \in V$$

Let Q_l be the relaxed LP formulation of Q and $\mathbf{Q}_l = \{x_h \colon h \in H\}$ be an optimal solution of Q_l. Since Q_l consists of n variables where $n = |H|$, solving Q_l takes $O(n^5)$ time [17]. Now we define the following sets.

$$V_1 = \left\{ v \in V \colon \sum_{h \in A(v)} x_h \geq \frac{1}{2} \right\}, V_2 = \left\{ v \in V \colon \sum_{h \in B(v)} x_h > \frac{1}{2} \right\}$$

$$H_1 = \bigcup_{v \in V_1} A(v), H_2 = \bigcup_{v \in V_2} B(v)$$

Based on these, we consider the following two integer programs Q' and Q''.

minimize $\sum_{h \in H_1} x'_h$		minimize $\sum_{h \in H_2} x''_h$	
subject to $\sum_{h \in A(v)} x'_h \geq 1, \forall v \in V_1$		subject to $\sum_{h \in B(v)} x''_h \geq 1, \forall v \in V_2$	
$x'_h \in \{0, 1\},$	$h \in H_1$	$x''_h \in \{0, 1\},$	$h \in H_2$
Q'		Q''	

Let Q'_l and Q''_l be the relaxed LP formulation of Q' and Q'' respectively. Clearly, the solutions of Q' and Q'' gives a feasible solution for Q. Hence $OPT(Q) \leq OPT(Q') + OPT(Q'')$. For each $x_h \in \mathbf{Q}_l$, define $y_h = \min\{1, 2x_h\}$ and define $\mathbf{Y}_l = \{y_h\}_{x_h \in \mathbf{Q}_l}$. Notice that $\mathbf{Y}_l$ gives a feasible solution to Q'_l and Q''_l. Therefore, $OPT(Q'_l) + OPT(Q''_l) \leq 2 \cdot OPT(Q_l)$. We have the following claim.

Claim. $OPT(Q') \leq 2 \cdot OPT(Q'_l)$ and $OPT(Q'') \leq 2 \cdot OPT(Q''_l)$.

To prove the first part, note that for each segment $v \in V_1$, $A(v)$ is non-empty and for each $h \in A(v)$, h intersects v above the line l (the straight line which intersects all segments in H). Since all segments in H_1 intersect the straight line l we can consider the horizontal segments in H_1 as leftward-directed rays and all vertical segments in V_1 lie above l. Hence, solving Q' is equivalent to solving an ILP formulation, say $\mathcal{E}$, of the problem of finding a minimum cardinality subset of leftward-directed rays in H_1 that intersects all vertical segments in the set V_1. Hence solving $\mathcal{E}$ is equivalent to solving an SSR instance with H_1 and V_1 as input. By Lemma 3, we have that

$$OPT(Q') = OPT(\mathcal{E}) \leq 2 \cdot OPT(\mathcal{E}_l) \leq 2 \cdot OPT(Q'_l)$$

where $\mathcal{E}_l$ is the relaxed LP formulation of $\mathcal{E}$. Hence we have proof of the first part. For the second part, using similar arguments as above, we can show that solving Q'' is equivalent to solving an SSR instance and therefore by Lemma 3, we have that $OPT(Q'') \leq 2 \cdot OPT(Q''_l)$. Hence the proof of the claim follows.

By Lemma 3, we can solve both Q' and Q'' in polynomial time. Let D' and D'' be solutions of Q' and Q'', respectively. Clearly, $D' \cup D''$ is a feasible solution to the $\mathcal{LVSC}(V, H)$ instance. Hence,

$$|D' \cup D''| \leq 4(OPT(Q'_l) + OPT(Q''_l)) \leq 8 \cdot OPT(Q_l)$$

Hence we have the proof of Lemma 1. Proof of Lemma 2 is similar as above and therefore will only appear in the extended version of the paper. We prove our main result in the next section using Lemmas 1 and 2.

4   Proof of Theorem 2

Let $\mathcal{R}$ be a stabbed rectangle overlap representation of a graph $G = (V, E)$ and l be the line that intersects all rectangles in $\mathcal{R}$. We shall also refer to l as the *cutting line*.

For a vertex $u \in V$, let R_u denote the rectangle corresponding to u in $\mathcal{R}$. Without loss of generality, we assume that the coordinates of all corner points of all the rectangles in $\mathcal{R}$ are distinct and that the cutting line passes through the origin at an angle in $[\frac{\pi}{2}, \pi)$ with the positive x-axis.

Each rectangle R_u consists of four *boundary segments* i.e. *left segment*, *top segment*, *right segment* and *bottom segment*. Without loss of generality, we assume that the cutting line intersects eactly two boundary segments of each rectangle in $\mathcal{R}$. For a vertex $u \in V$, let $N(u)$ denote the set of rectangles that overlaps with R_u in $\mathcal{R}$. Let $N'(u)$ be the set of rectangles in $N(u)$ having a boundary segment that intersects both the cutting line and some boundary segment of R_u that does not cut the boundary line. See Fig. 4(a) for an example. Now define $N''(u) = N(u) \setminus N'(u)$. We have the following observation.

Observation A. *For a rectangle $R_u \in \mathcal{R}$ and a rectangle $X \in N''(u)$, there is a boundary segment of R_u that intersects the cutting line and some boundary segment of X.*

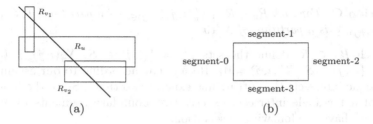

Fig. 4. (a) In this example $R_{v_1} \in N'(u)$ and $R_{v_2} \in N''(u)$. (b) Nomenclature for the four boundary segments of a rectangle.

Proof. Observe that X contains two boundary segments, say s_1 and s_2, such that none of s_1, s_2 intersects the cutting line. Since any two rectangles intersect exactly twice and since $X \in N''(u)$, R_u intersects both s_1 and s_2. If s_1 and s_2 belong to opposite sides of the cutting line, then both s_1 and s_2 are horizontal or both of them are vertical. In either case, R_u must have a boundary segment t that intersect both s_1, s_2 and the cutting line. Consider the case when both s_1 and s_2 lie below the cutting line. Then there exists $w \in \{s_1, s_2\}$ which is a vertical segment and $z \in \{s_1, s_2\} \setminus \{w\}$ which is a horizontal segment. Hence, R_u must have a horizontal boundary segment w' that intersects w and a vertical boundary segment z' that intersects z. If neither w' nor z' intersects the cutting line, then observe that the top-right corner of R_u must lie below the cutting line, implying that R_u does not intersect the cutting line. This is a contradiction. Similarly, the case when both s_1, s_2 lie above the cutting line also leads to a contradiction. □

We shall denote the left segment of a rectangle $R_u \in \mathcal{R}$ also as the *segment*-0 of R_u. Similarly *segment*-1, *segment*-2 and *segment*-3 of R_u shall refer to the top segment, the right segment and the bottom segment of R_u, respectively. See Fig. 4(b) for an illustration. Let $\mathcal{S} = \{(0,1),(0,3),(1,0),(1,2),(2,1),(2,3),(3,0),(3,2)\}$. Since no two horizontal segments or two vertical segments intersect, we have the following observation.

Observation B. *If two rectangles $R_u, R_v \in \mathcal{R}$ overlap there must be a pair $(i,j) \in \mathcal{S}$ such that segment-i of R_u intersects segment-j of R_v.*

Based on the above observation, we partition the sets $N'(u)$ and $N''(u)$ in the following way. For each rectangle $R_u \in \mathcal{R}$ and $(i,j) \in \mathcal{S}$, a rectangle $R_v \in N'(u)$ belongs to the set $X'_u(i,j)$ if and only if (i,j) is the smallest pair in the lexicographic order such that *(a)* segment-i of R_u intersects the segment-j of R_v and *(b)* segment-j of R_v intersects the cutting line.

Similarly, for each rectangle $R_u \in \mathcal{R}$ and $(i,j) \in \mathcal{S}$, a rectangle $R_v \in N''(u)$ belongs to the set $X''_u(i,j)$ if and only if (i,j) is the smallest pair in the lexicographic order such that *(a)* segment-i of R_u intersects the segment-j of R_v and *(b)* segment-i of R_u intersects the cutting line. The next observation follows from the above definitions.

Observation C. *For each $R_u \in \mathcal{R}$, $\{X'_u(i,j)\}_{(i,j)\in\mathcal{S}}$ is a partition of $N'(u)$ and $\{X''_u(i,j)\}_{(i,j)\in\mathcal{S}}$ is a partition of $N''(u)$.*

For each $R_u \in \mathcal{R}$, define the sets $\mathcal{S}'_u = \{(i,j) \in \mathcal{S}: X'_u(i,j) \neq \emptyset\}$ and $\mathcal{S}''_u(i,j) = \{(i,j) \in \mathcal{S}: X''_u(i,j) \neq \emptyset\}$. Recall that according to our assumption, each rectangle intersects the cutting line exactly two times. Since the boundary segment of a rectangle intersects exactly two boundary segments of another rectangle, we have the following observation.

Observation D. *For each $R_u \in \mathcal{R}$, $|\mathcal{S}'_u| \leq 4$ and $|\mathcal{S}''_u| \leq 4$.*

Let Q denote the following ILP formulation of the MDS problem on G and Q_l be the corresponding relaxed LP formulation.

$$
\begin{aligned}
&\text{minimize} \quad \sum_{R_v \in \mathcal{R}} x_v \\
&\text{subject to} \quad \sum_{(i,j)\in\mathcal{S}'_u} \sum_{R_v \in X'_u(i,j)} x_v + \sum_{(i,j)\in\mathcal{S}''_u} \sum_{R_v \in X''_u(i,j)} x_v \geq 1, \; \forall R_u \in \mathcal{R} \\
&\qquad\qquad x_v \in \{0,1\}, \qquad\qquad\qquad\qquad\qquad\qquad\qquad \forall R_v \in \mathcal{R}
\end{aligned}
$$
$$Q$$

Let $\mathbf{Q_l} = \{x_v : R_v \in \mathcal{R}\}$ be an optimal solution of Q_l. By Observation D, for each rectangle $R_u \in \mathcal{R}$, we have $|\mathcal{S}'_u| + |\mathcal{S}''_u| \leq 8$. Hence, there is a pair $(i,j) \in \mathcal{S}'_u \cup \mathcal{S}''_u$ such that either $\sum_{R_v \in X'_u(i,j)} x_v \geq \frac{1}{8}$ or $\sum_{R_v \in X''_u(i,j)} x_v \geq \frac{1}{8}$. For each pair $(i,j) \in \mathcal{S}$, define

$$
A'(i,j) = \left\{ R_u \in \mathcal{R}: (i,j) \in \mathcal{S}'_u, \sum_{R_v \in X'_u(i,j)} x_v \geq \frac{1}{8} \right\}
$$

$$
B'(i,j) = \bigcup_{R_u \in A'(i,j)} X'_u(i,j)
$$

$$
A''(i,j) = \left\{ R_u \in \mathcal{R}: (i,j) \in \mathcal{S}''_u, \sum_{R_v \in X''_u(i,j)} x_v \geq \frac{1}{8} \right\}
$$

$$
B''(i,j) = \bigcup_{R_u \in A''(i,j)} X''_u(i,j)
$$

For each $R_v \in \mathcal{R}$, let $\mathcal{T}_v = \{(i,j) \in \mathcal{S}: R_v \in B'(i,j) \text{ or } R_v \in B''(i,j)\}$.

Observation E. *For each $R_v \in \mathcal{R}$, we have that $|\mathcal{T}_v| \leq 12$.*

Based on these we have the following two ILP formulations for each pair $(i,j) \in \mathcal{S}$.

minimize $\displaystyle\sum_{R_v \in B'(i,j)} x'_v$	minimize $\displaystyle\sum_{R_v \in B''(i,j)} x''_v$
subject to $\displaystyle\sum_{R_v \in X'_u(i,j)} x'_v \geq 1, \forall R_u \in A'(i,j)$	subject to $\displaystyle\sum_{R_v \in X''_u(i,j)} x''_v \geq 1, \forall R_u \in A''(i,j)$
$\begin{array}{l} x'_v \in \{0,1\}, \\ Q'(i,j) \end{array} \qquad R_v \in B'(i,j)$	$\begin{array}{l} x''_v \in \{0,1\}, \\ Q''(i,j) \end{array} \qquad R_v \in B''(i,j)$

For each pair $(i,j) \in \mathcal{S}$, let $Q'_l(i,j)$ and $Q''_l(i,j)$ be the relaxed LP formulation of $Q'(i,j)$ and $Q''(i,j)$, respectively. Observe that $OPT(Q) \leq \displaystyle\sum_{(i,j)\in\mathcal{S}} (OPT(Q'(i,j)) + OPT(Q''(i,j)))$.

For each $x_v \in \mathbf{Q}_l$, define $y_v = \min\{1, 8x_v\}$ and $\mathbf{Y}_l = \{y_v\}_{x_v \in \mathbf{Q}_l}$. Due to Observations C and D, $\mathbf{Y}_l$ gives a feasible solution to $Q'_l(i,j)$ and $Q''_l(i,j)$ for all $(i,j) \in \mathcal{S}$. Therefore, $OPT(Q'_l(i,j)) \leq 8 \cdot OPT(Q_l)$ and $OPT(Q''_l(i,j)) \leq 8 \cdot OPT(Q_l)$ for all $(i,j) \in \mathcal{S}$. Now we have the following lemma.

Lemma 6. *For each $(i,j) \in \mathcal{S}$ there is a set $D'(i,j) \subseteq B'(i,j)$ such that $D'(i,j)$ gives a feasible solution of $Q'(i,j)$ and $|D'(i,j)| \leq 8 \cdot OPT(Q'_l(i,j))$.*

Proof. For any $(i,j) \in \mathcal{S}$, solving $Q'(i,j)$ is equivalent to finding a minimum cardinality subset D of $B'(i,j)$ such that each rectangle $R_u \in A'(i,j)$ overlaps a rectangle in $D \cap X'_u(i,j)$. Notice that, for each $R_u \in A'(i,j)$ the set $X'_u(i,j)$ is non-empty. Moreover for each $R_v \in X'_u(i,j)$, the segment-j of R_v intersects the cutting line and segment-i of R_u. Let $S = \{$segment-i of $R_u : R_u \in A'(i,j)\}$, $T = \{$segment-j of $R_v : R_v \in B'(i,j)\}$.

Solving $Q'(i,j)$ is equivalent to the problem finding a minimum cardinality subset D of T such that every segment in S intersect at least one segment in D. Moreover, every segment in T intersects the cutting line. Without loss of generality we can assume that S consists of vertical segments. Therefore T consists of horizontal segments all intersecting the cutting line. Hence solving $Q'(i,j)$ is equivalent to solving the $\mathcal{LVSC}(S,T)$ instance. Hence by Lemma 1, we have a feasible solution (say $D'(i,j)$) for $Q'(i,j)$ such that $|D'(i,j)| \leq 8 \cdot OPT(Q'_l(i,j))$. $\square$

Using similar arguments as in the proof of Lemma 6, we can prove that solving $Q''(i,j)$ is equivalent to solving an instance of the *local horizontal segment covering* problem. Then using Lemma 2 we can prove the following lemma.

Lemma 7. *For each $(i,j) \in \mathcal{S}$ there is a set $D''(i,j) \subseteq B''(i,j)$ such that $D''(i,j)$ gives a feasible solution of $Q''(i,j)$ and $|D''(i,j)| \leq 8 \cdot OPT(Q''_l(i,j))$.*

For   each   pair (i,j) $\in$ $\mathcal{S}$,
due to Lemmas 6 and 7, we have a feasible solution $D'(i,j)$ of $Q'(i,j)$ and a
feasible solution $D''(i,j)$ such that $|D'(i,j)| \leq 8 \cdot OPT(Q_l'(i,j))$ and $|D''(i,j)| \leq$
$8 \cdot OPT(Q_l''(i,j))$. Let D be the union of $D'(i,j)$'s and $D''(i,j)$ for all $(i,j) \in \mathcal{S}$.
Using Observation E we have

$$|D| = \sum_{(i,j) \in \mathcal{S}} (|D'(i,j)| + |D''(i,j)|)$$

$$\leq 8 \cdot \sum_{(i,j) \in \mathcal{S}} (OPT(Q_l'(i,j)) + OPT(Q_l''(i,j)))$$

$$\leq 768 \cdot OPT(Q_l) \leq 768 \cdot OPT(Q)$$

This completes the proof of Theorem 2. The next corollary follows from Theorem 2.

Corollary 1. *Let $\mathcal{R}$ be a stabbed rectangle intersection representation of a graph $G = (V, E)$ such that no two rectangles in $\mathcal{R}$ contain each other. There is an $O(|V|^5)$-time 768-approximation algorithm for the MDS problem on G.*

Acknowledgement. This research was partially funded by the IFCAM project "Applications of graph homomorphisms" (MA/IFCAM/18/39).

References

1. Bandyapadhyay, S., Maheshwari, A., Mehrabi, S., Suri, S.: Approximating dominating set on intersection graphs of rectangles and L-frames. In: MFCS, pp. 37:1–37:15 (2018)
2. de Berg, M., Cheong, O., van Kreveld, M., Overmars, M.: Computational Geometry: Algorithms and Applications. Springer, Heidelberg (2008). https://doi.org/10.1007/978-3-540-77974-2
3. Bousquet, N., Gonçalves, D., Mertzios, G.B., Paul, C., Sau, I., Thomassé, S.: Parameterized domination in circle graphs. Theory Comput. Syst. **54**(1), 45–72 (2014)
4. Catanzaro, D., et al.: Max point-tolerance graphs. Discret. Appl. Math. **216**, 84–97 (2017)
5. Chepoi, V., Felsner, S.: Approximating hitting sets of axis-parallel rectangles intersecting a monotone curve. Comput. Geom. **46**(9), 1036–1041 (2013)
6. Chlebík, M., Chlebíková, J.: Approximation hardness of dominating set problems in bounded degree graphs. Inf. Comput. **206**(11), 1264–1275 (2008)
7. Correa, J., Feuilloley, L., Pérez-Lantero, P., Soto, J.A.: Independent and hitting sets of rectangles intersecting a diagonal line: algorithms and complexity. Discrete Comput. Geom. **53**(2), 344–365 (2015)
8. Damian, M., Pemmaraju, S.V.: APX-hardness of domination problems in circle graphs. Inf. Process. Lett. **97**(6), 231–237 (2006)
9. Damian-Iordache, M., Pemmaraju, S.V.: A $(2+ \varepsilon)$-approximation scheme for minimum domination on circle graphs. J. Algorithms **42**(2), 255–276 (2002)

10. Erlebach, T., van Leeuwen, E.J.: Domination in geometric intersection graphs. In: Laber, E.S., Bornstein, C., Nogueira, L.T., Faria, L. (eds.) LATIN 2008. LNCS, vol. 4957, pp. 747–758. Springer, Heidelberg (2008). https://doi.org/10.1007/978-3-540-78773-0_64
11. Govindarajan, S., Raman, R., Ray, S., Basu Roy, A.: Packing and covering with non-piercing regions, In: ESA (2016)
12. Katz, M.J., Mitchell, J.S.B., Nir, Y.: Orthogonal segment stabbing. Comput. Geom.: Theory Appl. **30**(2), 197–205 (2005)
13. Keil, J.M., Mitchell, J.S.B., Pradhan, D., Vatshelle, M.: An algorithm for the maximum weight independent set problem on outerstring graphs. Comput. Geom. **60**, 19–25 (2017)
14. Khot, S., Regev, O.: Vertex cover might be hard to approximate to within 2-ε. J. Comput. Syst. Sci. **74**(3), 335–349 (2008)
15. Mudgal, A., Pandit, S.: Covering, hitting, piercing and packing rectangles intersecting an inclined line. In: Lu, Z., Kim, D., Wu, W., Li, W., Du, D.-Z. (eds.) COCOA 2015. LNCS, vol. 9486, pp. 126–137. Springer, Cham (2015). https://doi.org/10.1007/978-3-319-26626-8_10
16. Pandit, S.: Dominating set of rectangles intersecting a straight line. In: CCCG, pp. 144–149 (2017)
17. Tardos, E.: A strongly polynomial algorithm to solve combinatorial linear programs. Oper. Res. **34**(2), 250–256 (1986)

Minimizing the Cost of Batch Calibrations

Vincent Chau[1], Minming Li[2], Yinling Wang[3(✉)], Ruilong Zhang[2],
and Yingchao Zhao[4]

[1] Shenzhen Institutes of Advanced Technology, Chinese Academy of Sciences,
Shenzhen, China
[2] City University of Hong Kong, Hong Kong, China
[3] Dalian University of Technology, Dalian, China
yinling_wang@foxmail.com
[4] Caritas Institute of Higher Education, Hong Kong, China

Abstract. We study the scheduling problem with calibrations. We are
given a set of n jobs that need to be scheduled on a set of m machines.
However, a machine can schedule jobs only if a calibration has been performed beforehand and the machine is considered as valid during a fixed
time period of T, after which it must be recalibrated before running
more jobs. In this paper, we investigate the batch calibrations, calibrations occur in batch and at the same moment. It is then not possible
to perform any calibrations during a period of T. We consider different
cost function depending on the number of machines we calibrate at a
given time. Moreover, jobs have release time, deadline and unit processing time. The objective is to schedule all jobs with the minimum cost
of calibrations. We give a dynamic programming to solve the case with
arbitrary cost function. Then, we propose several faster approximation
algorithm for different cost function.

1 Introduction

The scheduling with calibrations was initially motivated from the Integrated
Stockpile Evaluation (ISE) program to test nuclear weapons periodically.
Because of sensitive application, the calibrations are expensive and need to be
performed at an appropriated moment. This motivation can be extended to the
scenarios where the machines need to be calibrated periodically to ensure the
quality of the products. Calibrations have applications in many areas, including
robotics [9,12,15], pharmaceuticals [3,7,14], and digital cameras [2,6,17].

Vincent Chau is supported by Shenzhen research grants (KQJSCX 20180330170311901,
JCYJ20180305180840138 and GGFW2017073114031767), NSFC (No. 61433012), Hong
Kong GRF 17210017 and Shenzhen Discipline Construction Project for Urban Com-
puting and Data Intelligence. Minming Li and Ruilong Zhang are supported by
a grant from Research Grants Council of the Hong Kong Special Administrative
Region, China (Project No. CityU 11268616). Yingchao Zhao is supported by Research
Grants Council of the Hong Kong Special Administrative Region, China (Project No.
UGC/FDS11/E03/16).

© Springer Nature Switzerland AG 2019
D.-Z. Du et al. (Eds.): COCOON 2019, LNCS 11653, pp. 78–89, 2019.
https://doi.org/10.1007/978-3-030-26176-4_7

The initial scheduling problem is as follows [8]. We are given a set of n jobs that need to be scheduled on a set of m identical machines. Each job j has a release time r_j, a deadline d_j and a processing time p_j. In this paper, we consider unit size jobs, i.e. $p_j = 1$, for all j. A job is scheduled if it is processed entirely inside its interval $[r_j, d_j)$. However, we can schedule jobs only if we perform calibrations beforehand. A calibration activates instantaneously the machine for a period of T time units and the machine can start to process jobs as soon as it is calibrated. After T time units, the machine cannot schedule any jobs unless we perform another calibration. The goal is to find a feasible schedule such that all jobs are scheduled with the minimum number of calibrations. We denote this problem by $\mathcal{O}$.

In some contexts, it may be possible to calibrate multiple machines simultaneously. The large expense of calibrations often involves experts or machinery that must be hired and transported to the machines before calibrations can be performed. As long as multiple machines are in the same location, the experts and machinery can be used to calibrate all of them without a significant increase in cost. After the experts leave, a minimum period is needed before they can come back. In this paper, we consider that the minimum period is T.

1.1   Related Works

Bender et al. [8] first proposed the theoretical framework of the scheduling problem with calibrations. In the seminal work, they studied the case where jobs have unit processing time, and proposed an optimal polynomial time algorithm when a single machine is available and a 2-approximation algorithm for the multiple machines case. Fineman et al. [13] considered a generalization where jobs have arbitrary processing time. They observed that minimizing calibrations for jobs with deadlines generalizes the well-known machine minimization problem when T is arbitrarily large, the problem is to minimize the number of machines. Chau et al. [10] worked on the case where jobs arrive in online fashion. They aim to find a tradeoff between the total flow-time (the elapsed time between the release time of a job and until its completion) and the total calibration costs. They gave several online approximation results on different settings of single or multiple machines for weighted or unweighted jobs and also a dynamic programming for the offline problem. Angel et al. [1] developed dynamic programming algorithms for generalizations where there are multiple kinds of calibrations (different lengths of calibration), or when calibrations are not instantaneous. They also give several properties that we will use throughout the paper. Wang [16] studied a variant with time slot cost, where scheduling a job incur a cost that depends on its starting time.

One scheduling problem, which has similitude with our problem, is that of minimizing the number of idle periods. In this problem, it is expensive to turn on a machine that is in idle state and it has the flexibility to keep the machine in active state, while the problem with calibrations has the notion of paying a cost for a fixed constant-sized period of activity. The problem of minimizing the number of idle periods has been proved to be polynomial time solvable when jobs have unit processing time in single machine case [4,5] as well as multiple machine case [11], and also when jobs have arbitrary processing time but on a single machine [5].

1.2 Our Problem

In this paper, we study a variant of the problem where calibrations can only occur simultaneously. In other words, at a given time t, we can decide to calibrate a fixed number $x \leq m$ of machines, and no calibration can occur in $[t, t + T)$. In the following, we refer to $\mathcal{B}$ for the *batch calibration scheduling problem*. We model the costs of the calibrations by a function $f(x)$ where x is the number of calibrations that occur in a given time. Typically, the cost function is a concave function, i.e. it is expensive to call an expert to calibrate the machines, but it is cheaper to calibrate an extra machine. Consider a scenario where it costs b to call an expert to come to the factory, and each calibration on a machine has a unit cost. The cost function can be modeled as $f(x) = x + b$ where x is the number of machines we calibrate at a given time and b is the travel cost of the expert.

1.3 Our Contributions and Organization of the Paper

We first show that the problem can be solved in polynomial time via dynamic programming for arbitrary cost function $f(x)$ in Sect. 3. We then propose several fast approximation algorithms:

- An optimal algorithm when $f(x) = b$ in Sect. 4,
- A 4-approximation algorithm when $f(x) = x$ in Sect. 5,
- A $(b + m)/(b + 1)$-approximation algorithm when $f(x) = x + b$ in Sect. 6.

Finally we conclude in Sect. 7.

2 Preliminaries

We first define a restricted set of times that a calibration can start as well as the starting time of jobs. We use the properties from [1]. It can be shown that in an optimal solution, there are a polynomial numbers of time steps in which a calibration can start. In fact, it is sufficient to consider that the starting time of calibrations can start at a distance at most n before a deadline of a job where n is the number of jobs. We denote the set of starting time of calibrations as Ψ.

Definition 1 (Definition 1 [1]). *Let* $\Psi := \bigcup_i \{d_i - n, d_i - n + 1, \ldots, d_i\}$.

Proposition 1 (Proposition 1 [1]). *There exists an optimal solution in which each calibration starts at a time in* Ψ.

Once the starting time of the calibrations are set, we are able to define a set of time steps in which jobs can be scheduled. In particular, jobs are scheduled as soon as the machine is calibrated, as well as right after their release time. We denote this set Φ.

Definition 2 (Definition 3 [1]). *Let* $\Phi := \{t + a \mid t \in \Psi, \ a = 0, \ldots, n\} \cup \bigcup_i \{r_i, r_i + 1, \ldots, r_i + n\}$.

Proposition 2 (Proposition 7 [1]). *There exists an optimal solution in which the starting times and completion times of jobs belong to Φ.*

Note that the above definitions and propositions hold for single machine as well as for the multiple machine case. Throughout the paper, we use s_j to denote the starting time of job j.

Definition 3. *A schedule S is defined by a set of calibrations and each calibration $c_i \in S$ is defined by a triplet (J_i, t_i, m_i), i.e. the jobs in $J_i \subseteq J$ are scheduled on machine m_i in the time interval $[t_i, t_i + T)$.*

Definition 4. *The EDF (Earliest Deadline First) policy is to schedule the job with the smallest deadline among the pending jobs.*

In the sequel, we consider that all schedules follow the EDF policy. When several machines are available at the same time slot, we consider each machine in increasing order of index while applying the EDF policy. Without loss of generality, we assume that jobs are sorted in non-decreasing order of deadlines, i.e., $d_1 \leq d_2 \leq \ldots \leq d_n$.

3   An Optimal Polynomial Time Algorithm for Arbitrary Cost Function

We inspire from the dynamic programming proposed by Demaine et al. in [11]. First, we define the dynamic programming table in Definition 5, then we enumerate the different cases to build the schedule.

Definition 5 *(See Fig. 1). Let $C(t_1, t_2, e, s, k, q, l_1, l_2)$ be the minimum cost of a schedule such that:*

- *jobs from $\{j \mid j \leq k, r_j \in [t_1, t_2)\}$ are scheduled in $[t_1, t_2)$*
- *the first batch of calibrations starts at time e with l_1 machines and ends after t_1 $(t_1 < e + T)$*
- *the last batch of calibrations starts at time s with l_2 machines and starts before t_2 $(s \leq t_2)$*
- *there are q reserved machines at time-slot $[t_2 - 1, t_2)$*
- *only the calibrations that start in $[t_1, t_2)$ are considered for the cost*
- *at least one batch of calibrations overlaps with $[t_1, t_2)$*

Optimality: To compute the schedule associated to $C(t_1, t_2, e, s, k, q, l_1, l_2)$, we consider the job k that has the largest deadline among the jobs in $\{1, 2, \ldots, k\}$. Suppose that job k is scheduled at time t' and we suppose that t' is maximal among all optimal solutions. We claim that jobs scheduled after k must be released after t'; otherwise, we could swap the scheduled times of k and of such a job, which is feasible because k has the latest deadline among the jobs,

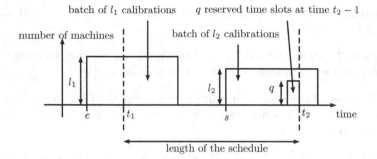

Fig. 1. Illustration of Definition 5. This corresponds to a schedule associated to $C(t_1, t_2, e, s, k, q, l_1, l_2)$.

thus we get a feasible schedule with same cost, but with a larger t', a contradiction. In this way, we reach two subproblems with intervals $[t_1, t' + 1)$ and $[t' + 1, t_2)$.

We distinguish 3 different cases to compute such a value (See Fig. 2 for an illustration of different cases): the job k is scheduled in the first batch of calibrations, in the last, or in the middle. Moreover, when dividing the problem into two subproblems, we need to consider whether a same calibration are in both subproblems. Formally, if a job k is scheduled at time t', the batch calibrations that allow us to schedule job k are either fully in the first subproblem (the batch calibrations ends at time $t' + 1$), or the batch calibrations start in the first subproblem and end in the second subproblem. Finally, we denote e' (resp. s' be the starting

We consider the 3 different cases as well as the sub-cases in the following. To simplify notation, we use C' to denote $C(t_1, t_2, e, s, k, q, l_1, l_2)$.

Case 1: job k is scheduled in the first calibration.
As mentioned previously, we have two subcases.
Case 1.1: job k is not scheduled at the last time slot of the calibration
$C' = \min_{t'} C(t_1, t' + 1, e, e, k - 1, 1, l_1, l_1) + C(t' + 1, t_2, e, s, k - 1, q, l_1, l_2)$ if
$t' + 1 < e + T$
Case 1.2: job k is scheduled at the last time slot of the calibration
$C' = C(t_1, t' + 1, e, e, k - 1, 1, l_1, l_1) + C(t' + 1, t_2, e', s, k - 1, q, l', l_2)$ if $t' + 1 = e + T$

Case 2: job k is not scheduled in the first calibration, and not in the last calibration.
Case 2.1: job k is not scheduled at the last time slot of the calibration
$C' = \min_{t', e', l'} C(t_1, t' + 1, e, e', k - 1, 1, l_1, l') + C(t' + 1, t_2, e', s, k - 1, q, l', l_2)$ if
$t' + 1 < e' + T$
Case 2.2: job k is scheduled at the last time slot of the calibration
$C' = \min_{t', e', s', l', l'_2} C(t_1, t' + 1, e, e', k - 1, 1, l_1, l') + C(t' + 1, t_2, s', s, k - 1, q, l'_2, l_2)$
if $t' + 1 = e' + T$

Case 3: job k is scheduled in the last calibration.
Case 3.1: job k is not scheduled at the last time slot of the schedule
$C' = \min_{t'} C(t_1, t' + 1, e, s, k - 1, 1, l_1, l_2) + C(t' + 1, t_2, s, s, k - 1, q, l_2, l_2)$ if $t' + 1 < t_2$
Case 3.2: job k is scheduled at the last time slot of the schedule
$C' = (t_1, t_2, e, s, k - 1, q + 1, l_1, l_2)$ if $t' + 1 = t_2$

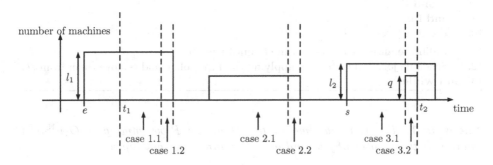

Fig. 2. Illustration of different cases of the Dynamic Programming

Initial Cases
For the initial cases, we distinguish two cases where only one batch of calibrations overlaps with the considered interval, or two batches of calibrations overlap with the interval. Moreover, we need to ensure that the time slot $[t_2 - 1, t_2)$ is covered by a batch of calibrations.

There is only one batch of calibrations
$C(t_1, t_2, e, e, 0, q, l_1, l_1) = f(l_1)$ if $t_1 \leq e < t_2 \leq e + T$ and $q \leq l_1$
$C(t_1, t_2, e, e, 0, 0, l_1, l_1) = f(l_1)$ if $t_1 \leq e < e + T < t_2$
$C(t_1, t_2, e, e, 0, q, l_1, l_1) = 0$ if $e < t_1 < t_2 \leq e + T$ and $q \leq l_1$
$C(t_1, t_2, e, e, 0, 0, l_1, l_1) = 0$ if $e < t_1 < e + T < t_2$

There are two batches of calibrations
$C(t_1, t_2, e, s, 0, q, l_1, l_2) = f(l_1) + f(l_2)$ if $t_1 \leq e$ and $e + T \leq s$ and $s < t_2 \leq s + T$ and $q \leq l_2$
$C(t_1, t_2, e, s, 0, 0, l_1, l_2) = f(l_1) + f(l_2)$ if $t_1 \leq e$ and $e + T \leq s$ and $s + T < t_2$ (and $q \leq l_2$)
$C(t_1, t_2, e, s, 0, q, l_1, l_2) = f(l_2)$ if $e < t_1$ and $e + T \leq s$ and $s < t_2 \leq s + T$ and $q \leq l_2$
$C(t_1, t_2, e, s, 0, 0, l_1, l_2) = f(l_2)$ if $e < t_1$ and $e + T \leq s$ and $s + T < t_2$ (and $q \leq l_2$)
$C(t_1, t_2, e, e, 0, q, l_1, l_1) = +\infty$ for all other cases.

Note that it is not necessary to consider 3 or more batches of calibrations in the initial case because no job will be assigned to the calibrations in the middle.

Algorithm 1. Multi-Lazy-Binning (MLB)

1: Jobs in $\mathcal{J}$ are sorted in non-decreasing order of deadline
2: **while** $\mathcal{J} \neq \emptyset$ **do**
3:    $t \leftarrow \max_{i \in \mathcal{J}} d_i$, $k \leftarrow 0$
4:    **for** $i \in \mathcal{J}$ **do**
5:       $t' \leftarrow d_i - \left\lceil \frac{|\{j \leq i, j \in \mathcal{J}\}|}{m} \right\rceil$
6:       **if** $t > t'$ **then**
7:          $t \leftarrow t'$, $k \leftarrow i$
8:       **end if**
9:    **end for**
10:    $u \leftarrow t + \left\lceil \frac{d_k - t}{T} \right\rceil \times T$
11:    Calibrate all m machines at time t (until time u)
12:    Schedule jobs from t to u by applying the EDF policy and remove them from $\mathcal{J}$.
13: **end while**

Theorem 1. *The running time of the Dynamic Programming is* $O(n^{15}m^5)$ *where* n *is the number of jobs and* m *is the number of machines.*

Proof. According to Proposition 1, the starting time of a calibration lies in Ψ while the starting time of a job lies in Φ, which both have size $O(n^2)$, in particular the values of e, s lie in Ψ and t_1, t_2, lie in Φ. k can take any value from 0 to n and q, l_1, l_2 are the number of machines which are less than m. Therefore the size of the table is $O(n^9m^3)$.

To compute a specific value of the table, we need to look at all the values of t' (the starting time of job k), and there are $O(n^2)$ values. By summing the running time of all cases, we need to look at the values of e', s', l', l'_2. So the running time is $O(n^6m^2)$. Thus the computing time of the dynamic programming is $O(n^{15}m^5)$. □

4 $f(x) = b$

In this section, we consider the case where calibrating has a constant cost and is independent on the number of machines we use. Therefore, it is never unprofitable to use all available machines. The idea is to use a similar algorithm to the Lazy-Binning algorithm in [8]. We delay each batch calibrations as late as possible until it is not possible and we calibrate all machines. A formal description can be found in Algorithm 1.

Theorem 2. *Algorithm 1 computes the minimum number of batch calibrations.*

Proof. Without loss of generality, we suppose that the optimal solution uses all machines in each batch calibrations.

We show how to transform our problem to an equivalent one using the algorithm proposed by Bender et al. [8].

Since we have m machines, we can simply consider that each initial time-slot becomes m time-slots in the new instance, i.e., for each time slot t in the initial problem, we divide it into m identical time slots denoted as $t_1, t_2, \ldots, t_m$. We consider that each job has a processing time of $1/m$ in order to fit into the new time slots. The release time and deadline remain the same. Finally, we add the constraint that the calibrations can only occur at some time t_1 (and not at time t_i for $i > 1$). So we have a new scheduling problem on a single machine. Since we calibrate all machines at a given time t, it means that we also calibrate all the consecutive time-slots $t_1, t_2, \ldots, t_m, \ldots, (t + T - 1)_1, \ldots, (t + T - 1)_m$. See Fig. 3 for an illustration. By using the algorithm in [8], we get the minimum number of calibrations (number of batches calibration for our problem). Thus, it is optimal for our problem.

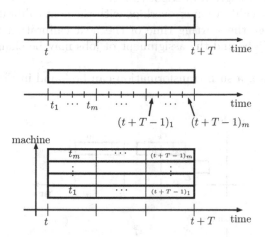

Fig. 3. Illustration of Theorem 2. We start from the initial scheduling problem with calibrations. Suppose that the machine is calibrated at time t, we divide each time slot into m equal time slots (second schedule). Then we make the correspondence between each new time slot of the second schedule with each time slot and the machine in the batch calibration scheduling problem (third schedule)

Now we show that with the resulting schedule, we can build a feasible solution for our problem.

Each new time slot corresponds to a machine, if a job is scheduled at the time slot t_i, it means this job is scheduled at time t on machine i in the initial problem.   □

5 $f(x) = x$

In this section, we propose a 4-approximation algorithm when the cost function is $f(x) = x$ in Algorithm 2. First, we show that we can transform any feasible solution for the initial scheduling problem (problem $\mathcal{O}$) into another feasible solution for the batch scheduling problem (problem $\mathcal{B}$) by loosing a factor of 2, i.e. by using at most twice more calibrations.

High Level Idea: Given a schedule, we find the first moment where two calibrations c_i, c_k on two different machines share a non-empty time interval and such that they do not start at the same time, i.e. we have $[t_i, t_i + T) \cap [t_k, t_k + T) \neq \emptyset$ and $m_i \neq m_k$. Without loss of generality, we assume that $t_i < t_k$. We modify the calibration c_k and create two new calibrations c_k^1 and c_k^2 such that:

- the jobs in J_k that were scheduled before (resp. after) $t_i + T$ are assigned to calibration c_k^1 (resp. c_k^2) and we denote the set of jobs J_k^1 (resp. J_k^2).
- $c_k^1 = (J_k^1, t_i, m_k)$
- $c_k^2 = (J_k^2, t_i + T, m_k)$

A formal description is given in Algorithm 2.

Note that if the calibration c_k^2 overlaps with the next calibration on the same machine, we change the starting time of the next calibration without modifying the scheduled jobs and the assignment of jobs may be changed accordingly (see Algorithm 3).

An illustration of a such transformation can be found in Fig. 4.

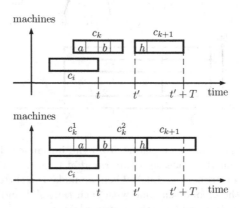

Fig. 4. Illustration of Algorithm 2. Calibrations c_i and c_k overlap and do not start at the same moment. We split the calibration c_k into two calibrations c_k^1 and c_k^2 such that the calibrations c_k^1 and c_i start at the same moment. Calibration c_k^2 starts right after calibration c_k^1 ends. Finally, if calibration c_k^2 overlaps with the next calibration, say c_{k+1}, we delay the calibration c_{k+1}, but the jobs remains unchanged: for example the job h did not change its starting time, but change the calibration in which it belongs to (from calibration c_{k+1} to c_k^2)

We denote the optimal solution of batch calibration problem as OPT_B, and the optimal solution of original calibration problem as OPT_O. Since the objective is to minimize the number of calibrations, the cost of a solution S can be written as $|S|$.

Lemma 1. *For any set of jobs, we have $|OPT_O| \leq |OPT_B|$ when the cost function of batch calibrations problem is $f(x) = x$.*

Algorithm 2. Batch calibrations

Require: Schedule S
1: **while** There exist two calibrations c_i and c_k that overlap on two different machines
    μ and μ' **do**
2:    Let denote $c_i = (J_i, t_i, \mu)$ and $c_k = (J_k, t_k, \mu')$ with $t_i \leq t_k < t_i + T$
3:    Find the first calibration that overlaps with a previous calibration on another
       machine
4:    $k^* \leftarrow \arg\min_k \{ t_i < t_k < t_i + T \}$
5:    $J_k^1 \leftarrow \{ j \mid j \in c_{k^*} \text{ and } s_j < t_k + T \}$
6:    $J_k^2 \leftarrow \{ j \mid j \in c_{k^*} \text{ and } s_j \geq t_k + T \}$
7:    $c_{k^*}^1 = (J_k^1, t_i, m_k)$
8:    $c_{k^*}^2 = (J_k^2, t_i + T, m_k)$
9:    $S \leftarrow \{ S \setminus c_{k^*} \} \cup c_{k^*}^1 \cup c_{k^*}^2$
10:   Modify S such that it becomes feasible (See Algorithm 3)
11: **end while**
12: **return** S

Algorithm 3. Feasibility

Require: Schedule S
1: **for** each machine $\mu \in [1, \ldots, m]$ **do**
2:    **while** there exists $c_i = (J_i, t_i, \mu)$ and $c_k = (J_k, t_k, \mu)$ such that $t_i \leq t_k < t_i + T$
      do
3:       $J \leftarrow \{ j \mid s_j \in [t_k, t_i + T), j \in J_k \}$ // Set of jobs scheduled in $[t_k, t_i + T)$
4:       $c_i \leftarrow (J_i \cup J, t_i, \mu)$
5:       $c_k \leftarrow (J_k \setminus J, t_i + T, \mu)$
6:       Update schedule S
7:    **end while**
8: **end for**
9: **return** S

Proof. The set of feasible solutions of problem $\mathcal{O}$ is denoted by $\mathcal{F}_\mathcal{O}$ while the set of feasible solutions of problem $\mathcal{B}$ is denoted by $\mathcal{F}_\mathcal{B}$. In fact, any feasible solution for the problem $\mathcal{B}$ is also a feasible solution for the problem $\mathcal{O}$, so we have $\mathcal{F}_b \subseteq \mathcal{F}_o$. Therefore, we have $\min_{S \in \mathcal{F}_\mathcal{O}} |S| \leq \min_{S' \in \mathcal{F}_\mathcal{B}} |S'|$ and the lemma follows. $\qquad\square$

We now prove that only the calibrations from the initial solution can be considered in Algorithm 2. In fact, for each calibration from the initial solution, we add an extra calibration.

Lemma 2. *Only calibrations from the initial solution can be doubled in Algorithm 2.*

Proof. It is sufficient to prove that the newly added calibrations will not be doubled, i.e. we will not create an additional calibration from a newly added calibration. Suppose that a calibration c_k is the first calibration to overlap with a former calibration c_i on another machine. From the construction of the calibrations c_k^1 and c_k^2, we know that the calibration c_k^1 and the calibration c_i start at the same moment. Since the calibration c_k was the first to overlap with another calibration, the calibration c_k^1 will never be considered because it starts before the calibration c_k.

Suppose that the algorithm considers the calibration c_k^2. It means that there is a calibration $c_{i'}$ overlaps with and starts before the calibration c_k^2. By construction, we know that the calibration c_k^2 starts right after the calibration c_i ends. So the calibration $c_{i'}$ also overlaps with calibration c_i. The algorithm should have chosen the calibration $c_{i'}$ instead of calibration c_k^2, so we have a contradiction.   □

Theorem 3. *Algorithm 2 is a 4-approximation algorithm.*

Proof. Algorithm 2 depends on the schedule given in the input. Let S (resp. ALG) be the schedule given as input (resp. output) in Algorithm 2. From Lemma 2, we know that each calibration in S can be doubled only once. So the solution returned by the algorithm cannot have more than $2|S|$ calibrations. Thus we have $|S| \leq |ALG| \leq 2|S|$. Bender et al. [8] proposed a 2-approximation algorithm for the problem $\mathcal{O}$, so $|OPT_\mathcal{O}| \leq |S| \leq 2|OPT_\mathcal{O}|$.

$$|ALG| \leq 2|S| \leq 4|OPT_\mathcal{O}|$$

Finally, from Lemma 1, we have $|OPT_\mathcal{O}| \leq |OPT_\mathcal{B}|$, so $|OPT_\mathcal{B}| \leq |ALG| \leq 4|OPT_\mathcal{O}| \leq 4|OPT_\mathcal{B}|$.   □

6 $f(x) = x + b$

We use the algorithm proposed in Sect. 4 and show that the approximation ratio depends on the cost function of batch calibrations.

Theorem 4. *The Multi-Lazy-Binning (MLB) algorithm (Algorithm 1) is a $(b + m)/(b + 1)$-approximation algorithm when the cost function is $f(x) = x + b$, where m is the number of machines.*

Proof. Let K^{MLB} (resp. $K^{OPT_\mathcal{B}}$) be the number of batches of calibrations in the schedule returned by Algorithm 1 (MLB) (resp. the optimal solution of problem $\mathcal{B}$ with $f(x) = x + b$). According to the Theorem 2, MLB uses the minimum number of batches, so $K^{MLB} \leq K^{OPT_\mathcal{B}}$. The cost of the Algorithm 1 is $(b + m)K^{MLB}$ because it calibrates all the machines at each batch, while the cost of the optimal solution is at least $(b + 1)K^{OPT_\mathcal{B}}$. Therefore, the approximation ratio ρ of the Algorithm 1 can be bounded as follows:

$$\rho \leq \frac{(b+m)K^{MLB}}{(b+1)K^{OPT_\mathcal{B}}} = \frac{bK^{MLB}}{(b+1)K^{OPT_\mathcal{B}}} + \frac{mK^{MLB}}{(b+1)K^{OPT_\mathcal{B}}} \leq \frac{b+m}{b+1}.$$ □

7   Conclusion

We show that the scheduling problem with batch calibrations can be solved in polynomial time. However, since the running time is high, we propose some fast approximation algorithms for several special case. A natural question is whether there exists a fast constant approximation algorithm for arbitrary cost function.

References

1. Angel, E., Bampis, E., Chau, V., Zissimopoulos, V.: On the complexity of minimizing the total calibration cost. In: Xiao, M., Rosamond, F. (eds.) FAW 2017. LNCS, vol. 10336, pp. 1–12. Springer, Cham (2017). https://doi.org/10.1007/978-3-319-59605-1_1
2. Baer, R.: Self-calibrating and/or self-testing camera module. US Patent App. 11/239,851, 30 September 2005
3. Bansal, S.K., et al.: Qualification of analytical instruments for use in the pharmaceutical industry: a scientific approach. Aaps Pharmscitech **5**(1), 151–158 (2004)
4. Baptiste, P.: Scheduling unit tasks to minimize the number of idle periods: a polynomial time algorithm for offline dynamic power management. In: SODA, pp. 364–367. ACM Press (2006)
5. Baptiste, P., Chrobak, M., Dürr, C.: Polynomial-time algorithms for minimum energy scheduling. ACM Trans. Algorithms **8**(3), 26:1–26:29 (2012)
6. Barton-Sweeney, A., Lymberopoulos, D., Savvides, A.: Sensor localization and camera calibration in distributed camera sensor networks. In: 3rd International Conference on 2006 Broadband Communications, Networks and Systems, BROADNETS 2006, pp. 1–10. IEEE (2006)
7. Beamex: Traceable and efficient calibrations in the process industry, October 2007. https://www.beamex.com/wp-content/uploads/2016/12/CalibrationWorld_2007-03-ENG.pdf
8. Bender, M.A., Bunde, D.P., Leung, V.J., McCauley, S., Phillips, C.A.: Efficient scheduling to minimize calibrations. In: SPAA, pp. 280–287. ACM (2013)
9. Bernhardt, R., Albright, S.: Robot Calibration. Springer, Heidelberg (1993)
10. Chau, V., Li, M., McCauley, S., Wang, K.: Minimizing total weighted flow time with calibrations. In: SPAA, pp. 67–76. ACM (2017)
11. Demaine, E.D., Ghodsi, M., Hajiaghayi, M.T., Sayedi-Roshkhar, A.S., Zadimoghaddam, M.: Scheduling to minimize gaps and power consumption. In: SPAA, pp. 46–54. ACM (2007)
12. Evans, R.C., Griffith, J.E., Grossman, D.D., Kutcher, M.M., Will, P.M.: Method and apparatus for calibrating a robot to compensate for inaccuracy of the robot. US Patent 4,362,977, 7 December 1982
13. Fineman, J.T., Sheridan, B.: Scheduling non-unit jobs to minimize calibrations. In: SPAA, pp. 161–170. ACM (2015)
14. Forina, M., Casolino, M.C., De la Pezuela Martínez, C.: Multivariate calibration: applications to pharmaceutical analysis. J. Pharm. Biomed. Anal. **18**(1), 21–33 (1998)
15. Nguyen, H.-N., Zhou, J., Kang, H.-J.: A new full pose measurement method for robot calibration. Sensors **13**(7), 9132–9147 (2013)
16. Wang, K.: Calibration scheduling with time slot cost. In: Tang, S., Du, D.-Z., Woodruff, D., Butenko, S. (eds.) AAIM 2018. LNCS, vol. 11343, pp. 136–148. Springer, Cham (2018). https://doi.org/10.1007/978-3-030-04618-7_12
17. Zhang, Z.: Method and system for calibrating digital cameras. US Patent 6,437,823, 20 August 2002

Variants of Homomorphism Polynomials Complete for Algebraic Complexity Classes

Prasad Chaugule$^{(\boxtimes)}$, Nutan Limaye, and Aditya Varre

Indian Institute of Technology, Bombay, India
{prasad,nutan}@cse.iitb.ac.in

Abstract. We present polynomial families complete for the well-studied algebraic complexity classes VF, VBP, VP, and VNP. The polynomial families are based on the homomorphism polynomials studied in the recent works of Durand et al. (2014) and Mahajan et al. [10]. We consider three different variants of graph homomorphisms, namely *injective homomorphisms*, *directed homomorphisms* and *injective directed homomorphisms* and obtain polynomial families complete for VF, VBP, VP, and VNP under each one of these. The polynomial families have the following properties:

- The polynomial families complete for VF, VBP, and VP are model independent, i.e. they do not use a particular instance of a formula, ABP or circuit for characterising VF, VBP, or VP, respectively.
- All the polynomial families are hard under p-projections.

Keywords: Algebraic complexity theory · Homomorphism

1 Introduction

Valiant [1] initiated the systematic study of the complexity of algebraic computation. There are many interesting computational problems which have an algebraic flavour, for example, determinant, rank computation, discrete log and matrix multiplication. These algebraic problems as well as many other problems, which do not prima facie have an algebraic flavour, can be reduced to the problem of computing certain polynomials. Valiant's work spurred the study of such polynomials and led to a classification of these polynomials as *easy to compute* and *possibly hard to compute*. His work also formalised the notion of a model of computation of polynomials.

He introduced various models of computation of polynomials. An *arithmetic circuit* is one such model of computation which has been well-studied. An arithmetic circuit is a DAG whose in-degree 0 nodes are labelled with variables ($X = \{x_1, \ldots, x_n\}$) or field constants (from, some field, say $\mathbb{F}$). All the other nodes are labelled with operators $+, \times$. Each such node computes a polynomial in a natural way. The circuit has an out-degree zero node, called the output gate. The circuit is said to compute the polynomial computed by its output gate. The size of the circuit is the number of gates in it.

© Springer Nature Switzerland AG 2019
D.-Z. Du et al. (Eds.): COCOON 2019, LNCS 11653, pp. 90–102, 2019.
https://doi.org/10.1007/978-3-030-26176-4_8

Any multivariate polynomial $p(X) \in \mathbb{F}[x_1, \ldots, x_n]$ is said to be *tractable* if its degree is at most poly(n) and there is a poly(n) sized circuit computing it. The class of such tractable polynomial families is called VP.

Many other models of computation have been considered in the literature such as arithmetic formulas and algebraic branching programs (ABPs). An *arithmetic formula* is a circuit in which the underlying DAG is a tree. The class of polynomial families computable by polynomial sized arithmetic formulas is called VF. The class of polynomial families computed by polynomial sized ABPs is called VBP.

Another important class of polynomials studied in the literature (and defined in [1]) is VNP. It is known that VF $\subseteq$ VBP $\subseteq$ VP $\subseteq$ VNP.

In [1], it was shown that *the Permanent* polynomial is *complete*[1] for the class VNP[2]. It was also shown that the syntactic cousin of the Permanent polynomial, namely *the Determinant* polynomial, is complete for VBP. For the longest time there were no natural polynomials which were known to be complete for VP.

Bürgisser in [2] proposed a candidate VP-complete polynomial which was obtained by converting a generic polynomial sized circuit into a VP-hard polynomial (similar to how the Circuit Value Problem is shown to be hard for P). Subsequently Raz in [3] gave a notion of a *universal circuit* and presented a VP-complete polynomial arising from the encoding of this circuit into a polynomial. More recently, Mengel [4] as well as Capelli et al. [5] proposed characterisations of polynomials computable in VP. In these and other related works, the VP-complete polynomial families were obtained using the structure of the underlying circuit. (See for instance [6] and references therein for other related work.)

In the Boolean setting, consider the following two problems: (1) $\{G = (V, E) \mid G$ has a hamiltonian cycle$\}$ and (2) $\{\langle M, x, 1^t \rangle \mid M$ accepts x in at most t steps on at least one non-deterministic branch$\}$. Both the problems are known to be NP-complete, but unlike the first problem, the second problem essentially codes the definition of NP into a decision problem. In that sense, the second problem is dependent on the model used to define NP, but the first one is independent of it. It is useful to have many problems like the first one that are NP-complete, as each such problem conveys a property of the class of NP-complete languages which is not conveyed by its definition.

In the Boolean world the study of NP-complete problems was initiated by the influential works of Cook and Levin [7,8]. Over the years we have discovered thousands of NP-complete problems. Similarly, many natural problems have also been shown to be P-complete. See for instance [9] which serves as a compendium of P-complete problems. Most of these problems are model independent.

---

[1] The hardness is shown with respect to p-projection reductions. We will define them formally in Sect. 2.

[2] Valiant [1] raised the question of whether the Permanent is computable in VP. This question is equivalent to asking whether VP= VNP, which is the algebraic analogue of the P vs. NP question.

In contrast, in the arithmetic world there is a paucity of circuit-description-independent VP-complete problems. Truly circuit-description-independent VP-complete polynomial families were introduced in the works of Durand et al. [6] and Mahajan et al. [10]. We extend their works by giving more such families of polynomials complete for VP. Along the way we obtain such polynomial families complete for VF, VBP, and VNP as well.

At the core of our paper are *homomorphism polynomials*, variants of which were introduced in [6] and [10]. (In fact, in [11,12], some variants of homomorphism polynomials were defined and they were studied in slightly different contexts.) Informally, a homomorphism polynomial is obtained by encoding a combinatorial problem of counting the number of homomorphisms from one graph to another as a polynomial. Say we have two graphs, G and H, then the problem of counting the number of a certain set of homomorphisms, say $\mathcal{H}$, from the graph G to H can be algebrised in many different ways. One such way is to represent the counting problem as the following polynomial. $f_{G,H,\mathcal{H}} = \sum_{\phi \in \mathcal{H}} \prod_{(u,v) \in E(G)} Y_{(\phi(u),\phi(v))}$, where $Y = \{Y_{(a,b)} \mid (a,b) \in E(H)\}$ and $\mathcal{H}$ is a set of homomorphisms from G to H^3.

Our work essentially builds on the ideas defined and discussed in the works of [6,10]. Although [10] and [13] provide homomorphism polynomial families complete for all the important algebraic complexity classes, [6] raises an interesting question, which remains unanswered even in [10] and [13].

The question is: do there exist homomorphism polynomial families complete for algebraic classes such as VP or VNP when $\mathcal{H}$ is restricted to only injective homomorphims (or only directed homomorphisms)? We explore this direction and answer this question positively. This helps us to obtain a complete picture of the work initiated in [6,10].

We consider three sets of homomorphisms, namely injective homomorphisms (denoted as $\mathcal{IH}$), directed homomorphisms (denoted as $\mathcal{DH}$) and injective directed homomorphisms (denoted as $\mathcal{IDH}$). Naturally, when we consider $\mathcal{DH}$ or $\mathcal{IDH}$, we assume that G and H are directed graphs. We then design pairs of classes of graphs which help us obtain polynomials such that they are complete for the following complexity classes: VF, VBP, VP, and VNP. Like in [6,10] our polynomials are also model independent, i.e. the graph classes we use can be defined without knowing anything about the exact structure of the formula, ABP, or the circuit. We also show the hardness in all the cases under more desirable p-projections.

Our constructions do not rely on tree-width/path-width bounded graphs (as the constructions of [10]) and also our upper bound proofs do not use techniques such as the result of Baur and Strassen (like the way it is used in [6]). We believe this makes some of our constructions and proofs slightly simpler as compared to the known constructions and proofs. Moreover, this makes our constructions and proofs conceptually different from those in [6,10].

---

^{3} Note that if we set all Y variables to 1, then this polynomial essentially counts the number of homomorphisms from G to H.

We are able to characterise all the well-studied arithmetic complexity classes using variants of homomorphism polynomials. This provides a unified way of giving characterisation for these classes. The table below (Table 1) shows the known results regarding the homomorphism polynomials prior to our work and also summarises the results in this paper.

There are two other definitions of homomorphism polynomials studied in the literature. One such variant from [6] defines homomorphism polynomials with additional X variables as follows: $\hat{f}_{G,H,\mathcal{H}}(Y) = \sum_{\phi \in \mathcal{H}}(\prod_{u \in V(G)} X_{\phi(u)}^{\alpha(u)})(\prod_{(u,v) \in E(G)} Y_{(\phi(u),\phi(v))})$, where $\alpha : V(G) \to \{0,1\}$. Yet another variant from [6,10] defines homomorphism polynomials using additional Z variables as $\tilde{\hat{f}}_{G,H,\mathcal{H}}(Y) = \sum_{\phi \in \mathcal{H}}(\prod_{u \in V(G)} Z_{u,\phi(u)})(\prod_{(u,v) \in E(G)} Y_{(\phi(u),\phi(v))})$. Since the overall goal in these works is to design polynomials complete for complexity classes, it is reasonable to compare our results with those in [6,10] without worrying about the different variations of homomorphism polynomials across the entries of Table 1[4].

Table 1. Comparison between our work and previous work. A cell containing the symbol ✓ represents the polynomial family designed in this paper

	VP		VBP and VF	VNP
	c-reductions	p-projections	p-projections	p-projections
InjDirHom	-	[6], ✓	✓	✓
InjHom	-	✓	✓	✓
DirHom	[6]	✓	[6], ✓	✓
Hom	[6]	[10]	[10]	[6,13]

The table gives the list of our results. We start with some notations and preliminaries in the following section. In Sect. 3 we present the details regarding VP-complete polynomial families. In Sect. 4 we present the results regarding VNP, VBP and VF-complete polynomial families.

2   Preliminaries

For any integer $n \in \mathbb{N}$, we use $[n]$ to denote the set $\{1, 2, \ldots, n\}$. For any set S, we use $|S|$ to denote the cardinality of the set.

Graph Theoretic Notions. A cycle graph on n nodes, denoted as $\mathcal{C}_n$, is a graph that has n nodes say $v_0, \ldots, v_{n-1}$, and n edges, namely $\{(v_{(i \bmod n)}, v_{(i+1 \bmod n)}) \mid 0 \leq i \leq n-1\}$. We assume that the cycle graph is undirected unless stated otherwise. A *spiked cycle graph* on $n+1$ ($n \geq 3$) nodes, denoted as

---

[4] We do not use c-reductions in this work. They are more general than p-reductions. The formal definition can be found in [6].

$\mathcal{S}_n$, is a cycle graph $\mathcal{C}_n$ with an additional edge (v, u), where u is an additional node which is not among $v_0, \ldots, v_{n-1}$, and $v \in \{v_0, \ldots, v_{n-1}\}$. We call the nodes $v_0, \ldots, v_{n-1}$ *the cycle nodes* and we call the additional node u *a spiked node.* For a graph G, a cycle graph $\mathcal{C}_n$ is said to be attached to a node v of G, if one of the nodes of $\mathcal{C}_n$ is identified with the node v. A spiked cycle graph $\mathcal{S}_n$ is said to be attached to a node v of G, if a node at distance 2 from the spiked node of $\mathcal{S}_n$ is identified with v (Fig. 1).

(a) Spiked cycle $\mathcal{S}_5$, spiked node u.   (b) $\mathcal{S}_5$ is attached to node v in G

Fig. 1. $\mathcal{S}_5$ attached to v in G. The distance between spiked node u and v is 2.

Algebraic Circuit Complexity Classes. Let $\mathbb{F}$ be any characteristic 0 field. From now on we will only work with characteristic 0 fields. Various algebraic circuit complexity classes have been studied in literature, see for instance [2] for the formal definitions of VF, VBP, VP, and VNP.

Projection Reductions. We say that a family of polynomials $\{f_n\}_{n \in \mathbb{N}}$ is a p-projection of another family of polynomials $\{g_n\}_{n \in \mathbb{N}}$ if there is a polynomially bounded function $m : \mathbb{N} \to \mathbb{N}$ such that for each $n \in \mathbb{N}$, f_n can be obtained from $g_{m(n)}$ by setting its variables to one of the variables of f_n or to field constants.

Normal Form Circuits and Formulas. In this section we present some notions regarding normal form circuits. We say that an arithmetic circuit is *multiplicatively disjoint* if the graphs corresponding to the subcircuits rooted at the children of any multiplication gate are vertex disjoint. We use a notion of a normal form of a circuit as defined in [6]. We recall the notion of universal circuit family in normal form as defined in [6]. (Universal circuit families were defined in [3].)

Definition 1 ([6]). *A family of universal circuits $\{D_n\}_{n \in \mathbb{N}}$ in normal form is a family of circuits such that for each $n \in \mathbb{N}$, D_n has the following properties:*

– *It is a layered circuit in which each $\times$ gate (+ gate) has fan-in 2 (unbounded fan-in resp.).*
– *Without loss of generality the output gate is a + gate. Moreover, the circuit has an alternating structure, i.e. the children of + ($\times$) gates are $\times$ (+, resp.) gates, unless the children are in-degree 1 gates, in which case they are input gates.*
– *The input gates have out-degree 1. They all appear on the same layer, i.e. the length of any input gate to output gate path is the same.*

- D_n is multiplicatively disjoint.
- Input gates are labelled by variables and no constants appear at the input gate.
- The depth of D_n is $2c\lceil \log n \rceil$, for some constant c. The number of variables, $v(n)$, and size of the circuit, $s(n)$, are both polynomial in n.
- The degree of the polynomial computed by the circuit is n.

We now recall a notion of a parse tree of a circuit [6,14].

Definition 2 ([14]). *The set of parse trees of a circuit C, $\mathcal{T}(C)$, is defined inductively based on the size of the circuit as follows.*

- *A circuit of size 1 has itself as its unique parse tree.*
- *If the circuit size is more than 1, then the output gate is either a $\times$ gate or a $+$ gate.*
  (i) if the output gate g of the circuit is a $\times$ gate with children g_1, g_2 and say C_{g_1}, C_{g_2} are the circuits rooted at g_1 and g_2 respectively, then the parse trees of C are obtained by taking a node disjoint copy of a parse tree of C_{g_1} and a parse tree of C_{g_2} along with the edges (g, g_1) and (g, g_2).
  (ii) if the output gate g of the circuit is a $+$ gate, then the parse trees of C are obtained by taking a parse tree of any one of the children of g, say h, and the edge (g, h).

Note that a parse tree computes a monomial. For a parse tree T, let f_T be the monomial computed by T. Given a circuit C (or a formula F), the polynomial computed by C (by F, resp.) is equal to $\sum_{T \in \mathcal{T}(C)} f_T$ ($\sum_{T \in \mathcal{T}(F)} f_T$, resp.).

We use the following fact about parse trees proved in [14].

Proposition 1 ([14]). *A circuit C is multiplicatively disjoint if and only if any parse tree of C is a subgraph of C. Moreover, a subgraph T of C is a parse tree if:*

- *T contains the output gate of C.*
- *If g is a $\times$ gate in T, with children g_1, g_2 then the edges (g, g_1) and (g, g_2) appear in T.*
- *If g is a $+$ gate in T, it has a unique child in T, which is one of the children of g in C.*
- *No edges other than those added by the above steps belong to C.*

Graph Homomorphism, Its Variants and Homomorphism Polynomials. Given two undirected graphs G and H, we say that $\phi : V(G) \to V(H)$ is *a homomorphism* from G to H if for any edge $(u, v) \in E(G)$, $(\phi(u), \phi(v)) \in E(H)$. The homomorphism is said to be *an injective homomorphism* if additionally for any node $a \in V(H)$, $|\phi^{-1}(a)| \leq 1$. Given two directed graphs G, H, $\phi : V(G) \to V(H)$ is said to be *a directed homomorphism* from G to H if for any directed edge $(u, v) \in E(G)$, $(\phi(u), \phi(v))$ is a directed edge in $E(H)$. A homomorphism is called an *injective directed homomorphism* if it is an injective as well as directed homomorphism.

Definition 3. *Let G, H be two undirected graphs. Let $Y = \{Y_{(a,b)} \mid (a,b) \in E(H)\}$ be a set of variables. Let $\mathcal{IH}$ be a set of injective homomorphisms from G to H. Then the injective homomorphism polynomial $f_{G,H,\mathcal{IH}}$ is defined as follows: $f_{G,H,\mathcal{IH}}(Y) = \sum_{\phi \in \mathcal{IH}} \prod_{(u,v) \in E(G)} Y_{(\phi(u),\phi(v))}$. If G, H are directed graphs and $\mathcal{DH}$ ($\mathcal{IDH}$) is a set of directed (injective directed, respectively) homomorphisms from G to H then $f_{G,H,\mathcal{DH}}(Y) = \sum_{\phi \in \mathcal{DH}} \prod_{(u,v) \in E(G)} Y_{(\phi(u),\phi(v))}$ ($f_{G,H,\mathcal{IDH}}(Y) = \sum_{\phi \in \mathcal{IDH}} \prod_{(u,v) \in E(G)} Y_{(\phi(u),\phi(v))}$, respectively) is said to be the directed (injective directed, respectively) homomorphism polynomial.*

3   Polynomial Families Complete for VP

In this section we present the details regarding VP-complete polynomial families. We start with some definitions of graph classes.

Definition 4 (Balanced Alternating-Unary-Binary tree). *A balanced alternating-unary-binary tree with k layers, denoted as AT_k, is a layered tree in which the layers are numbered from $1, \ldots, k$, where the layer containing the root node is numbered 1 and the layer containing the leaves is numbered k. The nodes on an even layer have exactly two children and the nodes on an odd layer have exactly one child. Figure 2(a) shows an example AT_k for $k = 5$.*

Definition 5 (Block tree). *Let $\mathsf{BT}_{k,s}$ denote an alternately-unary-binary block tree, which is a graph obtained from AT_k by making the following modifications: each node u of AT_k is converted into a block B_u consisting of s nodes. The block corresponding to the root node is called the root block. The blocks corresponding to the nodes on the even (odd) layers are called binary (unary, respectively) blocks. If v is a child of u in AT_k then B_v is said to be a child of B_u in $\mathsf{BT}_{k,s}$.*

After converting each node into a block of nodes, we add the following edges: say B is a unary block and block B' is its child, then for each node u in B and each node v in B' we add the edge (u,v). Moreover, if B is a binary block and B', B'' are its children, then we assume some ordering of the s nodes in these blocks. Say the nodes in B, B', B'' are $\{b_1, \ldots, b_s\}$, $\{b'_1, \ldots, b'_s\}$, and $\{b''_1, \ldots, b''_s\}$ respectively, then we add edges (b_i, b'_i) and (b_i, b''_i) for each $i \in [s]$. Figure 2(b) shows a block tree $\mathsf{BT}_{k,s}$, where $k = 5$ and $s = 3$.

Let $k_1 = 3 < k_2 < k_3$ be three distinct fixed odd numbers such that $k_3 > k_2 + 2$.

Definition 6 (Modified-Alternating-Unary-Binary tree). *We attach a spiked cycle $\mathcal{S}_{k_1}$ to the root of AT_k. We attach a spiked cycle $\mathcal{S}_{j \times k_2}$ ($\mathcal{S}_{j \times k_3}$ respectively) to each left child node (right child node respectively) in every odd layer $j > 1$. We call the graph thus obtained to be a modified alternating-unary-binary tree and denote it by MAT_k.*

Definition 7 (Modified Block tree). *We start with $\mathsf{BT}_{k,s}$ and make the following modifications: we keep only one node in the root block and delete all the other nodes from the root block. We then attach a spiked cycle $\mathcal{S}_{k_1}$ to the only*

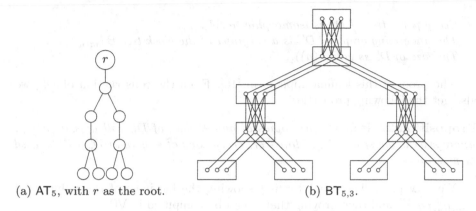

(a) AT_5, with r as the root.   (b) $BT_{5,3}$.

Fig. 2. Examples of AT_k and $BT_{k,s}$

node in root block. We attach a spiked cycle $S_{j \times k_2}$ ($S_{j \times k_3}$ respectively) to each left child node (right child node respectively) in every odd layer $j > 1$. We call the graph thus obtained a modified block tree, $MBT_{k,s}$.

We identify each node in graphs MAT_k, $MBT_{k,s}$ as either a *core node* or a *non core node*. We formally define this notion.

Definition 8 (Core nodes and Non-core nodes). *A* non-core node *is any node in* MAT_k *(or* $MBT_{k,s}$*) which was not already present in* AT_k *(or* $BT_{k,s}$ *respectively). Any node which is not a non-core node is a* core node.

3.1   Injective Homomorphisms

Consider the universal circuit family $\{D_n\}$ in normal form as in Definition 1. Let $m(n) = 2c\lceil \log n \rceil + 1$ be the number of layers in D_n and let $s(n)$ be its size.

Theorem 1. *The family* $f_{G_n, H_n, \mathcal{IH}}(Y)$ *is complete for class* VP *under* p*-projections, where* G_n *is* $MAT_{m(n)}$ *and* H_n *is* $MBT_{m(n),s(n)}$.

As the first step towards proving this theorem, we perform a few more updates to the normal form circuits we designed for polynomials in VP. From the definition of D_n, we know that any parse tree of D_n is isomorphic to $AT_{m(n)}$. From such a circuit D_n, we construct another circuit D'_n, which has all the properties that D_n has and additionally the underlying graph of D'_n is a subgraph of the block tree $BT_{m(n),s(n)}$, for $m(n), s(n)$ as mentioned above. Formally,

Lemma 1. *For every* $n \in \mathbb{N}$*, given any circuit* D_n *with* $m(n) = 2c\lceil \log n \rceil + 1$ *layers and size* $s(n)$ *in the normal form as in Definition 1, there is another circuit* D'_n *such that it has all the properties that the circuit* D_n *has and additionally it has the following properties:*

- *The polynomial computed by* D'_n *is the same as the polynomial computed by* D_n.

– *Every parse tree of D'_n is isomorphic to $AT_{m(n)}$.*
– *The underlying graph of D'_n is a subgraph of the block tree $BT_{m(n),s(n)}$.*
– *The size of D'_n is* poly($s(n)$).

The proof of this lemma appears in [15]. From the construction of D'_n, we also get the following properties.

Proposition 2. *At most $s(n)$ copies of any $+$ gate of D_n will appear in D'_n, where $s(n)$ is the size of D_n. Moreover, every copy of $+$ gate in D'_n will be used at most once.*

We now prove Theorem 1 by first showing the hardness of the polynomial $f_{G_n,H_n,\mathcal{IH}}(Y)$ and then proving that it can be computed in VP.

VP Hardness of $f_{G_n,H_n,\mathcal{IH}}(Y)$. We now show that if $f_n(X)$ is a polynomial computed in VP, then it is a p-projection of $f_{G_n,H_n,\mathcal{IH}}(Y)$. Let G_n, H_n be the source and target graphs defined in Theorem 1.

Let f_n be any polynomial in VP and D_n be the normal form universal circuit computing f_n with $m(n) = 2c\lceil \log n \rceil + 1$ layers and size $s(n)$. We convert this circuit into D'_n as specified at the start of this section. As observed earlier, it still computes the polynomial computed by D_n. Let $\mathcal{G}'_n$ be the underlying graph of the circuit D'_n. As D'_n is multiplicatively disjoint every parse tree of the circuit is a subgraph of $\mathcal{G}'_n$ and is of the form $AT_{m(n)}$.

If a spiked cycle is attached to a node v in layer ℓ of a layered graph then we will say that all the nodes of the cycle belong to the same layer ℓ.

Let $\phi : G_n \to H_n$ be any injective homomorphism. Let us use ϕ_i to denote the action of this homomorphism restricted to layer i on G_n. Let $\tilde{\phi}_i$ denote $\cup_{1 \le j \le i} \phi_i$, i.e. the action of ϕ up to layer i. We will prove the following lemma inductively.

Lemma 2. *Let ϕ be an injective homomorphism from G_n to H_n. For any $i \in [m(n)]$, $\tilde{\phi}_i(G_n)$ is simply a copy[5] of the graph MAT_i inside H_n with the following additional properties: (i) the root of MAT_i is mapped to the root of H_n. (ii) for any $i \in [m(n)]$, a core node u in layer i is mapped to a node in B_u at layer i.*

The lemma can be proved using induction on $i \in [m(n)]$ (See [15] for details). We will now show that using this lemma we are done. We saw that $\mathcal{G}'_n$ is embedded in $BT_{m,q}$ as a subgraph, where $m = 2c\lceil \log n \rceil + 1$ and $q = s(n)$.

We wish to set variables such that the monomial computed by each injective homomorphism is the same as the monomial computed by the corresponding parse tree. This can be achieved simply by setting variables as follows: Let e be an edge between two core nodes of H_n. If such an edge is not an edge in $\mathcal{G}'_n$ then set it to 0. (This carves out the graph $\mathcal{G}'_n$ inside H_n.) If such an edge is an edge associated with the leaf node, then locate the corresponding node in D'_n. It will be an input gate in D'_n. If the label of that input gate is x, then set this edge to

---

[5] It is a layer preserving isomorphic copy which maps the root node of MAT_i to the root of H_n.

x. If e is any other edge that appears in $\mathcal{G}'_n$, then set it to 1. (This allows for the circuit functionality to be realised along the edges of H_n.) Finally, suppose e is an edge between two non-core nodes (or between a core and a non-core node), i.e. along one of the attached cycles, then set it to 1. (This helps in suppressing the cycle edges in the final computation.)

This exactly computes the sum of all parse trees in the circuit D'_n, which shows that any polynomial computed in VP is also computed as a p-projection of $f_{G_n,H_n,\mathcal{IH}}(Y)$.

$f_{G_n,H_n,\mathcal{IH}(Y)}$ is in VP. The source graph G_n and target graph H_n are as described in the construction. We have already observed in Lemma 2 that all injective homomorphisms from G_n to H_n respect the layers. Therefore, it suffices to compute only such layer respecting homomorphisms.

Construction of the Circuit Computing $f_{G_n,H_n,\mathcal{IH}}(Y)$. The construction of the circuit, say C_n, is done from the bottom layer (i.e. from the leaves) to the top layer (i.e. to the root). For any *core node* $u \in V(\mathsf{MAT}_{m(n)})$ at layer ℓ of G_n and any *core node* a in block B_u at layer ℓ in H_n, we have a gate $\langle u, a \rangle$ in our circuit C_n at layer ℓ. Let us denote the sub-graph rooted at u in G_n by $G^{(u)}$ and that rooted at a in H_n to be $H^{(a)}$. Let $\mathcal{IH}_{(u,a)}$ be the set of injective homomorphism from sub-graph $G^{(u)}$ to $H^{(a)}$ where u is mapped to a. Let $f_{\langle u,a \rangle}$ be the polynomial computed at the gate $\langle u, a \rangle$.

We will describe the inductive construction of the circuit C_n starting with the leaves. We know that there is a spiked cycle $\mathcal{S}_{k_2 \times m(n)}$ or $\mathcal{S}_{k_3 \times m(n)}$ attached to each node at layer $m(n)$ in G_n[6]. For any spiked cycle $\mathcal{S}_k$ attached at a node x in H_n, let $\sigma^x_{\mathcal{S}_k}(Y)$ denote the monomial obtained by multiplying all the Y variables along the edges in $\mathcal{S}_k$ attached at x in H_n. Let u be a left (or right) child node in G_n at layer $m(n)$ and a be some node in B_u at layer $m(n)$ in H_n, then we set $\langle u, a \rangle = \sigma^a_{\mathcal{S}_{k_2 \times m(n)}}(Y)$ (or $\langle u, a \rangle = \sigma^a_{\mathcal{S}_{k_3 \times m(n)}}(Y)$, respectively).

Suppose we have a left (or right) child node, say u, at layer i in G_n which has only one child, say u' at layer $i+1$ in G_n. We know that there is a spiked cycle $\mathcal{S}_{k_2 \times i}$ (or $\mathcal{S}_{k_3 \times i}$ respectively) attached to u if it is the left (right respectively) child node. Let a be any node in B_u at layer i in H_n. Say a has t children, $a_1, \ldots, a_t$ in H_n. Inductively, we have gates $\langle u', a_\alpha \rangle$ for all $1 \le \alpha \le t$. We set

$$\langle u, a \rangle = \sum_{\alpha=1}^{t} \langle u', a_\alpha \rangle \times Y_{(a,a_\alpha)} \times \sigma^a_{\mathcal{S}_{k_2 \times i}}(Y) \quad \text{or} \tag{1}$$

$$\langle u, a \rangle = \sum_{\alpha=1}^{t} \langle u', a_\alpha \rangle \times Y_{(a,a_\alpha)} \times \sigma^a_{\mathcal{S}_{k_3 \times i}}(Y) \tag{2}$$

depending on whether u is a left child or a right child of its parent in G_n respectively. Suppose u in layer i in G_n has a left child u_1 and a right child u_2 in layer $i+1$. Let a be any node in the block B_u in H_n. Let a_1 and a_2 be the left child and right child of a in H_n respectively. It is easy to see that a_1 resides in the

---

[6] Recall that $m(n) = 2c\lceil \log n \rceil + 1$, which is odd. Also this is without loss of generality.

block B_{u_1} in H_n and a_2 resides in the block B_{u_2} in H_n. Inductively, we have gates $\langle u_1, a_1 \rangle$ and $\langle u_2, a_2 \rangle$. We set

$$\langle u, a \rangle = \langle u_1, a_1 \rangle \times Y_{(a, a_1)} \times \langle u_2, a_2 \rangle \times Y_{(a, a_2)} \tag{3}$$

This completes the description of C_n. The details regarding the correctness of C_n appears in [15].

3.2   Directed and Injective Directed Homomorphisms

We start with the definitions of the following directed graph classes.

Definition 9 (Directed Balanced Alternating-Unary-Binary tree, Directed Block tree). *Let AT_k^d and $BT_{k,s}^d$ denote the directed versions of AT_k and $BT_{k,s}$, respectively. The directions on the edges go from the root (block) towards the leaves (leaf blocks).*

Let $k_1' = 5 < k_2' < k_3' < k_4' \in \mathbb{N}$ be four distinct fixed mutually co-prime numbers.

Definition 10 (Modified Directed Alternating-Unary-Binary tree). *We attach a directed cycle $C_{k_1'}$ to the root of AT_k^d. We attach a directed cycle $C_{k_2'}$ to each node in every even layer in AT_k^d. We attach a directed cycle $C_{k_3'}$ ($C_{k_4'}$ respectively) to each left child node (right child node respectively) in every odd layer (except the root node at layer 1) in AT_k^d. We call the graph thus obtained to be a modified directed alternating-unary-binary tree, MAT_k^d.*

Definition 11 (Modified Directed Block tree). *We consider $BT_{k,s}^d$ and make the following modifications: we keep only one node in the root block node and delete all the other nodes from the root block node. We attach a directed cycle $C_{k_1'}$ to the only node in the root block of $BT_{k,s}^d$. We attach a directed cycle $C_{k_2'}$ to each node in every even layer in $BT_{k,s}^d$. We attach a directed cycle $C_{k_3'}$ ($C_{k_4'}$ respectively) to each left child node (right child node respectively) in every odd layer (except the root node at layer 1) in $BT_{k,s}^d$. We call the graph thus obtained to be a modified directed block tree and denote it by $MBT_{k,s}^d$.*

We identify each node in graphs MAT_k^d, $MBT_{k,s}^d$ as either a *core node* or a *non core node* just like we did in the undirected case. We have the following theorem (proof appears in [15]).

Theorem 2. *The families $f_{G_n, H_n, \mathcal{DH}}(Y)$, $f_{G_n, K_{p(n)}, \mathcal{DH}}(Y)$ and $f_{G_n, H_n, \mathcal{IDH}}(Y)$ are complete for class VP under p-projections where G_n is $MAT_{m(n)}^d$, H_n is $MAT_{m(n),s(n)}^d$ and $K_{p(n)}$ is the complete graph obtained from H_n by adding all directed edges between every pair of nodes of H_n, where $p(n)$ is the number of nodes of H_n.*

Remark 1. We get a VP-complete polynomial family when the right-hand-side graph is a complete graph and $\mathcal{H} = \mathcal{DH}$. It would be interesting to get this feature even when $\mathcal{H} = \mathcal{IH}$ or $\mathcal{IDH}$. It is easy to see that in the case of $\mathcal{H} = \mathcal{IH}$ or $\mathcal{IDH}$, if the right-hand-side graph is a complete graph, then the current proof can be easily modified to prove VP-hardness. On the other hand, the containment of these families in VP is not straightforward in this case. It is worth noting however that all our constructions ensure that the graphs are model independent in all three cases, i.e. when $\mathcal{H}$ equals $\mathcal{IH}, \mathcal{DH}$ and $\mathcal{IDH}$.

4 Polynomial Families Complete for **VNP**, **VBP** and **VF**

In this section, we present the results related to the homomorphism polynomial families complete for VNP, VBP and VF. At the core of our VNP-complete polynomials lies the Permanent polynomial, which is known to be VNP-complete. Whereas at the core of our VBP and VF polynomials is the $\mathrm{IMM}_{k,n}$ polynomial, which is defined to be the $(1,1)$th entry of the matrix arising from the multiplication of n matrices of dimension k. $\mathrm{IMM}_{k,n}$ is known to be VBP-complete for large k and VF-complete for $k = \mathcal{O}(1)$ (See [16] for more details). For every complexity class $\mathcal{C} \in \{\text{VNP}, \text{VBP}, \text{VF}\}$ and for any homomorphism class $\mathcal{H} \in \{\mathcal{IH}, \mathcal{DH}, \mathcal{IDH}\}$, we come up with a pair of graph families $\Pi_{\mathcal{H},\mathcal{C}}$ such that the homomorphism polynomial defined with respect to $\mathcal{H}$ and $\Pi_{\mathcal{H},\mathcal{C}}$ is complete for the class $\mathcal{C}$. Formally, we show the following. (See [15] for proof details.)

Theorem 3. *Let $\mathcal{C} \in \{VNP, VBP, VF\}$.*

There exist graph families $\{G_n^{\mathcal{C}}\}$ and $\{H_n^{\mathcal{C}}\}$ such that $f_{\{G_n^{\mathcal{C}}\}, \{H_n^{\mathcal{C}}\}, \mathcal{IH}}$ is complete for the class $\mathcal{C}$.

If $\mathcal{H} \in \{\mathcal{DH}, \mathcal{IDH}\}$, then there exist directed graph families $\{G_n^{\mathcal{C}}\}$ and $\{H_n^{\mathcal{C}}\}$ such that $f_{\{G_n^{\mathcal{C}}\}, \{H_n^{\mathcal{C}}\}, \mathcal{H}}$ is complete for the class $\mathcal{C}$.

References

1. Valiant, L.G.: Completeness classes in algebra. In: Proceedings of the Eleventh Annual ACM Symposium on Theory of Computing, STOC 1979, pp. 249–261 (1979)
2. Bürgisser, P.: Completeness and Reduction in Algebraic Complexity Theory, vol. 7. Springer, Heidelberg (2013). https://doi.org/10.1007/978-3-662-04179-6
3. Raz, R.: Elusive functions and lower bounds for arithmetic circuits. Theory Comput. **6**(1), 135–177 (2010)
4. Mengel, S.: Characterizing arithmetic circuit classes by constraint satisfaction problems. In: Aceto, L., Henzinger, M., Sgall, J. (eds.) ICALP 2011. LNCS, vol. 6755, pp. 700–711. Springer, Heidelberg (2011). https://doi.org/10.1007/978-3-642-22006-7_59
5. Capelli, F., Durand, A., Mengel, S.: The arithmetic complexity of tensor contraction. Theory Comput. Syst. **58**(4), 506–527 (2016)
6. Durand, A., Mahajan, M., Malod, G., de Rugy-Altherre, N., Saurabh, N.: Homomorphism polynomials complete for VP. In: LIPIcs-Leibniz International Proceedings in Informatics, vol. 29 (2014)

7. Cook, S.A.: The complexity of theorem-proving procedures. STOC **71**, 151–158 (1971)
8. Levin, L.A.: Universal search problems. Probl. Inf. Transm. **9**(3), 115–116 (1973). (in Russian)
9. Greenlaw, R., Hoover, H.J., Ruzzo, W.: A compendium of problems complete for P, vol. 11 (1992)
10. Mahajan, M., Saurabh, N.: Some complete and intermediate polynomials in algebraic complexity theory. Theory Comput. Syst. **62**(3), 622–652 (2018)
11. Rugy-Altherre, N.: A Dichotomy theorem for homomorphism polynomials. In: Rovan, B., Sassone, V., Widmayer, P. (eds.) MFCS 2012. LNCS, vol. 7464, pp. 308–322. Springer, Heidelberg (2012). https://doi.org/10.1007/978-3-642-32589-2_29
12. Engels, C.: Dichotomy theorems for homomorphism polynomials of graph classes. J. Graph Algorithms Appl. **20**(1), 3–22 (2016)
13. Saurabh, N.: Analysis of algebraic complexity classes and boolean functions. Ph.D. thesis (2017)
14. Malod, G., Portier, N.: Characterizing Valiant's algebraic complexity classes. In: Královič, R., Urzyczyn, P. (eds.) MFCS 2006. LNCS, vol. 4162, pp. 704–716. Springer, Heidelberg (2006). https://doi.org/10.1007/11821069_61
15. Chaugule, P., Limaye, N., Varre, A.: Variants of homomorphism polynomials complete for algebraic complexity classes, 31 July 2018
16. Shpilka, A., Yehudayoff, A., et al.: Arithmetic circuits: a survey of recent results and open questions. Found. Trends® Theor. Comput. Sci. **5**(3–4), 207–388 (2010)

Chance-Constrained Submodular Knapsack Problem

Junjie Chen[1] and Takanori Maehara[2(✉)]

[1] Department of Computer Science, Hong Kong Baptist University,
Kowloon Tong, Hong Kong, China
csjjchen@comp.hkbu.edu.hk
[2] Discrete Optimization Unit, RIKEN Center for Advanced Intelligence Project,
Nihonbashi, Tokyo, Japan
takanori.maehara@riken.jp

Abstract. In this study, we consider the *chance-constrained submodular knapsack problem*: Given a set of items whose sizes are random variables that follow probability distributions, and a nonnegative monotone submodular objective function, we are required to find a subset of items that maximizes the objective function subject to that the probability of total item size exceeding the knapsack capacity is at most a given threshold. This problem is a common generalization of the chance-constrained knapsack problem and submodular knapsack problem.

Specifically, we considered two cases: the item sizes follow normal distributions, and the item sizes follow arbitrary but known distributions. For the normal distribution case, we propose an algorithm that finds a solution that has an expected profit of at least $1 - e^{-1} - O(\epsilon)$ to the optimal. For the arbitrary distribution case, we propose an algorithm that finds a solution that has the same approximation factor but satisfies the relaxed version of the constraint, which relaxes both the knapsack capacity and overflow probability. Here, both algorithms are built on the same strategy: reduce the chance constraint to a multidimensional knapsack constraint by guessing parameters, and solve the reduced multidimensional knapsack constrained submodular maximization problem by the continuous relaxation and rounding method.

Keywords: Chance-constrained knapsack problem ·
Submodular maximization · Approximation algorithm

1 Introduction

1.1 Background and Motivation

The knapsack problem is one of the most fundamental combinatorial optimization problems: Given a set of items associated with sizes and profits, find a maximum profit subset of items whose total size is at most a given capacity of a

---

© Springer Nature Switzerland AG 2019
D.-Z. Du et al. (Eds.): COCOON 2019, LNCS 11653, pp. 103–114, 2019.
https://doi.org/10.1007/978-3-030-26176-4_9

knapsack. If the item sizes follow probability distributions, the problem is called a *stochastic knapsack problem*.

In this study, we consider the *chance-constrained knapsack problem* [7] (aka. *bounded overflow probability model* [2]), which is an important variant of the stochastic knapsack problem. In this variant, we are required to find a maximum-profit set of items subject to that the probability of the total item size violating the knapsack capacity is at most a given threshold ρ. Regarding the different probability distributions of item size, several results have been obtained. Goel and Indyk [5] studied the problem for the Poisson, exponential, and Bernoulli distributions and proposed PTASes and QPTASes. Goyal and Ravi [7] proposed a PTAS for the normal distribution case by reducing the problem to a two-dimensional knapsack problem. Recently, Shabtai, Raz, and Shavitt [17] proposed a FPTAS for the normal distribution case by utilizing dynamic programming and a polyhedral structure. Bhalgat, Goel, and Khanna [2] considered online version of the problem (i.e., the realized item size is immediately revealed when it is inserted to the knapsack), and proposed adaptive and non-adaptive policies that achieve PTASes for arbitrary-distribution case by slightly relaxing the overflow probability and knapsack capacity. For the same setting, Li and Yuan [13] also proposed an adaptive policy that achieves a PTAS.

As described above, there are many existing studies for the problem; however, all the existing models only consider linear objective functions. The purpose of this study is to generalize the objective function to a *monotone submodular function*. A (monotone) submodular function is an important class of discrete functions that naturally arises in combinatorial optimization [6,8], machine learning [9], and economics [15]. The problem of maximizing a monotone submodular function is NP-hard even under a cardinality constraint [16]; however, subject to several constraints, such as the cardinality constraint [16], knapsack constraint [10,18], matroid constraint [3,19], etc. [4,12,20], the problem admits polynomial time approximation algorithms. In particular, since the knapsack-constrained monotone submodular maximization problem has been applied to the online advertising problems [1,14] in which the cost (i.e., size) is uncertain, it would be beneficial to consider this generalization.

1.2 Contributions

Formally, our problem, *chance-constrained submodular knapsack problem*, is defined as follows. Let $N = \{1, \dots, n\}$ be a finite set of items. Each item $i \in N$ has stochastic size X_i, which is a random variable that follows a known distribution. Without loss of generality, we assume that the knapsack capacity is one. Let $\rho \in \mathbb{R}_{>0}$ be a given probability threshold, and $f \colon 2^N \to \mathbb{R}_{\geq 0}$ be a monotone submodular function. Then, our problem is formulated by

$$\underset{S \subseteq N}{\text{maximize}} \ f(S)$$

$$\text{subject to } \Pr\left(\sum_{i \in S} X_i > 1\right) \leq \rho. \tag{1}$$

For this problem, we obtain the following results.

Theorem 1. *Suppose that size X_i of item follows a normal distribution. Then, for any $\epsilon > 0$, there exists an algorithm that runs in polynomial-time and returns a feasible solution with the expected profit of at least $(1 - e^{-1} - O(\epsilon)) OPT$, where OPT is the optimal value.*

Theorem 2. *Suppose that size X_i of item follows an arbitrary but known distribution. Then, for any $\epsilon > 0$, there exists an algorithm that runs in polynomial-time and returns a solution with the expected profit of at least $(1 - e^{-1} - O(\epsilon)) OPT$ if relaxing the knapsack capacity to $1 + O(\epsilon)$ and the overflow probability to factor $\rho + O(\epsilon)$, where OPT is the optimal value.*

We prove these theorems by the same strategy: Convert the chance-constraint to a multiple deterministic knapsack constraint and apply an algorithm for multiple knapsack constrained monotone submodular maximization.

For the normal distribution case, we employ the technique of Goyal and Ravi [7] to achieve the first part: By guessing the expectation of the size of an optimal solution, the problem reduces to a two-dimensional knapsack problem, where the dimensions correspond to the mean and the variance. For the second part, instead of the LP relaxation used in Goyal and Ravi, we employ Kulik, Shachnai, and Tamir's approximation algorithm for submodular knapsack problem [10]. This method separates items into "light" items and "heavy" items, and solve the residual problem by a continuous greedy algorithm [20].

For the arbitrary distribution case, we employ the technique called the *Poisson approximation* established by Li and Yuan [13]. The key theorem of the Poisson approximation (Lemma 3 in this paper) shows that the size distribution of a solution to the problem is well-approximated by a *compound Poisson distribution*. Then, as same as the normal distribution case, the problem is reduced to a multiple knapsack problem, where the dimension of the knapsack depends on the accuracy of the distribution approximation. Then, we solve the multiple knapsack constrained submodular maximization problem by the continuous relaxation and rounding. It should be emphasized that the existing techniques for the stochastic knapsack problem with an arbitrary item size distribution [2, 13] cannot be used since these only considered the online version problem.

2   Preliminaries

Let N be a finite set. A function $f \colon 2^N \to \mathbb{R}$ is *normalized* if $f(\emptyset) = 0$. In the rest of the paper, we assume that the objective function f is normalized. A function $f \colon 2^N \to \mathbb{R}$ is *monotone* if for all $A \subseteq A' \subseteq N$, $f(A) \leq f(A')$. A function $f \colon 2^N \to \mathbb{R}$ is *submodular* if it satisfies the diminishing return property: For all $i \in N$ and $A \subseteq A' \subseteq N$, $f(A \cup \{i\}) - f(A) \geq f(A' \cup \{i\}) - f(A)$. For a subset $T \subseteq N$ and a function $f \colon 2^N \to \mathbb{R}$, the marginal function $f_T \colon 2^{N \setminus T} \to \mathbb{R}$ is defined by $f_T(S) := f(S \cup T) - f(S)$. If f is monotone submodular, f_T is also monotone submodular.

For a submodular function $f\colon 2^N \to \mathbb{R}$, the corresponding continuous relaxation, called the *multilinear extension* $F\colon [0,1]^N \to \mathbb{R}$, is defined as follows [3]:

$$F(x) = \mathbb{E}_{\hat{S}\sim x}[f(\hat{S})] = \sum_{S\subseteq N} f(S) \prod_{i\in S} x_i \prod_{j\in N\setminus S} (1 - x_j), \qquad (2)$$

where $\hat{S}$ is a random subset where each $i \in N$ is included to the subset with probability x_i independently. An important property of the multilinear extension of submodular function is that $F(x)$ is concave along any nonnegative direction.

The multilinear extension is widely used for submodular maximization problems. The following theorem shows that the multilinear extension is approximately maximized in polynomial time.

Theorem 3 (Feldman, Naor, and Schwartz [4]**).** *Let $f\colon 2^N \to \mathbb{R}_{\geq 0}$ be a monotone submodular function and $P \subseteq [0,1]^N$ be a downward-closed polytope with a separation oracle. For any $\epsilon > 0$, there exists a polynomial time algorithm that produces a fractional solution $x \in P$ such that $F(x) \geq (1 - e^{-1} - O(\epsilon)) \max_{x^* \in P} F(x^*)$.* $\square$

Evaluating the multilinear extension is #P-hard; however, we can arbitrary accurately estimate its value in polynomial time via random sampling [20].

Kulik, Shachnai, and Tamir [10] showed that a monotone submodular function is approximately maximized under a $d = O(1)$-dimensional knapsack constraint in polynomial time. Since our result heavily relies on their method, we briefly summarize their result. Consider a d-dimensional knapsack problem with item sizes $c_1, \ldots, c_n \in \mathbb{R}^d$. An item $i \in N$ is *light* if all the entries in c_i is at most ϵ^3 to the capacity; otherwise *heavy*. Note that an optimal solution contains at most d/ϵ^3 heavy items. If the problem contains no heavy items, a fractional solution is rounded to an integral solution by preserving the objective value.

Theorem 4 (Kulik, Shachnai, and Tamir [10]**).** *Let $f\colon 2^N \to \mathbb{R}_{\geq 0}$ be a monotone submodular function and $x \in [0,1]^N$ be a fractional solution to a d-dimensional knapsack constraint without heavy items. For any $\epsilon > 0$, there exists a polynomial time algorithm that produces a random subset S such that $\mathbb{E}[f(S)] \geq (1 - O(\epsilon))F(x)$.* $\square$

By finding an approximate continuous solution using Theorem 3, and rounding the solution using Theorem 4, we can obtain $(1 - e^{-1} - O(\epsilon))$-approximate solution to the d-dimensional knapsack constrained submodular maximization problem in polynomial time.

3   Normal Distribution Case

In this section, we propose an algorithm for the normal distribution case, i.e., for each item $i \in N$, its size X_i follows a normal distribution $\mathcal{N}(a_i, \sigma_i^2)$ with mean a_i and variance σ_i^2. We first show how to reformulate the chance constraint by a two-dimensional knapsack constraint. Then we propose an algorithm to solve the problem.

Algorithm 1. Algorithm for the normal distribution case.

Input: Given a item set N, where size of each item is normally distributed with mean a_i and variance σ_i^2. A monotone submodular set function f. An input parameter ϵ. Overflow probability ρ. $N_1 = \lfloor \log_{1+\epsilon} \min_{i \in N}\{a_i\} \rfloor$,

1: **for** each $T \subseteq N$ such that $|T| \le d\epsilon^{-3}$ **do**
2:     Compute a_T and b_T, $\mathcal{D} = \emptyset$
3:     Ensure $1 - a_T \ge 0$ and $(1 - a_T)^2 - b_T \ge 0$, otherwise terminate this iteration
4:     **for** each $\mu = a_T + (1 + \epsilon)^k$, $k = N_1, \ldots, 0$, such that $(1 - \mu)^2 - b_T \ge 0$ **do**
5:         Solve the two-dimensional knapsack problem whose residual capacity is $\bar{L} = (\mu - a_T, (1 - \mu)^2 - b_T)$ and return $\mathcal{D}$
6: Return the $S = \mathcal{D} \cup T$ with best $f(S)$

3.1 Problem Reformulation

The constraint in Eq. (1) can be equivalently written as $\Pr\left(\sum_{i \in S} X_i \le 1\right) \ge 1 - \rho$. Let $Z_S = \sum_{i \in S}(X_i - a_i)/\sqrt{\sum_{i \in S}\sigma_i^2}$. Then, the constraint is represented by $\Pr\left(Z_S \le (1 - \sum_{i \in S} a_i)/\sqrt{\sum_{i \in S}\sigma_i^2}\right) \ge 1 - \rho$. Since Z_S follows the standard normal distribution with mean zero and variance one, using the cumulative distribution function ϕ of the standard normal distribution, it can be further equivalently written as $\phi^{-1}(1 - \rho)\sqrt{\sum_{i \in S}\sigma_i^2} \le 1 - \sum_{i \in S} a_i$. Let S^* be an optimal solution and $\mu^* = \sum_{i \in S^*} a_i$. If we know the value of μ^*, the constraint is safely replaced by $\sum_{i \in S^*} b_i \le (1 - \mu^*)^2$, where $b_i = \phi^{-1}(1 - \rho)\sigma_i^2$. This inspires us to solve the following problem

$$\begin{aligned} \underset{S \subseteq N}{\text{maximize}} \quad & f(S) \\ \text{subject to} \quad & \sum_{i \in S} a_i \le \mu \\ & \sum_{i \in S} b_i \le (1 - \mu)^2 \end{aligned} \tag{3}$$

by guessing μ^* with μ. Given an estimated value μ, this problem is a deterministic two-dimensional submodular knapsack problem.

3.2 Proposed Algorithm

We propose an algorithm for the problem (Algorithm 1). In the algorithm, we first guess a subset $T \subseteq N$. Let $f_T(S) = f(S \cup T) - f(S)$, $a_T = \sum_{i \in T} a_i$, and $b_T = \sum_{i \in T} b_i^2$. Then, we guess μ by enumerating $\mu = (1 + \epsilon)^k + a_T$ for each k. Here, we obtain a residual knapsack problem whose objective function is f_T and the capacity is $(\mu - a_T, (1 - \mu)^2 - b_T)$. We define heavy items H and light items L according to this capacity (i.e., i is light if $a_i \le \epsilon^3(\mu - a_T)$ and $b_i \le \epsilon^3((1 - \mu)^2 - b_T))$. Then, we remove all the heavy items from the instance, and solve the problem using the multilinear relaxation and rounding (Theorems 3 and 4). We then output the best solution among the process.

We analyze the performance of this algorithm. Let S^* be an optimal solution and OPT be the optimal value of the original problem in Eq. (1).

Lemma 1. *Consider the case when $T \subseteq S^*$ and $\mu - a_T \leq \mu^* - a_T \leq (1+\epsilon)(\mu - a_T)$. Then, the solution obtained in line 6 of Algorithm 1 satisfies*

$$\mathbb{E}[f(D)] \geq (1 - e^{-1} - O(\epsilon))(OPT - f(T) - f_T(S^* \cap H)). \tag{4}$$

Proof. We define $\hat{x}$ by $\hat{x}_j = 1/(1 + \epsilon)$ if $j \in S^* \cap L$, otherwise $\hat{x}_j = 0$. Note that $\hat{x} = 1_{S^* \cap L}/(1 + \epsilon)$, where $1_{S^* \cap L}$ is the indicator vector of $S^* \cap L$. Then, $\hat{x}$ is a feasible solution to the residual knapsack problem with these parameters because

$$\sum_{j \in L} a_j \hat{x}_j = \sum_{j \in L \cap S^*} a_j \hat{x}_j = \sum_{j \in L \cap S^*} a_j \frac{1}{1 + \epsilon} \leq \frac{\mu^* - a_T}{1 + \epsilon} \leq \mu - a_T$$

where the last inequality comes from $\mu^* - a_T \leq (1 + \epsilon)(\mu - a_T)$, and

$$\sum_{j \in L} b_j \hat{x}_j = \sum_{j \in L \cap S^*} b_j \hat{x}_j = \sum_{j \in L \cap S^*} b_j \frac{1}{1 + \epsilon} \leq \frac{(1 - \mu^*) - b_T}{1 + \epsilon} \leq (1 - \mu)^2 - b_T$$

where the last inequality is due to $\mu \leq \mu^*$. The objective value at $\hat{x}$ is

$$F_T(\hat{x}) \geq \frac{1}{1 + \epsilon} F_T(1_{S^* \cap L}) = \frac{1}{1 + \epsilon} f_T(S^* \cap L), \tag{5}$$

because of the concavity of multilinear extension along any nonnegative direction. By the submodularity, we have

$$f_T(S^* \cap L) \geq OPT - f(T) - f_T(S^* \cap H) \tag{6}$$

Therefore, by Theorems 3 and 4, we obtain the lemma. □

Lemma 2. *Consider the case when $T \subseteq S^*$ and $\mu - a_T \leq \mu^* - a_T \leq (1+\epsilon)(\mu - a_T)$. The cardinality of $S^* \cap H$ is less than $(2 + \epsilon)\epsilon^{-3}$.*

Proof. If item j is heavy, either $a_j > \epsilon^3(\mu - a_T) \geq \epsilon^3/(1 + \epsilon)(\mu^* - a_T)$ or $b_j^2 > \epsilon^3((1 - \mu)^2 - b_T) \geq \epsilon^3/(1 + \epsilon)((1 - \mu^*)^2 - b_T)$ holds. Therefore, by the standard counting argument, there are at most $(2 + \epsilon)/\epsilon^3$ heavy items. □

Now we are ready to present the proof of Theorem 1.

Proof (Proof of Theorem 1). Let $S^* = \{o_1, o_2, \ldots, o_k\}$ be an optimal solution, where $\{o_i\}_{i=1}^k$ are sorted in the decreasing order in their marginal profits, that is $f_{S_{i-1}}(o_i) \geq f_{S_i}(o_{i+1})$ where $S_i = \{o_1, o_2, \ldots, o_i\}$. If $|S^*| \leq d\epsilon^{-3}e$, we can find the optimal solution by enumerating subset T. Therefore, in the following, we assume that $|S^*| > d\epsilon^{-3}e$ and T takes the first $d\epsilon^{-3}e$ items of S^*. Let $f(T) = \alpha OPT$ for some α. By submodularity and Lemma 2, we have

$$f_T(S^* \cap H) \leq \frac{f(T)}{d\epsilon^{-3}e} |S^* \cap H| \leq e^{-1}(1 + \epsilon)\alpha OPT$$

By combining with Lemma 1, we obtain

$$\alpha OPT + (1 - e^{-1} - O(\epsilon))(OPT - \alpha OPT - f_T(S^* \cap H))$$
$$\geq (1 - e^{-1} - O(\epsilon))OPT.$$

□

We analyze the complexity of our algorithm. There are at most $O(n^{de\epsilon^{-3}})$ different sets of T and $O(\log_{1+\epsilon} 1/a_{\min})$ different choices of μ. Thus, the algorithm solves the multidimensional submodular knapsack problem at most polynomially many times. The multidimensional submodular knapsack problem is solved in polynomial time by the continuous relaxation and rounding is performed in polynomial time [10]. Therefore, the complexity of algorithm is polynomial.

4 Arbitrary Distribution Case

In this section, we propose an algorithm for arbitrary distribution case. As same as the normal distribution case, we reformulate the chance-constraint by a multiple knapsack constraint. To handle arbitrary distribution, we apply the discretization technique, called the *Poisson approximation* [2,13].

4.1 Problem Reformulation

We first introduce the compound Poisson distribution [11,13]. Consider a K-dimensional nonnegative vector $V = (V_1, V_2, \ldots, V_K) \in \mathbb{R}^K_{\geq 0}$ and let $\lambda = \sum_k V_k$. Define $Y = \sum_{i=1}^M Y_i$ where $M \sim \text{Poisson}(\lambda)$ and Y_i's are independent random variables where $\Pr(Y_i = k) = V_k/\lambda$ and $\Pr(Y_i = 0) = 0$. Then, we say that Y follows a *compound Poisson distribution* corresponding to V. The most important property here is Le Cam's Poisson approximation theorem.

Lemma 3 (Le Cam's Poisson Approximation Theorem [11], rephrased in [13, Lemma 2.5]). *Let $X_1, X_2, \ldots$ be independent random variables taking integer values in $\{0, 1, \ldots, K\}$, and let $X = \sum_i X_i$. Let $\pi_i = \Pr(X_i \neq 0)$ and $V = (V_1, \ldots, V_K)$ for $V_k = \sum_i \Pr(X_i = k)$. Suppose $\lambda = \sum_i \pi_i = \sum_k V_k < \infty$. Let Y be the compound Poisson distribution corresponding to vector V. Then, the total variation distance between X and Y is bounded as follows:*

$$\Delta(X, Y) = \sum_{k \geq 0} |\Pr(X = k) - \Pr(Y = k)| \leq 2 \sum_i 2\pi_i^2. \qquad \square$$

This theorem implies that the sum of any bounded integer-valued random variable is approximated by a compound Poisson distribution. Intuitively, we apply this theorem to X_i and guess the signature (i.e., V) by enumeration. In the following, we discretize the values of X_i to apply the theorem.

Without loss of generality, we assume that the input size X_i of item i is in range $[0, 2]$. Otherwise, we put all the probability mass outside of the range to $\Pr(X_i = 2)$; this does not change the solution to the problem. We say that an item i is *big* if the expected size $\mathbb{E}[X_i]$ is greater than ϵ^{10}; otherwise, i is *small*. By Lemma 4 below, there are at most $3/\epsilon^{11}$ big items added into the knapsack.

Lemma 4 (A variation of [13, Lemma 2.1]). *Suppose each item i has a nonnegative size X_i randomly taken from $[0, 2]$. For any subset $S \subseteq N$ and any $0 < \epsilon < 1/2$, if $\Pr(X(S) > 1) \leq 1 - \epsilon$, the total expected size of S is $\mathbb{E}[X(S)] \leq 3/\epsilon$.* $\qquad \square$

Next, we discretize the value of X_i. For an item i, we say it realizes to *big size* if $X_i > \epsilon^4$, otherwise *small size*. If X_i realizes to small size, we define the corresponding discretized size $\widetilde{X}_i$ as a Bernoulli random variable that takes 0 or ϵ^4. We can find a value $0 \leq \delta \leq \epsilon^4$ such that $\Pr(X_i \geq \delta \mid X_i \leq \epsilon^4)\epsilon^4 = \mathbb{E}[X_i \mid X_i \leq \epsilon^4]$. That is to let the discretized size $\widetilde{X}_i = 0$ if $0 \leq X_i < \delta$; otherwise $\widetilde{X}_i = \epsilon^4$. Then, the probability $\Pr(\widetilde{X}_i = 0)$ and $\Pr(\widetilde{X}_i = \epsilon^4)$ are set accordingly such that $\mathbb{E}[\widetilde{X}_i \mid X_i \leq \epsilon^4] = \mathbb{E}[X_i \mid X_i \leq \epsilon^4]$. If X_i realizes to big size, we define the corresponding discretized size by $\widetilde{X}_i = \lfloor X_i/\epsilon^5 \rfloor \epsilon^5$. Note that there are $d = O(1/\epsilon^5)$ different discretized sizes. With the above discretization, we have the following lemma.

Lemma 5 (A variation of [13, Lemma 2.2]**).** *Let a set S of item where $\mathbb{E}[X(S)] \leq 3/\epsilon$, and let $0 \leq \beta \leq 1 + O(\epsilon)$. We have the following results*

- $\Pr(X(S) \leq \beta) \leq \Pr(\widetilde{X}(S) \leq \beta + O(\epsilon)) + O(\epsilon)$
- $\Pr(\widetilde{X}(S) \leq \beta) \leq \Pr(X(S) \leq \beta + O(\epsilon)) + O(\epsilon)$ □

The probability of item i after size discretization is denoted as $\widetilde{\pi}_i(s)$ where s ($s \neq 0$) is the discretized input size. We define a *signature* of an item as

$$\text{Sg}(i) = (\bar{\pi}_i(s_1), \bar{\pi}_i(s_2), \ldots, \bar{\pi}_i(s_d))$$

where $\bar{\pi}_i(s) = \lfloor \widetilde{\pi}_i(s)(n/\epsilon^6) \rfloor (\epsilon^6/n)$. Note that Sg is a $d = O(1/\epsilon^5)$ dimensional vector. For a subset $S \subseteq N$, the signature is defined as $\text{Sg}(S) = \sum_{i \in S} \text{Sg}(i)$. By Lemma 4, the number of different signature is at most $(3n/\epsilon^{11})^{O(1/\epsilon^5)} = n^{O(1/\epsilon^5)}$. Furthermore, let $\bar{X}_i$ be a random variable of input size for item i such that $\Pr(\bar{X}_i = s) = \bar{\pi}_b(s)$ and $\Pr(\bar{X}_i = 0)$ be the rest of probability mass.

The properties of signatures and their corresponding compound Poisson distributions are shown below.

Lemma 6 (Monotonicity [13, Lemma 2.6]**).** *Let S_1 and S_2 be two sets such that $\text{Sg}(S_1) \leq \text{Sg}(S_2)$ (element-wise). Then, their corresponding compound Poisson distribution variables Y_1 and Y_2 satisfy $\Pr(Y_1 > \beta) \leq \Pr(Y_2 > \beta)$ for any $\beta \geq 0$.*   □

Now we prove the following lemma.

Lemma 7. *Let S_1 and S_2 be a set of small items where $\mathbb{E}[\widetilde{X}(S_1)] \leq 3/\epsilon$ and $\mathbb{E}[\widetilde{X}(S_2)] \leq 3/\epsilon$. If $\text{Sg}(S_1) \leq \text{Sg}(S_2)$, for any $\beta \geq 0$, we have $\Pr(\widetilde{X}(S_1) > \beta) \leq \Pr(\widetilde{X}(S_2) > \beta) + O(\epsilon)$.*

Proof. Let us consider the total variation distance of two variables as $\triangle(X_1, X_2) = \sum_s |\Pr(X_1 = s) - \Pr(X_2 = s)|$. Since the probability distribution of item is discretized by ϵ^6/n and there are at most $O(1/\epsilon^5)$ different discrete sizes, we can conclude that $\triangle(\widetilde{X}_i, \bar{X}_i) \leq O(\epsilon/n)$.

Let Y be the compound Poisson distribution corresponding to $\text{Sg}(S)$. Then, by Le Cam's Poisson approximation theorem (Lemma 3), we have $\triangle(\bar{X}(S), Y) \leq$

$2\sum_{j\in S}\pi_j^2$, where $\pi_j = \Pr(\bar{X}_j \neq 0)$. Note that S_1 is a set of small items, and Y_1 is the corresponding compound Poisson random variable. We then have $\triangle(\bar{X}(S_1), Y_1) = \sum_{j\in S_1}(\Pr(\bar{X}_j \neq 0))^2 \leq O(\epsilon)$, because

$$\Pr(\bar{X}_i \neq 0) \leq \Pr(\tilde{X}_i \neq 0) = \Pr(\tilde{X}_i \geq \epsilon^4) \leq \mathbb{E}[\tilde{X}_i]/\epsilon^4 \leq \epsilon^6$$

where the last inequality is from the definition of small item $\mathbb{E}[\tilde{X}_b] \leq \epsilon^{10}$, and

$$\sum_{i\in S_1}\Pr(\bar{X}_i \neq 0) \leq \sum_{i\in S_1}\Pr(\tilde{X}_i \geq \epsilon^4) \leq \mathbb{E}[\tilde{X}(S_1)]/\epsilon^4 \leq 3/\epsilon^5,$$

where the last inequality comes from $\mathbb{E}[\tilde{X}(S_1)] \leq 3/\epsilon$. With similar analysis, we also have $\triangle(\bar{X}(S_2), Y_2) \leq O(\epsilon)$. By Lemma 6, we have that for any $\beta > 0$,

$$\begin{aligned}
\Pr(\tilde{X}(S_1) > \beta) &\leq \Pr(\bar{X}(S_1) > \beta) + \triangle(\bar{X}(S_1), \tilde{X}(S_1))\\
&\leq \Pr(Y_1 > \beta) + \triangle(\bar{X}(S_1), Y_1) + O(\epsilon)\\
&\leq \Pr(Y_2 > \beta) + O(\epsilon)\\
&\leq \Pr(\bar{X}(S_2) > \beta) + \triangle(\bar{X}(S_2), Y_2) + O(\epsilon)\\
&\leq \Pr(\tilde{X}(S_2) > \beta) + O(\epsilon)
\end{aligned}$$

$\square$

Corollary 1. *Given two sets S_1 and S_2 of small items where $\mathbb{E}[\tilde{X}(S_1)] \leq 3/\epsilon$ and $\mathbb{E}[\tilde{X}(S_2)] \leq 3/\epsilon$. If $\mathrm{Sg}(S_1) \leq \mathrm{Sg}(S_2)$ and any big item set B, we have $\Pr(\tilde{X}(S_1) + X(B) > \beta) \leq \Pr(\tilde{X}(S_2) + X(B) > \beta) + O(\epsilon)$ for any $\beta > 0$.* $\square$

Now we apply this corollary to reduce the chance constraint to a multidimensional knapsack constraint with discretization. We first guess a subset B of big items, which has only $n^{3/\epsilon^{11}}$ candidates, and consider the remaining problem that does not contain any big items. Since the remaining problem has no big items, the Poisson approximation preserves the probability within accuracy $O(\epsilon)$.

More precisely, let S^* be an optimal solution, and consider the case when $B \subseteq S^*$ is the big items in the optimal solution. By Corollary 1, we know that for any set S with $\mathrm{Sg}(S) \leq \mathrm{Sg}(S^*)$, $S \cup B$ is a feasible solution by differing probability at most $O(\epsilon)$. This inspires us to solve the following problem

$$\begin{aligned}
&\text{maximize } f_B(S)\\
&\text{subject to } \mathrm{Sg}(S) \leq \mathrm{Sg}
\end{aligned} \tag{7}$$

by guessing the signature of $\tilde{X}(S^* \setminus B)$ by Sg. Note that this problem is the d-dimensional knapsack constrained monotone submodular maximization problem. Therefore, as same as in Sect. 1, we can utilize the continuous relaxation and rounding to obtain a solution.

Algorithm 2. Algorithm for Arbitrary Distribution

Input: Given a item set N, where size of each item is realized from arbitrary distribution. A submodular set function f. An input parameter ϵ and dimension of knapsack d. Overflow probability ρ.

1: Apply size and probability discretization to all the items
2: **for** each subset of big items B such that $\mathbb{E}[X(B)] < 3/\epsilon$ **do**
3:     **for** each possible signature Sg **do**
4:         **for** each subset T with $|T| \leq d\epsilon^{-3}e$ **do**
5:             Divide the rest items into light item set L and heavy item set H
6:             Consider the optimization with only light item set L by removing H
7:             Compute a solution S by the continuous relaxation and rounding method
8:             Check the feasibility of set $S \cup B \cup T$.
9: Return the best possible set $S \cup B \cup T$.

4.2   Proposed Algorithm

We propose an algorithm for the problem (Algorithm 2). We first guess a subset $B \subseteq N$ of the big items of an optimal solution. Then, we solve the d-dimensional knapsack problem by the continuous relaxation and rounding method — we first guess the signatures Sg of the remaining problem and the subset $T \subseteq N \setminus B$ in the optimal solution. By removing the heavy items from the instance, we solve the residual knapsack problem by the continuous relaxation, and obtain a subset by the rounding. We then output the best solution among the process.

The analysis of the algorithm is given as follows.

Proof (Proof of Theorem 2). We consider the case when B is the set of big items in the optimal solution, Sg is the signature of $S^* \setminus B$, and $T \subseteq S^* \setminus B$ is the first $d\epsilon^{-3}e$ items in the decreasing order of the marginal profit as in the normal distribution case.

We first show that $S \cup T \cup B$ is feasible to the relaxed capacity and overflow probability. By Lemma 5, we have

$$\Pr(X(S \cup T \cup B) > 1 + O(\epsilon)) \leq \Pr(\widetilde{X}(S \cup T) + X(B) > 1 + O(\epsilon)) + O(\epsilon)$$

Furthermore, by Corollary 1 and Lemma 5,

$$\Pr(\widetilde{X}(S \cup T) + X(B) > 1 + O(\epsilon)) \leq \Pr(\widetilde{X}(S^* \setminus B) + X(B) > 1 + O(\epsilon)) + O(\epsilon)$$
$$\leq \Pr(X(S^*) > 1) + O(\epsilon).$$

Thus, $\Pr(X(S \cup T \cup B) > 1 + O(\epsilon)) \leq \rho + O(\epsilon)$.

We then prove the approximation guarantee. We put $f(B) = \beta\text{OPT}$ for some β and $OPT^* = (1 - \beta)OPT$. Let $f_B(T) = \alpha\text{OPT}$ for some α By the continuous relaxation and rounding method, we obtain a random solution S such that $\mathbb{E}[f_{B \cup T}(S)] \geq (1 - e^{-1} - O(\epsilon))F_{B \cup T}(\bar{x}^*)$, where $\bar{x}^*$ is the optimal fractional solution. We then know $F_{B \cup T}(\bar{x}^*) \geq f_{B \cup T}(L^*)$, where $L^* = L \cap (S^* \setminus (B \cup T))$. The expected profit to obtained by the algorithm is

$$\mathbb{E}[f_{B \cup T}(S)] + f(B) + f_B(T)$$
$$\geq \beta \mathrm{OPT} + \alpha \mathrm{OPT}^* + (1 - e^{-1} - O(\epsilon))(\mathrm{OPT}^* - \alpha \mathrm{OPT}^* - f_{B \cup T}((S^* \setminus (B \cup T)) \cap H))$$
$$\geq \beta \mathrm{OPT} + \alpha \mathrm{OPT}^* + (1 - e^{-1} - O(\epsilon))(\mathrm{OPT}^* - \alpha \mathrm{OPT}^* - e^{-1} \alpha \mathrm{OPT}^*)$$
$$\geq (1 - e^{-1} - O(\epsilon))\mathrm{OPT}$$

Here, the first inequality is due to Lemma 1, the second inequality comes from submodularity and the fact that the number of heavy items in the residual knapsack is less than $\epsilon^{-3}d$. □

At last, we analyze the complexity of Algorithm 2. The number of big item sets is bounded by $n^{O(1/\epsilon^{11})}$, the number of signatures is bounded by $n^{O(1/\epsilon^5)}$, and there are at most $O(n^{d\epsilon^{-3}}e)$ different set T. Thus, the algorithm solves the residual $d = O(1/\epsilon^5)$ dimensional knapsack problem polynomially many times. Solving such residual knapsack problem requires polynomial time; hence, the complexity of proposed algorithm is polynomial.

5 Conclusions

In this study, we considered the problem of maximizing monotone submodular function subject to a chance-constraint. Specifically, we proposed algorithms for two cases: the distributions of items follow normal distributions, and arbitrary but known distributions. For the normal distribution case, for any $\epsilon > 0$, the proposed algorithm runs in polynomial time and gives a solution whose expected profit is at least $(1 - e^{-1} - O(\epsilon))$ to the optimal. For the arbitrary distribution case, for any $\epsilon > 0$, the proposed algorithm runs in polynomial time and gives a solution whose expected profit is at least $(1 - e^{-1} - O(\epsilon))$ and violates the relaxed knapsack constraint with at most the relaxed overflow probability.

References

1. Alon, N., Gamzu, I., Tennenholtz, M.: Optimizing budget allocation among channels and influencers. In: Proceedings of the 21st International Conference on World Wide Web, pp. 381–388 (2012)
2. Bhalgat, A., Goel, A., Khanna, S.: Improved approximation results for stochastic knapsack problems. In: Proceedings of the 22nd Annual ACM-SIAM Symposium on Discrete Algorithms, pp. 1647–1665 (2011)
3. Calinescu, G., Chekuri, C., Pál, M., Vondrák, J.: Maximizing a submodular set function subject to a matroid constraint (extended abstract). In: Fischetti, M., Williamson, D.P. (eds.) IPCO 2007. LNCS, vol. 4513, pp. 182–196. Springer, Heidelberg (2007). https://doi.org/10.1007/978-3-540-72792-7_15
4. Feldman, M., Naor, J., Schwartz, R.: A unified continuous greedy algorithm for submodular maximization. In: Proceedings of the 52nd Annual IEEE Symposium on Foundations of Computer Science, pp. 570–579 (2011)

5. Goel, A., Indyk, P.: Stochastic load balancing and related problems. In: Proceedings of the 40th Annual IEEE Symposium on Foundations of Computer Science, pp. 579–586 (1999)
6. Goemans, M.X., Williamson, D.P.: Improved approximation algorithms for maximum cut and satisfiability problems using semidefinite programming. J. ACM **42**(6), 1115–1145 (1995)
7. Goyal, V., Ravi, R.: A PTAS for the chance-constrained knapsack problem with random item sizes. Oper. Res. Lett. **38**(3), 161–164 (2010)
8. Khot, S., Kindler, G., Mossel, E., O'Donnell, R.: Optimal inapproximability results for max-cut and other 2-variable CSPs? SIAM J. Comput. **37**(1), 319–357 (2007)
9. Krause, A., Cevher, V.: Submodular dictionary selection for sparse representation. In: Proceedings of the 27th International Conference on Machine Learning, pp. 567–574 (2010)
10. Kulik, A., Shachnai, H., Tamir, T.: Approximations for monotone and nonmonotone submodular maximization with knapsack constraints. Math. Oper. Res. **38**(4), 729–739 (2013)
11. Le Cam, L., et al.: An approximation theorem for the poisson binomial distribution. Pac. J. Math. **10**(4), 1181–1197 (1960)
12. Lee, J., Mirrokni, V.S., Nagarajan, V., Sviridenko, M.: Non-monotone submodular maximization under matroid and knapsack constraints. In: Proceedings of the 41st Annual ACM Symposium on Theory of Computing, pp. 323–332 (2009)
13. Li, J., Yuan, W.: Stochastic combinatorial optimization via poisson approximation. In: Proceedings of the 45th Annual ACM Symposium on Theory of Computing, pp. 971–980 (2013)
14. Maehara, T., Narita, A., Baba, J., Kawabata, T.: Optimal bidding strategy for brand advertising. In: Proceedings of the 27th International Joint Conference on Artificial Intelligence and the 23rd European Conference on Artificial Intelligence, pp. 424–432 (2018)
15. Murota, K.: Discrete convex analysis: a tool for economics and game theory. J. Mech. Inst. Des. **1**(1), 151–273 (2016)
16. Nemhauser, G.L., Wolsey, L.A., Fisher, M.L.: An analysis of approximations for maximizing submodular set functions–I. Math. Program. **14**(1), 265–294 (1978)
17. Shabtai, G., Raz, D., Shavitt, Y.: A relaxed FPTAS for chance-constrained knapsack. In: 29th International Symposium on Algorithms and Computation (2018)
18. Sviridenko, M.: A note on maximizing a submodular set function subject to a knapsack constraint. Oper. Res. Lett. **32**(1), 41–43 (2004)
19. Vondrák, J.: Optimal approximation for the submodular welfare problem in the value oracle model. In: Proceedings of the 40th Annual ACM Symposium on Theory of Computing, pp. 67–74 (2008)
20. Vondrák, J., Chekuri, C., Zenklusen, R.: Submodular function maximization via the multilinear relaxation and contention resolution schemes. In: Proceedings of the 43rd Annual ACM Symposium on Theory of Computing, pp. 783–792 (2011)

Approximation Hardness of Travelling Salesman via Weighted Amplifiers

Miroslav Chlebík[1] and Janka Chlebíková[2(✉)]

[1] Department of Mathematics, University of Sussex, Brighton, UK
m.chlebik@sussex.ac.uk
[2] School of Computing, University of Portsmouth, Portsmouth, UK
janka.chlebikova@port.ac.uk

Abstract. The expander graph constructions and their variants are the main tool used in gap preserving reductions to prove approximation lower bounds of combinatorial optimisation problems. In this paper we introduce the weighted amplifiers and weighted low occurrence of CON-STRAINT SATISFACTION problems as intermediate steps in the NP-hard gap reductions. Allowing the weights in intermediate problems is rather natural for the edge-weighted problems as TRAVELLING SALESMAN or STEINER TREE. We demonstrate the technique for TRAVELLING SALES-MAN and use the parametrised weighted amplifiers in the gap reductions to allow more flexibility in fine-tuning their expanding parameters. The purpose of this paper is to point out effectiveness of these ideas, rather than to optimise the expander's parameters. Nevertheless, we show that already slight improvement of known expander values modestly improve the current best approximation hardness value for TSP from $\frac{123}{122}$ ([9]) to $\frac{117}{116}$. This provides a new motivation for study of expanding properties of random graphs in order to improve approximation lower bounds of TSP and other edge-weighted optimisation problems.

1 Introduction

The TRAVELLING SALESMAN problem (TSP) is undoubtedly one of the most famous combinatorial optimisation problems. In its standard version, we are given an edge-weighted (undirected) graph and the goal is to find a closed tour with a minimum cost that visits each vertex at least once. This is equivalent to the GRAPHIC TRAVELLING SALESMAN problem where exactly one visit per vertex is allowed and the cost between any two vertices corresponds to their shortest path.

The shortest-path metric of the GRAPHIC TSP plays an important role in understanding of complexity for the METRIC TSP problem. The approximability of the METRIC TSP is a long-standing open problem, Christofides's approximation algorithm with ratio 1.5 [4] hasn't been improved for more than three decades. It is generally believed that the approximation ratio can be close to 4/3 due to known integrality gap for the Held-Karp LP relaxation [8].

© Springer Nature Switzerland AG 2019
D.-Z. Du et al. (Eds.): COCOON 2019, LNCS 11653, pp. 115–127, 2019.
https://doi.org/10.1007/978-3-030-26176-4_10

In the last decade, some significant progress has been done in the GRAPHIC TSP. Gharan et al. [6] made first breakthrough with an $(1.5 - \varepsilon)$-approximation algorithm where ε being of the order of 10^{-12}. Following that, Mömke and Svensson [11] obtained a significantly better approximation factor of $\frac{14(\sqrt{2}-1)}{12\sqrt{2}-13} \approx$ 1.461, which was improved further to $\frac{13}{9} \approx 1.444$ by Mucha [12]. To our best knowledge, currently the best known approximation ratio is 1.4 due to [15]. The overview about this recent development can also be found in [16].

However, there is still a significant gap between the ratio of the best approximation algorithm and the approximation ratio that provably can't be achieved unless P = NP. The first APX-hardness result showed the NP-hardness to approximate the TSP problem within $1 + \varepsilon$ without any explicit value for ε (Papadimitrious and Yannakakis, [13]). The first explicit value 5381/5380 was set by Engebretsen [5], further improved to 3813/3812 by Böckenhauser et al. [1] and 220/219 by Papadimitrious and Vempala [14]. The further progress in the reductions and amplifiers increased the threshold to 185/184 by Lampis [10] and to our best knowledge the currently best value is 123/122 by Karpinski et al. [9].

Main Contribution. The main novelty of this paper is using weighted amplifiers and weighted low occurrence of CONSTRAINT SATISFACTION problems (CSP) as intermediate steps in the NP-hard gap reductions to the TRAVELLING SALESMAN problem. Allowing the weights in intermediate problems to TSP (or the STEINER TREE problem) is rather natural, as the problems themselves are using edge weights. We demonstrate the technique for TSP and use the parametrised weighted amplifiers in the gap reductions to allow more flexibility in fine-tuning their expanding parameters. In this paper we don't aim to optimise the parameters of amplifiers that provably exist, but show that already slight improvement of known values modestly improve the hardness of approximation for TSP from the current best value $\frac{123}{122}$ [9] to the new value $\frac{117}{116}$. This provides a new motivation for study of expanding properties of random graphs in order to improve approximation lower bounds of TSP and other edge-weighted optimisation problems.

Preliminaries

All graphs in this paper are undirected and connected. Let $G = (V, E)$ be an edge-weighted graph with cost edge-function $c\colon E \to \mathbb{R}^+$. For an edge $e = \{u, v\} \in E$ we also use the notation uv as an shorthand. A tour in the graph G is an alternating sequence of vertices and edges, starting and ending at a vertex, where each vertex is incident with the previous and the following edge in the sequence. If a starting and ending vertex is the same, the tour is closed.

Any solution of TSP is a closed tour, hence an Eulerian multigraph (edges are taken with their multiplicities if they are used multiple times) spanning V. A quasi-tour T in G is a multiset of edges from E such that all vertices in G are balanced with respect to T (each vertex from V is incident with even number of edges from T, possibly 0); hence each such connected component in G is an Eulerian multigraph (or an isolated vertex).

MAX-E3-LIN-2

Our inapproximability results for the TRAVELLING SALESMAN problem use reductions from Håstad's NP-hard gap type result for MAX-E3-LIN-2, the MAX-IMUM SATISFIABILITY problem for linear equations modulo 2 with *exactly* 3 variables per equation [7] (more details can be found in Appendix). In fact, Håstad's tight inapproximability results can be stated in the form in which every variable occurs the same number of times in the system of equations, see e.g. [2].

Theorem 1. *For every $\varepsilon \in \left(0, \frac{1}{4}\right)$ and every fixed sufficiently large integer $k \geq k(\varepsilon)$, the following partial decision subproblem $Q(\varepsilon, k)$ of* MAX-E3-LIN-2 *is NP-hard: given an instance of* MAX-E3-LIN-2 *with m equations and exactly k occurrences of each variable, to decide if at least $(1 - \varepsilon)m$ or at most $(\frac{1}{2} + \varepsilon)m$ equations are satisfied by the optimal assignment.*

The results of such form were already used to prove the inapproximability results for other optimisation problems, e.g., the STEINER TREE problem [3].

For some optimisation problems it is more convenient to use reductions if all equations of MAX-E3-LIN-2 have the same right hand side. The NP-hard gap results in such a case can be easily enforced if we allow flipping some occurrences of variables, so also the literal $\overline{x}$ ($:= 1 - x$) can be used for a variable x. The canonical gap versions $Q_b(\varepsilon, 2k)$, for any fixed $b = 0$ or $b = 1$, of MAX-E3-LIN-2 are as follows:

THE $Q_b(\varepsilon, 2k)$ PROBLEM, $b \in \{0, 1\}$
Input: An instance of MAX-E3-LIN-2 with m equations of the form $x \oplus y \oplus z = b$, each variable occurring exactly k times as unnegated and k times negated.
Task: To decide if at least $(1 - \varepsilon)m$ or at most $(\frac{1}{2} + \varepsilon)m$ equations are satisfied by the optimal assignment.

The corresponding 'fixed occurrence' NP-hard gap result reads as follows (see [2] for the details of the following theorem):

Theorem 2. *For every $\varepsilon \in (0, \frac{1}{4})$ and every sufficiently large integer k, $k \geq k(\varepsilon)$, the partial decision subproblems $Q_0(\varepsilon, 2k)$ and $Q_1(\varepsilon, 2k)$ of* MAX-E3-LIN-2 *are NP-hard.*

Weighted Amplifiers

Amplifier graphs are useful in proving inapproximability results for CSPs in which every variable appears a bounded (and, typically, very low) number of times. Such CSPs are often used as intermediate steps in proving approximation hardness results for many combinatorial optimisation problems. For problems like TRAVELLING SALESMAN, or STEINER TREE which are based on edge weights, it is natural to consider the intermediate low degree CSPs with their edge weights as well.

For a graph $G = (V, E)$, a cut is a partition of V into two subsets U and $\overline{U} := V \setminus U$. The cut set $E(U, \overline{U})$ is defined as $E(U, \overline{U}) = \{uv \in E, u \in U \text{ and } v \in \overline{U}\}$ and the cut size as $|E(U, \overline{U})|$. If edges are weighted with $p \colon E \to \mathbb{R}^+$, then $p(E(U, \overline{U}))$ is weight of the cut set $E(U, \overline{U})$, hence $p(E(U, \overline{U})) = \sum_{uv \in E, u \in U, v \in \overline{U}} p(uv)$.

Definition 1. Let $G = (V, E)$ be a graph with edge weights $p \colon E \to \mathbb{R}^+$, and $D \subseteq V$, $|D| \geq 2$. We say that a weighted graph (G, p) is an amplifier for D if for every vertex set $A \subseteq V$

$$p(E(A, \overline{A})) \geq \min\{|D \cap A|, |D \cap \overline{A}|\}.$$

The vertices of the given set D are called the *contacts*, the rest of the vertices $(= V \setminus D)$ is the set of *checkers*. We say that an amplifier (G, p) for the set D is a d-regular amplifier if, additionally, all contacts have degree $(d - 1)$ and all checkers have degree d (in G).

In full generality, one could also allow distinct weights for vertices of D to replace the sizes $|D \cap A|$, $|D \cap \overline{A}|$ with their weighted version, but for our purposes the vertices of D are uniformly weighted each with weight 1.

2   Intermediate Weighted CSPs

In this section we extend the NP-hard gap results from a system of linear equations with exactly 3 variables to a low occurrence version of w-MAX-3-LIN-2, a weighted hybrid system of linear equations over $\mathbb{Z}_2$ with either 2 or 3 variables. Similarly to MAX-E3-LIN-2, the task of the w-MAX-3-LIN-2 problem is to find an assignment that maximizes weight of the satisfied equations in the hybrid system.

To prove the NP-hard gap results for the w-MAX-3-LIN-2 problem, we extend Håstad's results for MAX-E3-LIN-2 using the amplifiers defined in Sect. 1.

Reduction from $Q(\varepsilon, k)$ to w-MAX-3-LIN-2

Let $\varepsilon \in (0, \frac{1}{4})$, and $k > 0$ be an integer such that the problem $Q(\varepsilon, k)$ is NP-hard. Let an instance I of $Q(\varepsilon, k)$ be given, denote by $\nu(I)$ the set of variables of I, $\nu := |\nu(I)|$. Let's assume that $G = (V, E)$ with the edge weights $p \colon E \to \mathbb{R}^+$ be an amplifier for a set $D \subseteq V$ with $|D| = k$.

Now we describe a gap preserving reduction from $Q(\varepsilon, k)$ to the w-MAX-3-LIN-2 problem with an amplifier (G, p) as a parameter. The instance I of $Q(\varepsilon, k)$ is transformed to a weighted hybrid instance J of w-MAX-3-LIN-2.

- For each variable $x \in \nu(I)$ take a copy of the amplifier (G, p), let (G_x, p) denote that copy:
  - Inside (G_x, p) the vertices correspond to the variables in J and each edge vv' represents the equation $v \oplus v' = 0$ with weight $p(vv')$ in J.

  The contact vertices of (G_x, p) represent k occurrences of the variable x in the equations of I. Distinct occurrences of a variable x in I are represented by the distinct contact vertices in G_x.
- Every equation $x \oplus y \oplus z = b$ from I, $b \in \{0, 1\}$, also belongs to J with weight 1.

Remark 1. Observe that the above reduction from an instance I of $Q(\varepsilon, k)$ to an instance J of w-MAX-3-LIN-2 preserves the NP-hard gap of $Q(\varepsilon, k)$. Indeed, there is a simple dependence of an optimal value for J on that of I.

In the following we show that if we look at these problems as MINIMUM UNSATISFIABILITY problems, where OPT' is the corresponding minimum weight of unsatisfied equations over all assignments, then $\text{OPT}'(I) = \text{OPT}'(J)$. Clearly, any assignment to variables from $\nu(I)$ generate an assignment to variable of J in a natural way; the value of a variable $x \in \nu(I)$ is assigned to all variables of G_x. Such assignments to variables of J are called **standard**. Hence, obviously $\text{OPT}'(J) \leq \text{OPT}'(I)$.

The observation that the optimum $\text{OPT}'(J)$ is achieved on standard assignments is based on the amplifier's properties. Any assignment φ to the variables of J can be converted to a standard one in such a way that the weight of unsatisfied equations doesn't increase as follows: consider a variable x from $\nu(I)$. Assign to all variables in G_x the same value as it is assigned to the majority of contact vertices in G_x by the assignment φ. The fact that (G_x, p) is the amplifier ensures that the weight of unsatisfied equations in J doesn't increase. Now if we repeat the same operation for each variable from $\nu(I)$, one after another, the result will be a standard assignment without increase of the weight of unsatisfied equations in J. Consequently, $\text{OPT}'(J)$ is achieved on the standard assignments. But for every standard assignment the weight of unsatisfied equations of J is the same as the number of unsatisfied equations of I by that assignment, hence $\text{OPT}'(I) = \text{OPT}'(J)$.

Reduction from $Q_b(\varepsilon, 2k)$ to w-MAX-3-LIN-2

Now we slightly modify the previous reduction from $Q(\varepsilon, k)$ to deal with the instances of $Q_b(\varepsilon, 2k)$ for any fixed $b = 0$ or $b = 1$.

Let $\varepsilon \in (0, \frac{1}{4})$ and $k > 0$ be an integer such that $Q_b(\varepsilon, 2k)$ is NP-hard. Assume that $G = (V, E)$ with edge weights $p \colon E \to \mathbb{R}^+$ is an amplifier for a set $D \subseteq V$ with $|D| = 2k$. Let $\{V^u, V^n\}$ be a partition of V balanced in D, namely $|D \cap V^u| = |D \cap V^n| = k$. Denote further G^u and G^n the induced subgraph of G with the vertex sets V^u and V^n, respectively. In what follows we describe the reduction from $Q_b(\varepsilon, 2k)$ to w-MAX-3-LIN-2 parametrised by an amplifier (G, p) for $D \subseteq V$ with $|D| = 2k$ and with chosen balanced partition $\{V^u, V^n\}$ of V.

Let an instance I of $Q_b(\varepsilon, 2k)$ be given, $\nu(I)$ be the set of variables of I, $\nu = |\nu(I)|$.

- For each variable x from $\nu(I)$ take a copy of an amplifier (G, p), let G_x denote such a copy.
  - Any edge vv' inside either G_x^u or G_x^n represents the cycle equation $v \oplus v' = 0$ taken with weight $p(vv')$.
  - Any edge between $v \in V_x^u$ and $v' \in V_x^n$ in G_x represents the matching equation $v \oplus v' = 1$ taken with weight $p(vv')$.
- The contact vertices of G_x^u (resp. G_x^n) represent k occurrences of unnegated (resp. negated) variable x in the equations of I. Every equation $x \oplus y \oplus z = b$ from I, $b \in \{0, 1\}$, also belongs to J with weight 1.

This way we produce an instance J of the w-MAX-3-LIN-2 problem. Any assignment to variables from $\nu(I)$ generates an assignment to variables of J in a natural way: the value of a variable x is assigned to all variables of G_x^u, and the value opposite to x, $\bar{x} = 1 - x$, is assigned to all vertices of G_x^n. Such assignment to the variables of J is called **standard**. Similarly to the previous reduction, any assignment to variables of J can be converted to a standard one without increasing the weight of unsatisfied equations as it follows from properties of an amplifier.

3   The Weighted Bi-wheel Amplifiers

The previous reductions were based on a theoretical model of amplifiers with required properties, without proving their existence. In this section we introduce a class of weighted graphs with such expanding properties that generalise the bi-wheel amplifiers from [9]. Further we describe in the details the properties of the instances of the subproblem of w-MAX-3-LIN-2, called the Hybrid bi-wheel instances.

Definition 2. *Let an integer $k > 0$ and a rational number $\tau > 1$ be such that τk is an integer. The* **weighted $(2k, \tau)$-bi-wheel amplifier** $W_{k,\tau} = (V, E)$, $p \colon E \to \mathbb{R}^+$, *is a (weighted) 3-regular amplifier with a specific balanced partition constructed as follows: Take two disjoint cycles, each on τk vertices (connected in consecutive order), $V^u = \{1^u, 2^u, \ldots, (\tau k)^u\}$ and $V^n = \{1^n, 2^n, \ldots, (\tau k)^n\}$, respectively. Select the sets of k contacts $D^u \subseteq V^u$ and $D^n \subseteq V^n$ as $D^u = \{c_1^u, c_2^u, \ldots, c_k^u\}$, $D^n = \{c_1^n, c_2^n, \ldots, c_k^n\}$. The remaining vertices of both cycles, $V^u \setminus D^u$ and $V^n \setminus D^n$, are checkers.*

To complete the construction, consider a perfect matching between the checkers of these two cycles where each matching edge has one vertex in the first cycle $V^u \setminus D^u$ and another one in the second cycle $V^n \setminus D^n$.

We assume that in each cycle of the bi-wheel consecutive contacts are separated by a chain of several (at least 1) checkers. Hence, in particular, $\tau \geq 2$.

Remark 2. Let us denote by $\mathcal{C}^u$ ($\mathcal{C}^n$, resp.) the set of edges contained in the first (the second, resp.) cycle in $W_{k,\tau}$, so $\mathcal{C}^u = \{\{i^u, (i+1)^u\} : i = 1, 2, \ldots, \tau k\}$ and $\mathcal{C}^n = \{\{i^n, (i+1)^n\}\} : i = 1, 2, \ldots, \tau k\}$ (the vertex $\tau k + 1$ is the vertex 1), and by $\mathcal{M} \subseteq E$ the associated perfect matching on the set of checkers. Clearly, $|\mathcal{C}^u| = |\mathcal{C}^n| = \tau k$, $|\mathcal{M}| = |V^u \setminus X^u| = |V^n \setminus X^n| = (\tau - 1)k$.

In this paper we consider only bi-wheel amplifiers $(W_{k,\tau}, p)$ whose weights have uniform cycle weight p_c for all cycle edges of both $\mathcal{C}^u$ and $\mathcal{C}^n$, and another uniform matching weight p_m for all matching edges from $\mathcal{M}$.

Now we are ready to describe the specific properties of the Hybrid bi-wheel instances of w-MAX-3-LIN-2 based on a fixed $(2k, \tau)$-bi-wheel amplifier $W_{k,\tau}$ with weights p_c and p_m.

Theorem 3. *For every $\varepsilon \in (0, \frac{1}{4})$ and $b \in \{0, 1\}$ there exist instances of w-MAX-3-LIN-2, called Hybrid($W_{k,\tau}, p$), with the following properties:*

(i) each variable of the system equations Hybrid($W_{k,\tau}, p$) occurs exactly 3 times;
(ii) m equations are of the form $x \oplus y \oplus z = b$, each of weight 1;
(iii) $3\tau m$ equations are of the form $x \oplus y = 0$ each of weight p_c;
(iv) $\frac{3}{2}m(\tau - 1)$ equations are of the form $x \oplus y = 1$ each of weight p_m,

for which it is NP-hard to decide whether there is an assignment to the variables that leaves unsatisfied equations of weight at most εm, or every assignment to the variables leaves unsatisfied equations of weight at least $(0.5 - \varepsilon)m$.

The reduction from Hybrid($W_{k,\tau}, p$), presented later in Sect. 4, is a gap preserving reduction to TSP parametrised by a $(2k, \tau)$-bi-heel amplifier with cycle weights p_c and matching weights p_m. The trade-off between parameters p_c, p_m and τ is crucial for quality of approximation lower bounds.

Definition 3. *We call the triple (p_c, p_m, τ) admissible if for every k_0 there exists $k \geq k_0$ and a $(2k, \tau)$-bi-wheel that is an amplifier with cycle weights p_c and matching weights p_m.*

The bi-wheel amplifiers introduced by Berman and Karpinski [9] are based on the fact that the triple $(p_c = 1, p_m = 1, \tau = 7)$ is admissible. This leads to NP-hardness to approximate TSP to within any constant approximation ratio less than $\frac{123}{122}$. They also observed [9] that their proof (of amplification properties) doesn't seem to work with $\tau = 6$ instead $\tau = 7$. However, there is an opportunity for fine-tuning here if we allow non-integral τ. If, e.g., 90% of pairs of consecutive contacts in bi-wheel cycles are separated by 6 checkers, and 10% of such pairs are separated by a chain of 5 checkers only, then the proof of required amplification properties still works. The detailed explanation together with all computations for wheel amplifiers can be found in the paper [2]. The proof for bi-wheels is very similar, so along these lines one can argue that the triple $(p_c = 1, p_m = 1, \tau = 6.9)$ is admissible. This itself would (very modestly) improve on the lower approximation bound for TSP given in [9].

Introducing weighted amplifier graph constructions seems to have paid off even more compared to improvement of parameters for unweighted amplifiers. In this case we have more freedom in fine-tuning the approximation hardness lower bounds obtained in parametric way, if we can prove that bi-wheel amplifiers with certain parameters (p_c, p_m, τ) exist.

Let us explain trade-off between parameters (p_c, p_m, τ) of bi-wheels in a simple scenario with $p_m = 1$ fixed. Our contribution allows to use weighted amplifiers with $p_c < 1$ (strengthening of amplifiers) or with $p_c > 1$ (relaxing of amplifiers). One can achieve amplifiers with $p_c < 1$ by increasing τ from $\tau = 7$. On the other hand, to relax to $p_c > 1$ can be achieved with $\tau < 7$. These ideas indicate importance to better understand the exact trade-off between (p_c, p_m, τ) triples for bi-wheel amplifiers that provably exist.

In this paper we don't include too many new results on expanding properties of random graphs, we rather demonstrate effectiveness of weighted parametrised amplifiers and address the question of fine-tuning in (p_c, p_m, τ) triples for bi-wheel amplifiers. We sketch how these ideas will modestly improve known lower bounds for TSP if we allow bi-wheel with $p_c < 1$.

Theorem 4. *The triple $(p_c = \frac{1}{2}, \ p_m = 1, \tau = 11)$ is admissible, hence for every large enough $k \geq k_0$ there is a $(2k, 11)$-bi-wheel that is an amplifier with cycle weights $p_c = \frac{1}{2}$ and matching weights $p_m = 1$.*

4   Gap Preserving Reduction from Hybrid($W_{k,\tau}, p$) to TSP

In this section we describe a gap preserving reduction from the system of equations Hybrid($W_{k,\tau}, p$) to the TRAVELLING SALESMAN problem. In the reduction we suppose that all equations of Hybrid($W_{k,\tau}, p$) with three variables are of the form $x \oplus y \oplus z = 0$ to simplify a discussion later (hence Hybrid($W_{k,\tau}, p$) was obtained via reduction from $Q_0(\varepsilon, 2k)$). We also introduce a real parameter $\theta > 0$ set to $\theta = \frac{1}{\max\{1, \ p_m\}}$, in order to simultaneously capture different scenarios $p_m \leq 1$ and $p_m > 1$.

The gap preserving reduction is similar to the reduction presented in [9], the main difference is in using a parametrised weighted $(2k, \tau)$-bi-wheel amplifier $(W_{k,\tau}, p)$ introduced in Sect. 3. We use the concept of forced edges introduced by Lampis in [10] (used also in [9]). The idea is based on the observation that we are able to stipulate that some edges, called *forced* edges, are to be used at least once in any valid tour. It can be achieved by replacing such an edge with a path of many edges of the same total weight. With this trick we may assume without loss of generality that we can force some edges to be used at least once (see [9] for the details). If u and v are vertices that are connected by a forced edge e, we write $\{u, v\}_F$ or simply uv_F. The construction contains some forced edges, all other edges in the constructed graph are *unforced* edges with edge weight 1.

We start with an instance I of $Q_0(\varepsilon, 2k)$ with ν variables, m equations of the form $x \oplus y \oplus z = 0$ and use the reduction from Sect. 2 to create an instance J of Hybrid($W_{k,\tau}, p$). Using the same notation as in Theorem 3 we construct an instance $G[J]$ of TSP in the following way: for each copy $W_j := (W_{k,\tau}, p)$, $1 \leq j \leq \nu$, of a $(2k, \tau)$-bi-wheel we construct a subgraph of $G[J]$:

(i) each variable x of the bi-wheel W_j, corresponds to a vertex x in the subgraph,
(ii) for each cycle equation $x \oplus y = 0$, we create an unforced edge xy with weight 1.

Now we add the edges among the vertices of 'bi-wheel' subgraphs using two types of gadgets:

- a 3-*variable gadget* H^{3Q}:
  for each equation j, $1 \leq j \leq m$, of the form $x \oplus y \oplus z = 0$ we add a 3-variable gadget H_j^{3Q} connecting the contacts x, y, z, where each contact vertex x, y, and z is part of its own $(2k, \tau)$-bi-wheel. Each gadget H_j^{3Q} contains two new

vertices γ^l, γ^r for every vertex $\gamma \in \{x, y, z\}$ and two additional vertices e_j^l and e_j^r, see Fig. 1 how the vertices are connected. All edges $\{\gamma^\alpha, \gamma\}_F$ with $\alpha \in \{r, l\}$ and $\gamma \in \{x, y, z\}$ are forced edges with weight $w(\{\gamma^\alpha, \gamma\}_F) = 0.5 + p_c\theta$. All remaining edges of H_j^{3Q} are unforced with weight 1.

- a *matching gadget* H^{2M}:
  for each equation $x_t^u \oplus x_q^n = 1$ we add a matching gadget H^{2M} connecting the checkers x_t^u and x_q^n via two forced edges $\{x_t^u, x_q^n\}_F^1$ and $\{x_t^u, x_q^n\}_F^2$, each of the same weight $2p_c\theta$ (Fig. 2).

At the end of the construction, we add a new central vertex s that is connected to every gadget H_j^{3Q} with two forced edges $\{e_j^l, s\}_F$ and $\{e_j^r, s\}_F$, both with weight 0.5, $w(\{e_j^\alpha, s\}_F) = 0.5$ for both $\alpha \in \{r, l\}$.

Observe that the construction doesn't need gadgets for the cycle edges, the connections between the matching edge gadgets are sufficient to encode these constraints.

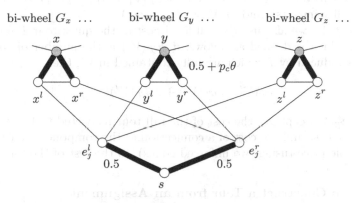

Fig. 1. An example of a 3-variable gadget H_j^{3Q} including the central vertex s, which is not part of the gadget. Thick lines represent forced edges.

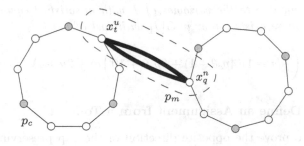

Fig. 2. A gadget H^{2M} inside the bi-wheel G_x for the equations $x_t^u \oplus x_q^n = 1$ contains only two forced edges, represented as thick lines.

Now in the following we describe in the details the properties of the gap preserving reduction from the $Hybrid(W_{k,\tau}, p)\}$ to the TRAVELLING SALESMAN problem.

Local Edge Cost. To count the cost $c(T)$ of a tour T, we use the local edge cost counting based on the ideas from [9]: the cost $w(uv)$ of any edge uv of T is split into two nonnegative parts, one attached to u and the second one to v. If an edge uv doesn't contain s then cost is split equally with contribution $0.5w(uv)$ for each vertex, but for edges of the form us, the full cost contributes to u, and none to s.

Let T be a multi-set of edges from E that defines a quasi-tour in $G[J](V, E)$. Then for a set $V' \subseteq V$, the local edge cost of V' is formally defined as

$$c_T(V') = \sum_{u \in V' \setminus \{s\}} \sum_{uv \in T} 0.5w(uv) + \sum_{e_j^\alpha \in V'} \sum_{e_j^\alpha s \in T} 0.5w(e_j^\alpha s).$$

Note that for two vertex sets V_1, V_2 we have $c_T(V_1 \cup V_2) \leq c_T(V_1) + c_T(V_2)$ (with equality for disjoint sets), and $c_T(V) = \sum_{e \in T} w(e)$.

In Subsect. 4.2 we also use the full local cost of the quasi-tour T for the set V', $c_T^*(V')$, which is defined as follows: if $\#_T(V')$ is the number of connected components induced by T which are fully contained in V', then

$$c_T^*(V') = c_T(V') + 2\#_T(V').$$

Intuitively, $c_T^*(V')$ captures the cost of the full tour restricted to V': it includes the local edge cost and the cost of a connection of the components on V' of the lowest possible price (using two unforced edges), to the rest of the tour.

4.1 How to Construct a Tour from an Assignment

Given an instance J of the $Hybrid(W_{k,\tau}, p)$ and an assignment φ to its variables, we describe a construction of a tour T in $G[J]$ with cost related to φ.

Lemma 1. *Let J be an instance of $Hybrid(W_{k,\tau}, p)$ from Theorem 3. If there exists an assignment φ to the variables of J with unsatisfied equations of total weight Δ, then there exists a tour in $G[J]$ with cost at most*

$$\left(\frac{3}{2}(\tau - 1)(4p_c\theta + 1) + 6p_c\theta + 10 \right)m + 2\nu + \Delta.$$

4.2 How to Define an Assignment from a Tour

Now we need to prove the opposite direction of the gap preserving reduction: given a tour in $G[J]$ the task is to define an assignment to the variables of the system equations I of $Hybrid(W_{k,\tau}, p)$ such that weight of unsatisfied equations is in a correlation with cost of a given tour.

Lemma 2. *If there is a tour in $G[J]$ with cost*

$$\left(\frac{3}{2}(\tau - 1)(4p_c\theta + 1) + 6p_c\theta + 10\right)m + \Delta - 2,$$

then there is an assignment to the instance J that leaves unsatisfied equations of weight at most $\Delta \cdot \max\{1,\ p_m\} = \frac{\Delta}{\theta}$, where $\theta = \frac{1}{\max\{1,\ p_m\}}$.

The high-level idea of the proof is to partition the vertex set of $G[J]$ into the gadget-based subgraphs similarly as in the proof of Lemma 1. For each such subgraph we give a lower bound on the local edge cost of any quasi-tour restricted to it, which in fact corresponds to cost of the tour constructed in Lemma 1. If a given quasi-tour behaves inside a gadget differently, its cost must be obviously higher. The difference between the tour's local edge cost and the lower bound is called the *credit* of the gadget. Based on the tour we define an assignment for J and show that the total sum of credits can be used to bound from above the weight of unsatisfied equation, where the total sum of credits is at most Δ.

Theorem 5. *If (p_c, p_m, τ) is an admissible triple then it is NP-hard to approximate the TRAVELLING SALESMAN problem to within any constant approximation ratio less than*

$$1 + \frac{1}{3(\tau - 1)(4p_c + \max\{1,\ p_m\}) + 12p_c + 20\max\{1,\ p_m\}}.$$

Proof. Let $\varepsilon \in (0, \frac{1}{4})$. Consider a $(2k, \tau)$-bi-wheel with large enough k, which is an amplifier with cycle weights p_c and matching weights p_m. We have instances of Hybrid($W_{k,\tau}, p$) with ν copies of a bi-wheel $(W_{k,\tau}, p)$, m equations of the form $x \oplus y \oplus z = 0$ each of weight 1, $3\tau m$ equations of the form $x \oplus y = 1$ each of weight p_m with the following NP-hard gap results: It is NP-hard to decide whether there is an assignment to the variables that leaves unsatisfied equations of weight at most εm, or every assignment to the variables leaves unsatisfied equations of weight at least $(0.5 - \varepsilon)m$. Due to Lemmas 1 and 2 we now know that for produced instances $G[J]$ of TSP it is NP-hard to decide whether there is a tour with cost at most $\left(\frac{3}{2}(\tau - 1)(4p_c\theta + 1) + 6p_c\theta + 10\right)m + 2\nu + \varepsilon m$, where $\theta = \frac{1}{\max\{1,\ p_m\}}$ or all tours have cost at least $\left(\frac{3}{2}(\tau - 1)(4p_c\theta + 1) + 6p_c\theta + 10\right)m + (0.5 - \varepsilon)m \cdot \theta - 2$.

The ratio between these two cases can get arbitrarily close to

$$1 + \frac{1}{3(\tau - 1)(4p_c + \max\{1, p_m\}) + 12p_c + 20\max\{1,\ p_m\}}$$

by appropriate choices of $\varepsilon > 0$ and large enough k. □

Therefore, using the constants of admissible triples from Theorem 4 we can conclude

Corollary 1. *It is NP-hard to approximate the TRAVELLING SALESMAN problem within any constant approximation ratio less than $\frac{117}{116}$.*

5 Conclusion

The methods of this paper provide a new motivation for the study of expanding properties of random graphs. As we have demonstrated, introducing the parametrised weighted amplifiers and weighted low occurrence CONSTRAINT SATISFACTION problems as intermediate steps in the NP-hard gap reductions, allows more flexibility in fine-tuning their expanding parameters. We show that already slight improvement of known expander values modestly improve the hardness of approximation for TSP from the current best value $\frac{123}{122}$ [9] to the new value $\frac{117}{116}$. The introduced method of weighted amplifiers (or expanders) can be of independent interest. Such technique could be used in the gap preserving reductions for other edge-weighted optimisation problems to improve their approximation hardness results.

References

1. Böckenhauer, H.-J., Hromkovič, J., Klasing, R., Seibert, S., Unger, W.: An improved lower bound on the approximability of metric TSP and approximation algorithms for the TSP with sharpened triangle inequality. In: Reichel, H., Tison, S. (eds.) STACS 2000. LNCS, vol. 1770, pp. 382–394. Springer, Heidelberg (2000). https://doi.org/10.1007/3-540-46541-3_32
2. Chlebík, M., Chlebíková, J.: Approximation hardness for small occurrence instances of NP-hard problems. In: Petreschi, R., Persiano, G., Silvestri, R. (eds.) CIAC 2003. LNCS, vol. 2653, pp. 152–164. Springer, Heidelberg (2003). https://doi.org/10.1007/3-540-44849-7_21
3. Chlebík, M., Chlebíková, J.: The Steiner tree problem on graphs: inapproximability results. Theor. Comput. Sci. **406**, 207–214 (2008)
4. Christofides, N.: Worst-case analysis of a new heuristic for the traveling salesman problem, Technical report, Graduate School of Industrial Administration, Carnegie-Mellon University, Pittsburgh PA (1976)
5. Engebretsen, L.: An explicit lower bound for TSP with distances one and two. Algorithmica **35**(4), 301–319 (2003). Preliminary version in STACS 1999
6. Gharan, O.S., Saberi, A., Singh, M.: A randomized rounding approach to the traveling salesman problem. In: IEEE 52nd Annual Symposium on Foundations of Computer Science, FOCS, pp. 550–559 (2011)
7. Håstad, J.: Some optimal inapproximability results. J. ACM **48**, 798–859 (2001)
8. Held, M., Karp, R.M.: The traveling-salesman problem and minimum spanning trees. Oper. Res. **18**, 1138–1162 (1970)
9. Karpinski, M., Lampis, M., Schmied, R.: New inapproximability bounds for TSP. J. Comput. Syst. Sci. **81**, 1665–1677 (2015)
10. Lampis, M.: Improved inapproximability for TSP. Theory Comput. **10**(9), 217–236 (2014)
11. Mömke, T., Svensson, O.: Approximating graphic TSP by matchings. In: IEEE 52nd Annual Symposium on Foundations of Computer Science, FOCS 2011, pp. 560–569 (2011)
12. Mucha, M.: 13/9-approximation for graphic TSP. Theory Comput. Syst. **55**(4), 640–657 (2014)

13. Papadimitriou, C.H., Yannakakis, M.: The Traveling Salesman Problem with distances one and two. Math. Oper. Res. **18**(1), 1–11 (1993)
14. Papadimitriou, C.H., Vempala, S.: On the approximability of the Traveling Salesman Problem. Combinatorica **26**(1), 101–120 (2006)
15. Sebö, A., Vygen, J.: Shorter tours by nicer ears: 7/5-approximation for the graph-TSP, 3/2 for the path version, and 4/3 for two-edge-connected subgraphs. Combinatorica **34**(5), 597–629 (2014)
16. Svensson, O.: Overview of new approaches for approximating TSP. In: Brandstädt, A., Jansen, K., Reischuk, R. (eds.) WG 2013. LNCS, vol. 8165, pp. 5–11. Springer, Heidelberg (2013). https://doi.org/10.1007/978-3-642-45043-3_2

Deleting to Structured Trees

Pratyush Dayal[ID] and Neeldhara Misra[(✉)][ID]

Indian Institute of Technology, Gandhinagar, India
{pdayal,neeldhara.m}@iitgn.ac.in
http://www.iitgn.ac.in

Abstract. We consider a natural variant of the well-known FEEDBACK
VERTEX SET problem, namely the problem of deleting a small subset
of vertices or edges to a full binary tree. This version of the problem is
motivated by real-world scenarios that are best modeled by full binary
trees. We establish that both the edge and vertex deletion variants of the
problem are NP-hard. This stands in contrast to the fact that deleting
edges to obtain a forest or a tree is equivalent to the problem of finding a
minimum cost spanning tree, which can be solved in polynomial time. We
also establish that both problems are FPT by the standard parameter.

Keywords: Full Binary Trees · Feedback Vertex Set · NP-hardness

1 Introduction

The FEEDBACK VERTEX SET (FVS) problem asks for a smallest subset S of
vertices in an undirected graph G to be removed such that the graph, $G \setminus S$,
becomes acyclic. This problem was one of the first problems shown to be NP-
complete [6], and has applications to problems that arise in several areas. These
applications include, but are not limited to, operating systems, database systems
and VLSI chip design. Consequently, the FVS problem has been widely studied
in the context of exact, parameterized and approximation algorithms.

Several variations of the FVS theme have also emerged over the years. For
instance, one line of work considers the task of "deleting to specialized forests",
such as forests of pathwidth one [3,8] or forests whose connected components
are stars of bounded degree [5]. In this case, the forests of pathwidth one turn
out to be graphs whose connected components are caterpillars.

Meanwhile, another line of work is the TREE DELETION SET (TDS) problem
that considers the issue of the connectivity of the structure after the solution
has been deleted. In particular, the TDS problem asks for a smallest subset of
vertices S such that $G \setminus S$ is a tree [7,10]. We remark that the NP-completeness
of this TDS problem follows from a general result of Yannakakis [12]. To state
this result, recall that a *property* π is a class of graphs, and we will say that π

The authors acknowledge funding support from IIT Gandhinagar for PD and NM, and
SERB (Grant No. MTR/2017/001033/MS) for NM.

D.-Z. Du et al. (Eds.): COCOON 2019, LNCS 11653, pp. 128–139, 2019.
https://doi.org/10.1007/978-3-030-26176-4_11

is satisfied by, or is true for, a graph G if $G \in \pi$. A property is said to be *non-trivial* if it is satisfied for at least one graph and false for at least one graph; it is *interesting* if the property is true for arbitrarily large graphs and is *hereditary on induced subgraphs* if the deletion of any node from a graph in π always results in a graph that is in π. The result in question states that the problem of finding a maximum connected subgraph satisfying a property π is NP-hard for any non-trivial and interesting property that is hereditary on induced subgraphs.

In this work, we pose a variation of FVS that is in the spirit of a combination of the variations that we have alluded to; here, however, we are looking for a connected object with additional structure. Specifically, we consider the problem of deleting to a full binary tree. We recall that a full binary tree is a tree that has exactly one vertex of degree two and no vertex of degree more than three. Consider the problem of optimally deleting to a full binary tree, posed in the language of the theorem of Yannakakis [12] stated above, which is to find a maximum connected subgraph that satisfies a certain property. Observe that the property in question could be defined as the property of not having cycles, having exactly one vertex of degree two and no vertex of degree more than three. Note that this property is not hereditary on induced subgraphs: in particular, the deletion of a leaf from a graph that has the property will lead to a violation of the property. In our first result, we explicitly establish the NP-hardness of this problem by reducing from a variant of the INDEPENDENT SET problem.

In addition, we also consider the edge deletion version of the question above. Recall that for a given connected graph on n vertices and m edges, deleting a smallest subset of edges to obtain a tree is straightforward: it is clear that we have to remove every edge that does not belong to a spanning tree, so the size of the solution is always $(m - (n - 1))$. In fact, this problem can be solved in polynomial time even when the edges have weights and we seek a subset of edges of smallest total weight, whose removal results in a tree. It is straightforward to see that any such solution is the complement of a maximum spanning tree and thus, can be found in polynomial time.

In a somewhat surprising twist, we show that the problem of deleting a subset of edges of minimum total weight to obtain a full binary tree is, in fact, NP-complete. To establish some intuition for why this is true, we briefly sketch a simple reduction from the problem of EXACT COVER BY 3-SETS to the closely related problem of deleting edges to obtain a full ternary tree.

A ternary tree is a tree where every non-leaf vertex, except the root, is exactly of degree four, while the root has degree three. Let $\mathcal{F} := \{S_1, \ldots, S_p\}$ be a family of sets of size three over the universe $\mathcal{U} := \{x_1, \ldots, x_q\}$. The goal is to find a subfamily of disjoint sets whose union is $\mathcal{U}$. We create a full ternary tree T with p leaves labeled $\{t_1, \ldots, t_p\}$, and set the weight of the edges of T to B, a quantity that we will specify later. Then, we introduce for every element x_i in the universe a vertex v_i that is adjacent to t_j, if and only if $x_i \in S_j$. The edges between the leaves of T and the vertices corresponding to the elements of $\mathcal{U}$ have unit weights. We also set $B = 3p - q + 1$. It is easy to verify that this graph has a solution of cost $3p - q$ if and only if the system $\mathcal{U}$ has an exact cover, as desired.

It turns out that establishing the hardness of the problem of deleting to full binary trees is non-trivial, and this is one of our main contributions. We reduce from a fairly restrained version of the SATISFIABILITY problem, the hardness of which is inspired by a reduction in [1] and is of independent interest. We note that we deal with the weighted versions of the problems considered, and we also fix a choice of root vertex as part of the input. Finally, we also note that both the problems we propose above are fixed-parameter tractable, when parameterized by the solution size. To this end, we describe a natural branching algorithm and remark that most preprocessing rules that work in a straightforward manner for FEEDBACK VERTEX SET fail when applied as-is to our problem. In particular, it is not trivial to delete degree-one vertices or short-circuit vertices of degree two.

We believe that the problem we propose and the study we undertake has considerable practical motivation. One of the applications of FVS and related problems is to understand noisy datasets. For example, let us say that we expect the data to have a certain structure, but errors in the measurement cause the data at hand not to have the properties expected by said structures. In this context, one approach will be to identify and eliminate the noise - for acyclic structures, that could translate identifying a FVS of small cost. Therefore, for scenarios where the data corresponds to full binary trees, for instances in the case of phylogenetic trees, the problem we present here will be a more relevant model.

2 Preliminaries

We follow standard notation and terminology from parameterized complexity [2] and graph theory [4]; we use $[n]$ to denote the set $\{1, 2, \ldots, n\}$. We now turn to the definitions of the problems that we consider.

FULL BINARY TREE DELETION BY VERTICES (FBT-DV)
Input: A graph $G = (V, E)$, a vertex $r \in V$, vertex weights $w : V \to \mathbb{R}^+$, and $k \in \mathbb{Z}^+$.
Question: Does G have a subset $S \subseteq V$ of total weight at most k such that $G \setminus S$ is a full binary tree rooted at r?

The problems of FULL BINARY TREE DELETION BY EDGES (FBT-DE), COMPLETE BINARY TREE DELETION (by edges or vertices) and BINARY TREE DELETION (by edges or vertices) can be defined analogously. Our focus in this contribution will be mainly on FBT-DV and FBT-DE.

The MULTI-COLORED INDEPENDENT SET problem is the following.

MULTI-COLORED INDEPENDENT SET (MCIS)
Input: A graph $G = (V, E)$ and a partition of $V = (V_1, \ldots, V_k)$ into k parts.
Parameter: k
Question: Does there exist a subset $S \subseteq V$ such that S is independent in G and for every $i \in [k]$, $|V_i \cap S| = 1$?

---

3 NP-hardness

In this section, we establish that the problems of deleting to full binary trees via vertices or edges are NP-complete. We first describe the hardness for the vertex-deletion variant.

Theorem 1. FBT-DV *is* NP-*complete.*

Proof. We reduce from MULTI-COLORED INDEPENDENT SET [2, Corollary 13.8]. Let (G, k) be an instance of MCIS where $G = (V, E)$ and further, let $V = (V_1, \ldots, V_k)$ denote the partition of the vertex set V. We assume, without loss of generality, that $|V_i| = n$ for all $i \in [k]$. Specifically, we denote the vertices of V_i by $\{v_1^i, \ldots, v_n^i\}$. We are now ready to describe the reduced instance of FBT-DV, which we denote by (H, ℓ).

To begin with, let H be a complete binary tree with $2nk$ leaves, where a complete binary tree is a full binary tree with 2^w vertices at distance w from the root for all $w \in [d - 1]$, where d the distance between the root and the leaf furthest away from the root. We denote these leaf vertices as:

$$\left(\cup_{1 \leqslant i \leqslant k} \{a_1^i, \ldots, a_n^i\} \right) \bigcup \left(\cup_{1 \leqslant i \leqslant k} \{b_1^i, \ldots, b_n^i\} \right),$$

where, for all $i \in [k]$ and $j \in [n]$, a_j^i and b_j^i are siblings, and their parent is denoted by p_j^i. We refer to this as the *backbone*, to which we will now add more vertices and edges.

For each $i \in [k]$ and $j \in [n]$, we now introduce a third child of p_j^i, which we denote by u_j^i. We refer to the u's as the *essential* vertices, while its siblings (the a's and the b's) are called *partners*. For all $1 \leqslant i \leqslant k$, we also introduce two *guards*, denoted by x_i and y_i, which are adjacent to all the essential vertices of type i, that is, all u_j^i for $j \in [n]$. Finally, we ensure that the graph induced on the essential vertices is a copy of G, more precisely, we have:

$$(u_i^r, u_j^s) \in E(H) \text{ if and only if } (v_i^r, v_j^s) \in E(G) \text{ for all } i \in [k] \text{ and } j \in [n].$$

We set $\ell = nk$. This completes the construction. We now turn to a proof of equivalence.

The Forward Direction. If $S \subseteq V$ is a multi-colored independent set, then consider the subset S^* given by all the essential vertices corresponding to $V \setminus S$, along with the partner vertices a_j^i for each (i, j), for which v_j^i belongs to S. It is easy to verify that the proposed set consists of nk vertices. Observe that the deletion of S^* leaves us with a full binary tree where each p_j^i now has two children - either two partner vertices (for vertices not in S) or one essential vertex along with one partner vertex (for vertices in S). Further, each pair of guards of type i now has an unique parent, which is the essential vertex corresponding to the vertex given by $S \cap V_i$. The essential vertices have degree exactly three because their only other neighbors in H were essential vertices corresponding to neighbors in G, but the presence of any such vertex in $H \setminus S^*$ will contradict the fact that S induces an independent set in G. This concludes the argument in the forward direction.

The Reverse Direction. Let S^* be a subset of $V(H)$ such that $H \setminus S^*$ is a full binary tree. We claim that $S^* \cap \{p_j^i \mid 1 \leqslant i \leqslant k \text{ and } 1 \leqslant j \leqslant n\} = \emptyset$, since the deletion of any parent of a partner vertex will result in the corresponding partner vertex becoming isolated in $H \setminus S^*$— which leads to a contradiction when we account for the budget constraint on S^*. Since all the parents of partner vertices survive and have degree four in H, it follows that at least one of its neighbors must belong to S^*. In particular, we claim that for every $i \in [k]$ and $j \in [n]$, $S^* \cap \{u_j^i, a_j^i, b_j^i\} \neq \emptyset$. Indeed, if this is not the case, then S^* contains the parent of p_i^j, and it is easy to verify that this leads to a situation where either $H \setminus S^*$ is disconnected or one of the guard vertices has degree two and is not the root, contradicting the assumption that $H \setminus S^*$ is a full binary tree.

From the discussion above, it is clear that S^* picks at least n vertices of type i for each $1 \leqslant i \leqslant k$, and combined with the fact that $|S^*| \leqslant nk$, we note that S^* does not contain any of the guard vertices. Our next claim is that for all $i \in [k]$, $G \setminus S^*$ contains at least one essential vertex of type i. If not, then S^* contains all the neighbors of the guards of type i, which makes them isolated in $G \setminus S^*$–a contradiction.

For each $1 \leqslant i \leqslant k$, consider the vertex in G corresponding to the essential vertex that is *not* chosen by S^* (in the event that there are multiple such vertices, we pick one arbitrarily). We denote this collection of vertices by S. We claim that S induces an independent set in G: indeed, if not, then any edge in $G[S]$ is also present in $H \setminus S^*$ and creates a cycle when combined with the unique path connecting its endpoints via the backbone, which is again a contradiction. This concludes the proof.   □

We now turn our attention to the edge-deletion variant. Here, we will find it convenient to reduce from a structured version of exact satisfiability, where the occurrences of the variables are bounded in frequency and also controlled in terms of how they appear. We will turn to a formal description in a moment, noting that here our reduction is similar to the one used to show that Linear-SAT is NP-complete [1].

Theorem 2. FBT-DE *is* NP-*complete.*

We first describe the version of SATISFIABILITY that we will reduce from. Our instance consists of $(4p + q)$ clauses which we will typically denote as follows:

$$\mathcal{C} = \{A_1, B_1, A_1', B_1', \cdots, A_p, B_p, A_p', B_p'\} \cup \{C_1, \cdots, C_q\}$$

We refer to the first $4p$ clauses as the *core* clauses, and the remaining clauses as the *auxiliary* clauses. The core clauses have two literals each, and also enjoy the following structure:

$$\forall i \in [p], A_i \cap B_i = \{x_i\} \text{ and } A_i' \cap B_i' = \{\overline{x}_i\}$$

We refer to the x_i's as the *main* variables and the remaining variables that appear among the core clauses as *shadow* variables. The shadow variables occur exactly once, and have negative polarity among the core clauses. Therefore, using $\ell(\cdot)$ to denote the set of literals occurring amongst a subset of clauses, we have:

$$\left| \ell \left(\bigcup_{i=1}^{p} \{A_i, B_i, A'_i, B'_i\} \right) \right| = 6p.$$

The auxiliary clauses have the property that they only contain the shadow variables, which occur exactly once amongst them with positive polarity. Also, every auxiliary clause contains exactly four literals. Note that this also implies, by a double-counting argument, that $q = p$. We say that a collection of clauses is a *chain* if it has all the properties described above. An instance of LINEAR NEAR-EXACT SATISFIABILITY (LNES) is the following: given a set of clauses that constitute a chain, is there an assignment τ of truth values to the variables such that exactly one literal in every core clause and two literals in every auxiliary clause evaluate to TRUE under τ?

For ease of discussion, given an assignment of truth values τ we often use the phrase "τ satisfies a literal" to mean that the literal in question evaluates to true under τ. For instance, the question from the previous paragraph seeks an assignment τ that satisfies exactly one literal in every core clause and two literals in every auxiliary clause. We also refer to such an assignment as a near-exact satisfying assignment. The following observation is a direct consequence of the definitions above.

Proposition 1. *Let $\mathcal{C}$ be a collection of clauses that form a chain. For any assignment of truth values τ, the main variables satisfy exactly two core clauses and the shadow variables satisfy either one core clause or one auxiliary clause.*

We first establish that LNES is NP-complete:

Lemma 1. LINEAR NEAR-EXACT SATISFIABILITY *is NP-complete.*

Proof. We reduce from $(2/2/4)$-SAT, which is the variant of SATISFIABILITY where every clause has four literals and every literal occurs exactly twice — in other words, every variable occurs in exactly two clauses with positive polarity and in exactly two clauses with negative polarity. The question is, if there exists an assignment τ of truth values to the variables under which exactly two literals in every clause evaluate to true. This problem is known to be NP-complete [11].

Let ϕ be a $(2/2/4)$-SAT instance over the variables $V = \{x_1, \ldots, x_n\}$ and clauses $\mathcal{C} = \{C_1, \ldots, C_m\}$. For every variable x_i, we introduce four new variables: p_i, r_i and q_i, s_i. We replace the two positive occurrences of x_i with p_i and r_i, and the two negated occurrences of x_i with q_i and s_i. We abuse notation and continue to use $\{C_1, \ldots, C_m\}$ to denote the modified clauses. Also, introduce the clauses: $A_i = (x_i, \overline{p_i}), B_i = (x_i, \overline{r_i}), A'_i = (\overline{x_i}, \overline{q_i}), B'_i = (\overline{x_i}, \overline{s_i})$, for all $1 \leqslant i \leqslant n$. Note that these collection of clauses form a chain, as required. We use ψ to refer to this formula. We now turn to the argument for equivalence.

In the forward direction, let τ be an assignment that sets exactly two literals of every clause in ϕ to true. Consider the assignment ζ given by:

$$\zeta(x_i) = \tau(x_i), \zeta(p_i) = \zeta(r_i) = \tau(x_i); \zeta(q_i) = \zeta(s_i) = 1 - \tau(x_i),$$

for all $1 \leqslant i \leqslant n$. It is straightforward to verify that ζ satisfies exactly one literal in every core clause and exactly two literals in every auxiliary clause.

In the reverse direction, let ζ be an assignment for the variables of ψ that satisfies exactly one literal in every core clause and exactly two literals in every auxiliary clause. Define τ as the restriction of ζ on the main variables. Let C be a clause in ϕ. To see that τ satisfies exactly two literals of C, note that the following:

$$\zeta(p_i) = \zeta(r_i) = \zeta(x_i) = \tau(x_i); \zeta(q_i) = \zeta(s_i) = 1 - \zeta(x_i) = 1 - \tau(x_i)$$

is forced by the requirement that ζ must satisfy exactly one literal in each core clause. Therefore, if τ satisfies more or less than two literals of any clause C, then that behavior will be reflected exactly in the auxiliary clause corresponding to C, which would contradict the assumed behavior of ζ. We make this explicit with an example for the sake of exposition. Let C from ϕ be the clause $(x_1, \overline{x}_3, \overline{x}_6, x_7)$, and let the clause constructed in ψ be (p_1, q_3, q_6, r_7). Suppose $\tau(x_1) = \tau(x_7) = \tau(x_6) = 1$ and $\tau(x_3) = 0$. Then we have $\zeta(p_1) = \zeta(r_7) = 1$ and $\zeta(q_6) = 0$, while $\zeta(q_3) = 1$. This demonstrates that ζ satisfies three literals in the auxiliary clause corresponding to C, in one-to-one correspondence with the literals that are satisfied by τ. This completes our argument.   □

We now turn to a proof of Theorem 2. The overall approach is the following. We will introduce a complete binary tree whose leaves will be used to represent variables using variable gadgets which will have obstructions that can be removed in a fixed number of ways, each of which corresponds to a "signal" for whether the variable is to be set to true or false. We will then introduce vertices corresponding to clauses that will be attached to the variable gadgets in such a way that they can only be "absorbed" into the rest of the structure precisely when exactly two of its literals are receiving a signal indicating that they are being satisfied.

The Shadow Variables. An instance of the gadget that we construct for the shadow variables is depicted in Fig. 1. We remark here that the notation used for the vertices here is to enable our discussion of how the gadget works and is not to be confused with the notation already used to denote the variables and clauses of the LNES instance.

The vertices p and q are called the *anchors* of the gadget, while the vertices x, y, a and b are called the *drivers* of the gadget. This is because, as we will see, the behavior of the edges incident to these vertices determines the fate of the variable—in terms of whether it "accepts" the vertex corresponding to the core clause or the auxiliary clause to which it belongs. We refer to the vertex u in the gadget as the *negative point of entry*, while the vertex v is called the *positive point of entry*.

We refer to the edges incident on the vertices x, y, a and b as *active* edges and the remaining edges (i.e, $(p, u), (p, v), (q, w), (q, c)$ and (w, z)) as *passive* edges. We say that a solution is *nice* if it does not contain any passive edges. We also say that an instance G of FBT-DE contains a clean copy of the gadget

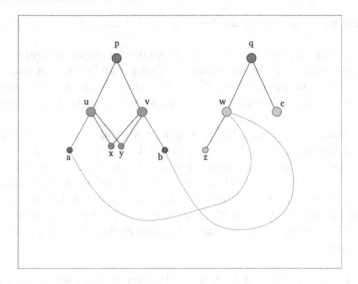

Fig. 1. The gadget corresponding to the shadow variables.

H if H appears in G as an induced subgraph, and further, $d_G(x) = d_G(y) = d_G(a) = d_G(b) = 2$ while $d_G(w) = 4$, $d_G(c) = 1$ and none of the vertices of H are chosen to be the target root vertex. We make the following observation about the behavior of this gadget.

Claim. Let H be a vertex gadget for a shadow variable as defined above. Let G be an instance of FBT-DE that contains a clean copy of H. Then, any nice solution S contains exactly four edges among the edges of H.

Proof. Let F denote the set of active edges in G. Since $d_G(x) = d_G(y) = d_G(a) = d_G(b) = 2$, we claim that any solution S must delete *exactly* four edges from F: in particular, S contains exactly one of the edges incident to each of the vertices. Indeed, if S deletes fewer edges than suggested then $G \setminus S$ contains a degree two vertex different from the root, which is a contradiction. On the other hand, if S contains more than four edges from F, then at least one of these four vertices is isolated in $G \setminus S$, which contradicts our assumption that $G \setminus S$ is connected. This clearly implies the claim, since all edges not considered are passive and a nice solution does not contain these edges by definition.   □

We now analyze the possible behaviors of a solution localized to the gadget in greater detail. We refer the reader to the full version of this paper for the figures associated with this explanation.

The possibilities (xv, yv, aw, bv) and (xu, yu, au, bw) do not arise because employing these deletions causes the entry point vertices to have degree four or more in $G \setminus S$. Further, since the solution S does not involve any of the passive edges, then we also rule out the following possibilities, since they all lead to a situation where the degree of w is four or more in $G \setminus S$:

▷ xv, yv, au, bv ▷ xu, yu, au, bv ▷ xu, yv, au, bv ▷ xv, yu, au, bv

Recalling that $d_G(w) = 4$ when H makes a clean appearance in G, we also safely rule out the possibilities: (xv, yv, aw, bw), (xu, yu, aw, bw), (xu, yv, aw, bw), (xv, yu, aw, bw). Note that they result in a situation where the degree of w is exactly two in $G \setminus S$ — since w is not the target root vertex, this is a contradiction as well.

Observe that, given a nice solution S, in all the valid scenarios possible, either $d_{H \setminus S}(u) = 2$ and $d_{H \setminus S}(v) = 3$, or $d_{H \setminus S}(u) = 3$ and $d_{H \setminus S}(v) = 2$. We say that a shadow variable gadget has a *negative signal* in solutions where $d_{H \setminus S}(u) = 2$. Similarly, we say that the gadget has a *positive signal* in the situations where $d_{H \setminus S}(v) = 2$. We refer to the edges $\{(v, x), (v, y), (u, a), (w, b)\}$ as the *negative witness* and the edges $\{(u, x), (u, y), (v, b), (w, a)\}$ as the *positive witness* for the shadow variable gadgets. This concludes the description of the gadget meant for shadow variables.

The Main Variables. We now turn our attention to the gadget corresponding to the main variables. Here, we find it convenient to incorporate vertices representing the core clauses that the main variables belong to also as a part of the gadget. The construction of the gadget is depicted in Fig. 2. As before, the notation used for the vertices here is to enable our discussion of how the gadget works. With the exception of A, B, A', B', which indeed are meaningfully associated with the analogously named core clauses, the notation is not to be confused with the notation already used to denote the variables and clauses of the LNES instance.

The edges (z, u) and (z, v) are the *passive* edges of this gadget, while the remaining edges are *active*. The vertex z is called the *anchor* of this gadget. As before, a solution is *nice* if it does not contain any of the passive edges. We say that an instance G of FBT-DE has a clean copy of H if H appears in G as an induced subgraph and, further, $d_G(p) = d_G(q) = d_G(u) = d_G(v) = 3$, $d_G(x) = d_G(y) = 2$, $d_G(B) = d_G(A') = 2$, $d_G(A) = d_G(B') = 3$, and none of the vertices of H are chosen to be the target root vertex.

We reflect briefly on the nature of a nice solution S in instances that have a clean copy of a main variable gadget H. First, since $d_G(x) = 2$ and x is not the target root vertex, we note that exactly one of (v, x) or (u, x) must belong to S. Suppose $(v, x) \in S$. The removal of (v, x) makes v a vertex of degree two, and since (z, v) is a passive edge and S is nice, $(v, q) \in S$ is forced. Along similar lines, we have that $(u, p) \notin S$. Now, we argue that $(q, B') \notin S$. Indeed, if $(q, B') \in S$, then A' has degree two from the deletions so far, and q is a degree-one vertex with A' as its sole neighbor. Recalling that A' is not the target root vertex, we are now forced to delete exactly one of the endpoints incident on A', but both possibilities lead us to a disconnected graph. Therefore, $(q, B') \notin S$. It is easy to see that this forces $(q, A') \in S$ and further, $(A, y) \in S$. A symmetric line of reasoning shows that if $(u, x) \in S$, then $(p, u), (p, B)$ and (B', y) are all in S as well. We refer the reader to the full version of this paper for the figures

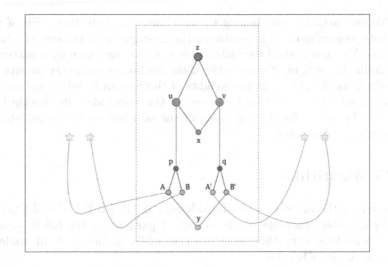

Fig. 2. The gadget corresponding to the main variables.

associated with this explanation. These two scenarios motivate the definitions of positive and negative signals that we now make explicit.

With respect to a nice solution S, we say that a main variable gadget has a *negative signal* if $d_{H \setminus S}(v) = 2$. Likewise, we say that the gadget has a *positive signal* if $d_{H \setminus S}(u) = 2$. We will also refer to the set of edges $\{(v, x), (v, q), (A', q), (A, y)\}$ as the *positive witness* of this gadget, and the *negative witness* is defined analogously.

We are now ready to discuss the overall construction. Let ϕ be an instance of LNES with clauses given by:

$$\mathcal{C} = \{A_1, B_1, A'_1, B'_1, \cdots, A_p, B_p, A'_p, B'_p\} \cup \{C_1, \cdots, C_p\},$$

where the main variable common to A_i and B_i is denoted by x_i and the auxiliary variables in these two clauses are denoted by p_i and r_i, while the auxiliary variables in the clauses A'_i and B'_i are denoted by q_i and s_i. We denote by $\mathcal{I}_\phi :=$ (G, r, w, k) the LNES instance that we will now construct based on ϕ.

First, we construct the smallest complete binary tree with at least $(9p)$ leaves and let v be the root of this tree. We refer to this tree as the *backbone* of G. Let the first $9p$ leaves of this tree be denoted by $\ell_1, \ldots, \ell_p; \alpha_1, \beta_1, \ldots, \alpha_{4p}, \beta_{4p}$. For each main variable x_i, let H_i be the corresponding gadget. We identify the anchor of H_i with ℓ_i. For each shadow variable, we introduce a gadget corresponding to it, and identify the first anchor vertex in the gadget with α_i and the second anchor vertex with β_i. For every core clause A_i, we add an edge between the vertex A in the gadget corresponding to A_i and the negative entry point in the gadget for the shadow variable contained in the clause A_i. We also do this in an analogous fashion for the core clauses A'_i, B_i and B'_i.

Finally, for each auxiliary clause C_i, we introduce two vertices ω_i and ω'_i. We connect these vertices with the positive entry point into all gadgets corresponding to the shadow variables that belong to the clause C_i. Note that each of these vertices have degree four. This completes the description of the construction of the graph G. We note that all the gadgets present in G are clean by construction. We now define the weight of every edge in the backbone and every passive edge in the gadgets as $(k + 1)$, while the weights of the remaining edges are set to be one. Finally, we set $k := 28p$ and let $r = v$—this concludes the description of the instance $\mathcal{J}_\phi$. We defer the argument of the equivalence of the instances to the full version of this paper.

4   FPT Algorithms

We observe that the problems considered here, namely FBT-DV and FBT-DE are fixed-parameter tractable by the standard parameter. We briefly describe a natural branching algorithm for FBT-DV while noting that an analogous argument works for FBT-DE.

Let $(G = (V, E), k, r, w)$ be an instance of FBT-DV. First, consider a vertex v, different from the designated root, that has four or more neighbors. Choose any four neighbors of v, say a, b, c, d, and branch on the set $\{v, a, b, c, d\}$ and we adjust the remaining budget by subtracting the respective weights of these vertices. The exhaustiveness of this branching rule follows immediately from the definition of a full binary tree. Along similar lines, we can also branch on the designated root along with three of its neighbors at a time, if the root has degree three or more, and also the closed neighborhoods of vertices of degree exactly two. We abort any branches where we have exhausted the budget.

We say that a graph with a designated root vertex is *nice* if it is connected, its maximum degree is three and the root is only vertex of degree two. Note that the depth of the branching thus far is bounded by $O(5^k)$, and we branch appropriately on disconnected instances, noting that only one of the components can "survive". Also, note that all the remaining instances are nice. If any of the remaining graphs are also acyclic, then we are already done.

If not, then we branch on these graphs further as follows. We pre-process vertices of degree three with a degree one neighbor by employing an appropriate short-circuiting rule. We can then start a breadth-first search (BFS) from the root vertex, noting that the depth of the BFS tree is at most $(\log_2 n + 1)$, since every internal vertex in this tree has at least two children. Therefore, we may infer that there exists a cycle of length $O(\log n)$, and we branch on all the vertices of this cycle other than the root vertex. If the deletion of a vertex on the cycle leads to a disconnected graph, then we abort the corresponding branch. Similarly, if the deletion of a vertex on the cycle creates vertices of degree two in the resulting graph, then we branch on the closed neighborhood of these degree two vertices and discard any disconnected graphs until we arrive at a nice instance, at which point we recurse in the fashion described here. The correctness of the overall algorithm follows from the exhaustiveness of the branching rules. The fact that the running time is FPT follows by a well-known argument [9].

Theorem 3. *The problems* FBT-DV *and* FBT-DE *are* FPT *with respect to solution size.*

References

1. Arkin, E.M., et al.: Choice is hard. In: Elbassioni, K., Makino, K. (eds.) ISAAC 2015. LNCS, vol. 9472, pp. 318–328. Springer, Heidelberg (2015). https://doi.org/10.1007/978-3-662-48971-0_28
2. Cygan, M., et al.: Parameterized Algorithms. Springer, Cham (2015). https://doi.org/10.1007/978-3-319-21275-3
3. Cygan, M., Pilipczuk, M., Pilipczuk, M., Wojtaszczyk, J.O.: An improved FPT algorithm and a quadratic kernel for pathwidth one vertex deletion. Algorithmica **64**(1), 170–188 (2012)
4. Diestel, R.: Graph Theory. Graduate Texts in Mathematics, vol. 173, 4th edn. Springer, New York (2012)
5. Ganian, R., Klute, F., Ordyniak, S.: On structural parameterizations of the bounded-degree vertex deletion problem. In: 35th Symposium on Theoretical Aspects of Computer Science, STACS, pp. 33:1–33:14 (2018)
6. Garey, M.R., Johnson, D.S.: Computers and Intractability: A Guide to the Theory of NP-Completeness. W. H. Freeman, New York (1979)
7. Giannopoulou, A.C., Lokshtanov, D., Saurabh, S., Suchý, O.: Tree deletion set has a polynomial kernel but no opt$^{o(1)}$ approximation. SIAM J. Discrete Math. **30**(3), 1371–1384 (2016)
8. Philip, G., Raman, V., Villanger, Y.: A quartic kernel for pathwidth-one vertex deletion. In: Thilikos, D.M. (ed.) WG 2010. LNCS, vol. 6410, pp. 196–207. Springer, Heidelberg (2010). https://doi.org/10.1007/978-3-642-16926-7_19
9. Raman, V., Saurabh, S., Subramanian, C.R.: Faster fixed parameter tractable algorithms for finding feedback vertex sets. ACM Trans. Algorithms **2**(3), 403–415 (2006)
10. Raman, V., Saurabh, S., Suchý, O.: An FPT algorithm for tree deletion set. J. Graph Algorithms Appl. **17**(6), 615–628 (2013)
11. Ratner, D., Warmuth, M.: Finding a shortest solution for the N × N extension of the 15-PUZZLE is intractable. In: Kehler, T.R.S. (ed.) Proceedings of the 5th National Conference on Artificial Intelligence, pp. 168–172
12. Yannakakis, M.: The effect of a connectivity requirement on the complexity of maximum subgraph problems. J. ACM **26**(4), 618–630 (1979)

Sensitivity, Affine Transforms and Quantum Communication Complexity

Krishnamoorthy Dinesh and Jayalal Sarma[✉]

Department of Computer Science and Engineering,
Indian Institute of Technology Madras, Chennai, India
{kdinesh,jayalal}@cse.iitm.ac.in

Abstract. In this paper, we study the Boolean function parameters sensitivity (s), block sensitivity (bs), and alternation (alt) under specially designed affine transforms and show several applications. For a function $f : \mathbb{F}_2^n \to \{0,1\}$, and $A = Mx + b$ for $M \in \mathbb{F}_2^{n \times n}$ and $b \in \mathbb{F}_2^n$, the result of the transformation g is defined as $\forall x \in \mathbb{F}_2^n, g(x) = f(Mx + b)$.

As a warm up, we study alternation under linear shifts (when M is restricted to be the identity matrix) called the *shift invariant alternation* (the smallest alternation that can be achieved for the Boolean function f by shifts, denoted by $\mathsf{salt}(f)$). By a result of Lin and Zhang [12], it follows that $\mathsf{bs}(f) \le O(\mathsf{salt}(f)^2 \mathsf{s}(f))$. Thus, to settle the SENSITIVITY CONJECTURE ($\forall\ f, \mathsf{bs}(f) \le \mathsf{poly}(\mathsf{s}(f))$), it suffices to argue that $\forall\ f, \mathsf{salt}(f) \le \mathsf{poly}(\mathsf{s}(f))$. However, we exhibit an explicit family of Boolean functions for which $\mathsf{salt}(f)$ is $2^{\Omega(\mathsf{s}(f))}$.

Going further, we use an affine transform A, such that the corresponding function g satisfies $\mathsf{bs}(f, 0^n) \le \mathsf{s}(g)$, to prove that for $F(x, y) \overset{\text{def}}{=} f(x \wedge y)$, the bounded error quantum communication complexity of F with prior entanglement, $Q^*_{1/3}(F)$ is $\Omega(\sqrt{\mathsf{bs}(f, 0^n)})$. Our proof builds on ideas from Sherstov [17] where we use specific properties of the above affine transformation. Using this, we show the following.
(a) For a fixed prime p and an ϵ, $0 < \epsilon < 1$, any Boolean function f that depends on all its inputs with $\deg_p(f) \le (1 - \epsilon) \log n$ must satisfy $Q^*_{1/3}(F) = \Omega\left(\frac{n^{\epsilon/2}}{\log n}\right)$. Here, $\deg_p(f)$ denotes the degree of the multilinear polynomial over $\mathbb{F}_p$ which agrees with f on Boolean inputs.
(b) For Boolean function f such that there exists primes p and q with $\deg_q(f) \ge \Omega(\deg_p(f)^\delta)$ for $\delta > 2$, the deterministic communication complexity - $\mathsf{D}(F)$ and $Q^*_{1/3}(F)$ are polynomially related. In particular, this holds when $\deg_p(f) = O(1)$. Thus, for this class of functions, this answers an open question (see [2]) about the relation between the two measures.
Restricting back to the linear setting, we construct linear transformation A, such that the corresponding function g satisfies, $\mathsf{alt}(f) \le 2\mathsf{s}(g) + 1$. Using this new relation, we exhibit Boolean functions f (other than the parity function) such that $\mathsf{s}(f)$ is $\Omega(\sqrt{\mathsf{sparsity}(f)})$ where $\mathsf{sparsity}(f)$ is the number of non-zero coefficients in the Fourier representation of f.

© Springer Nature Switzerland AG 2019
D.-Z. Du et al. (Eds.): COCOON 2019, LNCS 11653, pp. 140–152, 2019.
https://doi.org/10.1007/978-3-030-26176-4_12

Keywords: Affine transforms · Alternation · Degree · Sensitivity

1   Introduction

For a Boolean function $f : \{0,1\}^n \rightarrow \{0,1\}$, *sensitivity* of f on $x \in \{0,1\}^n$, is the maximum number of indices $i \in [n]$, such that $f(x \oplus e_i) \neq f(x)$ where $e_i \in \{0,1\}^n$ with exactly the ith bit as 1. The *sensitivity* of f (denoted by $s(f)$) is the maximum sensitivity of f over all inputs. A related parameter is the *block sensitivity* of f (denoted by $bs(f)$), where we allow disjoint blocks of indices to be flipped instead of a single bit. Another parameter is the deterministic *decision tree complexity* (denoted by $DT(f)$) which is the depth of an optimal decision tree computing the function f. The *certificate complexity* of f (denoted by $C(f)$) is the non-deterministic variant of the decision tree complexity. The parameter $s(f)$ was originally studied by Cook *et al.* [5] in connection with the CREW-PRAM model of computation. Subsequently, Nisan and Szegedy [13] introduced the parameters $bs(f)$ and $C(f)$ and conjectured that for any function $f : \{0,1\}^n \rightarrow \{0,1\}$, $bs(f) \leq poly(s(f))$ - known as the SENSITIVITY CONJECTURE. Later developments, which revealed several connections between sensitivity, block sensitivity and the other Boolean function parameters, demonstrated the fundamental nature of the conjecture (see [9] for a survey and several equivalent formulations of the conjecture).

Shi and Zhang [19] studied the parity complexity variants of $bs(f), C(f)$ and $DT(f)$ and observed that such variants have the property that they are invariant under arbitrary invertible linear transforms (over $\mathbb{F}_2^n$). They also showed existence of Boolean functions where under *all* invertible linear transforms of the function, the decision tree depth is linear while their parity variant of decision tree complexity is at most logarithmic in the input length.

Our Results: While the existing studies focus on understanding the Boolean function parameters under the effect of arbitrary invertible affine transforms, in this work, we study the relationship between the above parameters of Boolean functions $f : \mathbb{F}_2^n \rightarrow \{0,1\}$, under specific affine transformations over $\mathbb{F}_2^n$. More precisely, we explore the relationship of the above parameters for the function $g : \mathbb{F}_2^n \rightarrow \{0,1\}$ and f, where g is defined as $g(x) = f(Mx + b)$ for specific $M \in \mathbb{F}_2^{n \times n}$ and $b \in \mathbb{F}_2^n$ (where is M not necessarily invertible). We show the following results, and their corresponding applications, which we explain along with the context in which they are relevant.

Boolean Functions Under Shifts: We study the parameters when the transformation is very structured - namely the matrix M is the identity matrix and $b \in \mathbb{F}_2^n$ is a linear shift. More precisely, we study $f_b(x) \overset{\text{def}}{=} f(x + b)$ where b is the shift. Observe that all the parameters mentioned above are invariant under shifts. A Boolean function parameter which is neither shift invariant nor invariant under invertible linear transforms is the *alternation*, a measure of non-monotonicity of Boolean function (see Sect. 2 for a formal definition). To see this for the case of shifts, if we take f as the majority function on n bits, then there exists shifts $b \in \{0, 1\}^n$ where $\mathsf{alt}(f_b) = \Omega(n)$ while $\mathsf{alt}(f) = 1$.

A recent result related to SENSITIVITY CONJECTURE by Lin and Zhang [12] shows that $\mathsf{bs}(f) \leq O(\mathsf{s}(f)\mathsf{alt}(f)^2)$. This bound for $\mathsf{bs}(f)$, implies that to settle the SENSITIVITY CONJECTURE, it suffices to show that $\mathsf{alt}(f)$ is upper bounded by $\mathsf{poly}(\mathsf{s}(f))$ for all Boolean functions f. However, the authors [6] ruled this out, by exhibiting a family of functions where $\mathsf{alt}(f)$ is at least $2^{\Omega(\mathsf{s}(f))}$.

Observing that the parameters $\mathsf{s}(f), \mathsf{bs}(f)$ are invariant under shifts, we define a new quantity *shift-invariant alternation*, $\mathsf{salt}(f)$, which is the minimum alternation of any function g obtained from f upon shifting by a vector $b \in \{0, 1\}^n$ (Definition 1). By the aforementioned bound on $\mathsf{bs}(f)$ of Lin and Zhang [12], it is easy to observe that $\mathsf{bs}(f) \leq O(\mathsf{s}(f)\mathsf{salt}(f)^2)$. We also show that there exists a family of Boolean functions f with $\mathsf{bs}(f) = \Omega(\mathsf{s}(f)\mathsf{salt}(f))$ (Proposition 3).

It is conceivable that $\mathsf{salt}(f)$ is much smaller compared to $\mathsf{alt}(f)$ for a Boolean function f and hence that $\mathsf{salt}(f)$ can potentially be upper bounded by $\mathsf{poly}(\mathsf{s}(f))$ thereby settling the SENSITIVITY CONJECTURE. However, we rule this out by showing the following stronger gap, about the same family of functions demonstrated in [6].

Proposition 1. *There exists an explicit family of Boolean functions for which* $\mathsf{salt}(f)$ *is* $2^{\Omega(\mathsf{s}(f))}$.

Boolean Functions Under Affine Transformations: We now generalize our theme of study to the affine transforms over $\mathbb{F}_2^n$. In particular, we explore how to design affine transformations in such a way that block sensitivity of the original function (f) is upper bounded by the sensitivity of the new function (g). Let the *sensitivity* of f on a denoted as $\mathsf{s}(f, a) = |\{i \mid f(a \oplus e_i) \neq f(a), i \in [n]\}|$ and *block sensitivity* of f on a, denoted $\mathsf{bs}(f, a)$, be the maximum number of disjoint blocks $\{B_i \mid B_i \subseteq [n]\}$ such that $f(a \oplus e_{B_i}) \neq f(a)$.

Lemma 1. *For any* $f : \mathbb{F}_2^n \to \{-1, 1\}$ *and* $a \in \{0, 1\}^n$, *there exists an affine transform* $A : \mathbb{F}_2^n \to \mathbb{F}_2^n$ *such that for* $g(x) = f(A(x))$, *(a)* $\mathsf{bs}(f, a) \leq \mathsf{s}(g, 0^n)$, *and (b)* $g(x) = f((x_{i_1}, x_{i_2}, \ldots, x_{i_n}) \oplus a)$ *where* $i_1, \ldots, i_n \in [n]$ *are not necessarily distinct.*

The above transformation is used in Nisan and Szegedy (Lemma 7 of [13]) to show that $\mathsf{bs}(f) \leq 2\mathsf{deg}(f)^2$. Here, $\mathsf{deg}(f)$ is the degree of the multilinear polynomial over reals that agrees with f on Boolean inputs. We show another application of Lemma 1 in the context of two party quantum communication complexity, a model for which was introduced by Yao (we defer the details of the model to Sect. 2). The measure of interest in this setting is the communication cost for computing a function $F : \{0, 1\}^n \times \{0, 1\}^n \to \{0, 1\}$ which we denote as $Q_{1/3}^*(F)$. The corresponding analog in the classical setting is the bounded error randomized communication model where the two parties communicate with $0, 1$ bits and share an unbiased random source. We define $\mathsf{R}_{1/3}(F)$ as the minimum cost randomized protocol computing F with error at most $1/3$ and $\mathsf{D}(F)$ as the minimum cost deterministic protocol computing F. It can be shown that $Q_{1/3}^*(F) \leq \mathsf{R}_{1/3}(F) \leq \mathsf{D}(F)$.

One of the fundamental goals in quantum communication complexity is to see if there are functions where their randomized communication complexity is significantly larger than their quantum communication complexity. It has been conjectured by Shi and Zhu [18] that this is not the case in general (which they called the Log-Equivalence Conjecture). In this work, we are interested in the case when $F(x, y)$ is of the form $f(x \wedge y)$ where $f : \{0, 1\}^n \to \{0, 1\}$ and $x \wedge y$ is the string obtained by bitwise AND of x and y.

Question 1. *For $f : \{0, 1\}^n \to \{0, 1\}$, let $F : \{0, 1\}^n \times \{0, 1\}^n \to \{0, 1\}$ be defined as $F(x, y) = f(x \wedge y)$. Is it true that for any such F, $\mathsf{D}(F) \leq \mathrm{poly}(Q_{1/3}^*(F))$?*

Since $\mathsf{R}_{1/3}(F) \leq \mathsf{D}(F)$, answering the above question in positive would show that the classical randomized communication model is as powerful as the quantum communication model for the class of functions $F(x, y) = f(x \wedge y)$. This question for such restricted F has also been proposed by Klauck [10] as a first step towards answering the general question (see also [2]). In this direction, Razborov [15] showed that for the special case when f is symmetric, $F(x, y) = f(x \wedge y)$ satisfy $\mathsf{D}(F) \leq O(Q_{1/3}^*(F)^2)$. In the process, Razborov developed powerful techniques to obtain lower bounds on $Q_{1/3}^*(F)$ which were subsequently generalized by Sherstov [16], Shi and Zhu [18]. Subsequently, in a slightly different direction, Sherstov [17] showed that instead of computing $F(x, y) = f(x \wedge y)$ alone, if we consider F to be the problem of computing both of $F_1(x, y) = f(x \wedge y)$ and $F_2(x, y) = f(x \vee y)$, then $\mathsf{D}(F) = O(Q_{1/3}^*(F)^{12})$ for all Boolean functions f where $Q_{1/3}^*(F) = \max\left\{ Q_{1/3}^*(F_1), Q_{1/3}^*(F_2) \right\}$ and $\mathsf{D}(F) = \max\left\{ \mathsf{D}(F_1), \mathsf{D}(F_2) \right\}$. Using Lemma 1, we build on the ideas of Sherstov [17] and obtain a lower bound for $Q_{1/3}^*(F)$ where $F(x, y) = F_1(x, y) = f(x \wedge y)$.

Theorem 2. *Let $f : \{0, 1\}^n \to \{-1, 1\}$ and $F(x, y) = f(x \wedge y)$, then, $Q_{1/3}^*(F) = \Omega\left(\sqrt{\mathsf{bs}(f, 0^n)} \right)$.*

In this context, we make an important comparison with a result of Sherstov [17]. He proved that for $F'(x, y) = f_b(x \wedge y)$, where $b \in \{0, 1\}^n$ is the input on which $\mathsf{bs}(f, x)$ is maximum, $Q_{1/3}^*(F') = \Omega(\sqrt{\mathsf{bs}(f)}) \geq \Omega(\sqrt{\mathsf{bs}(f, 0^n)})$ (Corollary 4.5 of [17]). Notice that F and F' differ by a linear shift of f with b.[1] Moreover, $Q_{1/3}^*(F)$ can change drastically even under such (special) linear shifts of f. For example, consider $f = \wedge_n$. Since $\mathsf{bs}(f)$ is maximized at 1^n, $b = 1^n$. Hence, the function F' is the disjointness function for which $Q_{1/3}^*(F') = \Omega(\sqrt{n})$ [15] whereas, $Q_{1/3}^*(F) = O(1)$. The same counterexample also shows that $Q_{1/3}^*(F) = \Omega(\sqrt{\mathsf{bs}(f)})$ cannot hold for all f. Since the lower bounds shown on quantum

---

[1] More importantly, this b in Corollary 4.5 of [17] cannot be fixed to 0^n for all Boolean functions to conclude Theorem 2.

communication complexity are on different functions, Theorem 2 is incomparable with the result of Sherstov (Corollary 4.5 of [17]).

Using Theorem 2, for a prime p, we show that if f has small degree when expressed as a polynomial over $\mathbb{F}_p$ (denoted by $\deg_p(f)$), the quantum communication complexity of F is large.

Theorem 3. *Fix a prime p. Let $f : \{0,1\}^n \to \{-1,1\}$ where f depends on all the variables. Let $F(x,y) = f(x \wedge y)$. For any $0 < \epsilon < 1$ if $\deg_p(f) \leq (1-\epsilon)\log n$, then we have $Q^*_{1/3}(F) = \Omega\left(\frac{n^{\epsilon/2}}{\log n}\right)$.*

Observe that, though Theorem 2 does not answer Question 1 in positive for all functions, we could show a class of Boolean function for which $D(F)$ and $Q^*_{1/3}(F)$ are polynomially related. More specifically, we show this for the set of all Boolean functions f such that there exists two distinct primes p, q with $\deg_p(f)$ and $\deg_q(f)$ are sufficiently far apart.

Theorem 4. *Let $f : \{0,1\}^n \to \{-1,1\}$ with $F(x,y) = f(x \wedge y)$. Fix $0 < \epsilon < 1$. If there exists distinct primes p, q such that $\deg_q(f) = \Omega(\deg_p(f)^{\frac{2}{1-\epsilon}})$, then $D(F) = O(Q^*_{1/3}(F)^{2/\epsilon})$.*

By the result of Gopalan *et al.* (Theorem 1.2 of [7]), any Boolean function f with $\deg_p(f) = o(\log n)$ must have $\deg_q(f) = \Omega(n^{1-o(1)})$ thereby satisfying the condition of Theorem 4. Hence, for all such functions, Theorem 4 answers Question 1 in positive. Observe that the same can also be derived from Theorem 3.

We remark that Theorem 2 is to viewed as an *improvement over the method of using block sensitivity* (Corollary 4.5 of [17]) in proving quantum communication complexity lower bounds. Theorems 3 and 4 is obtained as a consequence of this improvement hence is an application of how Boolean function parameters change under structured affine transforms (Lemma 1).

Boolean Functions Under Linear Transforms: We now restrict our study to linear transforms. Again, the aim is to design special linear transforms which transforms the parameters of interest in this study. In particular, in this case, we show linear transforms for which we can upper bound the alternation of the original function in terms of the sensitivity of the resulting function. More precisely, we prove the following lemma:

Lemma 2. *For any $f : \mathbb{F}_2^n \to \{0,1\}$, there exists an invertible linear transform $L : \mathbb{F}_2^n \to \mathbb{F}_2^n$ such that for $g(x) = f(L(x))$, $\mathsf{alt}(f) \leq 2s(g) + 1$.*

We show an application of the above result in the context of sensitivity. Nisan and Szegedy [13] showed that for any Boolean function f, $s(f) \leq 2\deg(f)^2$. However, the situation is quite different for $\deg_2(f)$ - noticing that for f being parity on n variables, $\deg_2(f) = 1$ and $s(f) = n$ - the gap can even be unbounded. Though parity may appear as a corner case, there are other functions like the Boolean inner product function[2] IP_n whose $\mathbb{F}_2$-degree is constant while sensitivity is $\Omega(n)$ thereby ruling out the possibility that $s(f) \leq$

---

[2] $\mathsf{IP}_n(x_1, x_2, \ldots, x_n, y_1, y_2, \ldots, y_n) = \sum_i x_i y_i \mod 2$.

$\deg_2(f)^2$. It is known that if f is not the parity on n variables (or its negation), $\deg_2(f) \leq \log \mathsf{sparsity}(f)$ [1]. Hence, as a structural question about the
two parameters, we ask: for f other than the parity function, is it true that
$\mathsf{s}(f) \leq \mathsf{poly}(\log \mathsf{sparsity}(f))$. Observe that IP_n has high sparsity and hence does
not rule this out. We use Lemma 2, which is in the theme of studying alternation
and sensitivity in the context of linear transformations, to improve this gap and
show that there is a family of functions where this gap is exponential.

Theorem 5. *There exists a family of functions $\{g_k \mid k \in \mathbb{N}\}$ such that $\mathsf{s}(g_k) \geq \frac{\sqrt{\mathsf{sparsity}(g_k)}}{2} - 1$.*

2 Preliminaries

In this section, we define the notations used followed by description of the classical and quantum communication models.

Define $[n] = \{1, 2, \ldots, n\}$. For $S \subseteq [n]$, define $e_S \in \{0, 1\}^n$ to be the indicator
vector of the set S. For $x, y \in \{0, 1\}^n$, we denote $x \wedge y$ (resp. $x \oplus y$) $\in \{0, 1\}^n$
as the string obtained by bitwise AND (resp. XOR) of x and y. We use x_i to
denote the ith bit of x. We now define the Boolean function parameters we use.

Sensitivity, Block Sensitivity and Certificate Complexity: For a Boolean
function $f : \{0, 1\}^n \rightarrow \{0, 1\}$ and $a \in \{0, 1\}^n$, we define, (1) the *sensitivity* of
f on a as $\mathsf{s}(f, a) = |\{i \mid f(a \oplus e_i) \neq f(a), i \in [n]\}|$, (2) the *block sensitivity* of f
on a, $\mathsf{bs}(f, a)$ to be the maximum number of disjoint blocks $\{B_i \mid B_i \subseteq [n]\}$ such
that $f(a \oplus e_{B_i}) \neq f(a)$ and, (3) the *certificate complexity* of f on a, $\mathsf{C}(f, a)$ to
be the size of the smallest set $S \subseteq [n]$ such that fixing f according to a on the
location indexed by S causes the function to become constant. For $\phi \in \{\mathsf{s}, \mathsf{bs}, \mathsf{C}\}$,
we define $\phi(f) = \max_{a \in \{0,1\}^n} \phi(f, a)$ and are respectively called the sensitivity,
the block sensitivity and the certificate complexity of f. By definition, the three
parameters are shift invariant, by which we mean $\forall\, b \in \{0, 1\}^n$, $\phi(f_b) = \phi(f)$ for
$\phi \in \{\mathsf{s}, \mathsf{bs}, \mathsf{C}\}$ where $f_b(x) \overset{\text{def}}{=} f(x \oplus b)$. Also, it can be shown that $\mathsf{s}(f) \leq \mathsf{bs}(f) \leq \mathsf{C}(f)$.

Alternation: For $x, y \in \{0, 1\}^n$, define $x \prec y$ if $\forall i \in [n]$, $x_i \leq y_i$. We define
a *chain* $\mathcal{C}$ on $\{0, 1\}^n$ as $(0^n = x^{(0)}, x^{(1)}, \ldots, x^{(n-1)}, x^{(n)} = 1^n)$ such that for all
$i \in [n]$, $x^{(i)} \in \{0, 1\}^n$ and $x^{(i-1)} \prec x^{(i)}$. We define *alternation* of f for a chain
$\mathcal{C}$, denoted $\mathsf{alt}(f, \mathcal{C})$ as the number of times the value of f changes in the chain.
We define alternation of a function $\mathsf{alt}(f)$ as $\max_{\text{chain } \mathcal{C}} \mathsf{alt}(f, \mathcal{C})$.

Degree, Decision Tree Depth and Sparsity: Every Boolean function f can
be expressed uniquely as a multilinear polynomial $p(x)$ in $\mathbb{F}[x_1, \ldots, x_n]$ over any
field $\mathbb{F}$ such that $p(x) = f(x)\ \forall x \in \{0, 1\}^n$. Fix a prime p. We denote $\deg(f)$
(resp. $\deg_p(f)$) to be the degree of the multilinear polynomial computing f over
reals (resp. $\mathbb{F}_p$). A decision tree T is a rooted tree with non-leaf nodes labeled
by variables and leaf nodes labeled by Boolean values. For any input x, starting
from the root, there is natural path leading to a leaf based on the value of the

variable queried. Depth of a decision tree is the length of the longest path from root to any leaf. A decision tree T computes f if for every input x, there is a path leading to a leaf labeled $f(x)$. We define $\mathsf{DT}(f)$ as the depth of a decision tree computing f of minimum depth. It is known that for all Boolean functions f, $\deg_p(f) \leq \deg(f) \leq \mathsf{DT}(f) \leq \mathsf{bs}(f)^3$. For more details on $\mathsf{DT}(f)$ and other related parameters, see the survey by Buhrman, de Wolf [3] and Hatami et al. [9]. Sparsity of a Boolean function $f : \{0,1\}^n \to \{-1,1\}$ (denoted by $\mathsf{sparsity}(f)$) is the number of non-zero Fourier coefficients in the Fourier representation of f. For more details on this parameter, see [14].

Communication Models: We first describe the two party classical communication model, introduced by Yao. Given a function $F : \{0,1\}^n \times \{0,1\}^n \to \{0,1\}$, Alice is given an $x \in \{0,1\}^n$ and Bob is given $y \in \{0,1\}^n$. They can communicate with each other and their aim is to compute $F(x,y)$ while communicating minimum number of bits. We call the procedure employed by Alice and Bob to computing f as the *protocol*. We define $\mathsf{D}(F)$ as the minimum cost of a deterministic protocol computing F. For more details on classical communication complexity of Boolean functions, refer [11].

We now describe the quantum communication complexity, again introduced by Yao. Similar to the classical model, both the parties have to come up with a *quantum protocol* where they communicate qubits via a quantum channel and compute f while minimizing the number of qubits exchanged (which is the *cost* of the quantum protocol) in the process. In this model, we allow protocols to have prior entanglement. We define $Q^*_{1/3}(F)$ as the minimum cost quantum protocol computing F with prior entanglement. For a precise definition of the model and the protocol cost, see Sect. 2.6 of [17] (see also [15] for a detailed treatment). The important relation that we require is the following lower bound on $Q^*_{1/3}(F)$ due to Sherstov [17].

Corollary 1 (Corollary 4.5 of [17]). *Let $f : \{0,1\}^n \to \{-1,1\}$ be given. Then for some $z \in \{0,1\}^n$, the matrix $F' = [f_z(x \wedge y)]_{x,y} = [f(\ldots, (x_i \wedge y_i) \oplus z_i, \ldots)]_{x,y}$ obeys $Q^*_{1/3}(F') = \Omega(\sqrt{\mathsf{bs}(f)})$.*

3   Warm Up: Alternation Under Shifts

In this section, as a warm-up, we study sensitivity and alternation under linear shifts (when the matrix M is the identity matrix). We introduce a parameter, *shift-invariant alternation* (salt). We then show the existence of Boolean functions whose shift-invariant alternation is exponential in its sensitivity (Proposition 1) thereby ruling out the possibility that $\mathsf{salt}(f)$ can be upper bounded by a polynomial in $\mathsf{s}(f)$ for all Boolean functions f.

Recall from the introduction that the parameters s, bs and C are shift invariant while alt is not. We define a variant of alternation which is invariant under shifts.

Definition 1 (Shift-invariant Alternation). *For $f : \{0,1\}^n \to \{0,1\}$, the shift-invariant alternation (denoted by $\mathsf{salt}(f)$) is defined as $\min_{b \in \{0,1\}^n} \mathsf{alt}(f_b)$.*

A Family of Functions with $\mathsf{salt}(f) = \Omega(2^{s(f)})$**:** We now exhibit a family of functions $\mathcal{F}$ where for all $f \in \mathcal{F}$, $\mathsf{salt}(f) \geq 2^{s(f)}$ thereby ruling out the possibility that $\mathsf{salt}(f)$ can be upper bounded by a polynomial in $\mathsf{s}(f)$. The family $\mathcal{F}$ is the same class of Boolean functions for which alternation is at least exponential in sensitivity due to [6].

Definition 2 (Definition 1 from [6]. See also Proof of Lemma A.1 of [8]).
Consider the family defined as follows: $\mathcal{F} = \{f_k \mid f_k : \{0,1\}^{2^k-1} \to \{0,1\}, k \in \mathbb{N}\}$. *The Boolean function* f_k *is computed by a decision tree which is a full binary tree of depth k with 2^k leaves. A leaf node is labeled as 0 (resp. 1) if it is the left (resp. right) child of its parent. All the nodes (except the leaves) are labeled by a distinct variable.*

We remark that Gopalan *et al.* [8] demonstrates an exponential lower bound on tree sensitivity (introduced by them as a generalization of the parameter sensitivity) in terms of decision tree depth for the same family of functions in Definition 2. We remark that, in general, lower bound on tree sensitivity need not imply a lower bound on alternation. For instance, if we consider the Majority function[3] Maj_n, the tree sensitivity can be shown to be $\Omega(n)$ while alternation is 1.

The authors [6] have shown that for any $f \in \mathcal{F}$, there exists of a chain of large alternation in f. However, this is not sufficient to argue existence of a chain of large alternation under *every* linear shift. We now proceed to state an exponential lower bound on $\mathsf{salt}(f)$ in terms of $s(f)$ for all $f \in \mathcal{F}$. The proof is omitted due to lack of space.

Proposition 1. *For* $f_k \in \mathcal{F}$, $\mathsf{salt}(f_k) \geq 2^{\Omega(\mathsf{s}(f_k))}$.

A Family of Functions with $\mathsf{bs}(f) = \Omega(\mathsf{s}(f)\mathsf{salt}(f))$**:** Lin and Zhang [12] showed that for any Boolean function $f : \{0,1\}^n \to \{0,1\}$, $\mathsf{bs}(f) = O(\mathsf{alt}(f)^2\mathsf{s}(f))$. This immediately gives the following proposition.

Proposition 2. *For any* $f : \{0,1\}^n \to \{0,1\}$, $\mathsf{bs}(f) \leq O(\mathsf{salt}(f)^2\mathsf{s}(f))$.

We now exhibit a family of functions for which $\mathsf{bs}(f)$ is $\Omega(\mathsf{s}(f)\mathsf{salt}(f))$. Before proceeding, we show a tight composition result for alternation of Boolean functions when composed with OR_k (which is the k bit Boolean OR function). For functions $f_1, \ldots, f_k$ where each $f_i : \{0,1\}^n \to \{0,1\}$, define the function $OR_k \circ \overline{f} : \{0,1\}^{nk} \to \{0,1\}$ as $\vee_{i=1}^k f_i(x^{(i)})$ where for each $i \in [k]$, $x^{(i)} = (x_1^{(i)}, \ldots, x_n^{(i)}) \in \{0,1\}^n$ is input to the function f_i.

Lemma 3. *Consider k Boolean functions* $f_1, \ldots, f_k$ *where each* $f_i : \{0,1\}^n \to \{0,1\}$ *satisfy,* $f_i(0^n) = f_i(1^n) = 0$. *Then,* $\mathsf{alt}(OR_k \circ \overline{f}) = \sum_{i=1}^k \mathsf{alt}(f_i)$.

Using Lemma 3 we now argue the existence of a family of Boolean functions where $\mathsf{bs}(f) = \Omega(\mathsf{s}(f)\mathsf{salt}(f))$. The proof of the proposition has been omitted due to lack of space.

---

[3] $\mathsf{Maj}_n(x) = 1 \iff \sum_i x_i \geq \lceil n/2 \rceil$.

Proposition 3. *There exists a family of Boolean functions for which* $\mathsf{bs}(f) \geq \frac{\mathsf{s}(f) \cdot \mathsf{salt}(f)}{4}$.

4 Affine Transforms: Lower Bounds on Quantum Communication Complexity

In this section, we study the affine transformation in its full generality applied to block sensitivity and sensitivity, and use it to prove Theorems 3 and 4 from the introduction. We achieve this using affine transforms as our tool (Sect. 4.1), by which we derive a new lower bound for $Q^*_{1/3}(F)$ in terms of $\mathsf{bs}(f, 0^n)$ (Sect. 4.2). Using this and a lower bound on $\mathsf{bs}(f, 0^n)$ (Proposition 5), we show that for any Boolean function f, and any prime p, $Q^*_{1/3}(F) \geq \Omega\left(\frac{\sqrt{\mathsf{DT}(f)}}{\deg_p(f)}\right)$. This immediately implies that if there is a p such that $\deg_p(f)$ is constant, then $\mathsf{D}(F) \leq 2\mathsf{DT}(f) \leq O(Q^*_{1/3}(F)^2)$ thereby answering Question 1 in positive for such functions. We relax this requirement and show that if there exists distinct primes p and q for which $\deg_p(f)$ and $\deg_q(f)$ are not very close, then $\mathsf{D}(F) \leq \mathsf{poly}(Q^*_{1/3}(F))$ (Theorem 4).

4.1 Upper Bound for Block Sensitivity via Affine Transforms

In this section, we describe our main tool. Given an $f : \{0, 1\}^n \to \{0, 1\}$ and any $a \in \{0, 1\}^n$, we exhibit an affine transform $A : \mathbb{F}_2^n \to \mathbb{F}_2^n$ such that for $g(x) = f(Ax)$, $\mathsf{bs}(f, a) \leq \mathsf{s}(g, 0^n)$.

Before describing the affine transform, we note that a linear transform is already known to achieve a weaker bound of $\mathsf{bs}(f) \leq O(\mathsf{s}(g)^2)$ due to Sherstov [17].

Proposition 4 (Lemma 3.3 of [17]). *For any* $f : \mathbb{F}_2^n \to \{0, 1\}$, *there exists a linear transform* $L : \mathbb{F}_2^n \to \mathbb{F}_2^n$ *such that for* $g(x) = f(Lx)$, $\mathsf{bs}(f) = O(\mathsf{s}(g)^2)$.

We now prove Lemma 1 which describes an affine transform improving the bound on $\mathsf{bs}(f)$ in the above proposition to linear in $\mathsf{s}(g)$. This affine transform has already been used in Nisan and Szegedy (Lemma 7 of [13]) to show that $\mathsf{bs}(f) \leq 2\deg(f)^2$. Since the exact form of g is relevant in the subsequent arguments, we explicitly prove it here bringing out the structure of the affine transform that we require.

Proof (Proof of Lemma 1). Let $\mathsf{bs}(f, a) = k$ and $\{B_1, \ldots, B_k\}$ be the sensitive blocks on a. Since the blocks are disjoint, $\{B_i \mid i \in [k]\}$ viewed as vectors over $\mathbb{F}_2^n$ are linearly independent. Hence, there is a linear transform $L : \mathbb{F}_2^n \to \mathbb{F}_2^n$ such that $L(e_i) = B_i$ for $i \in [k]$.[4] Define $A(x) = L(x) \oplus a$. For $g(x) = f(A(x))$, $\mathsf{s}(g, 0^n) = |\{i \mid g(0^n) \neq g(0^n \oplus e_i), i \in [n]\}| = |\{i \mid f(a) \neq f(a \oplus L(e_i)), i \in [n]\}| = \mathsf{bs}(f, a)$ which completes the proof of main statement and **(a)**. Now, **(b)** holds as the sensitive blocks are disjoint. □

---

[4] For completeness of definition of L, for $i \notin [k]$, we define $L(e_i) = 0^n$.

4.2   From Block Sensitivity Lower Bound at 0^n to Quantum Communication Lower Bounds

We now prove Theorem 2, which lower bounds $Q^*_{1/3}(F)$ in terms of $\mathsf{bs}(f, 0^n)$.

Proof (of Theorem 2). We first state a weaker version of this result - Corollary 1 from Sect. 2. This result is based on a powerful method of proving quantum communication lower bounds due to Razborov [15] and Klauck [10], says that for a Boolean function $g : \{0,1\}^n \to \{-1,1\}$ with $G(x,y) = g(x \wedge y)$, if there exists an $z \in \{0,1\}^n$ such that $z_i = 0$ for $i \in [k]$ and $g(z \oplus e_1) = g(z \oplus e_2) = \dots = g(z \oplus e_k) \neq g(z)$, then $Q^*_{1/3}(G) = \Omega(\sqrt{k})$. This immediately implies that for any $g : \{0,1\}^n \to \{-1,1\}$, $Q^*_{1/3}(G) = \Omega\left(\sqrt{\mathsf{s}(g,0^n)}\right)$. Given an f, we now describe a $g : \{0,1\}^n \to \{-1,1\}$ such that $Q^*_{1/3}(F) \geq Q^*_{1/3}(G)$ and $Q^*_{1/3}(G) = \Omega(\sqrt{\mathsf{bs}(f,0^n)})$ as follows thereby completing the proof.

Applying Lemma 1 with $a = 0^n$ to f, we obtain $g(x) = f(x_{i_1}, x_{i_2}, \dots, x_{i_n})$. We note that F and G can be viewed as a $2^n \times 2^n$ matrix with (x,y)th entry being $f(x \wedge y)$ and $g(x \wedge y)$ respectively. By construction of g, using the observation that the matrix G appears as a submatrix of F, $Q^*_{1/3}(F) \geq Q^*_{1/3}(G)$. This observation is used in Sherstov (for instance, see proof of Theorem 5.1 of [17]) without giving details. For completeness, we give the details here. Let $S = \{i_1, \dots i_n\} \subseteq [n]$ of size k. For $j \in S$, let $B_j = \{t \mid i_t = j\}$. Hence g depends only on these k input variables of S and all the variables with indices in B_j are assigned the variable x_j. This implies that $g(x) = f(\oplus_{j \in S} x_j e_{B_j})$.

We now exhibit a submatrix of F containing G. Consider the submatrix of F with rows and columns restricted to $W = \{a_1 e_{B_1} \oplus a_2 e_{B_2} \oplus \dots a_k e_{B_k} \mid (a_1, a_2 \dots, a_k) \in \{0,1\}^k\}$. For $u, y \in W$, $F(u,y) = f(u \wedge y) = f((u_1 e_{B_1} \oplus \dots \oplus u_k e_{B_k}) \wedge (y_1 e_{B_1} \oplus \dots \oplus y_k e_{B_k}))$. But this equals $f(u_1 \wedge y_1 e_{B_1} \oplus \dots \oplus u_k \wedge y_k e_{B_k})$ since B_js are disjoint. Hence, by definition of g, $F(u,y) = f(u \wedge y) = g(u \wedge y)$.

For the g obtained, as reasoned before, we must have $Q^*_{1/3}(G) \geq \Omega(\sqrt{\mathsf{s}(g,0^n)})$. Hence, by **(a)** of Lemma 1, as $a = 0^n$, we have $Q^*_{1/3}(G) \geq \Omega(\sqrt{\mathsf{bs}(f,0^n)})$. □

4.3   Putting Them Together

We are now ready to prove Theorems 3 and 4. A critical component of our proof is the following stronger connection between $\mathsf{DT}(f)$ and $\mathsf{bs}(f, 0^n)$. Buhrman and de Wolf, in their survey [3], showed that $\mathsf{DT}(f) \leq \mathsf{bs}(f) \cdot \deg(f)^2$ where the proof is attributed to Noam Nisan and Roman Smolensky. The same proof can be adapted to show the following strengthening of their result.

Proposition 5. *For any $f : \{0,1\}^n \to \{0,1\}$, and any prime p, $\mathsf{DT}(f) \leq \mathsf{bs}(f, 0^n) \cdot \deg_p(f)^2$.*

We now give a proof of Theorems 3 and 4.

Proof (of Theorem 3). Applying Theorem 2 and Proposition 5, we have $Q_{1/3}^*(F) \geq \Omega\left(\frac{\sqrt{DT(f)}}{\deg_p(f)}\right)$. As observed in Gopalan *et al.* [7], by a modification to an argument in the proof of Nisan and Szegedy (Theorem 1 of [13]), it can be shown that $\deg(f) \geq \frac{n}{2^{\deg_p(f)}}$. Since, $DT(f) \geq \deg(f)$, we have $DT(f) \geq \frac{n}{2^{\deg_p(f)}}$. Hence, $Q_{1/3}^*(F) = \Omega\left(\frac{\sqrt{n}}{\deg_p(f)2^{\deg_p(f)/2}}\right) = \Omega\left(\frac{n^{\epsilon/2}}{(1-\epsilon)\log n}\right)$ where the last lower bound follows upon applying the bound on $\deg_p(f)$. □

As a demonstrative example, we show a weaker lower bound on quantum communication complexity with prior entanglement for the generalized inner product function $GIP_{n,k}(x,y) \stackrel{\text{def}}{=} \oplus_{i=1}^n \bigwedge_{j=1}^k (x_{ij} \wedge y_{ij})$ when $k = \frac{1}{2}\log n$. We remark that a lower bound of $\Omega(n)$ is known for the inner product function [4].

Note that $GIP_{n,k}$ can be expressed as $f \circ \wedge$, where $f(z) \stackrel{\text{def}}{=} \oplus_{i=1}^n \bigwedge_{j=1}^k z_{ij}$, with $\deg_2(f) = k$. Applying Theorem 3 with $\epsilon = 1/2$ and $p = 2$, we have $Q_{1/3}^*(GIP_{n,\frac{1}{2}\log n}) = \Omega\left(\frac{n^{1/4}}{\log n}\right)$. Though this bound is arguably weak, Theorem 3 gives a non-trivial lower bound for all those Boolean functions f with small $\deg_p(f)$ for some prime p.

Using the above results, we now prove Theorem 4.

Proof (of Theorem 4). Applying, Theorem 2 and Proposition 5, we get that for any prime t, $Q_{1/3}^*(F) \geq \Omega\left(\frac{\sqrt{DT(f)}}{\deg_t(f)}\right)$. By hypothesis, $\deg_p(f) \leq O(\deg_q(f)^{\frac{1-\epsilon}{2}}) \leq O(DT(f)^{\frac{1-\epsilon}{2}})$ implying that for $t = p$, $D(F) \leq 2DT(f) \leq O(Q_{1/3}^*(F)^{2/\epsilon})$ □

5   Linear Transforms: Sensitivity versus Sparsity

Continuing in the theme of affine transforms, in this section, we first establish an upper bound on alternation of a function in terms of sensitivity of the function after application of a suitable linear transform. Using this, we show the existence of a function whose sensitivity is asymptotically as large as square root of sparsity (see introduction for a motivation and discussion).

Proof (of Lemma 2). Let $0^n \prec x_1 \prec x_2 \ldots \prec x_n = 1^n$ be a chain $\mathcal{C}$ of maximum alternation in the Boolean hypercube of f. Since chain $\mathcal{C}$ has maximum alternation, there must be at least $(\text{alt}(f) - 1)/2$ many zeros and $(\text{alt}(f) - 1)/2$ many ones when the x_is are evaluated on f. Note that the set of n distinct inputs $x_1, x_2, \ldots, x_n$ seen as vectors in $\mathbb{F}_2^n$ are linearly independent and hence forms a basis of $\mathbb{F}_2^n$. Hence there exists a invertible linear transform $L : \mathbb{F}_2^n \to \mathbb{F}_2^n$ taking standard basis vectors to the these vectors, i.e. $L(e_i) = x_i$ for $i \in [n]$.

To prove the result, we now show that $s(g, 0^n) \geq \frac{\text{alt}(f)-1}{2}$. The neighbors of 0^n in the hypercube of g are $\{e_i \mid i \in [n]\}$ and each of them evaluates to $g(e_i) = f(L(e_i)) = f(x_i)$ for $i \in [n]$. Since there are at least $(\text{alt}(f) - 1)/2$ many zero and at least these many ones among x_is when evaluated by f, there must be at least $(\mathsf{alt}(f) - 1)/2$ many neighbors of 0^n which differ in evaluation with $g(0^n)$ (independent of the value of $g(0^n)$). Hence $\mathsf{s}(g) \geq s(g, 0^n) \geq \frac{\mathsf{alt}(f)-1}{2}$ which completes the proof. $\qquad\square$

We now describe the family of functions and argue an exponential gap between sensitivity and logarithm of sparsity, as stated in Theorem 5.

Proof (of Theorem 5). We remark that for the family of functions $f_k \in \mathcal{F}$ (Definition 2), $\mathsf{alt}(f_k) \geq 2^{(\log \mathsf{sparsity}(f_k))/2} - 1$ [6].

We now use this family $\mathcal{F}$ to describe the family of functions g_k. For every $f_k \in \mathcal{F}$, let $g_k(x) = f_k(L(x))$ such that $\mathsf{alt}(f_k) \leq 2\mathsf{s}(g_k) + 1$ as guaranteed by Lemma 2. Since, we have $\mathsf{alt}(f_k) \geq 2^{(\log \mathsf{sparsity}(f_k))/2} - 1$, it must be that $\mathsf{s}(g_k) \geq \frac{1}{2}(\mathsf{alt}(f_k) - 1) \geq \frac{1}{2}(2^{(\log \mathsf{sparsity}(f_k))/2} - 2) \geq \frac{\sqrt{\mathsf{sparsity}(f_k)}}{2} - 1$ As the parameter sparsity does not change under invertible linear transforms [14], $\mathsf{s}(g_k) \geq 0.5\sqrt{\mathsf{sparsity}(f_k)} - 1 = 0.5\sqrt{\mathsf{sparsity}(g_k)} - 1.$ $\qquad\square$

Acknowledgment. The authors would like to thank the anonymous reviewers for their constructive comments to this paper, specifically for pointing out an error in the earlier version of Theorem 2 by giving examples. See discussion after the Theorem 2 of this paper.

References

1. Bernasconi, A., Codenotti, B.: Spectral analysis of Boolean functions as a graph eigenvalue problem. IEEE Trans. Comput. **48**(3), 345–351 (1999)
2. Buhrman, H., de Wolf, R.: Communication complexity lower bounds by polynomials. In: Proceedings of the 16th Annual IEEE Conference on Computational Complexity, Chicago, Illinois, USA, 18–21 June 2001, pp. 120–130 (2001)
3. Buhrman, H., de Wolf, R.: Complexity measures and decision tree complexity: a survey. Theor. Comput. Sci. **288**(1), 21–43 (2002)
4. Cleve, R., van Dam, W., Nielsen, M., Tapp, A.: Quantum entanglement and the communication complexity of the inner product function. Theor. Comput. Sci. **486**, 11–19 (2013)
5. Cook, S.A., Dwork, C., Reischuk, R.: Upper and lower time bounds for parallel random access machines without simultaneous writes. SIAM J. Comput. **15**(1), 87–97 (1986)
6. Dinesh, K., Sarma, J.: Alternation, sparsity and sensitivity: combinatorial bounds and exponential gaps. In: Panda, B.S., Goswami, P.P. (eds.) CALDAM 2018. LNCS, vol. 10743, pp. 260–273. Springer, Cham (2018). https://doi.org/10.1007/978-3-319-74180-2_22
7. Gopalan, P., Lovett, S., Shpilka, A.: On the complexity of Boolean functions in different characteristics. In: Proceedings of the 24th Annual IEEE Conference on Computational Complexity, CCC 2009, Paris, France, 15–18 July 2009, pp. 173–183 (2009)
8. Gopalan, P., Servedio, R.A., Wigderson, A.: Degree and sensitivity: tails of two distributions. In: 31st Conference on Computational Complexity, CCC 2016, 29 May–1 June 2016, Tokyo, Japan, pp. 13:1–13:23 (2016)

9. Hatami, P., Kulkarni, R., Pankratov, D.: Variations on the Sensitivity Conjecture. Graduate Surveys, Theory of Computing Library, vol. 4 (2011)
10. Klauck, H.: Lower bounds for quantum communication complexity. SIAM J. Comput. **37**(1), 20–46 (2007)
11. Kushilevitz, E., Nisan, N.: Communication Complexity, 2nd edn. Cambridge University Press, Cambridge (2006)
12. Lin, C., Zhang, S.: Sensitivity conjecture and log-rank conjecture for functions with small alternating numbers. In: 44th International Colloquium on Automata, Languages, and Programming, ICALP 2017, 10–14 July 2017, Warsaw, Poland, pp. 51:1–51:13 (2017)
13. Nisan, N., Szegedy, M.: On the degree of Boolean functions as real polynomials. In: Proceedings of the 24th Annual ACM Symposium on Theory of Computing, STOC 1992, pp. 462–467. ACM, New York (1992)
14. O'Donnell, R.: Analysis of Boolean Functions. Cambridge University Press, Cambridge (2014)
15. Razborov, A.A.: Quantum communication complexity of symmetric predicates. Izv. Math. **67**(1), 145 (2003)
16. Sherstov, A.A.: The pattern matrix method for lower bounds on quantum communication. In: Proceedings of the 40th Annual ACM Symposium on Theory of Computing, Victoria, British Columbia, Canada, 17–20 May 2008, pp. 85–94 (2008)
17. Sherstov, A.A.: On quantum-classical equivalence for composed communication problems. Quantum Inf. Comput. **10**(5&6), 435–455 (2010)
18. Shi, Y., Zhu, Y.: Quantum communication complexity of block-composed functions. Quantum Inf. Comput. **9**(5), 444–460 (2009)
19. Zhang, Z., Shi, Y.: On the parity complexity measures of Boolean functions. Theor. Comput. Sci. **411**(26–28), 2612–2618 (2010)

On Exactly Learning Disjunctions
and DNFs Without Equivalence Queries

Ning Ding[✉]

School of Electronic Information and Electrical Engineering,
Shanghai Jiao Tong University, Shanghai 200240, China
dingning@sjtu.edu.cn

Abstract. In this paper we address the issue of exactly learning boolean functions. The notion of exact learning introduced by [2] endows a learner with access to oracles that can answer two types of queries: membership queries and equivalence queries, in which however, equivalence queries are unrealistically strong and cannot be really carried out. Thus we investigate exact learning without equivalence queries and provide some positive results of exactly learning disjunctions and DNFs as follows (without equivalence queries).

We present a general result for exactly properly learning disjunctions if probability mass of negative inputs and probabilities that all bits are assigned to 0 and 1 are all positive. Moreover, with at most n membership queries, we can reduce sample and time complexity.

We present a general result for exactly properly learning the class of s-DNFs with random examples, and obtain two concrete results under uniform distributions. First, the class of l-term s-DNFs with $l_1 \log 2l$-terms can be exactly learned using $O(2^{s+l_1} s \ln n)$ examples in time linear in $((\frac{2en}{s})^s, 2^{s+l_1} s \ln n)$. Second, if assume each literal appears in at most d terms, the class of l-term s-DNFs with $l_1 \log 4sd$-terms can be exactly learned using $O(2^{s+l_1} \cdot e^{\frac{l}{sd}} s \ln n)$ examples in time linear in $((\frac{2en}{s})^s, 2^{s+l_1} \cdot e^{\frac{l}{sd}} s \ln n)$.

1 Introduction

In computational learning theory one of main learning models is the PAC model [11], in which a boolean function class $\mathcal{C}$ is learnable if there is an efficient learner that, given many training random examples labeled by $C \in \mathcal{C}$, can output a hypothesis f satisfying $\Pr_x[C(x) \neq f(x)] < \epsilon$. In this model there have been many results for learning a large variety of boolean functions, such as [5,8,10,11].

Another model of exact learning introduced by [2] endows a learner with access to oracles answering two types of queries: membership queries and equivalence queries, and requires the output hypothesis should be functionally equivalent to the target function for all instances. On a membership query the learner can inquire any instance x and receive $C(x)$, and on an equivalence query the learner submits a hypothesis h and receives a differing-input x such that $h(x) \neq C(x)$ if h is not of same functionality as C. Under this definition, there

© Springer Nature Switzerland AG 2019
D.-Z. Du et al. (Eds.): COCOON 2019, LNCS 11653, pp. 153–165, 2019.
https://doi.org/10.1007/978-3-030-26176-4_13

are some results of learning DNFs such as [3,4,7,9]. It is noted that equivalence queries are unrealistically strong, underlying which the motivation is to draw PAC learning results since a function class that is exactly learnable under this definition is also PAC learnable.

Due to the unreality of equivalence queries, it is of great appeal to achieve exact learnability without equivalence queries (with random examples or membership queries). To our knowledge, there is no previous positive works towards this strong goal and thus we will try to touch this appealing problem.

Our Results. We provide some positive results of exactly learning disjunctions and DNFs (some are proper). Usually a learning result involves an accuracy ϵ and a confidence δ. Since we now deal with exact learning, there is no ϵ in all results in this paper and for simplicity we omit δ (by considering it as a constant) in this section and will specify it in the formal descriptions in later sections.

Exactly Learning Disjunctions. Exact learnability means that training examples can uniquely determine a target function. So there are two necessary conditions for achieving exact learning of disjunctions. First a set of training examples labeled by a disjunction should contain some of label 0, since otherwise there are many distinct disjunctions consistent with the examples. Precisely, let D denote any distribution over x and this necessary condition requires that for each target function C, $\Pr_{x \leftarrow D}[C(x) = 0]$ should be noticeable so that enough negative inputs of C can be sampled. Thus any set of disjunctions that is exactly learnable has to be related to D. For any number $0 < \mu_D \leq 1$, we use $\mathcal{C}_{\mu_D}$ to denote the class of all disjunctions C satisfying $\Pr_{x \leftarrow D}[C(x) = 0] \geq \mu_D$.

Second for many sampling of x from D, if some bit of x is always assigned to 0 or always to 1, there are at least two disjunctions consistent with the training examples. To see this assume C does not contain x_i and $x_i = 0$ in all given examples. Then both C and $C \vee x_i$ are consistent with the examples, which results in the impossibility of exact learning. So the second necessary condition is that each bit of x can be both assigned to 0 and 1 with positive probability under D. Let $0 < p_D < 1$ denote a upper bound for probabilities that all bits are assigned to 0 or to 1. Then our first result can be stated as follows.

Proposition 1. $\mathcal{C}_{\mu_D}$ can be exactly properly learned using $O(\frac{1}{1-p_D} \frac{1}{\mu_D} \ln n)$ random examples in time linear in $(n, \frac{1}{1-p_D} \frac{1}{\mu_D} \ln n)$ under D.

When restricted to learning s-disjunctions over x, each of which contains at most s literals, under uniform distributions U, in this scenario, $p_U = \frac{1}{2}, \mu_U \geq 2^{-s}$ and thus we have that s-disjunctions can be exactly properly learned using $O(2^s \ln n)$ examples in time linear in $(n, 2^s \ln n)$ under uniform distributions.

We then consider to get rid of p_D and reduce the sample and time complexity in Proposition 1 in the membership query model and have the second result.

Proposition 2. $\mathcal{C}_{\mu_D}$ can be exactly properly learned using $O(\frac{1}{\mu_D})$ random examples and at most n membership queries in time linear in $(n, \frac{1}{\mu_D})$ under D.

Exactly Learning DNFs. We then consider exact learning of DNFs. By viewing each term (i.e. a conjunction over x) of a target DNF as a variable, the DNF is a disjunction of all its terms. This shows we can employ Proposition 1 to learn it. We use $\mathcal{F}_{s,\mu_D}$ to denote the class of all s-DNFs C satisfying $\Pr_{x \leftarrow D}[C(x) = 0] \geq \mu_D > 0$, in which an s-DNF is a DNF in which each term contains at most s literals. Also there are totally less than $(\frac{2en}{s})^s$ terms of at most s-literals over x, and let $0 < p'_D < 1$ denote a upper bound for probabilities that all these terms are equal to 0 or 1. Then our general result for learning DNFs is as follows.

Proposition 3. $\mathcal{F}_{s,\mu_D}$ can be exactly properly learned using $O(\frac{1}{1-p'_D} \frac{1}{\mu_D} \cdot s \ln n)$ examples in time linear in $((\frac{2en}{s})^s, \frac{1}{1-p'_D} \frac{1}{\mu_D} \cdot s \ln n)$ under D.

Then we focus on uniform distributions U, under which $p'_U \leq 1 - 2^{-s}$. Then we would like to quantify μ_U. However, as we will show in general for an s-DNF C, the value of $\Pr_{x \leftarrow U}[C(x) = 0]$ can be quite arbitrary, ranging from 2^{-n} to $\frac{1}{2}$. So there is no meaningful quantity for μ_U which can be substituted in Proposition 3 even under uniform distributions. So we consider subclasses of s-DNFs, for which meaningful quantities for μ_U can be given.

The first one is the class of all l-term s-DNFs, each in which has at most $l_1 \log 2l$-terms and we denote it by $\mathcal{F}_{s,l,l_1}$. For instance, for $l = \mathrm{poly}(n)$, $\mathcal{F}_{s,l,l_1}$ consists of all l-term s-DNFs, each of which contains $l_1 O(\log n)$-terms. (When l_1 is not large, we can roughly say that $\mathcal{F}_{s,l,l_1}$ consists of all l-term s-DNFs, in each of which there are not many narrow terms.) So we will show for any $C \in \mathcal{F}_{s,l,l_1}$, either C is constant-1 or $\Pr[C(x) = 0] \geq 2^{-l_1-1}$, which brings $\mu_U \geq 2^{-l_1-1}$ for those non-constant-1 functions in $\mathcal{F}_{s,l,l_1}$. Then we have the following result.

Proposition 4. $\mathcal{F}_{s,l,l_1}$ can be exactly learned using $O(2^{s+l_1} s \ln n)$ examples in time linear in $((\frac{2en}{s})^s, 2^{s+l_1} s \ln n)$ under uniform distributions.

We then investigate whether this result can be improved. It can be seen that one way to improve it is to make it still hold while l_1 could denote the number of those terms containing literals fewer than $\log 2l$. We will show constant 2 in the bound $\log 2l$ is not essential and to make the argument sound, any one more than 1 works and however any one less than 1 cannot. That is, the bound cannot be reduced smaller, say $\log \frac{l}{10}$, $\log \frac{l}{\sqrt{n}}$.

Then we introduce a condition that each literal appears in at most d terms and call this d-appearance. We remark that this condition is mild since usually a literal will not appear in all terms. Then we employ the Local Lemma to deduce that any such l-term s-DNF C with $l_1 \log 4sd$-terms is either constant-1 or $\Pr[C(x) = 0] \geq 2^{-l_1} \cdot e^{-\frac{l_2}{sd}}$. This brings the following result.

Proposition 5. l-term s-DNFs with d-appearance and $l_1 \log 4sd$-terms can be exactly learned using $O(2^{s+l_1} \cdot e^{\frac{l}{sd}} s \ln n)$ examples in time linear in $((\frac{2en}{s})^s, 2^{s+l_1} \cdot e^{\frac{l}{sd}} s \ln n)$ under uniform distributions.

When $\frac{l}{sd}$ is a large constant, the sample and time complexity are of same magnitude as Proposition 4, while now the bound $\log 4sd$ can be less than $\log l$.

Organization. In Sect. 2 we present preliminaries. In Sects. 3 and 4 we present the results of learning disjunctions and DNFs respectively.

2   Preliminaries

We use $[n]$ to denote the integers in $[1, n]$. For any set S, we use $|S|$ to denote its cardinal number. For any two sets S_1, S_2, we use $S_1 + S_2$ to denote $S_1 \cup S_2$ if their intersection is empty.

A literal denotes a boolean variable or its negation. A disjunction of x is an OR of some literals of $(x_1, \cdots, x_n, \overline{x}_1, \cdots, \overline{x}_n)$, where $\overline{x}_i$ denotes the negation of x_i, equal to $1 - x_i$, $1 \le i \le n$. An s-disjunction is an OR of at most s literals.

A DNF is an OR of some conjunctions, each of which is an AND of some literals of $(x_1, \cdots, x_n, \overline{x}_1, \cdots, \overline{x}_n)$. An s-conjunction is an AND of at most s literals. We also call a conjunction a term. If a DNF contains at most l terms, we say it is a l-term DNF, and moreover, if each of all its conjunctions contains at most s literals, we say it is a l-term s-DNF. We also call s the width of the conjunctions and DNF.

Chernoff Bound. Let $X = \sum_{i=1}^{n} X_i$ where all X_i are independently distributed in $[0, 1]$. Then for any $0 < \epsilon < 1$, $\Pr[X < (1 - \epsilon)\mathbf{E}[X]] \le e^{-\frac{\epsilon^2}{2}\mathbf{E}[X]}$.

2.1   Exact Learning Without Equivalence Queries

Let $\mathcal{C}$ denote a class of boolean functions. In the random examples model, a labeled example is a pair $(x, f(x))$, where $x \in \{0, 1\}^n, f \in \mathcal{C}$. A training sample labeled by f is of the form $((x^1, f(x^1)), \cdots, (x^m, f(x^m)))$.

Definition 1 (Exact Learning). An algorithm L is called an exact learner for $\mathcal{C}$ under distribution D over $\{0, 1\}^n$, if it is given a training sample in which each x is sampled from D independently and its label is $f(x)$ for some unknown $f \in \mathcal{C}$, $\delta \in (0, 1)$, with probability at least $1 - \delta$, L outputs a function h such that $f(x) = h(x)$ for all $x \in \{0, 1\}^n$. If h is also in $\mathcal{C}$, we say L properly learns $\mathcal{C}$. We refer to δ as the confidence parameter.

If L is additionally given oracle access to f and it can inquire f with any input x and receives $f(x)$, we call L exactly learns $\mathcal{C}$ with membership queries.

3   Exactly Properly Learning Disjunctions

In Sect. 3.1 we present the exactly learning result for disjunctions with random examples. In Sect. 3.2 we present the result with membership queries.

3.1   Learning Disjunctions with Random Examples

Recall that [11] shows how to PAC properly learning disjunctions to any error ϵ (originally it aims at conjunctions but we adapt its idea to disjunctions). That is, initially set the unknown target function to be the OR of all literals. Then given many labeled examples, scan each one with label 0, in which if a literal is assigned to 1, then delete it from the OR representation. Lastly the remaining disjunction (which is an OR of literals including ones in the target function) is consistent with all examples with label 0. Since including more literals in the OR will not affect label 1, it is also consistent with all examples with label 1. Then using the VC theory to specify sample complexity and running this strategy, the output hypothesis is with desired accuracy.

We extend this strategy by further deleting those literals in the remaining representation but not in the target function. Assume each variable is assigned to 0 and to 1 with positive probability. If a literal appears with values both 0 and 1 among the examples with label 0, it cannot be in the target function, since one of its assignments 0/1 results in label 1. In contrast, a literal in the target function should always appear with value 0 among these examples.

Thus we have there are two necessary conditions for exact learning. Let C be the target function and D be the unknown distribution over x. The first condition is that the negative points of C have positive probability mass, i.e. $\Pr_{x \leftarrow D}[C(x) = 0] > 0$, and actually it should be at least noticeably more than 0 such that many examples with label 0 can be sampled. Otherwise, the given random examples are all with label 1 and thus there are more than one disjunction consistent with them and thus we definitely cannot exactly learn C. The second condition is that for $x \leftarrow D$, each literal should have positive probability mass on 0 and 1, since otherwise among all examples with label 0 there is no way to distinguish it from the very literals in C, resulting in the impossibility of exact learning. We specify the two conditions more precisely as follows.

Let C be a disjunction of some literals of $(x_1, \cdots, x_n, \overline{x}_1, \cdots, \overline{x}_n)$, where x_i denotes the ith bit of x and $\overline{x}_i$ denotes $1 - x_i$, $1 \le i \le n$. Let $S_1 \subset [n]$ consist of all those $i \in [n]$ satisfying x_i appears in C, and $S_2 \subset [n]$ consist of those $i \in [n]$ satisfying $\overline{x}_i$ appears in C. W.l.o.g. $S_1 \cap S_2 = \emptyset$. So $C(x) = \vee_{i \in S_1} x_i \vee \vee_{i \in S_2} \overline{x}_i$, where $\vee$ and $\bigvee$ denote OR. We introduce notations $\mu_D, \mathcal{C}_{\mu_D}$ and p_D as follows.

- For any $0 < \mu_D \le 1$, let $\mathcal{C}_{\mu_D}$ denote the class of all disjunctions C satisfying $\Pr_{x \leftarrow D}[C(x) = 0] \ge \mu_D$.
- Let $p_i = \Pr_{x \leftarrow D}[x_i = 1]$ and $1 - p_i = \Pr_{x \leftarrow D}[x_i = 0]$ for each i. Let p_D be $\max_{i \in [n]}(\max(p_i, 1 - p_i))$. We require $0 < p_D < 1$.

So when given random examples from D with $0 < p_D < 1$, what we consider to learn is $\mathcal{C}_{\mu_D}$. Let δ denote the confidence parameter. Let $(x^1, b^1), \cdots, (x^m, b^m)$ be m examples where $x^k \leftarrow D$ and $b^k = C(x^k)$, $k \in [m]$ and C is the unknown target disjunction. We first present the following claim.

Claim 1. In $m = O(\frac{1}{1-p_D} \frac{1}{\mu_D} \ln \frac{n}{\delta})$ examples (the constant in O is larger than 8), with probability $1 - O(\frac{\delta}{n})$, there are at least $O(\frac{1}{1-p_D} \ln \frac{n}{\delta})$ ones with label 0.

Proof. Let $n' = \frac{m\mu_D}{2}$. Define a random variable $\xi_k = 1$ if $C(x^k) = 0$ and $\xi_k = 0$ else for each k. Let $Y = \sum_{k=1}^m \xi_k$ and $\mathbf{E}[Y] \geq 2n'$. By the Chernoff bound, for any $0 < \epsilon' < 1$, $\Pr[Y < (1 - \epsilon')\mathbf{E}[Y]] < e^{-\frac{\epsilon'^2}{2}\mathbf{E}[Y]}$.

Choose $\epsilon' = \frac{1}{2}$. Then $\Pr[Y < n'] < e^{-\frac{1}{4}n'} \leq (\frac{\delta}{n})^{\frac{1}{1-p_D}} = O(\frac{\delta}{n})$. Since $n' = O(\frac{1}{1-p_D} \ln \frac{n}{\delta})$, the claim holds.   □

So in the following we take $m = O(\frac{1}{1-p_D} \frac{1}{\mu_D} \ln \frac{n}{\delta})$. By Claim 1, there are at least $n' = O(\frac{1}{1-p_D} \ln \frac{n}{\delta})$ examples with label 0. W.l.o.g. denote these examples by $(x^1, 0), \cdots, (x^{n'}, 0)$. Then we have the following fact.

Claim 2. Given $n' = O(\frac{1}{1-p_D} \ln \frac{n}{\delta})$ training examples $(x^k, 0)$ for $k \in [n']$, if $i \in S_1$ then $x_i^k = 0$ for all k (x_i^k denotes the ith bit of x^k), if $i \in S_2$ then $x_i^k = 1$ for all k, and if $i \notin S_1 + S_2$, then except for probability $\frac{\delta}{2n}$, there are two distinct $j, k \in [n']$ such that $x_i^j = 1$ and $x_i^k = 0$.

Proof. Since $C(x) = \bigvee_{i \in S_1} x_i \bigvee \bigvee_{i \in S_2} \overline{x}_i$, the results for $i \in S_1$ and $i \in S_2$ are obvious. Then we consider the case that $i \notin S_1 + S_2$. In this case values of x_i are independent of labels. Since each such bit x_i is valued with 0 or 1 in a random example with probability bounded by p_D, denoted $1 - \xi$ for some ξ, it is always valued with 0 or always valued with 1 in all the examples with probability $(1 - \xi)^{O(\frac{1}{1-p_D} \ln \frac{n}{\delta})} = (1 - \xi)^{O(\frac{1}{\xi} \ln \frac{n}{\delta})} = (\frac{\delta}{n})^{O(1)} \leq \frac{\delta}{2n}$. The clam holds.   □

By Claim 2 we have that for all $i \notin S_1 + S_2$, taking the union failure bounds, except for probability $\delta/2$, x_i can be valued both 0 and 1 in the examples. Thus by checking all values of x_i^k for $k \in [n'], i \in [n]$, we can determine whether i is in S_1 or S_2 or not in $S_1 + S_2$. That is, for each i: if the values of x_i^k are 0 for all k then $i \in S_1$; if the values of x_i^k are 1 for all k then $i \in S_2$; otherwise, i is not in $S_1 + S_2$. Lastly, we can recover C as $C(x) = \bigvee_{i \in S_1} x_i \bigvee \bigvee_{i \in S_2} \overline{x}_i$.

We remark that the total failure probability comes from Claim 2 and Claim 1 which is less than δ, and the learning strategy only needs to scan all examples to make decisions and thus runs in linear time. Thus we have the following result.

Proposition 6. For any distribution D with $0 < p_D < 1$, the class $\mathcal{C}_{\mu_D}$ with $\mu_D > 0$ can be exactly properly learned to any confidence δ using $O(\frac{1}{1-p_D} \frac{1}{\mu_D} \ln \frac{n}{\delta})$ examples in time linear in $(n, \frac{1}{1-p_D} \frac{1}{\mu_D} \ln \frac{n}{\delta})$ under D.

Then consider a specific case that D is the uniform distribution U and C is an s-disjunction over x for any $s \leq n$, i.e. $|S_1 + S_2| \leq s$. Thus $\mu_U \geq 2^{-s}$ and $p_U = \frac{1}{2}$ for the class of s-disjunctions. Thus we have the following corollary.

Corollary 1. The class of s-disjunctions can be exactly properly learned to any confidence δ using $O(2^s \ln \frac{n}{\delta})$ examples in time linear in $(n, 2^s \ln \frac{n}{\delta})$ under uniform distributions.

3.2   Learning with Membership Queries

Now we try to get rid of the parameter p_D and reduce the sample and time complexity in Proposition 6 in the membership query model. Since for $C \in \mathcal{C}_{\mu_D}$, for any given $\frac{1}{\mu_D} \ln \frac{1}{\delta}$ random examples, no one is labeled with 0 with probability $(1 - \mu_D)^{\frac{1}{\mu_D} \ln \frac{1}{\delta}} < \delta$. Thus with probability $1 - \delta$, there is at least an example with label 0. Denote it by $(x^j, 0)$.

Recall $C(x) = \bigvee_{i \in S_1} x_i \bigvee \bigvee_{i \in S_2} \overline{x}_i$. Exactly learning C is equivalent to deciding the membership of all i in S_1, S_2. For each i, if $x_i^j = 0$ then either $i \in S_1$ or $i \notin S_1 + S_2$, and if $x_i^j = 1$ then either $i \in S_2$ or $i \notin S_1 + S_2$.

Then membership queries can help for further deciding the membership of all i. That is, for any i belonging to the first case, the learner inquires the oracle with y where y equals x^j but with x_i^j flipped. If the oracle replies with 0, i cannot be in S_1 and thus does not belong to $S_1 + S_2$, and if it replies with 1, i must belong to S_1. Similarly, for any i belonging to the second case, the learner inquires the oracle with y where y equals x^j but with x_i^j flipped. If the oracle replies with 0, i cannot be in S_2 and thus does not belong to $S_1 + S_2$, and if it replies with 1, i must belong to S_2.

Thus by proposing at most n queries, the learner can decide membership of all i in S_1, S_2. So we have the following result.

Proposition 7. For any distribution D, the class $\mathcal{C}_{\mu_D}$ with $\mu_D > 0$ can be exactly properly learned to any confidence δ using $\frac{1}{\mu_D} \ln \frac{1}{\delta}$ examples and at most n membership queries in time linear in $(n, \frac{1}{\mu_D} \ln \frac{1}{\delta})$ under D.

4   Exactly Learning s-DNFs

In this section we present the results of exactly learning s-DNFs. In Sect. 4.1 we give a general result. In Sects. 4.2 and 4.3 we present two concrete results when s-DNFs satisfy mild conditions under uniform distributions.

4.1   Learning s-DNFs Under Any Distributions

Let $C : \{0, 1\}^n \to \{0, 1\}$ denote a l-term s-DNF. That is, C is an OR of at most l conjunctions of width at most s. Let $C_1, \cdots, C_l$ denote its l terms, each of which is a conjunction of at most s literals of x.

Note there are totally less than $\sum_{i=0}^{s} \binom{2n}{i} \leq (\frac{2en}{s})^s$ distinct s-conjunctions over x. We denote them by $C_1', \cdots, C_N'$ for $N \leq (\frac{2en}{s})^s$, and all C_i of C belong to them. Viewing each C_i' as a variable, C is a disjunction of l ones of $\{C_i' : i \in [N]\}$. So we can apply the result of learning disjunctions to exactly learn C.

Let D be any distribution over x. To apply Proposition 6, we need to refer to the counterparts of p_D, μ_D therein. That is, let μ_D be any number in $(0, 1]$ and $\mathcal{F}_{s,\mu_D}$ denote the class of all s-DNFs C satisfying $\Pr_{x \leftarrow D}[C(x) = 0] \geq \mu_D$. So for any D, what we attempt to learn is $\mathcal{F}_{s,\mu_D}$.

Similarly assume $\Pr_x[C'_i = 1] = p_i$ and $\Pr_x[C'_i = 0] = 1 - p_i$ for $i \in [N]$. Then let p'_D denote $\max_{i \in [N]}(\max(p_i, 1 - p_i))$. ($p'_D$ is the counterpart of p_D, since all C'_i are viewed as variables.)

Note that to apply Proposition 6 it requires $p'_D < 1$. For $p'_D < 1$, both $p_i, 1 - p_i$ should be less than 1 for all i, which means no C'_i is a constant. Since each C'_i is an AND of at most s literals of x, for C'_i not being a constant, it suffices to require that no bit of x can be determined by some other bits. In the statement of the result below, we just require that it should be satisfied $p'_D < 1$ for D instead of focusing on various specific cases of D satisfying $p'_D < 1$.

Lastly, the n in Proposition 6 is now changed to $N \le (\frac{2en}{s})^s$. Since $C \in \mathcal{F}_{s,\mu_D}$ can be viewed as a disjunction of l variables C_i out of N ones, we can exactly learn C by the proposition. Note that the learned C is an OR of some C'_i, indicating it is still an s-DNF, and it is exactly identical to the original C. So we have the following result.

Proposition 8. The class of s-DNFs $\mathcal{F}_{s,\mu_D}$ with $\mu_D > 0$ can be exactly properly learned to any confidence δ using $O(\frac{1}{1-p'_D} \frac{1}{\mu_D} \cdot s \ln \frac{n}{\delta})$ examples in time linear in $((\frac{2en}{s})^s, \frac{1}{1-p'_D} \frac{1}{\mu_D} \cdot s \ln \frac{n}{\delta})$ under D with $p'_D < 1$.

Also turn to the specific case that D is the uniform distribution U. Let us evaluate p'_U. For any s-conjunction C'_i, since C'_i contains at most s literals, $\Pr[C'_i = 1] \ge 2^{-s}, \Pr[C'_i = 0] \le 1 - 2^{-s}$. Thus $p'_U \le 1 - 2^{-s}$.

As for μ_U, in general for an s-DNF C, the value of $\Pr_x[C(x) = 0]$ can range broadly. For instance, if $C = x_i$ then $\Pr_x[C(x) = 0] = \frac{1}{2}$, and if $C = \bigvee_{i=1}^m x_i$ then $\Pr_x[C(x) = 0] = \frac{1}{2^m}$ for any $m \in [n]$. This shows we cannot provide a meaningful lower bound to substitute μ_U under uniform distributions.

Conversely, given any $2^{-n} < \mu_U < 1/2$, choose $m \approx \log \frac{1}{\mu_U}$ and thus any disjunction C over m literals, considered as a 1-DNF, admits $\Pr_x[C(x) = 0] \approx \mu_U$. This asserts the existence of s-DNFs $\mathcal{F}_{s,\mu_U}$ for quite arbitrary μ_U. So we have the following result which still needs to contain the parameter μ_U.

Proposition 9. The class of s-DNFs $\mathcal{F}_{s,\mu_U}$ with $\mu_U > 0$ can be exactly properly learned to any confidence δ using $O(2^s \frac{1}{\mu_U} s \ln \frac{n}{\delta})$ examples in time linear in $((\frac{2en}{s})^s, 2^s \frac{1}{\mu_U} s \ln \frac{n}{\delta})$ under uniform distributions.

In the following two subsections we will present concrete quantities for μ_U when s-DNFs satisfy some mild conditions.

4.2 Learning s-DNFs with Not Many Narrow Terms Under Uniform Distributions

We consider how to provide a lower bound for μ_U when C satisfies a mild condition. Recall $C = \bigvee_{i=1}^l C_i$ and each C_i is a conjunction of at most s literals and $l < (\frac{2en}{s})^s$. Let l_1 denote the number of those C_i which widths are at most $\log 2l$ (the constant 2 is not essential and any one more than 1 also works). Let $\mathcal{F}_{s,l,l_1}$

denote the class of all l-term s-DNF C with at most $l_1 \log 2l$-terms (when l_1 is not large, $\mathcal{F}_{s,l,l_1}$ consists of all C which do not have many narrow terms).

We estimate μ_U for $\mathcal{F}_{s,l,l_1}$. For each C in the class, first consider C is not constant-1. Divide all the conjunctions C_i of C to two parts in which one consists of those C_i's each of which contains at most $\log 2l$ literals and the other consists of those C_i's each of which contains more literals. Then for each C_i in the second part, $\Pr_x[C_i = 1] < 2^{-\log 2l} = \frac{1}{2l}$. Thus the total probability that at least one C_i in the second part is 1 for uniform x is less than $\frac{1}{2}$. Namely, all C_i in the second part are 0 with probability at least $\frac{1}{2}$. W.l.o.g. denote these C_i by $D_1, \cdots, D_{l_2}$ for some $l_2 \leq l$, and $\Pr_{x \leftarrow U}[D_i = 0, \forall i \in [l_2]] \geq \frac{1}{2}$. If the second part is empty, just consider $D_i = 0, \forall i \in [l_2]$ with probability 1.

Let $C_1, \cdots, C_{l_1}$ denote those C_i in the first part. Then we have the following claim that estimates the conditional probability that all $C_1, \cdots, C_{l_1}$ are 0 when all $D_1, \cdots, D_{l_2}$ are 0 for uniform x.

Claim 3. For any non-constant-1 $C \in \mathcal{F}_{s,l,l_1}$, if the first part of C is not empty, $\Pr_{x \leftarrow U}[C_j = 0, \forall j \in [l_1] | D_i = 0, \forall i \in [l_2]] \geq \frac{1}{2^{l_1}}$.

Proof. First we have the following formula (in which $[0]$ denotes the empty set).

$$\Pr_{x \leftarrow U}[C_j = 0, \forall j \in [l_1] | D_i = 0, \forall i \in [l_2]]$$
$$= \prod_{k=1}^{l_1} \Pr_{x \leftarrow U}[C_k = 0 | D_i = 0, \forall i \in [l_2], C_j = 0, \forall j \in [k-1]] \tag{1}$$

We consider $\Pr_{x \leftarrow U}[C_k = 0 | D_i = 0, \forall i \in [l_2], C_j = 0, \forall j \in [k-1]]$ for each k. If the condition that $D_i = 0, \forall i \in [l_2], C_j = 0, \forall j \in [k-1]$ can completely determine the value of C_k, then on this condition C_k is always 1 or always 0. Since C is not constant-1, C_k cannot be 1. In fact, when the conditional event occurs, if $C_k = 1$, then $C = 1$. When the event does not occur, there is one of $\{D_i, C_j, \forall i \in [l_2], \forall j \in [k-1]\}$ which is 1, which also ensures $C = 1$. Thus $C_k = 0$ with probability 1 on the occurrence of the conditional event.

Otherwise, that $D_i = 0, \forall i \in [l_2], C_j = 0, \forall j \in [k-1]$ cannot determine the value of C_k. This means there are at least two possibilities for the values of the involved literals in C_k. Since there is one which results in $C_k = 1$ (note $C_k = 1$ if and only if all involved literals are 1), we have $\Pr_{x \leftarrow U}[C_k = 0 | D_i = 0, \forall i \in [l_2], C_j = 0, \forall j \in [k-1]] \geq \frac{1}{2}$.

Combining the two cases, the value of Eq. 1 is at least $\prod_{k=1}^{l_1} \frac{1}{2} = \frac{1}{2^{l_1}}$. $\square$

Claim 4. For any C in $\mathcal{F}_{s,l,l_1}$, either C is constant-1 or $\Pr_{x \leftarrow U}[C(x) = 0] \geq 2^{-l_1-1}$.

Proof. If C is not constant-1, then when the first part of C is not empty, by Claim 3, $\Pr_{x \leftarrow U}[C = 0] = \Pr_{x \leftarrow U}[D_i = 0, \forall i \in [l_2]] \cdot \Pr_{x \leftarrow U}[C_j = 0, \forall j \in [l_1] | D_i = 0, \forall i \in [l_2]] \geq 2^{-l_1-1}$. When the first part is empty, $\Pr[C = 0] = \Pr[D_i = 0, \forall i \in [l_2]] \geq \frac{1}{2}$. The claim holds. $\square$

So $\mathcal{F}_{s,l,l_1}$ excluding constant-1 functions admits $\mu_U \geq 2^{-l_1-1}$. By Proposition 9, any non-constant-1 C in $\mathcal{F}_{s,l,l_1}$ can be exactly properly learned to any confidence δ using $O(2^{s+l_1}s\ln\frac{n}{\delta})$ examples in time linear in $((\frac{2en}{s})^s, 2^{s+l_1}s\ln\frac{n}{\delta})$ under uniform distributions.

On the other hand, if C is constant-1, for $O(2^{s+l_1}s\ln\frac{n}{\delta})$ random examples, they are always labeled with 1. When detecting this, we can output a constant-1 function as the output hypothesis.

Thus a learner can learn C as follows. When it finds all given random examples are of label 1, it outputs a constant-1 function and otherwise it adopts the strategy in Sect. 4.1 to obtain a hypothesis. If C is constant-1, then the learner learns always exactly. If C is not, these examples contain many with label 0 with probability $1-\delta$, and on this condition it learns C exactly as Proposition 9 shows. So (substituting δ with $\frac{\delta}{2}$) we have the following result.

Proposition 10. The class of $\mathcal{F}_{s,l,l_1}$ can be exactly learned to any confidence δ using $O(2^{s+l_1}s\ln\frac{n}{\delta})$ examples in time linear in $((\frac{2en}{s})^s, 2^{s+l_1}s\ln\frac{n}{\delta})$ under uniform distributions.

Although Proposition 10 does not contain l explicitly in sample and time complexity, it is actually embodied since $l < (\frac{2en}{s})^s$. Note the learned hypothesis may not be in $\mathcal{F}_{s,l,l_1}$. We then present concrete quantities of sample and time complexity for different magnitudes of s (consider constant l_1 for simplicity and the quantities when $l_1 = O(\log n)$ or $l_1 = n^\alpha$ for $\alpha < 1$ can be given similarly).

- If $s = \log^c n$ for some integer $c > 0$, the sample complexity and learning-time are now $O(2^{\log^c n}\log^c n\ln\frac{n}{\delta})$ and linear in $(n^{\log^c n}, 2^{\log^c n}\log^c n\ln\frac{n}{\delta})$.
- If $s = O(n)$, say $\frac{n}{16}$, the sample complexity is $O(2^{\frac{n}{16}}n\ln\frac{n}{\delta})$ and the learning-time is less than a linearity in $(2^{\frac{7}{16}n}, n\ln\frac{n}{\delta})$ since $(\frac{2en}{s})^s = (32e)^{n/16} < 2^{\frac{7}{16}n}$. This means that all $\frac{n}{16}$-DNFs with a constant number of terms of width $\log 2l$ can be exactly learned roughly using $O(2^{\frac{n}{16}})$ examples in time $O(2^{\frac{7}{16}n})$ (in contrast to exactly determining such DNFs via total 2^n inputs).
- Consider the most general case of $s = n$. Then the sample complexity is $O(2^n n\ln\frac{n}{\delta})$. Now N is the number of all distinct conjunctions over x ($2n$ literals), which is in the magnitude of $2^{1.5n}$. Thus the learning-time is approximately a linearity in $2^{1.5n}$. (To collect statistical information via random examples instead of via membership queries, it is probably reasonable to allow time complexity more than 2^n for the most general case.)

4.3   Learning Under Bounded-Appearance of All Literals

In Sect. 4.2 we presented the learning result for $\mathcal{F}_{s,l,l_1}$, in which each C is a l-term s-DNF with l_1 terms of width at most $\log 2l$. It can be seen that one way to improve this result is to make it still hold while l_1 is only restricted to be the number of those terms of width smaller than $\log 2l$. Namely, we could still learn C when the number of its narrower terms is not many (in contrast to learning C in Sect. 4.2 when the number of its narrow terms is not many).

As shown in Sect. 4.2, the constant 2 in the width bound $\log 2l$ is not essential and to make the argument sound, any one more than 1 works, but any one less than 1 cannot work. So the width bound cannot be reduced smaller, say $\log \frac{l}{10}$, $\log \frac{l}{\sqrt{n}}$. (Although this is not notable in terms of magnitude, we think whether the constant can be reduced to less than 1 is still of theoretical interest.)

Then our task is to obtain a smaller width bound. To do this, we introduce a more condition for C that each literal appears in at most d terms, $d < l$. We call this d-appearance for all literals. We remark that this condition is mild since usually a literal will not appear in all terms.

We also divide all terms C_i of C to two parts in which one consists of those C_i that involve at most $\log 4sd$ literals and the other consists of those C_i that involve more. (We will show $4sd$ can be less than l.) Then for each C_i in the second part, $\Pr_x[C_i = 1] < 2^{-\log 4sd} = \frac{1}{4sd}$. Note that now we cannot obtain the probability that all these C_i output 1 by counting the union probability $\frac{l_2}{4sd}$ for $l_2 = l - l_1$ since it can be larger than 1. Instead, we would like to apply The Local Lemma to evaluate this probability.

Still denote these C_i in the second part by $D_1, \cdots, D_{l_2}$. For convenience, let A_i denote the event that $D_i(x) = 1$, $i \in [l_2]$. Then the probability that all D_i are 0 can be represented as $\Pr[\bigwedge_{i=1}^{l_2} \overline{A_i}]$. (If the second part of C is empty, take $\Pr[\bigwedge_{i=1}^{l_2} \overline{A_i}] = 1$.) Then we introduce the Lemma.

Theorem 5 (The Local Lemma [6]). Let $A_1, \cdots, A_n$ be events in an arbitrary probability space. Suppose each event A_i is mutually independent of a set of all the other events A_j but at most d, and $\Pr[A_i] \leq p$ for all $1 \leq i \leq n$. If $ep(d+1) \leq 1$ then $\Pr[\bigwedge_{i=1}^{n} \overline{A_i}] \geq \prod_{i=1}^{n}(1 - \frac{1}{d+1})$.

The statement of the Lemma here is referred to [1] (Lemma 5.1.1 and Corollary 5.12 in Chapter 5). Returning to our setting, in general each D_i contains at most s literals, each appearing in at most d terms. So each A_i is mutually independent of all other A_j but at most sd ones. Also $\Pr[A_i] < \frac{1}{4sd}$ for all i and $e \cdot \frac{1}{4sd} \cdot (sd + 1) < 1$. Thus by the Lemma, we have the following claim.

Claim 6. $\Pr[\bigwedge_{i=1}^{l_2} \overline{A_i}] > e^{-\frac{l_2}{sd}} \geq e^{-\frac{l}{sd}}$.

Proof. If the second part of C is empty, $\Pr[\bigwedge_{i=1}^{l_2} \overline{A_i}] = 1$. Otherwise, it follows from The Local Lemma that $\Pr[\bigwedge_{i=1}^{l_2} \overline{A_i}] \geq \prod_{i=1}^{l_2}(1 - \frac{1}{sd+1}) > e^{-\frac{l_2}{sd}}$. $\square$

Let us compare Claim 6 to the result in Sect. 4.2 that $\Pr_{x \leftarrow U}[D_i = 0, \forall i \in [l_2]] \geq \frac{1}{2}$, in the notations of A_i, i.e. $\Pr[\bigwedge_{i=1}^{l_2} \overline{A_i}] > \frac{1}{2}$. In Sect. 4.2, the width cannot be less than $\log l$, while now the width $\log 4sd$ can be less than $\log l$, i.e. $4sd < l$, and actually $\Pr[\bigwedge_{i=1}^{l_2} \overline{A_i}]$ here can be more than a constant when $\frac{l}{sd}$ is any large constant or can be $e^{-\sqrt{n}}$ when $\frac{l}{sd} = \sqrt{n}$. When $l = \mathsf{poly}(n)$, d equals $O(\mathsf{poly}(n)/s) = \mathsf{poly}(n)$, which means each literal is allowed to appear in $\mathsf{poly}(n)$ terms and so the condition is quite mild.

Following the route in Sect. 4.2, we evaluate the probability that all C_i in the first part are 0 on the condition of $\bigwedge_{i=1}^{l_2} \overline{A_i}$ for uniform x. Still let $C_1, \cdots, C_{l_1}$ denote all ones in the first part. Then we have the following claim.

Claim 7. For any l-term s-DNF C with d-appearance and l_1 terms width at most $\log 4sd$, either C is constant-1 or $\Pr_{x \leftarrow U}[C(x) = 0] \geq 2^{-l_1} \cdot e^{-\frac{l}{sd}}$.

Proof. If C is not constant-1, using the argument of Claim 3, we have $\Pr_{x \leftarrow U}[C_j = 0, \forall j \in [l_1] | D_i = 0, \forall i \in [l_2]] \geq \frac{1}{2^{l_1}}$ (if the first part is not empty). In the notations A_i, $\Pr_{x \leftarrow U}[C_j = 0, \forall j \in [l_1] | \bigwedge_{i=1}^{l_2} \overline{A_i}] \geq \frac{1}{2^{l_1}}$. Combining it with Claim 6 and Claim 4 (if the first part is empty), this claim holds.   □

Then by the argument in Sect. 4.2 we have the following result.

Proposition 11. The class of l-term s-DNFs with d-appearance and $l_1 \log 4sd$-terms can be exactly learned to any confidence δ using $O(2^{s+l_1} \cdot e^{\frac{l}{sd}} s \ln \frac{n}{\delta})$ examples in time linear in $((\frac{2en}{s})^s, 2^{s+l_1} \cdot e^{\frac{l}{sd}} s \ln \frac{n}{\delta})$ under uniform distributions.

When $\frac{l}{sd}$ is a large constant, the sample and time complexity are of same magnitude as Proposition 10, while now the width $\log 4sd$ can be less than $\log l$.

Acknowledgments. We are grateful to the reviewers of COCOON 2019 for their useful comments. This work is supported by National Cryptography Development Fund of China (Grant No. MMJJ20170128).

References

1. Alon, N., Spencer, J.H.: The Probabilistic Method, 3rd edn. Wiley, Chichester (2007)
2. Angluin, D.: Queries and concept learning. Mach. Learn. **2**(4), 319–342 (1987). https://doi.org/10.1007/BF00116828
3. Beimel, A., Bergadano, F., Bshouty, N.H., Kushilevitz, E., Varricchio, S.: On the applications of multiplicity automata in learning. In: 37th Annual Symposium on Foundations of Computer Science, FOCS 1996, Burlington, Vermont, USA, 14–16 October 1996, pp. 349–358. IEEE Computer Society (1996). https://doi.org/10.1109/SFCS.1996.548494
4. Bshouty, N.H.: Simple learning algorithms using divide and conquer. Comput. Complex. **6**(2), 174–194 (1997). https://doi.org/10.1007/BF01262930
5. Ding, N., Ren, Y., Gu, D.: PAC learning depth-3 AC^0 circuits of bounded top fanin. In: International Conference on Algorithmic Learning Theory, ALT 2017. Proceedings of Machine Learning Research, PMLR, vol. 76, 15–17 October 2017, pp. 667–680. Kyoto University, Kyoto (2017). http://proceedings.mlr.press/v76/ding17a.html
6. Erdős, P., Lovász, L.: Problems and results on 3-chromatic hypergraphs and some related questions. Infinite and Finite Sets, pp. 609–628 (1975)
7. Hellerstein, L., Raghavan, V.: Exact learning of DNF formulas using DNF hypotheses. J. Comput. Syst. Sci. **70**(4), 435–470 (2005). https://doi.org/10.1016/j.jcss.2004.10.001
8. Klivans, A.R., Servedio, R.A.: Learning DNF in time $2^{\tilde{o}(n^{1/3})}$. J. Comput. Syst. Sci. **68**(2), 303–318 (2004). https://doi.org/10.1016/j.jcss.2003.07.007
9. Kushilevitz, E.: A simple algorithm for learning O (log n)-term DNF. Inf. Process. Lett. **61**(6), 289–292 (1997). https://doi.org/10.1016/S0020-0190(97)00026-4

10. O'Donnell, R., Servedio, R.A.: New degree bounds for polynomial threshold functions. In: Larmore, L.L., Goemans, M.X. (eds.) Proceedings of the 35th Annual ACM Symposium on Theory of Computing, San Diego, CA, USA, 9–11 June 2003, pp. 325–334. ACM (2003). https://doi.org/10.1145/780542.780592
11. Valiant, L.G.: A theory of the learnable. Commun. ACM **27**(11), 1134–1142 (1984)

Interactive Physical Zero-Knowledge Proof for Norinori

Jean-Guillaume Dumas[1][iD], Pascal Lafourcade[2][iD], Daiki Miyahara[3,5][(✉)], Takaaki Mizuki[4][iD], Tatsuya Sasaki[3], and Hideaki Sone[4]

[1] Univ. Grenoble Alpes, CNRS,
Grenoble INP - Institute of Engineering Univ. Grenoble Alpes,
LJK, 38000 Grenoble, France
[2] LIMOS, University Clermont Auvergne, CNRS UMR 6158,
Clermont-Ferrand, France
[3] Graduate School of Information Sciences, Tohoku University, Sendai, Japan
daiki.miyahara.q4@dc.tohoku.ac.jp
[4] Cyberscience Center, Tohoku University, Sendai, Japan
[5] National Institute of Advanced Industrial Science and Technology, Tokyo, Japan

Abstract. Norinori is a logic game similar to Sudoku. In Norinori, a grid of cells has to be filled with either black or white cells so that the given areas contain exactly two black cells, and every black cell shares an edge with exactly one other black cell. We propose a secure interactive physical algorithm, relying only on cards, to realize a zero-knowledge proof of knowledge for Norinori. It allows a player to show that he or she knows a solution without revealing it. For this, we show in particular that it is possible to physically prove that a particular element is present in a list, without revealing any other value in the list, and without revealing the actual position of that element in the list.

Keywords: Zero-knowledge proofs ·
Card-based secure two-party protocols · Puzzle · Norinori

1 Introduction

Sudoku, introduced under this name in 1986 by the Japanese puzzle company Nikoli, and similar games such as Akari, Takuzu, Makaro, and Norinori have gained immense popularity in recent years. Many of them have been proved to be NP-complete [1,5,9,10]. In 2007, Gradwohl, Naor, Pinkas, and Rothblum proposed the first physical zero-knowledge proof protocols for Sudoku [7]. Their protocol only use several cards and allow a prover to prove to a verifier that he knows a solutions of a Sudoku grid. More precisely, a zero-knowledge proof of knowledge is a protocol that allows a prover P to convince a verifier V that she knows the solution w of an instance of a computational problem. Such a protocol has the following properties:

Correctness. If P knows w, then P can convince V.

Extractability. If P does not know w, then P cannot convince V, except with some *small* probability (here will have *perfect* Extractability, that is the probability that V does not detect a wrong grid is zero). This implies the standard *soundness* property, which ensures that if there exists no solution of the puzzle, then the prover is not able to convince the verifier regardless of the prover's behavior.

Zero-Knowledge. V cannot obtain any information about w. To prove that a protocol satisfies the zero-knowledge property, it is sufficient to exhibit a probabilistic polynomial time algorithm $M(\mathcal{I})$, not knowing w, such that the outputs of the protocol and the outputs of $M(\mathcal{I})$ follow the same probability distribution.

Recently, a novel protocol for Sudoku using fewer cards and with no soundness error was then proposed [14]. Physical protocols for other games, such as Hanjie, Akari, Kakuro, KenKen, Takuzu, and Makaro, have been designed [2–4]. In this article, we propose the first interactive physical zero-knowledge proof protocol for *Norinori*.

Norinori: It is a pencil puzzle published in the famous puzzle magazine *Nikoli*. The puzzle instance is a rectangular grid of cells. The grid is partitioned into *rooms*, that is, areas surrounded by bold lines. The goal of the puzzle is to shade certain cells so that they become black, according to the following rules [12]:

1. *Room condition*: Each room must contain exactly two black cells.
2. *Pair condition*: The black cells come in pairs: each black cell must be adjacent to exactly one, and only one, other black cell.

In Fig. 1, we give a simple example of a Norinori game. It is easy to verify that both constraints are satisfied in the solution on the right part of the figure. We note that in a solution the number of black cells is exactly twice the number of rooms.

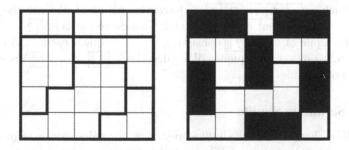

Fig. 1. Example of a Norinori grid and its solution.

Solving Norinori was shown to be NP-complete via a reduction from PLANAR 1-IN-3-SAT in [1]. Hence, it is possible to construct a cryptographic zero-knowledge proof of this game by using the generic cryptographic zero-knowledge proofs for all problems in NP given in [6]. However, this construction requires cryptographic primitives and is not very efficient.

Contributions: Our aim in this paper is to design an interactive physical protocol that uses only cards and envelopes for Nornori. This paper is not just another paper that proposed a physical zero-knowledge protocol for a Nikoli's game but we want to extend the physically verifiable set of functions that we are able to solve using only physical material. For instance, we know how to guarantee the presence of all numbers in some set without revealing their order [7], how to guarantee that two numbers are distinct without revealing their respective values [2], or how to prove that a number is the largest in a list, without revealing any value in the list [3]. In this paper, by providing a physical zero-knowledge proof for the Nikoli puzzle Norinori, we show in particular that it is possible to physically prove that a particular element is present in a list, without revealing any other value in the list, and without revealing the actual position of that element in the list.

Outline: In the next section, we introduce some notations and explain simple physical subprotocols that we use in our protocol for Norinori. In Sect. 3, we construct our zero-knowledge protocol for Norinori, before giving a security analysis in Sect. 4 and concluding in Sect. 5.

2   Preliminaries

In this section, we present some notations and introduce shuffling operations and a subprotocol that will be used later.

2.1   Physical Objects

The physical cards used in this paper are given in Table 1. We assume that the back sides of all cards are identical[1], and the face sides of all the cards of each type (such as ▢, ♣, ✓, S, 1, ...) are also identical. We use the notation $(c_1, c_2, \ldots, c_k)$ to represent a sequence of k face-down cards ? ? ... ? . We also consider a pile of cards ?▥ consisting of ℓ face-down cards and express it as a vector $\boldsymbol{p}$. Moreover, a sequence of k piles (?▥ , ?▥ , ... , ?▥) is expressed as a k-tuple of vectors $(\boldsymbol{p}_1, \boldsymbol{p}_2, \ldots, \boldsymbol{p}_k)$.

2.2   Pile-Scramble Shuffle

Pile-Scramble Shuffle [8] is a shuffle operation for piles: for a sequence of k piles $(\boldsymbol{p}_1, \boldsymbol{p}_2, \ldots, \boldsymbol{p}_k)$, applying a Pile-Scramble Shuffle results in $(\boldsymbol{p}_{\pi^{-1}(1)}, \boldsymbol{p}_{\pi^{-1}(2)}, \ldots, \boldsymbol{p}_{\pi^{-1}(k)})$, where $\pi \in S_k$ is a uniformly distributed random permutation, and S_k is the symmetric group of degree k. A Pile-Scramble Shuffle can be implemented by physical cases, for instance by using a big box where we place all the cases that contain the pile of cards and we blindly shuffle them.

---

[1] It means that the cards face down are indistinguishable from each other.

Table 1. Names of cards.

Face side	Back side	Name
☐	?	White card
♣	?	Black card
✓	?	Maker card
S	?	Starting card
1 2 3 4	?	Number card

2.3 Pile-Shifting Shuffle

Pile-Shifting Shuffle (which is also called Pile-Shifting Scramble) [13] cyclically shuffles piles of cards. That is, given a sequence of k piles, each of which consists of the same number of face-down cards, denoted by $(\boldsymbol{p}_1, \boldsymbol{p}_2, \ldots, \boldsymbol{p}_k)$, applying a Pile-Shifting Shuffle results in $(\boldsymbol{p}_{r+1}, \boldsymbol{p}_{r+2}, \ldots, \boldsymbol{p}_{r+k})$, where r is uniformly randomly generated from $\mathbb{Z}/k\mathbb{Z}$ and is unknown to everyone. To implement a Pile-Shifting Scramble, we use physical cases that can store a pile of cards, such as boxes or envelopes. One possible implementation is to place the different pile of cards in cases that are linked together in a circle in order to form a cycle. Then we just have to physically shuffle the cases, for instance by turning the circle.

2.4 Card Choosing Protocol

In this subsection, we describe the Card Choosing Protocol, which is immediately obtained by borrowing the idea behind the Chosen Cut [11]. This protocol is used as a subprotocol in our construction in Sect. 3.

Given a sequence of k face-down cards $(c_1, c_2, \ldots, c_k)$, the Card Choosing Protocol enables the prover P to choose a designated card secretly. More precisely, for some i such that $1 \leq i \leq k$, P can choose the i-th card without revealing any information about i to the verifier V. The protocol proceeds as follows:

1. P holds $k - 1$ white cards and one black card. Then, P puts them with their faces down below the cards $(c_1, c_2, \ldots, c_k)$, such that only the i-th card is the black card:

$$\boxed{?}\,\boxed{?}\,\cdots\,\boxed{?}\,\boxed{?}\,\boxed{?}\,\cdots\,\boxed{?}$$
$$c_1 \quad c_2 \qquad c_{i-1} \ c_i \ c_{i+1} \qquad c_k$$

$$\boxed{?}\,\boxed{?}\,\cdots\,\boxed{?}\,\boxed{?}\,\boxed{?}\,\cdots\,\boxed{?}$$
$$\boxed{\,}\,\boxed{\,}\qquad\boxed{\,}\,\boxed{♣}\,\boxed{\,}\qquad\boxed{\,}$$

2. Regarding cards in the same column as a pile, apply a Pile-Shifting Shuffle to the sequence of piles (which is denoted by $\langle\,\cdot\,|\ldots|\,\cdot\,\rangle$):

$$\left\langle \begin{matrix}\boxed{?}\\ \boxed{?}\end{matrix}\Big|\begin{matrix}\boxed{?}\\ \boxed{?}\end{matrix}\Big|\cdots\Big|\begin{matrix}\boxed{?}\\ \boxed{?}\end{matrix}\right\rangle \quad \rightarrow \quad \begin{matrix}\boxed{?}\,\boxed{?}\,\cdots\,\boxed{?}\\ \boxed{?}\,\boxed{?}\,\cdots\,\boxed{?}\end{matrix}$$
$$c_1 \quad c_2 \qquad c_k \qquad\qquad c_{r+1}\ c_{r+2}\quad c_{r+k}$$

where $r \in \mathbb{Z}/k\mathbb{Z}$ is a uniformly distributed random value.

3. Reveal all the cards in the second row. Then, one black card appears, and the card above the revealed black card is the i-th card:

$$\boxed{?}\,\boxed{?}\cdots\boxed{?}\,\underset{c_i}{\boxed{?}}\,\boxed{?}\cdots\boxed{?}$$

$$\square\,\square\cdots\square\,\boxed{\clubsuit}\,\square\cdots\square$$

Thus, P can show the designated card to V.

Because all the opened cards are shuffled in Step 2, V does not learn any information about the index i of the chosen card and is sure that only one card was designated.

3   Zero-Knowledge Proof for Norinori

We are now ready to describe our construction of a zero-knowledge proof for Norinori. For a puzzle instance of board size $m \times n$ including exactly t rooms, assume that the prover P knows the solution. Our protocol consists of three phases, namely:

1. Setup phase,
2. Pair Verification phase,
3. Room Verification phase.

It is important to perform the phases in this order. The Room Verification phase has to be the last one and the Setup phase the first phase.

Setup Phase: This phase has two steps:

1. P puts one face-down card on each cell according to the solution. That is, a black card is placed on every cell where a black square exists in the solution, and white cards are placed on all the other cells. For example, for the puzzle instance in Fig. 1, the cards are placed as follows:

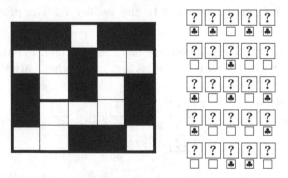

These cards are P's input, and we assume that P knows his or her input. This implies that if the cards conform to the solution, P knows the solution. Note that $2 \times t$ black cards and $m \times n - (2 \times t)$ white cards on a board of dimension $m \times n$ including exactly t rooms.

2. Next, remembering that m is the number of rows and n is the number of columns, V takes $2 \times m + 2 \times n + 3$ white cards and one starting card. Then, V puts these cards around the $m \times n$ "matrix" above to expand it to an $(m + 1) \times (n + 1)$ "matrix" as follows, where the starting card is placed at the top-left corner:

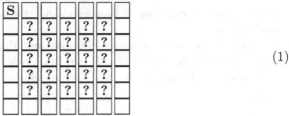

$$(1)$$

Pick each row from top to bottom to make a sequence of cards:

$$\overbrace{\qquad\qquad\qquad}^{m \times n + 2 \times m + 2 \times n + 3 \text{ cards}}$$

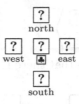

Notation: For any black card (whose location is known only to P) in the matrix, we define its four adjacent cards called *north*, *east*, *west*, and *south* cards:

```
        ?
      north
  ?   ?   ?
 west  ♣  east
        ?
      south
```

Pair Verification Phase: The verifier V verifies that P's input satisfies the Pair condition. If P has placed the cards correctly, every black card will have exactly one black card among its adjacent cards. The verification of the Pair condition is to guarantee this, and it proceeds as follows:

1. P selects a black card mentally from the card sequence, and performs the Card Choosing Protocol to choose that card:

```
 ?  ···  ?  ?  ?  ···  ?
 □  ···  □  ♣  □  ···  □
```

Then, V opens the card chosen by P to make sure that the chosen card is black:

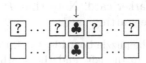

After confirming that the card is black, V replaces it by a marker card:

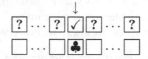

This operation prevents P from choosing the same black card again at later verifications.

2. V picks the four adjacent cards of the chosen black card. Because the sequence of cards was shifted cyclically, V can find such four cards, namely, the north, east, west, and south cards, by counting the distances[2]:

$$\boxed{?}\ \boxed{?}\ \boxed{?}\ \boxed{?}$$
west north east south

3. V puts number cards $\boxed{1}\boxed{2}\boxed{3}\boxed{4}$ in this order below these cards, and turns them over:

$$\begin{array}{cccc}\boxed{?}&\boxed{?}&\boxed{?}&\boxed{?}\\\boxed{1}&\boxed{2}&\boxed{3}&\boxed{4}\end{array}\quad\rightarrow\quad\begin{array}{cccc}\boxed{?}&\boxed{?}&\boxed{?}&\boxed{?}\\\boxed{?}&\boxed{?}&\boxed{?}&\boxed{?}\end{array}$$

4. P regards cards in the same column as a pile and applies a Pile-Scramble Shuffle to the sequence of piles:

$$\left\langle\begin{array}{c}\boxed{?}\\\boxed{?}\end{array}\begin{array}{c}\boxed{?}\\\boxed{?}\end{array}\cdots\begin{array}{c}\boxed{?}\\\boxed{?}\end{array}\right\rangle\quad\rightarrow\quad\begin{array}{cccc}\boxed{?}&\boxed{?}&\boxed{?}&\boxed{?}\\\boxed{?}&\boxed{?}&\boxed{?}&\boxed{?}\end{array}$$

5. V reveals the top row and checks that there is exactly one black or one marker card:

$$\begin{array}{cccc}\square&\clubsuit&\square&\square\\\boxed{?}&\boxed{?}&\boxed{?}&\boxed{?}\end{array}\quad\text{or}\quad\begin{array}{cccc}\square&\checkmark&\square&\square\\\boxed{?}&\boxed{?}&\boxed{?}&\boxed{?}\end{array}$$

After checking it, the cards in the top row are turned over.

6. P applies a Pile-Scramble Shuffle to the piles again, and reveals the bottom row:

$$\left\langle\begin{array}{c}\boxed{?}\\\boxed{?}\end{array}\begin{array}{c}\boxed{?}\\\boxed{?}\end{array}\cdots\begin{array}{c}\boxed{?}\\\boxed{?}\end{array}\right\rangle\rightarrow\begin{array}{cccc}\boxed{?}&\boxed{?}&\boxed{?}&\boxed{?}\\\boxed{?}&\boxed{?}&\boxed{?}&\boxed{?}\end{array}\rightarrow\begin{array}{cccc}\boxed{?}&\boxed{?}&\boxed{?}&\boxed{?}\\\boxed{2}&\boxed{3}&\boxed{4}&\boxed{1}\end{array}$$

Because the number cards indicate the original order, V can place each top card back in the original position in the card sequence:

$$\boxed{?}\boxed{?}\cdots\underset{\text{north}}{\boxed{?}}\cdots\underset{\text{west}}{\boxed{?}}\boxed{\checkmark}\underset{\text{east}}{\boxed{?}}\cdots\underset{\text{south}}{\boxed{?}}\cdots\boxed{?}$$

Turn over the face-up marker card. Note that P knows the locations of the other black cards and the starting card.

---

[2] For example, if the size of the puzzle board is 3×4, the north card is the fifth card away from the chosen card to the left, and the south card is the sixth card away from it to the right.

P and V repeat Steps 1 to 6 above $2 \times t - 1$ times more; recall that $2 \times t$ is the number of black cards[3]. Furthermore, as a targeted black card is replaced by a marker card, V is convinced that $2t$ black cards are verified.

Next, P and V place the card sequence back on the puzzle board, as follows:

1. P chooses the starting card in the card sequence by performing a Card Choosing Protocol:

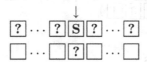

2. Shift the cards so that the starting card is leftmost:

$$\boxed{\text{S}}\,\boxed{?}\,\boxed{?}\,\boxed{?}\,\boxed{?}\,\boxed{?}\,\boxed{?} \cdots \boxed{?}$$

From the Card Choosing Protocols, the order of the cards is the same as the order of the card sequence generated in the Setup phase (although $2t$ black cards have been replaced by marker cards), and hence P and V can reconstruct the puzzle-board placement (1) by reversing the operations of the Setup phase.

Room Verification Phase: In this phase, V verifies the Room condition. To this end, the following is performed for each of the t rooms:

1. V picks all the cards from the room. Note that, regardless of the size of the targeted room, exactly two of the cards should be marker cards.
2. P shuffles the cards and reveals them.
3. V checks that exactly two marker cards appear.

If all phases have been passed, then the verifier accepts the proof by outputting 1.

Performance Analysis. Let us mention the performance of our protocol in terms of the numbers of shuffles and cards. The total number of shuffles is $7 \times t + 1$ (where t is the number of rooms), and the total number of required cards is $2 \times m \times n + 4 \times m + 4 \times n + 2 \times t + 12$, whose distribution is shown in Table 2 (where we have an $m \times n$ board).

---

[3] One might think that t times would suffice if any black card found in Step 5 was replaced by a marker card; however, this is not the case because we need to check that such a found black card also has exactly one black card among its adjacent cards.

Table 2. Numbers of cards required to execute our protocol.

Type of card	# of cards
White card (☐)	$2 \times m \times n + 4 \times m + 4 \times n - 2 \times t + 6$
Black card (♣)	$2 \times t + 1$
Marker card (✓)	$2 \times t$
Starting card (S)	1
Number card (1 2 3 4)	4

4   Security Analysis

We can easily see that our protocol satisfies the three properties of a zero-knowledge proof, as follows:

Correctness: If the prover P places the cards according to the solution in the Setup phase, P's input passes all verifications. Therefore, P who knows the solution, can always convince the verifier V.

Extractability: If P's input is invalid, V can detect it in the Verification phases. Therefore, if P does not know the solution, P cannot convince V.

Zero-knowledgeness: Since all the cards have been shuffled before they are opened, V learns nothing about the solution.

More precisely, we prove the following lemmas.

Lemma 1 (Correctness). *If the prover P has a solution for the Norinori puzzle, then P can always convince the verifier V (i.e., V outputs 1).*

Proof. We show that for a prover P with a solution, the verifier V never outputs 0. We look at the three phases:

Setup: In this phase the verifier just needs to check that the cards given by P correspond to size of the board. After that, V needs to place some extra cards and place all the cards in order to form a sequence. If P does not give the right number of cards, it is clear that he does not know the solution since he does not even know the puzzle itself.

Pair Verification: In this phase the goal of the verifier is to be convinced that all black cells come in pairs. If P knows a solution then he can place correctly the black cells. Hence the verifier never founds more than one marked or black card when he is opening the four adjacent cards of a black card. The only remaining uncertainty is whether P used exactly $2t$ black cards or more.

Room Verification: In this last phase, the verifier discovers exactly two black cards by room only if the prover knows a solution. As there are t rooms, this also proves that there were exactly $2t$ black cards to begin with.

Lemma 2 (Perfect Extractability). *If the prover does not know a solution for the Norinori puzzle, then the verifier V always rejects (i.e., V outputs 0) regardless of the prover P's behavior.*

Proof. It is important to notice that the order of our 3 phases is crucial and the fact that we can use the same cards for all the steps to ensure that we have no soundness error. In our proof, we consider two cases: the Pair condition is violated or the Room condition is violated.

Pair Condition: If the solution given by P does not satisfy the Pair condition, it means that some black cards do not come in pairs: some black cell are adjacent to more than one black card, or some black cells are isolated. We have three possible cases (modulo rotations and symmetries) that can occur:

1. There are 3 aligned black cards as follows: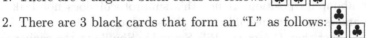

2. There are 3 black cards that form an "L" as follows:

3. A single black card is surrounded by white cards.

In the first two cases, when V opens the four adjacent cards of a black card, he detects that there is strictly more than one extra black card in the neighborhood. This shows that the prover does not have a solution that satisfies the Pair condition. Similarly, in case of an isolated black card, the neighborhood will show only white cards. Overall, the verifier is convinced after this phase if and only if all black cards come in pairs.

Room Condition: If the solution given by P does not satisfy the Room condition, it means that a room contains one or no black card, or that a room contains three or more black cards. Both situation are detected by the verifier that opens all the cards in all the rooms.

Lemma 3 (Zero-knowledge). *During an execution of our protocol, the verifier V learns nothing about P's Norinori solution.*

Proof. In order to prove this, we have to describe an efficient simulator that simulates any interaction between a cheating verifier and a real prover. The simulator does not have a correct solution, but it does have an advantage over the prover: when shuffling decks, it is allowed to swap the packets for different ones. We thus show how to construct a simulator for each challenge.

Setup: V learns only the size of the board when he receives P's inputs, which is a known data already publicly available in the puzzle grid.

Pair Condition: In this phase the verifier does not learn any information except the number of black cards, which is $2 \times t$. This is not a secret information since it is a known data already publicly available in the puzzle grid. It is sufficient to show that all distributions of opening values are simulated without knowing the prover's solution.

Now, from [8,11,13], respectively, we know that the subprotocols Pile-Scramble Shuffle, Pile-Shifting Shuffle and Card Choosing Protocol are zero-knowledge.

Note also, that during this verification phase, there is no need for the simulator to respect the Room condition. Thus a possible simulator is as follows: the simulator places at least one black card randomly on the grid. Then, for each pair verification, the simulator designates a black card and, at the Pile-Scramble step, P replaces the four adjacent cards by a single black and three white cards, e.g., as follows:

♣ □ □ □
1 2 3 4

For the first step, the Card Choosing Protocol ensures that V does not learn anything: indeed it reveals that there is at least one black card on the board, but nothing about its position. Hence this subprotocol does not leak any information and is indistinguishable from the simulator.

For Step 4, the fact that P uses the Pile-Scramble Shuffle to the sequence of piles implies that their order is uniformly distributed and then can be simulated without knowing the solution. Therefore the position of the adjacent black cell is not leaked either.

At Step 6, using the Pile-Scramble Shuffle, the simulator replaces the initial cards (not even switching the tested black card by a marked card, see in the next phase). Note that the fact that we can replace the verified cards in their initial position without leaking any information comes from the usage of Pile-Scramble Shuffle in Step 6 and the usage of Card Choosing Protocol in the next step. It allows us to use the same cards for the next verification[4].

Room Condition: This phase is similar to a room/row/column condition in Sudoku [7, Protocol 3]: since P shuffles all the cards within a room before revealing them, this is indistinguishable from a simulation putting randomly two marked cards among white ones. Thus V only learns that there were exactly two black cards in each rooms.

5   Conclusion

In this paper, we have designed an interactive zero-knowledge proof protocol for the famous puzzle, Norinori. Our protocol is quite simple and easy to implement by humans. By solving this Nikoli's puzzle, we also demonstrate that it is possible to physically prove that a particular element is present in a list, without revealing any other value in the list, and without relvealing the actual position of that element in the list. This functionality could be used to construct zero-knowledge proof protocols for other puzzles.

An interesting open problem is to design a more efficient protocol in terms of the numbers of cards and shuffles.

Acknowledgement. We thank the anonymous referees, whose comments have helped us to improve the presentation of the paper. This work was supported by JSPS KAKENHI Grant Number JP17K00001 and partly by the OpenDreamKit Horizon 2020 European Research Infrastructures project (#676541) and the Cyber@Alps French National Research Agency program (ANR-15-IDEX-02).

---

[4] This trick is inspired from [14] and allows us to also have no soundness error.

References

1. Biro, M., Schmidt, C.: Computational complexity and bounds for Norinori and LITS. In: EuroCG 2017, pp. 29–32 (2017)
2. Bultel, X., Dreier, J., Dumas, J.G., Lafourcade, P.: Physical zero-knowledge proofs for Akari, Takuzu, Kakuro and KenKen. In: Demaine, E.D., Grandoni, F. (eds.) FUN 2016. LIPIcs, vol. 49, pp. 8:1–8:20 (2016)
3. Bultel, X., et al.: Physical zero-knowledge proof for Makaro. In: Izumi, T., Kuznetsov, P. (eds.) SSS 2018. LNCS, vol. 11201, pp. 111–125. Springer, Cham (2018). https://doi.org/10.1007/978-3-030-03232-6_8
4. Chien, Y.-F., Hon, W.-K.: Cryptographic and physical zero-knowledge proof: from Sudoku to nonogram. In: Boldi, P., Gargano, L. (eds.) FUN 2010. LNCS, vol. 6099, pp. 102–112. Springer, Heidelberg (2010). https://doi.org/10.1007/978-3-642-13122-6_12
5. Demaine, E.D.: Playing games with algorithms: algorithmic combinatorial game theory. In: Sgall, J., Pultr, A., Kolman, P. (eds.) MFCS 2001. LNCS, vol. 2136, pp. 18–33. Springer, Heidelberg (2001). https://doi.org/10.1007/3-540-44683-4_3
6. Goldreich, O., Micali, S., Wigderson, A.: How to prove all NP statements in zero-knowledge and a methodology of cryptographic protocol design (Extended abstract). In: Odlyzko, A.M. (ed.) CRYPTO 1986. LNCS, vol. 263, pp. 171–185. Springer, Heidelberg (1987). https://doi.org/10.1007/3-540-47721-7_11
7. Gradwohl, R., Naor, M., Pinkas, B., Rothblum, G.N.: Cryptographic and physical zero-knowledge proof systems for solutions of Sudoku puzzles. In: Crescenzi, P., Prencipe, G., Pucci, G. (eds.) FUN 2007. LNCS, vol. 4475, pp. 166–182. Springer, Heidelberg (2007). https://doi.org/10.1007/978-3-540-72914-3_16
8. Ishikawa, R., Chida, E., Mizuki, T.: Efficient card-based protocols for generating a hidden random permutation without fixed points. In: Calude, C.S., Dinneen, M.J. (eds.) UCNC 2015. LNCS, vol. 9252, pp. 215–226. Springer, Cham (2015). https://doi.org/10.1007/978-3-319-21819-9_16
9. Iwamoto, C., Haruishi, M., Ibusuki, T.: Herugolf and Makaro are NP-complete. In: Ito, H., Leonardi, S., Pagli, L., Prencipe, G. (eds.) Fun with Algorithms 2018. LIPIcs, vol. 100, pp. 24:1–24:11. Schloss Dagstuhl - Leibniz-Zentrum fuer Informatik (2018)
10. Kendall, G., Parkes, A.J., Spoerer, K.: A survey of NP-complete puzzles. ICGA J. 31(1), 13–34 (2008)
11. Koch, A., Walzer, S.: Foundations for actively secure card-based cryptography. IACR Cryptology ePrint Archive 2017, 423 (2017)
12. Nikoli: Norinori. http://www.nikoli.co.jp/en/puzzles/norinori.html
13. Nishimura, A., Hayashi, Y.I., Mizuki, T., Sone, H.: Pile-shifting scramble for card-based protocols. IEICE Trans. Fundam. Electron. Commun. Comput. Sci. 101(9), 1494–1502 (2018)
14. Sasaki, T., Mizuki, T., Sone, H.: Card-based zero-knowledge proof for Sudoku. In: Ito, H., Leonardi, S., Pagli, L., Prencipe, G. (eds.) Fun with Algorithms 2018. LIPIcs, vol. 100, pp. 29:1–29:10. Schloss Dagstuhl - Leibniz-Zentrum fuer Informatik (2018)

On Proving Parameterized Size Lower Bounds for Multilinear Algebraic Models

Purnata Ghosal$^{(\boxtimes)}$ and B. V. Raghavendra Rao

Indian Institute of Technology Madras, Chennai 600036, India
purnatag@gmail.com, bvrr@iitm.ac.in

Abstract. We consider the problem of obtaining parameterized lower bounds for the size of arithmetic circuits computing polynomials with the degree of the polynomial as the parameter. In particular, we consider the following special classes of multilinear algebraic branching programs: (1) Read Once Oblivious Algebraic Branching Programs (ROABPs); (2) Strict interval branching programs; and (3) Sum of read once formulas with restricted ordering.

We obtain parameterized lower bounds (i.e., $n^{\Omega(t(k))}$ lower bound for some function t of k) on the size of the above models computing a multilinear polynomial that can be computed by a depth four circuit of size $g(k)n^{O(1)}$ for some computable function g.

Our proof is an adaptation of the existing techniques to the parameterized setting. The main challenge we address is the construction of hard parameterized polynomials. In fact, we show that there are polynomials computed by depth four circuits of small size (in the parameterized sense), but have high rank of the partial derivative matrix.

1 Introduction

Parameterized Complexity is a multi-dimensional study of computational problems which views the complexity of a problem in terms of both the input size and an additional parameter. This leads to a finer classification of computational problems, and a relaxed notion of tractability, given by a $f(k)\text{poly}(n)$ bound on the running time for decision problems with parameter k, known as *fixed-parameter tractability* or FPT. This was first studied by Downey and Fellows in their seminal work [8] where they developed parameterized complexity theory, and paved the way for extensive study of parameterized algorithms. The notion of intractability in parameterized complexity is captured by the W-hierarchy of classes [8].

Algebraic Complexity Theory is concerned with complexity of computing polynomials using elementary arithmetic operations such as addition and multiplication over an underlying ring or field. Valiant [25] formalized the notions of algebraic complexity theory and posed proving lower bound on the size of arithmetic circuits computing explicit polynomials as the primary challenge for the area. Following Valiant's work, there has been intense research efforts in the past

© Springer Nature Switzerland AG 2019
D.-Z. Du et al. (Eds.): COCOON 2019, LNCS 11653, pp. 178–192, 2019.
https://doi.org/10.1007/978-3-030-26176-4_15

four decades to prove lower bounds on the size of special classes of arithmetic circuits such as constant depth circuits, multilinear formula and non-commutative models. (See [23,24] for a survey.) Despite several techniques, the best known size lower bound for general arithmetic circuit is only super linear [3].

Given the lack of progress towards proving lower bounds on arithmetic circuits and the success of parameterized complexity theory in refining the notion of tractability, it is worthwhile exploring the feasibility of parameterizations of polynomials. Engels [9] initiated development of a parameterized theory for algebraic complexity classes and suggested suitable notions of tractability and reductions. The attempt in [9] was more at obtaining a complexity classification in the form of complete problems for a generic parameter. Müller [18] was the first to introduce parameterizations on polynomials in the context of designing parameterized algorithms problems on polynomials such as testing for identity of polynomials given as arithmetic circuits (ACIT). ACIT is one of the fundamental problems in algebraic complexity theory and has close connections to the circuit lower bound problem [15]. Müller studied parameters such as the number of variables in the polynomial, multiplication depth of the circuit computing the polynomial etc and obtained efficient randomized parameterized algorithms for ACIT. It may be noted that ACIT is non-trivial with these parameters since a polynomial can potentially have $n^{\Omega(k)}$ monomials where k is any of these parameters. In [6] Chauhan and Rao studied ACIT with degree as the parameter and obtained a randomness efficient parameterized algorithm for ACIT. It may be noted that polynomials with the degree bounded by the parameter are widely used in developing efficient parameterized algorithms [1,4,10] and in expressing properties of graphs [5]. For example, in [10] the polynomial representing homomorphisms between two graphs has indeed degree equal to the parameter, i.e. the number of vertices in the pattern graph. It may be noted that efficient computation of the polynomial defined in [10] by arithmetic formulas leads to space-efficient algorithms for detecting homomorphisms between graphs of bounded treewidth. In [12] the authors along with Prakash studied polynomials parameterized by the degree and showed limitations of an existing approach in obtaining deterministic parameterized algorithms for ACIT. In this article, we explore the possibility of obtaining parameterized lower bounds for the size of arithmetic circuits with degree as the parameter.

Let n be the number of variables and k be a parameter (e.g., degree of the polynomial). Throughout the article, $t(k)$ denotes a computable function that depends only on the parameter, e.g., $t(k) = 2^k, t(k) = 2^{2^k}, t(k) = \sqrt{k}$ etc. Any circuit is said to be of fpt size if the size of the circuit is bounded by $t(k)n^{O(1)}$ for some computable t. By a parameterized lower bound, we mean a lower bound of the form $n^{\Omega(t(k))}$ for a computable function t. It may be noted that the task of proving parameterized lower bounds is more challenging than classical lower bounds. In the case of degree as a parameter, most of the existing lower bounds of the form $n^{\Omega(\sqrt{k})}$ (e.g., [11,13,17]) for constant depth circuits are already parameterized lower bounds. In contrast, the lower bounds for other special classes such as multilinear formula [20,21] do not translate easily.

To understand the primary challenge in translating the results in [20,21] to the parameterized setting we need to delve a bit on the techniques used for proving lower bounds. Raz [21] used the notion of partial derivative matrix under a partition of variables with equal parts (See Sect. 2 for a detailed definition) as a measure of complexity. The idea is to show existence of such partitions where polynomials computed by a multilinear formula of small size will have small rank for the partial derivative matrix. Then for any polynomial that has large rank under every partition, a natural lower bound on the size follows. However, the analysis done in [21] or subsequent works [7,20,22] do not carry forward when parameterized by the degree. Similarly the construction of the polynomial family with high rank partial derivative matrix in [21] and subsequent works do not generalize to the parameterized setting.

In this article we address the challenge of translating lower bounds for the size of multilinear restrictions of arithmetic circuits to parameterized lower bounds.

Results. Our primary result is a parameterized family of polynomials (Theorem 1) such that under every equal sized bi-partition of the variables, the rank of the partial derivative matrix is the maximum possible up to a factor that depends only on the parameter. Further, we demonstrate a simple parameterized family of polynomials that can be written as a sum of three read once formulas such that under any partition of the variables into two equal parts, the rank of the partial derivative matrix is high (Theorem 2). As a consequence, we obtain parameterized lower bounds for the size of an ROABP (Theorem 3), a strict-interval ABP (Corollary 1) and sum of ROPs with restricted ordering of variables (Theorem 6) against the constructed hard polynomial.

Finally, we obtain a parameterized version of the separation between read-3 ABPs and ROABPs given in [16] (Theorem 4). This is done by constructing a parameterized variant of the hard polynomial given in [16] (Theorem 2).

2   Preliminaries

In this section we give all basic definitions related to arithmetic circuit. Let $\mathbb{F}$ denote a field. Most of the arguments in this article work for any $\mathbb{F}$. Let $X = \{x_1, \ldots, x_n\}$ denote the set of variables.

An *arithmetic circuit* is a model for computing polynomials using the basic operations $+$ and $\times$. An arithmetic circuit C is a directed acyclic graph, where every node (called a gate) has in-degree two or zero. The gates of in-degree zero are called input gates and are labeled from $X \cup \mathbb{F}$, where $X = \{x_1, \ldots, x_n\}$ is the set of variables called inputs and $\mathbb{F}$ is the underlying field. Internal gates of C are labeled by either $+$ or $\times$. Gates of out-degree zero are called output gates. Typically an arithmetic circuit will have a single output gate. Every gate in the circuit C is associated with a unique polynomial in $\mathbb{F}[X]$. The polynomial computed by the circuit is the polynomial associated at its output gate.

The complexity of arithmetic circuits is measured in terms of *size* and *depth*. Size is defined as the number of $+$ and $\times$ operations in the circuit. Depth of the circuit represents the length of the longest path from the output node (root) to an input node (leaf) of the circuit. Since a constant depth arithmetic circuit where fan-in of every gate is bounded by 2 (or even a constant) cannot even read all of the inputs, we assume unbounded fan-in in the case of constant depth circuits. Arithmetic circuits of constant depth have received wide attention [23].

An arithmetic circuit C is said to be *syntactic multilinear* if for every product gate $f = g \times h$ in C, the set of variables that appear under the sub-circuit rooted at g is disjoint from that of h. Naturally, a syntactic multilinear circuits computes a multilinear polynomial.

An arithmetic circuit where the underlying graph is a tree is known as *arithmetic formula*. An arithmetic formula is said to be read once (ROF for short) if every variable appears as a label in at most one leaf. Polynomials computed by ROFs are known as Read-once polynomials (ROPs for short). It may also be noted that ROFs are a proper subclass of syntactic multilinear formulas.

An algebraic branching program (ABP) P is a directed acyclic graph with a source vertex s of in-degree 0 and a sink vertex t of out-degree 0. The rest of the vertices can be divided into layers $L_1, L_2, \ldots, L_{r-1}$ between s and t, s being the only vertex in L_0, the first layer, and t being the only vertex in the last layer ℓ_r. Edges in P are between vertices of consecutive layers. Every edge e is labelled by either a constant from $\mathbb{F}$ or a variable from X. For a directed path ρ in P, let $w(\rho)$ denote the product of edge labels in ρ. For any pair of nodes u, v in P let $[u, v]_P$ denote the polynomial $\sum_{\rho \text{ is a } u \to v \text{ path}} w(\rho)$. The polynomial computed by P is $[s, t]_P$. The size of an ABP is the total number of nodes and edges in it and the depth of an ABP is the total number of layers in it excluding the layers containing s and t. Read-once ABPs are such that every input variable is read at most once along any path from s to t. Read-once Oblivious ABPs (ROABPs) are such that input variables are read at most once, in a fixed order, along any path from s to t, and any variable occur as a label in at most one layer of the program.

Let π be a permutation of the variables. An interval in π is a set of the form $\{\pi(i), \pi(i + 1), \ldots, \pi(j)\}$ for some $i < j$. Arvind and Raja [2] studied a restriction of multilinear ABPs called as interval ABPs where every node in the ABP computes a polynomial whose variable set forms an interval in $\{1, \ldots, n\}$. In this article, we consider a restriction of interval ABPs which we call as *strict interval* ABPs. A syntactically multilinear ABP P is said to be a π strict interval ABP, if for any pair of nodes (a, b) in P, the index set X_{ab} of the variables occurring on all paths from a to b is contained in some π interval I_{ab} in $[n]$ and for any node c the intervals I_{ab} and I_{bc} are non-overlapping.

For a polynomial $p \in X$, let $\mathsf{var}(p)$ denote the set of variables that p is dependent on and $\deg(p)$ denote its degree.

Partial Derivative Matrix of a Polynomial. Nisan [19] defined the partial derivative matrix of a polynomial and considered its rank as a complexity measure for non-commutative polynomials and proved exponential lower bounds for the size of non-commutative formulas. Raz [21] considered a variant of the partial derivative matrix and proved super polynomial size lower bounds for multilinear formulas. We describe the partial derivative matrix introduced by Raz [21] in more detail. Let $X = \{x_1, \ldots, x_n\}$ be the set of variables where n is even. A partition of X is an injective function $\varphi : X \to Y \cup Z$, where Y and Z are two disjoint sets of variables. A partition φ is said to be an equi-partition if $|Y| = |Z| = n/2$. In the remainder of the article, we assume that the number of variables n is an even number.

Definition 1 [21]. *Let $f \in \mathbb{F}[x_1, \ldots, x_n]$ be a polynomial of degree d, $\varphi : X \to Y \cup Z$ be a partition of the input variables of f. Then the partial derivative matrix of f with respect to φ, denoted by $M_{f\varphi}$ is a $2^{|Y|} \times 2^{|Z|}$ matrix where the rows are indexed by the set of all multilinear monomials μ in the variables Y, and columns indexed by the set of all multilinear monomials ν in variables in Z. For monomials μ and ν respectively in variables Y and Z, the entry $M_{f\varphi}(\mu, \nu)$ is the coefficient of the monomial $\mu\nu$ in f.*

For a multilinear polynomial $p \in \mathbb{F}[X]$ and an equi-partition φ, let $\mathsf{rank}_\varphi(p)$ be the rank of the matrix $M_{p\varphi}$ over $\mathbb{F}$. The following fundamental properties of the rank of a partial derivative matrix was given by Raz [21].

Lemma 1 [21].   *Let f_1 and f_2 be multilinear polynomials. Then $\mathsf{rank}_\varphi(f_1 + f_2) \leq \mathsf{rank}_\varphi(f_1) + \mathsf{rank}_\varphi(f_2)$ and if $\mathsf{var}(f_1) \cap \mathsf{var}(f_2) = \emptyset$ then $\mathsf{rank}_\varphi(f_1 f_2) = \mathsf{rank}_\varphi(f_1)\mathsf{rank}_\varphi(f_2)$.*

Lemma 2. *For any equi-partition $\varphi : X \to Y \cup Z$, and any multilinear polynomial p of degree k, we have $\mathsf{rank}_\varphi(p) \leq (k/2 + 1)\binom{n/2}{k/2}$.*

3   Construction of Hard Parameterized Polynomials

This section is devoted to the construction of two parameterized polynomial families $f = (f_{n,2k})_{k \geq 0}$ and $h = (h_{2n,k})$. The first family is computable by a depth four circuit of fpt (i.e., $t(k)n^{O(1)}$ for a computable t) size and the second family is a sum of three ROFs. Further, for any partition φ, $\mathsf{rank}_\varphi(f)$ is the maximum possible value up to a factor that depends only on the parameter and $\mathsf{rank}_\varphi(h)$ is the maximum possible value up to a constant factor in the exponent.

A Full Rank Polynomial: It may be noted that for a multilinear polynomial g of degree k in n variables, the maximum possible value of $\mathsf{rank}_\varphi(g)$ over all partitions φ is at most $(k/2 + 1)\binom{n/2}{k/2}$. Though it is possible to construct polynomials that achieve this bound under a fixed partition φ, it is not immediate if there is a polynomial g computed by small circuits that is full rank under every equipartition. In the following, we give the description of a multilinear polynomial of degree k that has rank $n^{k/2}/t(k)$ where t is a function that depends only on k. We assume that $2k|n$. Suppose $V_1 \cup \cdots \cup V_{2k} = X$ be a partition of the variable set $X = \{x_1, \ldots, x_n\}$ such that $|V_i| = |V_j|$ for $1 \leq i < j \leq 2k$. For convenience let $V_i = \{x_{i,1}, \ldots, x_{i,n/2k}\}$, where we assume a natural ordering among the variables.

Let $\mathcal{M}$ be the set of all possible perfect matchings on $G = K_{2k}$, the complete graph on $2k$ vertices. Let ζ_M for $M \in \mathcal{M}$, $\omega_{i,j}$ $1 \leq i < j \leq n/k$ be formal variables. Let $\mathbb{G}$ be any extension of $\mathbb{F}$ containing $\{\omega_{i,j}\} \cup \{\zeta_M \mid M \in \mathcal{M}\}$. We define a parameterized family of polynomial $f = (f_{n,2k})$, $f_{n,2k} \in \mathbb{G}[x_1, x_2, \ldots, x_n]$ as follows:

$$f(x_1, x_2, \ldots, x_n) = \sum_{M \in \mathcal{M}} \zeta_M \prod_{(i,j) \sim M} (1 + p(V_i \cup V_j)),$$

where p is a n/k variate quadratic multilinear polynomial defined as

$$p(v_1, \ldots, v_{n/k}) = \sum_{k < \ell} \omega_{k,\ell} v_k v_\ell.$$

Note that $f_{n,2k}$ is a degree $2k$ polynomial in n variables. When n and k are clear from the context, we use f to denote $f_{n,2k}$. Let $\mathbb{G} = \mathbb{F}(\{\zeta_M \mid M \in \mathcal{M}\} \cup \{\omega_{i,j} \mid 1 \leq i < j \leq n/k\})$, i.e the rational function field of the polynomial ring $\mathbb{F}[\{\zeta_M \mid M \in \mathcal{M}\} \cup \{\omega_{i,j} \mid 1 \leq i < j \leq n/k\}]$. In the remainder of the section, we argue that the polynomial family f defined above has almost full rank under every partition $\varphi : X \to Y \cup Z$, such that $|Y| = |Z| = |X|/2$.

Definition 2. *Consider a partition function* $\varphi : X \to Y \cup Z$ *such that* $|Y| = |Z|$ *and* $V \subseteq X$. *The set* V *is said to be* ℓ-*unbalanced with respect to* φ *if* $\frac{|X|}{2} - |\varphi(X) \cap Z| = \ell = |\varphi(X) \cap Y| - \frac{|X|}{2}$.

It may be noted that ℓ can be positive or negative accordingly as $|\varphi(X) \cap Y| > |\varphi(X) \cap Z|$ or otherwise. Our first observation is, even if the set $V = V_i \cup V_j$ is ℓ-unbalanced for $\ell < n/4k$, $i \leq q \leq [2k]$, $\mathrm{rank}_\varphi(p(V_i, V_j))$ remains large:

Lemma 3. *If* $V_i \cup V_j$ *is* ℓ *unbalanced with respect to a partition* $\varphi : X \to Y \cup Z$, *then* $\mathrm{rank}_\varphi(p(V_i, V_j)) = \Omega(n/2k - |\ell|)$.

Proof. Without loss of generality, suppose that $\ell > 0$, $V_i \cup V_j = \{v_1, \ldots, v_{n/k}\}$ and

$$\varphi(v_i) = \begin{cases} y_i & \text{if } i \leq n/2k + \ell \\ z_{i-(n/2k+\ell)} & \text{otherwise.} \end{cases}$$

Since p is a quadratic polynomial, the rows of $M_{p(V_i,V_j)^\varphi}$ are indexed by monomials $\emptyset, y_1, \ldots y_{n/2k+\ell}, y_i y_j, 1 \leq i < j \leq n/2k + \ell$ and columns are indexed by $\emptyset, z_1, \ldots, z_{n/2k-\ell}, z_i z_j, 1 \leq i < j \leq n/2k - \ell$. The rows and columns indexed by degree 2 monomials will have a rank of at most 2. Thus it is required to show that the submatrix of $M_{p(V_i,V_j)^\varphi}$ with rows indexed by $\emptyset, y_1, \ldots y_{n/2k+\ell}$ and columns indexed by $\emptyset, z_1, \ldots, z_{n/2k-\ell}$ has rank $\Omega(n/2k - |\ell|)$.

The $(y_i, z_j)^{\text{th}}$ entry of $M_{p(V_i,V_j)^\varphi}$ contains $\omega_{i,n/2k+j}$. By suitably substituting the variables $\omega_{i,n/2k+j}$ with values from $\mathbb{F}$, we see that the submatrix of $M_{p(V_i,V_j)^\varphi}$ restricted to rows and columns indexed respectively by $\emptyset, y_1, \ldots y_{n/2k+\ell}$ and $\emptyset, z_1, \ldots, z_{n/2k-\ell}$ has rank $\Omega(n/2k - |\ell|)$. $\square$

Theorem 1. *For the parameterized polynomial family* $f = (f_{n,2k})_{n,k\geq 0}$ *as above,*

$$\mathsf{rank}_\varphi(f_{n,2k}) = \Omega(\frac{n^k}{(2k)^{2k}})$$

for every equi-partition $\varphi : X \to Y \cup Z$ *and* $k > 3$.

Proof. Let φ be an equi-partition of X. Note that by the definition of f, it is enough to show that for all equi-partitions φ, there exists an optimal matching N such that $\mathsf{rank}_\varphi(f_N) = \Omega(\frac{n^k}{(2k)^{2k}})$, where $f_N = \prod_{(i,j)\in N} p(V_i, V_j)$. Since f_N is multilinear, it is enough to prove that $\forall (i,j) \in N$, $\mathsf{rank}_\varphi(p(V_i,V_j)) = \Omega(\frac{n}{k^2})$. Our argument is an iterative construction of the required matching.

We begin with some notations. Let $D(V_i) = |\varphi(V_i) \cap Y| - \frac{|V_i|}{2}$. Let $Y_i = \varphi(V_i) \cap Y$, $Z_i = \varphi(V_i) \cap Z$. We know $\forall i \in [2k]$, $|V_i| = \frac{n}{2k}$. So, $D(V_i) = |Y_i| - \frac{n}{4k}$ is the imbalance of φ. Since $0 \leq |Y_i| \leq |V_i| = \frac{n}{2k}$, $D(V_i) \in [\frac{-n}{4k}, \frac{n}{4k}]$.

Let $M \in \mathcal{M}$. For each edge $e = (i,j)$ in the matching M, we associate a weight with respect to φ: $\mathsf{wt}(e) = |D(V_i) + D(V_j)|$. The weight of the matching M, denoted by $\mathsf{wt}(M)$, is the sum of the weights of the edges in M, i.e., $\mathsf{wt}(M) = \sum_{e\in M} \mathsf{wt}(e)$. In the following, we give an iterative procedure, that given a matching M produces a matching N with the required properties. The procedure in each iteration, obtains a new matching of smaller weight than the given matching. The crucial observation then is, matchings that are weight optimal with respect to the procedure outlined below indeed have the required property. We say that a matching M is *good* with respect to φ, if $\forall\, e = (i,j) \in M$, $\mathsf{wt}(e) \leq n/2k - n/(2k(k-1))$. Note that if M is good then for every edge $(i,j) \in M$, we have $V_i \cup V_j$ is ℓ-unbalanced for some ℓ with $|\ell| \leq n/2k - n/(2k(k-1))$. Then, by Lemma 3 we have $\mathsf{rank}_\varphi(f_M) \geq n/(2k(k-1))$.

Suppose that the matching M is not good. Let $e = (i,j) \in M$ be an edge such that $\mathsf{wt}(e) > n/2k - n/2k(k-1)$. If there are multiple such edges, e is chosen such that $\mathsf{wt}(e)$ is the maximum, breaking ties arbitrarily. Note that we can assume that $D(V_i)$ and $D(V_j)$ are of the same sign, else we would have $\mathsf{wt}(e) \leq n/4k$. Without loss of generality, assume that both $D(V_i)$ and $D(V_j)$ to be non-negative, i.e., $\mathsf{wt}(e) = D(V_i) + D(V_j)$. Since φ is an equi-partition, we have

$$\sum_{m\in[2k]} D(V_m) = \sum_{m\in[2k]} \left(|Y_m| - \frac{n}{4k}\right) = 0 \implies \sum_{m\in[2k]\setminus\{i,j\}} D(V_m) = -\mathsf{wt}(e)$$

i.e., $$\sum_{e'\in M\setminus\{e\}} \mathsf{sgn}(e')\mathsf{wt}(e') < \frac{-n}{2k} + \frac{n}{2k(k-1)},$$

where $\mathsf{sgn}(e)$ is ± 1 depending on the sign of $\mathsf{wt}(e)$. By averaging, there is an $e_1 \in M$ such that $\mathsf{sgn}(e_1)\mathsf{wt}(e_1) < \frac{-n}{2k(k-1)} + \frac{n}{2k(k-1)^2}$. Suppose $e_1 = (i_1, j_1)$. Let $D(V_i) = a$, $D(V_j) = b$, $D(V_{i_1}) = c$, $D(V_{j_1}) = d$. Since $c + d < 0$, it must be that either $c < 0$ or $d < 0$. The new matching is constructed based on the values of a, b, c and d.

Case 1 Suppose $c, d < 0$. Then, $|a+b| + |c+d| > |a+c| + |b+d|$. We replace the edges (i, j) and (i_1, j_1) by $(i, i_1), (j, j_1)$ to get a new matching M'. We have $\mathsf{wt}(M') < \mathsf{wt}(M)$.

Case 2 Either $c \geq 0$ and $d < 0$ or $c < 0$ and $d \geq 0$. Without loss of generality, assume that $c \geq 0$ and $d < 0$. Suppose $c > \frac{n}{4k} - \frac{n}{2k(k-1)} + \frac{n}{2k(k-1)^2}$, then we have $d < \frac{-n}{2k(k-1)} + \frac{n}{2k(k-1)^2} - c < \frac{-n}{4k}$ which is impossible as $|d| \leq \frac{n}{4k}$. Therefore, we have $c \leq \frac{n}{4k} - \frac{n}{2k(k-1)} + \frac{n}{2k(k-1)^2}$. If $c > a, b$, then $a + b < 2c \leq \frac{n}{2k} - \frac{n}{k(k-1)} + \frac{n}{k(k-1)^2}$. For $k > 3$, this is impossible since $\mathsf{wt}(e) > \frac{n}{2k} - \frac{n}{2k(k-1)}$. We consider the following sub-cases:

Subcase (a) $a > c$. Then $a + b > c + b$, replace the edges (i, j) and (i_1, j_1) with the edges (i, j_1) and (i_1, j) to get the new matching M'.

Subcase (b) $b > c$. Then $a + b > a + c$, replace (i, j) and (i_1, j_1) with the edges (i, i_1) and (j, j_1) to get the new matching M'.

For the new matching M' obtained from M as above, we have one of the following properties:

- It has smaller total weight than M, i.e., $\mathsf{wt}(M') < \mathsf{wt}(M)$, or
- If M has a unique maximum weight edge, then the weight of any edge in M' is strictly smaller than that in M, i.e. $\max_{e' \in M'} \mathsf{wt}(e') < \mathsf{wt}(e)$, or
- The number of edges that have maximum weight in M' is strictly smaller than that in M, i.e., $|\{e'' \mid \mathsf{wt}(e'') = \max_{e' \in M'} \mathsf{wt}(e')\}| < |\{e'' \mid \mathsf{wt}(e'') = \max_{e' \in M} \mathsf{wt}(e')\}|$.

Since all of the invariants above are finite, by repeating the above procedure a finite number of times we get a matching $N \in \mathcal{M}$ such that any of the above steps are not applicable. That is, for every $e' \in N$, $\mathsf{wt}(e') \leq n/2k - n/2k(k-1)$.

Thus for every edge $(i, j) \in N$, we have $\mathsf{rank}_\varphi(p(V_i, V_j) = \Omega(n/2k(k-1))$ and $\mathsf{rank}_\varphi(f_N) = \Omega(n^k/(2k)^{2k})$. By the construction of the polynomial and Lemma 1, we have $\mathsf{rank}_\varphi(f) \geq \max_{M \in \mathcal{M}}\{\mathsf{rank}_\varphi(f_M)\} = \Omega(n^k/(2k)^{2k})$, as required. $\square$

A High Rank Sum of Three ROFs: In [16], Kayal et al. showed that there is a polynomial that can be written as sum of three ROFs such that any ROABP computing it requires exponential size. The lower bound proof in [16] is based on the construction of a polynomial using three edge disjoint perfect matchings on n vertices. We need a 3-regular mildly explicit family of expander graphs defined in [14]. Let $\mathcal{G} = (G(q))_{q>0, \text{ prime}}$ be a family of 3 regular expander graphs where a vertex x in $G(q)$ is connected to $x+1, x-1$ and x^{-1} where all of the operations are modulo q. When q is clear from the context, we denote $G(q)$ by G. Let G' be the double cover of G, i.e., $G' = (V_1, V_2, E')$ is the bipartite graph such that V_1, V_2 are copies of V and $u \in V_1, v \in V_2, (u, v) \in E' \iff (u, v) \in E$. It is known from [14] that the set of edges in E' can be viewed as the union of 3 edge disjoint perfect matchings. In [16], Kayal et al. construct a polynomial for each of these matchings and the hard polynomial is obtained by taking the sum of these three polynomials. This polynomial has degree $n/2$ and is unsuitable in the parameterized context.

We construct a polynomial h from G' similar to the one in [16], but having degree-k. Suppose $M_1 \cup M_2 \cup M_3 = E'$ be disjoint perfect matchings. We divide the n edges in each of the M_i into $\frac{k}{2}$ parts of $\frac{n}{k}$ edges each. Suppose $M_i = B_{i1} \cup B_{i2} \cup \cdots \cup B_{ik/2}$. The division is done arbitrarily. So, for each edge $(i, j) \in M$, we consider a monomial $x_i x_j$, and the final polynomial is the following:

$$h(x_1, \ldots, x_{2n}) = \sum_{i \in [3]} w_i \left(\prod_{j \in [\frac{k}{2}]} \sum_{(u,v) \in B_{ij}} x_u x_v \right),$$

where M_1, M_2 and M_3 are the edge-disjoint matchings such that $M_i = \prod_{j \in [\frac{k}{2}]} B_{ij}$, B_{ij} being the jth partition of edges in the matching M_i and w_1, w_2 and w_3 are formal variables. For a partition $\varphi : X \to Y \cup Z$, and an edge $(u, v) \in M_i$, (u, v) is said to be *bichromatic* with respect to φ if either $\varphi(x_u) \in Y$ and $\varphi(x_v) \in Z$ or $\varphi(x_u) \in Z$ and $\varphi(x_v) \in Y$. For a set of edges A over $\{x_1, \ldots, x_n\}$ let $\mathsf{be}_\varphi(A)$ be the number edges in A that are bichromatic with respect to φ. For a graph $G = (V, E)$, let $\mathsf{be}_\varphi(G)$ denote $\mathsf{be}_\varphi(E)$.

Let $\mathcal{D}$ denote the uniform distribution on the set of all partitions $\varphi : X \to Y \cup Z$ such that $|Y| = |Z|$. In the following we state the desired property of the polynomial h:

Theorem 2. *Let h be the polynomial defined as above. Then there is a constant $c > 0$ such that for every equi-partition φ of X, over the rational function field $\mathbb{F}(w_1, w_2, w_3)$*

$$\mathsf{rank}_\varphi(h) \geq \left(\frac{n}{k} \right)^{ck}.$$

Proof. Let $Y \subseteq X = \{x_1, \ldots, x_n\}$, $|Y| = \frac{n}{2}$ such that $\varphi : X \to Y \cup Z$. By the expander property of G (see [16]), the number of edges from Y to Z is lower bounded by $E(Y, Z) \geq \frac{(2 + 10^{-4})}{2} \cdot |Y| = \frac{(1+\epsilon)n}{2}$ for a fixed $\epsilon > 0$. (See [16] for details.)

Now, each perfect matching has $\frac{n}{2}$ edges, so the graph has $\frac{3n}{2}$ edges. By averaging, we get that there is a matching M_i, $1 \leq i \leq 3$ such that the number of bichromatic edges in M_i is at least $\frac{(1+\epsilon)n}{6}$. Without loss of generality, suppose $i = 1$. Let $h_1 = \prod_{j \in [\frac{k}{2}]} \sum_{(u,v) \in B_{1j}} x_u x_v$, i.e., the polynomial corresponding to M_1. Clearly, if the bichromatic edges in M_1 are distributed evenly across all sets in the partition $B_{11}, \ldots, B_{1k/2}$, $\mathsf{rank}_\varphi(h_1) = (((1 + \epsilon)/3k)n)^{k/2}$. However, this is not possible in general. Nevertheless, we get a smaller but good enough bound by a simple averaging argument. Let $\mathsf{be}_\varphi(M_i) = \sum_{j \in [\frac{k}{2}]} \mathsf{be}_\varphi(B_{ij})$. We have $\mathsf{be}_\varphi(M_1) \geq \frac{(1+\epsilon)n}{6}$. Let $\alpha = |\{j \mid \mathsf{be}_\varphi(B_{1,j}) \geq n/20k\}|$. Then

$$\frac{(1 + \epsilon)n}{6} \leq \mathsf{be}_\varphi(M_1) \leq \alpha \frac{n}{k} + (k/2 - \alpha) \frac{n}{20k}$$

$$\text{i.e., } \frac{(1 + \epsilon)n}{6} \leq \alpha \frac{n}{k} + (k/2 - \alpha) \frac{n}{20k}$$

$$\implies \alpha \geq \frac{(23 + 20\epsilon)}{114} k.$$

Note that $\text{rank}_\varphi(\sum_{(u,v)\in B_{1j}} x_u x_v) = \text{be}_\varphi(B_{1j})$ and hence we have $\text{rank}_\varphi(h_1) \geq (\frac{n}{20k})^\alpha = (\frac{n}{k})^{ck}$ for some constant $c > 0$ as required. $\qquad\square$

4   Lower Bounds

In this section we prove parameterized lower bounds for some special classes of syntactic multilinear ABPs. In particular, we prove lower bounds for the size of ROABPs, strict interval ABPs and a sum of restricted class of ROPs.

4.1   ROABP

In this section we prove a parameterized lower bound for the size of any ROABP computing the polynomials defined in Sect. 3. The lower bound argument follows from the fact that for any polynomial computed by an ROABP P, there exists an equi-partition φ of variables such that $\text{rank}_\varphi(P)$ is bounded by the size of the ROABP [19].

Theorem 3. *Any ROABP computing the polynomial family $f = (f_{n,2k})$ requires size $\Omega(n^k/(2k)^{2k})$.*

Proof. Let P be an ROABP of size S computing f. Consider an ordering from left to right of the variables occurring in the ROABP, $x_1, x_2, \ldots, x_n$. We can define the equi-partition $\varphi : X \to Y \cup Z$ such that,

$$\varphi(x_i) = \begin{cases} y_i, \text{ if } i \leq n/2 \\ z_{i-n/2} \text{ otherwise.} \end{cases}$$

Now, let L_i be a layer in P such that incoming edges to L_i are labelled with a linear polynomial in x_i. Then, we can represent f as

$$f(x_1, \ldots, x_n) = \sum_{j \in L_{n/2}} [s, v_j]_P \cdot [v_j, t]_P.$$

By definition of φ, for all $v_j \in L_{n/2}$, $\text{rank}_\varphi([s, v_j]_P \cdot [v_j, t]_P) = 1$.

Then, $\text{rank}_\varphi(f) \leq |L_{n/2}| \leq S$. By Theorem 1, $\text{rank}_\varphi(f) = \Omega(n^k/(2k)^{2k})$, therefore we have $S = \Omega(n^k/(2k)^{2k})$ as required. $\qquad\square$

Combining Theorem 3 with Theorem 2 we get:

Theorem 4. *An ROABP computing the family of polynomials h defined in Sect. 3 required size $n^{\Omega(k)}$.*

Proof. Follows from the proof of Theorem 3 that for any size S ROABP computing the polynomial h, there is an equi-partition φ such that $\text{rank}_\varphi(h) \leq S$. Then by Theorem 2, we have $S = n^{\Omega(k)}$ as required. $\qquad\square$

4.2   Strict Interval ABPs

In this section we prove a parameterized lower bound against the polynomial family f defined in Sect. 3 for the size of strict interval ABPs. Without loss of generality, assume that π is the identity permutation. Let P be a π strict-interval ABP computing the polynomial f. As a crucial ingredient in the lower bound proof, we show that using the standard divide and conquer approach, a strict-interval ABP can be transformed into a depth four circuit with $n^{\sqrt{k}}$ blow up in the size. To begin with, we need the following simple depth reduction for strict interval ABPs computing degree k polynomials. Proof is omitted.

Lemma 4. *Let P be a syntactic multilinear ABP of size S computing a homogeneous degree k polynomial g on n variables. Then there is a syntactic multilinear ABP P' of depth $k+1$ and size $O(S \cdot k)$ computing g such that:*

1. *Every node in the ith layer of P' computes a homogeneous degree i polynomial.*
2. *If P is strict interval then so is P'.*

Using Lemma 4 we can obtain a parameterized version of depth reduction to depth four circuits:

Lemma 5. *Let $g(x_1, \ldots, x_n)$ be a multilinear polynomial of degree k computed by a syntactic multilinear branching program P of size S. Then*

$$g(x_1, \ldots, x_n) = \sum_{i=1}^{T} \prod_{j=1}^{\sqrt{k}} f_{i,j} \tag{1}$$

for some $T = S^{O(\sqrt{k})}$ and $f_{i,j}$ is a degree $\sqrt{k}$ multilinear polynomial computed by a sub-program of P for $i \in \{1, \ldots, T\}, j \in \{1, \ldots, \sqrt{k}\}$.

Now, to prove the claimed lower bound for the size of strict interval ABPs, all we need is given a polynomial f computed by an strict interval ABP of size S, an equi-partition φ of X such that $\text{rank}_\varphi(f) \ll n^k$.

Lemma 6. *Let f be a polynomial computed by a strict interval ABP of size S. Then there is a partition φ such that $\text{rank}_\varphi(f) \leq S^{O(\sqrt{k})} n^{\sqrt{k}}$.*

Proof. Without loss of generality, assume that P is a strict interval ABP with respect to the identity permutation. Let $\varphi_{\text{mid}} : X \to Y \cup Z$ be the partition

$$\varphi_{\text{mid}}(x_i) = \begin{cases} y_i, & \text{if } i \leq n/2, \\ z_{i-n/2} & \text{otherwise.} \end{cases}$$

Consider the representation for f as in (1). Then for every $1 \leq i \leq T$, for all but one j, we have either $\varphi_{\text{mid}}(\text{var}([i_j, i_{j+1}])) \subseteq Y$ or $\varphi_{\text{mid}}(\text{var}([i_j, i_{j+1}])) \subseteq Z$. Therefore, $\text{rank}_{\varphi_{\text{mid}}}([s, i_1]_P \cdot \prod_{m=1}^{\sqrt{k}-2} [i_m, i_{m+1}]_P \cdot [i_{\sqrt{k}-1}, t]_P) \leq n^{\sqrt{k}}$, for every $i_j \in L_{j\sqrt{k}}$. By sub-additivity of rank_φ, we have $\text{rank}_\varphi(f) \leq S^{O(\sqrt{k})} n^{\sqrt{k}}$ for $\varphi = \varphi_{\text{mid}}$. $\square$

The required lower bound is immediate now.

Corollary 1. *Any strict-interval ABP computing the polynomial f has size $n^{\Omega(\sqrt{k})}$.*

4.3   Rank Bound for ROPs by Graph Representation

The reader might be tempted to believe that the lower bound arguments in the preceding sections might be applicable to more general models such as sum of ROFs and sum of ROABPs or even multilinear formulas. However, as we have seen in Sect. 3, there is a sum of three ROFs that has high rank under every partition. Thus our approach using rank_φ as a complexity measure is unlikely to yield lower bounds for even sum of ROFs, which is in contrast to the classical setting, where exponential lower bounds against models such as sum of ROFs and sum of ROABPs follow easily.

In this section, we develop a new method of analyzing rank of degree k polynomials computed by ROFs. Let $p \in \mathbb{F}[X]$ be the polynomial computed by a ROF Φ. We want to construct a graph $G_p = (X, E_p)$ corresponding to p so that $\mathsf{rank}_\varphi(p)$ can be related to certain parameters of the graph. A v in Φ is said to be a *maximal-degree-two gate* if v computes a degree two polynomial, and the parent of v computes a polynomial whose degree is strictly greater than two. Further, v is said to be a *maximal-degree-one* gate if v computes a linear form and the parent of v computes a polynomial of degree strictly greater than one. A gate v at depth 1 is said to be a *high degree gate* if the degree of the polynomial computed at v is strictly greater than two. Let V_2 denote the set of all maximal-degree-two gates in Φ, V_1 denote the set of all maximal-degree-one gates and V_0 denote the set of all high degree gates in Φ at depth one. Let $\mathsf{atomic}(\Phi) = V_0 \cup V_1 \cup V_2$. The following is a straightforward observation:

Observation 1. *Let Φ be an ROF and v be a maximal-degree-two gate in Φ. Then the polynomial Φ_v computed is of the form $\Phi_v = \sum_{i=1}^s \ell_{i_1} \ell_{i_2}$, where ℓ_{i_j} $1 \leq i \leq s$, $j \in \{1, 2\}$ are variable disjoint linear forms for some $s > 0$ such that each of the ℓ_{i_j} is dependent on at least one variable.*

For a linear form $\ell = \sum_{j=1}^r \alpha_{i_j} x_{i_j}$, let $\mathsf{path}(\ell)$ be the simple undirected path $(x_{i_1}, x_{i_2}), (x_{i_2}, x_{i_3}), \ldots, (x_{i_{r-1}}, x_{i_r})$. In the case when $r = 1$, $\mathsf{path}(\ell)$ is just single vertex. Similarly, for a subset $S \subseteq X$ of variables, let $\mathsf{path}(S)$ denote the path $(x_{i_1}, x_{i_2}), (x_{i_2}, x_{i_3}), \ldots, (x_{i_{r-1}}, x_{i_r})$ where $S = \{x_{i_1}, \ldots, x_{i_r}\}$, $i_1 < i_2 < \ldots < i_r$. For two variable disjoint linear forms ℓ and ℓ', let $\mathsf{path}(\ell\ell')$ be the path obtained by connecting the last vertex in $\mathsf{path}(\ell)$ to the first vertex of $\mathsf{path}(\ell')$ by a new edge. Now, we define a graph $G_p = (X, E_p)$ where vertices correspond to variables $x_u \in X$ and the set of edges E_p defined as follows. For each $v \in \mathsf{atomic}(\Phi)$ we add the following edges to E_p:

Case 1 $\Phi_v = \sum_{i=1}^r \ell_{i_1} \ell_{i_2}$ for some $r > 0$ add $\mathsf{path}(\ell_{i_1} \ell_{i_2})$ to G_p for every $1 \leq i \leq t$.

Case 2 $\Phi_v = \prod_{i \in S} x_i$ or $\Phi_v = \sum_{i \in S} c_i x_i$, where $S \subseteq X$, c_is are constants from $\mathbb{F}$, add $\mathsf{path}(S)$ to G_p.

It may be noted that the graph G_p is not unique as it depends on the given minimal ROF Φ computing f. In the following, we show that for a given partition φ, we bound the $\mathsf{rank}_\varphi(p)$ in terms of the number of bichromatic edges $\mathsf{be}_\varphi(G_p)$. We have:

Theorem 5. *Let $p \in \mathbb{F}[X_1,\ldots,X_n]$ be a multilinear polynomial of degree k computed by a ROF Φ. Then, for any equi-partition $\varphi : X \to Y \cup Z$, $\mathsf{rank}_\varphi(p) \le (4\mathsf{be}_\varphi(G_p))^{\frac{k}{2}}$.*

Proof. The proof is by induction on the structure of Φ. The base case is when the root gate of Φ is in $\mathsf{atomic}(\Phi)$. Consider a node $v \in \mathsf{atomic}(\Phi)$.

Case 1 $\Phi_v = \sum_{(i,j)\in S} x_i x_j$. If $\varphi(x_i)$, $\varphi(x_j)$ are not in the same partition, then each monomial $x_i x_j$ contributes 1 towards $\mathsf{rank}_\varphi(p)$. At the same time, the edge (x_i, x_j) added to E_p is bichromatic, so each monomial contributes 1 towards the measure $\mathsf{be}_\varphi(G_p)$ as well.

Case 2 $\Phi_v = \sum_{(a,b)\in T} \ell_a \ell_b$. If, for some $x_i, x_j \in \mathsf{var}(\ell_a)$, $\varphi(x_i), \varphi(x_j)$ are in different partitions, then the linear form ℓ_a contributes 2 towards $\mathsf{rank}_\varphi(\ell_a)$. If the same holds true for ℓ_b, then $\ell_a \ell_b$ would together contribute 4 towards $\mathsf{rank}_\varphi(p)$ and ≥ 2 towards the measure $\mathsf{be}_\varphi(G_p)$.

Case 3 $\Phi_v = \sum_{i\in W_1} c_i x_i$ or $\Phi_v = \prod_{i\in W_2} x_i$ for some $W_1, W_2 \subseteq X$. The first case has been considered already. For the second case, if $\exists x_a, x_b \in W_2$ such that $\varphi(x_a), \varphi(x_b)$ are in different partitions, the polynomial computed by the gate v will contribute a 1 towards $\mathsf{rank}_\varphi(p)$ and at least 1 towards $\mathsf{be}_\varphi(G_p)$, otherwise it contributes 0.

Thus we have verified that the statement is true when the root gate v of Φ is contained in $\mathsf{atomic}(\Phi)$. Suppose $p = p_1 \mathsf{op} p_2$ for $\mathsf{op} \in \{+, \times\}$ where p_1 and p_2 are variable disjoint and are computed by ROFs. By induction hypothesis, $\mathsf{rank}_\varphi(p_j) \le (4\mathsf{be}_\varphi(G_{p_j}))^{\frac{k_j}{2}}$ where $k_j = \deg(f_j)$. As $\mathsf{be}_\varphi(G_p) = \mathsf{be}_\varphi(G_{p_1}) + \mathsf{be}_\varphi(G_{p_2})$ and $k = k_1 + k_2$ ($\mathsf{op} = \times$) or $k = \max\{k_1, k_2\}$ ($\mathsf{op} = +$) we have, $\mathsf{rank}_\varphi(f) \le (4\mathsf{be}_\varphi(G_p))^{\frac{k}{2}}$ as required. $\qquad\square$

Recall that bisection of an undirected graph $G = (V, E)$ is a set $S \subseteq V$ such that $|S| = |V|/2$. The size of a bisection S is the number of edges across S and $\overline{S}$, i.e., $|\{(u, v) \mid (u, v) \in E, u \in S, v \notin S\}|$. The following is an immediate corollary to Theorem 5:

Theorem 6. *Let G be a graph on n vertices such that there is a bisection of G of size $n^{1-\epsilon}$. Suppose $p_1, \ldots, p_s$ be ROFs such that G_{p_i} is a sub-graph of G. Then, if $p = p_1 + \cdots + p_S$ we have $S = (n^{\Omega(k)}/t(k))$, where t is a computable function on k.*

Proof. Let $C = (S, \overline{S})$ be the cut and $\mathsf{size}(C)$ denote the number of edges across the cut. Define a partition $\varphi : X \to Y \cup Z$ as follows:

$$\varphi(x_i) \in \begin{cases} Y & \text{if } i \in S, \\ Z & \text{otherwise.} \end{cases}$$

Then by Theorem 5, $\text{rank}_\varphi(p_i) \leq \text{be}_\varphi(G_{p_i}))^{\frac{k}{2}}$. Since G_{p_i} is a sub-graph of G, we have $\text{be}_\varphi(G_p)) \leq \text{size}(C) \leq n^{1-\epsilon}$. Therefore, $\text{rank}_\varphi(p_i) \leq O_k(n^{(1-\epsilon)k/2})$. By sub-additivity, we have $\text{rank}_\varphi(f) \leq SO_k(n^{(1-\epsilon)k/2})$ where O_k is up to a factor that depends only on a function of k. By Theorem 1, we get $S = \Omega(n^{\epsilon k/2})$. $\square$

5   Conclusions

Our results demonstrate the challenges in translating classical arithmetic circuit lower bounds to the parameterized setting, when the degree of the polynomial is the parameter. We get a full rank polynomial that can be computed by depth four arithmetic circuits of fpt size, whereas in the classical setting, full rank polynomials cannot be computed by multilinear formulas of polynomial size [21]. This makes the task of proving parameterized lower bounds for algebraic computation much more challenging task. Given the application of polynomials whose degree is bound by a parameter in the design of efficient parameterized algorithms for many counting problems, we believe that this is a worthy research direction to pursue.

Further, we believe that our results are an indication that study of parameterized complexity of polynomials with degree as the parameter could possibly shed more light on the use of algebraic techniques in parameterized algorithms.

References

1. Amini, O., Fomin, F.V., Saurabh, S.: Counting subgraphs via homomorphisms. SIAM J. Discrete Math. **26**(2), 695–717 (2012)
2. Arvind, V., Raja, S.: Some lower bound results for set-multilinear arithmetic computations. Chicago J. Theor. Comput. Sci. (2016)
3. Baur, W., Strassen, V.: The complexity of partial derivatives. Theoret. Comput. Sci. **22**(3), 317–330 (1983)
4. Björklund, A.: Exact covers via determinants. In: STACS, pp. 95–106 (2010)
5. Björklund, A., Husfeldt, T., Taslaman, N.: Shortest cycle through specified elements. In: SODA, pp. 1747–1753 (2012)
6. Chauhan, A., Rao, B.V.R.: Parameterized analogues of probabilistic computation. In: Ganguly, S., Krishnamurti, R. (eds.) CALDAM 2015. LNCS, vol. 8959, pp. 181–192. Springer, Cham (2015). https://doi.org/10.1007/978-3-319-14974-5_18
7. Chillara, S., Engels, C., Limaye, N., Srinivasan, S.: A near-optimal depth-hierarchy theorem for small-depth multilinear circuits. In: FOCS (2018)
8. Downey, R.G., Fellows, M.R.: Fundamentals of Parameterized Complexity. Texts in Computer Science. Springer, London (2013). https://doi.org/10.1007/978-1-4471-5559-1
9. Engels, C.: Why are certain polynomials hard? A look at non-commutative, parameterized and homomorphism polynomials. Ph.D. thesis, Saarland University (2016)
10. Fomin, F.V., Lokshtanov, D., Raman, V., Saurabh, S., Rao, B.V.R.: Faster algorithms for finding and counting subgraphs. J. Comput. Syst. Sci. **78**(3), 698–706 (2012)
11. Fournier, H., Limaye, N., Malod, G., Srinivasan, S.: Lower bounds for depth 4 formulas computing iterated matrix multiplication. In: STOC, pp. 128–135 (2014)

12. Ghosal, P., Prakash, O., Rao, B.V.R.: On constant depth circuits parameterized by degree: identity testing and depth reduction. In: Cao, Y., Chen, J. (eds.) COCOON 2017. LNCS, vol. 10392, pp. 250–261. Springer, Cham (2017). https://doi.org/10.1007/978-3-319-62389-4_21
13. Gupta, A., Kamath, P., Kayal, N., Saptharishi, R.: Approaching the chasm at depth four. J. ACM (JACM) **61**(6), 33 (2014)
14. Hoory, S., Linial, N., Wigderson, A.: Expander graphs and their applications. Bull. Am. Math. Soc. **43**(4), 439–561 (2006)
15. Kabanets, V., Impagliazzo, R.: Derandomizing polynomial identity tests means proving circuit lower bounds. Comput. Complex. **13**(1–2), 1–46 (2004)
16. Kayal, N., Nair, V., Saha, C.: Separation between read-once oblivious algebraic branching programs (ROABPs) and multilinear depth three circuits. In: STACS, pp. 46:1–46:15 (2016)
17. Kumar, M., Saraf, S.: The limits of depth reduction for arithmetic formulas: it's all about the top fan-in. SIAM J. Comput. **44**(6), 1601–1625 (2015)
18. Müller, M.: Parameterized randomization. Ph.D. thesis, Albert-Ludwigs-Universität Freiburg im Breisgau (2008)
19. Nisan, N.: Lower bounds for non-commutative computation. In: STOC, pp. 410–418. ACM (1991)
20. Raz, R.: Separation of multilinear circuit and formula size. Theory Comput. **2**(6), 121–135 (2006)
21. Raz, R.: Multi-linear formulas for permanent and determinant are of super-polynomial size. J. ACM **56**(2), 8:1–8:17 (2009)
22. Raz, R., Yehudayoff, A.: Lower bounds and separations for constant depth multilinear circuits. Comput. Complex. **18**(2), 171–207 (2009)
23. Saptharishi, R.: A survey of lower bounds in arithmetic circuit complexity. Technical report (2019)
24. Shpilka, A., Yehudayoff, A.: Arithmetic circuits: a survey of recent results and open questions. Found. Trends Theor. Comput. Sci. **5**(3–4), 207–388 (2010)
25. Valiant, L.G.: The complexity of computing the permanent. Theor. Comput. Sci. **8**, 189–201 (1979)

Many-to-One Popular Matchings
with Two-Sided Preferences
and One-Sided Ties

Kavitha Gopal[1], Meghana Nasre[1], Prajakta Nimbhorkar[2,3],
and T. Pradeep Reddy[1(✉)]

[1] Indian Institute of Technology, Madras, Chennai, India
pradeeptenkayyagari@gmail.com
[2] Chennai Mathematical Institute, Chennai, India
[3] UMI ReLaX, Chennai, India

Abstract. We consider the problem of assigning applicants to posts
when each applicant has a strict preference ordering over a subset of
posts, and each post has all its neighbors in a single tie. That is, a
post is indifferent amongst all its neighbours. Each post has a capacity
denoting the maximum number of applicants that can be assigned to
it. An assignment M, referred to as a *matching*, is said to be *popular*,
if there is no other assignment M' such that the number of votes M'
gets compared to M is more than the number of votes M gets compared
to M'. Here votes are cast by applicants and posts for comparing M
and M'. An applicant a votes for M over M' if a gets a more preferred
partner in M than in M'. A post p votes for M over M' if p gets more
applicants assigned to it in M than in M'. The number of votes a post
p casts gives rise to two models. Let $M(p)$ denote the set of applicants p
gets in M. If $|M(p)| > |M'(p)|$, p can cast $|M(p)| - |M'(p)|$-many votes
in favor of M, or just one vote. The two models are referred to as the
multi-vote model and *one-vote model* in this paper.

We give a polynomial-time algorithm to determine the existence of a
popular matching in the multi-vote model, and to output one if it exists.
We give interesting connections between the two models. In particular,
we show that a matching that is popular in the one-vote model is also
popular in the multi-vote model, however the converse is not true. We
also give a polynomial-time algorithm to check if a given matching is
popular in the one-vote model, and if not, then output a more popular
matching.

1 Introduction

We consider the problem of assigning a set of applicants A to a set of posts P,
where applicants have preferences over posts. Formally, the input to our problem
is a bipartite graph $G = (A \cup P, E)$ with a capacity $c(p)$ associated with every
post p. A post can be assigned any number of applicants up to a maximum of
$c(p)$. Every edge has a rank associated with it; if edge (a, p) has rank i then p is

© Springer Nature Switzerland AG 2019
D.-Z. Du et al. (Eds.): COCOON 2019, LNCS 11653, pp. 193–205, 2019.
https://doi.org/10.1007/978-3-030-26176-4_16

the ith choice post for a. Applicant a prefers post p over p' if the rank of edge (a, p) is smaller than the rank of edge (a, p'). An assignment or a matching M in this setting is a subset of E such that each applicant has at most one edge incident to it in M and a post p has at most $c(p)$ edges incident to it in M.

We compare a matching M with another matching N based on the number of votes each of them gets compared to the other. An applicant a prefers a matching M to a matching N if either (i) a is matched in M and unmatched in N or (ii) a prefers $M(a)$ to $N(a)$; otherwise a is indifferent between M and N. Here, $M(a)$ denotes the post a is matched to in M. Similarly, a post p prefers a matching M to a matching N if and only if $|M(p)| > |N(p)|$. Thus posts are indifferent among all their neighbors.

There are two different models possible depending on the number of votes that a post casts – we denote them by the *multi-vote model* and the *one-vote model*. If $|M(p)| > |N(p)|$, in the multi-vote model, the post p gives $|M(p)| - |N(p)|$ votes to the matching M when comparing M and N. In contrast, in the one-vote model, the post gives only one vote to M when comparing M and N. Irrespective of the way in which a post votes, a matching M is more popular than a matching N, if M gets more votes than N. A matching M is popular if there exists no other matching more popular than M. Our goal is to compute a popular matching in the instance.

Cseh et al. [5] introduced this model of two-sided preferences (applicants and posts both have preferences) with one-sided ties (posts are indifferent amongst all the neighbours) in the one-to-one setting, where all posts have unit capacities. They show a polynomial-time algorithm for the popular matchings problem in this model. As mentioned in [5], the model of two-sided preferences and one-sided ties can be considered to be in between the one-sided preference list model (studied by Abraham et al. [1]) and the two-sided preferences model, well-known as the stable marriage model [8]. Since Cseh et al. study the problem in the one-to-one setting, the multi-vote and one-vote models coincide.

The many-to-one setting considered in our paper is a natural generalization motivated by practical scenarios like student course allocation. This setting allows courses to be indifferent between students, however, gives them vote(s) to distinguish between matchings. A multi-vote model is relevant when courses get votes proportional to their capacities, whereas the one-vote model is relevant when every course gets just one vote. The multi-vote model considered by us has been previously used in the context of popular matchings in the two-sided preference list model in [3,11,12].

Our main contribution in this paper is a polynomial-time algorithm for the many-to-one popular matchings problem in the two-sided preferences with one-sided ties model for the multi-vote model. We state our main result as Theorem 1 below. Throughout, we use m and n to denote the number of edges and vertices in G respectively. Let $\hat{C}$ denote the sum of capacities of all posts in G.

Theorem 1. *Given an instance $G = (A \cup P, E)$ of the many-to-one two-sided preferences with one-sided ties problem, there exists an $O(\sqrt{\hat{C}} n^2)$-time algorithm* *to decide the existence of a popular matching in G in the multi-vote model, and to compute one if it exists.*

We note that the algorithm for the many-to-one problem in multi-vote model closely follows the algorithm for the one-to-one problem [5]; however, subtle changes are required in the algorithm as well as for the proofs. We remark that even in the one-to-one setting there is no known characterization of popular matchings in this model. However, it is natural to ask, given a matching M in G, determine if M is popular. Our next theorem is an efficient algorithm to decide whether M is popular in the multi-vote model. The same algorithm can be used to decide whether M is popular in G in the one-to-one setting.

Theorem 2. *Given an instance $G = (A \cup P, E)$ of the many-to-one two-sided preferences with one-sided ties problem, there exists an $O(mn)$ time algorithm to decide whether a given matching M is popular in the multi-vote model in G.*

We remark that the standard technique of cloning (that is, making $c(p)$ copies of a post p) to reduce the many-to-one problem to the one-to-one setting does not work. Consider a simple example where there are three applicants $\{a_1, a_2, a_3\}$ and a single post p with $c(p) = 3$. It is clear that only the matching which matches all applicants to p is a popular matching. An approach of cloning will create three clones p_1, p_2, p_3, each with unit capacity, and set the preference of each applicant to be p_1 followed by p_2, and then followed by p_3. Note that the three copies of the post p are indifferent among all the three applicants. It is well-known that this instance does not admit a popular matching (see Abraham et al. [1] and Cseh et al. [5]).

Many-to-One Popular Matchings in the One-Vote Model: In light of Theorem 1, it is natural to determine the complexity of computing a popular matching in the one-vote model (if it exists). We leave this question open; however, we show interesting connections between the two models, and present an algorithm to check if a given matching is popular in the one-vote model.

The following example illustrates that an instance that admits a popular matching in the multi-vote model may not admit a popular matching in the one-vote model. Consider an instance with $A = \{a_1, \ldots, a_6\}$ and $P = \{p_1, p_2, p_3\}$, each post having a capacity of two. Applicants a_1, a_2 prefer p_1 followed by p_2 followed by p_3, whereas each of $a_3, \ldots, a_6$ prefers p_1 followed by p_2. It can be verified that the matching $M = \{(a_1, p_3), (a_2, p_3), (a_3, p_2), (a_4, p_2), (a_5, p_1), (a_6, p_1)\}$ is a popular matching in the multi-vote model. However, the instance does not admit a popular matching in the one-vote model. For any matching N, we can obtain a more popular matching N' by promoting the matched partners of p_3 (or unmatched applicants) to p_2, the matched partners of p_2 to p_1, and leaving the applicants in $N(p_1)$ unmatched. Thus four applicants prefer N' over N whereas two applicants (who are unmatched in N') prefer N over N'. The only possible post that gets fewer applicants in N' as compared to N is the post p_3 which gives one vote to N over N' due to the one-vote model. Thus, N' is more popular than N and since this happens for every matching, the instance does not admit a popular matching in the one-vote model. Interestingly, it turns out that any matching popular in the one-vote model is also popular in the multi-vote model. Our next result is an algorithm to decide whether a given matching is popular in the one-vote model.

Theorem 3. *Given an instance $G = (A \cup P, E)$ of the many-to-one two-sided preferences with one-sided ties problem, there exists an $O(mn^2)$ time algorithm to decide whether a given matching M is popular in the one-vote model in G.*

Background: A polynomial-time algorithm for the popular matchings problem in the one-sided preferences model is given by [1]. In the stable marriage setting, where both applicants and posts have strict preferences, a popular matching always exists and a maximum cardinality popular matching can be found in polynomial-time [10]. When ties are allowed in the stable marriage setting, the problem is NP-complete [2]. Cseh et al. [5] consider the problem in a restricted form of stable marriage setting where ties are allowed only on the posts' side, and show that the NP-completeness result still holds. They also show that a further restriction where every post has all its neighbours in a single tie makes the problem polynomial-time solvable. Subsequent to their work, Chang and Lai [4] consider the problem of deleting a set of applicants in an instance of the two-sided preferences with one-sided ties instance such that the resulting instance admits a popular matching.

2   Preliminaries

For a vertex $u \in V$, we let $nbr(u)$ denote the neighbours of u in G. If there is a different graph, say H under consideration, we explicitly use $nbr_H(u)$ to denote the neighbours of u in H. Given a matching M in the many-to-one setting, a post p is under-subscribed if $|M(p)| < c(p)$; if $|M(p)| = c(p)$ we say that p is fully-subscribed. For convenience, we will call an applicant unmatched in M as under-subscribed. We recall a well-known result from matching theory.

Dulmage-Mendelsohn Decomposition [6]: Any maximum matching M in a bipartite graph $G = (A \cup P, E)$ decomposes the vertex set $A \cup P$ into three pair-wise disjoint sets *even* ($\mathcal{E}$), *odd* ($\mathcal{O}$), and *unreachable* ($\mathcal{U}$) as defined below.

- **Even:** A vertex $v \in A \cup P$ is *even* if v has an even length alternating path w.r.t. M starting at an under-subscribed vertex. Any under-subscribed vertex is *even* since it has an zero length path starting at itself.
- **Odd:** A vertex $v \in A \cup P$ is *odd* if v has an odd length alternating path w.r.t. M starting at an under-subscribed vertex.
- **Unreachable:** A vertex $v \in A \cup P$ is *unreachable* in M if it is neither *even* nor *odd*.

It is known that these sets are *invariant* of the maximum matching M. Furthermore, a maximum matching in G does not contain any edge whose end points are $\mathcal{OO}$ or $\mathcal{OU}$. Finally, in the many-to-one setting G does not contain any edge whose end points are $\mathcal{EE}$. We refer the reader to [6,9,13] for a proof.

In several proofs we require to consider the symmetric difference $M \oplus M'$ of two matchings M and M'. In the many-to-one setting $M \oplus M'$ is a set of connected components. It is useful to decompose the component into a maximal collection of edge-disjoint alternating paths and cycles. We refer the reader to [11,12] for details of constructing such a decomposition.

Labeling of Edges w.r.t. a Matching: It is useful to label the edges not in the matching M with the vote of the applicant for that edge versus the matched edge. An edge $(a, p) \notin M$ gets the label $+1$ if either a is unmatched in M or a prefers p over $M(a)$. Otherwise the edge (a, p) gets the label -1. We remark that the labels on the edges do not capture the votes of the posts. Let $\Delta_m(M, M')$ (resp. $\Delta_1(M, M')$) denote the number of votes that M gets when compared to M' in the multi-vote model (resp. one-vote model). We write $M \succ_m M'$ if M is more popular than M' in the multi-vote model, that is $\Delta_m(M, M') > 0$. Analogously, we write that $M \succ_1 M'$ if M is more popular than M' in the one-vote model, that is $\Delta_1(M, M') > 0$.

3   Many-to-One Multi-vote Popular Matching

In this section, we describe a polynomial-time algorithm for the popular matching problem in multi-vote model. The input instance is a bipartite graph $G = (A \cup P, E)$ with $c(p)$ associated with every post $p \in P$.

Overview of the Algorithm: We associate two posts $f(a)$ and $s(a)$ with each applicant a, similar to [1] and [5]. Here $f(a)$ is the most preferred post of a. We call a post p an f-post if p is $f(a)$ for some applicant a; else we call p a non-f-post. To determine $s(a)$, we construct the graph G_1 on rank-1 edges of G. Let M_1 be a maximum matching of G_1. Decompose the vertices of G as *odd, even* and *unreachable* with respect to M_1 in G_1. For every applicant a, $s(a)$ is the most preferred *even* post of a in G_1 w.r.t. M_1. Note that $s(a)$ may not exist for some applicant. Let r_a be the rank of $s(a)$ in a's preference list, if $s(a)$ exists, otherwise define $r_a = \infty$. Algorithm 1 gives the detailed steps of our algorithm.

As in [5], we maintain three sets of posts X, Y, and Z. The set X is initialized to the set of all f-posts and Y is initialized to all non-f-posts. The set Z is initialized to be empty. In every iteration of the algorithm, we reconstruct the graph H. At the start of each iteration, the edge-set of H is empty. We add edges to H as described in lines 5 to 10 of Algorithm 1 and find a maximum matching M. We use M to partition the vertices as *odd, even* and *unreachable*. For every *even* post in Y, we freeze the rank-1 edges incident to it, if any, by storing them in a set S, and then move the post to Z after appropriately adjusting its capacity (lines 14 to 18). As mentioned in [5], the posts in Z are the *unwanted* posts. In the one-to-one setting [5] an f-post never moves to the set Z. In the many-to-one setting, however, an f-post may move to the set Z, hence we need the set S. We show that the edges in set S must belong to every popular matching of G, if one exists. We quit the while loop when there is no even post in Y.

Algorithm 1. Popular matching in many-to-one multi-vote model.

1: $A' = A$, $X =$ set of f-posts in G, $Y = P \setminus X$, $Z = \emptyset$, $S = \emptyset$.
2: **while** true **do**
3:     H is the empty graph on $A' \cup P$.
4:     **for** $a \in A' \setminus nbr(Z)$ **do**
5:         if $f(a) \in X$ then add the edge $(a, f(a))$ to H.
6:     **for** every $p \in X$ such that degree(p) in H is $< c(p)$ **do**
7:         delete p from X and add p to Y.
8:     **for** each $a \in A'$ **do**
9:         let p be a's most preferred post in Y.
10:         if the rank of p in a's preference list is $\leq r_a$, then add (a, p) to H.
11:     compute a max. matching M in H and a decomposition of vertices w.r.t M.
12:     **for** each *even* post $p \in Y$ **do**
13:         **if** p is an f-post **then**
14:             add all rank-1 edges incident on p to the set S.
15:             remove edges in S from G and delete the corr. applicants from A'.
16:             move p to Z and set $c(p)$ to $c(p)-$ # of rank-1 edges incident on p in M.
17:         **else if** p is a non-f-post **then**
18:             move p to Z.
19:     If all posts in Y are *odd* or *unreachable* then quit the while loop.
20: **for** each $a \in nbr(Z) \cap A'$ **do**
21:     add the edge (a, p) where p is a's most preferred post in Z.
22: **for** $a \in A'$ **do**, if $s(a)$ does not exist and $nbr(a) \subseteq X$, add unique last resort post $\ell(a)$. Add the edge $(a, \ell(a))$.
23: compute maximum matching M in H by augmentation.
24: **if** M matches all applicants in A' **then** return $M \cup S$ else return "No popular matching in multi-vote model".

At the end of the **while** loop, all the posts in X and Y are *odd* or *unreachable* in M. After the while loop, the algorithm adds edges to the posts in Z. For applicants a such that $s(a)$ does not exist and $nbr(a) \subseteq X$, we add a unique last resort post $\ell(a)$ and add the edge $(a, \ell(a))$. Note that the posts $\ell(a)$ belong to the set Y. Finally, we augment the matching using these additional edges and call the resultant matching M. If M matches all the applicants, we declare it to be popular in the multi-vote model, else we declare that G does not admit a popular matching.

Running Time: The main **while** loop of Algorithm 1 is executed at most $|P|$ times and in each iteration the most expensive step is to compute a many-to-one maximum cardinality matching in H. This can be done by Gabow's algorithm [7] in time $O(\sqrt{\hat{C}}n)$ time where $\hat{C}$ is the sum of capacities of all posts. Thus justifies the running time of Algorithm 1 in Theorem 1.

Properties of the Algorithm: Here we claim useful properties about Algorithm 1. All omitted proofs appear in the full-version. We first prove that a *high demand* f-post does not move to the set Z.

Lemma 1. *Let p be an f-post and k denote the number of applicants that treat p as its rank-1 post. If $k \geq c(p)$ then, p does not move to the set Z during the course of the algorithm.*

Below we provide justification for the applicants for whom we add dummy last-resort posts. Note that these applicants may be left unmatched in the matching output by our algorithm, which we claim to be popular.

Lemma 2. *Let a be an applicant such that $s(a)$ is defined. If a is left unmatched in a matching M, then M is not popular in the multi-vote model.*

Lemma 3. *Let a be an applicant such that $s(a)$ is not defined and $nbr(a) \nsubseteq X$ after the while loop of Algorithm 1. If a is left unmatched by M output by our algorithm, then M is not popular in the multi-vote model.*

3.1 Proof of Correctness

We prove that G has a popular matching if and only if the graph H constructed in Algorithm 1 has an A-complete matching (that matches all applicants). We prove the sufficiency here, the proofs of necessity appear in the full version.

Recall the labeling of unmatched edges with respect to a matching as described in Sect. 2. Let X, Y and Z be the sets at the end of Algorithm 1. We denote by $M(X)$ the set of applicants who are matched to posts in X. Similarly define $M(Y)$ and $M(Z)$. We use the following lemma from [5], see [5] for a proof.

Lemma 4 [5]. *If M is the matching output by Algorithm 1, then the labels on the edges satisfy the following properties.*

- *The edges in $(M(X) \times Y) \cup (M(Y) \times Z)$ gets -1 label.*
- *The edges in $M(L) \times L$ for $L = X, Y$, and Z gets -1 label.*
- *If an edge (a, p) gets labeled $+1$, then $(a, p) \in (M(Y) \times X) \cup (M(Z) \times (X \cup Y))$.*

Let M be the matching output by Algorithm 1. Let M' be any other matching in the graph G. We decompose $M \oplus M'$ into a collection of edge-disjoint alternating paths and cycles. For brevity, we denote the paths and cycles obtained after the decomposition by $M \oplus M'$.

Lemma 5. *Let M be the matching output by Algorithm 1 and M' be any matching in G.*

- *In any alternating cycle in $M \oplus M'$, the number of edges labeled -1 is at least the number of edges that are labeled $+1$.*
- *In any alternating path in $M \oplus M'$, the number of edges labeled $+1$ is at most 2 more than the number of edges that are labeled -1.*
- *In any even-length alternating path in $M \oplus M'$, the number of edges labeled -1 is at least the number of edges that are labeled $+1$.*

Proof. Let Q be a cycle in $M \oplus M'$. There are two cases depending on whether Q contains a post in Z.

- The cycle Q does not contain a post in Z. In this case, we claim that Q cannot have two consecutive M' edges labeled $+1$. Let $\langle a_1, p_1, a_2, p_2 \rangle$ be a path in the cycle Q. Here $M'(a_1) = p_1, M'(a_2) = p_2$ and $M(a_2) = p_1$. If the edge (a_1, p_1) is labeled $+1$, then edge (a_2, p_2) must be labeled -1. Let (a_1, p_1) be a $+1$ edge. By Lemma 4, $M(a_1) \in Y$ and $p_1 \in X$. Since $p_1 \in X$, (a_2, p_1) is a rank-1 edge. Therefore (a_2, p_2) is a -1 edge. Thus, in this case, for every $+1$ edge, there is a -1 edge in the cycle Q.
- If the cycle Q contains a post in Z, there can be two consecutive $+1$ edges. We will show that for every two consecutive $+1$ edges there exists two consecutive -1 edges. Let $\langle a_i, p_i, a_{i+1}, p_{i+1}, a_{i+2}, p_{i+2} \rangle$ be a path within Q as shown in the Fig. 1. The red edges are matched in M. From Lemma 4, the posts $p_i \in X$, $p_{i+1} \in Y$ and $p_{i+2} \in Z$. By the Step 4 of Algorithm 1, there is no edge between a_i and a vertex in the set Z. Since Q is a cycle, to reach p_{i+2}, we must use a post in Y and subsequently reach p_{i+2} in Z. Such a path must contain two consecutive -1 edges to reach the post p_{i+2}.

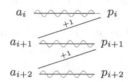

Fig. 1. A path in the alternating cycle Q in $M \oplus M'$. (Color figure online)

In any alternating path Q, there can be a sub-path with two consecutive $+1$ edges. However, the weight of the path Q cannot exceed 2. To see this, we note that, once we have two $+1$ edges in the path, we must be at an applicant $a \in M(X)$. An applicant in $M(X)$ does not have a $+1$ edge incident on it (because it is matched to its rank-1 post). Thus to have one more $+1$ edge in the path, it must traverse to vertex either in Y or in Z. Any path from an applicant in $M(X)$ to a post in Y (Z) is at least -1 (resp. -2). Thus the total weight of an alternating path cannot exceed 2.

Let $\langle p_1, a_1, p_2, a_2, \ldots, p_k \rangle$ be an even length alternating path. The post p_k and p_1 are under-subscribed in M and M' respectively. The post p_k must be in set Z as it is under-subscribed in M. If the post p_1 is in set Z, the number of edges labeled -1 are at least the number of edges labeled $+1$. Similarly, if the post $p_1 \in Y$ (X), then the number of edges labeled -1 at least one more (resp. at least two more) than the number of edges labeled $+1$. □

Lemma 6. *The matching M output by Algorithm 1 is multi-vote popular.*

Multi-vote versus One-Vote: We remark that, in the decomposition of $M \oplus M'$ into a collection of edge-disjoint paths and cycles, the same post p may appear as an end-point at multiple paths. However, since we are in the multi-vote model, this is not an issue, since the post casts multiple votes. We remark that Lemma 6 does not hold for the one-vote model.

4 Verifying Popularity

In this section we consider the problem of verifying whether a given matching M is popular in the multi-vote model. We consider the same question in the one-vote model. For the multi-vote model, we have already presented an algorithm to compute a popular matching if one exists. However, there is no known characterization of a popular matching for the two-sided preferences with one-sided ties model even in the one-to-one setting. Furthermore our verification algorithm in the multi-vote model is useful for the verification algorithm for the one-vote model. In fact, for the one-vote model, we assume that our input is a matching that is popular in the multi-vote model.

4.1 Multi-vote Model

To verify whether a given matching M is popular in the multi-vote model, we build a directed weighted graph in which we check for negative weight cycles. We describe the construction of the graph in Algorithm 2.

Algorithm 2. Verify whether a matching M is popular in the multi-vote model.

1: $H = (V, F)$ where $V = \{s, t\} \cup P \cup \{\ell(a) \mid a \in A\}$ and $F = \emptyset$.
2: For every matched post $p \in P$, add edge (s, p) with weight 1 to F.
3: For every under-subscribed post $p \in P$, add edge (p, t) with weight -1 to F.
4: **for** every matched applicant $a \in A$ **do**
5:     Add edge $(\ell(a), t)$ with weight 0 to F.
6:     Add edge $(M(a), \ell(a))$ with weight 1 to F.
7:     **for** every p in the preference list of a **do**
8:         **if** a prefers $M(a)$ over p **then**
9:             Add edges $(M(a), p)$ with weight 1 to F.
10:         **else**
11:             Add edges $(M(a), p)$ with weight -1 to F.
12: **for** every unmatched applicant $a \in A$ **do**
13:     Add edge $(s, \ell(a))$ with weight 0 to F.
14:     For every p in the preference list of a, add edges $(\ell(a), p)$ with weight -1 to F.
15: Add edge (t, s) with weight 0 to F.
16: **if** H contains a negative weight cycle **then**
17:     Output "M is not popular in the multi-vote model". Return.
18: **else**
19:     Output "M is popular in the multi-vote model". Return.

We prove the correctness of the construction using the following lemma.

Lemma 7. *A matching M in the many-to-one two-sided preferences with one-sided ties instance G is popular in the multi-vote model if and only if H constructed in Algorithm 2 has no negative weight cycle.*

We comment on the running time of the algorithm. Negative cycles can be detected using the Bellman-Ford Algorithm. H has $O(n)$ vertices and $O(m)$ edges. Thus verification of popularity in the multi-vote model can be done in $O(mn)$ time. This justifies Theorem 2.

4.2   One-Vote Model

Before we discuss our verification algorithm for one-vote model, we prove that a matching M that is one-vote popular is also multi-vote popular.

Lemma 8. *If G admits a popular matching M in the one-vote model then M is also a popular matching in the multi-vote model.*

Proof (Sketch). If M is one-vote popular but not multi-vote popular, then $M' \succ_m M$. We decompose $M \oplus M'$ into a collection of edge-disjoint paths and cycles. We show that, if for any path or cycle Q, $M \oplus Q \succ_m M$ then $M \oplus Q \succ_1 M$, contradicting that M is popular in the one-vote model.   □

We are given a matching M which is popular in the multi-vote model. Our goal is to verify whether M is popular in the one-vote model. We first label the posts similar to [5] as L_1, L_2, L_3 w.r.t. M as follows.

Labeling of Vertices w.r.t. a Multi-vote Popular Matching M:

1. Initialize $L_1 = L_2 = \emptyset$ and $L_3 = $ set of all posts under-subscribed in M.
2. For each 5-length alternating path $\rho = \langle a_i, p_i, a_{i+1}, p_{i+1}, a_{i+2}, p_{i+2} \rangle$ having two +1 edges (see Fig. 1), add p_i to L_1, p_{i+1} to L_2 and p_{i+2} to L_3.
3. Let $p \in P$ be a post that is not a part of any 5 length alternating path with two +1 edges. Repeat the following rules until there are no more posts to be added via below rules:
   - Suppose none of the applicants in $M(p)$ have a +1 edge incident on them and also one of the applicants in $M(p)$ is in $nbr(L_3)$: Add p to L_2.
   - If an applicant in $M(p)$ has a +1 edge to a post in L_2, then add p to L_3.
4. For each p such that no applicant in $M(p)$ has a +1 edge incident to it:
   - If $M(p) \cap nbr(L_3) = \emptyset$, then add p to L_1.
5. For each p not yet in $L_2 \cup L_3$ and there exists an applicant in $M(p)$ which has a +1 edge to a vertex in L_1:
   - Add p to L_2.

As observed in [5], the labeling of the vertices ensures that Lemma 4 holds when X, Y, Z are replaced by L_1, L_2, L_3 respectively. Using this labeling on the vertices and the fact that M is multi-vote popular, we show the following structure on any matching M' (if it exists) which is more popular than M in the one-vote model.

Lemma 9. *Let M be a popular matching in the multi-vote model. Let $M' \succ_1 M$ such that $M \oplus M' = C$ is edge minimal. Then, $C = Q_1 \cup Q_2$ where Q_1 and Q_2 are two edge disjoint alternating paths starting at a post $p \in L_3$ and ending at two vertices v_1 and v_2 belonging to $M(L_1) \cup L_3$ and $v_1 \neq v_2 \neq p$.*

Lemma 10 gives another observation on the set of paths in $C = M \oplus M'$. Define D (to be deleted) as the set of edges as defined by (1), (2) (3) below.

(1) All the -1 edges of the form $M(L_2) \times L_1$, $M(L_3) \times (L_1 \cup L_2)$.
(2) All the $+1$ edges of the form $M(L_3) \times L_1$.
(3) All the unmatched edges in the set $M(L_1) \times L_1$, $M(L_2) \times L_2$, and $M(L_3) \times L_3$.
    Note that all these edges are labeled -1.

Lemma 10. *Let M be a popular matching in the multi-vote model and assume that $M' \succ_1 M$. Let $M \oplus M' = C$, where C is a set of two edge disjoint paths Q_1 and Q_2 both of which start at a post $p \in L_3$. The paths Q_1 and Q_2 do not use the edges in set D.*

We are now ready to give our algorithm (Algorithm 3) to verify whether a matching M is popular in the one-vote model. The input is a matching M that is popular in the multi-vote model. The algorithm first labels the posts as L_1, L_2, L_3 w.r.t. M and deletes "unnecessary edges" in the first three steps. For each post $p \in L_3$, it constructs a flow network. Using the flow network, the goal is to explore $L_3 \leadsto M(L_1)$ paths and $L_3 \leadsto L_3$ paths. The deleted edges and the flow network construction ensures (i) every $L_3 \leadsto M(L_1)$ path has weight $+2$, (ii) every $L_3 \leadsto L_3$ path has weight 0 and (iii) the paths that we find are edge-disjoint and end at distinct posts.

Algorithm 3. Verify whether a matching M is popular in one-vote model.

Require: M is popular in the multi-vote model.
1: Delete all the -1 edges of the form $M(L_2) \times L_1$, $M(L_3) \times (L_1 \cup L_2)$.
2: Delete all the $+1$ edges of the form $M(L_3) \times L_1$.
3: Delete all the unmatched edges in the set $M(L_1) \times L_1$, $M(L_2) \times L_2$, and $M(L_3) \times L_3$.
    Note that all these edges are labeled -1.
4: Let G_M be the graph constructed above.
5: **for** each post $p \in L_3$ construct the flow network $H_M(p)$ as follows: **do**
6:     Add (s, p) edge with capacity $|M(p)|$.
7:     Direct all the matched edges in G_M from P to A.
8:     Direct all the unmatched edges in G_M from A to P.
9:     For each under-subscribed post $p' \neq p$, add an edge (p', t) with capacity 1.
10:     For each applicant $a \in M(L_1)$, add an edge (a, t) with capacity 1.
11:     Compute the max-flow from s to t in $H_M(p)$. If max-flow value is greater than 1, then output "M is not popular in one-vote model". Return.
12: Output "M is popular in one-vote model". Return.

Lemma 11. *Let M be a matching that is popular in the multi-vote model. Algorithm 3 correctly verifies whether M is popular in the one-vote model.*

Proof. Assume that Algorithm 3 returns "M is not popular in one-vote model". This is because, for some post $p \in L_3$, it found a flow of at least two in the flow network $H_M(p)$. Due to the construction of the flow network and the deletion of edges, this implies that there exists at least two edge disjoint paths from p which are of the form $L_3 \rightsquigarrow M(L_1)$ or $L_3 \rightsquigarrow L_3$. For an $L_3 \rightsquigarrow M(L_1)$ path, the weight is $+2$ and we lose the vote of the applicant at which the path ends. Thus we gain $+1$ along such a path. For an $L_3 \rightsquigarrow L_3$ path, the weight is 0 and we gain the vote of the post at which the path ends. Thus the gain along such a path is at least $+1$. Switching along at least two such paths gives us a matching M', which, when compared with M, gets at least two votes along the two or more edge-disjoint paths and loses the vote of p. Thus, $M' \succ_1 M$ and hence the output is correct.

Now let the algorithm return "M is popular in the one-vote model". By Lemmas 9 and 10 it is sufficient to explore paths starting at $p \in L_3$ which do not use any of the deleted edges. Thus if our algorithm did not find any such paths, M is popular in the one-vote model. □

Running Time: The algorithm computes a max-flow for each post $p \in L_3$, which takes $O(mn)$ time where m and n are respectively the number of edges and vertices in G_M. The number of vertices in L_3 are $O(n)$. Therefore the total time is $O(mn^2)$. This justifies Theorem 3.

References

1. Abraham, D.J., Irving, R.W., Kavitha, T., Mehlhorn, K.: Popular matchings. SIAM J. Comput. **37**(4), 1030–1045 (2007)
2. Biró, P., Irving, R.W., Manlove, D.F.: Popular matchings in the marriage and roommates problems. In: Calamoneri, T., Diaz, J. (eds.) CIAC 2010. LNCS, vol. 6078, pp. 97–108. Springer, Heidelberg (2010). https://doi.org/10.1007/978-3-642-13073-1_10
3. Brandl, F., Kavitha, T.: Popular matchings with multiple partners. In: Proceedings of 37th IARCS Annual Conference on Foundations of Software Technology and Theoretical Computer Science, pp. 19:1–19:15 (2017)
4. Chang, J., Lai, Y.: Popular condensation with two sided preferences and one sided ties. In: 2016 International Computer Symposium (ICS), pp. 22–27 (2016)
5. Cseh, Á., Huang, C.-C., Kavitha, T.: Popular matchings with two-sided preferences and one-sided ties. SIAM J. Discrete Math. **31**(4), 2348–2377 (2017)
6. Dulmage, A.L., Mendelsohn, N.S.: Coverings of bipartite graphs. Can. J. Math. **10**, 517–534 (1958)
7. Gabow, H.N.: An efficient reduction technique for degree-constrained subgraph and bidirected network flow problems. In: Proceedings of the Fifteenth Annual ACM Symposium on Theory of Computing, pp. 448–456 (1983)
8. Gale, D., Shapley, L.: College admissions and the stability of marriage. Am. Math. Mon. **69**, 9–14 (1962)

9. Irving, R.W., Kavitha, T., Mehlhorn, K., Michail, D., Paluch, K.E.: Rank-maximal matchings. ACM Trans. Algorithms **2**(4), 602–610 (2006)

10. Kavitha, T.: Popularity vs maximum cardinality in the stable marriage setting. In: Proceedings of the 23rd ACM-SIAM Symposium on Discrete Algorithms, pp. 123–134 (2012)

11. Nasre, M., Nimbhorkar, P.: Popular matchings with lower quotas. In: Proceedings of 37th IARCS Annual Conference on Foundations of Software Technology and Theoretical Computer Science, pp. 44:1–44:15 (2017)

12. Nasre, M., Rawat, A.: Popularity in the generalized hospital residents setting. In: Weil, P. (ed.) CSR 2017. LNCS, vol. 10304, pp. 245–259. Springer, Cham (2017). https://doi.org/10.1007/978-3-319-58747-9_22

13. Paluch, K.: Capacitated rank-maximal matchings. In: Spirakis, P.G., Serna, M. (eds.) CIAC 2013. LNCS, vol. 7878, pp. 324–335. Springer, Heidelberg (2013). https://doi.org/10.1007/978-3-642-38233-8_27

Feasibility Algorithms
for the Duplication-Loss Cost

Paweł Górecki[1]([✉]), Alexey Markin[2], and Oliver Eulenstein[2]

[1] Faculty of Mathematics, Informatics and Mechanics, University of Warsaw,
Warszawa, Poland
gorecki@mimuw.edu.pl
[2] Department of Computer Science, Iowa State University, Ames, USA
{amarkin,oeulenst}@iastate.edu

Abstract. Gene duplications are a dominant force in creating genetic
novelty, and studying their evolutionary history is benefiting various
research areas. The gene duplication model, which was introduced more
than 40 years ago, is widely used to infer duplication histories by resolv-
ing the discordance between the evolutionary history of a gene family
and the species tree through which this family has evolved. Today, for
many gene families lower bounds on the number of gene duplications that
have occurred along each edge of the species tree, called duplication scenarios, can be derived, for example from genome duplications. Recently,
the gene duplication model has been augmented to include duplication
scenarios and to address the question of whether such a scenario is feasible for a given gene family. Non-feasibility of a duplication scenario for a
gene family can provide a strong indication that this family might not be
well-resolved, and identifying well-resolved gene families is a challenging
task in evolutionary biology. However, genome duplications are often followed by episodes of gene losses, and lost genes can explain non-feasible
duplication scenarios. Here, we address this major shortcoming of the
augmented duplication model, by proposing a gene duplication model
that incorporates duplication-loss scenarios. We describe efficient algorithms that decide whether a duplication-loss scenario is feasible for a
gene family; and if so, compute a gene tree for the family that infers the
minimum duplication-loss events satisfying the scenario.

Keywords: Gene tree · Species tree · Duplication-loss ·
Reconciliation · Feasibility

1 Introduction

Gene duplication events are widely viewed as the dominant evolutionary source of
raw material for new genes creating genetic novelty in organisms [23, 26]. Duplication events give rise to operationally undistinguishable duplicate copies of a gene.
One of the duplicate copies can then evolve into a gene with a novel function, while
the other copy may maintain the original genes function. However, duplicate copies

© Springer Nature Switzerland AG 2019
D.-Z. Du et al. (Eds.): COCOON 2019, LNCS 11653, pp. 206–218, 2019.
https://doi.org/10.1007/978-3-030-26176-4_17

do not always establish new functions but result in a gene loss. The evolutionary histories by which gene duplication and subsequent loss occurred is full of complexities and present a primary research tool for studying how functional innovations of genes have evolved the way they are today. Potential applications of such studies are widespread and affecting a vast variety of fundamental biological research areas such as molecular biology, microbiology, and biotechnology [19]. For example, evolutionary histories of gene duplications provide a comprehensive way to describe the dynamics of gene family evolution [9,18] and are also a popular tool to differentiate between orthologous and paralogous genes [1,2], a primary task in the functional determination of genes [17].

The groundwork for inferring gene duplication and loss events has been laid by the gene duplication model. This model, which has been pioneered by Goodman et al. [13] nearly 40 years ago, has become a common choice for practitioners and is the focus of this work. The gene duplication model is taking a gene tree and a corresponding species tree, representing the evolutionary histories of genes and species respectively, where both trees are rooted and full binary. The gene tree represents the history of a set of genes belonging to the same family, and the species tree is the history of the species hosting the genes. There is often a discord between the gene tree and the species tree originating from complex histories of gene duplication and loss events, but discord can also arise from other evolutionary events, such as deep coalescence [27] or lateral gene transfer [20]. The gene duplication model is reconciling the gene tree with the species tree under the assumption that discordance originates from gene duplication and loss events. A reconciliation of the gene tree is an embedding of the gene tree into the species tree from which gene duplications and losses can be inferred. An example is depicted in Fig. 1 (Left hand box). Following the parsimony principle, the gene duplication model seeks a reconciliation that infers the minimum number of duplication and loss events. The interested reader is referred to [8,12] for a more detailed treatment of the gene duplication model.

Today, lower bounds on the number of gene duplications that have occurred along the edges of a species tree, called duplication scenarios, can be estimated, for example, from whole genome duplications [5,25]. Recently, the gene duplication model has been augmented to include duplication scenarios to address the question of whether a gene family is feasible under a given duplication scenario [22]. For example, naturally, every reconciliation of gene (family) tree infers a duplication scenario, and thus the gene family is *feasible*. However, not every duplication scenario can be inferred from reconciliations for a gene family, in which case the gene family is *non-feasible*. The *feasibility problem* is to decide whether a duplication scenario is feasible, which can be solved in linear time [28]. Non-feasibility of a duplication scenario for a gene family can provide a strong indication that this family might not be well-resolved, and identifying well-resolved gene families is a challenging task in evolutionary biology. However, a common criticism of the feasibility problem is that genome duplications are often followed by episodes of losses, which is not considered. This major drawback of the feasibility problem is leading to reconciliations that are non-feasible but can be explained by genes that have been lost. Estimates on losses are available for several gene families, e.g., [10]. Further, studying gene families is an ongoing and central effort in evolutionary biology that is likely to produce many more credible estimates on gene duplication and loss events.

Here, to overcome the drawback of the feasibility problem, we modify the augmented duplication model to also consider gene loss. Based on this modified model we introduce the *duplication-loss feasibility problem* that considers lower bounds on the overall number of duplication and loss events. To solve the duplication-loss feasibility problem, we describe a linear-time algorithm. Further, if the problem is feasible, we describe a polynomial-time algorithm that computes a gene tree that has a feasible reconciliation with the minimum overall duplication and loss count.

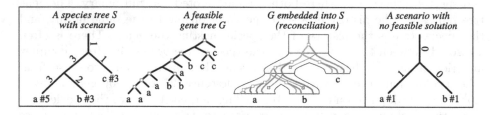

Fig. 1. LEFT HAND BOX. An example of a feasible duplication-loss scenario. The reconciliation of a gene tree G and its corresponding species tree S with three species a, b and c is shown. Through the embedded gene tree the following events are inferred: gene duplication (squares), loss (blue crosses), and speciation (others). The scenario for S is given: lower bounds, in green, on the number of duplication-loss events for each edge, and the number of gene copies sampled from each species, in blue with #. The provided reconciliation shows that the gene tree, or the corresponding gene family, is feasible for the given scenario. E.g., the reconciliation infers for the edge incident to leaf a three duplication events and one loss event, which is larger than the required lower bound of three for this edge. In addition, the gene tree G is under all feasible gene trees, a tree with the minimum overall number of duplication and loss events. RIGHT HAND BOX. A non-feasible duplication-loss scenario is depicted. (Color figure online)

The gene duplication model from Goodman et al. [13] infers gene duplication and subsequent loss events by reconciling possible discord between a gene tree its corresponding species tree under the assumption that the discord originates only from gene duplication and loss. An example of reconciliation is depicted in Fig. 1. Following the parsimony principle, for inferring the duplication-loss inference, the gene duplication model is using the unique reconciliation that results in the minimum overall number of inferred duplications [6,11,12,15]. This unique reconciliation can be specified by the *least common ancestor mapping*, which maps every gene in the gene tree to the most recent species in the species tree that possibly could have contained this gene, called its *host-species*. In the gene tree a gene that has a child with the same host-species is a *gene duplication*. Visually, we say that such gene duplication happened on the edge connecting the host-species to its ancestor.

The mapping and the gene duplications are linear-time computable [28]. There is a rich literature of extensions and variants of the gene duplication model, which can, in most cases, be efficiently computed [8,12]. A *gene loss* is an event that can be inferred from a reconciliation map as a gene lineage from species tree node s carrying on to only one child of s instead of both of them. Figure 1 indicates such events via blue crosses. For example, the pink gene lineage that appeared in the parent of species tree leaves a and b carries on only to the right child of this node and thus a loss is inferred within the other child (see a blue cross in leaf a). While computationally highly complex, probabilistic models for gene/species tree reconciliation, as well as gene sequence evolution, have also been developed [1,3].

Recently, the gene duplication model was augmented by also providing lower bounds on the number of gene duplications that have occurred along the edges of the species tree, termed a *duplication scenario* [22]. Using this augmented model the question of whether a duplication scenario is feasible for a given gene family could be decided in linear time. A scenario is *feasible* for a gene family, if there exists a gene tree for the family such reconciling this tree infers duplications that satisfy the lower bounds of the duplication scenario.

Contribution. To address the missing factor of gene losses in the previously explored duplication scenarios, we formulate a novel feasibility problem based on the standard duplication-loss model. This problem, referred to as *duplication-loss feasibility (DLF)*, takes a species tree annotated with the lower bounds on the number of events (duplications and losses) that have occurred along each species tree branch; and asks whether there exists a gene tree that satisfies these lower bounds. Note that such a gene tree must have for each species tree leaf s a fixed pre-determined number of leaves that map into s; otherwise, when the number of gene leaves is not fixed, the problem is trivial and not biologically relevant.

In spite of the added complexity of the employed reconciliation model, we demonstrate that this feasibility problem can be solved in linear time. The correctness of this algorithm follows from the important and non-trivial properties of the feasible gene trees, which we introduce in this work.

Further, we study a parsimonious modification of the feasibility problem, called *MinDLF*, that asks to find a gene tree that is not only feasible for the given scenario but also incurs the *minimum possible* number of events overall. To address this problem we present a dynamic programming formulation that solves MinDLF in $O(mn^3)$ time. Here n is the size of the species tree and m is the size of the gene tree.

2   Preliminaries

We recall needed basic definitions and introduce two feasibility problems.

2.1 Basic Definitions

We follow the basic definitions and notation from [14, 27]. A *(rooted phylogenetic) tree* $T = \langle V(T), E(T) \rangle$ is a connected acyclic graph where exactly one of its nodes has a degree of two (*root*), and the remaining nodes have a degree of one (*leaves*) or three. The nodes with a degree of at least two are called *internal*. $L(T)$ denotes the set of all leaves in T, and $|T| := |L(T)|$. By $\preceq$ we denote the partial order in a tree T where $a \preceq b$ if b is a node on the path between a and the root of T. Note, $a \prec b$ is equivalent to $a \preceq b$ and $a \neq b$. The least common ancestor (LCA) of a and b in T is denoted by $\mathsf{lca}(a, b)$. If a is not the root of T, then the *parent* of a, denoted by $\mathsf{par}(a)$, is the least node v such that $a \prec v$. The root of T is denoted by $\mathsf{root}(T)$. Nodes a and b with the same parent are called a *sibling* of each other. A sibling of a is denoted by $\mathsf{sib}(a)$. $T(a)$ denotes the maximal subtree of T rooted at node a.

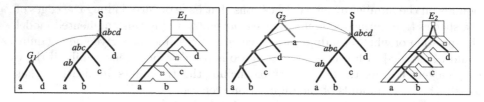

Fig. 2. Reconciling $G_1 = (a, d)$ and $G_2 = ((((a, b), c), d), a)$ with $S = (((a, b), c), d)$. E_1 and E_2 display embeddings of G_1 and G_2 into S respectively. There is one duplication (square) at the root of G_2 and the remaining internal nodes of both gene trees are speciation nodes (marked in red) In G_1, a leaf a visits nodes a, ab and abc in S, while in G_2, the top leaf a visits a, ab, abc and $abcd$. Further, there are two gene losses (crosses) assigned to b and c in S when reconciling with G_1 and three gene losses assigned to b, c and d in S for G_2, respectively. Here, $\mathrm{DL}(G_1, S) = 2$ and $\mathrm{DL}(G_2, S) = 4$. (Color figure online)

A *species tree* is a rooted tree whose leaves are called *species*. Here, we assume that the set of species is fixed. A *gene tree* is a rooted tree whose leaves (genes) are labeled by species. The labeling is interpreted as follows: a leaf is a gene and the label is the species from which the gene was sampled. Note, leaves in the gene tree have the same label when the respective genes are sampled from the same species.

In phylogenetic tree reconciliation a gene tree G is embedded into its corresponding species tree S. The embedding is determined by the *lca-mapping* $M: V(G) \to V(S)$ that is defined recursively as follows: (i) if g is a leaf then $M(g)$ is the label of g, i.e., g maps into its species (a leaf from S) – and (ii) $M(g) := \mathsf{lca}(M(g'), M(g''))$ if g has two children g' and g''. We write that a node g *maps to* s, which means $M(g) = s$.

An internal node g is called a *duplication* assigned to a node $M(g)$ if $M(g) = M(g')$ for a child g' of g. Every internal non-duplication node in G is called a *speciation (node)*. By δ_s^G and σ_s^G we denote the number of duplication and speciation nodes mapped to s, respectively. Note that $\sigma_s = 0$ if s is a leaf.

To model how a gene tree is embedded into its species tree, we first determine how genes (i.e., the nodes of G) correspond to the nodes of the species tree. Now, we define the set of species nodes *visited* by a given node $g \in V(G)$. First, if g is the root then g visits only $M(g)$. For a non-root node g, g visits every node s such that $M(g) \preceq s \prec M(\mathsf{par}(g))$ if $\mathsf{par}(g)$ is a speciation, and every node s such that $M(g) \preceq s \preceq M(\mathsf{par}(g))$, otherwise. Now, for each g, we assign one *gene loss* to a node s if both $\mathsf{sib}(s)$ and $\mathsf{par}(s)$ are visited by g. The number of gene losses assigned to a node s is denoted by λ_s^G. See an example in Fig. 2.

Finally, we define the *duplication-loss cost* function [24, 27], denoted $DL(G, S)$, as the total number of gene duplication and loss events assigned to the nodes of S. For a more detailed introduction to the model please refer to [15, 21, 24].

Note that here, for convenience, we assign duplication and loss events to species tree nodes; however, typically, we think of these events as occurring on the species tree edges that enter these nodes (i.e., parent edges). Therefore, Fig. 1 shows event counts assigned to edges rather than nodes.

2.2 Feasibility Problems

Here we define two feasibility problems. We assume that the species tree S is fixed. First, we need to define the notion of a scenario (see Fig. 1), which is a pair of functions that determine the number of events assigned to each node of the species tree and the number of genes present in each species. Formally, by the *event-count constraint* (on S) we denote any function $\gamma \colon V(S) \to \{0, 1, \dots\}$ and by *gene counts* (on S) we denote any non-zero function $\mathbf{g}^{\#} \colon L(S) \to \{0, 1, \dots\}$.

Problem 1 (DLF). Given: a species tree S, an event-count constraint γ, and gene counts $\mathbf{g}^{\#}$. Question: does exist a gene tree G such that (1) for each leaf s from S, G has exactly $\mathbf{g}^{\#}(s)$ leaves labeled s, and, (2) for each node s from S, $\gamma(s) \leq \delta_s^G + \lambda_s^G$.

A gene tree satisfying (1) and (2) will be called *feasible* (for S, γ, and $\mathbf{g}^{\#}$).

Problem 2 (MinDLF). Given: a species tree S, an event-count constraint γ, and gene counts $\mathbf{g}^{\#}$. Problem: reconstruct a feasible gene tree G with the minimal $DL(G, S)$ if such a tree exists.

3   Elementary Properties of Embeddings

We prove embedding properties that are elementary to solve the feasibility problems.

Figure 3 provides insight into the following lemma. Each colored edge in that figure represents a gene tree edge (or its part) and is also referred to as a gene *lineage*. Note that each duplication event creates one additional gene lineage "below" it. By β_s^G, we denote the number of non-duplication nodes that visit s.

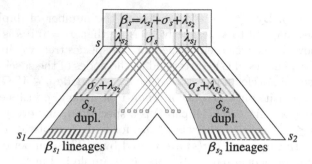

Fig. 3. A general situation of embedding gene tree lineages into an internal species tree node s with children s_1 and s_2. Note that the values of β's, σ's, δ's and λ's are uniquely determined by the lca-mapping between a given gene tree and its species tree. Gene losses are marked by blue crosses. See also Lemma 1. (Color figure online)

Lemma 1. *Let G be a gene tree and S be a species tree. For an internal node s from S with two children s_1 and s_2, we have[1]:*

$$\lambda_{s_i} = \beta_{\mathsf{sib}(s_i)} - \delta_{\mathsf{sib}(s_i)} - \sigma_s, \tag{1}$$

$$\beta_s = \beta_{s_1} + \beta_{s_2} - \delta_{s_1} - \delta_{s_2} - \sigma_s, \tag{2}$$

$$\delta_s \in \{0, 1, \ldots, \beta_s - 1\}, \tag{3}$$

$$\delta_s > 0 \Rightarrow \sigma_s > 0. \tag{4}$$

4   Algorithm for DLF

In this section we present the algorithm to solve DLF. First, we define a classical rooted subtree pruning and regrafting tree-rearrangement (rooted-SPR) and then two rules for transforming feasible gene trees. Having this, we limit our search for a feasible tree to a special case of gene trees. Finally, we present the algorithm for solving DLF.

The rooted-SPR operation is defined as follows [7]. Let g be a non-root node of gene tree G. For g' in G by $\mathsf{rSPR}(G, g, g')$ we denote the tree obtained from G by deleting the edge connecting g with its parent and

- if g' is not the root, by adding a new node v which subdivides the edge connecting g' with its parent, adding a new edge from v to g, and suppressing the non-root node of degree two or the root if its degree is one;
- otherwise, by adding a new root v and two edges connecting v with g and the previous root, and suppressing the non-root node of degree two.

Now we define two rules for transforming a feasible gene tree G.

---

[1] When it is clear from the context, the superscript G is omitted for clarity in γ, σ, δ and β.

Rule I The reducible expression (redex) here is a pair of two speciation nodes g and g' from G such that g is earlier in the prefix order than g'. Condition: $M(\mathsf{par}(g)) = M(g) = M(g')$. The rule transforms G into $G' = \mathsf{rSPR}(G, \mathsf{sib}(g), g')$.

Rule II The redex is a pair $\langle g, c \rangle$, where g is a non-root speciation node and c is a child of g. Condition: $M(\mathsf{par}(g)) \succ M(g)$. The rule transforms G into $G'' = \mathsf{rSPR}(G, c, \mathsf{root}(G))$.

Observe, that the requirement that g and g' are ordered in Rule I (any linear order is sufficient) is needed to ensure that any sequence of rule transformations is finite. Proofs of the next two lemmas follow from the definition of rules.

Lemma 2. *Let $\langle g, g' \rangle$ be a redex of Rule I in G. Then,*

- *the parent of g is a duplication in G,*
- *all event assignments are the same including β and σ, i.e., $\delta^G = \delta^{G'}$, $\lambda^G = \lambda^{G'}$, $\sigma^G = \sigma^{G'}$, and $\beta^G = \beta^{G'}$,*
- *and the number of duplications nodes above g that map into $M(g)$ (i.e., all duplications $d : d \succ g, M(d) = M(g)$) is decreased by one.*

Lemma 3. *Let $\langle g, c \rangle$ be a redex of Rule II. Let $M(g) = s$ and $r = M(\mathsf{root}(G))$. Then,*

- *for each $r \succ v \succeq s$, $\lambda^{G''}_{\mathsf{sib}(v)} = \lambda^G_{\mathsf{sib}(v)} + 1$ and $\beta^{G''}_v = \beta^G_v + 1$,*
- *for each child s' of s, $\lambda^{G''}_{s'} = \lambda^G_{s'} + 1$,*
- *$\delta^{G''}_r = \delta^G_r + 1$ and $\beta^{G''}_r = \beta^G_r + 1$,*
- *$\sigma^{G''}_s = \sigma^G_s - 1$,*
- *and the remaining values of the above functions are the same.*

Lemma 4. *If G is feasible then G' is feasible. The same holds for G''.*

Proof. It follows from Lemmas 2 and 3 that the transformation does not decrease the number of gene duplications and losses assigned to a species node.   □

Lemma 5. *If a given instance of DLF has a feasible solution, then there is a feasible gene tree in which (1) no two speciation nodes map into the same species tree node, and (2) a parent of each non-root speciation is a duplication mapped to the same node.*

Proof. Note that both conditions are satisfied if the tree has no redex of any rule. Thus, we transform G by Rules I and II until there is no redex left. It should be clear that the maximal sequence of reductions is finite. By Lemma 4, this procedure yields a feasible gene tree satisfying both properties.   □

This implies that the search for a feasible tree can be limited to the trees that satisfy the following conditions.

Algorithm 1. Solution to DLF

1: **Input:** A species tree S with $\gamma\colon V(S) \to \{0,1,\dots\}$ (events lower bounds) and $\mathbf{g}^{\#}\colon L(S) \to \{0,1,\dots\}$ (gene-count). For an internal node s of S, let s_1 and s_2 denote the children of s. **Output:** True iff there is a feasible tree for S, $\mathbf{g}^{\#}$, and γ.
2: **Function** $\mathsf{dlf}^*(s,\mathsf{d}_0,\mathsf{d}_1,\mathsf{d}_2)$:
3:     $b_1 := \mathsf{dlf}(s_1,\mathsf{d}_1); \quad b_2 := \mathsf{dlf}(s_2,\mathsf{d}_2)$
4:     **If** $b_1 = -\infty$ or $b_2 = -\infty$: **Return** $-\infty$. # *Non-feasible solution*
5:     **If** $\mathsf{d}_0 = 1$ and $(b_1 = 0$ or $b_2 = 0)$: **Return** $-\infty$. # *Speciation req.*
6:     **If** $(b_1 \le \mathsf{d}_1 \wedge \mathsf{d}_1 = 1)$ or $(b_2 \le \mathsf{d}_2 \wedge \mathsf{d}_2 = 1)$: **Return** $-\infty$ # *Child duplication req.*
7:     $e_1 := \mathsf{d}_1 + b_2 - \mathsf{d}_2 - \mathsf{d}_0; \quad e_2 := \mathsf{d}_2 + b_1 - \mathsf{d}_1 - \mathsf{d}_0;$ # *Initial DL counts at s_1 & s_2*
8:     $\Delta_1 := \mathsf{d}_1; \quad \Delta_2 := \mathsf{d}_2;$ # *Initial duplication counts at s_1 and s_2*
9:     **If** $b_1 - 2 \ge \gamma(s_1) - e_1 > 0$ and $\mathsf{d}_1 = 1$: $\Delta_1 := \Delta_1 + \gamma(s_1) - e_1$ # *Add dupl.*
10:     **Else If** $b_2 - 2 \ge \gamma(s_2) - e_2 > 0$ and $\mathsf{d}_2 = 1$: $\Delta_2 := \Delta_2 + \gamma(s_2) - e_2$
11:     **If** $\Delta_1 + b_2 - \Delta_2 - \mathsf{d}_0 < \gamma(s_1)$ or $\Delta_2 + b_1 - \Delta_1 - \mathsf{d}_0 < \gamma(s_2)$: **Return** $-\infty$
12:     **Return** $b_1 + b_2 - \mathsf{d}_0 - \Delta_1 - \Delta_2$
13: **Function** $\mathsf{dlf}(s,\mathsf{d})$:
14:     **If** s is a leaf: **Return** $\mathbf{g}^{\#}(s)$.
15:     **Return** $\max\{\mathsf{dlf}^*(s,\mathsf{d},0,0),\mathsf{dlf}^*(s,\mathsf{d},1,0),\mathsf{dlf}^*(s,\mathsf{d},0,1),\mathsf{dlf}^*(s,\mathsf{d},1,1)\}$.
16: # *Main body starts here*
17: **Let** r be the least common ancestor of all leaves s with $\mathbf{g}^{\#}(s) > 0$.
18: **For** every node s outside the subtree rooted at r: **If** $\gamma(s) > 0$: **Return False**
19: **If** $\mathsf{dlf}(r,1) \le \gamma(r)$: **Return False** # *Dupl. at the top node*
20: **Return True**

Lemma 6. *Assume that a given instance of DLF has a feasible solution G and let $r = M(\mathrm{root}(G))$. Then, there is a feasible gene tree in which, (1) for each internal node $s \ne r$ either $\delta_s \ge 1 = \sigma_s$ or $\delta_s = 0 = \sigma_s$, and (2) if r is not a leaf then $\sigma_r = 1$.*

We say that a gene tree is *s-feasible* for a node $s \in S$ if for each leaf l from $S(s)$, G has exactly $\mathbf{g}^{\#}(l)$ leaves labeled l, and for each node s' strictly below s, $\gamma(s') \le \delta_{s'}^G + \lambda_s^G$. Note that the last condition does not have to hold for s.

Lemma 7. *In Algorithm 1, for $\mathsf{d} \in \{0,1\}$ and a node s, $\mathsf{dlf}(s,\mathsf{d})$ returns the maximal value[2] of β_s in the set of all s-feasible trees under the assumption that a speciation node maps to s if and only if $\mathsf{d} = 1$.*

Lemma 8. *In Algorithm 1, for $\mathsf{d}_i \in \{0,1\}$ and an internal node s_0 with children s_1 and s_2, $\mathsf{dlf}^*(s_0,\mathsf{d}_0,\mathsf{d}_1,\mathsf{d}_2)$ returns the maximal value of β_s in the set of all s-feasible trees under the assumption that a speciation node maps to s_i if and only if $\mathsf{d}_i = 1$, for each i.*

Theorem 1 (Correctness). *Given a species tree S, event-count constraint γ and gene-counts $\mathbf{g}^{\#}$ Algorithm 1 returns true if and only if there is a feasible gene tree for S, γ and $\mathbf{g}^{\#}$.*

---

[2] We assume that the maximum is $-\infty$ if the set of s-feasible trees is empty.

Proof. Let r be the node as defined in line 17. First, outside the tree rooted at r there are no events allowed; therefore, the constraint must be 0, otherwise, no feasible solution exists. Next, by Theorem 1, if the returned value of $\mathsf{dlf}(s, \mathsf{d})$ is $-\infty$ then there is no feasible solution. Otherwise, there is an r-feasible gene tree. It remains to check whether the event-count constraint for r is satisfied. As $k = \mathsf{dlf}(s, \mathsf{d})$ is the maximal value of β_r for an r-feasible tree (by Lemma 7), we create $k - 1$ duplications at r (see Lemma 1 Eq. (3). Thus, the requirement is $k \geq \gamma(r)$ (see line 19). $\qquad\square$

Theorem 2 (Complexity). *Algorithm 1 requires $O(n)$ time and space, where n is the size of the species tree.*

Proof. There are three traversals of S: one for computing r in line 17, the second for checking the constraints outside $S(r)$ and the last one for the recursive calls of dlf and dlf^*. The memory is needed for the recursive calls and for a fixed number of local variables per each visited node s. $\qquad\square$

5   Algorithm for MinDLF

Here we propose a dynamic programming algorithm for solving our second problem. We start with several needed definitions, and then show the algorithm for the minimal cost inference. Finally, we describe the solution to MinDLF.

A non-root gene node g that visits a non-root species node s is called a *mid-lineage* of s if either g visits $\mathsf{par}(s)$ or the parent of g is a speciation that maps to $\mathsf{par}(s)$. It is not difficult to see the number of mid-lineages of s can be calculated as $\sigma_{\mathsf{par}(s)} + \lambda_{\mathsf{sib}(s)}$ or, equivalently, $\beta_s - \delta_s$ (see Lemma 1 or Fig. 3 and the gene lineages above duplications – they illustrate the lineages that we refer to as mid-lineages). The latter expression gives the number of mid-lineages for node $M(\mathsf{root}(G))$, which is 1.

For an s-feasible tree, *the partial DL cost* is the total number of events assigned to the nodes strictly below s (i.e., excluding s). We assume that the minimal partial cost is $+\infty$ if there is no s-feasible tree satisfying the corresponding conditions.

Our algorithm is composed of two dynamic programming formulas. The next two lemmas describe their properties.

Lemma 9. *For a node s, $P(s, b, p)$ is the minimal partial DL cost in the set of all s-feasible gene trees G such that (1) b is the number of non-duplication nodes that visit s (i.e., β_s^G), and, (2) for an internal node s, p is the lower bound for the number of speciation nodes mapped to s (i.e., $\sigma_s^G \geq p$).*

P^* has properties that are similar to P. However, in contrast, P^* incorporates the event-count constraint and the cost contribution of s.

Lemma 10. *For a node s, $P^*(s, b, l)$ is the minimal DL cost calculated as the number of events assigned to any node in $S(s)$ in the set of all s-feasible gene trees G such that the event-constraint is satisfied by s (i.e., $\gamma(s) \leq \lambda_s^G + \delta_s^G$), l is the number of losses assigned to s, and b is the number of mid-lineages of s.*

Algorithm 2. MinDLF cost

1: **Input:** The same as in Alg. 1. **Output:** The minimal DL cost of a feasible tree for S, $\mathbf{g}^{\#}$ and γ if exists and $-\infty$, otherwise.
2: **Let** $m = \sum_g \mathbf{g}^{\#}(g)$ and r be the LCA of all leaves s with $\mathbf{g}^{\#}(s) > 0$.
3: **For** every node $s \in V(S) \setminus V(S(r))$: **If** $\gamma(s) > 0$: **Return** $+\infty$
4: **Return** $P^*(r, 1, 0)$, where $P: V(S) \times \{0, 1, \dots, m\} \times \{0, 1\} \to \{0, 1, \dots\} \cup \{+\infty\}$ and $P^*: V(S) \times \{0, 1, \dots, m\}^2 \to \{0, 1, \dots\} \cup \{+\infty\}$ are defined below.
5: **For** any node s, if s is internal its children are denoted s_1 and s_2:

$$
P(s, b, p) = \begin{cases}
0 & s \text{ is a leaf and } \mathbf{g}^{\#}(s) = b, \\
+\infty & s \text{ is a leaf and } \mathbf{g}^{\#}(s) \neq b, \\
+\infty & s \text{ is internal, } p = 1 \text{ and } b = 0, \\
\min_{\substack{q \in \{p, \dots, b\} \\ l \in \{0, \dots, b-q\}}} P^*(s_1, b - l, l) + P^*(s_2, q + l, b - q - l) & \text{otherwise.}
\end{cases}
$$

6: **For** any node s, $P^*(s, 0, l) = \begin{cases} l & \gamma(s) \leq l \text{ and } \sum_{s' \in L(S(s))} \mathbf{g}^{\#}(s') = 0, \\ +\infty & \text{otherwise,} \end{cases}$

and, if $b > 0$, $P^*(s, b, l) = \min_{0 \leq d < m} \begin{cases} d + l + P(s, b + d, [\![d > 0]\!]) & \gamma(s) \leq l + d, \\ +\infty & \text{otherwise.} \end{cases}$

Here, for a predicate a, $[\![a]\!]$ is 1 if a is satisfied, and 0 otherwise.

Theorem 3 (Correctness). *The minimal cost of a feasible tree is given by $P^*(r, 1, 0)$, where r is the LCA of all leaves s with $\mathbf{g}^{\#}(s) > 0$.*

Proof. The proof follows immediately from Lemma 10 as r, being the map of every gene tree root satisfying the gene-count constraint, has exactly one mid-lineage and no losses. Additionally, the event-constraints has to be zero for any node outside $S(r)$ (see line 3). □

Theorem 4 (Complexity). *Algorithm 2 requires $O(|S|m^3)$ time and $O(|S|m^2)$ space, where $m = \sum_g \mathbf{g}^{\#}(g)$.*

Proof. The time complexity follows from the construction of P, where for each internal node from S and each $b \leq m$, we need two nested loops with up to m steps each to compute the minimum value in the last case of the definition in line 5. For the space complexity, we need $2|S|m$ and $|S|m^2$ memory to store P and P^*, respectively. □

Finally, we can solve MinDLF. Given the dynamic programming algorithm, we infer a feasible gene tree that minimizes the cost by using the standard back-tracking technique. For example, the gene tree G from Fig. 1 is the tree with the minimal DL cost among all feasible gene trees for S inferred from Algorithm 2 using backtracking. Note that $\text{DL}(G, S) = 11$, while the event-constraint gives the lower bound 10. In this case, there is no feasible gene tree with precisely 10 events.

6 Conclusion

Gene families play a significant role in the systematic analysis of protein function, diversity of multicellular organisms, and related areas [4,16,18]. Therefore, categorizing genes into credible families is an ongoing and central topic in phylogenetics. Recently, as part of this effort, the classic gene duplication model has been augmented by incorporating the knowledge of whole genome duplications [22]. A linear-time algorithm based on this model allowed then the refinement of gene families. While promising, the augmented model has the major drawback of not incorporating gene losses. Here, we are overcoming this shortcoming by including losses into the augmented model and providing efficient algorithms that support practitioners in carefully refining gene families.

Acknowledgments. The support was provided by the National Science Center grant 2017/27/B/ST6/02720 and the National Science Foundation under Grant No. 1617626.

References

1. Akerborg, O., Sennblad, B., Arvestad, L., Lagergren, J.: Simultaneous bayesian gene tree reconstruction and reconciliation analysis. Proc. Natl. Acad. Sci. U.S.A. **106**(14), 5714–5719 (2009)
2. Altenhoff, A.M., Dessimoz, C.: Inferring orthology and paralogy. In: Anisimova, M. (ed.) Evolutionary Genomics, pp. 259–279. Humana Press, Totowa (2012)
3. Arvestad, L., Berglund, A.C., Lagergren, J., Sennblad, B.: Bayesian gene/species tree reconciliation and orthology analysis using MCMC. Bioinformatics **19**(suppl1), 7–15 (2003)
4. Bininda-Emonds, O.R. (ed.): Phylogenetic Supertrees: Combining Information to Reveal the Tree of Life. Computational Biology, vol. 4. Springer, Dordrecht (2004). https://doi.org/10.1007/978-1-4020-2330-9
5. Blanc, G., Wolfe, K.H.: Widespread paleopolyploidy in model plant species inferred from age distributions of duplicate genes. Plant Cell **16**(7), 1667–1678 (2004)
6. Bonizzoni, P., Della Vedova, G., Dondi, R.: Reconciling a gene tree to a species tree under the duplication cost model. Theor. Comput. Sci. **347**, 36–53 (2005)
7. Bordewich, M., Semple, C.: On the computational complexity of the rooted subtree prune and regraft distance. Ann. Comb. **8**, 409–423 (2004)
8. Chauve, C., El-Mabrouk, N., Guéguen, L., Semeria, M., Tannier, E.: Duplication, rearrangement and reconciliation: a follow-up 13 years later. In: Chauve, C., El-Mabrouk, N., Tannier, E. (eds.) Models and Algorithms for Genome Evolution. Computational Biology, vol. 19, pp. 47–62. Springer, London (2013). https://doi.org/10.1007/978-1-4471-5298-9_4
9. Chen, K., Durand, D., Farach-Colton, M.: NOTUNG: a program for dating gene duplications and optimizing gene family trees. J. Comput. Biol. **7**(3–4), 429–447 (2000)
10. Dujon, B., et al.: Genome evolution in yeasts. Nature **430**, 35–44 (2004)
11. Eulenstein, O.: Vorhersage von Genduplikationen und deren Entwicklung in der Evolution. Ph.D. thesis, Rheinische Friedrich-Wilhelms-Universität Bonn, Germany (1998)

12. Eulenstein, O., Huzurbazar, S., Liberles, D.: Reconciling phylogenetic trees. In: Dittmar, L. (ed.) Evolution After Gene Duplication. Wiley, New York (2010)

13. Goodman, M., Czelusniak, J., Moore, G., Romero-Herrera, A., Matsuda, G.: Fitting the gene lineage into its species lineage. A parsimony strategy illustrated by cladograms constructed from globin sequences. Syst. Zool. **28**(2), 132–163 (1979)

14. Górecki, P., Eulenstein, O., Tiuryn, J.: Unrooted tree reconciliation: a unified approach. IEEE/ACM TCBB **10**(2), 522–536 (2013)

15. Górecki, P., Tiuryn, J.: DLS-trees: a model of evolutionary scenarios. Theor. Comput. Sci. **359**(1–3), 378–399 (2006)

16. Huson, D.H., Scornavacca, C.: A survey of combinatorial methods for phylogenetic networks. Genome Biol. Evol. **3**, 23–35 (2011)

17. Ihara, K., et al.: Evolution of the archaeal rhodopsins: evolution rate changes by gene duplication and functional differentiation. J. Mol. Biol. **285**(1), 163–174 (1999)

18. Kamneva, O.K., Knight, S.J., Liberles, D.A., Ward, N.L.: Analysis of genome content evolution in PVC bacterial super-phylum: assessment of candidate genes associated with cellular organization and lifestyle. Genome Biol. Evol. **4**(12), 1375–1390 (2012)

19. Kamneva, O.K., Ward, N.L.: Reconciliation approaches to determining HGT, duplications, and losses in gene trees. In: Michael Goodfellow, I.S., Chun, J. (eds.) New Approaches to Prokaryotic Systematics, Methods in Microbiology, Chap. 9, vol. 41, pp. 183–199. Academic Press, Cambridge (2014)

20. Koonin, E.V.: Orthologs, paralogs, and evolutionary genomics. Annu. Rev. Genet. **39**, 309–338 (2005)

21. Maddison, W.P.: Gene trees in species trees. Syst. Biol. **46**(3), 523–536 (1997)

22. Markin, A., Vadali, V.S.K.T., Eulenstein, O.: Solving the gene duplication feasibility problem in linear time. In: Wang, L., Zhu, D. (eds.) COCOON 2018. LNCS, vol. 10976, pp. 378–390. Springer, Cham (2018). https://doi.org/10.1007/978-3-319-94776-1_32

23. Ohno, S.: Evolution by Gene Duplication. Springer, Heidelberg (1970). https://doi.org/10.1007/978-3-642-86659-3

24. Page, R.: From gene to organismal phylogeny: reconciled trees and the gene tree/species tree problem. Mol. Phylogenet. Evol. **7**(2), 231–240 (1997)

25. Renny-Byfield, S., Wendel, J.F.: Doubling down on genomes: polyploidy and crop plants. Am. J. Bot. **101**(10), 1711–1725 (2014)

26. Thornton, K., Long, M.: Rapid divergence of gene duplicates on the Drosophila melanogaster X chromosome. Mol. Biol. Evol. **19**(6), 918–925 (2002)

27. Zhang, L.: From gene trees to species trees II: species tree inference by minimizing deep coalescence events. IEEE/ACM TCBB **8**, 1685–1691 (2011)

28. Zhang, L.: On a Mirkin-Muchnik-Smith conjecture for comparing molecular phylogenies. J. Comput. Biol. **4**(2), 177–187 (1997)

Imbalance, Cutwidth, and the Structure of Optimal Orderings

Jan Gorzny$^{(\boxtimes)}$ and Jonathan F. Buss

University of Waterloo, Waterloo, ON, Canada
{jgorzny,jonathan.buss}@uwaterloo.ca

Abstract. We show that both CUTWIDTH and IMBALANCE are fixed-parameter tractable when parameterized by the twin-cover number of the input graph. We further show that IMBALANCE is NP-complete for split graphs and therefore chordal graphs, but linear-time solvable for proper interval graphs, which equals the complexity of CUTWIDTH on these classes.

Both results follow from a new structural theorem, that every instance of CUTWIDTH or IMBALANCE has an optimal ordering of a restricted form.

Keywords: Imbalance · Cutwidth · Twin-cover · Vertex layout · Proper interval graph · Split graph

1 Background

The IMBALANCE problem was introduced by Biedl, et al. [2]. Given a linear layout of a graph G (i.e., an ordering of its vertices), the imbalance of each vertex is the absolute value of the difference in the sizes of its neighbourhood to its left and to its right. The imbalance of the layout is the sum of the imbalances of each vertex. An instance of the IMBALANCE problem consists of an n-vertex graph G and an integer k; a solution determines whether there exists a linear layout of G that has imbalance at most k (and, ideally, finds such a layout). This problem arises naturally for several graph-drawing applications, which require such an ordering [17, 18, 22, 27, 28]. (See the next section for a more formal definition of the problem.)

A related problem, CUTWIDTH, asks to determine the minimal size of a maximal cut; i.e., a set $\{ (v_i, v_j) \mid i \leq c < j, \ (v_i, v_j) \in E \}$, for some $c \geq 1, c < |V|$. Aside from the superficial resemblance of these two problems, Lokshtanov, et al. [21] showed that the imbalance of any graph is at least twice its cutwidth. We show a deeper relationship between the two problems; for many classes of graphs, a layout that optimizes imbalance also optimizes cutwidth. As a corollary, IMBALANCE and CUTWIDTH have the same complexity on these graphs.

The IMBALANCE problem was shown to be NP-complete for bipartite graphs with degree at most 6, for weighted trees [2], and for graphs of degree at most 4 [29]. Biedl et al. provide a pseudo-polynomial time algorithm for weighted trees,

© Springer Nature Switzerland AG 2019
D.-Z. Du et al. (Eds.): COCOON 2019, LNCS 11653, pp. 219–231, 2019.
https://doi.org/10.1007/978-3-030-26176-4_18

which runs in linear time on unweighted trees. Gaspers et al. [11] showed that IMBALANCE is equivalent to the GRAPH CLEANING problem, which yielded a $O(n^{\lfloor k/2 \rfloor}(n+m))$ time parametermized algorithm where k is the solution size. Lokshtanov, et al. [21] improved this result, and showed that IMBALANCE is fixed-parameter tractable (FPT) when parameterized by the solution size k by constructing an algorithm that runs in time $O(2^{O(k \log k)} \cdot n^{O(1)})$, or the treewidth of the graph $tw(G)$ and the maximum degree of the graph $\Delta(G)$ by constructing an algorithm that runs in time $O(\Delta(G)^{O(tw(G))} \cdot tw(G)! \cdot n)$ where $tw(G)$ is the treewidth of G. Fellows et al. showed that IMBALANCE is FPT when parameterized by the size of a minimum vertex cover of the graph [8]. Bakken [1] showed that IMBALANCE is FPT by the neighbourhood diversity of the graph.

CUTWIDTH has received more attention than IMBALANCE. CUTWIDTH is solvable in polynomial time on trees by Yannakakis [30]. Heggernes et al. showed that CUTWIDTH is NP-complete for split graphs, but has a linear time solution on proper interval graphs and threshold graphs [13]. Therefore, for all graph classes containing split graphs, e.g., chordal graphs, the problem is NP-complete. These tractable graph classes are all subsets of permutation graphs, and the complexity of CUTWIDTH on interval graphs, a subclass of both permutation and chordal graphs, was left as an open problem. CUTWIDTH was shown to be polynomially-solvable for superfragile graphs [20], a restricted subclass of interval graphs. Heggernes et al. later showed that CUTWIDTH has a linear time solution on proper interval bipartite graphs [14]. Thilikos et al. showed that CUTWIDTH is FPT when parameterized by the size of the solution [24,25], and Giannopoulous provided a simpler algorithm [12]. Fellows et al. [8] showed that CUTWIDTH is FPT when parameterized by the size of the minimum vertex cover of the graph $vc(G)$, which runs in time $O(2^{2^{O(vc(G))}} \cdot n^{O(1)})$. Cygan et al. [7] later improved this result and showed that CUTWIDTH parameterized by the size of a minimum vertex cover of the graph does not admit a polynomial kernel unless NP$\subseteq$coNP/poly.

We improve the fixed parameterized complexity results for these problems proved by [8]. We show that IMBALANCE and CUTWIDTH are FPT parameterized by the *twin-cover* number of the graph, a generalization of the vertex cover defined in the next section. A graph with bounded twin-cover number may have arbitrarily large vertex cover number [10], so that a bounded twin-cover number describes more graphs than the same bound on the vertex cover. This improvement is obtained by observing that sets of true twins (vertices with the same closed neighbourhood) must appear consecutively in some optimal ordering of each of these problems, a fact that may be useful in solving these problems on other classes of graphs. The observation allows the ILP approach used for graphs of bounded vertex cover to work as it did in [8]. We show that IMBALANCE is NP-complete for split graphs (a large sub-class of chordal graphs), while showing we can solve IMBALANCE on proper interval graphs in linear time. We also note that for graphs which are both k-trees and proper interval graphs, the cutwidth of G is exactly twice the imbalance of G. We also show that there is a polyno-mial time solution to IMBALANCE on superfragile graphs. We conjecture that for

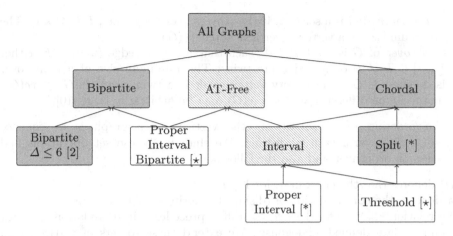

Fig. 1. The relationship between graph classes. IMBALANCE is NP-complete for shaded classes, unknown for hatched classes, and linear for the rest. Results that appear in this work are marked with [*]. Classes marked with [⋆] are conjectured results. An arrow from class A to class B indicates that class A is contained within class B.

proper interval bipartite graphs, as well as threshold graphs, there is an ordering which minimizes both the imbalance and the cutwidth of G simultaneously. Figure 1 illustrates the relationship between the classes described above.

2   Preliminaries

All graphs in this work are finite, undirected, and without multiple edges or loops. For a graph $G = (V, E)$, we will denote $n = |V|$ and $m = |E|$. The following definitions are standard.

A *complete graph* is a graph whose vertices are pairwise adjacent.

An *independent set* is a set $I \subseteq V$ of vertices with no edges among them (i.e., $(I \times I) \cap E = \emptyset$).

The *open neighbourhood* of a vertex v, denoted $N(v)$, is the set $\{u \in V \mid (v, u) \in E\}$ of vertices adjacent to v. The *closed neighbourhood* of a vertex, denoted $N[v]$, is the open neighbourhood of the vertex along with the vertex itself, i.e. $N[v] = N(v) \cup \{v\}$.

A vertex v is *simplicial* if $N(v)$ is a complete graph.

A set of vertices $X \subseteq V$ is a separator if $G(V - X)$ is disconnected. Such a set X is a *clique separator* if it is both a separator and a complete graph.

Two vertices u and v are *twins* if they have the same neighbours, except possibly for each other; that is, $N(u)\backslash\{v\} = N(v)\backslash\{u\}$. Equivalently, u and v are twins if each $w \in V \setminus \{u, v\}$ satisfies $(w, u) \in E$ if and only if $(w, v) \in E$. Two twins u and v are *true twins* they have an edge between them; i.e., $N[u] = N[v]$.

For a subset $S \subseteq V$, a vertex $v \in S$ is *universal to* S if $S \subseteq N(v)$. We will use $U(S)$ to denote the set of universal vertices for $S \subseteq V$, i.e., $U(S) = \{v \mid v \in S \text{ and } S \subseteq N(v)\}$.

A *vertex cover* of G is a set $C \subseteq V$ that covers every edge; i.e., $E \subseteq C \times V$. The minimum size of a vertex cover is denoted $vc(G)$.

A *twin cover* of G is a set $T \subseteq V$ such that for every edge $(u, v) \in E$, either $\{u, v\} \cap T \neq \emptyset$ or u and v are twins. The minimum size of a twin cover is denoted $tc(G)$. Since every vertex cover is a twin cover, $tc(G) \leq vc(G)$; however, the difference $vc(G) - tc(G)$ may be arbitrarily large [10].

An ordering of (a subset of) the vertices of a graph is a sequence $\langle v_1, v_2, \ldots, v_k \rangle$, with each v_i distinct. We shall freely use set operations and notation on orderings, and also the following.

$\sigma(i)$ denotes the ith vertex in σ, for $1 \leq i \leq |\sigma|$.

$\sigma\pi$ denotes the concatenation of (disjoint) orderings σ and π.

The relation $<_\sigma$ is defined by $u <_\sigma v$ iff u precedes v in σ. Relations $>_\sigma$, $\leq_\sigma$ and $\geq_\sigma$ are defined analogously. We extend these to sets of vertices: e.g., $x <_\sigma \{y, z\}$ iff $x <_\sigma y$ and $x <_\sigma z$.

For an element x of σ, $\sigma_{<x}$ denotes the ordering induced by σ on the set $\{y \in V \mid y <_\sigma x\}$. The orderings $\sigma_{\leq x}$, $\sigma_{>x}$, and $\sigma_{\geq x}$ are defined analogously. More generally, for a set $X \in V$, σ_X denotes the ordering induced by σ on X.

Definition 1 (Imbalance, cutwidth). *Let $G = (V, E)$ be a graph and σ an ordering of V.*

1. *For $v \in V$, let $pred_\sigma(v)$ and $succ_\sigma(v)$ respectively denote the number of neighbours of v that precede (resp. succeed) v in σ. That is, $pred_\sigma(v) = |\sigma_{<v} \cap N(v)|$ and $succ_\sigma(v) = |\sigma_{>v} \cap N(v)|$.*
   The imbalance *of v with respect to σ, denoted $\phi_\sigma(v)$, is $|succ_\sigma(v) - pred_\sigma(v)|$.*
   The cutwidth *after v with respect to σ, denoted $c_\sigma(v)$, is $c_\sigma(v) = |\{(x, y) \in E \mid x \leq_\sigma v \text{ and } v <_\sigma y\}|$.*
2. *The* imbalance *of σ is $im(\sigma) = \sum_{v \in \sigma} \phi_\sigma(v)$.*
   The cutwidth *of σ is $cw(\sigma) = \max_{v \in \sigma}\{c_\sigma(v)\}$.*
3. *$im(G)$, the* imbalance *of G, is the minimum of $im(\sigma)$ over all orderings σ of V.*
   $cw(G)$, the cutwidth *of G is the minimum of $cw(\sigma)$ over all orderings σ of V.*

We shall consider various classes of graphs.

A *chordal* graph is a graph with no induced cycle of size at least four. All classes considered here are sub-classes of the chordal graphs.

An *interval* graph is one in which each vertex v may be identified with an interval I_v of the real (or rational) line, such that $(u, v) \in E$ if and only if $I_u \cap I_v \neq \emptyset$ (i.e., two vertices are adjacent if and only if their respective intervals intersect). An interval graph is *proper* if no interval contains another; i.e., $I_u \not\subseteq I_v$ for every $u \neq v$. All interval graphs are chordal; a chordal graph is interval iff it contains no asteroidal triple [19]. An interval graph is proper iff it contains no claw [26].

A *split* graph is graph whose vertex set can be partitioned into a clique and an independent set. All split graphs are chordal [23].

3   True Twins, Minimal Imbalance, and Parameterization by Twin-Cover

We shall show that for each of the CUTWIDTH and IMBALANCE problems, there is an optimal order for which the sets of twins of a graph are grouped together. This enables the ILP approach from the parameterization of these problems by vertex cover to be applied to the more general parameter twin-cover.

Let Z be a clique in G, and let $Z \cup A \cup B$ be a partition of $V(G)$. We wish to consider layouts that have the form $\sigma_A \tau \sigma_B$, where τ orders Z. For $z \in Z$, we define its preferred position $p(z)$ in such a layout.

- If $|N(z) \cap A| > |Z| + |N(z) \cap B|$, let $p(z) = |A|$.
- If $|N(z) \cap A| + |Z| < |N(z) \cap B|$, let $p(z) = |A| + |Z| = |V| - |B|$.
- Otherwise, let $p(z) = |N(z)|/2$.

When $p(z)$ is an integer, placement of z at location $p(z)$ gives it its minimum possible imbalance in any layout of the considered form. When $p(z) = j + \frac{1}{2}$ (necessarily, $\deg(z)$ is odd), then either position j or $j + 1$ gives z its minimum inbalance $im(z) = 1$.

Lemma 1. *Let Z be a clique in G, let $Z \cup A \cup B$ be a partition of $V(G)$, and σ_A and σ_B arbitrary layouts of A and B, respectively. For $z \in Z$, define $p(z)$ as above.*

Suppose that σ is an ordering of Z with non-decreasing $p()$ values; i.e., whenever $x <_\sigma y$ we have $p(x) \leq p(y)$. Then the imbalance of $\sigma_A \sigma \sigma_B$ is the minimum possible imbalance over layouts of the form $\sigma_A \tau \sigma_B$.

Proof. Consider a layout $\sigma_A \tau \sigma_B$, and consider a series of exchanges of consecutive elements that ends at $\sigma_A \sigma \sigma_B$. Suppose that, at some step, $x \in Z$ and $y \in Z$ occur consecutively the layout, at positions q and $q + 1$. There are four cases.

1. $p(x) < p(y)$. The requirement on σ is that $x <_\sigma y$. No exchange is desired.
2. $p(x) \geq q + 1$. Exchanging x and y improves the imbalance of x; thus the total imbalance is no greater.
3. $p(y) \leq q$. Exchanging x and y improves the imbalance of y; thus the total imbalance is no greater.
4. $p(x) = p(y) = q + \frac{1}{2}$. Both x and y are indifferent to a swap; the imbalance is the same either way.

Therefore, the final order $\sigma_A \sigma \sigma_B$ has imbalance no more than the initial order $\sigma_A \tau \sigma_B$. Since τ was arbitrary, this must be the optimal imbalance.   □

Corollary 2. *Let Z be the connected union of two sets of twins, which occur consecutively an a layout. Then there is a layout of the same (or smaller) imbalance in which the two sets are separated from one another.*

Proof. The function $p()$ takes on at most two values; by Lemma 1, the imbalance is minimized by putting the twins with the lower value to the left. If $p()$ takes only one value, then any order of Z has the same total imbalance.   □

Theorem 3. *For any graph G, there exists an imbalance-optimal ordering σ of V such that each set of true twins appears consecutively in σ.*

Proof. In an ordering σ, let a *component* of a set of true twins M be a maximal subset of M that appears consecutively in σ. Suppose that σ has a separated set of true twins, and let M be the separated set of true twins of σ that has the rightmost vertex. Let U_1 and U_2 be the two rightmost components of a set of true twins in σ; let u be the rightmost vertex of U_1 and v be the leftmost vertex of U_2. Let $Y = \sigma_{>u} \cap \sigma_{<v} = y_1, \ldots, y_\ell$ be the vertices between u and v. (Since u and v are not adjacent, $\ell \geq 1$.) Let $Y_1 = y_1, \ldots, y_i$ be the maximal initial sequence of Y that forms a component of some set of true twins containing y_1, and let $Y_\ell = y_k \ldots, y_\ell$ be the maximal initial sequence of Y that forms a component of some set of true twins containing y_ℓ.

We may assume $y_\ell \in N(v)$ as otherwise we can swap U_2 left until it is immediately to the right of one of its neighbours. Similarly $y_1 \in N(u)$, as otherwise we can swap U_1 right until it is immediately to the left of one of its neighbours.

For $w \in \{u, v\}$, let L_w, R_w and B_w be the neighbours of w that lie respectively to the left of, to the right of, and between u and v; that is, $L_u = N(u) \cap \sigma_{<u}$, $L_v = N(v) \cap \sigma_{<u}$, $R_u = N(u) \cap \sigma_{>v}$, $R_v = N(v) \cap \sigma_{>v}$, $B_u = N(u) \cap Y$, and $B_v = N(v) \cap Y$. Since u and v are true twins, $L_u = L_v$, $B_u = B_v$, and $R_u = R_v$.

We wish to show that σ can be modified to an ordering π, either by shifting U_2 to the left (and shifting y_ℓ to the right), or by shifting U_1 to the right (and shifting y_1 to the left), such that $\pi < \sigma$ and $im(\pi) \leq im(\sigma)$.

We consider several cases.

Case 1: $Y_1 = Y_\ell$. In this case, we are done by Corollary 2, taking $Z = U_1 \cup U_2 \cup Y$.

Case 2: $Y_1 \neq Y_\ell$. In this case, let $Y' = N(u) \cap (Y \setminus (Y_1 \cup Y_\ell)) = N(v) \cap (Y \setminus (Y_1 \cup Y_\ell))$.
Case 2a: $|L_v| + 1 + |Y_1| + |Y'| \geq |R_v| + |Y_\ell|$.

Consider σ' obtained from σ by moving Y_ℓ to follow U_2. We will show that $im(\sigma') = im(\sigma)$, as required. Clearly, $\phi_{\sigma'}(x) = \phi_\sigma(x)$ for all vertices x except possibly $x \in Y_\ell$ and $x \in U_2$. Therefore it remains to be shown that $\phi_{\sigma'}(Y_\ell) + \phi_{\sigma'}(U_2) \leq \phi_\sigma(Y_\ell) + \phi_\sigma(U_2)$.

Observe that $|L_v| + 1 + |Y_1| + |Y'| = \operatorname{pred}_\sigma(v) - |Y_\ell|$ and $|R_v| = \operatorname{succ}_\sigma(v)$. Thus the assumption of the case yields

$$|L_v| + 1 + |Y_1| + |Y'| \geq |R_v| + |Y_\ell| \Rightarrow \operatorname{pred}_\sigma(v) - |Y_\ell| \geq \operatorname{succ}_\sigma(v) + |Y_\ell|$$
$$\Rightarrow \operatorname{pred}_\sigma(v) > \operatorname{pred}_\sigma(v) - |Y_\ell| \geq \operatorname{succ}_\sigma(v) + |Y_\ell|$$
$$\Rightarrow \operatorname{pred}_\sigma(U_2) > \operatorname{succ}_\sigma(U_2) + |U_2| - 1 + |Y_\ell| \geq \operatorname{succ}_\sigma(U_2) + |Y_\ell|.$$

where $|U_2| \geq 1$. Since v improves and has more neighbours to its left in σ, so does every vertex in U_2; therefore, $\phi_{\sigma'}(U_2) = \phi_\sigma(U_2) - 2 \cdot |Y_\ell| \cdot |U_2|$. Since $\phi_{\sigma'}(Y_\ell) \leq \phi_\sigma(Y_\ell) + 2 \cdot |U_2| \cdot |Y_\ell|$, $\phi_{\sigma'}(Y_\ell) + \phi_{\sigma'}(U_2) \leq \phi_\sigma(Y_\ell) + \phi_\sigma(U_2)$. We have moved U_2 $|Y_\ell|$ vertices closer to u.

Case 2b: $|L_v| + 1 + |Y_1| + |Y'| < |R_v| + |Y_\ell|$.

Case 2b(i): $|R_u| + 1 + |Y'| + |Y_\ell| \geq |L_u| + |Y_1|$. This case is analogous[*1] to Case 2a, except that U_1 is swapped with Y_1.

Case 2b(ii): $|R_u| + 1 + |Y'| + |Y_\ell| < |L_u| + |Y_1|$. From our case assumptions, $|L_u| + |Y_1| > |R_u| + 1 + |Y'| + |Y_\ell| > |L_u| + |Y_1|$, which is a contradiction; this case is impossible.

Therefore we can always move U_1 and U_2 closer together. This can be repeated for all sets of true twins in G.   □

Corollary 4. IMBALANCE *is fixed-parameter tractable when parameterized by the twin-cover number of the graph.*

Proof. For a graph G with a twin cover X of size $tc(G)$, the vertices in X cover of all edges except those between twins. Let $e = \{u, v\}$ be an uncovered edge; because u and v are twins, $N(v) = N(u)$; because e exists, $N[u] = N[v]$. Therefore, we can contract all true twins in G to get a graph G' with $vc(G') = tc(G)$. Now we can apply the fixed parameter algorithm for bounded vertex cover to get an optimal ordering, replacing all vertices obtained via a contraction with the corresponding set of true twins, as there is an optimal ordering with these vertices consecutive by Theorem 3, trying all possible orderings of the vertex cover of G', which has size $vc(G') = tc(G)$, as input to the ILP of [8].   □

A graph G is *superfragile* if G can be constructed with two operations: (1) adding a universal clique to a union of disjoint cliques, and (2) taking the union of disjoint superfragile graphs [20]. A superfragile graph is an interval graph (and thus chordal). Using Theorem 3, we have the following.

Corollary 5 (*). *If G is a superfragile graph, then $im(G)$ can be computed in $O(n^2)$ time.*

The proof of the next theorem is similar to the proof of Theorem 3, and yields the following corollary using the same approach as in Corollary 4.

Theorem 6 (*). *For any graph G, there exists an optimal-cutwidth ordering σ of V such that each set of true twins appears consecutively in σ.*

Corollary 7 (*). CUTWIDTH *is fixed-parameter tractable when parameterized by the twin-cover number of the graph.*

4   Split Graphs

In this section we show that the complexity of IMBALANCE equals the complexity of CUTWIDTH for split graphs using a similar reduction found in [13].

Lemma 8 (*). *Let σ be a layout of $V(G)$ for some graph. If $N(x) = \{u, v\}$ for some vertices u, v where $u <_\sigma v$, then there is an ordering σ' such that $\sigma' = \sigma$ except that x is between u and v in σ' and $im(\sigma') \leq im(\sigma)$.*

---

[1] The statements marked with a * are proved in the full version of the paper.

Theorem 9. *Imbalance minimization on split graphs is NP-complete.*

Proof. The proof is by reduction from imbalance minimization on general graphs, which is NP-complete [2]. Let (G, k) be an instance of the imbalance minimization problem; let $n = |V(G)|$ and $m = |E(G)|$. We create an instance G' of imbalance minimization on split graphs, with vertex set $V(G') = V(G) \cup K_n \cup I_{m+n(n-1)}$, where K_n is a clique and $I_{m+n(n-1)}$ an independent set. Each $v \in I_m$ corresponds to an edge $e = xy \in E(G)$, and we add edges between v and $x \in K_n$ and $y \in K_n$. Make each $v \in K_n$ adjacent to $n-1$ private neighbours $D_v = d_v^1, \ldots, d_v^{n-1}$ in the remainder of $I_{m+n(n-1)}$. The resulting instance is $(G', k + n(n-1))$.

Suppose that G has an ordering σ such that $im(\sigma) \leq k$. Let σ' be an ordering of K_n such that v_u is in the same position in σ' as u is in σ. For every $e = xy \in E(G)$ such that $x <_\sigma y$, place $v_{e=xy}$ between x and y in σ'. For each $v_x \in K_n$, place $|N(v_x) \cap K_n \cap \sigma_{>v_x}|$ vertices from D_x anywhere to the left of v_x; place the rest of D_x anywhere to the right of v_x. The resulting order σ' is an ordering with $im(\sigma') \leq k + n(n-1)$ as each vertex present in σ has the same imbalance (every new edge introduced by the clique is offset by a vertex of D_x), the new edge vertices are perfectly balanced, and there are $n(n-1)$ pendants each with imbalance 1.

Suppose instead that we have an optimal ordering σ' of G' such that $im(\sigma') \leq k + n(n-1)$. Observe that any vertex v_e, representing $e = xy$, can be placed between v_x and v_y without changing the imbalance of σ' by Lemma 8. Thus we may assume that every vertex v_e has imbalance 0. Since the $n(n-1)$ pendents necessarily contribute exactly $n(n-1)$ to the total imbalance of the graph, the vertices representing vertices of G must contribute at most k to the total imbalance of σ'. Therefore, we can get an ordering σ of G such that $im(\sigma) \leq k$ by taking the vertices of G as they appear in σ' as each v_x only has enough pendants to offset the edges introduced by the clique vertices, not the edge vertices, too. $\square$

5   Proper Interval Graphs

In this section, we show that IMBALANCE can be solved in linear time for proper interval graphs. An ordering σ of a graph G is called a *regular labelling* of G if for every edge $uv \in E(G)$ with $f(u) < f(v)$, $V(u, v) = \{x \in V(G)|f(u) \leq f(x) \leq f(v)\}$ is a clique of G. Such an order is also called a *Proper Interval* order (PI order). Regular orders of proper interval graphs are optimal for other linear layout problems, including CUTWIDTH [16].

Theorem 10 ([15]). *A graph G is a unit interval graph if and only if G has a regular labelling.*

Theorem 11 ([16]). *Let G be proper interval graph, and let σ be a regular labelling of G. Then $cw(\sigma) = cw(G)$.*

The results are easier to prove in the context of PI-orders which are obtained from Lexicographic Breadth First Search (LBFS); we therefore require the following established facts. Two vertices x, y are unrelated with respect to a vertex z if there exists an $(x-z)$-path P that misses y and a $(y-z)$-path Q that misses x. A vertex x is admissible if no pair of vertices is unrelated with respect to x.

Lemma 12 ([6]). *Let σ be an ordering of an interval graph generated by LBFS. Then $\sigma(n)$ is simplicial and admissible.*

The algorithm $LBFS^+$ is Lexicographic Breadth First Search (LBFS) with a tie-breaking rule. Let σ^R denote the reverse of σ; i.e., $\langle v_1, v_2, \ldots, v_k \rangle^R = \langle v_k, v_{k-1}, \ldots, v_1 \rangle$.

Theorem 13 ([4]). *Let G be a proper interval graph with PI-order σ. Then $LBFS^+(\sigma) = \sigma^R$.*

Corollary 14 (*). *For every regular ordering σ of a proper interval graph, both σ and σ^R are LBFS orderings. Therefore, $\sigma(1)$ and $\sigma(n)$ are simplicial and admissible vertices.*

The main result of this section is the next theorem.

Theorem 15. *Let G be a proper interval graph. If σ is a regular ordering of G, then $im(\sigma) = im(G)$.*

Proof. We establish the following stronger result, via induction.

Lemma 16. *Let G be a proper interval graph. Let σ be a regular ordering of a proper interval graph. Then, $im(\sigma) = im(G)$. Further, if $x = \sigma(1)$ and $y = \sigma(n)$, for any ordering σ', we have $\phi_{\sigma'}(N[x]) \geq \phi_\sigma(N[x])$, $\phi_{\sigma'}(N[y]) \geq \phi_\sigma(N[y])$, and also $\phi_{\sigma'}(M) \geq \phi_\sigma(M)$ for each set M which is either a clique-separator or a block of simplicial vertices of $G - (N[x] \cup N[y])$.*

The proof is by induction on ℓ, the number of maximal cliques of G. If $\ell = 1$, then G is a clique and the hypothesis holds as any ordering is optimal. If $\ell = 2$, i.e., G has two maximal cliques, then the result holds as well by Theorem 3 and the observation that $im(G)$ is minimized by placing the universal clique separating $U(G)$ in the centre of the ordering, as it is in σ.

If G has $\ell = 3$, partition $V(G)$ into disjoint sets $V = X \cup N(X) \cup M \cup N(Y) \cup Y$, where X is the set of twins of x, Y is the set of twins of y, and $M = V \setminus (X \cup N(X) \cup Y \cup N(Y))$. If $|M| > 0$, it must be that all $m \in M$ are twins (as otherwise there are more than 3 maximal cliques). Since each of the three parts of G is a set of true twins, by Theorem 3 we need only consider the ordering of the parts; the given one is minimal as it puts each set between its neighbours, except $N[X]$ and $N[Y]$, which only have one neighbouring set.

Assume the lemma holds for proper interval graphs with $\ell' \geq 3$, and consider a proper interval graph with $\ell = \ell' + 1 \geq 4$. Let σ be a regular ordering of G, $x = \sigma(1)$, and $y = \sigma(n)$. Observe that $\sigma = \{x\} \cdot N(x) \cdot M_1 \cdot \dots \cdot M_\ell \cdot N(y) \cdot \{y\}$. Consider $G' = G - N[x]$, and observe that G' has one fewer maximal clique ($N[x]$). Let $\sigma' = \sigma - N[x]$; by induction, $im(\sigma - N[x]) = im(G')$ as $\sigma - N[x]$ is still a regular ordering. Let $G'' = G - N[y]$; again G'' has one fewer maximal clique than G ($N[y]$). Let $\sigma'' = \sigma - N[y]$; by induction $im(\sigma - N[y]) = im(G'')$ as $\sigma - N[y]$ is still a regular ordering.

Since $\ell \geq 4$, there is a clique separator M_j such that if $m \in M_j$, $d(x, m) \geq 2$ and $d(m, y) \geq 2$: take the minimal clique separator that separates the rightmost $x' \in N(X)$ from the leftmost $y' \in N(y)$, where $N[x'] \neq N[x]$ and $N[y'] \neq N[y]$. (If no such separator existed, there would be at least two maximal cliques containing $N(x) \cap N(y)$, and taking a simplicial vertex from each along with a vertex in $N(y)$ and y itself would form an induced claw, a contradiction.)

Observe that $\sigma'_{\leq M_j}$ agrees with $\sigma_{\leq M_j}$ and $\sigma''_{\geq M_j}$ agrees with $\sigma_{\geq M_j}$. Since $M_i \not\subseteq N[x]$, $\phi_\sigma(M_i) = \phi_{\sigma''}(M_i)$ and by induction, $\phi_{\sigma''}(M_i)$ is minimum for all $M_i \subseteq \sigma''_{\geq M_j}$. Similarly, $M_i \not\subseteq N[y]$, $\phi_\sigma(M_i) = \phi_{\sigma'}(M_i)$ and by induction, $\phi_{\sigma'}(M_i)$ is minimum for all $M_i \subseteq \sigma''_{\leq M_j}$. Additionally, $N[x] \not\subseteq N[y]$, $\phi_\sigma(N[x]) = \phi_{\sigma'}(N[x])$ and by induction $\phi_{\sigma'}(N[x])$ is minimum. Lastly, $N[y] \not\subseteq N[x]$, $\phi_\sigma(N[y]) = \phi_{\sigma'}(N[y])$ and by induction $\phi_{\sigma'}(N[y])$ is minimum. Let σ^* be such that $im(\sigma^*) = im(G)$; then

$$\begin{aligned} im(G) = im(\sigma^*) &= \phi_{\sigma^*}(N[x]) + \phi_{\sigma^*}(N[y]) + \Sigma_{1 \leq i \leq \ell}\phi_{\sigma^*}(M_i) \\ &= \phi_{\sigma^*}(N[x]) + \Sigma_{1 \leq i \leq j}\phi_{\sigma^*}(M_i) + \Sigma_{j < i \leq \ell}\phi_{\sigma^*}(M_i) + \phi_{\sigma^*}(N[y]) \\ &\geq \phi_{\sigma'}(N[x]) + \Sigma_{1 \leq i \leq j}\phi_{\sigma'}(M_i) + \Sigma_{j < i \leq \ell}\phi_{\sigma''}(M_i) + \phi_{\sigma''}(N[y]) \\ &= im(\sigma) \geq im(G), \end{aligned}$$

as required.   □

The following theorem proves the next corollary and establishes the linear time complexity of IMBALANCE on proper interval graphs, as $LBFS^+$ can be implemented to run in linear time.

Theorem 17 ([5]). *A graph G is a proper interval graph if and only if the third $LBFS^+$ sweep on G is a PI-order.*

Corollary 18 (*). IMBALANCE *can be solved in linear time for proper interval graphs.*

5.1   Proper Interval ∩ k-Trees

A *k-tree* is a graph that can be obtained by starting with a $k + 1$ clique and repeatedly adding vertices so that each new vertex added has precisely k neighbours which formed a clique. All k-trees are chordal (e.g., [3]) and have no clique of size greater than $k + 1$. In this section, we show that proper interval graphs which are also k-trees have imbalance equal to exactly twice their cutwidth.

The main result for this section is a corollary of the following lemma.

Lemma 19 (*). *Let G be a proper interval graph which is also a k-tree. Let σ be a regular ordering of G. Then $2 \cdot cw(\sigma) \geq im(\sigma)$.*

Lemma 20 ([21]). *Let $G = (V, E)$ be a graph with treewidth $tw(G)$, path-width $pw(G)$, cutwidth $cw(G)$, and imbalance $im(G)$. Then $tw(G) \leq pw(G) \leq cw(G) \leq \frac{im(G)}{2}$. Furthermore, $\Delta(G) \leq im(G)$ where $\Delta(G)$ is the maximum degree of G.*

Corollary 21 (*). *Let G be a proper interval graph and a k-tree. Then $im(G) = 2 \cdot cw(G)$.*

6 Conclusion

We have shown that the complexity of IMBALANCE equals the complexity of CUTWIDTH on split graphs, proper interval graphs, and graphs of bounded twin-cover. We believe that IMBALANCE can be solved in linear time on proper inter-val *bipartite* graphs and *threshold*, but leave these classes as future work. Both problems are fixed-parameter tractable parameterized by the treewidth of the graph, but it is an open question as to whether these problems yield parameter-ized algorithms when parameterized by the *modular-width* of the graph [9]. The modular-width of a graph is a generalization of the twin-cover number (and the neighbourhood diversity) which is not comparable to treewidth. The complexity of both problems remains open for some restricted graph classes, like *cographs*, and *trivially perfect* graphs, which form a proper subset of cographs but is a superset of superfragile graphs.

Acknowledgements. The authors would like to thank Therese Biedl for helpful con-versations on these results, as well as the anonymous reviewers for their suggestions.

References

1. Bakken, O.R.: Arrangement problems parameterized by neighbourhood diversity. Master's thesis, The University of Bergen (2018)
2. Biedl, T., Chan, T., Ganjali, Y., Hajiaghayi, M.T., Wood, D.R.: Balanced vertex-orderings of graphs. Discrete Appl. Math. **148**(1), 27–48 (2005)
3. Brandstadt, A., Spinrad, J.P., et al.: Graph Classes: A Survey, vol. 3. SIAM, Philadelphia (1999)
4. Charbit, P., Habib, M., Mouatadid, L., Naserasr, R.: A new graph parameter to measure linearity. In: Gao, X., Du, H., Han, M. (eds.) COCOA 2017. LNCS, vol. 10628, pp. 154–168. Springer, Cham (2017). https://doi.org/10.1007/978-3-319-71147-8_11
5. Corneil, D.G.: A simple 3-sweep LBFS algorithm for the recognition of unit interval graphs. Discrete Appl. Math. **138**(3), 371–379 (2004)
6. Corneil, D.G., Olariu, S., Stewart, L.: The LBFS structure and recognition of interval graphs. SIAM J. Discrete Math. **23**(4), 1905–1953 (2009)

7. Cygan, M., Lokshtanov, D., Pilipczuk, M., Pilipczuk, M., Saurabh, S.: On cutwidth parameterized by vertex cover. Algorithmica **68**(4), 940–953 (2014)
8. Fellows, M.R., Lokshtanov, D., Misra, N., Rosamond, F.A., Saurabh, S.: Graph layout problems parameterized by vertex cover. In: Hong, S.-H., Nagamochi, H., Fukunaga, T. (eds.) ISAAC 2008. LNCS, vol. 5369, pp. 294–305. Springer, Heidelberg (2008). https://doi.org/10.1007/978-3-540-92182-0_28
9. Gajarský, J., Lampis, M., Ordyniak, S.: Parameterized algorithms for modularwidth. In: Gutin, G., Szeider, S. (eds.) IPEC 2013. LNCS, vol. 8246, pp. 163–176. Springer, Cham (2013). https://doi.org/10.1007/978-3-319-03898-8_15
10. Ganian, R.: Twin-cover: beyond vertex cover in parameterized algorithmics. In: Marx, D., Rossmanith, P. (eds.) IPEC 2011. LNCS, vol. 7112, pp. 259–271. Springer, Heidelberg (2012). https://doi.org/10.1007/978-3-642-28050-4_21
11. Gaspers, S., Messinger, M.-E., Nowakowski, R.J., Prałat, P.: Clean the graph before you draw it!. Inf. Process. Lett. **109**(10), 463–467 (2009)
12. Giannopoulou, A.C., Pilipczuk, M., Raymond, J.-F., Thilikos, D.M., Wrochna, M.: Cutwidth: obstructions and algorithmic aspects. arXiv preprint arXiv:1606.05975 (2016)
13. Heggernes, P., Lokshtanov, D., Mihai, R., Papadopoulos, C.: Cutwidth of split graphs, threshold graphs, and proper interval graphs. In: Broersma, H., Erlebach, T., Friedetzky, T., Paulusma, D. (eds.) WG 2008. LNCS, vol. 5344, pp. 218–229. Springer, Heidelberg (2008). https://doi.org/10.1007/978-3-540-92248-3_20
14. Heggernes, P., van 't Hof, P., Lokshtanov, D., Nederlof, J.: Computing the cutwidth of bipartite permutation graphs in linear time. In: Thilikos, D.M. (ed.) WG 2010. LNCS, vol. 6410, pp. 75–87. Springer, Heidelberg (2010). https://doi.org/10.1007/978-3-642-16926-7_9
15. Jinjiang, Y., Liying, K.: One characterization of unit interval graphs and its applications. J. Shijiazhuang Railway Inst. **7**(2), 50–54 (1994)
16. Jinjiang, Y., Sanming, Z.: Optimal labelling of unit interval graphs. Appl. Math. **10**(3), 337–344 (1995)
17. Kant, G.: Drawing planar graphs using the canonical ordering. Algorithmica **16**(1), 4–32 (1996)
18. Kant, G., He, X.: Regular edge labeling of 4-connected plane graphs and its applications in graph drawing problems. Theoret. Comput. Sci. **172**(1–2), 175–193 (1997)
19. Lekkerkerker, C., Boland, J.: Representation of a finite graph by a set of intervals on the real line. Fund. Math. **51**(1), 45–64 (1962)
20. Lilleeng, S.: A polynomial-time solvable case for the NP-hard problem cutwidth. Master's thesis, The University of Bergen (2014)
21. Lokshtanov, D., Misra, N., Saurabh, S.: Imbalance is fixed parameter tractable. Inf. Process. Lett. **113**(19–21), 714–718 (2013)
22. Papakostas, A., Tollis, I.G.: Algorithms for area-efficient orthogonal drawings. Comput. Geom. **9**(1–2), 83–110 (1998)
23. Stephane, F., Hammer, P.L.: Split graphs. In: Proceedings of the 8th Southeastern Conference on Combinatorics, Graph Theory and Computing, pp. 311–315 (1977)
24. Thilikos, D.M., Serna, M., Bodlaender, H.L.: Cutwidth I: a linear time fixed parameter algorithm. J. Algorithms **56**(1), 1–24 (2005)
25. Thilikos, D.M., Serna, M., Bodlaender, H.L.: Cutwidth II: algorithms for partial w-trees of bounded degree. J. Algorithms **56**(1), 25–49 (2005)
26. Wagner, G.: Eigenschaften der nerven homologische-einfactor familien in R^n. Ph.D. thesis, Universität Gottigen (1967)
27. Wood, D.R.: Optimal three-dimensional orthogonal graph drawing in the general position model. Theoret. Comput. Sci. **299**(1–3), 151–178 (2003)

28. Wood, D.R.: Minimising the number of bends and volume in 3-dimensional orthogonal graph drawings with a diagonal vertex layout. Algorithmica **39**(3), 235–253 (2004)
29. Wood, D.R., Kratochvil, J., Kára, J.: On the complexity of the balanced vertex ordering problem. Discrete Math. Theor. Comput. Sci. **9** (2007)
30. Yannakakis, M.: A polynomial algorithm for the min-cut linear arrangement of trees. J. ACM (JACM) **32**(4), 950–988 (1985)

Smaller Universal Targets for Homomorphisms of Edge-Colored Graphs

Grzegorz Guśpiel[(✉)] [ID]

Theoretical Computer Science Department, Faculty of Mathematics
and Computer Science, Jagiellonian University, Kraków, Poland
guspiel@tcs.uj.edu.pl

Abstract. For a graph G, the density of G, denoted $D(G)$, is the maximum ratio of the number of edges to the number of vertices ranging over all subgraphs of G. For a class $\mathcal{F}$ of graphs, the value $D(\mathcal{F})$ is the supremum of densities of graphs in $\mathcal{F}$. A k-edge-colored graph is a finite, simple graph with edges labeled by numbers $1, \ldots, k$. A function from the vertex set of one k-edge-colored graph to another is a homomorphism if the endpoints of any edge are mapped to two different vertices connected by an edge of the same color. Given a class $\mathcal{F}$ of graphs, a k-edge-colored graph $\mathbb{H}$ (not necessarily with the underlying graph in $\mathcal{F}$) is k-universal for $\mathcal{F}$ when any k-edge-colored graph with the underlying graph in $\mathcal{F}$ admits a homomorphism to $\mathbb{H}$. Such graphs are known to exist exactly for classes $\mathcal{F}$ of graphs with acyclic chromatic number bounded by a constant. The minimum number of vertices in a k-uniform graph for a class $\mathcal{F}$ is known to be $\Omega(k^{D(\mathcal{F})})$ and $O(k^{\lceil D(\mathcal{F}) \rceil})$. In this paper we close the gap by improving the upper bound to $O(k^{D(\mathcal{F})})$ for any rational $D(\mathcal{F})$.

Keywords: Universal graph · Homomorphism bound · Edge coloring · Graph density · Maximum average degree

1 Introduction

All graphs considered in this paper are finite, nonempty and contain no loops or multiple edges. By a class of graphs we mean a nonempty set of graphs closed under isomorphisms. For every positive integer k, the set $\{1, \ldots, k\}$ is denoted by $[k]$.

A *k-edge-colored graph* $\mathbb{G}$ is a pair (G, c), where G is a graph, called *an underlying graph* of $\mathbb{G}$, and c is a mapping from $E(G)$ to $[k]$, called a *k-edge-coloring* of $\mathbb{G}$. A *k-edge-colored graph over G* is a k-edge-colored graph with the underlying graph G.

Let $\mathbb{G}_1 = (G_1, c_1)$ and $\mathbb{G}_2 = (G_2, c_2)$ be two k-edge-colored graphs. A mapping $h : V(G_1) \to V(G_2)$ is a *homomorphism* of $\mathbb{G}_1$ to $\mathbb{G}_2$ if, for every two

Research was partially supported by MNiSW grant DI2013 000443.

D.-Z. Du et al. (Eds.): COCOON 2019, LNCS 11653, pp. 232–239, 2019.
https://doi.org/10.1007/978-3-030-26176-4_19

vertices u and v that are adjacent in G_1, $h(u)$ and $h(v)$ are adjacent in G_2 and $c_1(uv) = c_2(h(u)h(v))$. In other words, a homomorphism of $\mathbb{G}_1$ to $\mathbb{G}_2$ maps every colored edge in $\mathbb{G}_1$ into an edge of the same color in $\mathbb{G}_2$.

A k-edge-colored graph $\mathbb{H}$ is k-*universal*[1] for a class $\mathcal{F}$ of graphs if every k-edge-colored graph over any graph in $\mathcal{F}$ admits a homomorphism to $\mathbb{H}$. We denote by $\lambda_{\mathcal{F}}(k)$ the minimum possible number of vertices in a k-universal graph for $\mathcal{F}$. We set $\lambda_{\mathcal{F}}(k) = \infty$ if such a graph does not exist.

Observe that $\lambda_{\mathcal{F}}(1)$ is the maximum chromatic number of all graphs in $\mathcal{F}$. Although this parameter is of great importance in graph theory, this paper is focused on the behavior of $\lambda_{\mathcal{F}}(k)$ when k tends to infinity. In particular, the case $k = 1$ differs significantly from the case $k \geqslant 2$. Only the latter one is the subject of this paper.

The concept of finding a small k-universal graph for a certain class of graphs first arose in 1998, when Alon and Marshall [2] used it to obtain a result on Coxeter groups. They showed for the class of planar graphs $\mathcal{P}$ that $\lambda_{\mathcal{P}}(k)$ is between $k^3 + 3$ and $5k^4$. They also generalized their ideas for graphs with bounded acyclic chromatic number. An *acyclic coloring* of a graph G is an assignment of colors to the vertices of G such that adjacent vertices have different colors and every subgraph of G with vertices in at most 2 colors is acyclic. The *acyclic chromatic number* of a graph G, denoted $\chi_a(G)$, is the minimum number of colors in an acyclic coloring of G. For a class $\mathcal{F}$ of graphs, if the acyclic chromatic number of the graphs in $\mathcal{F}$ is bounded by a constant, we write $\chi_a(\mathcal{F}) = \max_{G \in \mathcal{F}} \chi_a(G)$ (and set $\chi_a(\mathcal{F}) = \infty$ otherwise). Alon and Marshall [2] showed that for every graph class $\mathcal{F}$ with $\chi_a(\mathcal{F}) = r < \infty$, we have $\lambda_{\mathcal{F}}(k) \leqslant rk^{r-1}$. Plugging in the famous result of Borodin [3] that $\chi_a(\mathcal{P}) \leqslant 5$ gives $\lambda_{\mathcal{P}}(k) \leqslant 5k^4$.

A concept similar to homomorphisms of edge colorings was considered by Raspaud and Sopena [9]. They show that for every oriented planar graph $\vec{G}$ there exists an oriented graph $\vec{H}$ on at most 80 vertices, such that $\vec{G}$ maps homomorphically to $\vec{H}$, where a homomorphism of an oriented graph $\vec{G}$ to an oriented graph $\vec{H}$ is a mapping $h : V(\vec{G}) \to V(\vec{H})$ such that for every directed edge $uv \in E(\vec{G})$, there is an edge from $h(u)$ to $h(v)$ in $\vec{H}$. This concept is also known under the name *oriented coloring*. A simple consequence of this result is that there exists a single graph $\vec{H}$ on at most 80 vertices, to which every oriented planar graph maps homomorphically. Later, Nešetřil and Raspaud [8] proved a theorem about *mixed graphs* (i.e. graphs with both oriented and unoriented colored edges) that implies both the results of Alon and Marshall and those of Raspaud and Sopena (see also [4]).

Universal graphs were recently analyzed further by Guśpiel and Gutowski [5]. We now shortly summarize their results. First, they show that for every $k \geqslant 2$, a class $\mathcal{F}$ of graphs admits a k-universal graph if and only if the acyclic chromatic number of graphs in $\mathcal{F}$ is bounded by a constant. In particular, this means that $\mathcal{F}$ either admits a k-universal graph for all $k \geqslant 2$ or for no $k \geqslant 2$. Next, they analyze the asymptotic behavior of $\lambda_{\mathcal{F}}(k)$. It happens that $\lambda_{\mathcal{F}}(k)$ can be much smaller than $O\left(k^{\chi_a(\mathcal{F})-1}\right)$ and is more closely related to the density of the graphs in $\mathcal{F}$.

---

[1] In literature, universal graphs are also called homomorphism bounds.

For a graph G, the *density* of G, denoted $D(G)$, is the maximum ratio of the number of edges to the number of vertices over all subgraphs of G. For a class $\mathcal{F}$ of graphs, its density $D(\mathcal{F})$ is the supremum of the densities of the graphs in $\mathcal{F}$. A simple argument gives $D(\mathcal{F}) \leqslant \chi_a(\mathcal{F}) - 1$ and there are examples of classes of graphs with bounded density and unbounded acyclic chromatic number. The main result of [5] is the following theorem:

Theorem 1 ([5]). *Let $\mathcal{F}$ be a class of graphs with $\chi_a(\mathcal{F}) = r < \infty$ and $\lceil D(\mathcal{F}) \rceil = d$. Then, for the constant $c = 8dr^4 \binom{8dr^4}{d}$, we have*

$$k^{D(\mathcal{F})} \leqslant \lambda_{\mathcal{F}}(k) \leqslant ck^{\lceil D(\mathcal{F}) \rceil},$$

for all $k \geqslant 2$.

Since $D(\mathcal{F}) \leqslant \chi_a(\mathcal{F}) - 1$, the above upper bound is asymptotically no worse than the one by Alon and Marshall (although the latter is usually much better for small k's). For any class of graphs with density being an integer, the bounds of Theorem 1 are asymptotically tight. For example, we get that $\lambda_{\mathcal{P}}(k) = \Theta(k^3)$.

In this paper, we continue the study of the asymptotics of $\lambda_{\mathcal{F}}(k)$. The paper [5] concludes with the question whether $\lambda_{\mathcal{F}}(k)$ is always $\Theta(k^{D(\mathcal{F})})$. Our next theorem confirms this hypothesis for $D(\mathcal{F})$ being a rational number and provides evidence that it should hold in general.

Theorem 2. *Let $\mathcal{F}$ be a class of graphs with $\chi_a(\mathcal{F}) = r < \infty$ and $D(\mathcal{F})$ bounded from above by a quotient s/t of natural numbers. Then, for the constant $c = 2^s (8r^4 s)^t \binom{8r^4 st}{s}$, we have*

$$k^{D(\mathcal{F})} \leqslant \lambda_{\mathcal{F}}(k) \leqslant ck^{s/t}, \tag{1}$$

for all $k \geqslant 2$.

The immediate consequences of Theorem 2 for a class $\mathcal{F}$ with $\chi_a(\mathcal{F}) < \infty$ are:

(1) $\lambda_{\mathcal{F}}(k) = \Theta(k^{D(\mathcal{F})})$, if $D(\mathcal{F})$ is a rational number,
(2) $\lambda_{\mathcal{F}}(k) = o(k^{D(\mathcal{F})+\varepsilon})$ for every $\varepsilon > 0$.

Our proof of Theorem 2 uses the techniques from [5] in a new way. In particular, the proof requires additional ideas that allow us to work with nonintegral graph densities. It is known that a graph with small density admits an orientation with small indegree. We employ a more fine-grained concept of fractional orientations to have a better control over the construction size. This technique seems to be quite interesting on its own and we hope that it can be used in other problems related to edge densities or other nonintegral graph parameters.

2   Universal Graph Construction

In this section, we collect several tools from [5] and apply them in a new way to obtain the main theorem. First, we introduce the terminology for orientations of graphs. An *orientation* of a graph G is an assignment of direction to each edge of G, which turns G into an oriented graph $\vec{G}$. For a vertex v, the number of edges oriented towards v is called the *indegree* of v. An orientation $\vec{G}$ of G is a *d-orientation* if every vertex of $\vec{G}$ has indegree at most d.

The paper [5] makes use of a simple observation, attributed to Hakimi [6], that for an integer d, a graph G admits a d-orientation if and only if $D(G) \leqslant d$. Not surprisingly, d may be relaxed to be a rational number s/t. First, we replace the original graph with a multigraph that contains t copies of each edge. Now, for a vertex v, instead of orienting at most d edges towards v in G, we orient at most s edges towards v in this multigraph. We formalize this idea in the following lemma:

Lemma 3. *For every graph G and a quotient s/t of natural numbers, we have $D(G) \leqslant s/t$ if and only if there exist orientations $\vec{G}_1, ..., \vec{G}_t$ of G such that for every $v \in V(G)$*

$$\sum_{i=1}^{t} \deg_{\vec{G}_i}^{in}(v) \leqslant s. \tag{2}$$

Proof. Assume that the orientations $\vec{G}_1, ..., \vec{G}_t$ of G have the property (2). Let H be a subgraph of G. For every $i = 1, ..., t$, we have

$$|E(H)| \leqslant \sum_{v \in V(H)} \deg_{\vec{G}_i}^{in}(v).$$

Summing up over the i's, we get

$$t \cdot |E(H)| \leqslant \sum_{v \in V(H)} \sum_{i=1}^{t} \deg_{\vec{G}_i}^{in}(v) \leqslant |V(H)| \cdot s.$$

Therefore, $|E(H)|/|V(H)| \leqslant s/t$.

For the other direction, we assume that $D(G) \leqslant s/t$ and apply Hall's Theorem to a bipartite graph B constructed as follows. The vertex set of B contains s copies of each vertex of G and t copies of each edge of G. For each $v \in V(G)$ and $e \in E(G)$ such that v is incident with e, we add an edge in B between every copy of v and every copy of e.

We show that B admits a matching of all edge copies by checking the Hall's condition for every set X of edge copies. Every such X induces a subgraph of G with at least $|X|/t$ edges and, according to the density bound, at least $|X|/s$ vertices. As every vertex of G has s copies in B, the set X is incident with at least $|X|$ vertices of B, and by Hall's Theorem the required matching indeed exists.

Now, given a matching in B, we construct the orientations $\vec{G}_1, ..., \vec{G}_t$ as follows: in the i-th orientation, every edge e is oriented towards the vertex whose copy is paired in the matching with the i-th copy of e. The bound on the sum of indegrees of $v \in V(G)$ in all orientations follows from the fact that there are s copies of v. □

Apart from the acyclic coloring, we use two other kinds of colorings. The first one, the *star coloring* of a graph, is an assignment of colors to the vertices of the graph such that:

(i) every two adjacent vertices get different colors,
(ii) every subsequent four vertices on any path in the graph get at least 3 different colors.

In other words, a star coloring is a proper coloring such that, for any two colors, every connected component in the graph induced by vertices of these two colors has at most one vertex of degree higher than one. Observe that any star coloring of G is an acyclic coloring of G. Conversely, Albertson, Chappell, Kierstead, Kündgen and Ramamurthi [1] showed that any acyclic coloring with r colors can be used to construct a star coloring with at most $2r^2 - r$ colors.

The other coloring we need is called out-coloring and is a technical tool from [5]. This concept appeared in almost the same form under the name in-coloring in [1], and earlier without name in Nešetřil and Ossona de Mendez [7]. Let $\vec{G}$ be an orientation of a graph G. We use the following notions: if uv is an edge of $\vec{G}$, then u is a *parent* of v; if uv and vw are edges of $\vec{G}$, then u is a *grandparent* of w. An *out-coloring* of an oriented graph is an assignment of colors to the vertices of the graph such that:

(i) every two adjacent vertices get different colors,
(ii) every two distinct parents of a single vertex get different colors,
(iii) a vertex gets different colors from any of its grandparents.

Out-colorings are closely related to star colorings (and acyclic colorings, in consequence). Indeed, every out-coloring of $\vec{G}$ is a star coloring of G. Moreover, [5] contains an easy construction of an out-coloring of $\vec{G}$ from a star coloring of G.

The upper bound of [5] is an explicit construction of a k-universal graph. Lemma 11 in [5] contains the details of the construction. We present it in a slightly modified form:

Lemma 4 ([5]). *For every $k \geqslant 2$, d and q there exists a k-edge-colored graph $\mathbb{H}$ on at most $q\binom{q}{d}k^d$ vertices such that for every k-edge-colored graph $\mathbb{G}$ with a d-orientation that admits an out-coloring with q colors there exists a homomorphism of $\mathbb{G}$ to $\mathbb{H}$.*

This lemma is the starting point of the proof of Theorem 2. In order to use it, we need to make an observation regarding the details of its proof. We now recall these details from [5].

The first step of the proof of Lemma 4 is the definition of $\mathbb{H}$. The vertex set of $\mathbb{H}$ is the set of all $(q+1)$-tuples of the form $(i, x_1, ..., x_q)$ such that $i \in [q]$, $x_j \in [k]$ for all $j \in [q]$ and among $x_1, ...x_q$ there are at most d values different from k. In the next step of the proof, a k-edge-colored graph $\mathbb{G} = (G, c_G)$ is fixed, together with a d-orientation $\vec{G}$ and an out-coloring f with at most q colors. Finally, a homomorphism h of $\mathbb{G}$ to $\mathbb{H}$ is defined. A vertex $u \in V(G)$ is mapped by h to the tuple $(f(u), x_1, ..., x_q) \in V(\mathbb{H})$, where for each $i \in [q]$

$$x_i = \begin{cases} c_G(up) & \text{if } u \text{ has a parent } p \text{ in } \vec{G} \text{ with } f(p) = i, \\ k & \text{otherwise.} \end{cases}$$

Notice that there are at most $q\binom{q}{i}k^i$ different tuples that a vertex with indegree i in $\vec{G}$ can be mapped to. Let V_i be the set of all such tuples, for every $i \leqslant d$. We obtain the following strengthening of Lemma 4:

Lemma 5. *For integers $k \geqslant 2$, d and q there exist a k-edge-colored graph $\mathbb{H}$ and sets $V_0 \subseteq ... \subseteq V_d = V(\mathbb{H})$ such that:*

(P1) for every k-edge-colored graph $\mathbb{G} = (G, c_G)$ and for every d-orientation $\vec{G}$ of G that admits an out-coloring with q colors there exists a homomorphism h of $\mathbb{G}$ to $\mathbb{H}$ such that for every vertex $v \in V(\mathbb{G})$ with indegree i in $\vec{G}$ we have $h(v) \in V_i$,
(P2) the size of each set V_i satisfies $|V_i| \leqslant q\binom{q}{i}k^i$.

After these preparations, we are ready to construct a k-universal graph on $O(k^{s/t})$ vertices.

Lemma 6. *Let $\mathcal{F}$ be a class of graphs with density bounded from above by a quotient s/t of natural numbers and let q be an integer such that every s-orientation of every graph in $\mathcal{F}$ admits an out-coloring with q colors. For any $k \geqslant 2$, the following holds:*

$$\lambda_{\mathcal{F}}(k) \leqslant 2^s q^t \binom{qt}{s} k^{s/t}.$$

Proof. Let $k_0 = \lceil k^{1/t} \rceil$ and let g be a one-to-one mapping from $[k]$ to $[k_0]^t$. After putting $k = k_0$ and $d = s$, Lemma 5 supplies us with a k_0-edge-colored graph $\mathbb{H}_0 = (H_0, c_{H_0})$ and a sequence $V_0 \subseteq ... \subseteq V_s$ of subsets of $V(H_0)$ with properties (P1) and (P2). We construct a k-universal graph $\mathbb{H} = (H, c_H)$ for $\mathcal{F}$ as follows. The vertex set of H is given by:

$$V(H) = \bigcup_{\substack{i_1,...,i_t \in \{0,...,s\} \\ i_1+...+i_t=s}} V_{i_1} \times ... \times V_{i_t}.$$

We note that $V(H)$ is a subset of $V(H_0)^t$ and the size of $V(H)$ is bounded by:

$$\sum_{\substack{i_1,...,i_t \in \{0,...,s\} \\ i_1+...+i_t=s}} \prod_{j=1}^{t} q\binom{q}{i_j}\lceil k^{1/t}\rceil^{i_j} \leqslant q^t\binom{qt}{s}\left(k^{1/t}+1\right)^s \leqslant q^t\binom{qt}{s}2^s k^{s/t}.$$

An edge between $(u_1, ..., u_t), (v_1, ..., v_t) \in V(H)$ exists if and only if $u_i v_i$ is an edge in H_0 for every $i \in [t]$ and $(c_{H_0}(u_1 v_1), ..., c_{H_0}(u_t v_t))$ is contained in the range of g. In such a case we put:

$$c_H((u_1, ..., u_t)(v_1, ..., v_t)) = g^{-1}((c_{H_0}(u_1 v_1), ..., c_{H_0}(u_t v_t))).$$

Now, let $\mathbb{G} = (G, c_G)$ be a k-edge-colored graph with $G \in \mathcal{F}$. To construct a homomorphism of $\mathbb{G}$ to $\mathbb{H}$, we start with defining $c_1, ..., c_t$ to be k_0-edge-colorings of G such that for every $uv \in E(G)$

$$g(c_G(uv)) = (c_1(uv), ..., c_t(uv)).$$

Next, let $\vec{G}_1, ..., \vec{G}_t$ be the s-orientations of G provided by Lemma 3. Applying Lemma 5 to the k_0-edge-colored graph (G, c_j) and the s-orientation $\vec{G}_j$, we get a homomorphism $h_j : (G, c_j) \to \mathbb{H}_0$. We want the mapping defined by

$$h(v) = (h_1(v), ..., h_t(v))$$

to be a homomorphism of $\mathbb{G}$ to $\mathbb{H}$.

To see that $h(v) \in V(H)$, note that the bound (2) of Lemma 3 together with property (P1) yields $h(v) \in V_{i_1} \times ... \times V_{i_t}$ for some $i_1, ..., i_t$ with $i_1 + ... + i_t = s$. To see that h preserves an edge $uv \in E(G)$, first note that each h_i preserves it, so that $h_i(u)h_i(v) \in E(H_0)$ and $c_i(uv) = c_{H_0}(h_i(u)h_i(v))$. The tuple

$$(c_{H_0}(h_1(u)h_1(v)), ..., c_{H_0}(h_t(u)h_t(v))) = (c_1(uv), ..., c_t(uv)) = g(c_G(uv))$$

is contained in the range of g, therefore the vertices

$$h(u) = (h_1(u), ..., h_t(u)), \qquad h(v) = (h_1(v), ..., h_t(v))$$

are connected by an edge in $\mathbb{H}$. Its color is given by

$$\begin{aligned}
c_H((h_1(u), ..., h_t(u))&(h_1(v), ..., h_t(v))) \\
&= g^{-1}((c_{H_0}(h_1(u)h_1(v)), ..., c_{H_0}(h_t(u)h_t(v)))) \\
&= g^{-1}((c_1(uv), ..., c_t(uv))) \\
&= c_G(uv),
\end{aligned}$$

which concludes the proof.   □

Now we are ready to prove our main theorem.

Proof of Theorem 2. Note that the lower bound in (1) appears already in Theorem 1. For the upper bound, first recall that one of the results in [1] says that every graph in $\mathcal{F}$ admits a star coloring with at most $2r^2 - r$ colors. By applying Lemma 10 from [5], we get that every s-orientation of every graph from $\mathcal{F}$ admits an out-coloring with at most $8r^4 s$ colors. We apply Lemma 6 with $q = 8r^4 s$ and obtain a k-universal graph for $\mathcal{F}$ of the desired size.   □

Acknowledgments. I would like to thank Grzegorz Gutowski and Prof. Paweł Idziak for their invaluable help with the preparation of this paper.

References

1. Albertson, M.O., Chappell, G.G., Kierstead, H.A., Kündgen, A., Ramamurthi, R.: Coloring with no 2-colored P4's. Electron. J. Comb. 11(1), R26 (2004)
2. Alon, N., Marshall, T.H.: Homomorphisms of edge-colored graphs and Coxeter groups. J. Algebraic Comb. 8(1), 5–13 (1998)
3. Borodin, O.V.: On acyclic colorings of planar graphs. Discrete Math. 25(3), 211–236 (1979)
4. Das, S., Nandi, S., Sen, S.: On chromatic number of colored mixed graphs. In: Gaur, D., Narayanaswamy, N.S. (eds.) CALDAM 2017. LNCS, vol. 10156, pp. 130–140. Springer, Cham (2017). https://doi.org/10.1007/978-3-319-53007-9_12
5. Guspiel, G., Gutowski, G.: Universal targets for homomorphisms of edge-colored graphs. J. Comb. Theory Ser. B 127, 53–64 (2017). https://doi.org/10.1016/j.jctb.2017.05.009
6. Hakimi, S.L.: On the degrees of the vertices of a directed graph. J. Franklin Inst. 279(4), 290–308 (1965)
7. Nešetřil, J., de Mendez, P.O.: Colorings and homomorphisms of minor closed classes. In: Aronov, B., Basu, S., Pach, J., Sharir, M. (eds.) Discrete and Computational Geometry. AC, vol. 25, pp. 651–664. Springer, Heidelberg (2003). https://doi.org/10.1007/978-3-642-55566-4_29
8. Nešetřil, J., Raspaud, A.: Colored homomorphisms of colored mixed graphs. J. Comb. Theory Ser. B 80(1), 147–155 (2000)
9. Raspaud, A., Sopena, E.: Good and semi-strong colorings of oriented planar graphs. Inf. Process. Lett. 51(4), 171–174 (1994). https://doi.org/10.1016/0020-0190(94)00088-3

A Simple Construction of Broadcast Graphs

Hovhannes A. Harutyunyan[1] and Zhiyuan Li[2(✉)]

[1] Department of Computer Science and Software Engineering,
Concordia University, Montreal, QC H3G 1M8, Canada
`haruty@cs.concordia.ca`
[2] Computer Science and Technology,
BNU-HKBU United International College, Zhuhai 0086-519000, China
`goliathli@uic.edu.hk`

Abstract. Broadcasting is one of the basic primitives of communication in usual networks. It is a process of information dissemination in which one informed node of the network, called the originator, distributes the message to all other nodes of the network by placing a series of calls along the communication lines. The network is modeled as a graph. The broadcast time of a given vertex is the minimum time required to broadcast a message from the originator to all other vertices of the graph. The broadcast time of a graph is the maximum time required to broadcast from any vertex in the graph. Many papers have investigated the construction of minimum broadcast graphs, the cheapest possible broadcast network architecture (having the fewest communication lines) in which broadcasting can be accomplished as fast as theoretically possible from any vertex. Since this problem is very difficult, numerous papers give sparse networks in which broadcasting can be completed in minimum time from any originator. In this paper, we improve the existing upper bounds on the number of edges by constructing sparser graphs and by presenting a minimum time broadcast algorithm from any originator.

1 Introduction

In today's world, due to massively parallel processing communication between different processors in a large network is of main concern. One of the main problems of information dissemination in large networks is called *broadcasting*. Broadcasting is the message dissemination problem in a connected network in which initially one informed node, called the originator, must distribute the message to all other nodes of the network. In this paper, we investigate the problem of designing sparse networks in which fast broadcasting is possible. Our study focuses on the classical broadcast model under the following assumptions.

- The process is split into discrete time units.
- Initially, only one node, called the *originator* has the information.
- In each time unit, every informed node can only call at most one uninformed neighbor.
- The process terminates when every node in the network is informed.

© Springer Nature Switzerland AG 2019
D.-Z. Du et al. (Eds.): COCOON 2019, LNCS 11653, pp. 240–253, 2019.
https://doi.org/10.1007/978-3-030-26176-4_20

A network can be modeled as a graph. The *broadcast scheme* from the originator v in the graph G is a sequence of parallel calls from v. Each call, expressed by a directed edge, defines the sender and the receiver. The broadcast scheme also defines a directed spanning tree of G rooted at v, which is the *broadcast tree* of originator v. The minimum number of time units required to broadcast from v in the graph G is the *broadcast time* of vertex v and denoted by $b(G, v)$. And the maximum broadcast time from any vertex in G is the broadcast time of G, denoted by $b(G) = max\{b(G, u), u \in V\}$.

Note that in each time unit, the number of informed vertices is at most doubled because every vertex can only inform at most one vertex in a single time unit. Thus, $b(G) \geq \lceil \log n \rceil$ for an arbitrary graph G.

A graph G on n vertices is called a *broadcast graph* if $b(G) = \lceil \log n \rceil$. A broadcast graph with the minimum number of edges is called a *minimum broadcast graph* (mbg). The broadcast function $B(n)$ denotes the number of edges in mbg on n vertices. From the application perspective, mbgs are the "cheapest" graphs with the fastest broadcasting. In this research area, there are two major topics.

- Given a graph G and a vertex v, determine the optimal broadcast scheme originating from v in G and the value of $b(G, v)$. It is called the *broadcast time* problem.
- Given an natural number n, construct a minimum broadcast graph on n vertices and determine the value of $B(n)$, called the *minimum broadcast graph* problem.

The broadcast time problem in general is NP-hard [22]. The minimum broadcast graph problem is even more difficult. The exact value of $B(n)$ is known only for $n \leq 15$ [7], $n = 17$ [18], $n = 18, 19$ [3,23], $n = 20, 21, 22$ [17], $n = 26$ [20,24], $n = 27, 28, 29, 58, 61$ [20], $n = 30, 31$ [3], $n = 63$ [16], $n = 127$ [9], and $n = 1023, 4095$ [21]. Knödel graphs [5,15], hypercubes [7], and recursive circulant graphs [19] give the exact value for $n = 2^m$. Knödel graphs also give the exact value for $n = 2^m - 2$ [5,14].

Determining the exact values of $B(n)$ is very difficult. Thus, the line of research in this area studies the construction of broadcast graphs (not necessarily minimum), which implies upper bounds on $B(n)$. The broadcast graph constructions include compounding method [1,2,10,12], ad-hoc constructions [8,12], vertex addition methods [3,9,11,13], and vertex deletion methods [3,12].

In this paper, we follow the compounding construction and improve the general upper bound on $B(n)$ for even $2^m - 2^{\frac{m+3}{2}} < n \leq 2^m - 8$. The previously best bound for this interval is $B(n) \leq \frac{1}{2}(m - 1)n$ [15]. Our improvement is no less than $\frac{n}{4}$. Section 2 reviews the important graphs being used in our construction. Section 3 gives a trivial compounding construction. Section 4 constructs our new graph. Section 5 proves that the newly constructed graph is a broadcast graph by demonstrating a broadcast algorithm from any originator. We also compare the new upper bound with the existing bounds.

2   Important Graphs

Definition 1. *A hypercube Q_k of dimension k, for any $k \in \mathbb{N}$ is a graph on 2^k vertices, where each vertex is represented by a binary string of length k and two vertices are adjacent if and only if the two strings have Hamming distance 1.*

In 1975, Knödel defined a class of broadcast graphs on even number of vertices [15]. We follow the equivalent definition given in [12,14].

Definition 2. *A Knödel graph $KG_n = (V, E)$ is defined for even values of n, where the vertex set is $V = \{v_0, v_1, v_2, ..., v_{n-1}\}$ and the edge set is $E = \{(v_x, v_y)| x + y \equiv 2^s - 1 \mod n, 1 \le s \le \lfloor \log n \rfloor\}$, where $0 \le x, y \le n - 1$.*

We recall the construction of a broadcast graph H from [12,14]. Let $s \ge 4$ and $1 \le t \le s - 2$. A graph H on $2^s - 2^t$ vertices consists of 2^{t-1} copies of a Knödel graph $KG_{2^{s-t+1}-2}$. In each copy, the vertices are labeled from 0 to $2^{s-t+1}-3$. All vertices with the same even label further form a hypercube Q_{t-1} of dimension $t - 1$ on 2^{t-1} vertices. Thus, H has two types of vertices: half of the vertices are of degree $s - t$, and the other half of the vertices are of degree $s - 1$. Thus, the number of edges of broadcast graph H is

$$\frac{1}{2}(\frac{n}{2}(s - t) + \frac{n}{2}(s - 1)) = \frac{n}{2}(s - \frac{t+1}{2})$$

For more details of the construction and the broadcast algorithm, see [12].

3   Trivial Compounding

Before constructing the new broadcast graph, we introduce a compounding based on two existing broadcast graphs with similar number of vertices (in the same interval of two consecutive powers of 2). This compounding is trivial, straight forward, and well-known in this research area.

Let $G_1 = (V_1, E_1)$ and $G_2 = (V_2, E_2)$ be two broadcast graphs on n_1 and n_2 vertices respectively, with $\lceil \log(n_1 + n_2) \rceil = \lceil \log \max(n_1, n_2) \rceil + 1$, and $n_1 \ge n_2$. To construct a broadcast graph $T = (V, E)$, we connect every $v \in V_2$ with a distinct vertex in V_1. Since $n_1 \ge n_2$, there are some vertices in graph G_1 without any neighbor in graph G_2. We connect each of these vertices with an arbitrary vertex in graph G_2. This construction is similar to the broadcast graph construction in [6] and [4]. It is clear that the number of vertices in T is $n = n_1 + n_2$ and the number of edges is $e = |E_1| + |E_2| + \max(n_1, n_2) = |E_1| + |E_2| + n_1$. This construction adds n_1 edges between G_1 and G_2, such that every vertex in G_1 is adjacent to at least one vertex in G_2 and every vertex in G_2 is also adjacent to at least one vertex in G_1. Two vertices v and u connected by this construction are *trivially adjacent*. And v is u's *trivial neighbor* or vice versa. We introduce a new notation for the trivial compounding.

Definition 3. *Let τ be a binary operator on two graphs G_1 and G_2, constructing the trivially compounded graph T described above.*

Compounding of graphs G_1 and G_2 results in the graph T.

$$T = G_1 \ \tau \ G_2$$

We assume that A_1 and A_2 are two broadcast algorithms for two different arbitrarily selected originators in graphs G_1 and G_2, respectively. Algorithm 1 is the broadcast algorithm originating from any vertex in graph T.

Algorithm 1. Broadcast algorithm γ

Input : Graph $T = (V, E)$ constructed above and an arbitrary originator $v \in V$.

Output: The broadcast time, $b(T, v)$.

```
1  begin
2  |   b(T, v) ⟵ 0;
3  |   Informed ⟵ {v};
4  |   Uninformed ⟵ V \ {v};                    /* Initialization */
5  |   if v ∈ G₁ then
6  |   |   u ⟵ v's neighbor in G₂;
7  |   |   Informed ⟵ Informed ∪ {u};
8  |   |   Uninformed ⟵ Uninformed \ {u};
9  |   |   b(T, v) ⟵ b(T, v) + 1;               /* First time unit */
10 |   |   b(T, v) ⟵ b(T, v) + max(A₁(G₁, v), A₂(G₂, u));
11 |   else
12 |   |   u ⟵ v's neighbor in G₁;
13 |   |   Informed ⟵ Informed ∪ {u};
14 |   |   Uninformed ⟵ Uninformed \ {u};
15 |   |   b(T, v) ⟵ b(T, v) + 1;               /* First time unit */
16 |   |   b(T, v) ⟵ b(T, v) + max(A₁(G₁, u), A₂(G₂, v));
17 |   end
18 end
```

Observation 1. *The graph T constructed above is a broadcast graph.*

Proof. We proof the observation by verifying the broadcast time $b(T, v) = \lceil \log n \rceil$ from an arbitrary originator v of graph T. $b(T, v)$ is increased by one on line 9 or 15. Then, the algorithm calls A_1 and A_2. By the assumption, G_1 and G_2 are broadcast graphs, A_1 and A_2 are broadcast algorithms and return $\lceil \log n_1 \rceil$ and $\lceil \log n_2 \rceil$ broadcast times respectively. Thus, the return value is $b(T, v) = 1 + \max(\lceil \log n_1 \rceil, \lceil \log n_2 \rceil)$ given by line 10 or 16.

Again by the assumption, the number of vertices in T is n and $\lceil \log n \rceil = \lceil \log(n_1 + n_2) \rceil = \lceil \log \max(n_1, n_2) \rceil + 1$, which is $b(T, v)$. Thus, T is a broadcast graph. $\qquad \square$

4   A New Graph Construction

We are now ready to construct a new broadcast graph, IBC on n vertices for even $2^{m-1} + 2 \leq n \leq 2^m - 2$, where $m \geq 5$. IBC stands for *imbalanced compounding* because in the trivial compounding, the two broadcast graphs used in our construction are of different orders. The strategy of the construction is simple. First, we construct a sequence of the broadcast graphs H described in Sect. 2 (also in [12,14]). The total number of vertices in all graphs H is equal to n. And the broadcast time of any H is exactly one time unit more than its nearest successor in the sequence. Then, the trivial compounding combines all graphs H to form a connected graph. We give the details of IBC construction below.

Since $2^{m-1} + 2 \leq n \leq 2^m - 2$ is an even integer, n has the form

$$n = 2^m - 2^{k_1} - 2^{k_2} - \cdots - 2^{k_p}$$

where $1 \leq k_p < k_{p-1} < \cdots < k_2 < k_1 \leq m - 2$ and $1 \leq p \leq m - 2$.

We further decompose n in order to use the broadcast graph H under the assumption $k_1 \leq m - 3$ and hence $2^m - 2^{m-2} + 2 \leq n \leq 2^m - 2$. Each H is on $n = 2^s - 2^t$ vertices, for some s and t, $s \geq 4$ and $1 \leq t \leq s - 2$.

$$n = (2^{m-1} - 2^{k_1}) + (2^{m-2} - 2^{k_2}) + \cdots + (2^{m-p+1} - 2^{k_{p-1}}) + (2^{m-p+1} - 2^{k_p}) \tag{1}$$

Note that if $p = 1$, the value of $n = 2^m - 2^{k_1}$ and Eq. (1) has only the first term. The new graph IBC is the same as the graph H. No trivial compounding is required. This obvious case is also excluded from the following discussions. So, the range of k_i and p is slightly different. And we further assume that n is even, $n = 2^m - 2^{k_1} - 2^{k_2} - \cdots - 2^{k_p}$, $m \geq 5$, $1 \leq k_p < k_{p-1} < \cdots < k_2 < k_1 \leq m - 3$, and $2 \leq p \leq k_1 \leq m - 3$ without any further specification in the rest of the paper.

Under the assumption, each term in Eq. (1) is the number of vertices in H.

$$n = \underbrace{2^{m-1} - 2^{k_1}}_{H_1} + \underbrace{2^{m-2} - 2^{k_2}}_{H_2} + \cdots + \underbrace{2^{m-p+1} - 2^{k_{p-1}}}_{H_{p-1}} + \underbrace{2^{m-p+1} - 2^{k_p}}_{H_p},$$

$$= \sum_{i=1}^{p-1} 2^{k_i-1}(2^{m-i-k_i+1} - 2) + 2^{k_p-1}(2^{m-p-k_p+2} - 2) \tag{2}$$

where H_i is the graph H on $2^{m-i} - 2^{k_i}$ vertices for $1 \leq i \leq p$. Then, we recursively define the graph

$$T_i = \begin{cases} H_p & \text{if } i = p; \\ H_i \ \tau \ T_{i+1} & \text{if } 1 \leq i \leq p - 1. \end{cases} \tag{3}$$

The new graph

$$IBC = T_1$$

Counting the exact number of edges in IBC is not trivial (Fig. 1).

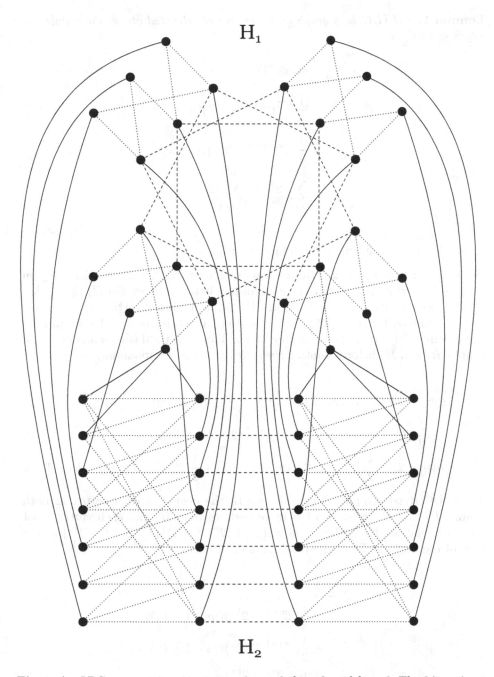

Fig. 1. An IBC on $n = 52$ vertices. $m = 6$, $p = 2$, $k_1 = 3$, and $k_2 = 2$. The 24 vertices on the top are in H_1, while the 28 vertices at the bottom belong to H_2. Dotted lines are the edges in Knödel graphs. Dashed lines are the compounded edges of H. And the solid lines are the trivial compounded edges.

Lemma 1. *Let IBC be a graph on n vertices constructed above, the number of edges in IBC is*

$$e = \frac{1}{2}(\frac{1}{2}n(m-1)$$

$$+ \frac{1}{2}\sum_{i=1}^{p-1}(2^{m-i} - 2^{k_i})(m - k_i)$$

$$+ \frac{1}{2}(2^{m-p+1} - 2^{k_p})(m - k_p)$$

$$+ \sum_{i=1}^{p-1}(2^{k_i} - \sum_{j=i+1}^{p} 2^j))$$

Proof. We prove the lemma by induction on p.

Base Case: When $p = 2$, IBC consists of H_1 on $2^{m-1} - 2^{k_1}$ vertices, H_2 on $2^{m-1} - 2^{k_2}$ vertices, and a trivial compounding between them. For H_1, $\frac{1}{2}(2^{m-1} - 2^{k_1})$ vertices are of degree $m - 1 - k_1$ and the same number of vertices are of degree $m - 2$. Similarly for H_2, $\frac{1}{2}(2^{m-1} - 2^{k_2})$ vertices are of degree $m - 1 - k_2$ and the same number of vertices are of degree $m - 2$. Thus, the total sum of all degrees of graphs H_1 and H_2 before counting the edges of trivial compounding is

$$\frac{1}{2}(2^{m-1} - 2^{k_1})(m - k_1 - 1) + \frac{1}{2}(2^{m-1} - 2^{k_1})(m - 2)$$

$$+ \frac{1}{2}(2^{m-1} - 2^{k_2})(m - k_2 - 1) + \frac{1}{2}(2^{m-1} - 2^{k_2})(m - 2)$$

$$= \frac{1}{2}n(m - 2) + \frac{1}{2}(2^{m-1} - 2^{k_1})(m - k_1 - 1) + \frac{1}{2}(2^{m-1} - 2^{k_2})(m - k_2 - 1)$$

Then the trivial compounding adds one to the degree of each vertex in both graphs H_1 and H_2. Since H_2 has more vertices than H_1 and each vertex must have a trivial neighbor, then, this will add $(2^{m-1} - 2^{k_2}) - (2^{m-1} - 2^{k_1})$ to the sum of the degrees. Thus, the total number of edges is

$$e = \frac{1}{2}(\frac{1}{2}n(m - 2)$$

$$+ \frac{1}{2}(2^{m-1} - 2^{k_1})(m - 1 - k_1)$$

$$+ \frac{1}{2}(2^{m-1} - 2^{k_2})(m - 1 - k_2)$$

$$+ n + 2^{k_1} - 2^{k_2})$$

$$= \frac{1}{2}(\frac{1}{2}n(m-2)$$

$$+ \frac{1}{2}(2^{m-1} - 2^{k_1})(m - 1 - k_1)$$

$$+ \frac{1}{2}(2^{m-1} - 2^{k_2})(m - 1 - k_2)$$

$$+ \frac{n}{2} + \frac{1}{2}(2^{m-1} - 2^{k_1}) + \frac{1}{2}(2^{m-1} - 2^{k_2}) + 2^{k_1} - 2^{k_2})$$

$$= \frac{1}{2}(\frac{1}{2}n(m-1)$$

$$+ \frac{1}{2}(2^{m-1} - 2^{k_1})(m - k_1)$$

$$+ \frac{1}{2}(2^{m-1} - 2^{k_2})(m - k_2)$$

$$+ 2^{k_1} - 2^{k_2})$$

Inductive Hypothesis: Assume that Lemma 1 is true for $p = r$, where $2 < r \le m - 3$, which means

$$e = \frac{1}{2}(\frac{1}{2}n(m-1)$$

$$+ \frac{1}{2}\sum_{i=1}^{r-1}(2^{m-i} - 2^{k_i})(m - k_i)$$

$$+ \frac{1}{2}(2^{m-r+1} - 2^{k_r})(m - k_r) \tag{4}$$

$$+ \sum_{i=1}^{r-1}(2^{k_i} - \sum_{j=i+1}^{r} 2^j)) \tag{5}$$

Inductive Step: When $p = r+1$, the last term of Eq. (2) is further decomposed into $2^{m-r} - 2^{k_r} + 2^{m-r} - 2^{k_{r+1}}$, representing H_p and H_{p+1}. After constructing the whole graph on n vertices, half of the vertices in H_r and H_{r+1} are of degree $m - 1$; the other half of the vertices in H_r are of degree $m - k_r$; and the other half of the vertices in H_{r+1} are of degree $m - k_{r+1}$. Thus, Eq. (4) is split into $\frac{1}{2}(2^{m-r} - 2^{k_r})(m - k_r) + \frac{1}{2}(2^{m-r} - 2^{k_{r+1}})(m - k_{r+1})$.

The trivial compounding further adds $\frac{1}{2}(2^{k_{r+1}} - 2^{k_r})$ edges to the compounding of H_r and H_{r+1}. And since each T_i contains H_{r+1}, each term in the summation on Eq. (5) has one additional term $-2^{k_{r+1}}$. Thus, when $p = r + 1$,

$$e = \frac{1}{2}(\frac{1}{2}n(m-1)$$

$$+ \frac{1}{2}\sum_{i=1}^{r}(2^{m-i} - 2^{k_i})(m - k_i)$$

$$+ \frac{1}{2}(2^{m-r} - 2^{k_{r+1}})(m - k_{r+1})$$

$$+ \sum_{i=1}^{r}(2^{k_i} - \sum_{j=i+1}^{r+1} 2^j))$$

which completes the proof. $\qquad\qquad\qquad\qquad\qquad\qquad\qquad\qquad\qquad\qquad$ $\square$

5   Broadcast Algorithm

Assume the broadcast algorithm for the graph H from any originator v in [12] is $C(H, v)$. Algorithm 2 is a $\lceil \log n \rceil$ time broadcast algorithm for the graph IBC. The If statement on line 4 is the termination condition for the recursion. When this statement is executed, the recursion calls $C(H, v)$ and halts. The cases in step 7 and 12 are distinguished because the vertex v can be either in H_{t+1} or T_{t+2}. If v is in H_{t+1}, u is the trivial neighbor in T_{t+2}. In the next iteration, the algorithm broadcasts from v in H_{t+1} and from u in T_{t+2}. And it is vice versa if v is in T_{t+2}.

Theorem 1. *Algorithm γ' completes the broadcast from any originator in graph IBC on n vertices in $\lceil \log n \rceil$ iterations and returns $b(IBC, v) = \lceil \log n \rceil$.*

Proof. We prove the theorem by induction on p, which is the number of additions in Eq. (2) (the number of trivial compoundings) plus 1, or the number of 0's minus 1 in the binary representation of n.

Base Case: When $p = 2$, IBC consists of H_1 on $2^{m-1} - 2^{k_1}$ vertices and H_2 on $2^{m-1} - 2^{k_2}$ vertices. Assume the originator v is in H_1 without loss of generality, the condition on line 8 is true and algorithm γ' executes line 9 to line 12 in the first iteration. Thus, v informs its trivial neighbor u in H_2 in the first time unit. Then, algorithm γ' calls $C(H_1, v)$ and $C(H_2, u)$, the broadcast algorithm in [12] from v and u in H_1 and H_2 respectively in the second iteration. Both of the algorithms return $m - 1$ broadcast time. Thus, the total broadcast time is $m - 1 + 1 = m = \lceil \log n \rceil$.

Inductive Hypothesis: Assume when $p = r$, γ' returns $b(IBC, v) = \lceil \log n \rceil$ for $2 < r \le m - 3$. This implies that in the last $(r-1)$-th iteration, line 5 returns $b(IBC, v) \longleftarrow C(H_r, v)$. Since H_r is on $2^{m-r+1} - 2^{k_r}$ vertices,

$$C(H_r, v) = m - r + 1 \qquad\qquad\qquad (6)$$

Algorithm 2. Broadcast algorithm γ'

Input : A graph $IBC = (V, E)$, an originator $v \in V$, and the broadcast
time t from the previous iteration, equals to 0 initially.

Output: Broadcast time $b(IBC, v)$

```
1  begin
2  │   Informed ⟵ {v};
3  │   Uninformed ⟵ V \ {v};
4  │   if v ∈ H_p then
5  │   │   b(IBC, v) ⟵ C(H_p, v);   /* The termination condition. And
   │   │   C(H, v) is the broadcast scheme on H from v given in
   │   │   [12]. */
6  │   end
7  │   if v ∈ H_{t+1} then
8  │   │   u ⟵ v's trivial neighbor in T_{t+2};
9  │   │   Informed ⟵ Informed ∪ {u};
10 │   │   Uninformed ⟵ Uninformed \ {u};
11 │   │   b(IBC, v) ⟵ t + 1 + max(γ'(T_{t+2}, u, t + 1), C(H_{t+1}, v));
12 │   else
13 │   │   u ⟵ v's trivial neighbor in H_{t+1};
14 │   │   Informed ⟵ Informed ∪ {u};
15 │   │   Uninformed ⟵ Uninformed \ {u};
16 │   │   b(IBC, v) ⟵ t + 1 + max(γ'(T_{t+2}, v, t + 1), C(H_{t+1}, u));
17 │   end
18 │   return b(IBC, v)
19 end
```

Inductive Step: When $p = r + 1$, the only difference is that the algorithm will
run one extra iteration. In the second last $(r - 1)$-th iteration, if the increment
on $b(IBC, v)$ is δ (assume $v \in T_{r+1}$ and $u \in H_r$ without loss of generality) line
16 will call the algorithm again and

$$\delta = 1 + max(\gamma'(T_{r+1}, u, r), C(H_r, v))$$

In the last iteration, the algorithm returns $C(H_{r+1}, u) = m - r$ because H_{r+1}
is on $2^{m-r} - 2^{k_{r+1}}$ vertices. And we know $C(H_r, v) = m - r$ since H_r is on
$2^{m-r} - 2^{k_r}$ vertices. Thus, $\delta = m - r + 1$ which is the same as the Eq. (6).
Therefore, the total broadcast time $b(IBC, v) = m = \lceil \log n \rceil$.   □

Summarizing Theorem 1 and Lemma 1, we obtain our upper bound on $B(n)$.

Theorem 2.

$$B(n) \leq \frac{1}{2}\left(\frac{1}{2}n(m-1)\right.$$

$$+ \frac{1}{2}\sum_{i=1}^{p-1}(2^{m-i} - 2^{k_i})(m - k_i)$$

$$+ \frac{1}{2}(2^{m-p+1} - 2^{k_p})(m - k_p)$$

$$+ 2^{k_p-1} - 2^{k_p}$$

$$+ \left.\sum_{i=1}^{p-1}\left(2^{k_i} - \sum_{j=i+1}^{p} 2^j\right)\right)$$

Let NB be the new upper bound given by Theorem 2 and

$$OB = \frac{1}{2}(m-1)n$$

in [15]. The comparison shows that NB is a better upper bound and

$$OB - NB \geq \frac{1}{2}(2^{m-1} - 2^{m-3} - 2^{m-3}) + \frac{1}{2}(2^{m-2} + 2^{m-3})$$

$$= 2^{m-2} + 2^{m-4}$$

The tedious calculation is given in the Appendix.

6   Conclusion

In this paper, we constructed a new broadcast graph based on the broadcast graph H defined in [12]. The new construction applies the trivial compounding on a sequence of broadcast graph H and improves the upper bound on $B(n)$ for $2^m - 2^{\frac{m+3}{2}} < n \leq 2^m - 8$. In the future, we can try to compound the sequence of H in a better way instead of just using the trivial compounding. This will reduce the number of edges because the trivial compounding is too simple and adds to many redundant edges.

Another work can also be done in the future is the comparison between the new upper bound NB and the bound OB' given by [10], which is the best bound on $B(n)$ when $2^{m-1} + 1 \leq n \leq 2^m - 2^{\frac{m+3}{2}}$. Currently, we only know that NB is better than OB' in some of the intervals, but the intervals are not continuous. Therefore, we can further investigate the comparison in the future.

Appendix: Comparison

By our assumption $k_p < \cdots < k_1 \leq m - 3$, k_i is strictly larger than k_{i+1}. Then in general $k_i \leq m - i - 2$ for $1 \leq i \leq p$. Similarly, since $1 \leq k_p < \cdots < k_1$, $p - i + 1 \leq k_i$ for $1 \leq i \leq p$. Thus, $p - i + 1 \leq k_i \leq m - i - 2$, where $1 \leq i \leq p$.

$OB - NB$

$$= \frac{1}{2}n(m-1) - \frac{1}{2}\left(\frac{1}{2}n(m-1) + \frac{1}{2}\sum_{i=1}^{p-1}(2^{m-i} - 2^{k_i})(m - k_i)\right.$$

$$+ \frac{1}{2}(2^{m-p+1} - 2^{k_p})(m - k_p) + \sum_{i=1}^{p-1}(2^{k_i} - \sum_{j=i+1}^{p} 2^{k_j}))$$

$$= \frac{1}{4}n(m-1) - \frac{1}{4}\sum_{i=1}^{p-1}(2^{m-i} - 2^{k_i})(m - k_i)$$

$$- \frac{1}{4}(2^{m-p+1} - 2^{k_p})(m - k_p) - \frac{1}{2}\sum_{i=1}^{p-1}(2^{k_i} - \sum_{j=i+1}^{p} 2^{k_j}))$$

$$= \frac{1}{4}n(m-1) - \frac{1}{4}\sum_{i=1}^{p-1}(2^{m-i} - 2^{k_i})(m - k_i)$$

$$- \frac{1}{4}(2^{m-p+1} - 2^{k_p})(m - k_p) - \frac{1}{2}(2^{k_1} - \sum_{i=2}^{p-1}(i-2)2^{k_i} - (p-1)2^{k_p})$$

By substituting $n = \sum_{i=1}^{p-1}(2^{m-i} - 2^{k_i}) + 2^{m-p+1} - 2^{k_p}$,

$$= \frac{1}{4}\sum_{i=1}^{p-1}(2^{m-i} - 2^{k_i})(m-1) + \frac{1}{4}(2^{m-p+1} - 2^{k_p})(m-1) - \frac{1}{4}\sum_{i=1}^{p-1}(2^{m-i} - 2^{k_i})(m - k_i)$$

$$- \frac{1}{4}(2^{m-p+1} - 2^{k_p})(m - k_p) - \frac{1}{2}(2^{k_1} - \sum_{i=2}^{p-1}(i-2)2^{k_i} - (p-1)2^{k_p})$$

$$= \frac{1}{4}\sum_{i=1}^{p-1}(2^{m-i} - 2^{k_i})(k_i + 1) + \frac{1}{4}(2^{m-p+1} - 2^{k_p})(k_p + 1)$$

$$- \frac{1}{2}2^{k_1} + \frac{1}{2}\sum_{i=2}^{p-1}(i-2)2^{k_i} + \frac{1}{2}(p-1)2^{k_p}$$

Since $k_i \geq 1$,

$$\geq \frac{1}{2} \sum_{i=1}^{p-1} (2^{m-i} - 2^{k_i}) + \frac{1}{2}(2^{m-p+1} - 2^{k_p})$$

$$- \frac{1}{2} 2^{k_1} + \frac{1}{2} \sum_{i=2}^{p-1} (i-2) 2^{k_i} + \frac{1}{2}(p-1) 2^{k_p}$$

$$= \frac{1}{2}(2^{m-1} - 2^{k_1} - 2^{k_1}) + \frac{1}{2} \sum_{i=2}^{p-1} (2^{m-i} - 2^{k_i}) + \frac{1}{2}(2^{m-p+1} - 2^{k_p})$$

$$+ \frac{1}{2} \sum_{i=2}^{p-1} (i-2) 2^{k_i} + \frac{1}{2}(p-1) 2^{k_p}$$

$$= \frac{1}{2}(2^{m-1} - 2^{k_1} - 2^{k_1}) + \frac{1}{2}(2^{m-2} + 2^{m-3} + 2^{m-p-2})$$

$$+ \frac{1}{2} \sum_{i=2}^{p-1} (i-2) 2^{k_i} + \frac{1}{2}(p-1) 2^{k_p}$$

$$\geq \frac{1}{2}(2^{m-1} - 2^{k_1} - 2^{k_1}) + \frac{1}{2}(2^{m-2} + 2^{m-3})$$

As $k_1 \leq m - 3$,

$$OB - NB \geq \frac{1}{2}(2^{m-1} - 2^{m-3} - 2^{m-3}) + \frac{1}{2}(2^{m-2} + 2^{m-3})$$

$$= 2^{m-2} + 2^{m-4}$$

Thus, our new upper bound is a better upper bound on $B(n)$ when $2^m - 2^{\frac{m+3}{2}} < n \leq 2^m - 8$.

References

1. Averbuch, A., Shabtai, R.H., Roditty, Y.: Efficient construction of broadcast graphs. Discrete Appl. Math. **171**, 9–14 (2014)
2. Bermond, J.-C., Fraigniaud, P., Peters, J.G.: Antepenultimate broadcasting. Networks **26**(3), 125–137 (1995)
3. Bermond, J.-C., Hell, P., Liestman, A.L., Peters, J.G.: Sparse broadcast graphs. Discrete Appl. Math. **36**(2), 97–130 (1992)
4. Chau, S.-C., Liestman, A.L.: Constructing minimal broadcast networks. J. Comb. Inf. Syst. Sci. **10**, 110–122 (1985)
5. Dinneen, M.J., Fellows, M.R., Faber, V.: Algebraic constructions of efficient broadcast networks. In: Mattson, H.F., Mora, T., Rao, T.R.N. (eds.) AAECC 1991. LNCS, vol. 539, pp. 152–158. Springer, Heidelberg (1991). https://doi.org/10.1007/3-540-54522-0_104
6. Farley, A.M.: Minimal broadcast networks. Networks **9**(4), 313–332 (1979)
7. Farley, A.M., Hedetniemi, S., Mitchell, S., Proskurowski, A.: Minimum broadcast graphs. Discrete Math. **25**(2), 189–193 (1979)

8. Grigni, M., Peleg, D.: Tight bounds on mimimum broadcast networks. SIAM J. Discrete Math. **4**(2), 207–222 (1991)
9. Harutyunyan, H.A.: An efficient vertex addition method for broadcast networks. Internet Math. **5**(3), 211–225 (2008)
10. Harutyunyan, H.A., Li, Z.: A new construction of broadcast graphs. In: Govindarajan, S., Maheshwari, A. (eds.) CALDAM 2016. LNCS, vol. 9602, pp. 201–211. Springer, Cham (2016). https://doi.org/10.1007/978-3-319-29221-2_17
11. Harutyunyan, H.A., Li, Z.: Broadcast graphs using new dimensional broadcast schemes for Knödel graphs. In: Gaur, D., Narayanaswamy, N.S. (eds.) CALDAM 2017. LNCS, vol. 10156, pp. 193–204. Springer, Cham (2017). https://doi.org/10.1007/978-3-319-53007-9_18
12. Harutyunyan, H.A., Liestman, A.L.: More broadcast graphs. Discrete Appl. Math. **98**(1), 81–102 (1999)
13. Harutyunyan, H.A., Liestman, A.L.: Upper bounds on the broadcast function using minimum dominating sets. Discrete Math. **312**(20), 2992–2996 (2012)
14. Khachatrian, L.H., Harutounian, H.S.: Construction of new classes of minimal broadcast networks. In: International Conference on Coding Theory, Dilijan, Armenia, pp. 69–77 (1990)
15. Knödel, W.: New gossips and telephones. Discrete Math. **13**(1), 95 (1975)
16. Labahn, R.: A minimum broadcast graph on 63 vertices. Discrete Appl. Math. **53**(1–3), 247–250 (1994)
17. Maheo, M., Saclé, J.-F.: Some minimum broadcast graphs. Discrete Appl. Math. **53**(1–3), 275–285 (1994)
18. Mitchell, S., Hedetniemi, S.: A census of minimum broadcast graphs. J. Comb. Inf. Syst. Sci. **5**, 141–151 (1980)
19. Park, J.-H., Chwa, K.-Y.: Recursive circulant: a new topology for multicomputer networks. In: International Symposium on Parallel Architectures, Algorithms and Networks (ISPAN 1994), pp. 73–80. IEEE (1994)
20. Saclé, J.-F.: Lower bounds for the size in four families of minimum broadcast graphs. Discrete Math. **150**(1–3), 359–369 (1996)
21. Shao, B.: On k-broadcasting in graphs. Ph.D. thesis, Concordia University (2006)
22. Slater, P.J., Cockayne, E.J., Hedetniemi, S.T.: Information dissemination in trees. SIAM J. Comput. **10**(4), 692–701 (1981)
23. Xiao, J., Wang, X.: A research on minimum broadcast graphs. Chin. J. Comput. **11**, 99–105 (1988)
24. Zhou, J., Zhang, K.: A minimum broadcast graph on 26 vertices. Appl. Math. Lett. **14**(8), 1023–1026 (2001)

No-Bend Orthogonal Drawings
and No-Bend Orthogonally Convex
Drawings of Planar Graphs
(Extended Abstract)

Md. Manzurul Hasan[1,2]([✉]) and Md. Saidur Rahman[1]

[1] Graph Drawing and Information Visualization Laboratory,
Department of Computer Science and Engineering (CSE),
Bangladesh University of Engineering and Technology (BUET), Dhaka, Bangladesh
saidurrahman@cse.buet.ac.bd
[2] Department of Computer Science,
American International University-Bangladesh (AIUB), Dhaka, Bangladesh
mhasan.cse00@gmail.com

Abstract. A plane graph is a planar graph with a fixed planar embedding in the plane. In an orthogonal drawing of a plane graph each vertex is drawn as a point and each edge is drawn as a sequence of vertical and horizontal line segments. A bend is a point at which the drawing of an edge changes its direction. A necessary and sufficient condition for a plane graph of maximum degree 3 to have a no-bend orthogonal drawing is known which leads to a linear-time algorithm to find such a drawing of a plane graph, if it exists. A planar graph G has a no-bend orthogonal drawing if any of the plane embeddings of G has a no-bend orthogonal drawing. Since a planar graph G of maximum degree 3 may have an exponential number of planar embeddings, determining whether G has a no-bend orthogonal drawing or not using the known algorithm for plane graphs takes exponential time. The best known algorithm takes $O(n^2)$ time for finding a no-bend orthogonal drawing of a biconnected planar graph of maximum degree 3. In this paper we give a linear-time algorithm to determine whether a biconnected planar graph G of maximum degree 3 has a no-bend orthogonal drawing or not and to find such a drawing of G, if it exists. We also give a necessary and sufficient condition for a biconnected planar graph G of maximum degree 3 to have a no-bend "orthogonally convex" drawing D; where any horizontal and vertical line segment connecting two points in a facial polygon P in D lies totally within P. Our condition leads to a linear-time algorithm for finding such a drawing, if it exists.

Keywords: Orthogonal drawings · Orthogonally convex drawings · Planar graphs

© Springer Nature Switzerland AG 2019
D.-Z. Du et al. (Eds.): COCOON 2019, LNCS 11653, pp. 254–265, 2019.
https://doi.org/10.1007/978-3-030-26176-4_21

1   Introduction

Automatic graph drawings have numerous applications in VLSI circuit layout, networks, computer architecture, circuit schematics etc. [6,12,14]. Among various drawing styles, "orthogonal drawings" of planar graphs have attracted much attention due to their practical applications, specially in circuit schematics, entity relationship diagrams, data flow diagrams etc. [3,9,13].

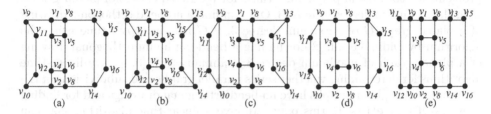

Fig. 1. (a)-(c) Three plane embeddings of a planar graph G which have no no-bend orthogonal drawing, (d) a planar embedding Γ of G which has a no-bend orthogonal drawing, and (e) a no-bend orthogonal drawing of Γ.

An *orthogonal drawing* of a planar graph G is a drawing of G in which each vertex is mapped to a *point*, each edge is drawn as a sequence of alternate horizontal and vertical line segments, and any two edges do not cross except at their common ends. A *bend* is a point where an edge changes its direction in a drawing. Figure 1(e) illustrates an orthogonal drawing without bends of the planar graph in Fig. 1(d). Every planar graph of the maximum degree four has an orthogonal drawing, but may need bends. Finding an orthogonal drawing of a planar graph of maximum degree four with the minimum number of bends is an NP-hard problem [5]. However, polynomial algorithms are known for finding bend-minimum orthogonal drawings of plane graphs (with fixed embedding) of maximum degree four and some restricted classes of planar graphs of maximum degree 3 [1,7–9,13]. Di Battista et al. [7] gave an $O(n^5 log n)$-time algorithm to find a bend-minimum orthogonal drawing of a planar graph G of maximum degree 3. Later Chang and Yen [1] gave an algorithm for bend-minimum orthogonal drawings of planar graphs of maximum degree 3 that runs in $O(n^{17/7})$ time. Recently Didimo et al. [8] gave an $O(n^2)$-time algorithm for bend-minimum orthogonal drawings of planar graphs of maximum degree 3.

An orthogonal drawing D of a plane graph is a *no-bend orthogonal drawing* if D has no bend. Not every biconnected plane graph has a no-bend orthogonal drawing. The biconnected plane graphs in Fig. 1(a)-(c) have no no-bend orthogonal drawing. Rahman et al. [11] gave a linear-time algorithm to determine whether a biconnected plane graph Γ with $\Delta \leq 3$ has a no-bend orthogonal drawing or not and to find a no-bend orthogonal drawing of Γ, if it exists, where Δ denotes the maximum degree of a graph.

A planar graph is said to have a no-bend orthogonal drawing if at least one of its plane embeddings has a no-bend orthogonal drawing. For the plane embeddings Γ_1, Γ_2, and Γ_3 in Figs. 1(a), 1(b), 1(c) respectively of a planar graph G there is no no-bend orthogonal drawing. But for the plane embedding in Fig. 1(d) of the same planar graph G, there exists a no-bend orthogonal drawing as illustrated in Fig. 1(e), and hence G has a no-bend orthogonal drawing. Since a planar graph G may have an exponential number of planar embeddings, finding no-bend orthogonal drawings of planar graphs is not a trivial problem. Similar problems are solvable in polynomial times for restricted classes of planar graphs with the maximum degree at most 3. One can find a no-bend orthogonal drawing of a planar graph G of maximum degree 3, if G has, in $O(n^2)$ time using the algorithm by Didimo et al. [8] for finding a bend-minimum orthogonal drawing of a planar graph of maximum degree 3. Rahman et al. [3,4] gave linear-time algorithms for determining whether series parallel graphs and subdivisions of triconnected cubic planar graphs have no-bend orthogonal drawings and for finding drawings if they exists. In this paper we give a linear-time algorithm that can check whether a biconnected planar graph of maximum degree 3 has a no-bend orthogonal drawing and can find a drawing if it exists.

In an orthogonal drawing of a plane graph Γ each inner face of Γ is drawn as rectilinear polygon. A rectilinear polygon P is called *orthogonally convex* if every horizontal or vertical segment connecting two points in P lies totally within P. An orthogonally convex drawing is an orthogonal drawing where each inner face is an orthogonally convex polygon. Chang and Yen [2] gave a necessary and sufficient condition for a biconnected plane graph of maximum degree 3 to have a no-bend orthogonally convex drawing and gave a linear-time algorithm to find such a drawing, if it exists. Hasan and Rahman [6] gave a necessary and sufficient condition for a subdivision of a triconnected cubic planar graph to have a no-bend orthogonally convex drawing and a linear-time algorithm to find a drawing if such one exists. In this paper we give a linear-time algorithm to determine whether a biconnected planar graph G of maximum degree 3 has a no-bend orthogonally convex drawing or not, and to find a no-bend orthogonally convex drawing of G, if it exists.

The rest of this paper is organized as follows. In Sect. 2, we give some terminologies and previous results. In Sect. 3, we describe a necessary and sufficient condition for a biconnected planar graph G of maximum degree 3 to have a no-bend orthogonal drawing which leads to a linear-time algorithm to find such a drawing, if it exists. In Sect. 4, we give a linear-time algorithm for no-bend orthogonally convex drawings of biconnected planar graphs of maximum degree 3. Finally, Sect. 5 concludes the paper.

2   Preliminaries

In this section we give some definitions and present some preliminary results.

Let $G = (V, E)$ be a connected simple graph with vertex set V and edge set E. The *degree* $d(v)$ of a vertex v is the number of neighbors of v in G. We call a vertex of degree k in G a k-vertex of G. We denote the maximum degree of a graph G by $\Delta(G)$ or simply by Δ. We sometimes denote a graph of maximum degree k by k-graph. A graph G is called cubic if $d(v) = 3$ for every vertex v. For $V' \subseteq V$, G - V' denotes a graph obtained from G by deleting all vertices in V' together with all edges incident to them. For a subgraph G' of G, we denote by $G - G'$ the graph obtained from G by deleting all vertices in G'.

Subdividing an edge (u, v) of a graph G is the operation of deleting the edge (u, v) and adding a path $u(= w_0), w_1, w_2, \ldots, w_k, v(= w_{k+1})$ passing through new vertices $w_1, w_2, \ldots, w_k$, $k \geq 1$, of degree 2. A graph G is called a *subdivision* of a graph G' if G is obtained from G' by subdividing some of the edges of G'. The *connectivity* $\kappa(G)$ of a graph G is the minimum number of vertices whose removal results in a disconnected graph or a single-vertex graph K_1. We say that G is *k-connected* if $\kappa(G) \geq k$. A subdivision of a triconnected cubic graph is biconnected, and the degree of any vertex is either 2 or 3. A drawing of a planar graph divides the plane into a set of connected regions, called *faces*. A subdivision of a triconnected cubic planar graph holds the following fact regarding faces [4].

Fact 1. *Let G be a subdivision of a triconnected cubic planar graph. Let Γ_1 and Γ_2 be two different arbitrary plane embeddings of G. Then every face in Γ_1 is a face in Γ_2 and vice versa.*

A *contour* of a face F is the cycle formed by vertices and edges along the boundary of F. Such a cycle is also called a *facial cycle*. The contour of the outer face F_o is denoted as C_o. If G is biconnected, all facial cycles are simple cycles. Let G be a planar graph, and Γ be an arbitrary plane embedding of G. The contour of a face of Γ is a cycle of G, and is simply called a face or a facial cycle of Γ. We denote by $F_o(\Gamma)$ the outer face of Γ. For a cycle C of Γ, we call the plane subgraph of Γ inside C (including C) the *inner subgraph* $\Gamma_I(C)$ *for* C, and call the plane subgraph of Γ outside C (including C) the *outer subgraph* $\Gamma_O(C)$ *for* C. Any face of Γ is either in $\Gamma_I(C)$ or in $\Gamma_O(C)$ for C. We call a face (or a cycle) inner if it is not F_o(or C_o). If an inner face or inner cycle contains a vertex on the boundary of F_o, then we call it *boundary face* or *boundary cycle*. If an inner face or inner cycle does not intersect with the boundary of F_o, then we call it *non-boundary face* or *non-boundary cycle*.

An edge which is incident to exactly one vertex of a cycle C and located outside C is called a *leg* of C. The vertex of C to which a leg is incident is called a *leg-vertex* of C. A cycle C in Γ is called a *k-legged cycle* of Γ if C has exactly k legs and there is no edge which joins two vertices on C and is located outside C. We say that cycles C and C' in Γ are *independent* if $\Gamma_I(C)$ and $\Gamma_I(C')$ have no common vertex. A set S of cycles is *independent* if every pair of cycles in S are independent. A k-legged cycle C in Γ is called a *regular k-legged cycle* if the plane graph $\Gamma - \Gamma_I(C)$ has a cycle. Similarly an edge of Γ which is incident to exactly one vertex of a cycle C in Γ and located inside C is called a *hand* of C. The vertex of C to which a hand is incident is called a *hand-vertex* of C. A cycle C is called a *k-handed cycle* if C has exactly k hands in Γ and there is no edge which joins two vertices on C and is located inside C. We call a k-handed cycle C a *regular k-handed cycle* if $\Gamma - \Gamma_O(C)$ contains a cycle. A path P on a k-legged cycle C such that P includes exactly two consecutive leg-vertices x and y of C, and x and y are the two endpoints of P is called a *contour path* of the cycle C. Therefore, each k-legged cycle has exactly k contour paths. Similarly a path P on a k-handed cycle C such that P includes exactly two consecutive hand vertices x and y of C, and x and y are the two endpoints of P is called a *contour path* of the cycle C. Thus each k-handed cycle has exactly k contour paths. If a contour path intersects (i.e., shares some edges with) the outer cycle, we call it *boundary contour path*. In fact, each boundary contour path is a sub path of C_o.

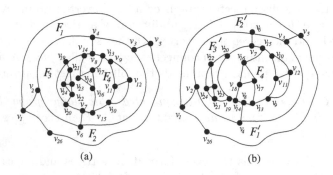

(a)   (b)

Fig. 2. Illustration of flipping operations and induced faces in a biconnected planar graph of maximum degree 3.

A planar graph G that is a subdivision of a triconnected cubic planar graph has a fixed number different planar embeddings. Let G be a biconnected planar graph of maximum degree 3, and Γ be an arbitrary plane embedding of G. Assume that G is not a subdivision of a triconnected cubic planar graph. If Γ does not have any regular 2-legged cycle, then Γ has at most two inner faces and it is trivial to make a decision whether G has a no-bend orthogonal drawing or not as well as a no-bend orthogonally convex drawing or not. We thus assume that Γ has a regular 2-legged cycle. In Γ for any 2-legged cycle or for any 2-handed cycle C, leg or hand vertices of C is a separation pair called a *flipping pair* like vertices (v_2, v_3) in Fig. 2. In a planar embedding Γ_1 of a biconnected planar graph G with $\Delta \leq 3$, flipping $\Gamma_I(C)$ of any 2-legged cycle C or flipping $\Gamma_O(C)$ of any 2-handed cycle C with respect to respective flipping pairs is called a *flipping operation* of cycle C that may create a new embedding Γ_2 as illustrated in Fig. 2. In a planar embedding of a planar graph shown in Fig. 2(a), three flippings are done simultaneously with respect to flipping pairs (v_2, v_3), (v_{19}, v_{20}), and (v_{21}, v_{22}), that creates a new planar embedding shown in Fig. 2(b). G may have an exponential number of different planar embeddings that are for each face of Γ as outer face additionally for flipping the 2-legged and 2-handed cycles with respect to the flipping pairs. The following Lemma can be observed for a biconnected planar graph with the maximum degree 3, whose proof is omitted.

Lemma 1. *Let G be a biconnected planar graph of maximum degree 3. Let Γ_1 and Γ_2 be two different arbitrary plane embeddings of G. Let F be a face in Γ_1 with the set V_f of all flipping pair vertices on F and let V' be the set of vertices on F which are neither on a 2-legged cycle nor on a 2-handed cycle. Then, there exists at least one face F' in Γ_2, that contains all the vertices of V_f and V'.*

We call a face F' that contains all the vertices of V_f and V' in mentioned Lemma 1 an *induced face* of F. The faces F_1', F_2', F_3', F_4 in Fig. 2(b) are induced faces for the faces F_1, F_2, F_3, F_4 in Figure 2(a), respectively.

Rahman et al. [11] gave a linear-time algorithm to check whether a biconnected plane graph Γ of $\Delta \leq 3$ has a no-bend orthogonal drawing and to find out a drawing if such one exists, as stated in the following lemma.

Lemma 2. *Assume that Γ is a biconnected plane graph with $\Delta \leq 3$. Γ has an orthogonal drawing without bends if and only if Γ satisfies the following conditions.*

(or1) There are four or more 2-vertices of Γ on $C_o(\Gamma)$,
(or2) every 2-legged cycle contains at least two 2-vertices of Γ, and
(or3) every 3-legged cycle contains at least one 2-vertex of Γ.

3   No-Bend Orthogonal Drawings

In this section we give a necessary and sufficient condition for a biconnected planar graph G of maximum degree 3 to have a no-bend orthogonal drawing. We also give an algorithm to draw such a drawing, if G has.

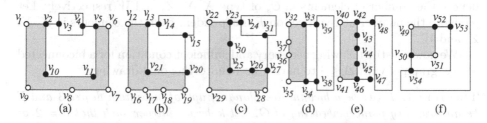

Fig. 3. Figure (a) illustrates X, Figures (b)-(c) illustrate Y, Figures (d)-(e) illustrate Z, and Figure (f) illustrates W types cycles, 2-vertices are represented by white circles.

Before stating our claims, we need to give some definitions.

Let F be a face in an arbitrary plane embedding Γ of G. We call a 2-legged cycle or a 2-handed cycle C in Γ of *type X* if C has leg or hand vertices on F such that one contour path of each cycle C in X contains four or more 2-vertices and other contour path contains at least two vertices except the leg or hand vertices. In Fig. 3(a), a cycle $v_1v_2v_{10}v_{11}v_5v_6v_7v_8v_9$ is such a cycle for F. We call a 2-legged cycle or a 2-handed cycle C in Γ of *type Y* if C has leg or hand vertices on F such that any cycle C in Y satisfies one of the following properties (a) and (b).

(a) One contour path contains four or more 2-vertices and the other contour path contains exactly one vertex except the leg or hand vertices as illustrated by the shaded region (Cycle $v_{12}v_{13}v_{21}v_{20}v_{19}v_{18}v_{17}v_{16}$) in Fig. 3(b).

(b) None of the two contour paths contains more that three 2-vertices but at least one contour path contains exactly three 2-vertices. If only one contour path contains exactly three 2-vertices, then the other one contains at least one vertex except the leg or hand vertices as illustrated by the shaded region (Cycle $v_{22}v_{23}v_{30}v_{25}v_{26}v_{27}v_{28}v_{29}$) in Fig. 3(c).

We call a 2-legged cycle or a 2-handed cycle C in Γ of *type Z* if C has leg or hand vertices on F such that any cycle C in Z satisfies one of the following properties (a) and (b).

(a) One contour path has three or more 2-vertices and another contour path contains no vertex except the leg or hand vertices as illustrated by the shaded region (Cycle $v_{32}v_{33}v_{34}v_{35}v_{36}v_{37}$) in Fig. 3(d).

(b) Both contour paths contain not more than two 2-vertices but at least one contour path contains exactly two 2-vertices as illustrated by the shaded region (Cycle $v_{40}v_{42}v_{43}v_{44}v_{45}v_{46}v_{41}$) in Fig. 3(e).

We call a 2-legged cycle or a 2-handed cycle C in Γ of *type W* if C has exactly one 2-vertex on each contour path such that their leg or hand vertices are on F, as illustrated by shaded region (Cycle $v_{49}v_{52}v_{51}v_{50}$) in Fig. 3(f).

Let C_s be a set of independent 2-legged and 2-handed cycles such that each element of C_s is one of the type X, Y, Z, and W. Let n_X, n_Y, n_Z, and n_W denote the number of elements in C_s of type X, Y, Z, and W respectively. Let p denote the number of 2-vertices on F independent from the elements in X, Y, Z, and W.

We now give the following necessary and sufficient condition for a biconnected planar graph G with $\Delta \leq 3$ to have a no-bend orthogonal drawing.

Theorem 1. *Let G be a biconnected planar graph of maximum degree 3 and Γ be an arbitrary plane embedding of G. Let k be an integer such that $k = 2$ or $k = 3$. Then G has a no bend orthogonal drawing if and only if Γ satisfies the following conditions $(i) - (iii)$.*

(i) *There exists at least one face F in Γ for which there is a set of independent 2-legged and 2-handed cycles such that $4(n_X) + 3(n_Y) + 2(n_Z) + n_W + p \geq 4$,*

(ii) *every k-handed cycle C for which $\Gamma_I(C)$ contains F, has at least $(4 - k)$ vertices of degree 2, and*

(iii) *every k-legged cycle C for which $\Gamma_O(C)$ contains F, has at least $(4 - k)$ vertices of degree 2.*

To prove Theorem 1 we need the following two lemmas, whose proofs are omitted.

Lemma 3. *Let G be a biconnected planar graph with $\Delta \leq 3$ and let Γ_1 and Γ_2 be two different arbitrary plane embeddings of G. Let F be the face of Γ_1 and let F' be the outer face in Γ_2, such that $F = F'$ or F' is an induced face of F in Γ_2. Assume that, C is a k-legged cycle in Γ_1 for $k = 2$ or 3, such that $\Gamma_I(C)$ contains F. If C is a k-legged cycle in Γ_1, then C is a k-handed cycle in Γ_2, and vice versa.*

Lemma 4. *Let G be a biconnected planar graph of $\Delta \leq 3$, and let Γ_1 and Γ_2 be two different arbitrary plane embeddings of G. Let F be the face of Γ_1 and let F' be the outer face in Γ_2, such that $F = F'$ or F' is an induced face of F in Γ_2. Assume that C is a k-legged cycle for $k = 2$ or 3 in Γ_1, such that $\Gamma_I(C)$ does not contain F. If C is a k-legged cycle in Γ_1, then C is a k-legged cycle in Γ_2. If C is a k-handed cycle in Γ_1, then C is a k-handed cycle in Γ_2.*

We now prove Theorem 1.

Outline of the Proof of Theorem 1

Necessity: Let G be a biconnected planar graph of maximum degree 3 and Γ be an arbitrary plane embedding of G. Assume that a planar embedding Γ' of G has a no-bend orthogonal drawing. Then Γ' satisfies Conditions $(or1) - (or3)$ of Lemma 2. Let F be the outer face of Γ'. We now show that in Γ there exists a face that satisfies Conditions (i) to (iii) of Theorem 1. We have two cases to consider.

Case 1: F exists in Γ.
In this case we show that the face F satisfies Conditions $(i) - (iii)$ of Theorem 1 in Γ.

(i) Since Γ' has a no-bend orthogonal drawing D, by Condition $(or1)$ of Lemma 2 the outer face F of Γ' has at least four 2-vertices. Among the 2-vertices at least four 2-vertices remain at corners of D whose angles are 270° in the outer face made by the contour of the outer face F. We consider such 2-vertices as *corner vertices*. Let C be a 2-legged cycle in D that has an edge on the outer face. By Condition $(or2)$ of Lemma 2, C has at least two 2-vertices. It can be observed that C is one of the type X, Y, Z and W. Let C_s be the set of independent 2-legged cycles which have edges on the outer face of D. If the elements of C_s contains four or more corner vertices, then one can show from the count of n_X, n_Y, n_Z, and n_W that F in Γ' satisfies Condition (i) in Theorem 1. If the elements in C_s have $t \leq 4$ corner vertices in D, then we can observe that the outer face F must have at least $4 - t$ corner vertices that are not on the elements in C_s. In this case $p \geq 4 - t$. Then one can easily observe that, F in Γ' satisfies the Condition (i) in Theorem 1. Since F exists in Γ, F in Γ satisfies the Condition (i) in Theorem 1.

(ii) Assume for a contradiction that, Γ has a 3-handed cycle C and $\Gamma_I(C)$ contains F but no 2-vertex. Then by Lemma 3, in Γ', C is a 3-legged cycle without any vertex of degree 2, a contradiction with the Condition $(or3)$ of Lemma 2. That is, every 3-handed cycle C for which $\Gamma_I(C)$ contains F has at least one vertex of degree 2. Similarly, every 2-handed cycle C for which

$\Gamma_I(C)$ contains F has at least two vertices of degree 2. Thus for $k = 2$ or $k = 3$, every k-handed cycle C for which $\Gamma_I(C)$ contains F has at least $4 - k$ vertices of degree 2.

(iii) Assume for a contradiction that Γ has a 3-legged cycle C such that $\Gamma_O(C)$ contains F and C does not contain any 2-vertex. Then by Lemma 4, in Γ', C is a 3-legged cycle without any vertex of degree 2, a contradiction with the Condition ($or3$) of Lemma 2. That is, every 3-legged cycle C for which $\Gamma_O(C)$ contains F has at least one vertex of degree 2. Similarly every 2-legged cycle C for which $\Gamma_O(C)$ contains F has at least two vertices of degree 2. Hence for $k = 2$ or $k = 3$, every k-legged cycle C for which $\Gamma_O(C)$ contains F has at least $4 - k$ vertices of degree 2.

Case 2: F does not exist in Γ.

In this case we show that an induced face $F'(\neq F)$ of F in Γ satisfies Conditions $(i) - (iii)$ of Theorem 1.

(i) Γ' has a no-bend orthogonal drawing whose outer face is F. F does not exist in Γ. By Lemma 1, there is an induced face F' for F in Γ. In Γ', let V_f be the set of all flipping pair vertices on F for the 2-legged cycles C_s, and V' be the set of vertices on F that are independent to the cycles C_s. By Lemma 1, in Γ', all the vertices of V_f and V' lie on F'. The difference one can see that, not all contour paths on F' remain same to all contour paths on F for C_s. Since F in Γ' satisfies Condition (i) of Theorem 1, F' in Γ satisfies the Condition (i) of Theorem 1.

(ii) Assume that Γ' has a 2-legged cycle C. By Condition ($or2$) Lemma 2, C has at least two vertices of degree 2. In Γ, if the induced face F' as above, resides in $\Gamma_I(C)$, then by Lemma 3, C is a 2-handed cycle in Γ, and the 2-handed cycle C has at least two vertices of degree 2. Thus we can say that, every 2-handed cycle C for which $\Gamma_I(C)$ contains F' has at least two vertices of degree 2. Similarly every 3-handed cycle C for which $\Gamma_I(C)$ contains F' has at least one vertex of degree 2. That is, every k-handed cycle C for which $\Gamma_I(C)$ contains F' has at least $4 - k$ vertices of degree 2.

(iii) Assume that Γ' has a 2-legged cycle C. By Condition ($or2$) Lemma 2, C has at least two vertices of degree 2. In Γ, if the induced face F' for F, resides in $\Gamma_O(C)$, then by Lemma 4, C is the 2-legged cycle in Γ, and the 2-legged cycle C has at least two vertices of degree 2. Every 2-legged cycle C for which $\Gamma_O(C)$ contains F' has at least two vertices of degree 2. Similarly, every 3-legged cycle C for which $\Gamma_O(C)$ contains F' has at least one vertex of degree 2. That is, every k-legged cycle C for which $\Gamma_O(C)$ contains F' has at least $4 - k$ vertices of degree 2.

Sufficiency: Assume that Γ has a face F satisfying Conditions $(i) - (iii)$ of Theorem 1. It is sufficient to show that G has a planar embedding which satisfies the condition in Lemma 2. Let Γ' be a planar embedding of G whose outer face is F. We have two cases to consider.

Case 1: Γ' has four 2-vertices at the outer face.

(i) Γ' has four vertices of degree 2 on the outer face, then it is obvious to the Condition $(or1)$ in Lemma 2.

(ii) Since Γ satisfies the Condition (ii) of Theorem 1, that is every 2-handed cycle C whose $\Gamma_I(C)$ contains F must have at least two 2-vertices in Γ. If such a cycle C exists in Γ, then C is a 2-legged cycle in Γ', that has at least two 2-vertices.

(iii) Since Γ satisfies the Condition (iii) of Theorem 1, that is every 3-handed cycle C whose $\Gamma_I(C)$ contains F must have at least one 2-vertex in Γ. If such a cycle C exists in Γ, then C is a 3-legged cycle in Γ', that has at least one 2-vertex.

Case 2: Γ' does not have four 2-vertices at the outer face.

Assume that Γ' does not have four 2-vertices on the outer face of Γ'. We can make another embedding Γ'' from Γ' by flipping any one or two independent 2-legged cycles C_k whose leg vertices are on the outer faces of both Γ' and Γ'' in such a such a way that $F_o(\Gamma'')$ contains at least four 2-vertices as illustrated in Fig. 1(d). We will now show that Γ'' satisfies the Conditions $(or1) - (or3)$ of Lemma 2

(i) We have already seen from the case 1 that, Γ' satisfies the Condition $(or2)$ of Lemma 2. Since Γ'' is obtained from Γ' by flipping any one or two independent 2-legged cycles C_k whose leg vertices are on the outer faces of both Γ' and Γ'', in such a such a way that $F_o(\Gamma'')$ contains four 2-vertices. It is obvious that Γ''' satisfies the Condition $(or1)$ of Lemma 2.

(ii) Γ'' is obtained from Γ' by flipping any one or two independent 2-legged cycles C_k, in such a way that they have edges at the outer faces in both Γ' and Γ''. It can be trivially observed that any 2-legged cycle in Γ'' is a 2-legged cycle in Γ'. It has already been proved in case 1 that every 2-legged cycle has at least two 2-vertices in Γ'. Hence Γ'' satisfies the Condition $(or2)$ of Lemma 2.

(iii) The only difference between Γ'' and Γ' is with respect to the 2-legged cycles that have edges on the outer faces in both Γ'' and Γ'. It can be trivially observed that any 3-legged cycle in Γ'' is a 3-legged cycle in Γ'. It has already been proved in Case 1 that every 3-legged cycle has at least one vertex of degree 2 in Γ'. Hence Γ'' satisfies the Condition $(or3)$ of Lemma 2.

Let G be a biconnected planar graph of maximum degree 3 and Γ be an arbitrary plane embedding of G. Traversing the contours of all faces in Γ, similar like approaches are described in $[2,4,10,12]$, one can check in linear time whether there exists a face F in Γ that satisfies the Conditions $(i) - (iii)$ in Theorem 1. If a face F in Γ satisfies the Conditions $(i) - (iii)$ of Theorem 1, then G has a planar embedding Γ' which satisfies the conditions in Lemma 2. In Γ' the outer face is F or an induced face F' of F. Hence using the drawing algorithm of Rahman et al. [11], one can also find a no-bend orthogonal drawing of the planar embedding Γ' of G in linear time. The whole process runs in linear time in total. Thus the following theorem holds.

Theorem 2. *Let G be a biconnected planar graph of maximum degree 3. Then one can determine in linear time whether G has a no-bend orthogonal drawing or not and find a drawing of G, if it exists.*

4   No-Bend Orthogonally Convex Drawings

In this section, we give a necessary and sufficient condition for a biconnected planar graph G of maximum degree 3 to have a no-bend orthogonally convex drawing, that can be checked in linear time. We also give a linear-time drawing algorithm if any drawing exists.

Let G be a biconnected planar graph of maximum degree 3, and Γ be an arbitrary plane embedding of G, and Γ has two or more independent regular 2-legged and 2-handed cycles. Let C_U be the set of all independent 2-legged and 2-handed cycles that have at least two 2-vertices on every cycle but all the 2-vertices of every cycle in C_U lie on exactly one contour path of the respective cycle. We add a dummy vertex d on the outer face of Γ, and connect the dummy vertex with all the leg and hand vertices of C_U, call the graph Γ^+. Let Γ_o^+ be a planar embedding of Γ^+, where the vertex d is on the outer face of Γ_o^+. Now we delete the vertex d from Γ_o^+, and call the obtained plane graph Γ_o. We can prove the following Theorem 3 whose proof is omitted.

Theorem 3. *Let G be a biconnected planar graph of maximum degree 3, and let Γ be an arbitrary plane embedding of G. Suppose that, Γ has two or more independent regular 2-legged and 2-handed cycles. G has a no bend orthogonally convex drawing if and only if Γ_o^+ can be obtained and the outer face of Γ_o satisfies the condition of a biconnected planar graph of maximum degree 3 for having a no-bend orthogonal drawing.*

By Theorem 2, we can determine whether Γ_o has a no-bend orthogonal drawing in linear time. Hence the following theorem holds.

Theorem 4. *Let G be a biconnected planar graph of maximum degree 3. Then one can determine in linear time whether G has a no-bend orthogonally convex drawing or not and can also find a drawing of G in linear time, if it exists.*

5   Conclusions

In this paper we have presented two different necessary and sufficient conditions for biconnected planar graphs G with $\Delta \leq 3$ to have no-bend orthogonal drawings and to have no-bend orthogonally convex drawings. We have also given linear-time algorithms for finding out no-bend orthogonal drawings and no-bend orthogonally convex drawings of planar graphs if they exist.

References

1. Chang, Y., Yen, H.: On bend-minimized orthogonal drawings of planar 3-graphs. In: Proceedings of 33rd International Symposium on Computational Geometry (SoCG 2017), pp. 29:1–29:15. Leibniz International Proceedings in Informatics (LIPICS), Schloss Dagstuhl-Leibniz-Zentrum für Informatik, Dagstuhl Publishing, Germany (2017)
2. Chang, Y., Yen, H.: On orthogonally convex drawings of plane graphs. Comput. Geom. Theory Appl. **62**, 34–51 (2017)
3. Rahman, M.S., Egi, N., Nishizeki, T.: No-bend orthogonal drawings of series-parallel graphs. In: Healy, P., Nikolov, N.S. (eds.) GD 2005. LNCS, vol. 3843, pp. 409–420. Springer, Heidelberg (2006). https://doi.org/10.1007/11618058_37
4. Rahman, M.S., Egi, N., Nishizeki, T.: No-bend orthogonal drawings of subdivisions of planar triconnected cubic graphs. IEICE Trans. Inf. Syst. **E88-D**(1), 23–30 (2005)
5. Garg, A., Tamassia, R.: On the computational complexity of upward and rectilinear planarity testing. SIAM J. Comput. **31**(2), 601–625 (2001)
6. Hasan, M.M., Rahman, M.S.: No-bend orthogonally convex drawings of subdivisions of planar triconnected cubic graphs. In: Proceedings of 3rd International Conference on Theoretical Computer Science and Discrete Mathematics (ICTCSDM 2018) (2018)
7. Battista, G.D., Liotta, G., Vargiu, F.: Spirality and optimal orthogonal drawings. SIAM J. Comput. **27**(6), 1764–1811 (1988)
8. Didimo, W., Liotta, G., Patrignani, M.: Bend-minimum orthogonal drawings in quadratic time. In: Biedl, T., Kerren, A. (eds.) GD 2018. LNCS, vol. 11282, pp. 481–494. Springer, Cham (2018). https://doi.org/10.1007/978-3-030-04414-5_34
9. Rahman, M.S., Nakano, S., Nishizeki, T.: A linear algorithm for bend optimal orthogonal drawings of triconnected cubic plane graphs. J. Graph Algorithms Appl. (JGAA) **3**(4), 31–62 (1999)
10. Rahman, M.S., Nakano, S., Nishizeki, T.: Box-rectangular drawings of plane graphs. J. Algorithms **37**, 363–398 (2000)
11. Rahman, M.S., Nishizeki, T., Naznin, M.: Orthogonal drawings of plane graphs without bends. J. Graph Algorithms Appl. (JGAA) **7**(4), 335–362 (2003)
12. Nishizeki, T., Rahman, M.S.: Planar Graph Drawing. World Scientific, Singapore (2004)
13. Rahman, M.S., Nishizeki, T.: Bend-minimum orthogonal drawings of plane 3-graphs. In: Goos, G., Hartmanis, J., van Leeuwen, J., Kučera, L. (eds.) WG 2002. LNCS, vol. 2573, pp. 367–378. Springer, Heidelberg (2002). https://doi.org/10.1007/3-540-36379-3_32
14. Tamassia, R.: Handbook of Graph Drawing and Visualization. Discrete Mathematics and Its Applications, 1st edition. Chapman & Hall/CRC (2016)

Branch-and-Cut Algorithms for Steiner Tree Problems with Privacy Conflicts

Alessandro Hill[1]([✉])[ID], Stefan Voß[2][ID], and Roberto Baldacci[3][ID]

[1] Department of Industrial & Manufacturing Engineering,
California Polytechnic State University, San Luis Obispo, USA
ahill29@calpoly.edu
[2] Institute of Information Systems (IWI),
University of Hamburg, Hamburg, Germany
stefan.voss@uni-hamburg.de
[3] Department of Electrical, Electronic, and Information Engineering
"Guglielmo Marconi", University of Bologna, Cesena, Italy
r.baldacci@unibo.it

Abstract. In this paper we propose two novel variants of the well-known Steiner tree problem in graphs that are motivated by applications in secure strategic telecommunication network design. Both network optimization models ask for a tree of minimal total edge cost that connects a pre-specified set of terminal nodes to a dedicated root node by optionally including intermediate Steiner nodes. Two types of privacy conflicts between pairs of conflicting terminals are considered: (I) The path from the root to a terminal must not include the conflicting terminal, and (II) conflicting terminals have to be on separate branches of the tree. We develop non-compact integer programming formulations and elaborate branch-and-cut algorithms. We incorporate problem specific valid inequalities that are crucial in order to solve these problems, and establish dominance relationships between these cuts and the induced polyhedra. The effectiveness of the cutting planes with respect to the dual bound and the performance of the exact algorithm are assessed on a diverse set of SteinLib-based test instances.

Keywords: Steiner tree problem · Integer programming ·
Branch-and-cut · Information privacy conflicts · Telecommunications

1 Introduction

In the strategic design of telecommunication infrastructure, the protection of data privacy is an important concern. The physical network structure is crucial in order to prevent from unwanted information leaks. A common requirement of information senders, and receivers, is that specific third party network participants must not be able to tap the network path used to route privacy-sensitive information packages. Protective measures such as data depersonalization, encryption technologies, and protocol changes are often not sufficient in order to prevent from attacks such as eavesdropping, traffic analysis, and camouflage marketing. Applications can be found when connecting facilities of high-tech industry, military, or government agencies to backbone networks and so forth.

© Springer Nature Switzerland AG 2019
D.-Z. Du et al. (Eds.): COCOON 2019, LNCS 11653, pp. 266–278, 2019.
https://doi.org/10.1007/978-3-030-26176-4_22

In this work, we address the case of centralized networks with minimal connectivity requirements. That is, we design tree topologies in which exactly one path exists between two customers in the network. These structures are commonly modeled using rooted Steiner trees in which given customers, called *terminals*, are sought to be connected to a central distributor. Intermediate nodes, called *Steiner nodes*, may be used in order to minimize the overall network cost.

We present topological concepts that allow the a priori specification of privacy conflicts between pairs of conflicting customers. More precisely, the following two types of restrictions are incorporated into the strategic network design model. In a first *path-critical* model, we embed the customer requirement that a conflicting customer is not allowed to be physically situated on its unique connecting path to the distributor. Furthermore, we suggest a *branch-critical* model in which the two paths connecting two customers that are in a privacy conflict to the distributor have to be disjoint. The latter requirement is more restrictive than the former since it forbids both customers from being on a common network branch. In both cases we assume that the distributor is considered neutral and serves either as source, destination, or both types of transmission.

The Steiner tree problem (STP) was introduced in Dreyfus and Wagner [1971] and is one of the prominent NP-hard combinatorial optimization problems listed by Karp [1972]. The most efficient exact approaches are based on branch-and-cut Gamrath et al. [2017]; Koch and Martin [1998]. Catalogues of mathematical programming formulations for the STP are given in Goemans and Myung [1993] and Polzin and Daneshmand [2001]. An overview on existing heuristic approaches is given in Duin and Voß [1994]. Several STP variants have been studied in the literature (e.g., Di Puglia Pugliese et al. [2016]; Johnson et al. [2000]; Leggieri et al. [2014]; Voß [1999]).

In this paper, we devise two exact algorithms based on integer programming for the novel models with complicating privacy side constraints. Our methods are built upon cut-set formulations. In order to strengthen the obtained dual bounds that are obtained from the solution of the linear relaxations during the branch-and-bound method, we develop effective problem-specific cutting planes. Our computational analysis shows the efficiency of our techniques for a broad range of privacy conflict densities on diverse classes of test instances derived from the literature. Our main contributions are as follows:

1. Introduction of two novel variants of the Steiner tree problem that incorporate customer privacy conflicts.
2. Introduction of cut-set based integer-programming formulations and of problem-specific valid inequalities.
3. Theoretical analysis of polyhedral formulation strengths through projection results.
4. Development of branch-and-cut algorithms and computational evaluation of algorithmic performance for a diverse set of test instances.

The optimization models are formally presented in Sect. 2, followed by our mathematical formulations in Sect. 3. Section 4 is devoted to the cutting plane techniques and we provide the results of our computational evaluation in Sect. 5, before closing with a conclusion in Sect. 6.

2 Privacy-Oriented Optimization Models

Before recalling the definition of the Steiner tree problem we introduce some notation. We refer to the vertex set of a (directed) graph G by $V[G]$, and to its edge (arc) set by $E[G]$ ($A[G]$), respectively. For a (directed) tree B, we define the graph representing the unique (directed) path connecting node i to j by $P_{i,j}(B)$.

Let T be the set of *terminals* (or *customer nodes*), W the set of *Steiner nodes*, r the *root node*, V the set of all nodes (i.e., $V = W \mathbin{\dot{\cup}} T \mathbin{\dot{\cup}} \{r\}$), and $V' = V \setminus \{r\}$. The set $E \subseteq 2^V$ contains all possible edges. We denote the non-negative cost of an edge $e = \{i, j\} \in E$ by c_e. The classical *Steiner tree problem in graphs* (STP) consists of finding a tree B of minimal total edge costs $c(B) = \sum_{e \in E[B]} c_e$, such that $T \cup \{r\} \subseteq V[B] \subseteq V$ and $E[B] \subseteq E$. Note that the STP is often defined without the root node, which is equivalent to replacing r by an additional terminal node. However, the root plays an important role when considering privacy conflict requirements.

2.1 STP with Path-Privacy Conflicts

Consider an instance of the STP and $C_P \subseteq 2^T$ a set of *path-privacy conflicts*. Then the *STP with path-privacy conflicts* (STP+CP) asks for a solution B such that for two conflicting customers in $\{i, j\} \in C_P$, we have $i \notin V[P_{r,j}(B)]$ and $j \notin V[P_{r,i}(B)]$. In other words, the conflicting terminals cannot both be on any unique path from the root r to any vertex in the tree. In Fig. 1 an optimal solution for an STP instance with 8 customers, 4 Steiner nodes and $E = 2^V$ is depicted (left), along with an optimal solution for a corresponding STP+CP instance with $C_P = \{\{1,2\}, \{3,4\}, \{6,8\}\}$ (center).

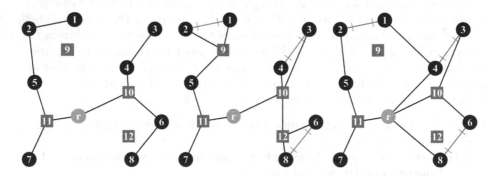

Fig. 1. Optimal solutions for the classical STP (left), the STP+CP (center) and the STP+CB (right) with eight terminals and three privacy conflicts.

2.2 STP with Branch-Privacy Conflicts

Consider an instance of the STP and $C_B \subseteq 2^T$ a set of *branch-privacy conflicts*. Then the *STP with branch-privacy conflicts* (STP+CB) asks for a solution B such that, for two conflicting customers in $\{i, j\} \in C_B$, we have $V[P_{r,i}(B)] \cap V[P_{r,j}(B)] = \emptyset$. In other words, the two unique paths in B that connect i and j to r have to be node disjoint. In this case, the customers only fully trust the root node regarding the information routing. Every other node in the network is considered a potentially insecure point of transmission. Note that in the STP+CB, no two conflicting terminals are allowed to be contained in the same connected component when removing r from the Steiner tree. Note that a branch-privacy constraint for terminals i and j dominates a corresponding path-privacy constraint. That is, the former requirement is stronger since a feasible solution for the STP+CB with $\{i, j\} \in C_B$ satisfies the requirement associated with $\{i, j\} \in C_P$, but not vice versa. The impact of branch privacy conflicts ($C_B = \{\{1, 2\}, \{3, 4\}, \{6, 8\}\}$) is illustrated by the optimal STP+CB solution in Fig. 1 (right). The STP+CB is related to the hop-constrained Steiner tree problem (STP+H) (see, for instance, Voß [1999]). In contrast to the STP+H, where a fixed given number of hops, or edges, must not be exceeded in order to reach a terminal from r, the STP+CB does not allow two conflicting terminals to be connected in $G \setminus r$ via paths of any length. Both the STP+CP and the STP+CB are NP-hard since they reduce to the STP when $C_P = C_B = \emptyset$.

3 Integer Programming Formulations

In this section, we develop mathematical formulations for the STP+CP and the STP+CB based on directed trees, or arborescences, that are rooted in r. Our formulations are cut-set based and therewith non-compact. Binary *edge variables* are used for each edge $e \in E$ in order to encode the undirected Steiner tree: x_e takes value one if e is installed in the network and zero otherwise. We recall that there is a one-to-one correspondence between the (undirected) trees rooted in r and the arborescences with root r. To incorporate the arborescence structure in our formulations, let A be the set of arcs obtained from all possible orientations of edges in E; i.e., $A = \{(i, j) \in V^2 : \{i, j\} \in E\}$. For each arc $a \in A$, we define a binary *arc variable* x_a that is equal to one if a is used in the network and zero otherwise. The arborescence representation allows us to derive stronger formulations for the privacy-oriented models, which is also known to apply to the STP per se (Polzin and Daneshmand 2001). Let a^- denote the tail node i of an arc $a = (i, j)$ and a^+ its head node j. We define $\delta^-(U)$ ($\delta^+(U)$) to be the set of arcs in A with head (tail) in U and tail (head) in $V \setminus U$ for $U \subseteq V$.

The following non-compact cut-set based formulation for the STP was suggested by Wong [1984].

$$(F) \quad \min \quad \sum_{e \in E} c_e y_e \tag{1a}$$

$$s.t. \quad x_{i,j} + x_{j,i} = y_{i,j} \qquad (\{i,j\} \in E) \tag{1b}$$

$$\sum_{a \in \delta^-(i)} x_a = 1 \qquad (i \in T) \tag{1c}$$

$$\sum_{a \in \delta^-(S)} x_a \geq 1 \qquad (S \cap T \neq \emptyset, S \subseteq V') \tag{1d}$$

$$x_a \in \{0,1\} \qquad (a \in A) \tag{1e}$$

$$y_e \in \{0,1\} \qquad (e \in E) \tag{1f}$$

Linking inequalities (1b) ensure that if an arc between two nodes is chosen to be in the directed Steiner tree, the corresponding edge is also selected. Additionally, they forbid directed subtours of length 2. Node *in-degree* inequalities (1c) force the sum of arcs entering each terminal to one. Inequalities (1d) are called *Steiner cut constraints* and ensure the 1-connectivity within the solution network. Note that edge variables y_e will take integer values if arc variables are integer feasible. It is known that the formulation (F) is equally strong as a multi-commodity flow formulation, but computationally superior when implemented carefully (Koch and Martin 1998). Stronger flow-based formulations are known to yield even higher solve times for the LP-relaxations (e.g., common-flow formulations (Polzin and Daneshmand 2001)) and are not considered to be of practical use.

In the following, we extend formulation (F) to obtain formulations for the STP+CP and the STP+CB. To this end, let $\mathcal{P}_{i,j}(G \setminus r)$ denote the set of directed paths from i to j in $G \setminus r$ $(i \neq j)$. Then path-privacy requirements can be integrated into formulation (F) by adding the following *path-privacy inequalities*.

$$\sum_{a \in A[P]} x_a \leq |A[P]| - 1 \qquad (P \in \mathcal{P}_{i,j}, \{i,j\} \in C_P) \tag{2}$$

Inequalities (2) enforce that not all arcs of a directed path (not including r) that connects i and j can be selected in a solution simultaneously. Note that we obtain two inequalities for each conflict by considering paths starting at i and j separately. These cover inequalities are related to inequalities used for the STP+H in Costa et al. [2009] in order to forbid integer-feasible solutions, and they have also been separated in the fractional case in Hill and Schwarze [2018] for another STP variant. We denote the formulation for the STP+CP obtained by adding the exponential number of model inequalities (2) to (F) by (F_{CP}). More generally, we use $(F_{(I_1),...,(I_k)})$ for the formulation obtained from (F) by adding inequalities $(I_1), \ldots, (I_k)$; hence, for instance, $(F_{CP}) = (F_{(2)})$.

The STP+CB can be formulated by adding the following *branch-privacy inequalities* to (F).

$$\sum_{a \in \delta^-(S)} x_a \geq 2 \qquad (S \subseteq V' : \{i,j\} \subseteq S, \{i,j\} \in C_B) \tag{3}$$

Inequalities (3) force two conflicting terminals to be connected from r by two disjoint directed paths. They are a special case of capacitated connectivity inequalities known for vehicle routing problems, which ensure that the vehicle capacity is respected on each tour. These general versions have also been used to formulate capacitated minimum spanning tree problems. The in-degree of a Steiner node may exceed 1 when imposing inequalities (3). To see this, consider a STP+CB instance with $T = \{4, 5\}$, $W = \{1, 2, 3\}$ and $C_B = \{T\}$. Then $B = (T \cup W \cup \{r\}, \{\{r, 1\}, \{r, 2\}, \{1, 3\}, \{2, 3\}, \{3, 4\}, \{3, 5\}\})$ is an optimal solution for (F) with inequalities (3) if $c_e = 0$ for $e \in E(B)$ and $c_e = 1$ otherwise. Thus, the following *Steiner in-degree inequalities* are necessary.

$$\sum_{a \in \delta^-(i)} x_a \leq 1 \qquad (i \in W) \tag{4}$$

Note that inequalities (4) are also needed for the STP in the case of zero edge weights. Zero-cost Steiner subtours may also appear which can be identified and removed efficiently a posteriori in an integer-feasible solution. We denote the formulation obtained from (F) by adding inequalities (3) and inequalities (4) by (F_{CB}). We note that the use of arc variables in our formulations above technically allows us to consider directed variants of the STP+CP and the STP+CB, too.

4   Cutting Planes

In this section we develop valid inequalities that tighten formulations (F_{CP}) and (F_{CB}) and we establish relationships based on the achieved polyhedral strength. Furthermore, we describe techniques needed to ensure computational efficiency in practice. We use $P(D)$ to denote the polyhedron described by a formulation D when replacing binary variables by continuous variables with interval domain $[0, 1]$; e.g., $P(F) = \{(x, y) \in [0, 1]^{|A|} \times [0, 1]^{|E|} : (1b), (1c), (1d)\}$

4.1   Valid Inequalities

We first focus on valid inequalities for the STP, followed by cuts for the STP+CP and the STP+CB. Due to the dominance relation between the STP+CP and the STP+CB described in Sect. 2, any valid inequality for (F_{CP}) is also valid for (F_{CB}) when assuming that $C_B = C_P$ which we will formally specify in a theorem at the end of this subsection.

STP The following flow-balance inequalities are known to be valid for formulation (F) (Polzin and Daneshmand 2001).

$$\sum_{a \in \delta^-(i)} x_a \leq \sum_{a \in \delta^+(i)} x_a \qquad (i \in W) \tag{5}$$

Inequalities (5) ensure that the number of arcs leaving a Steiner node is at least the number of entering arcs. Formulation $(F_{(5)})$ is denoted $F + FB$ in Polzin and

Daneshmand [2001]. Moreover, the Steiner node flow can be further balanced by

$$x_{i,j} \leq \sum_{(k,i)\in\delta^-(i):k\neq j} x_a \qquad ((i,j) \in A, i \in W). \qquad (6)$$

Note that model inequalities (4) in (F_{CB}) are valid for (F_{CP}). We refer to $(F_{(4),(5),(6)})$ as (F^+) and we assume that $\delta^-(r) = \emptyset$. In the following, we establish polyhedral relationships between formulations. Following Polzin and Daneshmand [2001], we say that a formulation D is strictly stronger than formulation D' if $P(D) \subseteq P(D')$ and there exists a problem instance such that $P(D) \subset P(D')$; we denote this relation by $D \succ D'$.

STP+CP Inequalities (2) can be lifted by considering variables of forward and backward arcs on a path from i to j ($\{i,j\} \in C_P$). Note that it is necessary to maintain the path direction property in order to avoid forbidding both nodes being on different branches. Therefore, we impose the path direction by only allowing backward arcs that are not incident with i as follows.

$$\sum_{(i,l)\in A[P]} x_{i,l} + \sum_{(k,l)\in A[P]:k\neq i} (x_{k,l} + x_{l,k}) \leq |A[P]| - 1 \qquad (P \in \mathcal{P}_{i,j}, \{i,j\} \in C_P) \quad (7)$$

Lemma 1. $(F_{(7)}^+) \succ (F_{(2)}^+)$.

Proof. $P(F_{(2)}^+) \supseteq P(F_{(7)}^+)$: $\{a \in A[P]\} \subseteq \{(i,l) \in A[P]\} \cup \{(k,l) \in A[P] : k \neq i\}$. $P(F_{(7)}^+) \neq P(F_{(2)}^+)$: Let $C_P = \{\{1,2\},\{2,3\}\}$. Inequalites (7) are violated by the fractional solution for $(F_{(2)}^+)$ in Fig. 2 (left) for $P = (\{1,2,4\},\{(1,4),(4,2)\})$. $\square$

Instead of focusing on forbidding paths that connect conflicting terminals $\{i,j\} \in C_P$, we can also directly enforce alternative connectivity with respect to r. The following *path-privacy connectivity inequalities* impose a directed path from r to i in the network that is obtained after removing j from G.

$$\sum_{a\in\delta^-(S):a^-\neq j} x_a \geq 1 \qquad (i \in S, S \subseteq V' \setminus \{j\}, \{i,j\} \in C_P) \qquad (8)$$

The number of cut sets is exponential in $|V|$. For each S and each conflict in C_P, we obtain two inequalities depending on the choice of j. Inequalities (8) are stronger than inequalities (7) which is expressed by the following lemma.

Lemma 2. $(F_{(8)}^+) \succ (F_{(7)}^+)$.

Proof. Rewrite (7) as $-(\sum_{(i,l)\in A[P]} x_{i,l} + \sum_{(k,l)\in A[P]:k\neq i}(x_{k,l} + x_{l,k})) + |A[P]| \geq 1$; since $\sum_{(i,l)\in A[P]} x_{i,l} + \sum_{(k,l)\in A[P]:k\neq i}(x_{k,l} + x_{l,k}) \leq |A[P]| - \sum_{a\in\delta^-((V[P]\setminus\{i\})\setminus A[P])} x_a$ (using (1c) and (4)), $\sum_{a\in\delta^-(V[P]\setminus\{i\})\setminus A[P])} x_a \geq 1$ implies (7); having $\{\mathcal{P}_{i,j} \setminus i : \{i,j\} \in C_P\} = \{S \subseteq V' \setminus \{j\} : \{i,j\} \in C_B\}$, $P(F_{(8)}^+) \subseteq P(F_{(7)}^+)$ holds. $P(F_{(8)}^+) \neq P(F_{(7)}^+)$: Inequalities (8) are violated for $S = \{2,4,5\}$ and $\{i = 2, j = 3\} \in C_P$ by the solution for $(F_{(7)}^+)$ given in Fig. 2 (right) since $x_{r,2} + x_{r,4} + x_{r,5} + x_{1,2} + x_{1,4} + x_{1,5} = 0.5 < 1$. $\square$

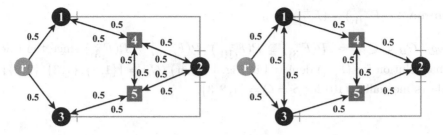

Fig. 2. Fractional solutions with arc-variable values for formulation $(F_{(2)}^+)$ that violate inequalities (7) (left), and for formulation $(F_{(7)}^+)$ that violate inequalities (8) (right).

Let $C_P(i) \subset T$ denote the set of terminals that are in path-privacy conflict with $i \in T$. Then inequalities (8) can be generalized by requiring root-connectivity for each terminal $i \in T$ after removing all conflicting terminals in $C_P(i)$.

$$\sum_{a \in \delta^-(S): a^- \notin C_P(i)} x_a \geq 1 \qquad (i \in S, S \subseteq V' \setminus C_P(i), i \in T) \qquad (9)$$

Inequalities (9) collapse to inequalities (1d) in the case that $C_P(i) = \emptyset$.

Lemma 3. $(F_{(9)}^+) \succ (F_{(8)}^+)$.

Proof. $P(F_{(8)}^+) \supseteq P(F_{(9)}^+)$ since $\{j\} \subseteq C_P(i)$. $P(F_{(8)}^+) \neq P(F_{(9)}^+)$: The weight of a minimal $(r,2)$-cut after removing either node 1 or node 3 in Fig. 3 (left) equals 1. However, the cut weight after removing both terminals in $C_P(2)$ equals 0.5. Hence, a corresponding cut with $S = \{2, 4, 5\}$ is violated for $i = 2$. $\square$

STP+CB Let H_B denote the conflict graph with respect to the conflict relation given by C_B; i.e., $G_B = (T, C_B)$. Let $\mathcal{Q}$ denote the set of all maximal cliques in G_C. Then, for $Q \in \mathcal{Q}$, the number of branches emerging from r in order to connect terminals in Q must be at least $|Q|$. Therefore, the following *clique connectivity inequalities* can be used to strengthen (F_{CB}^+).

$$\sum_{a \in \delta^-(Q)} x_a \geq |Q| \qquad (Q \in \mathcal{Q}, S \subseteq V' : Q \subseteq S) \qquad (10)$$

Note that Inequalities (10) imply the installation of at least as many root tree branches in a solution as the cardinality of a largest conflict clique. Thus, the following root out-degree inequality is a special case of inequalities (10) when $S = V'$ and $Q = \text{argmax}_{Q' \in \mathcal{Q}}\{|Q'|\}$.

$$\sum_{a \in \delta^+(\{r\})} x_a \geq \max_{Q \in \mathcal{Q}} |Q| \qquad (11)$$

Lemma 4. $(F^+_{(10)}) \succ (F^+_{(3)})$.

Proof. $C_B \subseteq \mathcal{Q} \implies P(F^+_{(3)}) \supseteq P(F^+_{(10)})$. $P(F^+_{(10)}) \neq P(F^+_{(3)})$ since the fractional solution for $(F^+_{(3)})$ depicted in Fig. 3 (right) $(C_B = \{\{1,2\},\{1,3\},\{2,3\}\})$ violates inequality (10) for $S = Q = \{1,2,3\}$. $\qquad\qquad\square$

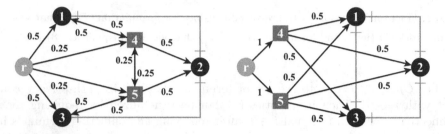

Fig. 3. Fractional solutions with arc-variable values for formulation $(F^+_{(8)})$ that violate inequalities (9) (left), and for formulation $(F^+_{(3)})$ that violate inequalities (10) (right).

Theorem 1. *Assume that $C_P = C_B$. Then it holds that*
$$(F^+_{(10)}) \succ (F^+_{(3)}) \succ (F^+_{(9)}) \succ (F^+_{(8)}) \succ (F^+_{(7)}) \succ (F^+_{(2)}) \succ (F^+) \succ (F).$$

Proof. Follows from Lemmata 1-4, relation between STP+CP and STP+CB $((F^+_{(3)}) \succ (F^+_{(9)}))$, and Polzin and Daneshmand [2001]. $\qquad\qquad\square$

4.2 Cut Generation

It is well-known that branch-and-cut algorithms for the STP that are based on formulation (F) suffer from an extensive number of model cuts (1d) in practice. To overcome this computational challenge, Koch and Martin [1998] suggest several techniques that aim at generating multiple diverse and at the same time effective connectivity cuts. In the following, we describe our approach which incorporates the ideas of the authors. Note that these techniques do not improve the dual bounds obtained by formulation (F) and solely speed up convergence of the cutting plane method.

- We use the following slightly different, but equivalent, version of inequalities (1d) which enforces connectivity for each terminal node separately. This may result in the separation of multiple cuts.

$$\sum_{a \in \delta^-(S)} x_a \geq 1 \qquad (i \in S, S \subseteq V', i \in T) \qquad (12)$$

- Forward and backward cuts: When separating inequalities (12), we compute two minimal cuts: The $r - i$ cut and the $i - r$ cut which are of equal weight but may differ in terms of cut sets.
- Steiner node connectivity: Similar to cut inequalities (12), we derive the following connectivity cuts for each Steiner node which ensure connectivity dependent on whether the node is used in the tree or not.

$$\sum_{a\in\delta^-(S)} x_a \geq \sum_{a\in\delta^-(\{i\})} x_a \quad (i \in S, S \subseteq V', i \in W) \tag{13}$$

- Cut ranking: We limit ourselves to the 15 most violated connectivity cuts. Moreover, we add at most 15 conflict cuts of each type in each round.
- Connected component connectivity. Whenever the support graph w.r.t. y is disconnected then we add dynamically connectivity inequalities (1d) in which S contains the component nodes (especially during the first rounds).
- Certainly, two conflicting terminals cannot be adjacent in any feasible solution for the STP+CP and the STP+CB, which leads to the following edge variable fixing cuts: $y_{i,j} = 0$ ($\{i,j\} \in (C_P \cup C_B)$). Using inequalities (1c), the latter can be rewritten as $\sum_{a\in\delta^-(\{i,j\})} x_a \geq 2$. They can be expressed by two model inequalities (8) obtained from $S = \{i\}$ and $S = \{j\}$, and are therefore redundant for (F_{CP}). Note that they correspond to inequalities (3) when $S = \{i,j\}$. We add these $|C_P|$ (or $|C_B|$) cuts in advance.
- Flow-balance inequalities (5) are added to the initial formulation, but inequalities (6) are separated dynamically.
- Only the theoretically strongest inequalities are separated during the branch-and-cut algorithm.

The separation problem for connectivity inequalities (8), (9), (12), (13) corresponds to finding a cut of minimal weight in the support graph. Capacitated connectivity inequalities (3), (10), and model inequalities (1d) can be separated similarly on an auxiliary network. For inequalities (10), we explicitly compute the set of all maximal cliques $\mathcal{Q}$ using the Bron-Kerbosch algorithm. Path cover inequalities (2) and (7) can be separated using truncated enumeration as suggested in Hill and Schwarze [2018].

5 Computational Study

To evaluate the performance of our approaches we perform a computational analysis on a diverse set of instances[1]. We use IBM ILOG CPLEX 12.80 as branch-and-bound framework (generic cuts disabled) and run our algorithm on an Intel Core i7-7600 2.80 GHz machine with 16 GB RAM. We derive 390 instances with up to 100 nodes from 30 base STP instances from Beasley [1989][2] (Steinb$k$, $k \in \{3,4,5,16,17,18\}$) and the SteinLib[3] (Koch et al. 2001)

---

[1] The test instances can be obtained from the authors upon request.
[2] http://people.brunel.ac.uk/~mastjjb/jeb/orlib/steininfo.html.
[3] http://steinlib.zib.de.

(from test sets PUC, ES20FST, P4Z, and P4E). We randomly draw pairs of conflict nodes from T using a uniform distribution with three different seeds. For $\gamma \in \{0, 0.25, 0.5, 0.75, 1.0\}$, we incrementally generate $\lceil \gamma|T| \rceil$ conflicts for sparse instances ($|E[G]| < 0.5\binom{n}{2}$) and $\lceil \gamma|T|(|T|-1)/2 \rceil$ conflicts for dense instances. That is, for a base instance and a seed, the conflicts obtained from γ_2 contain all conflicts for γ_1 if $\gamma_1 \leq \gamma_2$. The node of highest degree in G with lowest index is assigned to be the root node r.

In the following, we provide an analysis of the effectiveness of the techniques suggested in Sect. 4 at the root node of the branch-and-bound tree and within the branch-and-cut method. In our experiments, we do not see an advantage in separating STP+CP cuts when solving an STP+CB instance. We force each cut to stay in subsequent linear programs (LPs) since letting CPLEX manage the cuts tends to increase the solution times. Cut separation time did not exceed 10% of the total run time. Table 1 summarizes the results obtained for both models. The first three columns indicate the density of G (**Type**), the privacy conflict rate (γ), and the number of instances in this category (**#**). Columns * contain the number of instances that could be solved to optimality for the corresponding formulation. The cumulative run time in seconds is given in column $\mathbf{t(s)}$. Columns $\mathbf{\Delta_0^{LB}}$ and $\mathbf{\Delta^{LB}}$ state the average relative improvement of the lower bound compared to the base formulation at the root node and after at most one hour of branch-and-cut, respectively.

Table 1. Root node cutting plane impact and branch-and-cut effectiveness for STP+CP and STP+CB by density of G and different privacy conflict rates γ.

Instances			STP+CP						STP+CB						
Type	γ	#	$(\mathbf{F_{CP}^+})$		$(\mathbf{F_{(9)}^+})$				$(\mathbf{F_{CB}^+})$		$(\mathbf{F_{(10)}^+})$				
			*	t(s)	$\mathbf{\Delta_0^{LB}}$	*	t(s)	$\mathbf{\Delta^{LB}}$	*	*	t(s)	$\mathbf{\Delta_0^{LB}}$	*	t(s)	$\mathbf{\Delta^{LB}}$	*
Sparse	0	13	6	79	0	6	66	1.8	**8**	6	65	0	6	69	1.8	**8**
	0.25	39	26	284	0	26	285	1.7	**36**	16	62	0	16	66	1	**16**
	0.5	39	24	348	0	24	325	1.4	**33**	15	99	0.1	16	102	0.8	**16**
	0.75	39	27	408	0	27	335	1.5	**34**	10	138	0.5	16	111	1.6	**16**
	1	39	27	460	0	27	321	1.5	**34**	8	228	0.9	15	114	2	**15**
	All	169	110	1579	0	110	1332	1.5	**145**	55	592	0.3	69	462	1.4	**71**
Dense	0	17	**5**	159	0	**5**	157	0	**5**	17	176	0	17	157	0	**17**
	0.25	51	38	2534	0.4	46	1003	0.4	**51**	17	3089	6.4	21	1948	7.9	**21**
	0.5	51	41	4151	0.3	47	924	0.4	**51**	10	4825	19	20	3798	20.2	**20**
	0.75	51	40	5392	0.4	51	860	0.4	**51**	7	5828	46.6	25	4490	47.4	**25**
	1	51	39	6124	0.6	48	842	0.7	**51**	6	7094	199.9	51	24	199.9	**51**
	All	221	163	18360	0.4	197	3786	0.4	**209**	57	21012	62.8	134	10417	63.6	**134**
All	All	390	273	19939	0.2	307	5118	0.9	**354**	112	21604	35.7	203	10879	36.6	**205**

It can be seen that the number of problems that are solved at the root node increases notably (STP+CP: 12.5%; STP+CB: 81.3%). Furthermore, corresponding LPs can be solved significantly faster for the stronger formulations (STP+CP: 74.3%; STP+CB: 49.6%). Except for sparse STP+CP instances, the dual bounds are raised. Most notably, for dense STP+CB instances we observe an average increase of 62.8%, compared to an overall average increase of 18%. Note that optimality gaps can be closed at the root node for 97.1% of all dense full-conflict instances. Moreover, our cutting plane techniques resulted in an overall reduction of the number of cut separation rounds (78.5%). Finally, the branch-and-cut algorithms (without any initial upper bound provided) solve 71.7% of the instances to optimality. Note that without inequalities (4)-(6) and the improvements described in Subsection 4.2, the STP base instances cannot be solved efficiently at all.

6 Conclusion

We studied two new Steiner tree problem variants that find application in strategic telecommunication network planning. Customer privacy requirements were incorporated by enforcing additional topological constraints in cost-optimal networks. In order to solve the resulting combinatorial problems we developed mathematical formulations based on integer programming which were used in a branch-and-cut algorithm. We developed cutting plane techniques that are essential regarding computational efficiency. Polyhedral relationships between the cuts theoretically support their importance. A computational analysis was conducted using a diverse set of problem instances. We could show that our algorithmic enhancements significantly improve the obtained results compared to our initial approaches. Finally, we could compute optimal networks for more than 70% of the test instances. We see potential for further strengthening of our algorithms in the development of effective heuristic primal methods in order to obtain stronger upper bounds.

References

Beasley, J.E.: An SST-based algorithm for the Steiner problem in graphs. Networks **19**(1), 1–16 (1989)

Costa, A.M., Cordeau, J.-F., Laporte, G.: Models and branch-and-cut algorithms for the Steiner tree problem with revenues, budget and hop constraints. Networks **53**(2), 141–159 (2009)

Di Puglia Pugliese, L., Gaudioso, M., Guerriero, F., Miglionico, G.: An algorithm to find the link constrained Steiner tree in undirected graphs. In: Greuel, G.-M., Koch, T., Paule, P., Sommese, A. (eds.) ICMS 2016. LNCS, vol. 9725, pp. 492–497. Springer, Cham (2016). https://doi.org/10.1007/978-3-319-42432-3_63

Dreyfus, S.E., Wagner, R.A.: The Steiner problem in graphs. Networks **1**(3), 195–207 (1971)

Duin, C., Voß, S.: Steiner tree heuristics—a survey. In: Operations Research Proceedings 1993. ORP, vol. 1993, pp. 485–496. Springer, Heidelberg (1994). https://doi.org/10.1007/978-3-642-78910-6_160

Gamrath, G., Koch, T., Maher, S.J., Rehfeldt, D., Shinano, Y.: SCIP-Jack—a solver for STP and variants with parallelization extensions. Math. Program. Comput. **9**(2), 231–296 (2017)

Goemans, M.X., Myung, Y.-S.: A catalog of Steiner tree formulations. Networks **23**(1), 19–28 (1993)

Hill, A., Schwarze, S.: Exact algorithms for bi-objective ring tree problems with reliability measures. Comput. Oper. Res. **94**, 38–51 (2018)

Johnson, D.S., Minkoff, M., Phillips, S.: The prize collecting Steiner tree problem: theory and practice. In: Proceedings of the Eleventh Annual ACM-SIAM Symposium on Discrete Algorithms, pp. 760–769. Society for Industrial and Applied Mathematics (2000)

Karp, R.M.: Reducibility among combinatorial problems. In: Miller, R.E., Thatcher, J.W., Bohlinger, J.D. (eds.) Complexity of Computer Computations. IRSS, pp. 85–103. Springer, Boston (1972)

Koch, T., Martin, A.: Solving Steiner tree problems in graphs to optimality. Networks **32**(3), 207–232 (1998)

Koch, T., Martin, A., Voß, S.: SteinLib: an updated library on steiner tree problems in graphs. In: Cheng, X.Z., Du, D.Z. (eds.) Steiner Trees in Industry. Combinatorial Optimization, vol. 11, pp. 285–325. Springer, Boston (2001). https://doi.org/10.1007/978-1-4613-0255-1_9

Leggieri, V., Haouari, M., Triki, C.: The Steiner tree problem with delays: a compact formulation and reduction procedures. Discrete Appl. Math. **164**, 178–190 (2014)

Polzin, T., Daneshmand, S.V.: A comparison of Steiner tree relaxations. Discrete Appl. Math. **112**(1), 241–261 (2001)

Voß, S.: The Steiner tree problem with hop constraints. Ann. Oper. Res. **86**, 321–345 (1999)

Wong, R.T.: A dual ascent approach for Steiner tree problems on a directed graph. Math. Program. **28**(3), 271–287 (1984)

3D Path Network Planning: Using a Global Optimization Heuristic for Mine Water-Inrush Evacuation

Yi Hong[1]($\boxtimes$), Deying Li[3], Qiang Wu[4], and Hua Xu[2]

[1] School of Information Science and Technology, Beijing Forestry University, Beijing 100083, People's Republic of China
hongyi1003@hotmail.com
[2] Information Engineering College, Beijing Institute of Petrochemical Technology, Beijing 102617, People's Republic of China
[3] School of Information, Renmin University of China, Beijing 100872, People's Republic of China
[4] National Engineering Research Center of Coal Mine Water Hazard Controlling, China University of Mining And Technology, Beijing 100083, People's Republic of China

Abstract. The evacuation planning for mine water-inrush is of great importance for personal and property security, and the research of this field in 3D scenarios can provide intuition vision for the geographic space and contribute to the evacuation plan and implementation. In this paper, based on 3D mine model, we address a multi-objective optimization problem for evacuation path planning in mine water-inrush scenario, namely the *global-optimized multi-path finding* problem, which aims to minimize the global evacuation time-consuming and balance the evacuation loads of the emergency exits. Based on the auxiliary graph transformation, we propose a 3-phase heuristic referred to the classical problem, Minimum Weighted Set Cover. We finally conduct extensive experiments to evaluate the performance of the proposed algorithm, whose results indicate the heuristic outperform the existing alternatives in terms of the utilization as well as timeliness.

Keywords: Mine water-inrush · 3D model ·
Multi-objective optimization · Evacuation path planning

1 Introduction

With the rise of various hazards and risk raised by the global technology development, disaster avoiding and evacuation planning have drawn much attention due to its huge potential applications. The research on evacuation planning primarily contains risk management [1], escape routes planning [2,3], emotion mechanism of escape behavior [4,5] and escape virtual simulation [6,7].

Escape routes/paths planning is one of the most important problems in evacuation planning for accidents, which is to compute the optimal evacuation paths to minimize the escaping time. The problem in underground mine scenarios has

© Springer Nature Switzerland AG 2019
D.-Z. Du et al. (Eds.): COCOON 2019, LNCS 11653, pp. 279–290, 2019.
https://doi.org/10.1007/978-3-030-26176-4_23

been paid much attention all these years which focused on the cases of fire and explosion [8,9]. For the escape routes finding problem of coal mines in water-inrush case, the existing research concentrated on 2D mine environment and can be classified into two kinds by the pair of number of the involved evacuators and that of the emergency exits: one is *one-source to one-destination model* (finding the escape route from one evacuator to one exit) and the other one is *one-source to multi-destination* (constructing multi-path from one evacuator to several exits). Both of the two kinds has been effectively solved based on the classical algorithms for solving the shortest path problem, the Dijkstra algorithm [10,11], the Floyd-Warshall algorithm [12], the first k shortest paths [13–15], and dynamic programming [16]. For *one-source to one-destination*, in [13,14], the authors studied the first k shortest paths problem whose goal is to find another $k − 1$ suboptimal paths such that there is an alternative when the shortest one is impassable because of the water spread or congested on account of traffic load. Furthermore, many efforts are made to investigate on *one-source to multi-destination* in terms of modified or hybrid classical algorithms. In [15], the authors introduced a new variant of k shortest paths problem in a time-schedule network with constraints on arcs and they solved the problem based on the modified Dijkstra algorithm to find shortest paths and enumerate all paths. In [16], the authors considered a path networks with vulnerable links and they proposed a framework based on dynamic programming based on known link disruption probabilities and knowledge of transition probabilities for recovering.

It can be found that a 2D mine laneway model which cannot provide intuition vision for the geographic space and may effect the guideline of evacuation work. Furthermore, in the practical evacuation process, the mine personnel with a certain number locate in different places which may create the congestion on the feasible exits. But the above existing related literatures ignored the critical issue or cannot formally solve it in the problem. In this paper, based on a **3D geological mine model**, we introduce an evacuation routes planning problem for the scenarios with a set of mining personnel and a set of feasible escaping exits (**multi-source to multi-destination**), with the consideration of **the traffic load/utilization of the exits** in the solution. The objective of the problem is to find multi-path connecting each personnel position and its most suitable exit for the goal of global optimization. The list of our contributions is as follows.

(i) We introduce a new evacuation path planning problem in 3D mine water-inrush, the *global-optimized multi-path finding* (GMF) problem, and prove its NP-hardness.

(ii) We propose a global-optimized strategy to determine the evacuation paths starting at the personnel initial location such that the time consumed by the latest escaped mine worker is minimized and the escaping load of each exit can be balanced.

(iii) We conduct the simulations and evaluate the performance of the proposed algorithm in terms of utilization and timeliness.

The rest of the paper is organized as follows. Section 2 analyzes modeling of 3D mine laneway model and formulates problem definitions and its hardness proof. Section 3 introduces the framework of the 3-phase solution for GMF problem and describes the related algorithms. Simulation results and corresponding discussions are given in Sect. 4. Section 5 concludes this paper.

2 Modeling of Mine Network Using the Graph Concept and Problem Formulation

2.1 The Spatial Hierarchy Analysis of Mine Network

In this paper, we consider a 3D laneway model in mine water-inrush scenarios, $G = (V, E)$, where $V = V(G) = \{v_1, \cdots, v_n\}$ is the set of predetermined observation vertices, $E = E(G)$ is the set of bidirectional edges, and for each $v_i \in V$, $v_i = (ID_i, x_i, y_i, z_i)$ and for each $e_k \in E$, $e_k = (ID_i, ID_j)$ $(1 \leq i < j \leq n)$. For any vertex subset $V' \subseteq V$, $G[V']$ is the subgraph of G induced by V'. Similarly, $G[E']$ is the subgraph of G induced by an edge subset $E' \subseteq E$. Important notations in the vertex set of 3D connected graph G are as follows: (**a.**) **Water-bursting nodes** where water-inrush occurred; (**b.**) **Source nodes** where the personnel located at the water-inrush moment and will start to escape from; (**c.**) **Destination nodes in mine laneway** which are safe for personnel evacuation, e.g. the pithead, the throat of a mine, or mine refuge; (**d.**) **Impassable nodes** which cannot be passable for evacuation, i.e. silting-up obstacles or submerged objects raised by water-inrush. Note that the obstacles considered here are a part and do not represent all possible types.

In practice of mining, the laneways are constructed in system structure, i.e. there are several relatively complete and independent components, which are relevant to each other via bend channels (similar to the corridors in a building). Such a mining sub-element of laneways is deployed in certain range of height, e.g., $(-700\,\text{m}, -500\,\text{m})$. Thus a 3D laneway model G can be regarded as an equivalent multi-layer 3D model via decomposing G into several relatively independent subgraph $G_1, G_2, ..., G_L$ based on the spatial geometry speciality, i.e., G_1 is composed of the vertices with z-axis value in the range of $(-600\,\text{m}, -500\,\text{m})$ and those in the range of $(-700\,\text{m}, -600\,\text{m})$ belong to G_2. Thus a given 3D laneway model G can be reformulated into $G = \{< G_1, G_2, E_{1,2} >, < G_2, G_3, E_{2,3} >, ..., < G_{L-1}, G_L, E_{L-1,L} >\}$. The endpoints in all $E_{l-1,l}$ sets $(2 \leq l \leq L)$ are called as **turning points**.

2.2 The Traffic Capacity Analysis of Mine Network

When the water-inrush happens, the laneway is the main carrier of the trapped workers. And the traffic capacity of the laneway is affected by the following primary influence factors: Laneway types related to the section shape, slope (β_1), possibility influenced by the obstacles and the vehicle of the escape persons(β_2), water height (β_3), and geometric length $(length(e_k))$. Here the other influence factors like wind velocity are assumed to be negligible. Then the traffic capacity of laneway can be expressed in term of the equivalent length/edge weight and the calculation of the edge weight is according to the equation: $weight(e_k) = length(e_k) \cdot \beta_1(e_k)\beta_2(e_k)\beta_3(e_k), 1 \leq k \leq |E|$. Thus the mine laneway can be further modeled as a 3D connected and edge-weighted graph $G = (V, E, W)$, and G_l can be rewrote as $G_l = (V_l, E_l, W_l)$, where $1 \leq l \leq L$.

2.3   Problem Definitions and Hardness Results

In this paper, we consider the global evacuation planning problem in mine water-inrush, which is the overall construction of evacuation-paths for all the personnel in mine model. The problem is investigated from two points of view: the one is for each mine worker, to decide its globally ideal escape exit and the escaping path; the other one is for each escape exit, to assign relatively balanced escape count, i.e. to balance the loads of all the exits in order to take full advantage of them. Our problem in mine water-inrush evacuation is defined as follows.

Definition 1 (Global-optimized Multi-path Finding (GMF) Problem).
Given a 3D mine laneway model, a 3D connected and edge-weighted graph $G = (V, E, W)$, and two vertex subsets, $S \subseteq V$ is the set of source nodes (personnel locations), $D \subseteq V$ is the set of destination nodes (escape exits), the global-optimized multi-path finding problem is to find the evacuation paths from nodes in S to the nodes in D such that the whole evacuation delay can be minimized and the evacuation efficiency can be maximized.

Here the *whole evacuation delay* is defined as the elapsed time consumed by the latest trapped worker. And the *evacuation efficiency* is for the evacuation exits (the nodes in D) and is represented by the balance performance among the personnel evacuation to all the exits, i.e. maximizing the evacuation efficiency is equivalent to minimizing the maximal difference between the average personnel evacuation times consumed by the mine workers to any pair of exits.

Based on two important and classical NP-hard problems, **Set Cover** and **Minimum Weighted Set Cover** problems, the hardness proof of our problem is given as follows.

Theorem 1. *GMF Problem in Mine Water-inrush Evacuation is NP-hard.*

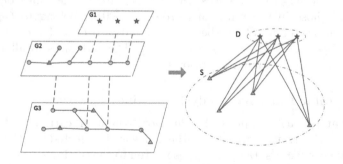

Fig. 1. Bipartite graph transformation of laneway model

Proof. To proof the hardness of GMF Problem, we consider a special case of it: the lengths of the paths between all pair of a root and a leaf are equivalent, $\max_{\forall d_i \in D, \forall s_i \in S} length(d_i, s_j)$, is the same value.

Based on the mathematical formulation of our problem, the forest $\mathcal{P}$ to be found has to meet the three conditions except (iii) in this case for the reason that (iii) is a default satisfying condition, i.e. the path lengths between all pair of roots and leaves are equal. Thus via transforming these isometric paths into one-hop edges as shown in Fig. 1, for each node $d_i \in D$, d_i can be used to represent accessible nodes in S, i.e. $S(d_i) \subseteq S$. Thus the problem becomes to finding a conditional (condition (iv)) set cover in the bipartite graph $G^*[D, S]$, i.e. finding a sub-collection $\mathcal{C} \subseteq \mathcal{F}$ (here $\mathcal{F} = \{S(d_1), S(d_2), ..., S(d_{|D|})\}$) such that $\bigcup_{d_i \in D} S(d_i) = S$.

This special version of Global-optimized Multi-path Finding Problem is equivalent to Minimum Weighted Set Cover Problem, which is proven to be NP-hard [17]. Therefore, GMF Problem is NP-hard in general.   □

3   Global-Optimized Evacuation Paths Algorithm

In this section, we propose the global-optimized algorithm for GMF Problem, which performs three distinct phases to compute evacuation paths from multiple sources to their most suitable destinations: **(1.)** Transform $G = (V, E, W)$ into a new 2D graph $G^* = (V^*, E^*, W^*)$ via extraction process; **(2.)** Find a forest $\mathcal{P}^*$ in G^* with the root set of $\mathcal{P}$ is a subset of D^* and the leaf set of $\mathcal{P}^*$ is S^* such that the whole path length between a root and a leaf can be minimized and the leaf numbers of roots can be balanced; **(3.)** Recover the evacuation network/forest $\mathcal{P}$ in G based on $\mathcal{P}^*$.

3.1   Auxiliary Graph G^* Induction

In the first phase, we construct a 2D logical graph G^* for the election process of shortest paths from the original graph G, whose details are as follows.

(i) The vertex set V^*: For each $v_i \in V$, $v_i = (ID_i, x_i, y_i, z_i)$, a corresponding node is generated in V^*, i.e. $V^* = \{v_i = (ID_i)|\forall v_i \in V, 1 \le i \le n\}$. Note that the new node in V^* need not change any properties of the original node v_i, thus we continue to adopt v_i denote the new node. The same procedure may be easily adapted to obtain D^* and S^*.

(ii) The edge set and edge-weight set E^* and W^*: For each $e_k \in E$, $e_k = (ID_i, ID_j)$, a corresponding edge is generated in E^{temp} and the edge-weight is remained, i.e. $E^{temp} = \{e_k = (ID_i, ID_j)|\forall e_k \in E, 1 \le k \le |E|\}$ and $W^{temp} = \{weight(e_k)|\forall e_k \in E^{temp}, 1 \le k \le |E^{temp}|\}$.

Based on the temporary graph $G^{temp} = (V^*, E^{temp}, W^{temp})$ and D^*, S^*, we operation an **extraction process** to obtain the set of candidate shortest paths P and E^*, W^*. For each pair (d_i, s_j), where $\forall d_i \in D^*$ and $\forall s_j \in S^*$, we find a candidate shortest path, which will be joined into P and all the edges on it (and their corresponding weights) constitute E^* (and W^*). The finding process is stated as follows:

a. Locate d_i, s_j's **layer numbers:** $d_i \in G_{l_i}^{temp}$ and $s_j \in G_{l_j}^{temp}$. Here according to the properties of the laneway structure, the assumption that d_i's location is not lower than the position of s_j (i.e. $l_i \leq l_j$) can be established.

b. If $l_i = l_j$: Dijkstra Algorithm can be executed on $(G_{l_i}^{temp}, d_i, s_j)$ to obtain $SP(d_i, s_j)$.

c. If $l_i < l_j$: Firstly determine the turning points in E_{l_j-1,l_j}^{temp} for s_j in its own layer$G_{l_j}^{temp}$; Secondly calculate the shortest paths between s_j to these turning points and choose the shortest one as the segment of $SP(d_i, s_j)$; Thirdly record the destination of the shortest segment as t_j, and utilize the similar way to find the second segment of $SP(d_i, s_j)$ in $G_{l_j-1}^{temp}$. Repeat the process until the last segment with the destination d_i is found in the layer $G_{l_i}^{temp}$.

3.2 MWSC-based Strategy to Construct Global Optimized Multi-path in G^*

This phase is based on MWSC algorithm and computing global optimized multi-path in G^* between the nodes in D^* and the nodes in S^*, as shown in Algorithm 1. In the phase, we firstly sort all the paths of P in the non-decreasing order of their lengths to obtain an ordered path set $P_O = \{path_1, path_2, \cdots, path_q\}$ as shown in step 2 in Algorithm 1. According to the ordered path lengths in P_O, we construct a global optimal forest/multi-path to minimize the longest shortest path in the solved forest, which is a binary search process as shown by steps 3–13. The binary search is starting from $Mid = \lfloor \frac{1+q}{2} \rfloor$, and in each iteration, we temporarily remove all the paths $path \in P$ whose length satisfies $length(path_{Mid}) \leq length(path)$, and check if the subgraph of G^* induced by P without those edges can still construct a MWSC which is verified by function Greedy-Forest$(G^*[P], D^*, S^*)$ (steps 15–27 in Algorithm 1). If a MWSC can be constructed, we update P and obtain a MWSC $Forest$ on the subgraph of G^* induced by the updated P, and proceed with the decreased Mid. Otherwise, we keep those edges in P and proceed with the increased Mid.

To balance the leaf numbers among all the roots of the solved forest, the criteria of greedy selection of Greedy-Forest$(G^*[P], D^*, S^*)$ is the difference between the leaf number of each root and the average leaf number of all the roots $avrNL$. Note that in G^*, each node d_i in D^* can be regarded as a subset of S^*, which is composed of its reachable nodes in S^*.

3.3 Paths Reduction for G from the Multi-path in G^*

Based on the forest $\mathcal{P}^*$ in G^* with the root set D^* and the leaf set S^*, we transfer $\mathcal{P}^*$ into an equivalent forest $\mathcal{P}$ from D to S in original graph G in this phase. In the logical graph, the association information for vertex set V has been restored and the coordinate information for V has been masked, which has not changed any association properties of the original node. Thus the phase is to restore the masked information of forest $\mathcal{P}^*$ for the original graph G as follows.

Algorithm 1. Global Optimized Multi-path Computation Algorithm (G^*, P, D^*, S^*)

1: Set $\mathcal{P}^* \leftarrow \emptyset$, $avrNL(= \frac{\sum_{d_i \in D^*} numofleaf(d_i)}{|D^*|})$, $Min = Max = Mid = 0$.
2: Sort the paths in P according to their weights in the non-decreasing order and store the order in $P_O = \{path_1, path_2, \cdots, path_q\}$, and $Min = 1$, $Max = q$
3: **while** $Min \leq Max$ **do**
4:     $Mid = \lfloor \frac{Min+Max}{2} \rfloor$, $P_{temp} = \{path | path \in P \text{ and } length(path_{Mid}) \leq length(path)\}$
5:     $P \leftarrow P \setminus P_{temp}$, $d_i \leftarrow d_i \setminus \{s_j \in P_{temp}\}$ $(\forall d_i \in P_{temp})$, $D^* \leftarrow D^* \setminus \{d_i | \forall d_i \in D^* \text{ and } |d_i| = 0\}$
6:     Apply the greedy forest construction function **Greedy-Forest**$(G^*[P], D^*, S^*)$ to find whether a forest can be constructed in $G^*[P]$ rooted at D^* and with leaves in S^*
7:     **if** **Greedy-Forest**$(G^*[P], D^*, S^*)$ returns **False then**
8:         $Min = Mid$
9:         $P \leftarrow P \bigcup P_{temp}$, $d_i \leftarrow d_i \bigcup \{s_j \in P_{temp}\}$ $(\forall d_i \in P_{temp})$, $D^* \leftarrow D^* \bigcup \{d_i | \forall d_i \in D^* \cap P_{temp}\}$
10:     **else**
11:         $Max = Mid$, $\mathcal{P}^* = $ **Greedy-Forest**$(G^*[P], D^*, S^*)$
12:     **end if**
13: **end while**
14: Return $\mathcal{P}^*$.
15: **function** GREEDY-FOREST$((G^*[P], D^*, S^*))$
16:     $\mathcal{S} \leftarrow S^*$, $\mathcal{D} \leftarrow D^*$, $Forest \leftarrow \emptyset$
17:     **while** $\mathcal{D} \neq \emptyset$ **do**
18:         Select a element $d_i \in \mathcal{D}$ that minimizes $|\frac{numofleaf(d_i)-avrNL}{|d_i \cap \mathcal{S}|}|$
19:         **if** $numofleaf(d_i) > avrNL$ **then**
20:             $d_i \leftarrow \{$The first $avrNL$ nearest leaf nodes in $d_i\}$
21:         **end if**
22:         $d_i \leftarrow d_i \cap \mathcal{S}$, $\mathcal{D} \leftarrow \mathcal{D} - \{d_i\}$, $\mathcal{S} \leftarrow \mathcal{S} - d_i$, $Forest \leftarrow Forest \bigcup \{d_i\}$
23:     **end while**
24:     **if** $\mathcal{S} \neq \emptyset$ **then**   Return False
25:     **else**   Return $Forest$
26:     **end if**
27: **end function**

(i) The vertex set of forest $\mathcal{P}$, $V[\mathcal{P}]$: For each $v_i = (ID_i) \in V[\mathcal{P}^*]$, $V[\mathcal{P}] \leftarrow V[\mathcal{P}] \bigcup \{v_i = (ID_i, x_i, y_i, z_i)\}$. The same restoring procedure is adapted to obtain D and S as well.

(ii) The edge set of $\mathcal{P}$, $E[\mathcal{P}]$: For each $e_k \in E[\mathcal{P}^*]$, $E[\mathcal{P}] \leftarrow E[\mathcal{P}] \bigcup \{e_k\}$.

4 Simulation and Discussion

4.1 Simulation Plan

In this simulation, we will investigate the influence of the following two important parameters on our algorithm: the number of source nodes, m and the number of impassable nodes, o. In particular, we will consider two group of setting for each performance factor: (i) o is fixed as 25 and m varies from 10 to 40 by the step of 5; (ii) m is fixed as 25 and o varies from 10 to 40 by the step of 5. For each parameter setting, we run 50 instances and compute their average for evaluation.

Since the problem is NP-hard, it is unlikely for us to compare an output of the algorithms with an optimal solution. Instead, we employ the *theoretical bound* of the problem for comparison, which is the performance factor generated by the shortest path(SP) algorithm for each source nodes. And we study the average characteristic behavior of the proposed algorithm and evaluate their average performance in terms of *the utilization factor, the average length factor* and *the global length factor*.

(i) *The utilization factor* reflects the utilizing balance of each destination, which can be calculated as dividing the sum of the difference between the number of the escaping persons of each escape exit and the average escaping number of all the escape exits by the number of effective destinations, i.e. $\frac{\sum_{d_i \in D^*} |numofleaf(d_i) - avrNL|}{|D^*|}$. For a path-finding algorithm, the lower the utilization factor, the better the performance of the path-finding algorithm.

(ii) *The average length factor* stands for the whole shortest property of the planned escaping paths, which is valued as the result of dividing the total sum of all the differences between each planned path and the shortest path with the same source and destination by the number of escaping paths, i.e. $\frac{\sum_{\forall d_i \in D, \forall s_j \in S} (length(d_i, s_j) - SP(d_i, s_j))}{\sum_{\forall d_i \in D, \forall s_j \in S} SP(d_i, s_j)}$, where p is the number of obtained paths by our algorithm, $length(d_i, s_j)$ is the path length between d_i and s_j and $SP(d_i, s_j)$ is the length of shortest path between them.

(iii) *The global length factor* stands for the global shortest property of the planned escaping paths, which is valued as the result of the difference between the length of the longest path generated by our algorithm and the length of the longest one among the shortest paths, i.e. $\frac{|\max_{\forall d_i \in D, \forall s_j \in S} length(d_i, s_j) - \max_{\forall d_i \in D, \forall s_j \in S} SP(d_i, s_j)|}{\max_{\forall d_i \in D, \forall s_j \in S} SP(d_i, s_j)}$. For the SP algorithm, the average and global length factors are both 0 which is the lower bound of all the path-finding algorithms.

4.2 Performance of Proposed Algorithms for GMF Problem

Firstly, we focus on the impact of the number of source nodes, m and the number of impassable nodes, o on the *the utilization factor* in Fig. 2. Overall in the two figures, the utilization factors of our algorithm with the variation of m and o are both obviously lower than those of the SP algorithm. In Fig. 2(a), we can

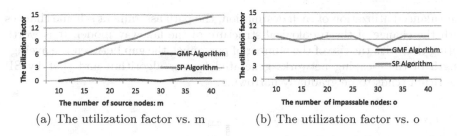

(a) The utilization factor vs. m   (b) The utilization factor vs. o

Fig. 2. Performance on the utilization factor

observe that with the increasing of m, the factor of our algorithm stays below that of the SP algorithm and its fluctuation is within the stable and tolerable range, i.e. $[0, 0.7]$. Furthermore, we also find that the number of personnel has no appreciable impact on the utilization factor of our algorithm, while the poorer performance on the utilization factor of the SP algorithm is more apparent with the growth of the personnel number. Next, we study how the utilization factor is affected by the number of impassable nodes o. As shown in Fig. 2(b), we can learn that the factor of our algorithm has no obvious change as o grows, while the factor of the SP algorithm presents obvious ups and downs with the growth of o. The reason is that the variation of the number of impassable nodes has direct influence on the topological structure of the laneway model, and the utilization factor gained from our algorithm can be guaranteed into a reasonable range for what topological structure the laneway model has.

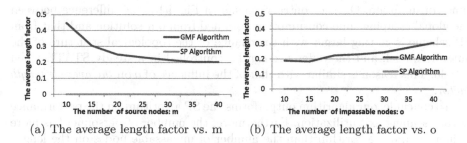

(a) The average length factor vs. m   (b) The average length factor vs. o

Fig. 3. Performance on the average length factor

Secondly, we investigate the impact of the number of source nodes, m and the number of impassable nodes, o on the *the average length factor*. From Fig. 3(a) and (b), we can observe that the number of source nodes m has more influence on this factor of our solutions than the number of impassable nodes o. Reasons are as follow: Firstly, the increasing number of obstacles or collapse may lead to the impassability of initial feasible segments. Thus it may raises that the passable segments become longer to some extent, and then it causes the average length of the generated paths becomes larger as well. Secondly, due to the consideration of the balanced utilization of each destination, the path assignment scheme changes with the change of the topological structure of the laneway model. It is worth noting that with the rise of the number of personnel, the gap between the average length of the escaping networks generated by our algorithm and that by SP algorithm tends to be gradually narrowing.

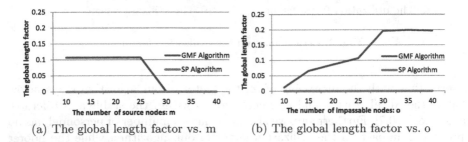

(a) The global length factor vs. m   (b) The global length factor vs. o

Fig. 4. Performance on the global length factor

Thirdly, comparing with the variation trend of *the average length factor*, the trend of *the global length factor* becomes more apparent as m and o increases. Seen from Fig. 4(a), the difference between the global escaping time-consumption obtained from our solution and that from SP scheme is getting smaller with the rising of the personnel number. And when m is larger than 30, the difference is getting 0 which indicates our strategy can enhance the utilization efficiency of the escaping exits with the minimization guarantee of the global escaping time-consumption. On the contrary, as shown Fig. 4(b), the difference between the global escaping time-consumption obtained from our solution and that from SP scheme is getting larger with the change of the topological structure of the laneway model, and the difference can be controlled in a steady STATE when o ≥ 30. This reason is similar with that of the influence of o on *the average length factor*.

To conclude, our algorithm outperforms the SP algorithm on the performance of the escaping exit utilization. Furthermore, the number of personnel has more influence on our algorithm than the number of impassable nodes on the length factor. Although there exists a gap on the global and average path length of the generated escaping network between our algorithm and the SP algorithm, with the rise of the number of personnel, the gap is tending to be gradually narrowing. Therefore, our strategy can enhance the utilization efficiency of the escaping exits and guarantee a tolerable range of the global escaping time-consumption.

5   Conclusions

In this paper, we have proposed a new problem the *Global-optimized Multi-path Finding (GMF) Problem in Mine Water-inrush Evacuation*, which has two optimization objects, maximizing exit utilization and minimizing the global escaping delay. Based on the classical problem, Minimum Weighted Set Cover (MWSC) Problem, we proposed a 3-phase heuristic algorithm for GMF problem. Based on our simulation results, our algorithm outperforms the shortest path algorithm on one of the most important performance factors, the utilization efficiency of the escaping exits. And our solution can guarantee that the deviation of the global escaping delay from the shortest delay is in a reasonable range.

Future Works. Despite our through studies conducted in this paper, we believe there are lots of room for further investigation. In this paper, we consider one of the multi-objective optimization problems on mine emergency application and we can further take other multi-objective evacuation plan problems, i.e. maximizing the utilization of the prioritized exits' utilization and minimizing the whole escaping delay. In addition, we plan to investigate intelligent evacuation multi-path plan problems in more application scenarios.

Acknowledgments. This research was supported in part by General Project of Science and Technology Plan of Beijing Municipal Education Commission (KM201910017006), Program of Beijing Excellent Talents Training for Young Scholar (2016000020124G056). Prof. Wu and Prof. Xu were supported in part by China National Scientific and Technical Support Program (2016YFC0801801), China National Natural Science Foundation (41430318, 41272276, 41572222, 41602262), Beijing Natural Science Foundation (8162036) and STATE Key Laboratory of Coal Resources and Safe Mining.

References

1. Faisal, K., Samith, R., Salim, A.: Methods and models in process safety and risk management: past, present and future. Process Saf. Environ. Prot. **98**, 116–147 (2015)
2. Özdamar, L., Demir, O.: A hierarchical clustering and routing procedure for large scale disaster relief logistics planning. Transp. Res. Part E Logistics Transp. Rev. **48**(3), 591–602 (2012)
3. Lahmar, M., Assavapokee, T., Ardekani, S.A.: A dynamic transportation planning support system for hurricane evacuation. In: Intelligent Transportation Systems Conference (ITSC 2006), pp. 612–617 (2006)
4. Wang, L., Wang, Y., Cao, Q., Li, X., Li, J., Wu, X.: A framework for human error risk analysis of coal mine emergency evacuation in China. J. Loss Prev. Process Ind. **30**, 113–123 (2014)
5. Radianti, J., Granmo, O., Sarshar, P., Goodwin, M., Dugdale, J., Gonzalez, J.J.: A spatio-temporal probabilistic model of hazard- and crowd dynamics for evacuation planning in disasters. Appl. Intell. **42**(1), 3–23 (2015)
6. Ji, J., Zhang, J., Chen, J., Wu, S.: Computer simulation of evacuation in underground coal mines. Min. Sci. Technol. **20**, 0677–0681 (2010)
7. Adjiski, V., Mirakovski, D., Despodov, Z., Mijalkovski, S.: Simulation and optimization of evacuation routes in case of fire in underground mines. J. Sustain. Min. **14**(3), 1–11 (2015)
8. Zhang, S., Wu, Z., Zhang, R., Kang, J.: Dynamic numerical simulation of coal mine fire for escape capsule installation. Saf. Sci. **50**, 600–606 (2012)

9. Goodwin, M., Granmo, O., Radianti, J.: Escape planning in realistic fire scenarios with Ant Colony Optimisation. Appl. Intell. **42**, 24–35 (2015)
10. Evans, J., Minieka, E.: Optimization Algorithms for Networks and Graphs, 2nd edn. Marcel Dekker, New York (1992)
11. Cherkassky, B.V., Goldberg, A.V., Radzik, T.: Shortest paths algorithms: theory and experimental evaluation. Math. Program. **73**(2), 129–174 (1996)
12. Jalali, S.E., Noroozi, M.: Determination of the optimal escape routes of underground mine networks in emergency cases. Saf. Sci. **47**, 1077–1082 (2009)
13. Martins, E., Pascoal, M., Santos, J.: A new algorithm for ranking loopless paths. Technical report, Univ. de Coimbra (1997)
14. Eppstein, D.: Finding the k shortest paths. SIAM J. Comput. **28**(2), 652–673 (1998)
15. Jin, W., Chen, S., Jiang, H.: Finding the K shortest paths in a time-schedule network with constraints on arcs. Comput. Oper. Res. **40**, 2975–2982 (2013)
16. Sever, D., Dellaert, N., Woensel, T., Kok, T.: Dynamic shortest path problems: hybrid routing policies considering network disruptions. Comput. Oper. Res. **40**(2013), 2852–2863 (2013)
17. Garey, M.R., Johnson, D.S.: Strong NP-completeness results: motivation, examples, and implications. J. ACM **25**, 499–508 (1978)

Max-Min 3-Dispersion Problems

Takashi Horiyama[1], Shin-ichi Nakano[2(✉)], Toshiki Saitoh[3], Koki Suetsugu[4],
Akira Suzuki[5], Ryuhei Uehara[6], Takeaki Uno[7], and Kunihiro Wasa[7]

[1] Saitama University, Saitama, Japan
[2] Gunma University, Maebashi, Japan
nakano@cs.gunma-u.ac.jp
[3] Kyushu Institute of Technology, Kitakyushu, Japan
[4] Kyoto University, Kyoto, Japan
[5] Tohoku University, Sendai, Japan
[6] JAIST, Nomi, Japan
[7] National Institute of Informatics, Tokyo, Japan

Abstract. Given a set P of n points on which facilities can be placed
and an integer k, we want to place k facilities on some points so that the
minimum distance between facilities is maximized. The problem is called
the k-dispersion problem. In this paper we consider the 3-dispersion problem when P is a set of points on a plane. Note that the 2-dispersion
problem corresponds to the diameter problem. We give an $O(n)$ time
algorithm to solve the 3-dispersion problem in the L_∞ metric, and an
$O(n)$ time algorithm to solve the 3-dispersion problem in the L_1 metric. Also we give an $O(n^2 \log n)$ time algorithm to solve the 3-dispersion
problem in the L_2 metric.

Keywords: Dispersion problem · Facility location

1 Introduction

The facility location problem and many of its variants have been studied [9,10].
Typically, given a set of points on which facilities can be placed and an integer
k, we want to place k facilities on some points so that a designated function on
distance is minimized. By contrast in the *dispersion problem*, we want to place
facilities so that a designated function on distance is maximized.

The intuition of the problem is as follows. Assume that we are planning to
open several chain stores in a city. We wish to locate the stores mutually far
away from each other to avoid self-competition. So we wish to find k points so
that the minimum distance between them is maximized. See more applications,
including *result diversification*, in [7,17,18].

Now we define the *max-min k-dispersion problem*. Given a set P of n possible
points, a distance function d for each pair of points (we assume that d is a
symmetric nonnegative function satisfying $d(p,p) = 0$ for all $p \in P$), and an
integer k with $k \ll n$, we wish to find a subset $S \subset P$ with $|S| = k$ such that

D.-Z. Du et al. (Eds.): COCOON 2019, LNCS 11653, pp. 291–300, 2019.
https://doi.org/10.1007/978-3-030-26176-4_24

the cost $cost(S) = \min_{\{u,v\} \subset S}\{d(u,v)\}$ is maximized. Such a set S is called a k-dispersion of P. This is the max-min version of the k-dispersion problem [17,19]. For the max-sum version see [4–8,12,15,17], and for a variety of related problems see [4,8]. The max-min k-dispersion problem is NP-hard even when the triangle inequality is satisfied [11,19]. An exponential time exact algorithm for the max-min k-dispersion problem is known [2]. The running time is $O(n^{\omega k/3} \log n)$, where $\omega < 2.373$ is the matrix multiplication exponent.

A geometric version of the problem in D-dimensional space can be solved in $O(kn)$ time for $D = 1$ (if the order of points in P on the line is given) and is NP-hard for $D = 2$ [19]. The running time for $D = 1$ was improved to $O(n \log \log n)$ [3] (if the order of points in P on the line is given) by the sorted matrix search method [13] (see a good survey for the sorted matrix search method in [1, Section 3.3]), then $O(n)$ [2] by a reduction to the path partitioning problem [13]. Ravi et al. [17] proved that the max-min k-dispersion problem cannot be approximated within any constant factor in polynomial time, and cannot be approximated within a factor of two in polynomial time when the distance satisfies the triangle inequality, unless P = NP. They also gave a polynomial-time algorithm with approximation ratio two when the triangle inequality is satisfied.

In this paper we consider the case $k = 3$, namely the max-min 3-dispersion problem, when a set P of n points lie on a plane. Note that the 2-dispersion of P corresponds to the diameter of P, and one can compute it in $O(n \log n)$ time [14].

We first study the case where d is the L_∞ metric. We give an algorithm to compute the 3-dispersion of P in $O(n)$ time.

Then we study the case where d is the L_1 metric. We show that a similar algorithm can compute the 3-dispersion of P in $O(n)$ time.

Finally we study the case where d is the L_2 metric. We give an algorithm to compute the 3-dispersion of P in $O(n^2 \log n)$ time. By slightly improving the algorithm we can also compute the 3-dispersion of P in dimension D in $O(Dn^2 + Tn \log n)$ time where T is the time to compute the diameter of n points in dimension D.

The remainder of this paper is organized as follows. Section 2 gives an $O(n)$ time algorithm to solve the 3-dispersion problem if d is the L_∞ metric. Section 3 gives an $O(n)$ time algorithm to solve the 3-dispersion problem if d is the L_1 metric. Section 4 gives an $O(n^2 \log n)$ time algorithm to solve the 3-dispersion problem if d is the L_2 metric. Finally Sect. 5 is a conclusion.

2   3-Dispersion in L_∞ Metric

In this section we give an $O(n)$ time algorithm to solve the 3-dispersion problem if P is a set of n points on a plane and d is the L_∞ metric.

Let $P = \{p_1, p_2, \cdots, p_n\}$ and assume $x(p_1) \geq x(p_2) \geq \cdots \geq x(p_n)$. Let $S = \{p_a, p_b, p_c\}$ be a 3-dispersion of P. We say that a pair (p_u, p_v) in P is type-H if $d(p_u, p_v) = |x(p_u) - x(p_v)|$, and type-V otherwise. We have the following four cases for S. Let $E = \{(p_a, p_b), (p_b, p_c), (p_c, p_a)\}$.

Case 1: All three pairs in E are *type-H*.

Case 2: Two pairs in E are *type-H* and one pair in E is *type-V*.

Case 3: Two pairs in E are *type-V* and one pair in E is *type-H*.

Case 4: All three pairs in E are *type-V*.

Our algorithm computes three points having the maximum cost for each case, then chooses the maximum one among those four solutions, as a 3-dispersion. Now we consider how to compute a 3-dispersion S restricted for each case.

Case 1: All three pairs in E are *type-H*.

The solution consists of (1) the leftmost point p_n in P, (2) the rightmost point p_1 in P, and (3) the point p_m in P which is the closest points to the midpoint between p_1 and p_n.

One can find p_n and p_1 in $O(n)$ time, then find p_m in $O(n)$ time.

Thus we can compute a 3-dispersion $S = \{p_1, p_m, p_n\}$ of P in $O(n)$ time if S is Case 1.

Case 2: Two pairs in E are *type-H* and one pair in E is *type-V*.

Let $S = \{p_a, p_b, p_c\}$ be a 3-dispersion with $x(p_a) \leq x(p_b) \leq x(p_c)$, and assume that S is Case 2.

Either (p_a, p_b) or (p_b, p_c) is *type-V*. Note that if (p_a, p_c) is *type-V* then either (p_a, p_b) or (p_b, p_c) is also *type-V*, a contradiction. Assume (p_b, p_c) is *type-V*. The other case is symmetrical. Let $P_i = \{p_1, p_2, \cdots, p_i\}$ be the subset of P consisting of the rightmost i points in P. We have the following two lemmas.

Lemma 1. *There is a 3-dispersion $S = \{p_a, p_b, p_c\}$ such that p_a is the leftmost point p_n in P.*

Proof. Otherwise let S' be $\{p_n, p_b, p_c\}$, which is derived from S by replacing p_a with the leftmost points p_n in P. Since (p_a, p_b) and (p_a, p_c) are *type-H*, $cost(S') \geq cost(S)$ holds. If $cost(S') = cost(S)$ then the claim is satisfied. If $cost(S') > cost(S)$ then S is not a 3-dispersion, a contradiction. (Note that S' may not be Case 2.)

Lemma 2. *There is a 3-dispersion $S = \{p_a, p_b, p_c\}$ satisfying the following. Let $p_i = p_b$. If $y(p_b) \leq y(p_c)$ then p_b is a lowest point in P_i and p_c is a highest point in P_i. If $y(p_b) > y(p_c)$ then p_b is a highest point in P_i and p_c is a lowest point in P_i.*

Proof. Assume that $y(p_b) \leq y(p_c)$ but p_b is not the lowest point p_ℓ in P_i. Then let S' be $\{p_a, p_\ell, p_c\}$, which is derived from S by replacing p_b with p_ℓ. Since (p_b, p_c) is *type-V*, now $cost(S') \geq cost(S)$ holds. If $cost(S') = cost(S)$ then the claim is satisfied for some $P_{b'}$ with $b' < b$. If $cost(S') > cost(S)$ then S is not a 3-dispersion, a contradiction. (Note that S' may not be Case 2.)

Similar for the other case.

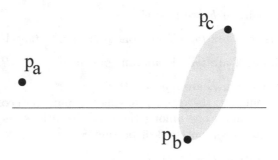

Fig. 1. An illustration for Case 2.

Let h_i be the highest point in P_i and ℓ_i the lowest point in P_i.

If $\{p_a, p_b, p_c\}$ is a 3-dispersion of P, $x(p_a) \leq x(p_b) \leq x(p_c)$, (p_b, p_c) is *type-V* and $\{p_a, p_b, p_c\}$ is Case 2, then $p_a = p_n$, and $\{p_b, p_c\} = \{h_i, \ell_i\}$ for $p_b = p_i$. See Fig. 1.

We first compute $\min\{d(p_a, p_i), d(h_i, \ell_i)\}$ for each i, then choose the maximum one among them satisfying $p_i \in \{h_i, \ell_i\}$. It corresponds to a 3-dispersion $\{p_a, p_b, p_c\}$ of P, since $d(p_a, p_i) = \min\{d(p_a, p_b), d(p_a, p_c)\}$ and $d(h_i, \ell_i) = d(p_b, p_c)$.

We can compute a 3-dispersion $\{p_a, p_b, p_c\}$ by binary search as follows.

First we sort P by x-coordinates in $O(n \log n)$ time.

By scanning P from right to left we can compute the highest point h_i and the lowest point ℓ_i in each P_i with $1 \leq i \leq n$ in $O(n)$ time in total. We also compute $d(p_n, p_i)$ for each i with $0 < i < n$ in $O(n)$ time in total.

Now we compute $\max_i \min\{d(p_a, p_i), d(h_i, \ell_i)\}$. Clearly $d(p_a = p_n, p_i)$ is monotonically decreasing with respect to i and $d(h_i, \ell_i)$ is monotonically increasing with respect to i. Then by binary search we can compute the optimal i with $S = \{p_a = p_n, h_i, \ell_i\}$ having the maximum cost $\min\{d(p_n, p_i), d(h_i, \ell_i)\}$ in $\log n$ stages. Each stage of the binary search requires $O(1)$ time.

Thus we can compute a 3-dispersion S in $O(n \log n)$ time in total if S is Case 2.

Case 3: Two pairs in E are *type-V* and one pair in E is *type-H*.

Similar to Case 2. Swap x and y axes.

Case 4: All three pairs in E are *type-V*.

Similar to Case 1. Swap x and y axes.

We have the following lemma.

Lemma 3. *If P is a set of n points on a plane and d is the L_∞ metric then one can solve the max-min 3-dispersion problem in $O(n \log n)$ time.*

We can improve the running time to $O(n)$ by removing the sort in Cases 2 and 3. The binary search proceeds as follows. In the j-th stage we have a set I of points having consecutive x-coordinates containing optimal p_i and $|I| = n/2^{j-1}$. We find the median $p_{j'}$ in I in $O(n/2^{j-1})$ time by the linear-time median-finding algorithm. Then find the highest point h_j and the lowest point ℓ_j in the right half points of I consisting of $n/2^j$ points in $O(n/2^j)$ time. By the two points h_j and ℓ_j above and the h_i, ℓ_i of a suitable preceding stage i we can compute the $h_{j'}, \ell_{j'}$ of current $P_{j'}$ in constant time. Depending on the result of $d(p_a, p_{j'}) < d(h_{j'}, \ell_{j'})$ we proceed to the next stage with suitable parameters.

We have the following theorem.

Theorem 1. *If P is a set of n points on a plane and d is the L_∞ metric then one can solve the max-min 3-dispersion problem in $O(n)$ time.*

We cannot simply generalize the algorithm to 3-dimension since there is an example in which any 3-dispersion has no points with an extreme coordinate value. Let $P = \{(1,0,0), (0,1,0), (0,0,1), (-1,0,0), (0,-1,0), (0,0,-1), (0.9,-0.9,0), (0,0.9,-0.9), (-0.9,0,0.9)\}$ then the 3-dispersion of P is $\{(0.9,-0.9,0), (0,0.9,-0.9), (-0.9,0,0.9)\}$, and none of which has an extreme coordinate value.

We now give an $O(n \log n)$ time algorithm to solve the 3-dispersion problem in 3-dimension.

We say a pair (p_u, p_v) in P is *type-X* if $d(p_u, p_v) = |x(p_u) - x(p_v)|$, *type-Y* if $d(p_u, p_v) = |y(p_u) - y(p_v)|$ and $|y(p_u) - y(p_v)| > |x(p_u) - x(p_v)|$, *type-Z* otherwise.

Let $P = \{p_1, p_2, \cdots, p_n\}$ be the set of points in 3-dimensional space, and let $S = \{p_a, p_b, p_c\}$ be the 3-dispersion of P.

We have 3^3 cases for S. Let $E = \{(p_a, p_b), (p_b, p_c), (p_c, p_a)\}$.

Case 1: (p_a, p_b) is *type-X*, (p_a, p_c) is *type-Y* and (p_b, p_c) is *type-Z*.

We have eight subcases for S depending on the order of p_a, p_b, p_c on each coordinate.

Case 1(a): $x(p_a) \leq x(p_b)$, $y(p_a) \leq y(p_c)$, and $z(p_b) \leq z(p_c)$. (The other cases are similar so omitted.)

Fix p_a. We wish to compute three points $\{p_a, p_b, p_c\}$ with the maximum d satisfying $\min\{x(p_b) - x(p_a), y(p_c) - y(p_a)\} = d$, and $z(p_c) - z(p_b) \geq d$. Each candidate value for d is the distance between p_a and a point in P, so the number of such values is at most n. By binary search we are going to find the maximum such d. We now need some definitions.

We sort the points with their x-coordinates in $O(n \log n)$ time. Similarly sort the points with their y-coordinates. Assume that $P = \{p_1, p_2, \cdots, p_n\} = \{p_1', p_2', \cdots, p_n'\}$, $x(p_1) \geq x(p_2) \geq \cdots \geq x(p_n)$, and $y(p_1') \geq y(p_2') \geq \cdots \geq y(p_n')$.

Let $B_i = \{p | x(p) \geq x(p_i)\}$ and $C_j = \{p | y(p) \geq y(p_j)\}$. We compute a table T_B as a preprocessing step so that $T_B(i) = \min\{z(p) | p \in B_i\}$. Similarly we compute a table T_C as a preprocessing step so that $T_C(j) = \max\{z(p) | p \in C_j\}$. We need $O(n)$ time for these tables.

We maintain the set Pb of candidates for p_b. Initially we set $Pb = \{p | x(p) > x(p_a)\}$. Similarly we maintain the set Pc of candidates for p_c. Initially we set $Pc = \{p | y(p) > y(p_a)\}$.

The binary search proceeds the following way.

Let $p_{b'}$ be the point in Pb having the median of x-coordinate, and $p'_{c'}$ be the point in Pc having the median of y-coordinate. Let $d' = \min\{x(p_{b'}) - x(p_a), y(p'_{c'}) - y(p_a)\}$. Assume $d' = x(p_{b'}) - x(p_a)$. (The other case is similar.) If $T_C(c') - T_B(b') \geq d'$ then there exists (p_a, p_b, p_c) satisfying (1) $\min\{x(p_b) - x(p_a), y(p_c) - y(p_a)\} \geq d'$, and (2) $z(p_c) - z(p_b) \geq d'$. Now $b' \geq b$ and $c' \geq c$ hold. In this case we can halve the size of the two candidate sets Pb and Pc, respectively. If $T_C(c') - T_B(b') < d'$ then there is no $\{p_a, p_b, p_c\}$ with cost d', so $b' < b$ holds for a 3-dispersion $\{p_a, p_b, p_c\}$, and we can halve the size of the candidate set Pb.

Thus by binary search we can find the maximum d in at most $2 \log n$ stages. In each stage we can compute $T_C(c') - T_B(b')$ in $O(1)$ time.

We need to compute above for each possible p_a. Thus the running time for **Case 1** is $O(n \log n)$.

The other cases are similar so omitted. Since the number of cases is a constant the total running time is $O(n \log n)$.

3   3-Dispersion in L_1

In this section we give an $O(n)$ time algorithm to solve the 3-dispersion problem when P is a set of n points on a plane and d is the L_1 metric.

We consider four coordinate systems each of which is derived from the original coordinate system by rotating 45, 135, 225 or 315° clockwise around the origin. We can observe that there is a 3-dispersion of P containing a point having extreme x-coordinate in one of those four coordinate systems. Note that each coordinate system has two extreme points for x-coordinates. We only explain 45° case with the point having the minimum x-coordinate. Other cases are similar.

Let x' and y' be the coordinates of the rotated coordinate system. Let $P = \{p_1, p_2, \cdots, p_n\}$ and assume $x'(p_1) \geq x'(p_2) \geq \cdots \geq x'(p_n)$. Let $S = \{p_a, p_b, p_c\}$ be the 3-dispersion of P with $x'(p_a) \leq x'(p_b) \leq x'(p_c)$ and $p_a = p_n$ is the point having the minimum x'-coordinate.

We say two points (p_u, p_v) with $x'(p_u) \leq x'(p_v)$ in P are *type-U* (upward) if $y(p_u) \leq y(p_v)$, and *type-D* (downward) otherwise.

We compute the optimal i^*, which is the i with the maximum $\min\{d(p_a, p_i), diam(P_i)\}$, where p_i is the i-th farthest point from p_a in the L_1 metric and $diam(P_i)$ is the diameter of P_i where $P_i = \{p_1, p_2, \cdots p_i\}$ is the subset of P consisting of the i farthest points from $p_n = p_a$ in P. Let p_b and p_c be the points corresponding to the diameter of P_i. See Fig. 2. If p_b and p_c are *type-U* then p_b is the point with the minimum $y'(p_b)$ in P_i and p_c is the point with the maximum $y'(p_b)$ in P_i. If p_b and p_c are *type-D* then p_b is the point p_i with the minimum $x'(p_b)$ in P_i and p_c is the point p_1 with the maximum $x'(p_c)$ in P_i.

Similar to the L_∞ metric case, we can compute a max-min 3-dispersion of P in $O(n)$ time, by binary search with the linear-time median-finding algorithm.

Now we have the following theorem.

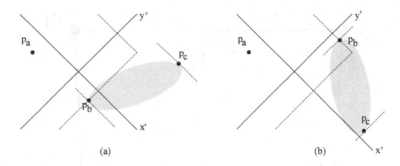

Fig. 2. Illustrations for the diameter of P_i in the L_1 metric with (a) *type-U* and (b) *type-D*.

Theorem 2. *If P is a set of n points on a plane and d is the L_1 metric then one can solve the max-min 3-dispersion problem in $O(n)$ time.*

4   3-Dispersion in L_2 Metric

In this section we design an $O(n^2 \log n)$ time algorithm to solve the 3-dispersion problem when P is a set of n points on a plane and d is the L_2 metric.

Let $S = \{p_a, p_b, p_c\}$ be a 3-dispersion of P, and assume that $d(p_b, p_c)$ is the shortest one among $\{d(p_a, p_b), d(p_b, p_c), d(p_c, p_a)\}$, $d(p_a, p_b) \le d(p_a, p_c)$ and p_b is the i-th farthest point from p_a in P. Let $P_i = \{p_1, p_2, \cdots, p_i\}$ be the subset of P consisting of the i farthest points from p_a. We have the following lemma. Let $diam(P)$ be the diameter of P.

Lemma 4. $d(p_b, p_c) = diam(P_i)$.

Proof. Otherwise there are $p_{b'}, p_{c'} \in P_i$ with $diam(P_i) = d(p_{b'}, p_{c'})$. Let $S' = \{p_a, p_{b'}, p_{c'}\}$. Now $d(p_a, p_b) \le d(p_a, p_{b'})$, $d(p_a, p_b) \le d(p_a, p_{c'})$, and $d(p_b, p_c) < d(p_{b'}, p_{c'})$ hold. Thus $cost(S) < cost(S')$, a contradiction.

Thus if we compute i maximizing $\min\{d(p_a, p_i), diam(P_i)\}$ for each p_a, then choose the maximum one, it corresponds to a 3-dispersion of P.

For a fixed p_a we can compute the optimal i^* with the maximum $\min\{d(p_a, p_{i^*}), diam(P_{i^*})\}$ by binary search, as follows.

Clearly $d(p_a, p_i)$ is monotonically decreasing with respect to i and $diam(P_i)$ is monotonically increasing with respect to i.

First we sort the points in P by the distance from p_a. Then we are going to find the optimal i^*. First set $I = [1, n]$.

In the j-th stage we check for the median i in I consisting of $n/2^{j-1}$ numbers containing i^*, whether $d(p_a, p_i) < diam(P_i)$ or not. For the check we first compute the convex hull C_j of P_i by constructing the convex hull of suitable $n/2^j$ points in $O((n/2^j) \log n)$ time, then possibly merging it to suitable $C_{j'}$ with $j' < j$ in $O(n)$ time. Using C_j we can compute $diam(P_i)$ in $O(n)$ time.

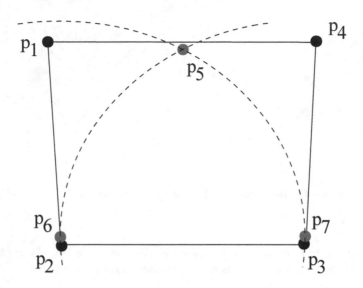

Fig. 3. A 3-dispersion may contain no corner points of the convex hull of P.

Depending on the result of $d(p_a, p_i) < diam(P_i)$ we proceed to the next stage with suitable parameters.

Since the number of stages is at most $\log n$, the total running time for a fixed p_a is $O(n \log n)$.

We have the following theorem.

Theorem 3. *If P is a set of n points on a plane and d is the L_2 metric then one can solve the max-min 3-dispersion problem in $O(n^2 \log n)$ time.*

If any P has a 3-dispersion with at least one point on the corner points of the convex hull of P then we can check p_a only for the corner points of the convex hull of P. However there is a counterexample. See Fig. 3, in which $\{p_5, p_6, p_7\}$ is a 3-dispersion. Note that the dotted circles have centers at p_6 and p_7 with radius $d(p_5, p_6) = d(p_6, p_7) = d(p_7, p_5)$ and so $d(p_6, p_7) > d(p_2, p_7) = d(p_3, p_6)$ holds.

By generalizing the algorithm to dimension D we have the following theorem.

Theorem 4. *If P is a set of n points in dimension D and d is the L_2 metric then one can solve the max-min 3-dispersion problem in $O(Dn^2 + Tn \log n)$ time, where T is the time to compute the diameter of n points in dimension D.*

Note that for a fixed p_a we need to compute medians in $O(Dn) + O(Dn/2) + O(Dn/2^2) + \cdots = O(Dn)$ time.

One can compute the diameter of n points in 3-dimension in $O(n \log n)$ time [16]. So we can compute a 3-dispersion of n points in 3-dimension in $O(n^2 \log^2 n)$ time, which is faster than the $O(n^{\omega k/3} \log n)$ time algorithm [2] for $k = 3$. The diameter of n points in dimension D can be computed in time

$O(n^{2-a(k)}(\log n)^{1-a(k)})$, where $a(k) = 2^{-(k+1)}$ [20]. So we can compute the 3-dispersion of n points in dimension D in $o(n^3)$ time for any D.

5 Conclusion

In this paper we have designed some algorithms to solve the 3-dispersion problem for a set of points on a plane. We have designed $O(n)$ time algorithms to solve the 3-dispersion problem when d is the L_∞ metric or the L_1 metric. Also we have designed an $O(n^2 \log n)$ time algorithm to solve the 3-dispersion problem when d is the L_2 metric.

Given a set P of points on a plane the diameter problem finds a pair (u, v) of points in P with the maximum $d(u, v)$. Thus the diameter corresponds to the 2-dispersion of P. There is a linear time reduction from the diameter problem to the 3-dispersion problem with the L_2 metric, as follows. Given P we append a dummy point p' so that it is far enough from P. Then a 3-dispersion of $P \cup \{p'\}$ always contains p' and the other two points correspond to the diameter of P. It is known that any algorithm to solve the diameter problem requires $\Omega(n \log n)$ time [14]. Thus any algorithm to solve the 3-dispersion problem with the L_2 metric requires $\Omega(n \log n)$ time. Therefore there is a chance to either design a faster algorithm to solve the 3-dispersion problem with the L_2 metric, or show a greater lower bound.

For a set P of points in a metric space we can compute the 3-dispersion of P as follows. By replacing $(+, \cdot)$ to $(\max, \min)$ in the matrix multiplication algorithm we can compute $\max_c\{\min\{d(p_a, p_c), d(p_b, p_c)\}\}$ for each $p_a, p_b \in P$ in $O(n^\omega)$ time. So we can compute $\min\{d(p_a, p_b), \max_c\{\min\{d(p_a, p_c), d(p_b, p_c)\}\}\}$ for each $p_a, p_b \in P$ in $O(n^\omega)$ time, then choose the maximum one among them as a 3-dispersion. Thus we can compute a 3-dispersion of n points in a metric space in $O(n^\omega)$ time, where $\omega < 2.373$.

Acknolwedgement. Shin-ichi Nakano was supported by JST CREST Grant Number JPMJCR1402. Takeaki Uno was supported by JST CREST Grant Number JPMJCR1401. Kunihiro Wasa was supported by JSPS KAKENHI Grant Number 19K20350 and JST CREST Grant Number JPMJCR1401.

References

1. Agarwal, P., Sharir, M.: Efficient algorithms for geometric optimization. ACM Comput. Surv. **30**, 412–458 (1998)
2. Akagi, T., et al.: Exact algorithms for the max-min dispersion problem. In: Chen, J., Lu, P. (eds.) FAW 2018. LNCS, vol. 10823, pp. 263–272. Springer, Cham (2018). https://doi.org/10.1007/978-3-319-78455-7_20
3. Akagi, T., Nakano, S.: Dispersion on the line. IPSJ SIG Technical reports, 2016-AL-158-3 (2016)
4. Baur, C., Fekete, S.P.: Approximation of geometric dispersion problems. In: Jansen, K., Rolim, J. (eds.) APPROX 1998. LNCS, vol. 1444, pp. 63–75. Springer, Heidelberg (1998). https://doi.org/10.1007/BFb0053964

5. Birnbaum, B., Goldman, K.J.: An improved analysis for a greedy remote-clique algorithm using factor-revealing LPs. Algorithmica **50**, 42–59 (2009)
6. Cevallos, A., Eisenbrand, F., Zenklusen, R.: Max-sum diversity via convex programming. In: Proceedings of SoCG 2016, pp. 26:1–26:14 (2016)
7. Cevallos, A., Eisenbrand, F., Zenklusen, R.: Local search for max-sum diversification. In: Proceedings of SODA 2017, pp. 130–142 (2017)
8. Chandra, B., Halldorsson, M.M.: Approximation algorithms for dispersion problems. J. Algorithms **38**, 438–465 (2001)
9. Drezner, Z.: Facility location: A Survey of Applications and Methods. Springer, New York (1995)
10. Drezner, Z., Hamacher, H.W.: Facility Location: Applications and Theory. Springer, Heidelberg (2004)
11. Erkut, E.: The discrete p-dispersion problem. Eur. J. Oper. Res. **46**, 48–60 (1990)
12. Fekete, S.P., Meijer, H.: Maximum dispersion and geometric maximum weight cliques. Algorithmica **38**, 501–511 (2004)
13. Frederickson, G.: Optimal algorithms for tree partitioning. In: Proceedings of SODA 1991, pp. 168–177 (1991)
14. Preparata, F.P., Shamos, M.I.: Computational Geometry: An Introduction. Springer, New York (1985). https://doi.org/10.1007/978-1-4612-1098-6
15. Rubinstein, S., Hassin, R., Tamir, A.: Approximation algorithms for maximum dispersion. Oper. Res. Lett. **21**, 133–137 (1997)
16. Ramos, E.A.: Deterministic algorithms for 3-D diameter and some 2-D lower envelopes. In: Proceedings of Symposium on Computational Geometry, pp. 290–299 (2000)
17. Rosenkrantz, D.J., Ravi, S.S., Tayi, G.K.: Heuristic and special case algorithms for dispersion problems. Oper. Res. **42**, 299–310 (1994)
18. Sydow, M.: Approximation guarantees for max sum and max min facility dispersion with parameterised triangle inequality and applications in result diversification. Mathematica Applicanda **42**, 241–257 (2014)
19. Wang, D.W., Kuo, Y.-S.: A study on two geometric location problems. Inf. Process. Lett. **28**, 281–286 (1988)
20. Yao, A.C.: On constructing minimum spanning trees in k-dimensional spaces and related problems. SIAM J. Comput. **11**, 721–736 (1982)

Matching Cut in Graphs with Large Minimum Degree

Sun-Yuan Hsieh[1], Hoang-Oanh Le[2], Van Bang Le[3], and Sheng-Lung Peng[4(✉)]

[1] Department of Computer Science and Information Engineering, National Cheng Kung University, No. 1, University Road, Tainan 70101, Taiwan
hsiehsy@mail.ncku.edu.tw
[2] Berlin, Germany
LeHoangOanh@web.de
[3] Institut für Informatik, Universität Rostock, Rostock, Germany
van-bang.le@uni-rostock.de
[4] Department of Computer Science and Information Engineering, National Dong Hwa University, Hualien 974, Taiwan
slpeng@mail.ndhu.edu.tw

Abstract. In a graph, a matching cut is an edge cut that is a matching. MATCHING CUT is the problem of deciding whether or not a given graph has a matching cut, which is known to be NP-complete. While MATCHING CUT is trivial for graphs with minimum degree at most one, it is NP-complete on graphs with minimum degree two.

In this paper, we show that, for any given constant $\epsilon > 0$, MATCHING CUT is NP-complete in the class of n-vertex (bipartite) graphs with minimum degree $\delta > n^{1-\epsilon}$. We give an exact branching algorithm to solve MATCHING CUT for graphs with minimum degree $\delta \geq 3$ in time $O^*(\lambda^n)$, where λ is the positive root of the polynomial $x^{\delta+1} - x^{\delta} - 1$. This is a very fast exact exponential time algorithm for MATCHING CUT on graphs with large minimum degree; for instance, the running time is $O^*(1.0099^n)$ on graphs with minimum degree $\delta \geq 469$. Complementing our hardness results, we show that, for any fixed constant $1 < c < 4$, MATCHING CUT is solvable in polynomial time for graphs with very large minimum degree $\delta \geq \frac{1}{c}n$.

1 Introduction

In a graph $G = (V, E)$, a *cut* is a partition $V = X \dot\cup Y$ of the vertex set into disjoint, non-empty sets X and Y, written (X, Y). The set of all edges in G having an endvertex in X and the other endvertex in Y, also written (X, Y), is called the *edge cut* of the cut (X, Y). A *matching cut* is an edge cut that is a (possibly empty) matching. Another way to define matching cuts is as follows [5,8]. A partition $V = X \dot\cup Y$ of the vertex set of the graph $G = (V, E)$ into disjoint, non-empty sets X and Y, is a matching cut if and only if each vertex in X has at most one neighbor in Y and each vertex in Y has at most one neighbor in X.

© Springer Nature Switzerland AG 2019
D.-Z. Du et al. (Eds.): COCOON 2019, LNCS 11653, pp. 301–312, 2019.
https://doi.org/10.1007/978-3-030-26176-4_25

Graham [8] studied matching cuts in graphs in connection to a number theory problem called cube-numbering. In [6], Farley and Proskurowski studied matching cuts in the context of network applications. Patrignani and Pizzonia [14] pointed out an application of matching cuts in graph drawing. Matching cuts have been used by Araújo et al. [1] in studying good edge-labellings in the context of WDM (Wavelength Division Multiplexing) networks.

Not every graph has a matching cut; the MATCHING CUT problem is the problem of deciding whether or not a given graph has a matching cut:

MATCHING CUT
Instance: A graph $G = (V, E)$.
Question: Does G have a matching cut?

Obviously, disconnected graphs and graphs having a bridge have a matching cut. In particular, MATCHING CUT is trivial for graphs with minimum degree at most one. It is known, however, MATCHING CUT is NP-complete on (bipartite) graphs with minimum degree two and on (bipartite) graphs with minimum degree three (see [12]). This paper considers the computational complexity of the MATCHING CUT problem in graphs of large minimum degree.

1.1 Previous Results

Graphs admitting a matching cut were first discussed by Graham in [8] under the name *decomposable graphs*. The first complexity and algorithmic results for MATCHING CUT have been obtained by Chvátal, who proved in [5] that MATCHING CUT is NP-complete, even when restricted to graphs of maximum degree four, and is solvable in polynomial time for graphs with maximum degree at most three. These results triggered a lot of research on the computational complexity of MATCHING CUT in graphs with additional structural assumptions; see [3,4,10,12–14]. In particular, the NP-hardness of MATCHING CUT has been further strengthened for planar graphs of maximum degree four [3] and bipartite graphs of minimum degree three and maximum degree four [12]. Moreover, it follows from Bonsma's result [3] and a simple reduction observed by Moshi [13] that MATCHING CUT remains NP-complete on bipartite planar graphs of minimum degree two and maximum degree eight. Recently, Le and Le [11] proved that MATCHING CUT is NP-complete on graphs of diameter d for any fixed integer $d \geq 3$ and on bipartite graphs of diameter d for any fixed integer $d \geq 4$.

On the positive side, some polynomially solvable cases have been identified: graphs of maximum degree 3 [5], graphs without chordless cycle of length at least five (including chordal and chordal bipartite graphs) [13], claw-free graphs (including line graphs) [3], graphs of diameter 2 [4,11], bipartite graphs of diameter at most 3 [11]. In [10], the cases of graphs of maximum degree 3 and of claw-free graphs have been extended to a larger class in which MATCHING CUT is still solvable in polynomial time.

Parameterized and exact exponential algorithms for MATCHING CUT on graphs without any restriction have been considered in [2,9,10]. Kratsch and

Le [10] provided the first exact branching algorithm for MATCHING CUT running in time $O^*(1.4143^n)^1$ and noted that MATCHING CUT cannot be solved for n-vertex graphs in $2^{o(n)}$ time, if the Exponential Time Hypothesis (ETH) is true. Recently, the running time has been improved in [9] to $O^*(1.3803^n)$. We note that the running time analysis of these branching algorithms highly depends on the minimum degree ≥ 2 of the input graphs. This fact was the main motivation of the present work.

1.2 Our Contributions

We prove that MATCHING CUT is NP-complete, even when restricted to (bipartite) n-vertex graphs of minimum degree at least $n^{1-\epsilon}$ for any fixed constant $\epsilon > 0$. This hardness result is somehow surprising, as, intuitively, MATCHING CUT would be easy for very dense graphs, graphs with very large minimum degree. We show that the exact branching algorithms in [9,10] solving MATCHING CUT can be simplified for graphs with large minimum degree δ, which results in a running time $O^*(\lambda^n)$, where λ is the positive root of the polynomial $x^{\delta+1} - x^\delta - 1$. For $\delta \geq 4$, this running time is $O^*(1.3248^n)$, improving upon the previous best known $O^*(1.3803^n)$-time algorithm [9]. It is quite interesting to note that, for $\delta \geq 469$, our exact algorithm solves an NP-complete problem in time $O^*(1.0099^n)$! Complementing our hardness results, we show that MATCHING CUT can be solved in polynomial time in graphs with minimum degree at least $\frac{1}{c}n$, where $1 < c < 4$ is a fixed constant.

1.3 Notation and Terminology

Let $G = (V, E)$ be a graph with vertex set $V(G) = V$ and edge set $E(G) = E$. An *independent set* (a *clique*) in G is a set of pairwise non-adjacent (adjacent) vertices. The neighborhood of a vertex v in G, denoted $N_G(v)$, is the set of all vertices in G adjacent to v; the closed neighborhood of v is $N_G[v] = N_G(v) \cup \{v\}$. If the context is clear, we simply write $N(v)$ and $N[v]$. Set $\deg(v) = |N(v)|$, the degree of the vertex v, and $\delta(G) = \min\{\deg(v) \mid v \in V(G)\}$, the minimum degree of G. For a subset $W \subseteq V$, the W-*neighbors* of a vertex v are the vertices in $N(v) \cap W$, $G[W]$ is the subgraph of G induced by W, and $G - W$ stands for $G[V \setminus W]$. The complete graph and the path on n vertices is denoted by K_n and P_n, respectively; K_3 is also called a *triangle*. The complete bipartite graph with one color class of size p and the other of size q is denoted by $K_{p,q}$.

Given a graph $G = (V, E)$ and a partition $V = X \dot\cup Y$, it can be decided in linear time if (X, Y) is a matching cut of G. This is because (X, Y) is a matching cut of G if and only if the bipartite subgraph $B_G(X, Y)$ of G with the color classes X and Y and edge set (X, Y) is P_3-free. That is, (X, Y) is a matching cut of G if and only if the non-trivial connected components of the bipartite graph $B_G(X, Y)$ are edges. We say that $S \subset V$ is *monochromatic* if, for any matching cut (X, Y) of G (if any), $S \subseteq X$ or else $S \subseteq Y$. Observe that

---

[1] Throughout the paper we use the O^* notation which suppresses polynomial factors.

(the vertex set of) any K_n with $n \geq 3$, and any $K_{p,q}$ with $p \geq 2, q \geq 3$, in G is monochromatic. A *bridge* in a graph is an edge whose deletion increases the number of the connected components. Since disconnected graphs and graphs having a bridge have a matching cut, we may assume that all graphs considered are connected and 2-edge connected.

When an algorithm branches on the current instance of size n into r sub-problems of sizes at most $n - t_1, n - t_2, \ldots, n - t_r$, then $(t_1, t_2, \ldots, t_r)$ is called the *branching vector* of this branching, and the unique positive root of the characteristic polynomial $x^n - x^{n-t_1} - x^{n-t_2} - \cdots - x^{n-t_r}$, denoted $\tau(t_1, t_2, \ldots, t_r)$, is called its *branching number*. The running time of a branching algorithm is $O^*(\lambda^n)$, where $\lambda = \max_i \lambda_i$ and λ_i is the branching number of branching rule i, and the maximum is taken over all branching rules. We refer to [7] for more details on exact branching algorithms.

1.4 Structure of the Paper

The paper is organized as follows. In Sect. 2 we show that MATCHING CUT is NP-complete when restricted to graphs with minimum degree $\delta \geq n^{1-\epsilon}$, for any fixed constant $\epsilon > 0$. The same result holds also for bipartite graphs. In Sect. 3 we describe the fast exact branching algorithm solving MATCHING CUT for graphs with large minimum degree. In Sect. 4 we show that MATCHING CUT can be solved in polynomial time for graphs with very large minimum degree at least $\frac{1}{c}n$ for any fixed constant $1 < c < 4$. We conclude the paper with Sect. 5.

2 Hardness Results

If G has a vertex v with $\deg(v) \leq 1$, then $(\{v\}, V(G) \setminus \{v\})$ is clearly a matching cut of G. It is known that MATCHING CUT is NP-complete for graphs with minimum degree two. It is quite easy to see that, for any constant $c \geq 2$, MATCHING CUT is NP-complete for graphs with minimum degree c: Given a graph $G = (V, E)$, let G' be obtained from G and $|V|$ cliques Q_v, $v \in V$, each of size c by joining edges between v and all vertices in Q_v. Then, G' has minimum degree c and G has a matching cut if and only if G' has a matching cut (this is because the cliques $Q_v \cup \{v\}$, $v \in V$, are monochromatic).

Our main result in this section says that MATCHING CUT remains NP-complete on graphs with very large minimum degree, even on such bipartite graphs. This is somehow surprising as one might expect that it would be easy to check if a very dense graph has a matching cut.

Theorem 1. *Let $\epsilon > 0$. MATCHING CUT is NP-complete on n-vertex graphs with minimum degree at least $n^{1-\epsilon}$. Moreover, MATCHING CUT remains NP-complete on n-vertex bipartite graphs with minimum degree at least $n^{1-\epsilon}$.*

Proof. Given an n-vertex graph $G = (V, E)$ with $\delta = \delta(G) \geq 2$ and $\epsilon > 0$, let $G' = (V', E')$ be obtained from $t = \lceil n^{\frac{1-\epsilon}{\epsilon}} \rceil$ copies of G by making all t copies of any vertex $v \in V$ to a clique. More precisely, writing $V = \{v_1, \ldots, v_n\}$ we have; see also Fig. 1.

- $V' = V^1 \cup \ldots \cup V^t$, where
    - $V^s = \{v_1^s, \ldots, v_n^s\}$ is the sth copy of the vertex set V, $1 \le s \le t$.
- $E' = E^1 \cup \ldots \cup E^t \cup F_1 \cup \ldots \cup F_n$, where
    - $E^s = \{v_i^s v_j^s \mid 1 \le i, j \le n, v_i v_j \in E\}$ is the sth copy of the edge set E, $1 \le s \le t$, and
    - $F_i = \{v_i^s v_i^{s'} \mid 1 \le s, s' \le t, s \neq s'\}$, $1 \le i \le n$.

Thus, $G'[V^s] = (V^s, E^s)$, $1 \le s \le t$, are the t copies of G and for each $1 \le i \le n$, the set $Q_i = \{v_i^1, \ldots, v_i^t\}$ of the t copies of $v_i \in V$ induces a clique in G' (with the edge set F_i).

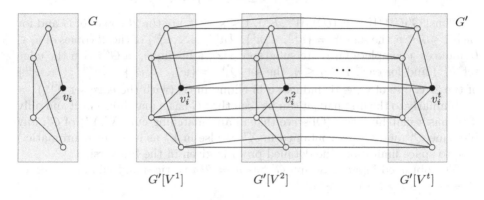

Fig. 1. The graphs G and G'.

Clearly, given $\epsilon > 0$ and G, G' can be constructed in polynomial time $t \cdot O(nt^2) = O(n^{3\frac{1-\epsilon}{\epsilon}+1})$. Moreover, G has a matching cut if and only if G' has a matching cut: first, if (X, Y) is a matching cut of G, then (X', Y') with

$$X' = \bigcup_{1 \le i \le n \,:\, v_i \in X} Q_i \quad \text{and} \quad Y' = \bigcup_{1 \le i \le n \,:\, v_i \in Y} Q_i$$

obviously, by definition of G', is a matching cut of G'. Second, if (X', Y') is a matching cut of G', then for each $1 \le i \le n$, the clique Q_i is contained in X' or else in Y' (as Q_i is monochromatic). Hence, (X, Y) with

$$X = \{v_i \mid 1 \le i \le n, Q_i \subseteq X'\} \text{ and } Y = \{v_i \mid 1 \le i \le n, Q_i \subseteq Y'\}$$

is a matching cut of G.

Now, G' has $n' = tn$ vertices. Hence $t \ge n^{\frac{1-\epsilon}{\epsilon}} = (n'/t)^{\frac{1-\epsilon}{\epsilon}}$, implying $t \ge (n')^{1-\epsilon}$. Therefore, G has minimum degree $\delta' = \delta + t - 1 > t \ge (n')^{1-\epsilon}$. Thus, the first part of the theorem is proved.

For the second part, let $\epsilon > 0$ and let $G = (V, E)$ be a bipartite graph with bipartition $V = U \cup W$ and minimum degree $\delta = \delta(G) \ge 2$. The previous construction can be modified to obtain a desired bipartite graph $G' = (U' \cup W', E')$ consisting of $2t$ copies of G as follows. Write $U = \{u_1, \ldots, u_p\}$, $W = \{w_1, \ldots, w_q\}$. Thus, $n = p + q$ is the vertex number of G.

- Let $t = \lceil (2n)^{\frac{1-\epsilon}{\epsilon}} \rceil$.
- $U' = U^1 \cup \ldots \cup U^t \cup W^{t+1} \cup \ldots \cup W^{2t}$, and
- $W' = W^1 \cup \ldots \cup W^t \cup U^{t+1} \cup \ldots \cup U^{2t}$, where
  - $U^s = \{u_1^s, \ldots, u_p^s\}$ is the sth copy of the color class U, $1 \le s \le 2t$,
  - $W^s = \{w_1^s, \ldots, w_q^s\}$ is the sth copy of the color class W, $1 \le s \le 2t$.
- $E' = E^1 \cup \ldots \cup E^{2t} \cup F_1^U \cup \ldots \cup F_p^U \cup F_1^W \cup \ldots \cup F_q^W$, where
  - $E^s = \{u_i^s w_j^s \mid 1 \le i \le p, 1 \le j \le q, u_i w_j \in E\}$ is the sth copy of the edge set E, $1 \le s \le 2t$,
  - $F_i^U = \{u_i^s u_i^{s'} \mid 1 \le s \le t, t+1 \le s' \le 2t\}$, $1 \le i \le p$, and
  - $F_j^W = \{w_j^s w_j^{s'} \mid 1 \le s \le t, t+1 \le s' \le 2t\}$, $1 \le j \le q$.

Thus, $G'[U^s \cup W^s] = (U^s \cup W^s, E^s)$, $1 \le s \le 2t$, are the $2t$ copies of G and for each $1 \le i \le p$, the set $Q_i^U = \{u_i^1, \ldots, u_i^t\} \cup \{u_i^{t+1}, \ldots, u_i^{2t}\}$ of the $2t$ copies of $u_i \in U$ induces a complete bipartite subgraph, viz., a bi-clique, in G' (with the edge set F_i^U) and for each $1 \le j \le q$, the set $Q_j^W = \{w_j^1, \ldots, w_j^t\} \cup \{w_j^{t+1}, \ldots, w_j^{2t}\}$ of the $2t$ copies of $w_j \in W$ induces a bi-clique in G' (with the edge set F_j^W).

Similarly to the first part, it can be seen that G has a matching cut if and only if G' has a matching cut. (Observe that in any matching cut (X', Y') of G', any bi-clique Q_i^U and Q_j^W is contained in X' or else in Y' as it is monochromatic.) Due to space limitation, the detailed proof is given in the full version.

The obtained bipartite graph G' has $n' = 2tn$ vertices and minimum degree $\delta' = \delta + t > t \ge (2n)^{\frac{1-\epsilon}{\epsilon}} \ge (n')^{1-\epsilon}$.   □

3   A Very Fast Exact Exponential Time Algorithm for Graphs with Large Minimum Degree

In [10], the first exact algorithm solving MATCHING CUT for n-vertex graphs has the running time $O^*(1.4143^n)$. This branching algorithm has been improved to $O^*(1.3803^n)$ in [9]. We remark that the branching steps, hence the time complexity of the proposed algorithms, strongly depend on the minimum degree $\delta \ge 2$ of the input graphs. In this section, we observe that the algorithms proposed in [9,10] can be simplified for input graphs with minimum degree $\delta \ge 3$ that results in an exact algorithm with running time $O^*(\lambda^n)$, where λ is the positive root of the polynomial $x^{\delta+1} - x^\delta - 1 = 0$. For graphs with minimum degree $\delta \ge 4$, this running time is $O^*(1.3248^n)$, improved upon the mentioned $O^*(1.3803)$-time algorithm. For graphs with large minimum degree $\delta \ge 469$, the running time is $O^*(1.0099^n)$; see also the Table 1 below. Recall that, by Theorem 1, MATCHING CUT remains NP-complete on graphs with minimum degree $\delta \ge c$ for arbitrary large constant c. The idea of our algorithm is as follows [10]. If the input graph $G = (V, E)$ has a matching cut (X, Y), then some vertex a is contained in X and some vertex b is contained in Y. The algorithm is a branching algorithm and will be executed for all possible pairs $a, b \in V$, hence $O(n^2)$ times. To do this set $A := \{a\}$, $B := \{b\}$, and $F := V \setminus \{a, b\}$ and call the branching algorithm. At each stage of the algorithm, A and/or B will be extended or it will be determined that there is no matching cut separating A and B, that is a matching cut (X, Y)

Table 1. Roots λ (rounded up) of $x^{\delta+1} - x^{\delta} - 1 = 0$ for some concrete minimum degree δ.

δ	3	4	25	199	469	6165	7747
λ	1.3803	1.3248	1.0976	1.0199	1.0099	1.0011	1.0009

with $A \subseteq X$ and $B \subseteq Y$. We describe our algorithm by a list of reduction and branching rules given in preference order, i.e., in an execution of the algorithm on any instance of a subproblem one always applies the first rule applicable to the instance, which could be a reduction or a branching rule. A reduction rule produces one instance/subproblem while a branching rule results in at least two instances/subproblems, with different extensions of A and B. Note that G has a matching cut that separates A from B if and only if in at least one recursive branch, extensions A' of A and B' of B are obtained such that G has a matching cut that separates A' from B'. Typically a rule assigns one or more free vertices, vertices of F, either to A or to B and removes them from F, that is, we always have $F := V \setminus (A \cup B)$.

The algorithm first applies Reduction Rules (R1)–(R4) given in [10]; the correctness of these rules is easy to see.

(R1) If a vertex in A has two B-neighbors, or a vertex in B has two A-neighbors then STOP: "G has no matching cut separating A, B".
  If $v \in F$, $|N(v) \cap A| \geq 2$ and $|N(v) \cap B| \geq 2$ then STOP: "G has no matching cut separating A, B".
  If there is an edge xy in G such that $x \in A$ and $y \in B$ and $N(x) \cap N(y) \cap F \neq \emptyset$ then STOP: "G has no matching cut separating A, B".
(R2) If $v \in F$ and $|N(v) \cap A| \geq 2$ then $A := A \cup \{v\}$.
  If $v \in F$ and $|N(v) \cap B| \geq 2$ then $B := B \cup \{v\}$.
(R3) If $v \in A$ has two adjacent F-neighbors w_1, w_2 then $A := A \cup \{w_1, w_2\}$.
  If $v \in B$ has two adjacent F-neighbors w_3, w_4 then $B := B \cup \{w_3, w_4\}$.
(R4) If there is an edge xy in G such that $x \in A$ and $y \in B$ then add $N(x) \cap F$ to A (if $N(x) \cap F \neq \emptyset$), and add $N(y) \cap F$ to B (if $N(y) \cap F \neq \emptyset$).

If none of these reduction rules can be applied then

- the A, B-edges of G form a matching cut in $G[A \cup B] = G - F$ due to (R1),
- every vertex in F is adjacent to at most one vertex in A and at most one vertex in B due to (R2),
- the neighbors in F of any vertex in A and the neighbors in F of any vertex in B form an independent set due to (R3), and
- every vertex in A adjacent to a vertex in B has no neighbor in F and every vertex in B adjacent to a vertex in A has no neighbor in F.

Clearly these properties hold for the instance (G, A, B) if none of the Rules (R1)–(R4) can be applied.

In contrast to the previous two algorithms in [9,10], our algorithm consists of only two branching rules, which are based on the following fact. See also Fig. 2.

Lemma 1. *Let $a \in A$ be a vertex with two (non-adjacent) neighbors u and v in F. Then*

(B1) *G has a matching cut separating A and B if and only if G has a matching cut separating $A \cup \{u\}$ and $B \cup (N[v] \cap F)$ or a matching cut separating $A \cup \{v\}$ and B.*

(B2) *If $N(v) \cap B \neq \emptyset$, then G has a matching cut separating A and B if and only if G has a matching cut separating $A \cup \{u\}$ and $B \cup (N[v] \cap F)$ or a matching cut separating $A \cup (N[v] \cap F)$ and B.*

Proof. (B1): Let (X, Y) be a matching cut of G such that $A \subseteq X$ and $B \subseteq Y$. If $v \in X$, then (X, Y) clearly separates $A \cup \{v\}$ and B. If $v \in Y$, then, as v is adjacent to $a \in A \subseteq X$, u must belong to X and $N[v] \cap F$ must belong to Y. Thus, (X, Y) is a matching cut separating $A \cup \{u\}$ and $B \cup (N[v] \cap F)$. Conversely, it is clear that any matching cut separating $A \cup \{u\}$ and $B \cup (N[v] \cap F)$, or $A \cup \{v\}$ and B is particularly a matching cut separating A and B.

(B2): Assume $N(v) \cap B \neq \emptyset$, and let (X, Y) be a matching cut of G such that $A \subseteq X$ and $B \subseteq Y$. By (B1), it remains to consider the case where $v \in X$. In this case, as v has a neighbor in $B \subseteq Y$, $N[v] \cap F$ must belong to X. Thus, (X, Y) is a matching cut separating $A \cup \{u\}$ and $B \cup (N[v] \cap F)$ or separating $A \cup (N[v] \cap F)$ and B. □

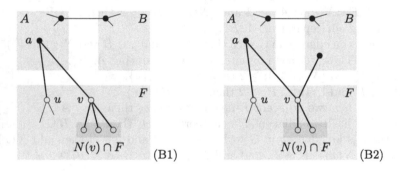

Fig. 2. When the branching rules (B1) and (B2) are applicable.

To determine the branching vectors which correspond to our branching rules, we set the size of an instance (G, A, B) as its number of free vertices, i.e., $|V(G)| - |A| - |B|$.

Let $a \in A$ with $|N(a) \cap F| \geq 2$, and choose two neighbors $u \neq v$ in F of a. Due to Reduction Rule (R3), u and v are non-adjacent. Due to Reduction Rule (R2), $N(v) \cap A = \{a\}$.

- If $N(v) \cap B = \emptyset$, we branch into two subproblems according to (B1): First, add u to A and add $N[v] \cap F = N[v] \setminus \{a\}$ to B. Second, add v to A. Hence the branching vector is $(\delta + 1, 1)$.

– If $N(v) \cap B \neq \emptyset$, we branch into two subproblems according to (B2): First, add u to A and add $N[v] \cap F$ to B. Second, add $N[v] \cap F$ to A. Due to Reduction Rule (R2), we have $|N(v) \cap B| = 1$, hence the branching vector is $(\delta, \delta - 1)$.

Note that, for $\delta \geq 3$, the branching number of the branching vector $(\delta, \delta - 1)$ is smaller than the branching number of the branching vector $(\delta + 1, 1)$. Consequently, the running time of our algorithm for n-vertex graphs with minimum degree $\delta \geq 3$ is $O^*(\lambda^n)$, where λ is the branching number of the branching vector $(\delta + 1, 1)$, that is, λ is the positive root of $x^{\delta+1} - x^\delta - 1 = 0$.

It remains to show that if none of the reduction rules and none of the branching rules is applicable to an instance (G, A, B) then the graph G has a matching cut (X, Y) such that $A \subseteq X$ and $B \subseteq Y$.

In fact, in this case, $(A, V(G) \setminus A)$ is a matching cut of G: First, every vertex in $V(G) \setminus A$ has at most one neighbor in A; this is clear because none of Reduction Rules (R1)–(R4) is applicable. Second, if some vertex $a \in A$ has at least two neighbors outside A, then no such neighbor is in B (as Reduction Rule (R4) is not applicable). So, a has two non-adjacent neighbors in $F = V(G) \setminus (A \cup B)$, and Branching Rule (B1) or (B2) is applicable, a contradiction. Thus, every vertex in A has at most one neighbor outside A. Therefore, $(A, V(G) \setminus A)$ is a matching cut of G as claimed.

To conclude, we obtain the following result:

Theorem 2. MATCHING CUT *can be solved in time* $O^*(\lambda^n)$ *for n-vertex graphs with minimum degree $\delta \geq 3$, where λ is the positive root of the polynomial* $x^{\delta+1} - x^\delta - 1$.

Perhaps, it is important to note that our algorithm labels the vertices of the input graph G by either A or B but never changes the graph G. Hence, the minimum degree δ of G remains unchanged in all subproblems. This is not the case by other minimum degree-based branching algorithms, like those computing the minimum independent set; cf. [7].

4   Polynomial-Time Solvable Cases: Graphs with Very Large Minimum Degree

In this section, we show that MATCHING CUT can be solved in polynomial time on graphs with large enough minimum degree. By Theorem 1, one can hope for polynomial-time results in terms of minimum degree only for graphs with $\delta \geq \frac{1}{c} n$ for some fixed constant $c > 1$. Examples of such dense graphs include Dirac graphs and Ore graphs. Graphs G with n vertices and $\delta(G) \geq \frac{1}{2} n$ are called *Dirac graphs*. *Ore graphs* are those graphs such that, for any two non-adjacent vertices x and y, $\deg(x) + \deg(y) \geq n$. Dirac graphs are particularly Ore graphs but not vice versa. Dirac and Ore graphs are well-studied classical graph classes, especially in the longest path and cycle community.

Our polynomial-time result is the following theorem.

Theorem 3. *Let* $0 < c_1 < 2$ *and* $c_2 \geq 0$ *be two arbitrary fixed constants.* MATCHING CUT *is polynomial-time solvable for n-vertex graphs, in which any two non-adjacent vertices* x *and* y *satisfy* $\deg(x) + \deg(y) \geq \frac{1}{c_1}n - c_2$.

Proof. Our strategy is to decide in polynomial time, given two disjoint vertex subsets A and B of the input graph G, if G admits a matching cut separating A and B, i.e., A is one part and B is in the other part of the matching cut.

First our algorithm applies Reduction Rules (R1)–(R4) mentioned in Sect. 3 to the current instance (in the order of the rules). In addition, we need one new reduction rule; recall from Sect. 3 that $F = V(G) \setminus (A \cup B)$.

(R5) If there are vertices $u, v \in F$ with a common neighbor in A and $|N(u) \cap N(v) \cap F| \geq 2$, then $A := A \cup \{u, v\} \cup (N(u) \cap N(v) \cap F)$.

Reduction Rule (R5) is safe: Assume that (X, Y) is a matching cut of G with $A \subseteq X$ and $B \subseteq Y$. Since $|N(u) \cap N(v) \cap F| \geq 2$, $u, v, N(u) \cap N(v) \cap F$ and the common neighbor $a \in A$ of u and v are contained in a monochromatic $K_{2,q}$ for some $q \geq 3$. Hence $u, v, N(u) \cap N(v) \cap F$ all must belong to X as $a \in A \subseteq X$. Thus, (X, Y) is a matching cut separating $A \cup \{u, v\} \cup (N(u) \cap N(v) \cap F)$ and B. The other direction is obvious: any matching cut of G separating $A \cup \{u, v\} \cup (N(u) \cap N(v) \cap F)$ and B clearly separates A and B, too. We have seen that Reduction Rule (R5) is correct.

Now, let $G = (V, E)$ be an n-vertex graph satisfying the condition in the theorem, and recall that we may assume that G is 2-edge connected. Hence every matching cut (X, Y) of G, if any, must contain at least two edges. Our algorithm will check, for each choice of two edges $a_1b_1, a_2b_2 \in E$, if G has a matching cut containing a_1b_1 and a_2b_2. To do this we start with $A := \{a_1, a_2\}$, $B := \{b_1, b_2\}$ and apply rules (R1)–(R5) as long as possible. If (R1) is applicable, then clearly G has no matching cut containing a_1b_1 and a_2b_2. So let us assume that (R1) was never applied and none of (R2), (R3), (R4) and (R5) is applicable.

Then, as (R1) was never applied, b_1 is the only neighbor in B of a_1 and a_2 is the only neighbor in A of b_2. As (R4) is not applicable, $|A| \geq |N[a_1] \setminus \{b_1\}| = \deg(a_1)$ and $|B| \geq |N[b_2] \setminus \{a_2\}| = \deg(b_2)$. Since a_1 and b_2 are non-adjacent, we therefore have

$$|A| + |B| \geq \frac{1}{c_1}n - c_2. \tag{1}$$

It follows from $|F| = n - |A| - |B|$ and (1) that

$$|F| \leq \left(1 - \frac{1}{c_1}\right)n + c_2. \tag{2}$$

Now, as (R2) is not applicable, every vertex in F has at most one neighbor in A (and at most one neighbor in B). So, if every vertex in A has at most one neighbor in F, then $(A, B \cup F)$ is a matching cut of G separating A and B.

Thus, it remains the case that some vertex $v \in A$ has two neighbors x, y in F. Then x and y are non-adjacent as (R3) is not applicable. Recall that v is the only neighbor in A of x and of y, and each of x and y has at most one neighbor in B. Now, as (R5) is not applicable, $|N(x) \cap N(y) \cap F| \leq 1$, and so we have $|F| \geq |N[x]| + |N[y]| - 4 = \deg(x) + \deg(y) - 2$. Hence

$$|F| \geq \frac{1}{c_1}n + c_2 - 2. \tag{3}$$

By (2) and (3), $\frac{1}{c_1}n + c_2 - 2 \leq \left(1 - \frac{1}{c_1}\right)n + c_2$. Since $0 < c_1 < 2$, we therefore have $n \leq \frac{2c_1}{2-c_1}$. Thus, G has bounded size. Hence, we can determine if G has a matching cut or not in constant time.

Observe that rule (R1) can be applied in constant time and applies at most once. Each of the other rules applies at most n times because it removes at least one vertex from F. Thus, the running time for the application of the rules is roughly $n \cdot O(m) = O(nm)$. Since we have at most $O(m^2)$ many choices for the edges a_1b_1, a_2b_2, we conclude that it can be decided in time $O(n \cdot m^3)$ if G has a matching cut. $\square$

Consider a graph G with minimum degree $\delta \geq \frac{1}{c}n$, where $1 < c < 4$ is a given constant. Then, for every two vertices x and y in G, $\deg(x) + \deg(y) \geq 2\delta \geq \frac{1}{c'}n$ with $c' = c/2 < 2$. Hence Theorem 3 leads to the following result:

Corollary 1. *Let $1 < c < 4$ be a fixed constant.* MATCHING CUT *is in polynomial-time solvable for n-vertex graphs with minimum degree at least $\frac{1}{c}n$.*

4.1 Dirac and Ore Graphs

In this subsection, we show that there is a good characterization for Dirac graphs and Ore graphs having a matching cut. It turns out that Ore graphs having a matching cut are particularly $\frac{1}{2}n$-regular Dirac graphs. Ore graphs having a matching cut can be characterized as follows.

Theorem 4. *An Ore graph G has a matching cut if and only if G is a co-bipartite graph with a bipartition into cliques A and B such that every vertex in A is adjacent to exactly one vertex in B and vice versa.*

Due to space limitation, we leave the proof of Theorem 4 for the full version. Theorem 4 implies that MATCHING CUT can be solved in quadratic time on Ore graphs, hence also on Dirac graphs.

5 Conclusion

We showed that MATCHING CUT remains NP-complete for graphs with unbounded minimum degree. For n-vertex graphs with minimum degree $\delta \geq 3$, however, we provided an exact branching algorithm to solve MATCHING CUT in time $O^*(\lambda^n)$ time, where λ is the positive root of the polynomial $x^{\delta+1} - x^{\delta} - 1$. For large minimum degree $\delta \geq 469$, the running time is $O^*(1.0099^n)$, and for

$\delta \geq 7747$, the running time is $O^*(1.0009^n)$. So, one might ask the question: is such an NP-hard problem really hard?

For graphs with very large minimum degree, MATCHING CUT becomes easier. We showed that, for any fixed constant $1 < c < 4$, MATCHING CUT is polynomial-time solvable for graphs with minimum degree at least $\frac{1}{c}n$. We leave the question open if MATCHING CUT can be solved in polynomial time for graphs with minimum degree at least $\frac{1}{c}n$, where $c \geq 4$ is any fixed constant.

References

1. Araújo, J., Cohen, N., Giroire, F., Havet, F.: Good edge-labelling of graphs. Discrete Appl. Math. **160**(18), 2502–2513 (2012)
2. Aravind, N.R., Kalyanasundaram, S., Kare, A.S.: On structural parameterizations of the matching cut problem. In: Gao, X., Du, H., Han, M. (eds.) COCOA 2017, Part II. LNCS, vol. 10628, pp. 475–482. Springer, Cham (2017). https://doi.org/10.1007/978-3-319-71147-8_34
3. Bonsma, P.S.: The complexity of the matching-cut problem for planar graphs and other graph classes. J. Graph Theory **62**(2), 109–126 (2009)
4. Borowiecki, M., Jesse-Józefczyk, K.: Matching cutsets in graphs of diameter 2. Theor. Comput. Sci. **407**(1–3), 574–582 (2008)
5. Chvátal, V.: Recognizing decomposable graphs. J. Graph Theory **8**(1), 51–53 (1984)
6. Farley, A.M., Proskurowski, A.: Networks immune to isolated line failures. Networks **12**(4), 393–403 (1982)
7. Fomin, F.V., Kratsch, D.: Exact Exponential Algorithms. Springer, Heidelberg (2010). https://doi.org/10.1007/978-3-642-16533-7
8. Graham, R.L.: On primitive graphs and optimal vertex assignments. Ann. N. Y. Acad. Sci. **175**(1), 170–186 (1970)
9. Komusiewicz, C., Kratsch, D., Le, V.B.: Matching cut: kernelization, single-exponential time FPT, and exact exponential algorithms. In: 13th International Symposium on Parameterized and Exact Computation, IPEC 2018, 20–24 August 2018, Helsinki, pp. 19:1–19:13 (2018)
10. Kratsch, D., Le, V.B.: Algorithms solving the matching cut problem. Theor. Comput. Sci. **609**, 328–335 (2016)
11. Le, H.-O., Le, V.B.: A complexity dichotomy for matching cut in (bipartite) graphs of fixed diameter. Theor. Comput. Sci. **770**, 69–78 (2019)
12. Le, V.B., Randerath, B.: On stable cutsets in line graphs. Theor. Comput. Sci. **1–3**(301), 463–475 (2003)
13. Moshi, A.M.: Matching cutsets in graphs. J. Graph Theory **13**(5), 527–536 (1989)
14. Patrignani, M., Pizzonia, M.: The complexity of the matching-cut problem. In: Brandstädt, A., Le, V.B. (eds.) WG 2001. LNCS, vol. 2204, pp. 284–295. Springer, Heidelberg (2001). https://doi.org/10.1007/3-540-45477-2_26

Incremental Optimization of Independent Sets Under the Reconfiguration Framework

Takehiro Ito[1], Haruka Mizuta[1(✉)], Naomi Nishimura[2], and Akira Suzuki[1]

[1] Tohoku University, Sendai, Japan
takehiro@ecei.tohoku.ac.jp, haruka.mizuta.s4@dc.tohoku.ac.jp,
a.suzuki@ecei.tohoku.ac.jp
[2] University of Waterloo, Waterloo, Canada
nishi@uwaterloo.ca

Abstract. Suppose that we are given an independent set I_0 of a graph G, and an integer $l \geq 0$. Then, we are asked to find an independent set of G having the maximum size among independent sets that are reachable from I_0 by either adding or removing a single vertex at a time such that all intermediate independent sets are of size at least l. We show that this problem is PSPACE-hard even for bounded pathwidth graphs, and remains NP-hard for planar graphs. On the other hand, we give a linear-time algorithm to solve the problem for chordal graphs. We also study the parameterized complexity of the problem with respect to the following three parameters: the degeneracy d of an input graph, a lower bound l on the size of independent sets, and a lower bound s on the size of a solution reachable from I_0. We show that the problem is fixed-parameter intractable when only one of d, l, and s is taken as a parameter. On the other hand, we give a fixed-parameter algorithm when parameterized by $s + d$; this result implies that the problem parameterized only by s is fixed-parameter tractable for planar graphs, and for bounded treewidth graphs.

1 Introduction

Recently, the reconfiguration framework [10] has been intensively applied to a variety of search problems (See, e.g., surveys [8,19]). For example, the INDEPENDENT SET RECONFIGURATION problem is one of the most well-studied reconfiguration problems [2,3,9,11–13,15–17,23]. For a graph G, a vertex subset $I \subseteq V(G)$ is an *independent set* of G if no two vertices in I are adjacent in G. Suppose that we are given two independent sets I_0 and I_r of G, and imagine that a token

Research by the first, second and fourth authors is partially supported by JST CREST Grant Number JPMJCR1402, and JSPS KAKENHI Grant Numbers JP17K12636, JP18H04091, JP19J10042 and JP19K11814, Japan. Research by Naomi Nishimura is partially supported by the Natural Science and Engineering Research Council of Canada.

D.-Z. Du et al. (Eds.): COCOON 2019, LNCS 11653, pp. 313–324, 2019.
https://doi.org/10.1007/978-3-030-26176-4_26

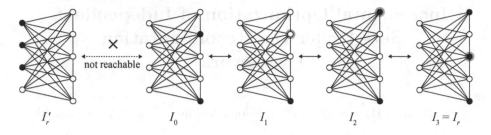

Fig. 1. A sequence $\langle I_0, I_1, I_2, I_3 \rangle$ of independent sets under the TAR rule for the lower bound $l = 1$, where the vertices in independent sets are colored with black .

(coin) is placed on each vertex in I_0. Then, for an integer lower bound $l \geq 0$, INDEPENDENT SET RECONFIGURATION *under the* TAR *rule* is the problem of determining whether we can transform I_0 into I_r via independent sets of size at least l such that each intermediate independent set can be obtained from the previous one by either adding or removing a single token.[1] In the example of Fig. 1, I_0 can be transformed into $I_r = I_3$ via the sequence $\langle I_0, I_1, I_2, I_3 \rangle$, but not into I_r', when $l = 1$.

Like this problem, many reconfiguration problems have the following basic structure: we are given two feasible solutions of an original search problem, and are asked to determine whether we can transform one into the other by repeatedly applying a specified reconfiguration rule while maintaining feasibility. These kinds of reconfiguration problems model several "dynamic" situations of systems, where we wish to find a step-by-step transformation from the current configuration of a system into a more desirable one.

However, it is not easy to obtain a more desirable configuration for an input of a reconfiguration problem, because many original search problems are NP-hard. Furthermore, there may exist (possibly, exponentially many) desirable configurations; even if we cannot reach a given target from the current configuration, there may exist another desirable configuration which is reachable. Recall the example of Fig. 1, where both I_r and I_r' have the same size three (which is larger than that of the current independent set I_0), but I_0 can reach only I_r.

Our Problem

In this paper, we propose a new variant of reconfiguration which asks for a more desirable configuration that is reachable from the current one. As the first example of this new variant, we consider INDEPENDENT SET RECONFIGURATION because it is one of the most well-studied reconfiguration problems.

Suppose that we are given a graph G, an integer lower bound $l \geq 0$, and an independent set I_0 of G. Then, we are asked to find an independent set I_{sol} of G such that $|I_{\text{sol}}|$ is maximized and I_0 can be transformed into I_{sol} under the

---

[1] TAR stands for Token Addition and Removal, and there are two other well-studied reconfiguration rules called TS (Token Sliding) and TJ (Token Jumping) [13]. We omit the details in this paper.

TAR rule for the lower bound l. We call this problem the *optimization variant* of INDEPENDENT SET RECONFIGURATION (denoted by OPT-ISR). To avoid confusion, we call the standard INDEPENDENT SET RECONFIGURATION problem the *reachability variant* (denoted by REACH-ISR).

Note that I_{sol} is not always a maximum independent set of the graph G. For example, the graph in Fig. 1 has a unique maximum independent set of size four (consisting of the vertices on the left side), but I_0 cannot be transformed into it. Indeed, $I_{\mathsf{sol}} = I_3$ for this example when $l = 1$.

Related Results

Although OPT-ISR is being introduced in this paper, some previous results for REACH-ISR are related in the sense that they can be converted into results for OPT-ISR. We present such results here.

Ito et al. [10] showed that REACH-ISR under the TAR rule is PSPACE-complete. On the other hand, Kamiński et al. [13] proved that any two independent sets of size at least $l + 1$ are reachable under the TAR rule with the lower bound l for even-hole-free graphs.

REACH-ISR has been studied well from the viewpoint of fixed-parameter (in)tractability. Mouawad et al. [17] showed that REACH-ISR under the TAR rule is W[1]-hard when parameterized by the lower bound l and the length of a desired sequence (i.e., the number of token additions and removals). Lokshtanov et al. [16] gave a fixed-parameter algorithm to solve REACH-ISR under the TAR rule when parameterized by the lower bound l and the degeneracy d of an input graph.

From our problem setting, one may be reminded of the concept of local search algorithms [1,20]. To the best of our knowledge, known results for this concept do not have direct relations to our problem, because they are usually evaluated experimentally. In addition, note that our problem assumes that an initial independent set I_0 is given as an input. In contrast, a local search algorithm is allowed to choose the initial solution, sometimes randomly.

Our Contributions

In this paper, we study OPT-ISR from the viewpoints of polynomial-time solvability and fixed-parameter (in)tractability.

We first study the polynomial-time solvability of OPT-ISR with respect to graph classes, as summarized in Fig. 2. More specifically, we show that OPT-ISR is PSPACE-hard even for bounded pathwidth graphs, and remains NP-hard even for planar graphs. On the other hand, we give a linear-time algorithm to solve the problem for chordal graphs. We note that our algorithm indeed works in polynomial time for even-hole-free graphs (which form a larger graph class than that of chordal graphs) if the problem of finding a maximum independent set is solvable in polynomial time for even-hole-free graphs; currently, its complexity status is unknown.

We next study the fixed-parameter (in)tractability of OPT-ISR, as summarized in Table 1. In this paper, we consider mainly the following three parameters: the degeneracy d of an input graph, a lower bound l on the size of independent

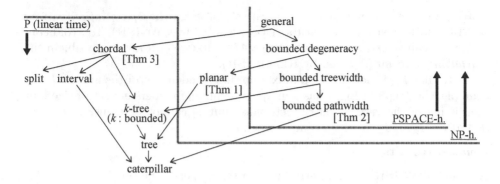

Fig. 2. Our results with respect to graph classes.

Table 1. Our results with respect to parameters.

	(no parameter)	Lower bound l	Solution size s
(no parameter)	–	NP-hard for fixed l (i.e., no FPT, no XP) [Corollary 1]	W[1]-hard, **XP** (i.e., no FPT) [Theorems 4 and 5]
Degeneracy d	PSPACE-hard for fixed d (i.e., no FPT, no XP) [Theorem 2]	NP-hard for fixed $d + l$ (i.e., no FPT, no XP) [Corollary 1]	**FPT** [Theorem 6]

sets, and the size s of a solution reachable from a given independent set I_0. As shown in Table 1, we completely analyze the fixed-parameter (in)tractability of the problem according to these three parameters; details are explained below.

We first consider the problem parameterized by a single parameter. We show that the problem is fixed-parameter intractable when only one of d, l, and s is taken as a parameter. In particular, we prove that OPT-ISR is PSPACE-hard for a fixed constant d and remains NP-hard for a fixed constant l, and hence the problem does not admit even an XP algorithm for each single parameter d or l under the assumption that P $\neq$ PSPACE or P $\neq$ NP. On the other hand, OPT-ISR is W[1]-hard for s, and admits an XP algorithm with respect to s.

We thus consider the problem taking two parameters. However, the problem still remains NP-hard for a fixed constant $d + l$, and hence it does not admit even an XP algorithm for $d + l$ under the assumption that P $\neq$ NP. Note that the combination of l and s is meaningless, since $l + s \leq 2s$. On the other hand, we give a fixed-parameter algorithm when parameterized by $s + d$; this result implies that OPT-ISR parameterized only by s is fixed-parameter tractable for planar graphs, and for bounded treewidth graphs.

We omit proofs for the claims marked with ($*$) due to the page limitation.

2   Preliminaries

In this paper, we consider only simple graphs, without loss of generality. For a graph G, we denote by $V(G)$ and $E(G)$ the vertex set and edge set of G, respectively. For a vertex $v \in V(G)$, let $N_G(v) = \{w \in V(G) : vw \in E(G)\}$, and let $N_G[v] = N_G(v) \cup \{v\}$. The set $N_G(v)$ is called the (open) neighborhood of v in G, while $N_G[v]$ is called the closed neighborhood of v in G. For a graph G and a vertex subset $S \subseteq V(G)$, $G[S]$ denotes the subgraph of G induced by S, that is, $V(G[S]) = S$ and $E(G[S]) = \{vw \in E(G) : v, w \in S\}$. For a vertex subset $V' \subseteq V(G)$, we simply write $G \setminus V'$ to denote $G[V(G) \setminus V']$. We denote by $A \triangle B$ the symmetric difference between two sets A and B, that is, $A \triangle B = (A \setminus B) \cup (B \setminus A)$.

We now formally define our problem OPT-ISR. For an integer $l \geq 0$ and two independent sets I_p and I_q of a graph G such that $|I_p| \geq l$ and $|I_q| \geq l$, a sequence $\mathcal{I} = \langle I_1, I_2, \ldots, I_\ell \rangle$ of independent sets of G is called a reconfiguration sequence between I_p and I_q under the TAR rule if $\mathcal{I}$ satisfies the following three conditions:

(a) $I_1 = I_p$ and $I_\ell = I_q$;
(b) I_i is an independent set of size at least l for each $i \in \{1, 2, \ldots, \ell\}$; and
(c) $|I_i \triangle I_{i+1}| = 1$ for each $i \in \{1, 2, \ldots, \ell - 1\}$.

To emphasize the lower bound l on the size of any independent set, we sometimes write TAR(l) instead of TAR. Note that any reconfiguration sequence is reversible, that is, $\langle I_\ell, I_{\ell-1}, \ldots, I_1 \rangle$ is a reconfiguration sequence between I_q and I_p under the TAR(l) rule. We say that two independent sets I_p and I_q are reachable under the TAR(l) rule if there exists a reconfiguration sequence between I_p and I_q under the TAR(l) rule. We write $I_p \overset{l}{\leftrightsquigarrow} I_q$ if I_p and I_q are reachable under the TAR(l) rule.

Our problem aims to optimize a given independent set under the TAR rule. Specifically, the optimization variant of INDEPENDENT SET RECONFIGURATION (OPT-ISR for short) is defined as follows:

> **Input:** A graph G, an integer $l \geq 0$, and an independent set I_0 of G such that $|I_0| \geq l$.
>
> **Task:** Find an independent set I_{sol} of G such that $I_0 \overset{l}{\leftrightsquigarrow} I_{\mathsf{sol}}$ and $|I_{\mathsf{sol}}|$ is maximized.

We denote by a triple (G, l, I_0) an instance of OPT-ISR, and call a desired independent set I_{sol} of G a solution to (G, l, I_0). Note that a given independent set I_0 may itself be a solution. OPT-ISR simply outputs a solution to (G, l, I_0), and does not require the specification of an actual reconfiguration sequence from I_0 to the solution.

We close this section with noting the following observation which says that OPT-ISR for an instance $(G, 0, I_0)$ is equivalent to finding a maximum independent set of G.

Lemma 1 *(∗). Every maximum independent set I_{max} of a graph G is a solution to an instance $(G, 0, I_0)$ of* OPT-ISR, *where I_0 is any independent set of G.*

3   Polynomial-Time Solvability

In this section, we study the polynomial-time solvability of OPT-ISR.

3.1   Hardness Results

Lemma 1 implies that results for the MAXIMUM INDEPENDENT SET problem can be applied to OPT-ISR for $l = 0$. For example, we have the following theorem, because MAXIMUM INDEPENDENT SET remains NP-hard for planar graphs [7].

Theorem 1. OPT-ISR *is NP-hard for planar graphs and $l = 0$, where l is a lower bound on the size of independent sets.*

For an integer $d \geq 0$, a graph G is d-*degenerate* if every induced subgraph of G has a vertex of degree at most d [14]. The *degeneracy* of G is the minimum integer d such that G is d-degenerate. It is known that the degeneracy of any planar graph is at most five [14], and hence we have the following corollary.

Corollary 1. OPT-ISR *is NP-hard for 5-degenerate graphs and $l = 0$, where l is a lower bound on the size of independent sets.*

This corollary implies that OPT-ISR admits neither a fixed-parameter algorithm nor an XP algorithm when parameterized by $d + l$ under the assumption that P $\neq$ NP, where d is the degeneracy of an input graph and l is a lower bound on the size of independent sets. We will discuss the fixed parameter (in)tractability of OPT-ISR more deeply in Sect. 4.

We then show that OPT-ISR is PSPACE-hard even if the pathwidth of an input graph is bounded by a constant. We first define the pathwidth of a graph, as follows [21]. A *path-decomposition* of a graph G is a sequence $\langle X_1, X_2, \ldots, X_t \rangle$ of vertex subsets of $V(G)$ such that

(a) for each vertex u of G, there exists a subset X_i such that $u \in X_i$;
(b) for each edge vw of G, there exists a subset X_j such that $v, w \in X_j$; and
(c) for any three indices a, b, c such that $a < b < c$, $X_a \cap X_c \subseteq X_b$ holds.

The *pathwidth* of G is the minimum value p such that there exists a path-decomposition $\langle X_1, X_2, \ldots, X_t \rangle$ of G for which $|X_i| \leq p + 1$ holds for all $i \in \{1, 2, \ldots, t\}$. A *bounded pathwidth graph* is a graph whose pathwidth is bounded by a fixed constant.

The following theorem is the main result of this subsection.

Theorem 2 *(∗). OPT-ISR is PSPACE-hard for bounded pathwidth graphs.*

3.2   Linear-Time Algorithm for Chordal Graphs

A graph G is *chordal* if every induced cycle in G is of length three [4]. The main result of this subsection is the following theorem.

Theorem 3. OPT-ISR *is solvable in linear time for chordal graphs.*

This theorem can be obtained from the following lemma; we note that a maximum independent set I_{max} of a chordal graph can be found in linear time [6], and the maximality of a given independent set can be checked in linear time.

Lemma 2. *Let* (G, l, I_0) *be an instance of* OPT-ISR *such that* G *is a chordal graph, and let* I_{max} *be any maximum independent set of* G. *Then, a solution* I_{sol} *to* (G, l, I_0) *can be obtained as follows:*

$$I_{sol} = \begin{cases} I_0 & \text{if } I_0 \text{ is a maximal independent set of } G \text{ and } |I_0| = l; \\ I_{max} & \text{otherwise.} \end{cases}$$

Proof. We first consider the case where I_0 is a maximal independent set of G and $|I_0| = l$. In this case, we cannot remove any vertex from I_0 because $|I_0| = l$. Furthermore, since I_0 is maximal, we cannot add any vertex in $V(G) \setminus I_0$ to I_0 while maintaining independence. Therefore, G has no independent set I' ($\neq I_0$) which is reachable from I_0, and hence $I_{sol} = I_0$.

We then consider the other case, that is, I_0 is not a maximal independent set of G or $|I_0| > l$. Observe that it suffices to consider the only case where $|I_0| > l$ holds; if $|I_0| = l$ and I_0 is not maximal, then we can obtain an independent set I_0'' of G such that $|I_0''| = l + 1$ and $I_0 \overset{l}{\leftrightsquigarrow} I_0''$ by adding some vertex in $V(G) \setminus I_0$. To prove $I_{sol} = I_{max}$, we below show that $I_0 \overset{l}{\leftrightsquigarrow} I_{max}$ holds if $|I_0| > l$.

Let $I_0' \subseteq I_0$ be any independent set of size $l+1$. Then, $I_0 \overset{l}{\leftrightsquigarrow} I_0'$ holds, because we can obtain I_0' from I_0 by removing vertices in $I_0 \setminus I_0'$ one by one. Similarly, let $I' \subseteq I_{max}$ be any independent set of size $l + 1$; we know that $I' \overset{l}{\leftrightsquigarrow} I_{max}$ holds. Kamiński et al. [13] proved that any two independent sets of the same size $l + 1$ are reachable under the TAR(l) rule for even-hole-free graphs. Since any chordal graph is even-hole free, we thus have $I_0' \overset{l}{\leftrightsquigarrow} I'$. Therefore, we have $I_0 \overset{l}{\leftrightsquigarrow} I_0' \overset{l}{\leftrightsquigarrow} I' \overset{l}{\leftrightsquigarrow} I_{max}$, and hence we can conclude that $I_0 \overset{l}{\leftrightsquigarrow} I_{max}$ holds as claimed.   □

We note that Lemma 2 indeed holds for even-hole-free graphs, which contain all chordal graphs. However, the complexity status of the MAXIMUM INDEPENDENT SET problem is unknown for even-hole-free graphs, and hence we do not know if we can obtain I_{max} in polynomial time.

4   Fixed Parameter Tractability

In this section, we study the fixed parameter (in)tractability of OPT-ISR. We take the solution size of OPT-ISR as the parameter. More formally, for an instance (G, l, I_0), the problem OPT-ISR *parameterized by solution size s* asks whether G has an independent set I such that $|I| \geq s$ and $I_0 \overset{l}{\rightsquigarrow} I$. We may assume that $s > l$; otherwise we are dealing with a yes-instance because I_0 itself is a solution. We sometimes denote by a 4-tuple (G, l, I_0, s) an instance of OPT-ISR parameterized by solution size s.

4.1   Single Parameter: Solution Size

We first give an observation that can be obtained from INDEPENDENT SET. Because INDEPENDENT SET is W[1]-hard when parameterized by solution size s [18], Lemma 1 implies the following theorem.

Theorem 4. OPT-ISR *is W[1]-hard when parameterized by solution size s.*

This theorem implies that OPT-ISR admits no fixed-parameter algorithm with respect to solution size s under the assumption that FPT $\neq$ W[1]. However, it admits an XP algorithm with respect to s, as in the following theorem.

Theorem 5. (*). OPT-ISR *parameterized by solution size s can be solved in time $O(s^3 n^{2s})$, where n is the number of vertices in a given graph.*

4.2   Two Parameters: Solution Size and Degeneracy

As we have shown in Theorem 4, OPT-ISR admits no fixed-parameter algorithm when parameterized by the single parameter of solution size s under the assumption that FPT $\neq$ W[1]. In addition, Theorem 2 implies that the problem remains PSPACE-hard even if the degeneracy d of an input graph is bounded by a constant, and hence OPT-ISR does not admit even an XP algorithm with respect to the single parameter d under the assumption that P $\neq$ PSPACE. In this subsection, we take these two parameters, and develop a fixed-parameter algorithm as in the following theorem.

Theorem 6. OPT-ISR *admits a fixed-parameter algorithm when parameterized by $s + d$, where s is the solution size and d is the degeneracy of an input graph.*

Before proving the theorem, we note the following corollary which holds for planar graphs, and for bounded treewidth graphs. Recall that OPT-ISR is intractable (from the viewpoint of polynomial-time solvability) for these graphs, as shown in Theorems 1 and 2.

Corollary 2. OPT-ISR *parameterized by solution size s is fixed-parameter tractable for planar graphs, and for bounded treewidth graphs.*

Proof. Recall that the degeneracy of any planar graph is at most five. It is known that the degeneracy of a graph is at most the treewidth of the graph. Thus, the corollary follows from Theorem 6.   □

Outline of Algorithm

As a proof of Theorem 6, we give such an algorithm. We first explain our idea and the outline of the algorithm. Our idea is to extend a fixed-parameter algorithm for REACH-ISR when parameterized by $l + d$ [16].

Consider the case where an input graph G consists of only a fixed-parameter number of vertices, that is, $|V(G)|$ can be bounded by some function of $s+d$. Then, we apply Theorem 5 to the instance and obtain the answer in fixed-parameter time (Lemma 4). We here use the fact (stated by Lokshtanov et al. [16, Proposition 2]) that a d-degenerate graph consists of a small number of vertices if it has a small number of low-degree vertices (Lemma 3).

Therefore, it suffices to consider the case where an input graph has many low-degree vertices. In this case, we will kernelize the instance: we will show that there always exists a low-degree vertex which can be removed from an input graph without changing the answer (yes or no) to the instance. Our kernelization has two stages. In the first stage, we focus on "twins" (two vertices that have the same closed neighborhoods), and prove that one of them can be removed without changing the answer (Lemma 5). The second stage will be executed only when the first stage cannot kernelize the instance into a sufficiently small size. The second stage is a bit involved, and makes use of the Sunflower Lemma by Erdös and Rado [5].

Graphs Having a Small Number of Low-Degree Vertices

We now give our algorithm. Suppose that (G, l, I_0, s) is an instance of OPT-ISR parameterized by solution size such that G is a d-degenerate graph. We assume that $|I_0| < s$; otherwise (G, l, I_0, s) is a yes-instance because I_0 itself is a solution.

We start with noting the following property for d-degenerate graphs, which is a little bit stronger claim than that of Lokshtanov et al. [16, Proposition 2]; however, the proof is almost the same as that of [16].

Lemma 3. *Suppose that a graph G is d-degenerate, and let $D \subseteq V(G)$ be the set of all vertices of degree at most $2d$ in G. Then, $|V(G)| \leq (2d + 1)|D|$.*

Proof. Suppose for a contradiction that $|V(G)| = (2d+1)|D| + c$ holds for some integer $c \geq 1$. Then, $|V(G) \setminus D| = 2d|D| + c$, and hence we have

$$|E(G)| = \frac{1}{2} \sum_{v \in V(G)} |N_G(v)| \geq \frac{1}{2} \sum_{v \in V(G) \setminus D} (2d + 1)$$

$$= \frac{1}{2}(2d + 1)(2d|D| + c) = d|V(G)| + \frac{1}{2}c > d|V(G)|.$$

This contradicts the fact that $|E(G)| \leq d|V(G)|$ holds for any d-degenerate graph G [14]. □

Let $D = \{v \in V(G) : |N_G(v)| \leq 2d\}$, and let $D' = D \setminus I_0$. We introduce a function $f(s, d)$ which depends on only s and d; more specifically, let $f(s, d) = (2d + 1)!((2s + d + 1) - 1)^{2d+1}$. We now consider the case where G has only a fixed-parameter number of vertices of degree at most $2d$.

Lemma 4. *If* $|D'| \leq f(s,d)$, *then* OPT-ISR *can be solved in fixed-parameter time with respect to* s *and* d.

Proof. Since $D' = D \setminus I_0$ and $|I_0| < s$, we have $|D| \leq |D'| + |I_0| < f(s,d) + s$. By Lemma 3 we thus have $|V(G)| \leq (2d+1)|D| < (2d+1)(f(s,d)+s)$. Therefore, $|V(G)|$ depends only on s and d. Then, this lemma follows from Theorem 5. $\square$

First Stage of Kernelization

We now consider the remaining case, that is, $|D'| > f(s,d)$ holds. The first stage of our kernelization focuses on "twins," two vertices having the same closed neighborhoods, and removes one of them without changing the answer.

Lemma 5. *Suppose that there exist two vertices* b_i *and* b_j *in* D' *such that* $N_G[b_i] = N_G[b_j]$. *Then,* (G, l, I_0, s) *is a yes-instance if and only if* $(G \setminus \{b_i\}, l, I_0, s)$ *is.*

Proof. We note that $b_i \notin I_0$ and $b_j \notin I_0$, because $D' = D \setminus I_0$. Then, the if direction clearly holds, and hence we prove the only-if direction. Suppose that (G, l, I_0, s) is a yes-instance, and hence G has an independent set I_{sol} such that $|I_{\mathsf{sol}}| \geq s$ and $I_0 \overset{l}{\leadsto} I_{\mathsf{sol}}$. Then, there exists a reconfiguration sequence $\mathcal{I} = \langle I_0, I_1, \ldots, I_\ell = I_{\mathsf{sol}} \rangle$. Since $N_G[b_i] = N_G[b_j]$, we know that b_i and b_j are adjacent in G and hence no independent set of G contains b_i and b_j at the same time. We now consider a new sequence $\mathcal{I}' = \langle I_0', I_1', \ldots, I_\ell' \rangle$ defined as follows: for each $x \in \{0, 1, \ldots, \ell\}$, let

$$I_x' = \begin{cases} I_x & \text{if } b_i \notin I_x; \\ (I_x \setminus \{b_i\}) \cup \{b_j\} & \text{otherwise.} \end{cases}$$

Since each I_x, $x \in \{0, 1, \ldots, \ell\}$, is an independent set of G and $N_G[b_i] = N_G[b_j]$, each I_x' forms an independent set of G. In addition, since $|I_{x-1} \triangle I_x| = 1$ for all $x \in \{1, 2, \ldots, \ell\}$, we have $|I_{x-1}' \triangle I_x'| = 1$. Therefore, $\mathcal{I}'$ is a reconfiguration sequence such that no independent set in $\mathcal{I}'$ contains b_i. Since $|I_\ell'| = |I_\ell| = |I_{\mathsf{sol}}| \geq s$, we can conclude that $(G \setminus \{b_i\}, l, I_0, s)$ is a yes-instance. $\square$

We repeatedly apply Lemma 5 to a given graph, and redefine G as the resulting graph; we also redefine D and D' according to the resulting graph G. Then, any two vertices b_i and b_j in D' satisfy $N_G[b_i] \neq N_G[b_j]$. If $|D'| \leq f(s,d)$, then we have completed our kernelization; recall Lemma 3. Otherwise, we will execute the second stage of our kernelization described below.

Second Stage of Kernelization

In the second stage of the kernelization, we use the classical result of Erdös and Rado [5], known as the *Sunflower Lemma*. We first define some terms used in the lemma. Let $P_1, P_2, \ldots, P_p$ be p non-empty sets over a universe U, and let $C \subseteq U$ which may be an empty set. Then, the family $\{P_1, P_2, \ldots, P_p\}$ is called a *sunflower* with a *core* C if $P_i \setminus C \neq \emptyset$ holds for each $i \in \{1, 2, \ldots, p\}$, and $P_i \cap P_j = C$ holds for each $i, j \in \{1, 2, \ldots, p\}$ satisfying $i \neq j$. The set $P_i \setminus C$ is called a *petal* of the sunflower. Note that a family of pairwise disjoint sets always forms a sunflower (with an empty core). Then, the following lemma holds.

Lemma 6 *(Erdős and Rado [5]). Let $\mathcal{A}$ be a family of sets (without duplicates) over a universe U such that each set in $\mathcal{A}$ is of size at most t. If $|\mathcal{A}| > t!(p-1)^t$, then there exists a family $\mathcal{S} \subseteq \mathcal{A}$ which forms a sunflower having p petals. Furthermore, $\mathcal{S}$ can be computed in time polynomial in $|\mathcal{A}|$, $|U|$, and p.*

We now explain the second stage of our kernelization, and make use of Lemma 6. Let $b_1, b_2, \ldots, b_{|D'|}$ denote the vertices in D', and let $\mathcal{A} = \{N_G[b_1], N_G[b_2], \ldots, N_G[b_{|D'|}]\}$ be the set of closed neighborhoods of all vertices in D'. In the second stage, recall that $N_G[b_i] \neq N_G[b_j]$ holds for any two vertices b_i and b_j in D', and hence no two sets in $\mathcal{A}$ are identical. We set $U = \bigcup_{b_i \in D'} N_G[b_i]$. Since each $b_i \in D'$ is of degree at most $2d$ in G, each $N_G[b_i] \in \mathcal{A}$ is of size at most $2d + 1$. Notice that $|\mathcal{A}| = |D'| > f(s, d) = (2d+1)!((2s+d+1)-1)^{2d+1}$. Therefore, we can apply Lemma 6 to the family $\mathcal{A}$ by setting $t = 2d + 1$ and $p = 2s+d+1$, and obtain a sunflower $\mathcal{S} \subseteq \mathcal{A}$ with a core C and p petals in time polynomial in $|\mathcal{A}|$, $|U|$, and $p = 2s+d+1$. Notice that $|\mathcal{A}| \leq n$ and $|U| \leq n$, and hence we can obtain $\mathcal{S}$ in time polynomial in n. Let $S = \{b_1', b_2', \ldots, b_p'\} \subseteq D'$ be the set of p vertices whose closed neighborhoods correspond to the sunflower $\mathcal{S}$, that is, $\mathcal{S} = \{N_G[b_1'], N_G[b_2'], \ldots, N_G[b_p']\} \subseteq \mathcal{A}$. We finally obtain the following lemma, as the second stage of the kernelization.

Lemma 7 *(*). Let b_q' be any vertex in S. Then, (G, l, I_0, s) is a yes-instance if and only if $(G \setminus \{b_q'\}, l, I_0, s)$ is.*

We can repeatedly apply Lemma 7 to G until the resulting graph has the corresponding vertex subset D' such that $|D'| \leq f(s, d)$. Then, by Lemma 4 we have completed our kernelization. This completes the proof of Theorem 6.

5 Conclusion

In this paper, we have introduced a new variant of reconfiguration, and studied OPT-ISR as the first example of the variant. As shown in Fig. 2 and Table 1, we have studied the problem from the viewpoints of polynomial-time solvability and the fixed-parameter (in)tractability, and shown several interesting contrasts among graph classes and parameters. In particular, we gave a complete analysis of the fixed-parameter (in)tractability with respect to the three parameters.

Acknowledgments. We thank Tatsuhiko Hatanaka for his insightful suggestions. We are grateful to Tesshu Hanaka, Benjamin Moore, Vijay Subramanya, and Krishna Vaidyanathan for valuable discussions with them.

References

1. Andrade, D., Resende, M., Werneck, R.: Fast local search for the maximum independent set problem. J. Heuristics **18**(4), 525–547 (2012)
2. Bonamy, M., Bousquet, N.: Token sliding on chordal graphs. In: Bodlaender, H.L., Woeginger, G.J. (eds.) WG 2017. LNCS, vol. 10520, pp. 127–139. Springer, Cham (2017). https://doi.org/10.1007/978-3-319-68705-6_10

3. Bousquet, N., Mary, A., Parreau, A.: Token jumping in minor-closed classes. In: Klasing, R., Zeitoun, M. (eds.) FCT 2017. LNCS, vol. 10472, pp. 136–149. Springer, Heidelberg (2017). https://doi.org/10.1007/978-3-662-55751-8_12
4. Brandstädt, A., Le, V., Spinrad, J.: Graph Classes: A Survey. SIAM, Philadelphia (1999)
5. Erdös, P., Rado, R.: Intersection theorems for systems of sets. J. Lond. Math. Soc. **35**, 85–90 (1960)
6. Frank, A.: Some polynomial algorithms for certain graphs and hypergraphs. In: Proceedings of BCC 1975, pp. 211–226 (1975)
7. Garey, M., Johnson, D.: Computers and Intractability: A Guide to the Theory of NP-Completeness. Freeman, San Francisco (1979)
8. van den Heuvel, J.: The complexity of change. In: Surveys in Combinatorics 2013. London Mathematical Society Lecture Note Series, vol. 409, pp. 127–160. Cambridge University Press (2013)
9. Hoang, D., Uehara, R.: Sliding tokens on a cactus. In: Proceedings of ISAAC 2016, pp. 37:1–37:26 (2016)
10. Ito, T., et al.: On the complexity of reconfiguration problems. Theoret. Comput. Sci. **412**(12–14), 1054–1065 (2011)
11. Ito, T., Kamiński, M., Ono, H.: Fixed-parameter tractability of token jumping on planar graphs. In: Ahn, H.-K., Shin, C.-S. (eds.) ISAAC 2014. LNCS, vol. 8889, pp. 208–219. Springer, Cham (2014). https://doi.org/10.1007/978-3-319-13075-0_17
12. Ito, T., Kamiński, M., Ono, H., Suzuki, A., Uehara, R., Yamanaka, K.: On the parameterized complexity for token jumping on graphs. In: Gopal, T.V., Agrawal, M., Li, A., Cooper, S.B. (eds.) TAMC 2014. LNCS, vol. 8402, pp. 341–351. Springer, Cham (2014). https://doi.org/10.1007/978-3-319-06089-7_24
13. Kamiński, M., Medvedev, P., Milanič, M.: Complexity of independent set reconfigurability problems. Theoret. Comput. Sci. **439**, 9–15 (2012)
14. Lick, D., White, A.: k-degenerate graphs. Can. J. Math. **22**, 1082–1096 (1970)
15. Lokshtanov, D., Mouawad, A.: The complexity of independent set reconfiguration on bipartite graphs. ACM Trans. Algorithms **15**(1), 7:1–7:19 (2019)
16. Lokshtanov, D., Mouawad, A., Panolan, F., Ramanujan, M., Saurabh, S.: Reconfiguration on sparse graphs. J. Comput. Syst. Sci. **95**, 122–131 (2018)
17. Mouawad, A., Nishimura, N., Raman, V., Simjour, N., Suzuki, A.: On the parameterized complexity of reconfiguration problems. Algorithmica **78**(1), 274–297 (2017)
18. Niedermeier, R.: Invitation to Fixed-Parameter Algorithms. Oxford University Press, Oxford (2006)
19. Nishimura, N.: Introduction to reconfiguration. Algorithms **11**(4), 52 (2018)
20. Pullan, W., Hoos, H.: Dynamic local search for the maximum clique problem. J. Artif. Intell. Res. **25**(1), 159–185 (2006)
21. Robertson, N., Seymour, P.: Graph minors. I. Excluding a forest. J. Comb. Theory Ser. B **35**, 39–61 (1983)
22. Schiex, T.: A note on CSP graph parameters. Technical report 1999/03, French National Institute for Agricultural Research (INRA) (1999)
23. Wrochna, M.: Reconfiguration in bounded bandwidth and tree-depth. J. Comput. Syst. Sci. **93**, 1–10 (2018)

Deconstructing Parameterized Hardness of Fair Vertex Deletion Problems

Ashwin Jacob[✉], Venkatesh Raman, and Vibha Sahlot

The Institute of Mathematical Sciences, HBNI, Chennai, India
{ajacob,vraman}@imsc.res.in, sahlotvibha@gmail.com

Abstract. For a graph G and a positive integer d, a set $S \subseteq V(G)$ is a fair set with the fairness factor d if for every vertex in G, at most d of its neighbors are in the S. In the Π-VERTEX DELETION problem, the aim is to find in a given graph a set S of minimum size such that $G \setminus S$ satisfies the property Π. We look at Π-FAIR VERTEX DELETION problem where we also want S to be a fair set.

It is known that the general Π-FAIR VERTEX DELETION problem where Π is expressible by a first order formula and is also given as input is $W[1]$-hard when parameterized by treewidth. Our first observation is that if we also parameterize by the fairness constant d, then Π-FAIR VERTEX DELETION is FPT (fixed-parameter tractable) even if Π-VERTEX DELETION can be expressed by a Monadic Second Order (MSO) formula. As a corollary we get an FPT algorithm for FAIR FEEDBACK VERTEX SET (FFVS) and FAIR VERTEX COVER (FVC) parameterized by solution size.

We then do a deep dive on FVC and more generally Π-FAIR VERTEX DELETION problems parameterized by solution size k when Π is characterized by a finite set of forbidden graphs. We show that these problems are FPT and develop a polynomial kernel when d is a constant. While the FPT algorithms use the standard branching technique, the fairness constraint introduces challenges to design a polynomial kernel. En route, we give a polynomial kernel for a special instance of MIN-ONES-SAT and FAIR q-HITTING SET, a generalization of q-hitting set, with a fairness constraint on an underlying graph structure on the universe. These could be of independent interest.

To complement our FPT results, we show that FAIR SET and FAIR INDEPENDENT SET problems are $W[1]$-hard even in 3-degenerate graphs when the fairness factor is 1. We also show that FVC is polynomial-time solvable when $d = 1$ or 2, and NP-hard for $d \geq 3$, and that FFVS is NP-hard for all $d \geq 1$.

Keywords: Vertex deletion problems · Parameterized complexity · Kernelization · W-hardness

1 Introduction and Motivation

Vertex deletion problems are extensively studied in multiple paradigms of algorithms and complexity. Given a graph class Π, in Π-VERTEX DELETION, the

© Springer Nature Switzerland AG 2019
D.-Z. Du et al. (Eds.): COCOON 2019, LNCS 11653, pp. 325–337, 2019.
https://doi.org/10.1007/978-3-030-26176-4_27

objective is to find in a given graph, the minimum number of vertices whose deletion results in a graph in graph class Π. Many well-studied problems like VERTEX COVER and FEEDBACK VERTEX SET can be modeled as Π-VERTEX DELETION. For example, Π is the class of edgeless graphs in the case of VERTEX COVER and the class of acyclic graphs in the case of FEEDBACK VERTEX SET. The result by Lewis and Yannakakis showed that for the hereditary graph classes that are not trivial, Π-VERTEX DELETION is NP-complete [13].

Recently, there is an interest on these vertex deletion problems where along with optimizing the vertex deletion set, we want the deletion set to be fair in the sense that it does not have too many vertices from the neighborhood of any vertex [9,10,15].

We formalize this below. Given a graph $G = (V, E)$ and a positive integer d, a set $S \subseteq V$ is *fair* if it contains at most d vertices from the neighborhood of each vertex. That is, for each vertex $v \in G$, $|N(v) \cap S| \le d$. We call d the *fairness factor* of S. Given a set $S \subseteq V(G)$, checking whether S is a fair set can be done in polynomial time by going over the neighborhoods of all the vertices in G. If for some vertex v, $|N(v) \cap S| > d$, we say that the fairness constraint for v with respect to S is violated.

Π-Fair Vertex Deletion

We define Π-FAIR VERTEX DELETION problem as follows:

Π-FAIR VERTEX DELETION
Input: A graph $G = (V, E)$ and $k, d \in \mathbb{N}$.
Question: Does there exist a set $S \subseteq V(G)$ of at most k vertices such that $G[V - S]$ belongs to a hereditary graph class Π and for each vertex $v \in V(G)$, $|N(v) \cap S| \le d$?

Using Π-FAIR VERTEX DELETION, we can define FAIR VERTEX COVER (FVC) and FAIR FEEDBACK VERTEX SET (FFVS) where Π is the class of edgeless graphs and acyclic graphs respectively.

1.1   Previous Work and Deconstructing Hardness

In [15], the general problem Π-FAIR VERTEX DELETION where Π can be expressed by a first order formula ϕ, also given as input, is shown to be hard for the parameterized complexity class $W[1]$, when parameterized by the sum of treedepth and the size of the minimum feedback vertex set in the graph. In the same paper, an FPT algorithm is presented when Π can be expressed by a Monadic Second Order (MSO) logic formula given as input, parameterized by the neighborhood diversity of the graph. Knop et al. [10] showed that Π-FAIR VERTEX DELETION where Π can be expressed by a formula with one free variable is FPT parameterized by the twin cover number of the graph. They also showed that FAIR VERTEX COVER is $W[1]$-hard when parameterized by both treedepth and feedback vertex set of the input graph. Fair edge deletion problems were also studied in [11,14].

Since Π-FAIR VERTEX DELETION problems like FVC are $W[1]$-hard parameterized by the minimum feedback vertex set of the graph, they are $W[1]$-hard parameterized by the treewidth t of the graph as the latter is a smaller parameter. But we note that in the reduction used in the proofs, the fairness factor d is in $\Omega(n)$. Inspired by the parameter ecology program of Fellows et al. [7], we feel that, in these problems it is more natural to consider d also as a parameter.

1.2   Our Results

We first show that Π-FAIR VERTEX DELETION is FPT parameterized by $t + d$ if Π-VERTEX DELETION can be expressed even by an MSO logic formula of constant length. This includes the fair version of a huge class of well-studied vertex deletion problems like FEEDBACK VERTEX SET, ODD CYCLE TRANSVERSAL, CLUSTER VERTEX DELETION. Unfortunately, the FPT algorithms obtained above have a large exponential dependence on k in the running time. Hence we study two classic problems FAIR VERTEX COVER and FAIR FEEDBACK VERTEX SET and give dynamic programming algorithms with much better running time parameterized by $t + d$. As a corollary, we get an FPT algorithm for FFVS parameterized by solution size as well. We remark that the standard reduction rules of FFVS don't seem to have an easy implementation to maintain the fairness constraint, and hence we do not know of other FPT algorithms for FFVS parameterized by solution size.

We then take a closer look at FAIR VERTEX COVER and more generally Π-FAIR VERTEX DELETION where the graph class Π has a finite forbidden set $\mathcal{F}$. We call the forbidden set $\mathcal{F}$ a q-forbidden set if the vertex set of each graph in $\mathcal{F}$ is of size at most q. In a graph G, we say a subset $S \subseteq V(G)$ hits all graphs in $\mathcal{F}$ when for all induced subgraphs H of G such that H is isomorphic to some member in $\mathcal{F}$, $S \cap V(H) \neq \emptyset$.

FAIR q-FORBIDDEN SET VERTEX DELETION
Input: Given a graph $G = (V, E)$, a q-forbidden set $\mathcal{F}$ where $q, k, d \in \mathbb{N}$.
Question: Does there exist a subset $S \subseteq V(G)$ of at most k vertices such that S hits all the occurrences of graphs in family $\mathcal{F}$ in G and for each vertex $v \in V(G)$, $|N(v) \cap S| \leq d$?

We define a hitting set variant with fairness and use it to solve FAIR q-FORBIDDEN SET VERTEX DELETION.

FAIR q-HITTING SET
Input: Given a graph $G = (V, E)$, a family $\mathcal{F}$ of subsets of $V(G)$ of size at most q with $|\mathcal{F}| = m$ and $|V(G)| = n$ where $q, k, d, n, m \in \mathbb{N}$.
Question: Does there exist a subset $S \subseteq V(G)$ of at most k vertices such that S hits all the sets in $\mathcal{F}$ and for each vertex $v \in V(G)$, $|N(v) \cap S| \leq d$?

The problem is different from the normal q-HITTING SET as there is a graph on the universe elements and the problem requires to pick a fair set from the universe. We give an $\mathcal{O}^{\star}(q^k)$ [1]FPT algorithm for FAIR q-HITTING SET parameterized by solution size k. As FAIR q-FORBIDDEN SET VERTEX DELETION can be reduced to FAIR q-HITTING SET, we also have an FPT algorithm for the former.

The more challenging task is to design a polynomial kernel for FAIR q-HITTING SET as here again the standard reduction rules for q-Hitting Set do not keep track of the fairness constraint. We obtain our kernel by reducing the problem instance to a MIN-ONES-SAT instance formula where all the clauses are constant-sized and each clause is either monotone or anti-monotone. In MIN-ONES-SAT, we are given a CNF formula and we need to find out the satisfying assignment for the formula such that minimum number of variables set to true. We show that this variant of MIN-ONES-SAT has a polynomial kernel parameterized by number of variables set to true. The latter result is of independent interest as in contrast, the general MIN-ONES-SAT problem even when all the clauses are of size 3 does not have a polynomial kernel [12] under same parameterization, unless NP $\subseteq$ coNP/poly. The FVC problem can also be thought of as a special instance of FAIR q-FORBIDDEN SET VERTEX DELETION with the q-Forbidden Set being the single edge K_2. For FVC we obtain a much better kernel through a different technique.

Finally we complement our FPT results with some hardness results. We show that FAIR SET (defined below) and FAIR INDEPENDENT SET are $W[1]$-hard even in 3-degenerate graphs, with fairness factor 1 when parameterized by solution size.

Fair Set

FAIR SET
Input: A graph $G = (V, E)$ and $k, d \in \mathbb{N}$.
Question: Does there exist a set $S \subseteq V(G)$ of at least k vertices such that for each vertex $v \in V(G)$, $|N(v) \cap S| \leq d$?

We note that the FAIR SET can be seen as a special case of (σ, ρ) DOMINATING SET [17] where the sets σ and $\rho = \{0, 1, \ldots, d\}$. The (σ, ρ) DOMINATING SET is shown to be FPT when parameterized by treewidth when the sets σ and ρ are finite or cofinite [18] with running time $\mathcal{O}^{*}((st)^{s-2}s^t)$ where t is the treewidth of the graph and s is the number of states in the sets. As the number of states in the FAIR SET can be shown to be $d + 1$, the problem is FPT parameterized by sum of treewidth t and fairness factor d with running time $\mathcal{O}^{*}(((d+1)t)^{d-1}(d+1)^t)$.

A FAIR DOMINATING SET is a dominating set which is also a fair set. A problem very closely related to FAIR DOMINATING SET named $[1, j]$ DOMINATING SET has been studied recently [1] where it is shown to be $W[1]$-hard in graphs of degeneracy $j+1$ and FPT in nowhere dense graphs parameterized by the solution size. We note that in the $W[1]$-hardness reduction, the $[1, j]$ dominating set is also a fair dominating set with fairness j and the FPT algorithm for nowhere dense graphs can be easily extended to FAIR DOMINATING SET.

---

[1] $\mathcal{O}^*$ notation ignores polynomial factors.

Concerning the solvability of FVC and FFVS for specific values of d, we note that FVC and FFVS are NP-hard when $d = 3$ and $d = 4$ respectively as VERTEX COVER is NP-hard in cubic graphs [8] and FEEDBACK VERTEX SET is NP-hard on graphs with degree at most 4 [16]. For FVC, we show that the problem is polynomial-time solvable when $d = 1$ or 2. For FFVS, we complete the picture by showing that the problem is NP-hard even when $d \in \{1, 2, 3\}$.

We end this section with the following observation, which essentially says that when the parameter is the solution size, Π-FAIR VERTEX DELETION problems are interesting only when $d \leq k$.

Observation 1. *When $d \geq k$, the Π-FAIR VERTEX DELETION problem turns into the standard Π-VERTEX DELETION problem as every vertex has at most $k \leq d$ neighbors in the solution. Hence when $d \geq k$, Π-FAIR VERTEX DELETION is FPT when parameterized by solution size k whenever the corresponding Π-VERTEX DELETION problem (without the fairness constraint) is FPT when parameterized by solution size.*

2 Preliminaries

Given a graph $G = (V, E)$, V and E denote its vertex-set and edge-set, respectively. For a graph $G = (V, E)$, we define $G[V']$ to be the graph induced on vertex set $V' \subseteq V$. For sets V and V, we denote $V \setminus V'$ as $V - V'$. If V' is singleton, say $\{a\}$, then we denote $V - \{a\}$ as $V - a$. Next, $G \setminus V'$ denotes the graph induced on vertex set $V - V'$. For a vertex $v \in V$, $G - v$ denotes graph induced on vertex set $V - v$. For a vertex $v \in V$, the open neighborhood of the vertex is denoted by $N(v)$ and closed neighborhood of the vertex is denoted by $N[v]$, that is $N(v) = \{u | (u, v) \in E(G) \text{ and } u \neq v\}$ and $N[v] = N(v) \cup \{v\}$. We use standard notation and terminology from the book of [6] for graph-related terms which are not explicitly defined here.

Definition 1 (Degeneracy). *A graph $G = (V, E)$ is said to be d'-degenerate if there exists an ordering of V such that each vertex has at most d' neighbors to its right in the ordering. The minimum such possible d' a graph can have is called its degeneracy.*

The basic definitions related to parameterized complexity, Monadic Second Order logic, tree decomposition, nice tree decomposition, treewidth, cluster vertex deletion set, sunflower and Sunflower Lemma can be found in [4].

3 FAIR VERTEX DELETION Parameterized by Treewidth + Fairness Factor

Theorem 1 (Courcelle's theorem [3]). *Given an MSO formula ϕ, an n-vertex graph G and a tree decomposition of G of width t, there exists an algorithm that verifies whether ϕ is satisfied in G in time $f(|\phi|, t) \cdot n$ for some computable function f and $|\phi|$ denoting the length of encoding of ϕ as a string.*

Theorem 2. *The Π-FAIR VERTEX DELETION is FPT parameterized by the sum of treewidth and fairness factor if the corresponding Π-VERTEX DELETION problem can be expressed by an MSO formula of constant length.*

Proof. Let ϕ_1 be the constant length formula expressing the Π-VERTEX DELETION. We can express the fairness property with fairness factor d by the following MSO logic formula ϕ_2

$$\phi_2 := \exists S \; \forall u \in V \; \nexists v_1, \ldots, v_{d+1} \in S \text{ such that } \{(u, v_1), \ldots, (u, v_{d+1})\} \in E \quad (1)$$

It can be seen that the length of the formula ϕ_2 is linear in d. Let ϕ_1' and ϕ_2' be the formula strings obtained after removing the prefix $\exists S$ from ϕ_1 and ϕ_2 respectively. The Π-FAIR VERTEX DELETION problem can be expressed by MSO formula $\phi = \exists S(\phi_1' \wedge \phi_2')$ which is of length linear in d. Hence by Courcelle's theorem, the result follows. □

Most of the Π-VERTEX DELETION problems like VERTEX COVER and FEEDBACK VERTEX SET can be expressed by a constant length MSO formula. Hence the above theorem gives FPT algorithms for these problems. But the running time of these FPT algorithms have huge exponents. So we focus on FVC and FFVS and give better FPT algorithms using dynamic programming on tree decompositions.

Theorem 3. $(\star)$ [2] *FVC can be solved in running time $\mathcal{O}^*(2^\omega (d+1)^{3\omega})$ on graphs of treewidth ω if a tree decomposition of width ω is given as input.*

In FAIR VERTEX COVER problem on a graph having a cluster vertex deletion size k, if a fair solution exists, then the treewidth of the graph can be proved to be at most $k + d$. This gives the following corollary.

Corollary 1. $(\star)$ *FVC is FPT parameterized by the sum of cluster vertex deletion size and fairness factor.*

Theorem 4. $(\star)$ *FFVS can be solved in running time $\mathcal{O}^*((\omega(d+1))^{2\omega})$ on graphs of treewidth ω if a tree decomposition of width ω is given as input.*

Corollary 2. $(\star)$ *FFVS is FPT parameterized by solution size k.*

Corollary 3. $(\star)$ *FFVS is FPT parameterized by cluster vertex deletion size.*

We use the Cut and Count Technique introduced by Cygan et al. [5] to get an improved algorithm for FFVS.

Theorem 5. $(\star)$ *Given a tree decomposition of the graph G of width ω, there is a randomized algorithm running in $\mathcal{O}^*(3^\omega (d+1)^{3\omega})$ time that either states that there exists an FFVS of size at most k or that it could not verify this hypothesis. If there indeed exists such a set, the algorithm will return "unable to verify" with probability at most $1/2$.*

---

[2] The proofs of Theorems, Lemmas and safeness of Reduction Rules marked $\star$ with some omitted details can be found in full version of the paper.

4   FAIR q-HITTING SET Parameterized by Solution Size

4.1   FPT Algorithm

We first give a $\mathcal{O}^{\star}(q^k)$ algorithm for FAIR q-HITTING SET via the standard branching technique.

Theorem 6. ($\star$) FAIR q-HITTING SET *can be solved in* $\mathcal{O}^{\star}(q^k)$ *time.*

We can give a parameter preserving reduction for FAIR q-FORBIDDEN SET VERTEX DELETION to FAIR q-HITTING SET as follows. Given the finite forbidden set $\mathcal{F}$, we can construct a family $\mathcal{F}'$ containing at most q-sized subsets $S \subseteq V(G)$ such that $G[S]$ is isomorphic to a member of $\mathcal{F}$. Even though $\mathcal{F}'$ is not explicitly given, since q is a constant, we can find the family of induced subgraphs of G that are isomorphic to at least an element of the q-forbidden set $\mathcal{F}$ in $\mathcal{O}(n^q)$ time. Hence if there exists a member H of $\mathcal{F}'$ which is not hit, we can find it in polynomial time and one of the vertices of the corresponding subgraph H has to go in the solution. This gives rise to the following corollary.

Corollary 4. FAIR q-FORBIDDEN SET VERTEX DELETION *is* FPT *parameterized by solution size* k *with running time* $\mathcal{O}^{\star}(q^k)$.

Since FVC is a special case of FAIR q-FORBIDDEN SET VERTEX DELETION with the forbidden set being an edge, we have the following corollary.

Corollary 5. *FVC can be solved in* $\mathcal{O}^{\star}(2^k)$ *running time.*

4.2   Polynomial Kernel

We noticed that the standard Sunflower Lemma based reduction rules used to obtain a kernel for q-HITTING SET do not work when we bring fairness constraints. Hence we take a different approach by casting the problem as a special case of the well studied MIN ONES-SAT problem.

A clause of a formula in the conjunctive normal form (CNF) is *monotone* if all its literals are positive. If all its literals are negative, we call it *anti-monotone*. Let us define the following problem.

MIN-ONES-MONOTONE/ANTI-MONOTONE l-SAT
Input: A CNF formula ϕ on n variables and m clauses such that all the clauses have at most l variables and are either monotone or anti-monotone.
Question: Does there exist an assignment A with at most k variables set to true that satisfies ϕ?

Theorem 7. MIN-ONES-MONOTONE/ANTI-MONOTONE ℓ-SAT *parameterized by* k *has a polynomial kernel for constant* ℓ.

Proof. Let $\phi = \phi_1 \wedge \phi_2$ where ϕ_1 is the conjunction of all the monotone clauses in ϕ and ϕ_2 is the conjunction of all the anti-monotone clauses in ϕ. Let a and b be the maximum size of the clauses in ϕ_1 and ϕ_2, respectively. We have the following reduction rules.

Reduction Rule 1: Since ϕ_1 is a monotone formula, it can be treated as a set system with the universe being the variables and the family of sets being the monotone clauses. Suppose that in this corresponding set system of the formula ϕ_1, there exists a sunflower $\mathcal{S} = \{C_1, \ldots, C_{k+1}\}$ with the core set of positive literals $Y = \{x_{y_1}, \ldots, x_{y_p}\}$. If Y is empty, then we return NO. Else in the formula ϕ we remove all the clauses C_i with $i \in [k+1]$ and add the clause $C_Y = x_{y_1} \vee \ldots \vee x_{y_q}$. If $|Y| = 1$, we set the corresponding variable to true in all the clauses in the formula.

Claim. ($\star$) Reduction Rule 1 is safe and can be performed in polynomial time.

Reduction Rule 2 ($\star$): Let x_v be a variable that is present only in negative form in ϕ, then delete all the clauses containing x_v.

Claim. If reduction rules 1 and 2 are not applicable, then ϕ has $\mathcal{O}((a! k^a \cdot a^2)^b)$ clauses.

Proof. Look at the number N of monotone clauses of size $a' \in [a]$ in ϕ. If $N > a'! k^{a'}$, then by Sunflower Lemma there exists a sunflower in ϕ. Hence we can apply Reduction Rule 1 which is not the case. Hence the number of monotone clauses is at most $\sum_{a'=1}^{a} a'! k^{a'} \leq a! k^a \cdot a$. Let M be the set of variables appearing in these clauses. We have $|M| \leq a! k^a \cdot a^2$.

Now we look at the anti-monotone clauses of ϕ. There are at most $\binom{M}{b}$ clauses here containing only variables in M. In the rest of the clauses there exists a variable that is occurring only in the anti-monotone clauses of ϕ and hence only in negative form in ϕ. Reduction rule 2 would remove all such clauses.

Hence the number of clauses in ϕ is bounded by $a! k^a \cdot a + \binom{a! k^a \cdot a^2}{b} = \mathcal{O}((a! k^a \cdot a^2)^b)$. $\qquad\square$

When $a, b \leq l$ are constants, the input size is polynomial in k. $\qquad\square$

Theorem 8. *There is a polynomial kernel for* FAIR q-HITTING SET *parameterized by the solution size k when both q and the fairness factor d are constants.*

Proof. Let $l = max\{q, d+1\}$. Given an instance $(G, \mathcal{F}, k)$ of FAIR q-HITTING SET, we construct an instance of MIN-ONES-MONOTONE/ANTI-MONOTONE l-SAT which is the formula ϕ as follows:

For each vertex $u \in V$, we have a variable x_u. We define two types of clauses:

- Monotone clauses: For each set $S \in \mathcal{F}$ with $S = \{u_1, \ldots, u_q\}$, we define the clause $C_S = x_{u_1} \vee \ldots \vee x_{u_q}$.

- Anti-monotone clauses: For each vertex $u \in V(G)$, we look at the open neighborhood set $N(u)$. For all sets $D \subseteq N(u)$ of size $d+1$, say $D = \{v_1, \ldots, v_{d+1}\}$, we construct a clause $C_D = \overline{x}_{v_1} \vee \ldots \vee \overline{x}_{v_{d+1}}$.

If we look at any $(d+1)$-sized set from the open neighbourhood of any vertex $u \in V$, at least one of the vertices cannot be in the solution as otherwise it violates the fairness of vertex u. This is captured by the anti-monotone clauses C_D.

The formula ϕ is the conjunction of all the clauses C_S and C_D. Note that since $|N(u)| \leq n-1$, the number of clauses is $\mathcal{O}(m+n^{d+1} \cdot n)$ which is polynomial in n and m for a constant d. Let the formula formed by the conjunction of all the monotone clauses be ϕ_1 and by all the anti-monotone clauses be ϕ_2. Note that here $a = q$ and $b = d+1$.

Claim. $(\star)$ There is a solution of size k for FAIR q-HITTING SET instance $(G, \mathcal{F}, k)$ if and only if there is a solution of size k for MIN-ONES-MONOTONE/ANTI-MONOTONE l-SAT instance ϕ.

From Theorem 7, we have formula ϕ' of size $\mathcal{O}((q!k^q \cdot q^2)^{d+1})$. Since both the problems are NP-complete, there is a polynomial time reduction from MIN-ONES-MONOTONE/ANTI-MONOTONE SAT back to FAIR q-HITTING SET. We use this reduction on ϕ' to get a FAIR q-HITTING SET instance of size $|\phi'|^{\mathcal{O}(1)} = \mathcal{O}(((q!k^q \cdot q^2)^{d+1})^{\mathcal{O}(1)})$ which is polynomial in k when q and d are constants. $\square$

As FVC is a special case of FAIR q-HITTING SET with $q = 2$, we have

Corollary 6. *FVC parameterized by solution size k has a kernel of size $\mathcal{O}(k^{\mathcal{O}(d)})$.*

4.3 Improved Kernel for FVC Parameterized by Solution Size

For FVC, we observe that we can get an improved kernel by modifying the classical Buss kernel [2] for VERTEX COVER.

We apply the following reduction rules in sequence, only once.

Reduction Rule 1. *Delete all isolated vertices in (G, k).*

The above reduction rule is safe as isolated vertices do not cover any edge.

Reduction Rule 2. $(\star)$ *Let H be the set of vertices in G having degree greater than d. If $|H| > k$ or H is not a fair set, then return NO. Else delete all the isolated vertices in $G[V - H]$. Add $d+1$ many pendant vertices adjacent to each $v \in H$. Return the resulting graph as (G', k).*

Reduction Rule 3. $(\star)$ *Let (G, k) be an input instance on which reduction rules 1 and 2 are not applicable. Let H be the set of vertices in G having degree greater than d. If there are more than $k \cdot d$ edges in $G[V - H]$, then return NO.*

Reduction Rule 4. $(\star)$ *Let (G, k) be an input instance on which reduction rules 1 and 2 are not applicable. If G has more than $3kd + 2k$ vertices or $2k^2 d + kd + k + k^2 + kd^2$ edges, then return NO.*

The reduction rules lead to the following theorem.

Theorem 9. $(\star)$ *There exists a kernel for FVC with $\mathcal{O}(kd)$ vertices and $\mathcal{O}(k^2d)$ edges.*

5   Hardness Results

5.1   W[1]-Hardness for Fair Set

Theorem 10. FAIR SET *with $d = 1$ is $W[1]$-hard when parameterized by solution size k for graphs with degeneracy three.*

Proof. We give a reduction from the MULTICOLORED INDEPENDENT SET problem known to be $W[1]$-hard [4] defined as follows.

MULTICOLORED INDEPENDENT SET
Input: A graph $G = (V, E)$ and partition of $(V_1, \ldots, V_k)$ of V for $k \in \mathbb{N}$.
Question: Does there exist a set $S \subseteq V$ of k vertices such that S forms an independent set and for each vertex V_i, $|V_i \cap S| = 1$?

Let $(G, V_1, \ldots, V_k)$ be the MULTICOLORED INDEPENDENT SET instance. Without loss of generality, assume that $G[V_i]$ is an independent set for all $i \in [k]$. We construct an instance $(G', k + 2, 1)$ of FAIR SET with $d = 1$ as follows:

We start constructing G' with the same vertex set of G. For each class V_i, we introduce a vertex v_i and make it adjacent to all the vertices in V_i. For each edge $e_j = (u, v) \in E(G)$, we add a vertex e_j in G' and add edges (u, e_j) and (e_j, v). We also add a vertex s adjacent to all edge vertices e_j for $j \in [m]$ and the vertices $v_1, v_2, \ldots, v_k$. Finally we add vertices t adjacent to s and t' adjacent to t.

Claim. $(\star)$ $(G, V_1, \ldots, V_k)$ is a yes instance for MULTICOLORED INDEPENDENT SET if and only if $(G', k + 2, 1)$ is a yes instance for FAIR SET.

Now we look at the degeneracy of the graph G'. We give the degeneracy order where we first put edge vertices $e_1, \ldots, e_m$, then all the vertices in $V_1, \ldots, V_k$ and then vertices $v_1, \ldots, v_k, s, t, t'$ in that order. It can be verified that the degeneracy of the graph is 3 from this order.   □

A fair independent set is an independent set which is also a fair set. The reduction in Theorem 10 can be slightly modified to give a $W[1]$-hardness result for FAIR INDEPENDENT SET problem.

Theorem 11. *The FAIR INDEPENDENT SET problem is $W[1]$-hard parameterized by solution size k for graphs with degeneracy three.*

5.2   NP-Hardness of FVC and FFVS

Theorem 12. ($\star$) *FVC is polynomial-time solvable when $d = 1, 2$ and is NP-hard when $d \geq 3$.*

Proof. Since vertex cover is NP-hard on subcubic graphs [8], we know that FVC is NP-hard for graph with $d \geq 3$.

Case $d = 1$: Let S denote the set of all vertices of the graph G with degree more than 1. All the vertices $v \in S$ have to go into the solution of FAIR VERTEX COVER as otherwise the fairness of v is violated. Hence if S is not a fair set, return NO. Since the vertices in $G \setminus S$ has degree at most 1, it consists of isolated vertices and edges. Note that the endpoints of the isolated edges have no neighbors to S as well. Create set T by picking an arbitrary vertex from every isolated edge. We an easily see that $T \cup S$ is the minimum sized FAIR VERTEX COVER in G.

Case $d = 2$: Let S denote the set of all vertices of the graph G with degree more than 2. All the vertices $v \in S$ have to go into the solution of FAIR VERTEX COVER as otherwise the fairness of v is violated. Hence if S is not a fair set, return NO. Since the vertices in $G \setminus S$ has degree at most two, it has isolated vertices, paths and cycles. Note that since all the vertices of a cycle has degree two, there is no edge from any vertex in cycle to S. Hence we can arbitrarily pick alternate vertices of the cycles in $G \setminus S$ into the solution.

Look at a path $P = (v_1, v_2, \ldots, v_l)$ in $G \setminus S$ with $l > 2$. Since all the internal vertices of P have degree two, there are no edges from any such vertex to S. If l is odd, picking all vertices v_{2i} which are internal vertices will cover all the edges of P. If l is even, picking all vertices v_{2i} expect v_l and the vertex v_{l-1} will cover all the edges of P without violating any fairness constraint. Hence all the edges of G are covered except isolated edges in $G \setminus S$. Look at any isolated edge (u, v). If u or v does not have any edges to S, then pick the corresponding vertex into the solution. Hence assume that both u and v have edges to S. Since u and v has degree at most two, they have a unique neighbor in S, n_u and n_v respectively. Note that $G[S]$ is has degree at most two as otherwise fairness is violated for some vertex. If n_u or n_v is a degree two vertex in $G[S]$, then the corresponding vertex u or v cannot go into the solution as fairness of n_u or n_v is violated. Hence both n_u and n_v are either isolated vertices in $G[S]$ or endpoints of some path in $G[S]$.

Let A denote the isolated edges (u, v) in $G \setminus S$ remaining to be covered and B denote set of vertices n_u where u is an endpoint of these edges. The problem reduces to picking exactly one endpoint of each isolated edge in $G \setminus S$ such that for all vertices $v \in B$, the number of vertices pick is at most one or two depending on whether v is an endpoint of a path or an isolated edge respectively. To solve this problem, we construct a bipartite graph $H = (A, B')$ from G with $B' = B \cup I$ where I is a copy of all the isolated vertices in $G[S]$ which are present in B. We add edges (a, b) for $a \in A$ and $b \in B'$ when $(a_0, b_0) \in E$ where a_0 is one of the endpoints of the edge a and $b_0 \in B$ is the corresponding original vertex in S.

Claim. (⋆) There is a matching saturating A in H if and only if there is a solution for FAIR VERTEX COVER in G.

Hence we have a polynomial-time algorithm for FVS with d $= 2$ since we can find the above matching if it exists in polynomial time.   □

Since FEEDBACK VERTEX SET is NP-hard on graphs with maximum degree 4 [16], we know that FFVS is NP-hard when $d \geq 4$. We complete the picture with the following theorem where we give a reduction from 3-SAT.

Theorem 13. (⋆) *FFVS is* NP-*hard for* $d \in \{1, 2, 3\}$.

6   Conclusion

We initiated a systematic study on various $\varPi$-FAIR VERTEX DELETION problems under various parameterizations. An open problem is to give a polynomial kernel for FFVS parameterized by solution size. Also finding FPT algorithms for other $\varPi$-FAIR VERTEX DELETION problems like FAIR ODD CYCLE TRANSVERSAL remains open.

References

1. Alambardar Meybodi, M., Fomin, F., Mouawad, A.E., Panolan, F.: On the parameterized complexity of [1, j]-domination problems. In: FSTTCS 2018 (2018)
2. Buss, J.F., Goldsmith, J.: Nondeterminism within P. In: Choffrut, C., Jantzen, M. (eds.) STACS 1991. LNCS, vol. 480, pp. 348–359. Springer, Heidelberg (1991). https://doi.org/10.1007/BFb0020811
3. Courcelle, B.: The monadic second-order logic of graphs. I. Recognizable sets of finite graphs. Inf. Comput. **85**(1), 12–75 (1990)
4. Cygan, M., et al.: Parameterized Algorithms. Springer, Cham (2015). https://doi.org/10.1007/978-3-319-21275-3
5. Cygan, M., Nederlof, J., Pilipczuk, M., Pilipczuk, M., van Rooij, J.M., Wojtaszczyk, J.O.: Solving connectivity problems parameterized by treewidth in single exponential time. In: FOCS, pp. 150–159. IEEE (2011)
6. Diestel, R.: Graph Theory. Springer, Heidelberg (2005)
7. Fellows, M.R., Jansen, B.M., Rosamond, F.: Towards fully multivariate algorithmics: parameter ecology and the deconstruction of computational complexity. Eur. J. Comb. **34**(3), 541–566 (2013)
8. Garey, M.R., Johnson, D.S., Stockmeyer, L.: Some simplified NP-complete graph problems. TCS **1**(3), 237–267 (1976)
9. Knop, D., Koutecký, M., Masařík, T., Toufar, T.: Simplified algorithmic metatheorems beyond MSO: treewidth and neighborhood diversity. In: Bodlaender, H.L., Woeginger, G.J. (eds.) WG 2017. LNCS, vol. 10520, pp. 344–357. Springer, Cham (2017). https://doi.org/10.1007/978-3-319-68705-6_26
10. Knop, D., Masařík, T., Toufar, T.: Parameterized complexity of fair deletion problems II. CoRR abs/1803.06878 (2018)
11. Kolman, P., Lidický, B., Sereni, J.S.: On fair edge deletion problems. Manuscript (2009). http://kam.mff.cuni.cz/kolman/papers/kls09.pdf

12. Kratsch, S., Wahlström, M.: Two edge modification problems without polynomial kernels. Discrete Optim. **10**(3), 193–199 (2013)
13. Lewis, J.M., Yannakakis, M.: The node-deletion problem for hereditary properties is NP-complete. JCSS **20**(2), 219–230 (1980)
14. Lin, L., Sahni, S.: Fair edge deletion problems. IEEE Trans. Comput. **38**(5), 756–761 (1989)
15. Masařík, T., Toufar, T.: Parameterized complexity of fair deletion problems. In: Gopal, T.V., Jäger, G., Steila, S. (eds.) TAMC 2017. LNCS, vol. 10185, pp. 628–642. Springer, Cham (2017). https://doi.org/10.1007/978-3-319-55911-7_45
16. Rizzi, R.: Minimum weakly fundamental cycle bases are hard to find. Algorithmica **53**(3), 402–424 (2009)
17. Telle, J.A.: Complexity of domination-type problems in graphs. Nord. J. Comput. **1**(1), 157–171 (1994)
18. van Rooij, J.M.M., Bodlaender, H.L., Rossmanith, P.: Dynamic programming on tree decompositions using generalised fast subset convolution. In: Fiat, A., Sanders, P. (eds.) ESA 2009. LNCS, vol. 5757, pp. 566–577. Springer, Heidelberg (2009). https://doi.org/10.1007/978-3-642-04128-0_51

On 1-Factorizations of Bipartite Kneser Graphs

Kai Jin$^{(\boxtimes)}$ (iD)

The Hong Kong University of Science and Technology, Kowloon, Hong Kong SAR
cscjjk@gmail.com

Abstract. It is a challenging open problem to construct an explicit 1-factorization of the bipartite Kneser graph $H(v,t)$, which contains as vertices all t-element and $(v-t)$-element subsets of $[v] := \{1, \ldots, v\}$ and an edge between any two vertices when one is a subset of the other. In this paper, we propose a new framework for designing such 1-factorizations, by which we solve a nontrivial case where $t = 2$ and v is an odd prime power. We also revisit two classic constructions for the case $v = 2t + 1$—the *lexical factorization* and *modular factorization*. We provide their simplified definitions and study their inner structures. As a result, an optimal algorithm is designed for computing the lexical factorizations. (An analogous algorithm for the modular factorization is trivial.)

Keywords: Graph Theory: 1-factorization · Modular factorization · Lexical factorization · Bipartite Kneser graph · Perpendicular array

1 Introduction

The *bipartite Kneser graph* $H(v,t)$ $(t < v/2)$ has as vertices all t-element and $(v-t)$-element subsets of $[v] := \{1, \ldots, v\}$ and an edge between any two vertices when one is a subset of the other. Because it is regular and bipartite, each bipartite Kneser graph admits a 1-factorization due to Hall's Marriage Theorem [7]. (A 1-factor of a graph G is a subgraph in which each node of G has degree 1, and a 1-factorization of G partitions the edges of G into disjoint 1-factors.) For the special case $v = 2t + 1$, the graph $H(2t + 1, t)$ is also known as the *middle level graph* and it admits two explicit 1-factorizations – the *lexical factorization* [10] (see Subsect. 1.2) and *modular factorization* [4] (see Sect. 4). However, to the best of our knowledge, for decades it remains a challenging open problem to design explicit 1-factorizations for the general bipartite Kneser graphs.

In this paper, we propose a natural framework to attack the open problem (Sect. 2). It attempts to find a special kind of 1-factorizations called resolvable 1-factorizations. We noticed that the lexical and modular factorizations and any 1-factorization of $H(2t+1, t)$ are resolvable. We also checked (by a C++ program) that there are no resolvable 1-factorization for $(v, t) = (6, 2)$. Therefore, we can only expect for solving part of the open problem using this framework.

© Springer Nature Switzerland AG 2019
D.-Z. Du et al. (Eds.): COCOON 2019, LNCS 11653, pp. 338–349, 2019.
https://doi.org/10.1007/978-3-030-26176-4_28

As our main result, Theorem 1 states that finding a resolvable 1-factorization of $H(v,t)$ is equivalent to designing a special type of combinatorial designs, called perpendicular arrays [2,3]. In particular, $\mathsf{CPA}(t, t + d, 2t + d)$, where $d = v - 2t$. According to this theorem and by using the known perpendicular arrays found in [17,18], we obtain the first resolvable 1-factorizations of $H(v,t)$ when $t = 2$ and v is an odd prime power or when $(v, t) \in \{8,3\}, \{32,3\}$. On the other direction, we use the lexical and modular factorizations to obtain the first explicit constructions of $\mathsf{CPA}(t, t + 1, 2t + 1)$, which are known to be existed in [12].

In addition to the construction of the new factorizations, we conduct a comprehensive study of the two previously known factorizations of the middle level graph, which serves as part of an ongoing effort to solve the general case.

In Sect. 3, we unveil an inner structure of the lexical factorization, which leads to not only the first constructive proof for the fact that the lexical factorization is well-defined, but also an optimal algorithm for the following problem: Given i and a t-element subset A, find the unique A' such that (A, A') belongs to the i-th 1-factor of the lexical factorization. The case $i = t + 1$ of this problem was studied in [16]. For $i \leq t$ it becomes more difficult and a trivial algorithm takes $O(v^2)$ time in the RAM model (where an atomic operation on a word accounts $O(1)$ time). We improve it to optimal $O(v)$ time in this paper. (An $O(v)$ time algorithm for this problem on modular factorization is trivial.)

In Sect. 4, we propose an intuitive definition of the modular factorization, which establishes an interesting connection between this factorization and the *inversion number of permutations* (see section 5.3 of [11]). As it is simpler than the original definition in most aspects, a few existing results about the modular factorization become more transparent with this new definition.

Also, we prove properties called *variation laws* for the known 1-factorizations.

We will see the alternative definitions, inner structure, and variation laws are important for understanding the existing 1-factorizations. They have not been reported in literature and obtaining them requires nontrivial analysis.

1.1 Motivation and Related Work

A 1-factor of the bipartite Kneser graph is also known as an antipodal matching in the subset lattice. It is strongly related to the *set inclusion matrix* introduced in [19], which has connections to t-design in coding theory (see [1,5] and the references within). See [16] for its another application in coding theory.

The 1-factorization problem of the middle level graph was motivated by the *middle level conjecture*, which states that all the middle level graphs are Hamiltonian. It was hoped that people can find two 1-factors which form a Hamiltonian cycle [10]. Yet after extensive studies for thirty years the conjecture itself was settled by Mütze [13]; see also [6] for a recent and shorter proof and see [14] for an optimal algorithm for computing such a Hamiltonian cycle. Moreover, Mütze and Su [15] settles the Hamiltonian problem for all the bipartite Kneser graphs.

We give new applications of the 1-factorizations of $H(v,t)$ in hat-guessing games. We show that an optimal strategy in the unique-supply hat-guessing games [8] can be designed from a 1-factorization of $H(v,t)$. To make the strategy easy to play, such a 1-factorization must be simple or at least admit an explicit construction. The details can be found in the full version of this paper [9].

1.2 Preliminaries

The *subset lattice* is the family of all subsets of $[v]$, partially ordered by inclusion. Let $\mathcal{P}_t$ denote the t-th layer of this subset lattice, whose members are the t-element subsets of $[v]$. Throughout the paper, denote $d = v - 2t$. Let the words clockwise and counterclockwise be abbreviated as CW and CCW respectively.

A representation of the edges of $H(v, t)$. We identify each edge (A, A') of $H(v, t)$ by a permutation ρ of t $\bigcirc$'s, t $\triangle$'s, and d $\times$'s: the (positions of) t '$\bigcirc$'s indicate the t elements in A; the t '$\triangle$'s indicate the t elements that are **not** in A' (recall that A' has $v - t$ elements); and the '$\times$'s indicate those in $A' - A$. We do not distinguish the edges with their corresponding permutations.

Denote $[t\bigcirc, t\triangle, d\times]$ as the multiset of $2t + d$ characters with t '$\bigcirc$'s, t '$\triangle$'s, and d '$\times$'s. Giving a 1-factorization of $H(v, t)$ is equivalent to giving a **labeling function** f from the $\binom{2t+d}{t,t,d}$ permutations of $[t\bigcirc, t\triangle, d\times]$ to $1, \ldots, \binom{t+d}{d}$ so that

(a) $f(\rho) \neq f(\sigma)$ for those pairs ρ, σ who admit the same positions for t $\bigcirc$'s; and
(b) $f(\rho) \neq f(\sigma)$ for those pairs ρ, σ who admit the same positions for t $\triangle$'s.

If (a) and (b) hold, for fixed i, all edges labeled by i constitute a 1-factor, denoted by $F_{f,i}$, and $F_{f,1}, \ldots, F_{f,\binom{t+d}{d}}$ constitute a 1-factorization of $H(v, t)$.

An example of the labelling function that satisfies (a) and (b) is given in [10]:

The lexical factorization [10]. Let $\rho = (\rho_1, \ldots, \rho_{2t+1})$ be any permutation of $[t\bigcirc, t\triangle, 1\times]$. Arrange $\rho_1, \ldots, \rho_{2t+1}$ in a cycle in CW order. For any ρ_j that equals $\bigcirc$, it is *positive* if there are strictly more $\bigcirc$'s than $\triangle$'s in the interval that starts from the unique $\times$ and ends at ρ_j in CW order. The number of positive $\bigcirc$'s modular $(t + 1)$ is defined to be $f_{\mathsf{LEX}}(\rho)$ (here, we restrict the remainder to $[t + 1]$ by mapping 0 to $t + 1$). See Fig. 1. It is proved in [10] that f_{LEX} satisfies the above conditions (a) and (b). We provide in Sect. 3 a more direct proof for this. The lexical factorization is $\{\mathcal{L}_1, \ldots, \mathcal{L}_{t+1}\}$, where $\mathcal{L}_i = F_{f_{\mathsf{LEX}},i}$.

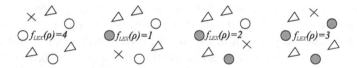

Fig. 1. Illustration of the definition of f_{LEX}. In the graph, the solid circles indicate positive $\bigcirc$'s. Note that the positions of $\bigcirc$'s are identical in all the permutations drawn here. As we see, the four permutations are mapped to different numbers under f_{LEX}.

Note: The original definition [10] of $f_{\mathsf{LEX}}(\rho)$ actually calculates the number of nonnegative $\triangle$'s rather than positive $\bigcirc$'s. For $\rho_j = \triangle$, it is said *nonnegative* if there the number of $\bigcirc$'s is no less than the number of $\triangle$'s in the interval that starts from the unique $\times$ and ends at ρ_j in CW order. Nevertheless, it is clear that the number of nonnegative $\triangle$'s is the same as the number of positive $\bigcirc$'s.

Note: The original definition [10] use $\mathcal{L}_0$ to denote $\mathcal{L}_{t+1}$. In this paper, however, we choose $\mathcal{L}_{t+1}$ instead of $\mathcal{L}_0$ to make it consistent with the case $d > 1$.

2   Construct "resolvable" 1-Factorizations of $H(v, t)$

This section introduces resolvable 1-factorizations of $H(v, t)$ and constructs some of them using combinatorial designs called perpendicular arrays (defined below).

Definition 1. *Assume* $\{g_A \mid A \in \mathcal{P}_t\}$ *is a group of functions where* g_A *is a bijection from* $A^C = [v] - A$ *to* $[t + d]$ *for every* $A \in \mathcal{P}_t$. *Assume* γ *is a bijection from the set of all d-element subsets of $[t+d]$ to* $1, \ldots, \binom{t+d}{d}$. *We define a labeling function* $f_{\gamma,g}$ *on the edges of $H(v, t)$ as follows:* $f_{\gamma,g}(A, A') := \gamma(g_A(A' - A))$.

Note 1: Throughout, we use $g_A(X)$ to denote $\bigcup_{x \in X} g_A(x)$ for any $X \subseteq A^C$.

Note 2: Function $f_{\gamma,g}$ satisfies condition (a) trivially. Yet in most cases it does not satisfy condition (b) and hence does not define a 1-factorization of $H(v, t)$.

Definition 2. *Let f be the labelling function of a 1-factorization of $H(v, t)$. We say f is* resolvable *if there are* $\{g_A \mid A \in \mathcal{P}_t\}$ *and γ as mentioned in Definition 1 such that* $f(A, A') \equiv \gamma(g_A(A' - A))$. *In this case, we call g_A's for $A \in \mathcal{P}_t$ the* resolved functions *of f and we say the 1-factorization defined by f is* resolvable.

Remark 1. Among other merits which make the resolvable 1-factorizations more interesting than the general ones, a resolvable 1-factorization takes only $(t + d)$ over $\binom{t+d}{d}$ fraction of storing space comparing to a general 1-factorization.

The following two lemmas are easy; proofs are given in the full version [9].

Lemma 1 *1. Any 1-factorization of $H(v = 4, t = 1)$ is resolvable.*
2. Any 1-factorization of $H(2t + 1, t)$, including the lexical factorization and modular factorization, is resolvable. (This claim is actually trivial.)
3. No 1-factorization of $H(6, 2)$ is resolvable. (Will be proved by a program.)

As shown by Lemma 1, there could be $H(v, t)$'s without a resolvable 1-factorization, hence we are not always able to design a resolvable 1-factorization of $H(v, t)$. Nevertheless, the first two claims of Lemma 1 and the results given in the rest part of this section point out that for several cases we can do so.

Lemma 2. *Given* $\{g_A \mid A \in \mathcal{P}_t\}$ *and γ as above, the following are equivalent:*

(i) function $f_{\gamma,g}(A, A')$ satisfies condition (b); and
(ii) When $A_1 \neq A_2$ and $(A_1, A'_1), (A_2, A'_2)$ are two edges in $H(v, t)$, then $g_{A_1}(A'_1 - A_1) = g_{A_2}(A'_2 - A_2)$ implies that $A'_1 \neq A'_2$.

By Lemma 2, it is independent with the choice of γ whether $f_{\gamma,g}(A, A')$ defines a 1-factorization of $H(v,t)$. Therefore, if we want to design a resolvable 1-factorization, the difficulty lies in and only lies in designing $\{g_A \mid A \in \mathcal{P}_t\}$.

By Lemma 2 and Definition 1, $H(v,t)$ has a resolvable factorization if and only if there exist resolved functions $\{g_A \mid A \in \mathcal{P}_t\}$ such that for $A_1 \neq A_2$, $g_{A_1}(A_1' - A_1) = g_{A_2}(A_2' - A_2)$ implies $A_1' \neq A_2'$. The following theorem shows that finding such functions is equivalent to designing some perpendicular arrays.

A *perpendicular array* [2,3] with parameters t, k, v, denoted by $\mathsf{PA}(t, k, v)$, is a $\binom{v}{t} \times k$ matrix over $[v]$, where each row has k distinct numbers and each set of t columns contain each t-element subset of $[v]$ as a row exactly once.

For $d \geq 0$, a $\mathsf{PA}(t, t+d, 2t+d)$ is *complete*, hence denoted by $\mathsf{CPA}(t, t+d, 2t+d)$, if each $(t+d)$-element subset of $[2t+d]$ is also contained in exactly one row.

Theorem 1. *$H(2t+d, t)$ has a resolvable 1-factorization $\Leftrightarrow \exists \mathsf{CPA}(t, t+d, 2t+d)$.*

Proof. $\Rightarrow$: Assume f is the labeling function of a resolvable 1-factorization of $H(v = 2t + d, t)$. Then, $f(A, A') \equiv \gamma(g_A(A' - A))$ for some resolved functions $\{g_A \mid A \in \mathcal{P}_t\}$ and a bijection γ as mentioned in Definition 1.

We construct a matrix M over $[v]$ as follows. For each $A \in \mathcal{P}_t$, we build a row $(a_1^{(A)}, \ldots, a_{t+d}^{(A)})$ in M, where $a_i^{(A)} = g_A^{-1}(i)$ (which belongs to A^C and thus belongs to $[v]$). As $\mathcal{P}_t$ has $\binom{v}{t}$ elements, the size of matrix M is $\binom{v}{t}$ by $k = t + d$.

We now verify that M is a $\mathsf{PA}(t, t+d, 2t+d)$. First, since g_A^{-1} is bijective, $a_1^{(A)}, \ldots, a_{t+d}^{(A)}$ are distinct and so each row of M contains $k = t + d$ distinct numbers. Next, for any t columns $i_1, \ldots, i_t$, we show that

$$\{a_{i_1}^{(A_1)}, \ldots, a_{i_t}^{(A_1)}\} \neq \{a_{i_1}^{(A_2)}, \ldots, a_{i_t}^{(A_2)}\} \tag{1}$$

for any distinct $A_1, A_2 \in \mathcal{P}_t$. Assume $\{j_1, \ldots, j_d\} = [t + d] - \{i_1, \ldots, i_t\}$.

Let $A_1' = A_1 \uplus \{g_{A_1}^{-1}(j_1), \ldots, g_{A_1}^{-1}(j_d)\}$ and $A_2' = A_2 \uplus \{g_{A_2}^{-1}(j_1), \ldots, g_{A_2}^{-1}(j_d)\}$. Clearly, $g_{A_1}(A_1' - A_1) = \{j_1, \ldots, j_d\} = g_{A_2}(A_2' - A_2)$, thus $A_1' \neq A_2'$ by Lemma 2. Thus $[v] - A_1' \neq [v] - A_2'$. Because $\{j_1, \ldots, j_d\} \uplus \{i_1, \ldots, i_t\} = [t + d]$, we know $A_1^C = \{g_{A_1}^{-1}(j_1), \ldots, g_{A_1}^{-1}(j_d)\} \uplus \{g_{A_1}^{-1}(i_1), \ldots, g_{A_1}^{-1}(i_t)\}$, which implies that $[v] - A_1' = \{g_{A_1}^{-1}(i_1), \ldots, g_{A_1}^{-1}(i_t)\}$. Similarly, $[v] - A_2' = \{g_{A_2}^{-1}(i_1), \ldots, g_{A_2}^{-1}(i_t)\}$. Altogether, $\{g_{A_1}^{-1}(i_1), \ldots, g_{A_1}^{-1}(i_t)\} \neq \{g_{A_2}^{-1}(i_1), \ldots, g_{A_2}^{-1}(i_t)\}$, i.e., (1) holds.

Next, we argue that M is a $\mathsf{CPA}(t, t+d, 2t+d)$. This reduces to proving that each row of M forms a distinct $(t+d)$-element subset of $[2t+d]$, which follows from the fact that the row constructed from A is a permutation of A^C.

$\Leftarrow$: Assume M is a $\mathsf{CPA}(t, t + d, 2t + d)$. First, we construct $\{g_A \mid A \in \mathcal{P}_k\}$. For any row $(a_1, \ldots, a_{t+d})$ of M, assuming that $A^C = \{a_1, \ldots, a_{t+d}\}$, define $g_A(a_i) = i$ for $i \in [t + d]$. Obviously, each g_A for $A \in \mathcal{P}_k$ is defined exactly once.

Below we verify that when $A_1 \neq A_2$, equality $g_{A_1}(A_1' - A_1) = g_{A_2}(A_2' - A_2)$ would imply $A_1' \neq A_2'$. According to Lemma 2, this further implies that for any γ as mentioned in Definition 1, $f_{\gamma,g}(A, A')$ is a labeling function satisfying conditions (a) and (b), and hence $H(2t + d, t)$ has a resolvable 1-factorization.

Suppose to the opposite that $g_{A_1}(A' - A_1) = g_{A_2}(A' - A_2) = \{j_1, \ldots, j_d\}$. Assume $\{i_1, \ldots, i_t\} = [t+d] - \{j_1, \ldots, j_d\}$. Because $g_{A_1}(A' - A_1) = \{j_1, \ldots, j_d\}$, we know $g_{A_1}([v] - A') = \{i_1, \ldots, i_t\}$, so $[v] - A' = \{g_{A_1}^{-1}(i_1), \ldots, g_{A_1}^{-1}(i_t)\}$. Similarly, because $g_{A_2}(A' - A_2) = \{j_1, \ldots, j_d\}$, we get $[v] - A' = \{g_{A_2}^{-1}(i_1), \ldots, g_{A_2}^{-1}(i_t)\}$. Moreover, because M is a $\mathsf{PA}(t, t+d, 2t+d)$ where $\{g_{A_1}^{-1}(i_1), \ldots, g_{A_1}^{-1}(i_t)\}$ and $\{g_{A_2}^{-1}(i_1), \ldots, g_{A_2}^{-1}(i_t)\}$ appear in two rows of M in the columns indexed by $i_1, \ldots, i_t$, these two sets are distinct. Thus $[v] - A' \neq [v] - A'$. Contradiction. □

2.1 Applications of Theorem 1

Lemma 3 *1. For $t = 1$, there is always a $\mathsf{PA}(t, 2t+d, 2t+d)$. (trivial)*
2. [17] For $t = 2$ and an odd prime power $2t+d$, there is a $\mathsf{PA}(t, 2t+d, 2t+d)$.
3. [18] For $t = 3$ and $2t+d \in \{8, 32\}$, there is a $\mathsf{PA}(t, 2t+d, 2t+d)$.

The following lemma is trivial; proof can be found in full version [9].

Lemma 4. *Any $t+d$ columns of a $\mathsf{PA}(t, 2t+d, 2t+d)$ form a $\mathsf{CPA}(t, t+d, 2t+d)$.*

Lemma 4 points out a way to construct a $\mathsf{CPA}(t, t+d, 2t+d)$. Yet it is unknown whether every $\mathsf{CPA}(t, t+d, 2t+d)$ can be constructed this way. We conjecture so. If so, finding resolvable 1-factorizations reduces to finding $\mathsf{PA}(t, 2t+d, 2t+d)$'s.

The following is a corollary of Lemmas 3, 4, and Theorem 1.

Corollary 1. *Graph $H(2t+d, t)$ has a resolvable 1-factorization when 1. $t = 1$, or 2. ($t = 2$ and $2t+d$ is an odd prime power), or 3. ($t = 3$ and $2t+d \in \{8, 32\}$).*

The constructions of $\mathsf{PA}(t, 2t+d, 2t+d)$ for those pairs of (t, d) discussed in Lemma 3 are explicit and quite simple (see [17, 18]). Also, our construction of the resolvable 1-factorization of $H(2t+d, t)$ using a $\mathsf{CPA}(t, t+d, 2t+d)$ is extremely simple (as shown in the proof of Theorem 1). As a result, the resolvable 1-factorizations of $H(2t+d, t)$ mentioned in this corollary are explicit and simple.

Perpendicular arrays have not been studied extensively in literature. In addition to the existence results mentioned in Lemma 3, there do exist $\mathsf{PA}(3, 5, 5)$ and $\mathsf{PA}(t, t+1, 2t+1)$ ($t \geq 1$) and some other perpendicular arrays. Yet the construction of $\mathsf{PA}(t, t+1, 2t+1)$ (in [12]) is not explicit and thus not too useful (regarding that we are only interested in explicit factorizations of $H(2t+1, t)$). A $\mathsf{PA}(3, 5, 5)$ is also useless to us since $5 < 2 \times 3$. Because a $\mathsf{CPA}(t, t+d, 2t+d)$ automatically implies a resolvable 1-factorization of $H(v, t)$, we hope that our results motivate more study on the perpendicular arrays in the future.

Another application of Theorem 1—construction of $\mathsf{CPA}(t, t+1, 2t+1)$. As shown in Lemma 1, the lexical and modular factorization of $H(2t+1, t)$ are both resolvable. The resolved functions of f_{LEX} and f_{MOD} will be demonstrated in the next sections. Using these resolved functions and applying the proof of Theorem 1, we can easily construct two $\mathsf{CPA}(t, t+1, 2t+1)$s. Therefore, as byproducts, we obtain (the first) explicit constructions of (complete) $\mathsf{PA}(t, t+1, 2t+1)$ (note that [12] only showed the existence of $\mathsf{PA}(t, t+1, 2t+1)$).

3   Revisit the Lexical Factorization

Recall f_{LEX} in Subsect. 1.2, which is a labeling function of $H(2t+1,t)$. In this section, we first give $\{g_A\}$ of γ so that $f_{\mathsf{LEX}} = f_{\gamma,g}$. Based on this formula we then show that f_{LEX} satisfies (a) and (b) and thus that it indeed defines a 1-factorization. Moreover, by applying $f_{\mathsf{LEX}} = f_{\gamma,g}$, we design optimal algorithms for solving two fundamental computational problems about this factorization (P1 and P2 below). Finally, we introduce a group of *variation laws* of f_{LEX}.

P1.   Given $A \in \mathcal{P}_t$ and $i \in \{1,\ldots,t+1\}$, how do we find the unique A' so that $(A, A') \in \mathcal{L}_i$? In other words, given number i and the positions of $\bigcirc$'s in ρ and suppose $f_{\mathsf{LEX}}(\rho) = i$, how do we determine the position of $\times$ in ρ?

P2.   Given a $A' \in \mathcal{P}_{t+1}$ and $i \in \{1,\ldots,t+1\}$, how do we find the unique A so that $(A, A') \in \mathcal{L}_i$? In other words, given number i and the positions of $\triangle$'s in ρ and suppose $f_{\mathsf{LEX}}(\rho) = i$, how do we determine the position of $\times$ in ρ?

3.1   Preliminary Lemmas

The two lemmas given in this subsection are trivial; proofs can be found in [9].

Given $S = (s_1,\ldots,s_v)$, the j-th $(0 \le j < v)$ *cyclic-shift* of S is $S^{(j)} := (s_{1+j},\ldots,s_{v+j})$, where subscripts are taken modulo v (and restricted to $[v]$).

Lemma 5. *Given any sequence S of t of right parentheses ')' and $t + 1$ left parentheses '('. There exists a unique cyclic-shift $S^{(j)}$ of S whose first $2t$ parentheses are paired up when parenthesized, and we can compute j in $O(t)$ time.*

Example 1. Assume $t = 9$, $S = (_1(_2)_3)_4)_5(_6(_7(_8)_9)_{10}(_{11})_{12})_{13}(_{14}(_{15})_{16}(_{17}(_{18})_{19}$.
The unique cyclic-shift in which the first $2t$ parentheses are paired up is:

$$
S^{(14)} = \left(\right)_{15} \, {}_{16}\left(_{17}\left(_{18}\right)_{19}\left(_1(_2)_3\right)_4\right)_5\left(_6\left(_7(_8)_9\right)_{10}\left(_{11}\right)_{12}\right)_{13}\left(_{14}\right.
$$

Definition 3. *Given $S = (s_1,\ldots,s_{2t+1})$, t of which are ')' and $t+1$ are '('. It is said* canonical *if its first $2t$ parentheses are paired up when parenthesized.*

Definition 4. (Indices of the $2t+1$ parentheses). *For any canonical parentheses sequence S, we index the $t+1$ left parentheses in S by $0,\ldots,t$ according to the following rule:* **The smaller the depth, the less the index; and index from right to left for those under the same depth.** *Here, depth is defined in the standard way; it is the number of pairs of matched parentheses that cover the fixed parenthesis. Moreover,* **we index the t right parentheses in such a way that any two paired parentheses have the same index.**

For $S^{(14)}$ above, the depth and index are shown below (index on the right).

This definition of indices is crucial to the next lemma and the entire section. For convenience, denote by $\mathsf{depth}(s_i), \mathsf{index}(s_i)$ the depth and index of s_i.

Lemma 6. *When S is canonical, for any $s_l = ($ and $s_r =)$, there are more $)$'s than $($'s in the cyclic interval $\{s_{l+1}, \ldots, s_r\}$ if and only if $\mathsf{index}(s_l) \geq \mathsf{index}(s_r)$.*

3.2   Finding Resolved Functions $\{g_A\}$ of f_{LEX}

Parenthesis representation. We can represent any $A \subseteq [v]$ by a sequence of parentheses $S = (s_1, \ldots, s_v)$ where $s_x = ')'$ if $x \in A$ and $s_x = '('$ if $x \notin A$. For example, $A = \{3, 4, 5, 9, 10, 12, 13, 16, 19\}$ is represented by the S given in Example 1 above. Notice that if $A \in \mathcal{P}_t$, its associate sequence S contains t ')'s.

Definition 5. *Fix $A \in \mathcal{P}_t$ and let S denote its parentheses sequence. We abuse $\mathsf{index}(s_x)$ to mean the index of s_x in the unique canonical cyclic-shift of S (uniqueness is by Lemma 5). For any $x \in A^C$ (hence $s_x = '('$), define $g_A(x) := \mathsf{index}(s_x) \bmod (t+1)(\in [t+1])$ (restrict to $[t+1]$ by mapping 0 to $t + 1$).*

Because left parentheses have distinct indices, g_A is a bijection as required.

Theorem 2. *Let γ be the natural bijection from all the 1-element subsets of $[t+1]$ to $[t+1]$, which maps $\{x\}$ to x. Define $\{g_A \mid A \in \mathcal{P}_t\}$ as in Definition 5. Then, $f_{\mathsf{LEX}} = f_{\gamma,g}$. In other words, $f_{\mathsf{LEX}}(A, A \cup \{x\}) \equiv g_A(x)$ $(x \in A^C)$.*

Proof. Build the parentheses sequence $S = (s_1, \ldots, s_{2t+1})$ of A and the permutation $\rho = (\rho_1, \ldots, \rho_{2t+1})$ of $[t\bigcirc, t\triangle, 1\times]$ corresponding to edge $(A, A\cup\{x\})$. Recall that $f_{\mathsf{LEX}}(A, A \cup \{x\}) := p \bmod (t+1) \in [t+1]$, where p is the size of $P = \{\rho_r = \bigcirc \mid$ there are more $\bigcirc$s than $\triangle$s in the cyclic interval $(\rho_{x+1}, \ldots, \rho_r)\}$. Observe that S can be constructed from ρ by replacing $\bigcirc, \triangle, \times$ to $')', '(', '('$. So, $\{s_r = ')' \mid$ there are more $')'$'s than $'('$s in cyclic interval $(s_{x+1}, \ldots, s_r)\}$, which equals $\{s_r = ')' \mid \mathsf{index}(s_x) \geq \mathsf{index}(s_r)\}$ by Lemma 6 (indices refer to those in the canonical cyclic-shift of S), has the same size as P, so $\mathsf{index}(s_x) = p$. Further by Definition 5, $g_A(x) = \mathsf{index}(s_x) \bmod (t+1)(\in [t+1]) = f_{\mathsf{LEX}}(A, A \cup \{x\})$. □

Theorem 3. f_{LEX} *satisfies conditions (a) and (b).*

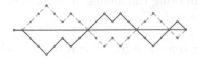

Fig. 2. $f_{\mathsf{LEX}}(\rho^*) \bmod (t+1) + f_{\mathsf{LEX}}(\rho) \bmod (t+1) = t$. The dashed line indicates ρ^*.

Proof. Because f_{LEX} equals $f_{\gamma,g}$, applying Note 2 below Definition 1, this labeling function satisfies condition(a). Below we prove that it also satisfies condition (b).

Define the *dual* of ρ, denoted by ρ^*, to be another permutation of $[t\bigcirc, t\triangle, 1\times]$ which is constructed from ρ by swapping the $\triangle$'s with $\bigcirc$'s. As illustrated in Fig. 2, we have (i): $f_{\mathsf{LEX}}(\rho^*) \bmod (t+1) + f_{\mathsf{LEX}}(\rho) \bmod (t+1) = t$ for any ρ.

Consider $t + 1$ distinct permutations $\rho^0, \dots, \rho^t$ sharing the same positions of $\triangle$'s. Then, $(\rho^0)^*, \dots, (\rho^t)^*$ share the same positions of $\bigcirc$'s. Using condition (a), $f_{\mathsf{LEX}}((\rho^0)^*), \dots, f_{\mathsf{LEX}}((\rho^t)^*)$ are distinct. So $t - f_{\mathsf{LEX}}((\rho^0)^*) \bmod (t+1), \dots, t - f_{\mathsf{LEX}}((\rho^t)^*) \bmod (t+1)$ are distinct. So $f_{\mathsf{LEX}}(\rho^0) \bmod (t+1), \dots, f_{\mathsf{LEX}}(\rho^t) \bmod (t+1)$ are distinct by (i), i.e., $f_{\mathsf{LEX}}(\rho^0), \dots, f_{\mathsf{LEX}}(\rho^t)$ are distinct. Thus (b) holds. $\square$

Remark 2. In the original proof of Theorem 3 in [10], it proves the existence of bijections g_A's ($A \in \mathcal{P}_t$) such that $f_{\mathsf{LEX}} = f_{\gamma,g}$, yet how to define such g_A's is neither explicitly given, nor implicitly given. As we have seen in Definition 4, giving this definition is not easy, even though the definition of f_{LEX} is known.

There are two advantages of having explicit $\{g_A\}$. First, the ideas we used in defining g_A could be useful in finding resolvable 1-factorizations for the case $v > 2t+1$. Second, to solve P1 and P2 (in the next subsection), it seems necessary to have an explicit definition of $\{g_A\}$ for the efficiency of computation.

3.3   Linear Time Algorithms for P1 and P2

Problem P1 admits a trivial $O(t^2)$ time solution as follows. Given the positions of $\bigcirc$'s in ρ and the number i, we can enumerate the position of the unique $\times$ among the remaining $t + 1$ positions and compute $f_{\mathsf{LEX}}(\rho)$ in $O(t)$ time, until that the computed value is i. Problem P2 can be solved symmetrically.

Applying the results in Subsect. 3.2, we can solve P1 much more efficiently. Briefly, using those indices of parentheses in Definition 4, we can compute $f_{\mathsf{LEX}}()$ for all permutations $\rho^0, \dots, \rho^t$ in which the positions of $\bigcirc$'s are as given altogether, and then find ρ^j so that $f_{\mathsf{LEX}}(\rho^j) = i$. See the details in Algorithm 1.

Input: A set $A \in \mathcal{P}_t$ and a number $i \in [t + 1]$.
Output: The set $A' = A \cup \{z\}$ so that $(A, A') \in \mathcal{L}_i$.
        (Integer z indicates the position of $\times$ so that $f_{\mathsf{LEX}}(\rho) = i$.)
1 Compute the parentheses sequence S of A.
2 Compute the unique j so that the first $2t$ parentheses are paired up in $S^{(j)}$.
3 Compute the indices of all parentheses in $S' = S^{(j)}$ according to Definition 4.
4 Find $s'_{z-j} =$ '(' in S with index $(i \bmod (t+1))$ and output $A' = A \cup \{z\}$.

Algorithm 1. Computing the unique A' such that $(A, A') \in \mathcal{L}_i$.

Theorem 4 *1. Given a canonical S', we can compute the indices of all parentheses in S' in $O(t)$ time. Therefore, Algorithm 1 solves P1 in $O(t)$ time.*

2. An instance (A', i) of P2 reduces to the instance $([v] - A', j)$ of P1, where $i \bmod (t+1) + j \bmod (t+1) = t$. Thus P2 can be solved in $O(t)$ time.

The proof of Theorem 4 is trivial and is omitted due to space limits.

3.4   Variation Laws of f_{LEX}

We prove some *variations laws* of f_{LEX} as summarized in Lemma 7, which are comparable to the laws of modular factorization given below in Lemma 9.

Lemma 7 (Variation laws of f_{LEX})

1. $f_{\mathsf{LEX}}(\rho) \neq t+1 \Leftrightarrow$ *there is a CW-balanced* $\triangle \Leftrightarrow$ *there is a CCW-balanced* $\bigcirc$.
2. $f_{\mathsf{LEX}}(\rho) \neq t \Leftrightarrow$ *there is a CW-balanced* $\bigcirc \Leftrightarrow$ *there is a CCW-balanced* $\triangle$.
3. *When* $f_{\mathsf{LEX}}(\rho) \neq t+1$, *let* $\rho^{\times \rightarrowtail \triangle}$ $(\rho^{\bigcirc \leftarrowtail \times})$ *be constructed from* ρ *by swapping* $\times$ *with the CW first CW-balanced* $\triangle$ *(the CCW first CCW-balanced* $\bigcirc$*).*

   Then,

   $$f_{\mathsf{LEX}}(\rho^{\times \rightarrowtail \triangle}) = f_{\mathsf{LEX}}(\rho^{\bigcirc \leftarrowtail \times}) = (f_{\mathsf{LEX}}(\rho) - 1) \bmod (t+1)(\in [t+1]).$$

4. *When* $f_{\mathsf{LEX}}(\rho) \neq t$, *let* $\rho^{\times \rightarrowtail \bigcirc}$ $(\rho^{\triangle \leftarrowtail \times})$ *be constructed from* ρ *by swapping* $\times$ *with the CW first CW-balanced* $\bigcirc$ *(the CCW first CCW-balanced* $\triangle$*).* *Then,*

   $$f_{\mathsf{LEX}}(\rho^{\times \rightarrowtail \bigcirc}) = f_{\mathsf{LEX}}(\rho^{\triangle \leftarrowtail \times}) = (f_{\mathsf{LEX}}(\rho) + 1) \bmod (t+1)(\in [t+1]).$$

4   Revisit the Modular Factorization

This section presents a new and simpler definition of the modular factorization. When a number modulo $t+1$ in this section, the remainder is restricted to $[t+1]$.

The modular factorization [4]. The modular factorization was originally given by $t+1$ 1-factors $\mathcal{M}_1, \ldots, \mathcal{M}_{t+1}$ where $\mathcal{M}_i$ was defined as follows. Consider $A \in \mathcal{P}_t$. Let ΣA indicate the sum of elements in A. Let $y = (\Sigma A + i) \bmod (t+1)(\in [t+1])$. Then, $\mathcal{M}_i(A) := A \cup \{z\}$, where z is the y-th **largest** element in $[v] - A$.

Take $t = 3, v = 7$, and $A = \{2, 4, 6\}$ for example:

For $i = 1$, we have $y = 13 = 1 \pmod 4$ and $z = 7$. So $\mathcal{M}_1(A) = \{2, 4, 6, 7\}$.
For $i = 2$, we have $y = 14 = 2 \pmod 4$ and $z = 5$. So $\mathcal{M}_2(A) = \{2, 4, 5, 6\}$.
For $i = 3$, we have $y = 15 = 3 \pmod 4$ and $z = 3$. So $\mathcal{M}_3(A) = \{2, 3, 4, 6\}$.
For $i = 4$, we have $y = 16 = 4 \pmod 4$ and $z = 1$. So $\mathcal{M}_4(A) = \{1, 2, 4, 6\}$.

Note 1. It is proved in [4] that $\mathcal{M}_i$ is a 1-factor for each i $(1 \leq i \leq t+1)$. Moreover, it is obvious that all the 1-factors $\mathcal{M}_1, \ldots, \mathcal{M}_{t+1}$ are pairwise-disjoint.

Note 2. The origins of modular factorization are murky, said by the authors of [4], who credited it to Robinson, who asked if it is the same as the lexical one.

Note 3. Assume $\mathcal{M}_i(A) = A'$. We can compute A from i and A' symmetrically. Let $x = (\Sigma A' + i) \bmod (t+1)(\in [t+1])$ where $\Sigma A'$ indicates the sum of elements in A'. Then $A = A' - \{z\}$, where z is the x-th **smallest** element in A' [4]. So, the problems on modular factorizations analogous to P1 and P2 are easy to solve.

The original definition of the modular factorization above does not explicitly give its labeling function. Such a labeling function will be needed in analyzing

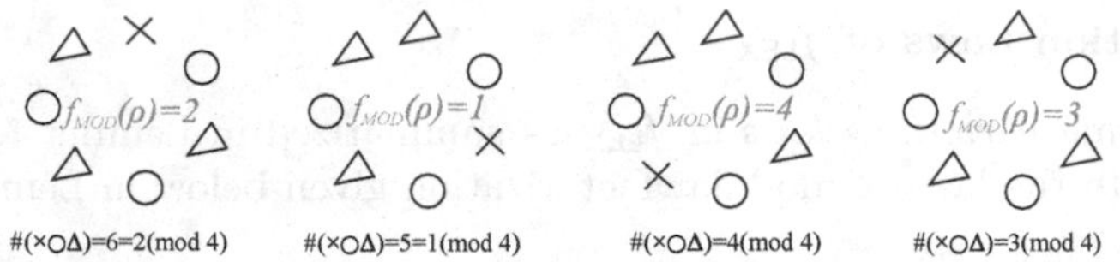

$f_{MOD}(\rho)=2$ $f_{MOD}(\rho)=1$ $f_{MOD}(\rho)=4$ $f_{MOD}(\rho)=3$

#(×○△)=6≡2(mod 4)   #(×○△)=5≡1(mod 4)   #(×○△)=4(mod 4)   #(×○△)=3(mod 4)

Fig. 3. Illustration of the definition of f_{MOD}. The four permutations drawn here share the same positions of ○'s, and they are mapped to different numbers under f_{MOD}.

the variation laws of the above modular factorization in Lemma 9 below and hence we state it Lemma 8. However, our definition of the modular factorization is **not** given by Lemma 8. The proof of Lemma 8 can be found in [9].

Consider any permutation $\rho = (\rho_1, \ldots, \rho_{2t+1})$ of $[t○, t△, 1×]$. For each $i \in [2t+1]$, the *position* of ρ_i is i. Let $O_1^\rho, \ldots, O_t^\rho$ be the positions of t ○'s in ρ and $T_1^\rho, \ldots, T_t^\rho$ the positions t △'s. Denote by $\operatorname{rank}_△^○(\rho)$ the rank of × when enumerating all △'s and × in ρ from ρ_{2t+1} back to ρ_1. So, $\operatorname{rank}_△^○(\rho) - 1$ is the number of △'s with positions larger than the position of ×. Denote by $\operatorname{rank}_○^○(\rho)$ the rank of × when enumerating all ○'s and × in ρ from ρ_{2t+1} back to ρ_1.

Lemma 8. *The labeling function of* $\{\mathcal{M}_1, \ldots, \mathcal{M}_{t+1}\}$ *is given by* f_{mod}*, where*

$$f_{mod}(\rho) := \operatorname{rank}_△^○(\rho) - \Sigma_{j=1}^t O_j^\rho \pmod{t+1}(\in [t+1]), \ or$$

$$f_{mod}(\rho) := 1 + \Sigma_{j=1}^t T_j^\rho - \operatorname{rank}_○^○(\rho) \pmod{t+1}(\in [t+1]).$$

We now introduce a labeling function f_{MOD} and proves that $f_{MOD} \equiv f_{mod} + C$ for some constant C. Thus we give an alternative yet equivalent definition of the modular factorization, which is $\{F_{f_{MOD},1}, \ldots, F_{f_{MOD},t+1}\}$.

Definition 6. *Assume* $\rho = (\rho_1, \ldots, \rho_{2t+1})$ *is any permutation of* $[t○, t△, 1×]$. *Arrange* $\rho_1, \ldots, \rho_{2t+1}$ *in CW order. We count* **the number of tuples** $(×, ○, △)$ *which are located in CW order within this cycle of characters (positions may be inconsecutive) (such a tuple is an inversion when we cut the sequence at* ×*). Taken modulo* $(t+1)$*, the remainder, restricted to* $[t+1]$*, is* $f_{MOD}(\rho)$*. See Fig. 3.*

By Definition 6, we establish an interesting connection between the modular factorization and the **inversion number of permutations** (Sect. 5.3 of [11]).

Let $\rho^{×\to△}$ be constructed from ρ, which swaps × with its CW next △.
Let $\rho^{×\to○}$ be constructed from ρ, which swaps × with its CW next ○.
Let $\rho^{△\leftarrow×}$ be constructed from ρ, which swaps × with its CCW next △.
Let $\rho^{○\leftarrow×}$ be constructed from ρ, which swaps × with its CCW next ○.

Lemma 9. (Variation laws of f_{mod} **and** f_{MOD}**).**

$$f_{MOD}(\rho^{×\to△}) = f_{MOD}(\rho^{○\leftarrow×}) = f_{MOD}(\rho) - 1 \pmod{t+1}, \tag{2}$$

$$f_{MOD}(\rho^{×\to○}) = f_{MOD}(\rho^{△\leftarrow×}) = f_{MOD}(\rho) + 1 \pmod{t+1}. \tag{3}$$

$$f_{mod}(\rho^{×\to△}) = f_{mod}(\rho^{○\leftarrow×}) = f_{mod}(\rho) - 1, \pmod{t+1} \tag{4}$$

$$f_{mod}(\rho^{×\to○}) = f_{mod}(\rho^{△\leftarrow×}) = f_{mod}(\rho) + 1. \pmod{t+1} \tag{5}$$

Proof of Lemma 9 is given in [9]. Its corollary below is trivial; proof omitted.

Corollary 2. *Because f_{mod} and f_{MOD} have the same variation law, there is a constant C so that $f_{\mathsf{MOD}} \equiv f_{\mathsf{mod}} + C$. Specifically,* $\begin{cases} C = 0, & t \text{ is even;} \\ C = (t+1)/2, & t \text{ is odd.} \end{cases}$

At last, we point out that the resolved functions of f_{mod} or f_{MOD} can easily be deduced according to the original definition of modular factorization.

References

1. Bapat, R.: Moore-Penrose inverse of set inclusion matrices. Linear Algebra Appl. **318**(1), 35–44 (2000)
2. Bierbrauer, J., Edel, Y.: Theory of perpendicular arrays. J. Comb. Des. **2**(6), 375–406 (1994)
3. Colbourn, C., Dinitz, J. (eds.): CRC Handbook of Combinatorial Designs, 2nd edn. CRC Press Inc., Boca Raton (2007)
4. Duffus, D., Kierstead, H., Snevily, H.: An explicit 1-factorization in the middle of the Boolean lattice. J. Comb. Theory Ser. A **65**(2), 334–342 (1994)
5. Ghorbani, E., Khosrovshahi, G., Maysoori, C., Mohammad-Noori, M.: Inclusion matrices and chains. J. Comb. Theory Ser. A **115**(5), 878–887 (2008)
6. Gregor, P., Mütze, T., Nummenpalo, J.: A short proof of the middle levels theorem. CoRR abs/1710.08249 (2018)
7. Hall, P.: On representatives of subsets. J. Lond. Math. Soc. **s1–10**(1), 26–30 (1935)
8. Jin, K., Jin, C., Gu, Z.: Cooperation via codes in restricted hat guessing games. In: International Conference on Autonomous Agents and Multiagent Systems (2019)
9. Jin, K.: On 1-factorizations of bipartite Kneser graphs. CoRR abs/1704.08852 (2017)
10. Kierstead, H., Trotter, W.: Explicit matchings in the middle levels of the Boolean lattice. Order **5**(2), 163–171 (1988)
11. Kleinberg, J., Tardos, E.: Algorithm Design. Addison-Wesley Longman Publishing Co., Inc., Boston (2005)
12. Kramer, E., Wu, Q., Magliveras, S., Trung, T.: Some perpendicular arrays for arbitrarily large t. Discrete Math. **96**(2), 101–110 (1991)
13. Mütze, T.: Proof of the middle levels conjecture. Proc. Lond. Math. Soc. **112**(4), 677 (2016)
14. Mütze, T., Nummenpalo, J.: A constant-time algorithm for middle levels Gray codes. In: Proceedings of the 28th Annual ACM-SIAM Symposium on Discrete Algorithms, pp. 2238–2253. Society for Industrial and Applied Mathematics (2017)
15. Mütze, T., Su, P.: Bipartite Kneser graphs are Hamiltonian. Combinatorica **37**(6), 1207–1219 (2017)
16. Ordentlich, E., Roth, R.: Low complexity two-dimensional weight-constrained codes. IEEE Trans. Inf. Theory **58**(6), 3892–3899 (2012)
17. Rao, C.: Combinatorial arrangements analogous to orthogonal arrays. Sankhyā Indian J. Stat. Ser. A **23**(3), 283–286 (1961)
18. Stinson, D., Teirlinck, L.: A construction for authentication/secrecy codes from 3-homogeneous permutation groups. Eur. J. Comb. **11**(1), 73–79 (1990)
19. Wilson, R.: Incidence matrices of t-designs. Linear Algebra Appl. **46**, 73–82 (1982)

An Optimal Algorithm for 2-Bounded Delay Buffer Management with Lookahead

Koji M. Kobayashi[(⊠)]

The University of Tokyo, Tokyo, Japan
kojikoba@mi.u-tokyo.ac.jp

Abstract. The bounded delay buffer management problem, which was proposed by Kesselman et al. (STOC 2001 and SIAM Journal on Computing 33(3), 2004), is an online problem focusing on buffer management of a switch supporting Quality of Service (QoS). The problem definition is as follows: Packets arrive to a buffer over time and each packet is specified by the *release time*, *deadline* and *value*. An algorithm can transmit at most one packet from the buffer at each integer time and can gain its value as the *profit* if transmitting a packet by its deadline after its release time. The objective of this problem is to maximize the gained profit. We say that an instance of the problem is s-bounded if for any packet, an algorithm has at most s chances to transmit it. For any $s \geq 2$, Hajek (CISS 2001) showed that the competitive ratio of any deterministic algorithm is at least $(1 + \sqrt{5})/2 \geq 1.618$. Very recently, Veselý et al. (SODA 2019) designed an online algorithm matching the lower bound.

Böhm et al. (ISAAC 2016 and Theoretical Computer Science, 2019) introduced the *lookahead* ability to an online algorithm, that is the algorithm can gain information about future arriving packets, and showed that for $s = 2$, there is an algorithm which achieves the competitive ratio of $(-1 + \sqrt{13})/2 \leq 1.303$. Also, they showed that the competitive ratio of any deterministic algorithm is at least $(1 + \sqrt{17})/4 \geq 1.280$. In this paper, for the 2-bounded model with lookahead, we design an algorithm with a matching competitive ratio of $(1 + \sqrt{17})/4$.

1 Introduction

The online buffer management problem proposed by Aiello et al. [1] formulates the management of buffers to store arriving packets in a network switch with Quality of Service (QoS) support as an online problem. This problem has received much attention among online problems and has been studied for the last fifteen years, which leads to developing various variants of this problem (see comprehensive surveys [17, 26]). Kesselman et al. [23] proposed the *bounded delay buffer management* problem as one of the variants, whose definition is as follows: Packets arrive to a buffer over time. A packet p is specified by the *release time* $r(p)$, *value* $v(p)$ and *deadline* $d(p)$. An algorithm is allowed to transfer at most one packet at each integer time. If the algorithm transmits a packet between

© Springer Nature Switzerland AG 2019
D.-Z. Du et al. (Eds.): COCOON 2019, LNCS 11653, pp. 350–362, 2019.
https://doi.org/10.1007/978-3-030-26176-4_29

its release time and deadline, it can gain its value as the *profit*. The objective of this problem is to maximize the gained profit. The performance of an online algorithm for this problem is evaluated using *competitive analysis* [11,27]. If for any problem instance, the profit of an optimal offline algorithm OPT is at most c times that of an online algorithm A, then we say that the competitive ratio of A is at most c. We call a problem instance the *s-bounded instance* (or *s-bounded delay buffer management* problem) in which for any packet p, $d(p) - r(p) + 1 \leq s$. For any $s \geq 2$, Hajek [19] showed that the competitive ratio of any deterministic algorithm is at least $(1 + \sqrt{5})/2 \geq 1.618$. Very recently, Veselý et al. [28] designed an online algorithm matching the lower bound.

There is much research among online problems to reduce the competitive ratio of an online algorithm for the original problems by adding extra abilities to the algorithm. One of the major methods is called the *lookahead* ability, with which an online algorithm can obtain information about arriving packets in the near future. This ability is introduced to various online problems: the bin packing problem [18], the paging problem [2,12], the list update problem [3], the scheduling problem [25] and so on. Then, Böhm et al. [9,10] introduced the lookahead ability to the bounded delay buffer management problem, that is, they gave an online algorithm for this problem an ability to obtain the information about future arriving packets and analyzed its performance.

Previous Results and Our Results. Böhm et al. [9,10] studied the 2-bounded bounded delay buffer management problem with lookahead. They designed a deterministic algorithm whose competitive ratio is at most $(-1 + \sqrt{13})/2 \leq 1.303$. Also, they proved that the competitive ratio of any deterministic algorithm is at least $(1 + \sqrt{17})/4 \geq 1.280$.

In this paper, we show an online algorithm matching their lower bound for this problem, that is, its competitive ratio is exactly $(1 + \sqrt{17})/4$. Since the original bounded delay buffer management problem has been solved completely by Veselý et al. [28] just recently, the bounded delay buffer management problem with lookahead is one of the most important variants which should be solved among several variants of this problem. Our result will help to develop an optimal algorithm for s-bounded instances.

Related Results. In the full version [10] of the paper [9], Böhm et al. studied lower bounds on the competitive ratios of online algorithms with more generalized lookahead. Specifically, the lookahead ability in [9] at a time t enables an online algorithm to obtain the information about packets p such that $r(p) = t+1$. In [10], for a positive integer ℓ, they considered the case where the ability at a time t enables an online algorithm to obtain the information about packets p such that $r(p) \leq t + \ell$. They showed that a lower bound of any deterministic algorithm is $\frac{1+\sqrt{5+8\ell+4\ell^2}}{2\ell+2}$. Moreover, they proved that for any $\ell \geq 1$, a lower bound of any randomized online algorithm is 1.25.

As mentioned above, for the s-bounded delay model *without* lookahead, Hajek [19] showed that the competitive ratio of any deterministic algorithm is at least $(1 + \sqrt{5})/2 \geq 1.618$ in the case of $s \geq 2$. Independently, this bound was also shown in [4,13,29]. Several deterministic algorithms have been developed [5,9,10,14,16,23] and very recently, Veselý et al. [28] designed an optimal online algorithm. Moreover, in the case where an algorithm must decide which packet to transmit on the basis of the current buffer situation, called the *memoryless* case, some results were shown [5,14,16]. The *agreeable deadline* variant has also been studied. In this variant, the larger the release times of packets are, the larger their deadlines are. Specifically, for any packets p and p', $d(p) \leq d(p')$ if $r(p) < r(p')$. The lower bound of $(1 + \sqrt{5})/2$ by Hajek [19] is applicable to this variant. Li et al. [21,24] displayed an optimal algorithm, whose competitive ratio matches the lower bound. The case in which for any packet p, $d(p) - r(p) + 1 = s$ has also been studied, called the *s-uniform delay* variant, which is a specialized variant of the agreeable deadline variant. The current best upper bound for this variant is $(1 + \sqrt{5})/2$ [21,24]. Also, in the case of $s = 2$, Chrobak et al. [15] designed an optimal online algorithm whose competitive ratio is 1.377 [15].

The research on randomized algorithms for the bounded delay buffer management problem has also been conducted extensively [5,6,13,14,20–22]. In the case in which s is general, the current best upper and lower bounds are $e/(e-1) \leq 1.582$ [5,14,22] and $5/4 = 1.25$ [13], respectively, against an oblivious adversary were shown. Upper and lower bounds of $e/(e-1)$ [6,22] and $4/3 \geq 1.333$ [6], respectively, against an adaptive adversary were shown. For any fixed s, lower bounds are the same with the bounds in the case in which s is general while upper bounds are $1/(1 - (1 - \frac{1}{s})^s)$ [22] against the both adversaries.

A generalization of the bounded delay buffer management problem has been studied, called the *weighted item collection* problem [7,8,22]. In this problem, an online algorithm does not know the deadline of each packet but knows the relative order of the deadlines of packets. Many other variants of the buffer management problem have been studied extensively (see e.g. [17,26]).

2   Model Description

We formally give the definition of the 2-bounded delay buffer management problem with lookahead, which is addressed in this paper. An *input* of this problem is a sequence of phases. Time begins with zero and a phase occurs at an integer time. Each phase consists of three subphases. The first occurring subphase is the *arrival subphase*. At an arrival subphase, arbitrarily many packets can arrive to a buffer. The buffer has no capacity limit and hence, all arriving packets can always be accepted to the buffer. A packet p is characterized by the *release time*, *deadline* and *value*, denoted by $r(p)$, $d(p)$ and $v(p)$ respectively. Arrival times and deadlines are non-negative integers and values are positive reals. $d(p) - r(p) \leq 1$ holds because we focus on 2-bounded instances. The second subphase is the *transmission subphase*. At a transmission subphase, an algorithm can transmit at most one packet from its buffer if any packet. At the transmission subphase at a time t, the algorithm can obtain the information about packets arriving at time $t + 1$ using the lookahead ability. The third subphase is the *expiration subphase*. At an expiration subphase, a packet which has reached its deadline is discarded from its buffer. That is, at the expiration subphase at a time t, all the packets p in the buffer such that $d(p) = t$ are discarded.

The *profit* of an algorithm is the sum of the values of packets transmitted by the algorithm. The objective of this problem is to maximize the gained profit. Let $V_A(\sigma)$ denote the profit of an algorithm A for an input σ. Let OPT be an optimal offline algorithm. We say that the competitive ratio of an online algorithm ON is at most c if for any input σ, $V_{OPT}(\sigma) \le V_{ON}(\sigma)c$.

3   Matching Upper Bound

3.1   Notation and Definitions

We give definitions before defining our algorithm COMPAREWITHPARTIALOPT (CP). For any integer time t and any algorithm A, $B_A(t)$ denotes the set of packets in A's buffer immediately before the arrival subphase at time t. That is, each packet p in the set is not transmitted before t such that $t > r(p)$ and $t \le d(p)$. Let us define an offline algorithm PO, which stands for a Partial OPT, which stores all the packets in the buffer of CP at a time and woks optimally given a subinput from the time. For integer times $t, t' \ge t$ and $t'' \in \{t', t'+1\}$ and an input σ, let $PO(t, t', t'')$ be an offline algorithm such that the set of packets in $PO(t, t', t'')$'s buffer immediately before the arrival subphase at time t is equal to that of $B_{CP}(t)$'s, and if the subinput of σ during time $[t, t']$ is given to $PO(t, t', t'')$, that is, packets p such that $r(p) \in [t, t']$ arrive to $PO(t, t', t'')$'s buffer during time $[t, t']$, then $PO(t, t', t'')$ is allowed to transmit $t'' - t + 1$ packets only from time t to t'' inclusive, that is, at $t'' - t + 1$ transmission subphases, and chooses the packets whose total profit is maximized. If there exist packets with the same value in $PO(t, t', t'')$'s buffer, $PO(t, t', t'')$ follows a fixed tie breaking rule. Also, $P(t, t', t'')$ denotes the set of $t'' - t + 1$ packets transmitted by $PO(t, t', t'')$ during time $[t, t'']$. Note that for any t and $t' \ge t$, the following relations hold because of the optimality of packets transmitted by $PO(t, t', t'')$ during time $[t, t'']$:

$$P(t, t', t') \subseteq P(t, t'+1, t'+1) \qquad (1)$$
$$P(t, t', t') \subseteq P(t, t', t'+1) \qquad (2)$$

and

$$P(t+1, t', t') \subseteq P(t, t', t'). \qquad (3)$$

We define for any t and $i \geq 0$,

$$\{m_i(t)\} = P(t, t+i, t+i) \backslash P(t, t+i-1, t+i-1)$$

and any $i \geq 1$,

$$\{q_i(t)\} = P(t, t+i, t+i+1) \backslash P(t, t+i, t+i).$$

Also, we define

$$P(t, t-1, t-1) = \varnothing.$$

If $m_i(t)$ $(q_i(t))$ does not exist, that is, the above equality is the empty set, then we assume that a packet whose value is 0 is given. This assumption is used to make the description of CP simpler and does not affect the performance of CP. Moreover, if there exists a packet p such that both $v(p) = v(q_1(t))$ holds and either $r(p) = t$ or $p \in B_{CP}(t)$, that is, CP can transmit p at t, then $q_0(t)$ denotes p. Also, we define $m_{01}(t) \in \arg\max\{m_0(t), m_1(t)\}$. Let $V(t, t', t'')$ denote the total value of packets in $P(t, t', t'')$. That is, $V(t, t', t'') = \sum_{p \in P(t,t',t'')} v(p)$. We describe each value in the algorithm definition for ease of presentation as follows: $m_i = m_i(t)$, $m_{01} = m_{01}(t)$, $q_i = q_i(t)$ and, $R = (1 + \sqrt{17})/4$.

3.2   Idea Behind Algorithm Design

In this section, we explain the idea behind designing our algorithm CP for better understanding. Suppose that CP decides which packet to transmit at a time t. Let us assume that at t, the buffer of CP stores all the packets in the buffer of OPT at t. We guarantee that this assumption holds at a time satisfying some conditions in a lemma of the full version of this paper. Due to page limitations, we omit all of the proofs in this paper. The full version of this paper is available at https://arxiv.org/abs/1807.00121. If this assumption holds, CP is able to *detect* two packets OPT transmits at times t and $t + 1$. To detect here means that CP calculates which packets OPT transmits at these times cause the worst situation with respect to the profit ratio. Let V be the maximum total value of two packets which OPT transmits at t and $t + 1$. CP chooses packets p and p' at t and $t + 1$, respectively, from packets which are revealed to CP at t such that $V \leq R(v(p) + v(p'))$ holds. Note that p' may arrive at $t + 1$. Although both CP's and OPT's buffers have the same packets at some time, the optimal choice depends on the instance, which in turn depends on CP's choice and thus CP might make a non-optimal choice in general. and hence, CP does not always transmit the packets as the ones which OPT transmits although they have the same packets at t. If CP could choose packets for each $t = 0, 2, 4, \ldots$ to satisfy the above inequality, we could prove that the competitive ratio of CP is at most R. However, this is impossible. For example, suppose that packets p_0, p_1 and p_2 are given at time 0 such that $d(p_0) = 0$, $d(p_1) = 1$ and $d(p_2) = 2$ and no other packets are given further. Also, suppose that CP transmits p_1 and p_2 at times 0 and 1, respectively and OPT transmits p_0, p_1 and p_2 at times 0, 1 and 2, respectively. In this case, CP does not transmit any packet at time 2 and thus, we cannot prove the above inequality.

Thus, the length of a time interval which CP uses to evaluate its competitive ratio is not fixed (such as 2 mentioned above) but variable as follows. Let us assume again that at a time t, the buffer of CP stores all the packets in the buffer of OPT at t. Also, suppose that CP decides which packet to transmit at a time $t'(\geq t)$ (the fact that $t' \leq t+2$ holds will be shown later by the definition of CP). By this assumption, CP can detect $t'-t+2$ packets transmitted by OPT during the time $[t, t'+1]$ (in some special cases, CP can detect $t'-t+3$ packets transmitted by OPT during $[t, t'+2]$). Let V' be the maximum total value of packets which OPT transmits during this time interval. CP chooses packets p and p' at t' and $t'+1$, respectively, considering the total value U of packets which CP already transmitted during time $[t, t'-1]$ such that $V' \leq R(U+v(p)+v(p'))$ holds. For example, suppose that packets $q_0(0), m_0(0)$ and $m_1(0)$ are given at time 0 whose values satisfy the execution conditions of Case 1.2.3.1 in CP. If CP transmits $m_0(0)$ and $m_1(0)$ at times 0 and 1, respectively, then CP can detect that OPT transmits $q_0(0), m_0(0)$ and $m_1(0)$ at times 0, 1 and 2, respectively. In this case, $t = t' = 0$ holds, and the above inequality holds by the condition of Case 1.2.3.1.

The sequences of packets which OPT transmits during $[t, t'+1]$ (or $[t, t'+2]$) are classified into three categories according to a packet p' which CP transmits at $t'+1$ (this fact will be proved in some lemmas of the full version of this paper): (a) Packets which are given during $[t, t'+1]$ satisfy some conditions and OPT transmits specific packets whose total value is $V(t, t'+1, t'+1)$. (b) If the deadline of p' is $t'+1$, then the total value of packets which OPT transmits during $[t, t'+1]$ is at most $V(t, t'+1, t'+1)$. (c) If the deadline of p' is $t'+2$, then the total value of packets which OPT transmits during $[t, t'+2]$ is at most $V(t, t'+1, t'+2)$. Please refer to Table 1. 'Case' column shows the names of cases executed by CP at t. 'Type' column shows the categories of packet sequences transmitted by OPT during $[t, t'+1]$ (or $[t, t'+2]$). 't' and '$t+1$' in 'CP' column show packets which CP transmits at t and $t+1$, respectively. Similarly, 't', '$t+1$' and '$t+2$' in 'OPT' column show packets which CP detects at time t that OPT transmits at t, $t+1$ and $t+2$, respectively. 'Value' column shows the total value of the packets detected by CP. For example, packets detected at Case 1.2.3.1 are classified into (c). CP transmits $m_0(t)$ and $m_1(t)$ at times 0 and 1, respectively. CP can detect that OPT transmits $q_0(t)(= q_1(t)), m_0(t)$ and $m_1(t)$ at times 0, 1, and 2, respectively, and the total value of these packets is $V(t, t+1, t+2)$. On the other hand, suppose that packets satisfying the condition of Case 1.2.3.2 are given. In this case, at time t, if CP decides which packet to transmit at $t+1$, then a situation in which the above inequality does not hold can occur whichever packet which arrives at or before $t+1$ CP chooses. Note that if this condition is satisfied, then *this situation occurs not only for CP but also any online algorithm*, which causes the definition of CP lengthy. Hence, CP chooses $q_0(t)$ as a packet for the transmission subphase at t, and decides which packet to transmit for the transmission subphase of time $t+2$ after making sure of packets at $t+1$ with lookahead. That is, CP executes Step 2 at the transmission subphase at $t+1$ to choose packets which CP transmits at $t+1$ and $t+2$.

Similarly to the case at time t, CP chooses packets at $t+1$ considering the value $U = v(q_0(t))$ of the packet $q_0(t)$ which CP transmitted at t so that the above inequality holds. Please refer to the row of Case 2.2.1 in Table 2, which is described in the same manner as the previous one. Suppose that packets are given satisfying the conditions of Case 2.2.1. If CP transmits $m_{01}(t)$ and $m_2(t)$ at times $t+1$ and $t+2$, respectively, then CP can detect that OPT transmits $m_0(t), m_1(t)$ and $m_2(t)$ at times t, $t+1$ and $t+2$, respectively. These packets are classified into (b) and the above inequality holds because of the condition of Case 2.2.1. Unfortunately, suppose that packets satisfying the condition of Case 2.2.2.3 are given. In this case, at $t+1$, if CP decides which packet to transmit at $t+2$, then a situation in which the above inequality does not hold can also occur. Note that if the conditions of Cases 1.2.3.2 and 2.2.2.3 are satisfied at times t and $t+1$, respectively, then *this situation occurs for any online algorithm*. Thus, CP chooses $m_0(t)$ as a packet for $t+1$, and executes Step 3 at $t+2$ to choose packets which CP transmits at $t+2$ and $t+3$. Fortunately, as Table 3 shows, at time $t+2$, if CP chooses packets to transmit at $t+2$ and $t+3$ appropriately, then the above inequality holds at any of Cases 3.1 - 3.2.3. Moreover, we will prove that the buffer of CP stores all the packets in the buffer of OPT at $t'+2$ in a lemma of the full version (there exists some exception for a packet sequence classified into (c)). Hence, in the next step, we can regard time $t'+2$ as a new base time, which was time t in the above discussion, and evaluate the profit ratio for each time interval recursively. In this way, CP is designed so that at each time interval $[t, t'+1]$ (or $[t, t'+2]$), the corresponding profit ratio is at most R, that is, its competitive ratio is at most R.

Table 1. Packet prediction at Step 1 at time t

CP				OPT			
Case	Type	t	$t+1$	Value	t	$t+1$	$t+2$
1.1	a, b	m_0		$V(t,t,t)$	m_0		
1.2.1	b	m_1	m_0	$V(t,t+1,t+1)$	m_1	m_0	
1.2.2	b	m_0	m_1	$V(t,t+1,t+1)$	m_0	m_1	
1.2.3.1	c	m_0	m_1	$V(t,t+1,t+2)$	q_0	m_0	m_1
1.2.3.2		q_0	(Step 2)				

m_i and q_i denote $m_i(t)$ and $q_i(t)$ for ease of presentation. $q_0 = q_1$ by definition.

Table 2. Packet prediction at Step 2 at time $t+1$

CP					OPT				
Case	Type	t	$t+1$	$t+2$	Value	t	$t+1$	$t+2$	$t+3$
2.1	b	q_0	m_0	m_1	$V(t, t+2, t+2)$	m_0	m_1	m_2	
2.2.1	b		m_{01}	m_2	$V(t, t+2, t+2)$	m_0	m_1	m_2	
2.2.2.1	a		m_{01}		$V(t, t+1, t+1)$	m_0	m_1		
2.2.2.2	c		m_{01}	m_2	$V(t, t+2, t+3)$	q_0	m_0	m_1	m_2
2.2.2.3			m_0	(Step 3)					

m_i, m_{01} and q_i denote $m_i(t), m_{01}(t)$ and $q_i(t)$. $q_0 = q_1$ by definition.

Table 3. Packet prediction at Step 3 at time $t+2$

CP						OPT					
Case	Type	t	$t+1$	$t+2$	$t+3$	Value	t	$t+1$	$t+2$	$t+3$	$t+4$
3.1	b	q_0	m_0	m_1	m_2	$V(t, t+3, t+3)$	m_0	m_1	m_2	m_3	
3.2.1	b			m_2	m_3	$V(t, t+3, t+3)$	m_0	m_1	m_2	m_3	
3.2.2	a			m_2		$V(t, t+2, t+2)$	m_0	m_1	m_2		
3.2.3	c			m_2	m_3	$V(t, t+3, t+4)$	q_0	m_0	m_1	m_2	m_3

m_i, m_{01} and q_i denote $m_i(t), m_{01}(t)$ and $q_i(t)$. $q_0 = q_1$ by definition.

3.3 Algorithm

The executions of CP are divided into *stages*. Each stage consists of a single transmission subphase, two consecutive transmission subphases, three consecutive transmission subphases or four consecutive transmission subphases.

CP uses the internal variable s_t for holding the name of a packet which CP transmits at a time t. $s_{t'} = $ null holds at first for any integer t'. CP uses the constant tmp1 (tmp2) if at time t $(t+1)$, CP cannot decide which packet to transmit at $t+1$ $(t+2)$ in Case 1.2.3.2 (2.2.2.3). On the other hand, once the name of a packet is set to s_{t+1} at time t, CP certainly transmits the packet at $t+1$. It is applied to s_{t+2} (s_{t+3}) which is set at $t+1$ $(t+2)$.

COMPAREWITHPARTIALOPT (CP)

Initialize: For any integer time t', $s_{t'}$:=null.

Suppose that a stage starts at a time t.

Step 1 (the transmission subphase at t):

Execute the following cases (Case 1.1 - 1.2.3.2) and transmit the packet s_t. If s_{t+1} =null after this transmission (i.e., Case 1.1 is executed), then finish the stage.

Case 1.1 ($d(m_0) = t$ or q_0 does not exist): $s_t := m_0$.

Case 1.2 ($d(m_0) \neq t$):

Case 1.2.1 ($d(m_1) = t$): $s_t := m_1$ and $s_{t+1} := m_0$.

Case 1.2.2 ($d(m_1) = t + 1$): $s_t := m_0$ and $s_{t+1} := m_1$.

Case 1.2.3 ($d(m_1) \neq t + 1$):

Case 1.2.3.1 $\frac{V(t,t+1,t+2)}{v(m_0)+v(m_1)} \leq R$): $s_t := m_0$ and $s_{t+1} := m_1$.

Case 1.2.3.2 $\frac{V(t,t+1,t+2)}{v(m_0)+v(m_1)} > R$): $s_t := q_0$ and $s_{t+1} := $ tmp1.

Step 2 (the transmission subphase at $t + 1$):

If s_{t+1} =tmp1, then execute the following cases (Case 2.1 - 2.2.2.3). Transmit the packet s_{t+1}. If s_{t+2} =null after this transmission (i.e., Case 2.2.2.1 is executed), then finish the stage.

Case 2.1 ($\frac{V(t,t+2,t+2)}{v(q_0)+v(m_0)+v(m_1)} \leq R$): $s_{t+1} := m_0$ and $s_{t+2} := m_1$.

Case 2.2 ($\frac{V(t,t+2,t+2)}{v(q_0)+v(m_0)+v(m_1)} > R$):

Case 2.2.1 ($d(m_2) = t + 2$): $s_{t+1} := m_{01}$ and $s_{t+2} := m_2$.

Case 2.2.2 ($d(m_2) \neq t + 2$):

Case 2.2.2.1 ($v(q_2) \neq v(q_1)$): $s_{t+1} := m_{01}$.

Case 2.2.2.2 ($v(q_2) = v(q_1)$ and $\frac{V(t,t+2,t+3)}{v(q_0)+v(m_{01})+v(m_2)} \leq R$): $s_{t+1} := m_{01}$ and $s_{t+2} := m_2$.

Case 2.2.2.3 ($v(q_2) = v(q_1)$ and $\frac{V(t,t+2,t+3)}{v(q_0)+v(m_{01})+v(m_2)} > R$): $s_{t+1} := m_0$ and $s_{t+2} := $ tmp2.

Step 3 (the transmission subphase at $t + 2$):

If s_{t+2} =tmp2, then execute the following cases (Case 3.1 - 3.2.3). Transmit the packet s_{t+2}. If s_{t+3} =null after this transmission (i.e., Case 3.2.2 is executed), then finish the stage.

Case 3.1 ($\frac{V(t,t+3,t+3)}{v(q_0)+v(m_0)+v(m_1)+v(m_2)} \leq R$): $s_{t+2} := m_1$ and $s_{t+3} := m_2$.

Case 3.2 ($\frac{V(t,t+3,t+3)}{v(q_0)+v(m_0)+v(m_1)+v(m_2)} > R$):

Case 3.2.1 ($d(m_3) = t + 3$): $s_{t+2} := m_2$ and $s_{t+3} := m_3$.

Case 3.2.2 ($d(m_3) \neq t + 3$ and $v(q_3) \neq v(q_1)$): $s_{t+2} := m_2$.

Case 3.2.3 ($d(m_3) \neq t + 3$ and $v(q_3) = v(q_1)$): $s_{t+2} := m_2$ and $s_{t+3} := m_3$.

Step 4 (the transmission subphase at $t + 3$): Transmit s_{t+3} and finish the stage.

3.4   Overview of the Analysis

For ease of analysis, we assume that if CP does not store any packet in its buffer at the transmission subphase at a time t, no packets arrive at or after time $t + 1$ any more, that is, the input is over. Note that CP stores no packets but OPT may store one at t, that is, transmit it then. Since we consider a 2-bounded instance, the buffers of OPT and CP are both empty after the expiration subphase at t. This situation is equal to the one before the first packet arrives at time 0 and by the definition of CP, we regard a time at which the buffers are empty as time 0. Therefore, this assumption does not affect the performance of CP.

Consider a given input σ. Let k denote the number of stages after σ is over. Let τ be the last time at which CP transmits a packet. We partition the time sequence $[0, \tau]$ into k sequences T_i $(i = 1, \ldots, k)$ disjointly such that T_i consists of times at which the executions of the ith stage are done. Specifically, if $T_i = [t_i, t_i']$, then $t_i \leq t_i'$, $t_1 = 0$, $t_k' = \tau$ and for any $j = 2, \ldots, k$, $t_j = t_{j-1}' + 1$. The size of each T_i depends on times at which CP does the executions of the ith stage, that is, which case CP executes at each time: Suppose that $T_i = [t, t']$, in which t and t' are integer times.

- If Case 1.1 is executed at t, then $t' = t$.
- If Case 1.2.1, 1.2.2 or 1.2.3.1 is executed at t, then $t' = t + 1$.
- If Case 1.2.3.2 is executed at t and Case 2.1, 2.2.1 or 2.2.2.2 at $t + 1$, then $t' = t + 2$.
- If Cases 1.2.3.2 and 2.2.2.1 are executed at t and $t + 1$, respectively, then $t' = t + 1$.
- If Cases 1.2.3.2 and 2.2.2.3 are executed at t and $t + 1$, respectively, and Case 3.1, 3.2.1 or 3.2.3 is executed at $t + 2$, then $t' = t + 3$.
- If Cases 1.2.3.2 and 2.2.2.3 are executed at t and $t + 1$, respectively, and Case 3.2.2 is executed at $t + 2$, then $t' = t + 2$.

For a time t, a packet whose release time is t and deadline is $t + 1$ is called a 2_t-packet. If for a time t, CP transmits a 2_t-packet p at t and OPT transmits p at $t + 1$, then we call the time $t + 1$ an extra time (e-time, for short). On the other hand, for each $i = 1, \ldots, k$, let us define T_i', which is formally defined later, each of which is a subsequence of the time sequence $[0, \tau']$, in which τ' is the last time at which OPT transmits a packet. They are not always disjoint differently from T_i. To analyze the performance of CP, for each $i \in [1, k]$, we will compare the total value of packets transmitted by CP during the time T_i with that by OPT during the time T_i'. T_i' is defined as follows: For $T_i = [t, t']$ in which t and $t'(\geq t)$ are integer times, we define $T_i' = [t, \hat{t}']$, in which if $t' + 1$ is an e-time, then $\hat{t}' = t' + 1$. Otherwise, $\hat{t}' = t'$. We give the lemma about T_i'.

Lemma 1. *A time in $[0, \tau']$ is contained in some T_i'.*

For any i, we define an offline algorithm OPT_i to bound the value of packets transmitted by OPT during time $T_i' = [t, t']$, in which t and $t'(\geq t)$ are integer times. Roughly speaking, if t is not an e-time, then OPT_i transmits the same packet as a packet OPT transmits during T_i'. If t is an e-time, then OPT_i

transmits the same packet as a packet OPT transmits during T_i' except for t. However, OPT_{i-1} transmits the same packet as OPT at t.

First, let us define packets in the buffer of OPT_i for $T_i = [t, t']$. If $t = 1$, $B_{OPT_i}(t) = B_{OPT}(t)$. If $t \geq 2$ and t is not an e-time, then $B_{OPT_i}(t) = B_{OPT}(t)$. If $t \geq 2$ and t is an e-time, then $B_{OPT_i}(t) = B_{OPT}(t) \setminus \{p\}$, in which p is the 2_{t-1}-packet which OPT transmits at t. Then, for $T_i = [t, t']$ and $T_i' = [t, \hat{t}']$, we define OPT_i as follows: The subinput of σ during time T_i' is given to OPT_i, that is, packets p such that $r(p) \in T_i'$ arrive to OPT_i's buffer during time T_i' according to their release times. Then, OPT_i is allowed to transmit $\hat{t}' - t + 1$ packets only from time t to $\hat{t}'$ inclusive, that is, at $\hat{t}' - t + 1$ transmission subphases, and chooses the packets whose total profit is maximized. If there exist packets with the same value in OPT_i's buffer, OPT_i follows the same tie breaking rule as OPT.

We use $PO(t, t', \hat{t}')$ to define CP and can evaluate the profit of CP using the profit of $PO(t, t', \hat{t}')$ during T_i. On the other hand, we bound the profit of OPT using that of OPT_i during T_i'. Then, we evaluate the relations between the profit of $PO(t, t', \hat{t}')$ and that of OPT_i. For any $i \in [1, k]$, let V_i denote the total value of packets transmitted by CP during T_i. By definition, $V_{CP}(\sigma) = \sum_{i=1}^{k} V_i$. On the other hand, Lemma 1 indicates that a packet which OPT transmits is transmitted at a time in some $T_{i'}$ by either $OPT_{i'}$ or $OPT_{i'-1}$. Also, by the definition of OPT_i, if t is not an e-time, OPT_i transmits a packet at t whose value is at least that transmitted by OPT. If t is an e-time, then OPT_{i-1} transmits a packet at t whose value is at least that transmitted by OPT and OPT_i may also transmit a packet. That is, the total value of packets transmitted by OPT_i over all $i \in [1, k]$ is at least that of OPT. For any $i \in [1, k]$, let V_i' denote the total value of packets transmitted by OPT_i during T_i'. Hence, $V_{OPT}(\sigma) \leq \sum_{i=1}^{k} V_i'$. Since

$$\frac{V_{OPT}(\sigma)}{V_{CP}(\sigma)} \leq \frac{\sum_{i=1}^{k} V_i'}{\sum_{i=1}^{k} V_i} \leq \max_{i \in [1,k]} \left\{ \frac{V_i'}{V_i} \right\},$$

we will prove the following lemma:

Lemma 2. *For any* $i \in [1, k]$, $V_i'/V_i \leq (1 + \sqrt{17})/4$.

Therefore, we have the following theorem:

Theorem 1. *The competitive ratio of* CP *is at most* $(1 + \sqrt{17})/4$.

Acknowledgments. This work was supported by JSPS KAKENHI Grant Number 19K11819.

References

1. Aiello, W., Mansour, Y., Rajagopalan, S., Rosén, A.: Competitive queue policies for differentiated services. J. Algorithms **55**(2), 113–141 (2005)
2. Albers, S.: On the influence of lookahead in competitive paging algorithms. Algorithmica **18**(3), 283–305 (1997)
3. Albers, S.: A competitive analysis of the list update problem with lookahead. Theoret. Comput. Sci. **197**(1–2), 95–109 (1998)
4. Andelman, N., Mansour, Y., Zhu, A.: Competitive queueing policies for QoS switches. In: Proceedings of the 14th ACM-SIAM Symposium on Discrete Algorithms, pp. 761–770 (2003)
5. Bartal, Y., et al.: Online competitive algorithms for maximizing weighted throughput of unit jobs. In: Diekert, V., Habib, M. (eds.) STACS 2004. LNCS, vol. 2996, pp. 187–198. Springer, Heidelberg (2004). https://doi.org/10.1007/978-3-540-24749-4_17
6. Bienkowski, M., Chrobak, M., Jeż, Ł.: Randomized competitive algorithms for online buffer management in the adaptive adversary model. Theoret. Comput. Sci. **412**(39), 5121–5131 (2011)
7. Bienkowski, M., et al.: Collecting weighted items from a dynamic queue. Algorithmica **65**(1), 60–94 (2013)
8. Bienkowski, M., et al.: A Φ-competitive algorithm for collecting items with increasing weights from a dynamic queue. Theoret. Comput. Sci. **475**, 92–102 (2013)
9. Böhm, M., Chrobak, M., Jeż, Ł., Li, F., Sgall, J., Veselý, P.: Online packet scheduling with bounded delay and lookahead. In: Proceedings of the 27th International Symposium on Algorithms and Computation, pp. 21:1–21:13 (2016)
10. Böhm, M., Chrobak, M., Jeż, Ł., Li, F., Sgall, J., Veselý, P.: Online packet scheduling with bounded delay and lookahead. Theoret. Comput. Sci. **776**, 95–113 (2019)
11. Borodin, A., El-Yaniv, R.: Online Computation and Competitive Analysis. Cambridge University Press, Cambridge (1998)
12. Breslauer, D.: On competitive on-line paging with lookahead. Theoret. Comput. Sci. **209**(1–2), 365–375 (1998)
13. Chin, F.Y.L., Fung, S.P.Y.: Online scheduling for partial job values: does time-sharing or randomization help? Algorithmica **37**, 149–164 (2003)
14. Chin, F.Y.L., Chrobak, M., Fung, S.P.Y., Jawor, W., Sgall, J., Tichý, T.: Online competitive algorithms for maximizing weighted throughput of unit jobs. J. Discrete Algorithms **4**(2), 255–276 (2006)
15. Chrobak, M., Jawor, W., Sgall, J., Tichý, T.: Improved online algorithms for buffer management in QoS switches. ACM Trans. Algorithms **3**(4), 50:1–50:19 (2007)
16. Englert, M., Westermann, M.: Considering suppressed packets improves buffer management in quality of service switches. SIAM J. Comput. **41**(5), 1166–1192 (2012)
17. Goldwasser, M.: A survey of buffer management policies for packet switches. ACM SIGACT News **41**(1), 100–128 (2010)
18. Grove, E.F.: Online bin packing with lookahead. In: Proceedings of the 6th ACM-SIAM Symposium on Discrete Algorithms, pp. 430–436 (1995)
19. Hajek, B.: On the competitiveness of online scheduling of unit-length packets with hard deadlines in slotted time. In: Proceedings of the 35th Conference on Information Sciences and Systems, pp. 434–438 (2001)
20. Jeż, Ł.: Randomized algorithm for agreeable deadlines packet scheduling. In: Proceedings of the 27th Symposium on Theoretical Aspects of Computer Science, pp. 489–500 (2010)

21. Jeż, Ł., Li, F., Sethuraman, J., Stein, C.: Online scheduling of packets with agreeable deadlines. ACM Trans. Algorithms **9**(1), 5:1–5:11 (2012)
22. Jeż, Ł.: A universal randomized packet scheduling algorithm. Algorithmica **67**(4), 498–515 (2013)
23. Kesselman, A., Lotker, Z., Mansour, Y., Patt-Shamir, B., Schieber, B., Sviridenko, M.: Buffer overflow management in QoS switches. SIAM J. Comput. **33**(3), 563–583 (2004)
24. Li, F., Sethuraman, J., Stein, C.: An optimal online algorithm for packet scheduling with agreeable deadlines. In: Proceedings of the 16th ACM-SIAM Symposium on Discrete Algorithms, pp. 801–802 (2005)
25. Motwani, R., Saraswat, V., Torng, E.: Online scheduling with lookahead: multipass assembly lines. INFORMS J. Comput. **10**(3), 331–340 (1998)
26. Nikolenko, S.I., Kogan, K.: Single and Multiple Buffer Processing. In: Kao, M.Y. (ed.) Encyclopedia of Algorithms. Springer, New York (2016). https://doi.org/10.1007/978-1-4939-2864-4_535
27. Sleator, D., Tarjan, R.: Amortized efficiency of list update and paging rules. Commun. ACM **28**(2), 202–208 (1985)
28. Veselý, P., Chrobak, M., Jeż, Ł., Sgall, J.: A ϕ-competitive algorithm for scheduling packets with deadlines. In: Proceedings of the 25th ACM-SIAM Symposium on Discrete Algorithms, pp. 202–208 (2019)
29. Zhu, A.: Analysis of queueing policies in QoS switches. J. Algorithms **53**, 123–142 (2004)

Reoptimization of Path Vertex Cover Problem

Mehul Kumar⬭, Amit Kumar$^{(\boxtimes)}$⬭, and C. Pandu Rangan

Department of Computer Science and Engineering,
Indian Institute of Technology Madras, Chennai, India
{mehul,amitkr,rangan}@cse.iitm.ac.in

Abstract. Most optimization problems are notoriously hard. Considerable efforts must be spent in obtaining an optimal solution to certain instances that we encounter in the real world scenarios. Often it turns out that input instances get modified locally in some small ways due to changes in the application world. The natural question here is, given an optimal solution for an old instance I_O, can we construct an optimal solution for the new instance I_N, where I_N is the instance I_O with some local modifications. Reoptimization of NP-hard optimization problem precisely addresses this concern. It turns out that for some reoptimization versions of the NP-hard problems, we may only hope to obtain an approximate solution to a new instance. In this paper, we specifically study the reoptimization of path vertex cover problem. The objective in *k-path* vertex cover problem is to compute a minimum subset S of the vertices in a graph G such that after removal of S from G there is no path with k vertices in the graph. We show that when a constant number of vertices are inserted, reoptimizing unweighted *k-path* vertex cover problem admits a PTAS. For weighted 3-*path* vertex cover problem, we show that when a constant number of vertices are inserted, the reoptimization algorithm achieves an approximation factor of 1.5, hence an improvement from known 2-approximation algorithm for the optimization version. We provide reoptimization algorithm for weighted *k-path* vertex cover problem ($k \geq 4$) on bounded degree graphs, which is also an NP-hard problem. Given a ρ-approximation algorithm for *k-path* vertex cover problem on bounded degree graphs, we show that it can be reoptimized within an approximation factor of $(2 - \frac{1}{\rho})$ under constant number of vertex insertions.

Keywords: Reoptimization · Approximation algorithms · Path vertex cover

1 Introduction

Most combinatorial optimization problems are NP-hard. Efficient algorithms to find an optimal solution for such problems are not known. By efficient, we mean running in time polynomial in the input size. Hence, we resort to approximation

© Springer Nature Switzerland AG 2019
D.-Z. Du et al. (Eds.): COCOON 2019, LNCS 11653, pp. 363–374, 2019.
https://doi.org/10.1007/978-3-030-26176-4_30

algorithms which aim to efficiently provide a near-optimal solution. For minimization problems, a ρ-approximation algorithm ($\rho > 1$) efficiently outputs a solution of cost at most ρ times the optimum, where ρ is called the approximation ratio. A family of $(1+\epsilon)$ approximation algorithms ($\forall \epsilon > 0$) with polynomial running times is called a polynomial time approximation scheme (PTAS).

In many practical applications, the problem instance can arise from small perturbations in the previous instance of an optimization problem. A naive approach is to work on the new problem instance from scratch using known ρ-approximation algorithm. But, with some prior knowledge of the solution for old instance, can we perform better? The computational paradigm of reoptimization addresses this question.

We consider the case where one has devoted a substantial amount of time to obtain an exact solution for the NP-hard optimization problem. Now, the goal is to reoptimize the solution whenever the modified instance is known. A reoptimization problem $Reopt(\pi)$ can be built over any optimization problem π. An input instance for $Reopt(\pi)$ is a triple $(I_N, I_O, OPT(I_O))$, where I_O is an old instance, I_N is a modified instance and $OPT(I_O)$ is an optimal solution for π on I_O.

Suppose I_N is a hard instance obtained via some perturbations in I_O and assume that we have an optimal solution of I_O. The natural question is, can we find an optimal solution of I_N? For most of the cases this may not be the case and it is not difficult to show that, if the optimization problem is *path vertex cover* and the perturbation is a single vertex insertion, then even possessing $OPT(I_O)$ does not help to find an optimal solution for I_N efficiently, unless $P = NP$. Hence, the objective of an efficient algorithm for $Reopt(\pi)$ is to either achieve a better approximation ratio or improve the running time of the known approximation algorithm. In this paper, the optimization problem we consider for reoptimization is the path vertex cover problem. This problem has its applications in traffic control and secure communication in wireless networks [7]. We briefly explain the optimization problem below:

A path of order k in a graph is a simple path containing k vertices. For a given graph $G = (V, E)$, $S \subseteq V$ is a feasible k-*path* vertex cover iff every path of order k in G contains at least one vertex from S. The problem of finding a feasible k-path vertex cover on a graph is known as k-path vertex cover problem (k-$PVCP$). This problem has two variants: weighted and unweighted. The goal in unweighted k-$PVCP$ is to find a feasible subset of minimum cardinality whereas in weighted k-$PVCP$, the objective is to find minimum weighted subset of vertices that covers all the paths of order k or more.

2   Related Work and Contributions

For any fixed integer $k \geq 2$, the k-*path* vertex cover problem (k-$PVCP$) is known to be NP-complete for an arbitrary graph G and also it's NP-hard to approximate it within a factor of 1.3606, unless $P = NP$ [2]. However, unweighted and weighted k-*path* vertex cover problems on trees have polynomial time algorithms [2,3]. The problem has been studied in [6] as k-*path* traversal problem which presents a $\log(k)$-approximation algorithm for the unweighted version. For $k = 2$, the k-*PVCP* corresponds to the conventional vertex cover problem. The 3-*PVCP* is a dual problem to the dissociation number of the graph. Dissociation number is the maximum cardinality of a subset of vertices that induce a subgraph with maximum degree at most 1. [8] provides a 2-approximation algorithm for weighted 3-*PVCP* and there is a 3-approximation algorithm for 4-*PVCP* [4].

For the reoptimization version, G. Ausiello et al. present an algorithm for reoptimizing unweighted vertex cover problem. Following the approach in [5], Sect. 4 shows that reoptimization of unweighted k-*PVCP* admits a PTAS under the constant number of vertex insertions. Also respectively extending it to the case where old solution is an approximated solution on I_O. In Sect. 5, we extend the reoptimization paradigm for weighted vertex cover problem in [5] to weighted k-*PVCP*. Using the results from Sect. 5, we show in Sect. 6 that weighted 3-*PVCP* can be reoptimized with an approximation factor of 1.5 under constant number of vertex insertions. In Sect. 7, we present an algorithm for reoptimization version of weighted k-*PVCP* ($k \geq 4$) on bounded degree graphs under constant number of vertex insertions. For a given ρ-approximation algorithm for weighted k-*PVCP* ($k \geq 4$), the algorithm achieves an approximation ratio of $(2 - \frac{1}{\rho})$ for such graphs.

3   Preliminaries

In this paper, the graphs we consider are simple undirected graphs. A graph G is a pair of sets (V, E), where V is the set of vertices and E is the set of edges formed by unordered pairs of distinct vertices in V. For a vertex $v \in V$, we denote the set of neighbours of v in G by $N_G(v)$, where $N_G(v) = \{u \in V \mid (u, v) \in E\}$. For any $S \subseteq V$, we define $N_G(S)$ to be the neighbouring set of S in G, where $N_G(S) \subseteq (V - S)$ and $\forall u \in N_G(S) \, \exists v \in S$ such that $(u, v) \in E$. For any $S \subseteq V$, we use $G[S]$ to represent the subgraph induced on the vertex set S in G. Let $V(G)$ and $E(G)$ denote the vertex set and edge set of G respectively. A degree of a vertex is the number of edges incident on it. We use $\Delta(G)$ to denote the maximum degree of the vertices in graph G. In the case of weighted graphs, with every vertex we associate a positive weight function $f : V \to \mathbb{R}^+$. For any $v \in V$, let $w(v)$ be the weight of the vertex and for any subset $S \subseteq V$, the weight of the subset $w(S)$ is $\sum_{v \in S} w(v)$. Size of a graph is defined as the number of vertices in it. A constant-size graph is a graph where number of vertices are constant and independent of input parameters of the algorithm. Two graphs are said to be disjoint if they do not share any common vertices.

Let $G = (V, E)$ and $G_A = (V_A, E_A)$ be two graphs where $V \cap V_A = \emptyset$. Given a set of attachment edges $E^a \subseteq (V \times V_A)$, insertion of G_A into G yields the undirected graph $G' = (V', E')$, where $V' = V \cup V_A$ and $E' = E \cup E_A \cup E^a$. Thus, a constant number of vertex insertions can be realized as a constant-size graph insertion. We define a vertex insertion in G as a special case of graph insertion where the inserted graph G_A is a single vertex $v \notin V[G]$. In general, we denote $OPT(G)$ as the optimal solution and $ALG(G)$ as the solution output by an algorithm for the corresponding problem on G.

Let π denote the optimization problem and $Reopt(\pi)$ is the reoptimization version of it. The π we consider in this paper is the k-path vertex cover problem. The input instance of $Reopt(\pi)$ is a triple $(G_O, G_N, OPT(G_O))$, where G_O is the old graph, G_N is the new graph and $OPT(G_O)$ is an optimal solution for π on G_O. Let $A_\rho(\pi)$ be a known ρ-approximation algorithm for π.

Lemma 1. *Minimum unweighted k-path vertex cover problem on graphs with maximum degree $\Delta(G) \geq 3$ is NP-complete.*

Proof. k-path vertex cover is in NP as enumerating over all paths of order k would verify a k-path vertex cover instance, where runtime of verification is $O(n^k)$. We will show it is NP-hard by reducing vertex cover problem for cubic graphs to it, which is known to be NP-complete [9]. Applying the same reduction given in Theorem 1 of [2], for the input instance of a cubic graph G we get the reduced graph instance G'. Since the proof of reduction given in Theorem 2 of [2] is independent of the $\Delta(G)$ and $\Delta(G') = \Delta(G) + 1 = 4$, hence the reduction implies NP-hardness for k-path vertex cover on bounded degree graphs too.

Corollary 1. *Minimum weighted k-path vertex cover for bounded degree graphs is NP-hard.*

4   Reoptimization of Unweighted k-$PVCP$

Let π be unweighted k-$PVCP$. We consider the reoptimization version $Reopt(\pi)$ where a constant-size graph $G_A = (V_A, E_A)$ is inserted to the old graph $G_O = (V_O, E_O)$ to yield the new graph $G_N = (V_N, E_N)$. Let $|V_A| = c$. For a given ϵ, we design an algorithm $Unwtd$-$kpath$ for $Reopt(\pi)$ that outputs $ALG(G_N)$ as a solution.

Algorithm 1. $Unwtd$-$kpath(G_O, G_N, OPT(G_O), \epsilon)$

1: $V_A = V(G_N) - V(G_O)$
2: $c = |V_A|$
3: $m = \lceil c/\epsilon \rceil$
4: $S_1 = V(G_N)$
5: **for each subset** X **of** $V(G_N)$ **where** $|X| \leq m$ **do**
6:     **if** $(X$ **covers all** k-paths **in** G_N **and** $|X| < |S_1|)$
7:         $S_1 = X$
8: $S_2 = OPT(G_O) \cup V_A$
9: $ALG(G_N) = min(|S_1|, |S_2|)$
10: **return** $ALG(G_N)$

Theorem 1. *$Unwtd$-$kpath$ for $Reopt(\pi)$ under constant-size graph insertion admits a PTAS.*

Proof. Since $OPT(G_N) \cap V(G_O)$ and $OPT(G_O) \cup V_A$ is a feasible k-*path* vertex cover on G_O and G_N respectively, we get

$$|OPT(G_O)| \leq |OPT(G_N)| \leq |OPT(G_O)| + c \quad \cdots \quad (1)$$

If $OPT(G_N)$ has size at most m, it would have been found in step 7 of *Unwtd-kpath*. We know,

$$|ALG(G_N)| \leq |OPT(G_O)| + c = |S_2|$$

and S_2 is picked when $|OPT(G_N)| \geq m \geq \frac{c}{\epsilon}$. Thus, approximation factor for $ALG(G_N)$ using inequality (1) and above observation is,

$$\frac{|ALG(G_N)|}{|OPT(G_N)|} \leq \frac{|OPT(G_O)| + c}{|OPT(G_N)|} \leq \frac{|OPT(G_N)| + c}{|OPT(G_N)|} \leq 1 + \epsilon$$

Further, we analyze the runtime. Enumerating all possible k-*paths* in a graph of n vertices takes $O(n^k)$ time. Thus for a given set X, we can decide in polynomial time whether all paths of order k are covered by the set. There are $O(n^m)$ subsets of size at most m, where $n = |V_N|$. The runtime of Algorithm 1 is $O(n^m \cdot n^k) = O(n^{\frac{c}{\epsilon}} \cdot n^k)$, and hence a valid PTAS. Note that the runtime can be improved by using color coding algorithm for finding a k-*path* [1], which runs in $O(2^k n^{O(1)})$ time.

Substituting $|OPT(G_O)|$ by an α-approximate solution in the proof, we get the following corollary:

Corollary 2. *Given an α-approximate solution for G_O, there is $(\alpha + \epsilon)$-approximate solution for π on G_N, where G_N is obtained via a constant-size graph insertion to G_O*

5 Subroutine for Reoptimization of Weighted k-*PVCP*

Let π_k be weighted k-*PVCP*. $A_\rho(\pi_k)$ be a known ρ-approximation algorithm for π_k. In reoptimization version of the problem $Reopt(\pi_k)$, a new graph G_N is obtained by inserting a graph G_A to G_O.

Definition. A family $\mathcal{F} = \{F \mid F \subseteq V_N\}$ is called a **good** family if it satisfies the following two properties:

- **Property 1:** $\exists \ F \in \mathcal{F}$ such that $F \subseteq OPT(G_N)$ and,
- **Property 2:** $\forall \ F \in \mathcal{F}$, F covers all the k-*paths* which contains at-least one vertex from $V(G_A)$ in graph G_N.

We provide below a generic algorithm that works on the good family $\mathcal{F}$. This family of sets will be constructed in different ways for different problems. The details are provided in the respective sections.

An algorithm for $Reopt(\pi_k)$ constructs the good family $\mathcal{F}$ and feeds it to the subroutine *Construct-Sol*. The algorithm *Construct-Sol* iteratively prepares a solution S_i for each set $F_i \in \mathcal{F}$. The inputs to the algorithm *Construct-Sol* are: modified graph G_N, inserted graph G_A, old optimal solution $OPT(G_O)$), a good family $\mathcal{F}$ and $A_\rho(\pi_k)$.

Algorithm 2. $Construct\text{-}Sol(G_N, G_A, OPT(G_O), \mathcal{F}, A_\rho(\pi_k))$

1: **for** $i = 1$ **to** $|\mathcal{F}|$ **do**
2:     $S_i^1 = OPT(G_O) \cup F_i$
3:     $G' = G_N[(V_N - V(G_A)) - F_i]$
4:     Run $A_\rho(\pi_k)$ on G' and denote the output set as S_i^2
5:     $S_i^2 = S_i^2 \cup F_i$
6:     $S_i = minWeight(S_i^1, S_i^2)$
7: $ALG(G_N) = minWeight(S_1, S_2, \ldots, S_{|\mathcal{F}|})$
8: **return** $ALG(G_N)$

Lemma 2. *If $OPT(G)$ is an optimal solution for weighted k-PVCP for G, then for any $S \subseteq OPT(G)$, $w(OPT(G[V - S])) \leq w(OPT(G)) - w(S)$.*

Proof. If F is a feasible k-path cover for $G[V]$, then for any $V^* \subseteq V$, $F \cap V^*$ is a feasible k-path cover for $G[V^*]$.

$OPT(G) - S$ is a feasible solution for $G[V - S]$ because $(V - S) \cap OPT(G) = OPT(G) - S$. Since $S \subseteq OPT(G)$, $w(OPT(G) - S) = w(OPT(G)) - w(S)$. Hence, $w(OPT(G[V - S])) \leq w(OPT(G)) - w(S)$.

Theorem 2. *The algorithm Construct-Sol outputs a solution $ALG(G_N)$ with an approximation factor of $(2 - \frac{1}{\rho})$, running in $O(|V(G_N)|^2 \cdot |\mathcal{F}| \cdot T(A_\rho(\pi_k), G_N))$ steps, where ρ is the approximation factor of a known $A_\rho(\pi_k)$.*

Proof. A graph G_A is inserted to G_O to yield the new graph G_N. By property 1 of the good family $\mathcal{F}$, the optimal solution for G_N must include at least one set in $\mathcal{F} = \{F_1, \ldots, F_\psi\}$, where $\psi = |\mathcal{F}|$. Thus, at least one $S_i(1 \leq i \leq \psi)$ is prepared by the subroutine.

Let $OPT(G_N)_i$ be the optimal solution which includes F_i and not $(V(G_A) - F_i)$. We prepare ψ number of solutions for the graph G_N.

S_i^1 is a feasible k-path cover for G_N, where feasibility follows from property 2 of the family. We can write the following inequalities:

$$w(OPT(G_O)) \leq w(OPT(G_N)_i)$$
$$w(S_i^1) = w(OPT(G_O) \cup F_i) \leq w(OPT(G_O)) + w(F_i)$$

From above two inequalities,

$$w(S_i^1) \leq w(OPT(G_N)_i) + w(F_i) \quad \cdots (2)$$

Another solution S_i^2 is prepared. From Lemma 2 and construction of S_i^2, we can write the following inequality:

$$w(S_i^2) \leq \rho(w(OPT(G_N)_i) - w(F_i)) + w(F_i) \quad \cdots (3)$$

Since $\rho > 1$, adding $(\rho - 1) \times (1)$ and (2), we get

$$(\rho - 1)w(S_i^1) + w(S_i^2) \leq (2\rho - 1)(w(OPT(G_N)_i))$$

Minimum weighted subset between S_i^1 and S_i^2 is chosen to be S_i. Then,

$$(\rho - 1)w(S_i) + w(S_i) \leq (2\rho - 1)(w(OPT(G_N)_i))$$

$$\implies \forall i \in [1, \psi], w(S_i) \leq (2 - \frac{1}{\rho})(w(OPT(G_N)_i)) \quad \cdots (4)$$

We have prepared a set of ψ number of solutions that is, $\{S_1, S_2, \cdots, S_\psi\}$. By definition of $OPT(G_N)_i$ and property 2 of good family $\mathcal{F}$, we get that, if $F_i \subseteq OPT(G_N)$, then $w(OPT(G_N)_i) = w(OPT(G_N))$. By the property 1 of $\mathcal{F}$, there exists an F_i such that $F_i \subseteq OPT(G_N)_i$. Hence, the following inequality for such an i holds true:

$$w(S_i) \leq \left(2 - \frac{1}{\rho}\right)(w(OPT(G_N)_i)) = \left(2 - \frac{1}{\rho}\right)(w(OPT(G_N)))$$

We know, $\forall i \in [1, \psi], w(ALG(G_N)) \leq w(S_i)$. So,

$$w(ALG(G_N)) \leq \left(2 - \frac{1}{\rho}\right)(w(OPT(G_N)))$$

Thus, algorithm *Construct-Sol* outputs a solution with an approximation factor of $(2 - \frac{1}{\rho})$. Note that step 3 of the algorithm takes $O(|V_N|^2)$ time. Moreover, if the running time of $A_\rho(\pi_k)$ on input graph G_N is $T(A_\rho(\pi_k), G_N)$, then the running time of algorithm *Construct-Sol* is $O(|V(G_N)|^2 \cdot \psi \cdot T(A_\rho(\pi_k), G_N))$.

6   Reoptimization of Weighted 3-*PVCP*

Let π_3 be weighted 3-*PVCP*. A constant-size graph $G_A = (V_A, E_A)$ is inserted to G_O to yield the new graph $G_N = (V_N, E_N)$. Let $T(A_2(\pi_3), G_N)$ denote the runtime of 2-approximation algorithm [8] for π_3. Let $|V_N| = n$.

Algorithm 3. $Wtd\text{-}3path(G_N, G_A, OPT(G_O), A_2(\pi_3))$

1: $\mathcal{F} = \emptyset$
2: **for all** $X \subseteq (V_A = V(G_A))$ and X is a 3-path cover for G_A **do**
3:     V_I = Set of isolated v in $G_A[V_A - X]$ and $N_{G_N}(v) \cap V(G_O) \neq \emptyset$
4:     E_I = Set of isolated (u, v) in $G_A[V_A - X]$ and $N_{G_N}(\{u, v\}) \cap V(G_O) \neq \emptyset$
5:     **for all** $(u, v) \in E_I$ **do**
6:         $X = X \cup N_{G_O}(\{u, v\})$
7:     $Y = N_{G_N}(V_A) - X$
8:     **for all** $Y' \subseteq Y, |Y'| \leq |V_I|$ **do**
9:         **if** $(Y - Y')$ is a 3-*path* cover for $G_N[V_I \cup Y]$
10:             $X' = X \cup (Y - Y')$
11:             $X' = X' \cup N_{G_O}(Y')$
12:             $\mathcal{F} = \mathcal{F} \cup \{X'\}$
13: **return** $Construct\text{-}Sol(G_N, G_A, OPT(G_O), \mathcal{F}, A_2(\pi_3))$

Theorem 3. *Algorithm Wtd-3path is a 1.5 approximation for Reopt(π_3) under constant-size graph insertion.*

Proof. The algorithm works in 3 phases to construct the good family $\mathcal{F}$. The algorithm prepares a subset X for each feasible 3-*path* cover X for G_A because the optimal solution for G_N must contain one subset among the feasible X's. In the first phase, if an edge in $G_A[V_A - X]$ has a neighbour in G_O, it must be included in X.

Let $Y = N_{G_N}(V_A) - X$. In the second phase, $G_N[V_I \cup Y]$ is made free from all 3-*paths* by removing a feasible subset $Y - Y'$. In the third phase, the neighbours of the vertices in Y' are included in X' because such a neighbour will form 3-*path* with the vertices in Y' and V_I.

Since we consider all feasible subsets X and feasible $Y - Y'$ for the corresponding X, the constructed family $\mathcal{F}$ satisfies both the properties of good family. Thus from Theorem 2 and $A_2(\pi_3)$, we get the desired approximation. Further, we analyze the running time. Let $c = |V(G_A)|$. The maximum cardinality of Y is n. Then, the steps 3 to 12 in the algorithm run in $O(n^3 n^{c+1})$ because $|Y'| \leq c$. Thus $|\mathcal{F}| \in O(c^2 \cdot 2^c \cdot n^{c+4})$. Hence the algorithm Wtd-3path runs in $O(n^{c+6} \cdot 2^c \cdot T(A_2(\pi_3), G))$.

7   Reoptimization of Weighted k-$PVCP$ ($k \geq 4$) for Bounded Degree Graphs

A graph free from 2-*paths* contains only isolated vertices. A graph that does not have any 3-*path* contains isolated vertices and isolated edges. But, in the case of graphs that are free from k-*paths* ($k \geq 4$), star graph is a possible component. As the number of subsets needed to be considered for preparation of $\mathcal{F}$ would be exponential in the vertex degree, we restrict the reoptimization of weighted k-$PVCP$ ($k \geq 4$) to bounded degree graphs. Lemma 1 shows that the problem on bounded degree graphs is NP-complete.

The local modification which we consider for reoptimization is constant-size graph insertion. Let $G_O = (V_O, E_O)$ be the old graph. Given G_O, constant-size graph $G_A = (V_A, E_A)$ and attachment edges E^a, the new graph $G_N = (V_N, E_N)$ is obtained. Let $|V_N| = n$ and $|V_A| = c$. Let the maximum degree of the graph G_N be Δ. We use $P_k(G, V')$ to denote the collection of k-*paths* in graph G containing at least one vertex from $V' \cap V(G)$. For a set of vertices V', a graph is said to be V'-connected graph if every connected component in the graph contains at least one vertex from V'. Let $A_\rho(\pi_k)$ be a ρ-approximation algorithm for weighted k-$PVCP$ (π_k) and $T(A_\rho(\pi_k), G_N)$ be the running time of $A_\rho(\pi_k)$ on G_N.

Definition. We define a variation of BFS on a graph G, where traversal starts by enqueuing a set of vertices V' instead of a single root vertex. Initially, all the vertices in V' are at the same level and unvisited. Now the unvisited nodes are explored in breadth first manner. In this variation, we obtain the BFS forest for the input (G, V'), where the vertices of V' are at level 1 and the subsequent levels signify the order in which the vertices are explored.

Consider the BFS forest obtained from V_A in G_N. We use L_i to denote the set of vertices at level i $(i \geq 0)$ of the BFS forest. Let $S_j = \bigcup_{i=0}^{j} L_i$, where $L_0 = \emptyset$ and $L_1 = V_A$. Then $L_i = N_{G_N}(S_{i-1})$ for $i \geq 2$. Note that this BFS forest has $|V_A|$ number of disjoint BFS trees, where the trees have distinct root vertices from V_A.

Lemma 3. *In a BFS forest obtained after performing BFS traversal from a set of vertices $V_A \subseteq V$ in a graph $G = (V, E)$ having no k-paths, the number of vertices at each level is at most $|V_A|\Delta(\Delta - 1)^{\lceil \frac{k-5}{2} \rceil}$.*

Proof. Consider the case when BFS is performed from a single vertex set $V_A = \{v_1\}$ to obtain a BFS tree. For any level i, $|L_i| \leq \Delta(\Delta - 1)^{i-2}$. Thus the statement holds true for $i \leq \lceil \frac{k-1}{2} \rceil$. For the case when $i > \lceil \frac{k-1}{2} \rceil$, let $j = i - \lceil \frac{k-3}{2} \rceil$. We claim that there exists a vertex v in L_j such that v is a common ancestor for all the vertices in L_i. Assume to contrary that the claim is false. If $|L_i| = 1$ the claim is trivially true. Otherwise we have two distinct vertices v_x and $v_y \in L_i$ such that they have the lowest common ancestor in $L_{j'}$, where $1 \leq j' \leq j - 1 = i - \lceil \frac{k-1}{2} \rceil$. This imposes a path $\langle v_i, \cdots v, \cdots v_j \rangle$ of order $\lceil \frac{k-1}{2} \rceil + 1 + \lceil \frac{k-1}{2} \rceil \geq k$. But it contradicts the fact that G has no paths of order k or more. Hence $|L_i| \leq \Delta(\Delta - 1)^{i-j-1} = \Delta(\Delta - 1)^{\lceil \frac{k-5}{2} \rceil}$.

Now, when BFS is performed for the case when $|V_A| > 1$, the BFS forest obtained has $|V_A|$ number of disjoint BFS trees where each tree satisfies the above argument. Hence the number of vertices in each level in the BFS forest is at most $|V_A|\Delta(\Delta - 1)^{\lceil \frac{k-5}{2} \rceil}$

Algorithm 4. $Construct\text{-}F(X, V, L, level, V_A, G_N, \mathcal{F}, k)$

1: $\mathcal{F} = \mathcal{F} \cup \{X \cup L\}$
2: **if** $level \geq k - 1$
3:     **return** $\mathcal{F}$
4: $b = (|V_A|\Delta(\Delta - 1)^{\lceil \frac{k-5}{2} \rceil})$
5: **for each non-empty subset** V' of L and $|V'| \leq b$ **do**
6:     **if** $G_N[V \cup V']$ **is a k-path free** V_A**-connected graph**
7:         $X'' = X \cup (L - V')$
8:         $V'' = V \cup V'$
9:         $L'' = N_{G_N}(V'') - X''$
10:         $\mathcal{F} = Construct\text{-}(X'', V'', L'', level + 1, V_A, G_N, \mathcal{F}, k)$
11: **return** $\mathcal{F}$

Theorem 4. *Algorithm Wtd-kpath is a $(2 - \frac{1}{\rho})$ approximation for $Reopt(\pi_k)$ under graph insertion and runs in $O(n^{O(1)} \cdot 2^{k(\Delta+1)b} \cdot T(A_\rho(\pi_k), G_N))$, where $b = |V_A|\Delta(\Delta - 1)^{\lceil \frac{k-5}{2} \rceil}$ and $((\Delta + 1)b) \in O(\log n)$.*

Algorithm 5. $Wtd\text{-}kpath(G_N, G_A, OPT(G_O), A_\rho(\pi_k), k)$

1: Initialization: $\mathcal{F} = \emptyset$, $level = 1$, $X = \emptyset$ and $V = \emptyset$.
2: $\mathcal{F} = Construct\text{-}F(X, V, V(G_A), level, V(G_A), G_N, \mathcal{F}, k)$
3: $ALG(G_N) = Construct\text{-}Sol(G_N, G_A, OPT(G_O), \mathcal{F}, A_\rho(\pi_k))$
4: **return** $ALG(G_N)$

Proof. We first prove that for every call to the function
$Construct\text{-}F(X, V, L, level, V_A, G_N, \mathcal{F}, k)$, the following invariants on V, X and
L are maintained:

- $V \subseteq S_{level}$ and $G_N[V]$ is a k-path free V_A-connected graph.
- X is the set of neighbours of V in $G_N[S_{level}]$
- L is the set of neighbours of V in graph G_N that are also in $L_{level+1}$, i.e.
  $L = N_{G_N}(V) - S_{level}$.

The above invariants trivially hold true during the first call to the function
$Construct\text{-}F$. Assuming the invariants to be true during a call to $Construct\text{-}F$, we show that the subsequent recursive calls maintain the invariants. Note
that the parameter '$level$' is incremented to $level + 1$ during the recursive call.
$G_N[V \cup V'] = G_N[V'']$ is a k-path free V_A-connected subgraph in G_N. Also,
$V'' \subseteq S_{level+1}$ because $V \subseteq S_{level}$ and $L \subseteq L_{level+1}$. The invariance property of
X and L implies $X'' = X \cup (L - V')$ is the set of neighbours of V'' in $G_N[S_{level+1}]$.
From previous observation about X'' and V'', we get that $L'' = N_{G_N}(V'') - X''$ is
the set of neighbours of V'' in G_N which are also in $L_{level+2}$. Thus, the invariants
are maintained.

Note that X covers all the paths in $P_k(G_N[S_{level}], V_A)$. $X \cup L$ is a k-path
cover for G_N because the paths in $P_k(G_N, V_A) - P_k(G_N[S_{level}], V_A)$ contain at
least one vertex from L. Thus, $\{X \cup L\}$ is included in $\mathcal{F}$ to satisfy property 2 of
good family.

By Lemma 3, it is sufficient to consider non empty subsets V' of size at most
$(|V_A| \Delta (\Delta - 1)^{\lceil \frac{k-5}{2} \rceil})$ from subsequent level L to construct V''. For each recursive
call, the case when V' or L is empty is handled in the step 1. A V_A-connected
graph that has no $k\text{-}paths$ will have a maximum level of $k - 1$ in the BFS forest.
The algorithm explores all feasible subsets V' for each $level \leq k - 1$. Thus
the property 1 of good family holds true for $\mathcal{F}$, because the family includes all
possibilities for $\{X \cup L\}$ that covers $P_k(G_N, V_A)$. Thus, the constructed family
$\mathcal{F}$ is indeed a good family.

Let $RT(l)$ be the running time of the function $Construct\text{-}F$, where l is the
parameter '$level$'. Let $C = \Sigma_{i=1}^{i=b} \binom{\Delta b}{i}$. Observe that $|L| \leq (\Delta \cdot b)$ due to the
construction of L'' in the previous recursion. As we are choosing sets of size at
most b from L, we get the recursion $RT(l) = O(n^{O(1)} \cdot C^k \cdot RT(l+1))$ for $1 \leq l \leq$
$(k - 1)$ and $RT(k) = O(n^{O(1)})$. Thus step 6 in $Wtd\text{-}kpath$ runs in $O(n^{O(1)} \cdot C^k)$
time. In each function call, $|\mathcal{F}|$ is incremented by one element. Thus, $|\mathcal{F}| \leq 2^{kb}$
because $|V''| \leq b$ for each level. Note that $C \leq 2^{b\Delta}$. Hence using Theorem 2,
the algorithm $Weighted\text{-}kpath$ runs in $O(n^{O(1)} \cdot C^k \cdot 2^{kb} \cdot T(A_\rho(\pi_k), G_N)) =$
$O(n^{O(1)} \cdot 2^{k(\Delta+1)b} \cdot T(A_\rho(\pi_k), G_N))$ and achieves the desired approximation.

Using 3-*approximation* algorithm for weighted 4-*PVCP* [4] and Theorem 4, we get the following corollary:

Corollary 3. *Algorithm Wtd-kpath is a* $(\frac{5}{3})$*-approximation for Reopt*(π_4) *under constant-size graph insertion, where* $\Delta \in O(1)$ *and* $A_\rho(\pi_4)$ *is* $A_3(\pi_4)$.

It is not hard to show that a greedy algorithm for π_k will output a n-approximation. Also reducing π_k to minimum weighted subset selection problem and using the Theorem 15.3 in [10], we get a k-approximation algorithm for π_k. Hence, we get the following corollaries:

Corollary 4. *Algorithm Wtd-kpath is a* $(2 - \frac{1}{n})$*-approximation for Reopt*(π_k) *under constant-size graph insertion, where* $\Delta \in O(1)$ *and* $A_\rho(\pi_k)$ *is* $A_n(\pi_k)$.

Corollary 5. *Algorithm Wtd-kpath is a* $(2 - \frac{1}{k})$*-approximation for Reopt*(π_k) *under constant-size graph insertion, where* $\Delta \in O(1)$ *and* $A_\rho(\pi_k)$ *is* $A_k(\pi_k)$.

Note that the algorithm only explores the vertices till level $k - 1$ that is, the vertices in the set S_{k-1}. Thus, $|\mathcal{F}|$ is at most $2^{|S_{k-1}|}$. Therefore, the algorithm will also run efficiently for the scenarios where the graph G_A is attached to a '*sparse*' part of G_O, that is for $|S_{k-1}| \in O(\log n)$.

Corollary 6. *Algorithm Wtd-kpath is a* $(2 - \frac{1}{\rho})$*-approximation for Reopt*(π_k) *under graph insertion, where* $|S_{k-1}| \in O(\log n)$.

8   Concluding Remarks

In this paper, we have given a PTAS for reoptimization of unweighted k-*PVCP* under constant number of vertex insertions. When constant-size graph is inserted to the old graph, we have presented 1.5-approximation algorithm for reoptimization of weighted 3-*PVCP*. Restricting our inputs to bounded degree graphs, we have presented a $(2 - \frac{1}{k})$-approximation for reoptimization of weighted k-*PVCP* under constant-size graph insertion. For the reasons we mentioned in Sect. 7, our technique for reoptimization of weighted k-*PVCP* ($k \geq 4$) cannot be extended to arbitrary graphs. Hence, reoptimization of weighted k-*PVCP* ($k \geq 4$) for arbitrary graphs under constant number of vertex insertions is an intriguing open problem.

Acknowledgment. We thank Narayanaswamy N S for enlightening discussions on the problem.

References

1. Alon, N., Yuster, R., Zwick, U.: Color-coding: a new method for finding simple paths, cycles and other small subgraphs within large graphs. In: Proceedings of the Twenty-Sixth Annual ACM Symposium on Theory of Computing, Montréal, Québec, Canada, 23–25 May 1994, pp. 326–335 (1994). https://doi.org/10.1145/195058.195179

2. Bresar, B., Kardos, F., Katrenic, J., Semanisin, G.: Minimum k-path vertex cover. Discrete Appl. Math. **159**(12), 1189–1195 (2011). https://doi.org/10.1016/j.dam. 2011.04.008
3. Bresar, B., Krivos-Bellus, R., Semanisin, G., Sparl, P.: On the weighted k-path vertex cover problem. Discrete Appl. Math. **177**, 14–18 (2014). https://doi.org/ 10.1016/j.dam.2014.05.042
4. Camby, E., Cardinal, J., Chapelle, M., Fiorini, S., Joret, G.: A primal-dual 3-approximation algorithm for hitting 4-vertex paths. In: 9th International Colloquium on Graph Theory and Combinatorics, ICGT, p. 61 (2014)
5. Escoffier, B., Bonifaci, V., Ausiello, G.: Complexity and approximation in reoptimization, February 2011. https://doi.org/10.1142/9781848162778_0004
6. Lee, E.: Partitioning a graph into small pieces with applications to path transversal. In: Proceedings of the Twenty-Eighth Annual ACM-SIAM Symposium on Discrete Algorithms, SODA 2017, pp. 1546–1558 (2017). https://doi.org/10.1137/1. 9781611974782.101
7. Novotný, M.: Design and analysis of a generalized canvas protocol. In: Samarati, P., Tunstall, M., Posegga, J., Markantonakis, K., Sauveron, D. (eds.) WISTP 2010. LNCS, vol. 6033, pp. 106–121. Springer, Heidelberg (2010). https://doi.org/10. 1007/978-3-642-12368-9_8
8. Tu, J., Zhou, W.: A primal-dual approximation algorithm for the vertex cover p_3 problem. Theor. Comput. Sci. **412**(50), 7044–7048 (2011). https://doi.org/10. 1016/j.tcs.2011.09.013
9. Uehara, R.: NP-complete problems on a 3-connected cubic planar graph and their applications (1996)
10. Vazirani, V.V.: Approximation Algorithms. Springer, Heidelberg (2001). https:// doi.org/10.1007/978-3-662-04565-7

A Simple Local Search Gives a PTAS for the Feedback Vertex Set Problem in Minor-Free Graphs

Hung Le[1(⊠)] and Baigong Zheng[2]

[1] Department of Computer Science,
University of Victoria, Victoria, BC, Canada
hungle@uvic.ca
[2] School of Electrical Engineering and Computer Science,
Oregon State University, Corvallis, OR, USA
zhengb@oregonstate.edu

Abstract. We show that a simple local search gives a PTAS for the Feedback Vertex Set (FVS) problem in minor-free graphs. An efficient PTAS in minor-free graphs was known for this problem by Fomin, Lokshtanov, Raman and Sauraubh [13]. However, their algorithm is a combination of many advanced algorithmic tools such as contraction decomposition framework introduced by Demaine and Hajiaghayi [10], Courcelle's theorem [9] and the Robertson and Seymour decomposition [29]. In stark contrast, our local search algorithm is very simple and easy to implement. It keeps exchanging a constant number of vertices to improve the current solution until a local optimum is reached. Our main contribution is to show that the local optimum only differs the global optimum by $(1 + \epsilon)$ factor.

Keywords: Feedback vertex set · PTAS · Local search · Minor-free graphs

1 Introduction

Given an undirected graph, the *Feedback Vertex Set* (FVS) problem asks for a minimum set of vertices whose removal makes the graph acyclic. This problem arises in a variety of applications, including deadlock resolution, circuit testing, artificial intelligence, and analysis of manufacturing processes [12]. Due to its importance, the problem has been studied for a long time. It is one of Karp's 21 NP-complete problems [18] and is still NP-hard even in planar graphs [30]. It is one of the two problems that motivates the development of the seminal contraction decomposition framework for designing polynomial time approximation schemes[1] (PTASes) for many optimization problems in planar graphs [10].

---

[1] A polynomial-time approximation scheme for a minimization problem is an algorithm that, given a fixed constant $\epsilon > 0$, runs in polynomial time and returns a solution within $1 + \epsilon$ of optimal.

© Springer Nature Switzerland AG 2019
D.-Z. Du et al. (Eds.): COCOON 2019, LNCS 11653, pp. 375–386, 2019.
https://doi.org/10.1007/978-3-030-26176-4_31

In general graphs, the current best approximation ratio for the FVS problem is 2 due to Becker and Geiger [4] and Bafna, Berman and Fujito [3]. For some special classes of graphs, better approximation algorithms are known. Kleinberg and Kumar [19] gave the first PTAS for the FVS problem in planar graphs, followed by an efficient PTAS[2] by Demaine and Hajiaghayi [10] which is generalizable to bounded genus graphs and single-crossing-minor-free graphs. Recently, Cohen-Addad et al. [8] gave a PTAS for the weighted version of this problem in bounded-genus graphs. By generalizing the contraction decomposition of Demaine and Hajiaghayi to minor-free graphs, Fomin, Lokshtanov, Raman and Sauraubh [13] obtained a PTAS for the FVS problem in this class of graphs. A graph is H-*minor-free*, or simply minor-free, if it excludes some fixed graph H as a minor. We note that the class of minor-free graphs are vastly bigger than planar graphs and bounded-genus graphs. A typical example is the complete bipartite graph $K_{3,n}$ which has unbounded genus but is K_5-minor-free. In Sect. 5, we show that in some sense, minor-free graphs are the limit for which we are still able to obtain a PTAS for this problem.

A common theme in all known algorithms is complication in both implementation and analysis. The algorithm of Kleinberg and Kumar [19] is obtained by recursively applying the planar separator theorem by Lipton and Tarjan [22] and analyzing several special cases. The algorithm by Demaine and Hajiaghayi [13] employs the primal-dual relationship of planar graphs to decompose the graphs into several bounded treewidth instances, then applies dynamic programming (DP) to solve the FVS problem on bounded treewidth graphs. DP on bounded treewidth graphs is a very strong algorithmic tool. However, the implementation details typically are quite complicated. Additionally, the NP-hardness complexity of finding a tree decomposition of minimum width in planar graphs is still a long standing open problem. The algorithm of Cohen-Addad et al. [8] for bounded-genus graphs is not simpler and has worst running time; however, it can work with node-weighted graphs. Given the complicated nature of the algorithms for planar and bounded-genus graphs, it is not surprising that the technical level of the algorithm by Fomin, Lokshtanov, Raman and Sauraubh [13] for minor-free graphs is much higher. It uses advanced tools such as Courcelle's theorem [9] and the Robertson and Seymour decomposition [29]. We note that the decomposition of Robertson and Seymour was built through a series of papers which span 20 years with several hundred pages [28,29]. Thus, even understanding Robertson and Seymour decomposition is a real challenge, let alone implementing it. All of this motivates our current work.

We show that a simple local search algorithm gives a PTAS for the FVS problem in minor-free graphs. The algorithm is depicted in Algorithm 1. Intuitively, the local search algorithm starts with an arbitrary solution for the problem and tries to change a constant number (depending on ϵ) of vertices in the current solution to obtain a better solution. The algorithm outputs the current solution when it cannot obtain a better solution in this way.

---

[2] A PTAS is efficient if the running time is of the form $2^{\text{poly}(1/\epsilon)} n^{O(1)}$.

Local search is among the most successful heuristics in combinatorial optimization, partly due to its simplicity. It has been applied to scheduling, graph coloring, graph partitioning, Hopfield networks; we refer readers to the monograph by Michiels, Aarts and Korst [23] for more details. However, one of the hardest questions regarding local search is the performance guarantee. We provide an answer this question for the FVS problem. The analysis of our algorithm is simple, but non-trivial: it only uses two well-known properties of H-minor-free graphs as black boxes, namely sparsity and separability, and can be described in about four pages. A key ingredient in our analysis is the introduction of Steiner vertices into the construction of exchange graphs which is different from all previous works [5,6,25]; we defer further details of this discussion to Subsect. 1.1.

Algorithm 1. LOCALSEARCH(G, ϵ)

1: $S \leftarrow$ an arbitrary solution of G
2: $c \leftarrow$ a constant depending on ϵ
3: **while** there is a solution S' such that $|S \setminus S'| \leq c$, $|S' \setminus S| \leq c$ and $|S'| < |S|$ **do**
4:     $S \leftarrow S'$
5: output S

Theorem 1. *For any fixed $\epsilon > 0$, there is a local search algorithm that finds an $(1 + \epsilon)$-approximate solution for the FVS problem in H-minor-free graphs with running time $O(n^c)$ where $c = \frac{\mathrm{poly}(|V(H)|)}{\epsilon^2}$.*

Beside simplicity, our algorithm has two other interesting properties. First, to run the algorithm, we do not need to know beforehand whether the graph under consideration is minor-free or not; it will give a PTAS in case the graph is minor-free. All known algorithms discussed above need to test topological properties of the graph, such as planarity, genus-boundedness or minor-freeness, to be able to decide whether the algorithms are applicable. Except for planarity, other testings are quite expensive [17,24]. Second, the dependency of the exponent of the running time in our algorithm on the size of the minor is poly$(|V(H)|)$, or $O(|V(H)|^{3/2})$ precisely while the constant behind the big-O in the running time of the algorithm by Fomin, Lokshtanov, Raman and Sauraubh [13] is a tower function of $|V(H)|$. Even when $|V(H)| = 5$, the constant is still bigger than the size of the universe [17].

Perhaps the only drawback of our result is the running time dependency on ϵ, which is roughly $n^{O(\frac{1}{\epsilon^2})}$. However, our result should be seen as the first step toward theoretically understanding of the power of local search for the FVS problem: as long as we are willing to pay for computational time, we are guaranteed to get better approximation ratio. For APX-hard problems, such as the FVS problem, there is a limit to which, if one increases the neighborhood size, the gain in approximation is zero or negligible. Thus, a natural question is: when the input has some structural properties, would it be possible to obtain better approximation ration when the neighborhood size increases? A yes answer to this question would be quite significant in practice because real instances typically have some structural properties and the local search algorithm does not need to test such properties. Our Theorem 1 provides a yes answer to this question, when the structure of the input is minor-free. Also, in practice, one often runs local search with $c = 4$ or 5 (c is in line 2 of Algorithm 1.). It will be interesting to know, even in planar graphs, when $c = 4$ or 5, what is the approximation guarantee we can obtain? Indeed, there have been some recent work [2,26] toward this direction for optimization problems admitting local search PTASes (with the same running time as our algorithm in Theorem 1). Our Theorem 1 says that there has to be a constant c such that when we apply local search to planar graphs with c, we will beat the best known 2-approximation algorithm for general graphs [3,4]. We leave the problem of determining the exact constant c as an open problem for future research. Finally, we would like to point out that local search was experimentally applied to the FVS problem with good results [27,31]. In a certain sense, our result helps justifying for them.

To complement our positive result, we provide several negative results. The work of Har-Peled and Quanrud [16] shows that local search provides PTASes for several problems, including vertex cover, independent set, dominating set and connected dominating set, in graphs with polynomial expansion (all of these problems are known to have PTASes in minor-free graphs.). Minor-free graphs are a special case of graphs with polynomial expansion. Thus, their work gives a hope that local search can be used to generalize known PTASes for optimization problems from minor-free graphs to graphs of polynomial expansion. However, our first negative result refuses this hypothesis. By a simple reduction, we show that the FVS problem is APX-hard in 1-planar graphs. Note that 1-planar graphs also a special case of graphs of polynomial expansion. Second, we show that two closely related variants of the FVS problem, namely: *odd cycle transversal* and *subset feedback vertex set*, do not have such simple local search PTASes, even in planar graphs. We remark that these two problems are not known to have PTASes in planar graphs.

1.1   Our Analysis Technique

To better put our technique into context, we briefly discuss previous work. Chan and Har-Peled [6] and Mustafa and Ray [25] independently showed that a simple local search gives PTASes for many geometric problems. Cabello and Gajser [5] observed that the same local search can be used to design PTASes for the maximum independent set, the minimum vertex cover and minimum dominating set problems in minor-free graphs. Cohen-Addad, Klein and Mathieu [7] showed that local search yields PTASes for k-means, k-median and uniform uncapacitated facility location in minor-free graphs. In analyzing local search algorithms, one typically relies on an *exchange graph* constructed from the optimal solution[3] O and the local search solution L. For independent set and vertex cover, the exchange graph is the subgraph induced by $O \cup L$, and for other problems, the

---

[3] For k-means and k-median, the exchange graph is constructed from L and a nearly optimal solution O', which is obtained by removing some vertices of O.

exchange graph is obtained by contracting each vertex of $V(G) \setminus (O \cup L)$ to a nearest vertex in $O \cup L$. Then local properties of these problems naturally appear in the exchange graphs: if we consider a small neighborhood R in the exchange graph and replace the vertices of L in R with the vertices of O in R and its the boundary, the resulting vertex set is still a feasible solution. By decomposing the exchange graph into small neighborhoods, we can bound the size of L by the size of O and the total size of the boundaries of these neighborhoods.

However, the FVS problem does not have such local properties and hence, just simply deleting vertices and contracting edges do not give us an exchange graph. This is because for a cycle C in the original graph, the vertex of L that covers C may be inside of some neighborhood but the vertex of O that covers C may be outside of that neighborhood. One may try to argue the boundary of the neighborhood could cover C. But unfortunately, the boundary may not be helpful since the crossing vertices of C and the boundary may not be in both solutions and then they may be deleted or contracted to other vertices.

To solve this problem, we construct an exchange graph with the following property: for any cycle C of the original graph, in our exchange graph, there is (i) a vertex in $O \cap L \cap C$, or (ii) an edge between a vertex in $O \cap C$ and a vertex in $L \cap C$, or (iii) another cycle C' such that vertices in C' is a subset of vertices in C and $C' \cap (O \cup L) = C \cap (O \cup L)$. Property (i) and/or (ii) are typically achieved in previous analyses [5,7] by vertex deletion or edge contraction. It is property (iii) that is specific to our problem and is a main challenge. To additionally achieve this property, we need to introduce vertices, called *Steiner vertices*, that are not in both solutions, into the exchange graph. Meanwhile, we need to guarantee that the number of such vertices is linear to the size of $O \cup L$. The linear size bound is essential to the correctness of our algorithm and we prove this size bound by a structural lemma (Lemma 1) which may be of independent interest.

In summary, this is the first time Steiner vertices are proved useful in analyzing local search due to the non-local nature of the FVS problem. Given that many optimization problems, such as minor covering and packing problems [13], exhibit the same non-local properties, we believe that our technique is useful in studying the local search algorithm for these problems as well.

2 Preliminaries

For a graph G, we denote the vertex set and the edge set of G by $V(G)$ and $E(G)$, respectively. For a subgraph H of G, the *boundary* of H is the set of vertices that are in H but have at least one incident edge that is not in H. We denote by $int(H)$ the set of vertices of H that are not in the boundary of H. The *degree* of a vertex is the number of its incident edges.

A graph H is a *minor* of G if H can be obtained from G by a sequence of vertex deletions, edge deletions and edge contractions. G is H-*minor-free*, if G does not contain a fixed graph H as a minor. We sometimes call H-minor-free graphs *minor-free graphs* when the order of H is not relevant. It is well known [20,21] that H-minor-free graph is sparse; an H-minor-free graphs with n vertices has at most $O(\sigma_H n)$ edges where $\sigma_H = |V(H)|\sqrt{\log |V(H)|}$.

A *balanced separator* of a graph is a set of vertices whose removal partitions the graph roughly in half. A separator theorem typically provides bounds for the size of each part and the size of the balanced separator. Usually, the size of the balanced separator is sublinear w.r.t. the size of the graph. Separator theorems have been found for planar graphs [22], bounded-genus graphs [15], and minor-free graphs [1].

An *r-division* is a decomposition of a graph, which was first introduced by Frederickson [14] for planar graphs to speed up planar shortest path algorithms.

Definition 1. *For an integer r, an r-division of a graph G is a collection of edge-disjoint subgraphs of G, called* regions, *with the following properties:*

1. *Each region contains at most r vertices and each vertex is contained in at least one region.*
2. *The number of regions is at most $c_{\text{div}} \frac{n}{r}$.*
3. *The number of boundary vertices, summed over all regions, is at most $c_{\text{div}} \frac{n}{\sqrt{r}}$.*

where c_{div} is a constant.

We say a graph is *r-divisible* if it has an r-division. A graph is *divisible* if it is r-divisible for every r. Given any r and a planar graph G, Frederickson [14] gave a construction for the r-division of G that only relies on the planar separator theorem [22]. It is straightforward to extend the construction to any family of graphs with balanced separators of sublinear size. Since H-minor-free graphs are known to have balanced separators [1], H-minor-free graphs are divisible with $c_{\text{div}} = \text{poly}(|V(H)|)$.

3   Exchange Graphs Imply PTASes by Local Search

In this section, we show that if for a minor-free graph G, we can construct another graph, called *exchange graph*, such that it is divisible, then Algorithm 1 is a PTAS for the FVS problem. Let O be an optimal solution of the FVS problem and L be the output of the local search algorithm. We say a vertex u a *solution vertex* if $u \in O \cup L$ and a *Steiner vertex* otherwise. Unlike prior works [5,16], we allow *Steiner vertices* in our exchange graphs.

Definition 2. *A graph Ex is an exchange graph for optimal solution O and local solution L of the FVS problem in a graph G if it satisfies the following properties:*

(1) $L \cup O \subseteq V(\text{Ex}) \subseteq V(G)$.
(2) $|V(\text{Ex})| \leq c_{\text{ex}}(|L| + |O|)$ *for some constant c_{ex}.*
(3) *For every cycle C of G, there is (3a) a vertex of C in $O \cap L$ or (3b) an edge $uv \in E(\text{Ex})$ between a vertex $u \in L \cap C$ and a vertex $v \in O \cap C$ or (3c) a cycle C' of Ex such that $V(C') \subseteq V(C)$ and $C \cap (O \cup L) = C' \cap (O \cup L)$.*

We now prove Theorem 1 given that we can construct a *divisible exchange graph* for G. The details of the construction will be given in Sect. 4.

Proof of Theorem 1. We set the constant c in line 2 of Algorithm 1 to be $1/\delta^2$ where $\delta = \frac{\epsilon}{2c_{\mathrm{div}}c_{\mathrm{ex}}(2+\epsilon)} = O(\frac{\epsilon}{c_{\mathrm{div}}c_{\mathrm{ex}}})$. Note that c_{div} and c_{ex} are constants in Definition 1 and Definition 2, respectively. Since in each iteration, the size of the solution is reduced by at least one, there are at most n iterations. Since each iteration can be implemented in $n^{O(c)}$ time by enumerating all possibilities, the total running time is $n^{O(c)} = n^{O(1/\epsilon^2)}$. We now show that the output L has size at most $(1+\epsilon)|O|$.

Let Ex be a divisible exchange graph for O and L. We find an r-division of Ex for $r = c = \lceil 1/\delta^2 \rceil$. Let B be the multi-set containing all the boundary vertices in the r-division. By the third property in Definition 1, $|B|$ is at most $c_{\mathrm{div}}\frac{|V(\mathrm{Ex})|}{\sqrt{r}}$. By the second property in Definition 2, $|V(\mathrm{Ex})| \leq c_{\mathrm{ex}}(|O| + |L|)$. Thus, $|B| \leq c_{\mathrm{div}}c_{\mathrm{ex}}\delta(|O| + |L|)$. In the following, we will show that:

$$|L| \leq |O| + 2|B| \tag{1}$$

If so, we have:

$$|L| \leq |O| + 2c_{\mathrm{div}}c_{\mathrm{ex}}\delta(|O| + |L|) = |O| + \frac{\epsilon}{2+\epsilon}(|O| + |L|)$$

that implies $|L| \leq (1+\epsilon)|O|$.

To prove Eq. (1), we study some properties of Ex. For any region R_i of the r-division, let B_i be the boundary of R_i and $M_i = (\mathrm{L} \setminus R_i) \cup (\mathrm{O} \cap R_i) \cup B_i$.

Claim. M_i is a feedback vertex set of G.

Proof. For a contradiction, assume that there is a cycle C of G that is not covered by M_i. Then C does not contain any vertex of $\mathrm{L} \setminus R_i$, $\mathrm{O} \cap R_i$ and B_i. So C can only be covered by some vertices of $(\mathrm{L} \setminus \mathrm{O}) \cap int(R_i)$ and some vertices of $\mathrm{O} \setminus (\mathrm{L} \cup R_i)$. This implies that C does not contain any vertex of $\mathrm{O} \cap \mathrm{L}$ and there is no edge in Ex between $C \cap \mathrm{O}$ and $C \cap \mathrm{L}$. By the third property of exchange graph, there must be a cycle C' in Ex such that $V(C') \subseteq V(C)$ and $C \cap (\mathrm{O} \cup \mathrm{L}) = C' \cap (\mathrm{O} \cup \mathrm{L})$. Let u be the vertex of $(\mathrm{L} \setminus \mathrm{O}) \cap int(R_i)$ in C and v be the vertex of $\mathrm{O} \setminus (\mathrm{L} \cup R_i)$ in C. Then cycle C' contains both u and v, which implies C' crosses the boundary of R_i, that is $C' \cap B_i \neq \emptyset$. Let w be a vertex in $C' \cap B_i$, then w also belongs to C in G. This implies M_i contains a vertex of C, a contradiction. □

By the construction of M_i, we know the difference between L and M_i is bounded by the size of the region R_i, that is r. Recall that $c = r = 1/\delta^2$. Since L is the output of Algorithm 1, it cannot be improved by changing at most r vertices. Thus, we have $|L| \leq |M_i|$. By the construction of M_i, this implies

$$|L \cap R_i| \leq |M_i \cap R_i| \leq |O \cap int(R_i)| + |B_i|.$$

Thus, we have:

$$|L \cap int(R_i)| \leq |L \cap R_i| \leq |O \cap int(R_i)| + |B_i|.$$

Since $int(R_i)$ and $int(R_j)$ are vertex-disjoint for any two distinct i and j, by summing over all regions in the r-division, we get

$$|L| - |B| \le \sum_i |L \cap int(R_i)| \le \sum_i (|O \cap int(R_i)| + |B_i|) \le |O| + |B|.$$

This proves Eq. (1). $\qquad\qquad\qquad\qquad\qquad\qquad\qquad\qquad\qquad\qquad\qquad\qquad$ □

4   Exchange Graph Construction

Recall that $\sigma_H = |V(H)|\sqrt{\log |V(H)|}$ is the sparsity of H-minor-free graphs. In this section, we will show that H-minor-free graphs have divisible exchange graphs for the FVS problem with $c_{\mathrm{ex}} = O(\sigma_H)$. We construct the exchange graph in three steps:

Step 1 We delete all edges in G that are incident to vertices of $O \cap L$. We then remove all components that do not contain any solution vertex. Note that the removed components are acyclic.

Step 2 Let $v \in V(G) \setminus (O \cup L)$ be a non-solution vertex of degree at most 2. Recall that isolated vertices are removed in Step 1. If v has degree 1, we simply remove v from G. If v has degree 2, we remove v from G and add an edge between two neighbors of v in G. We can view this step in terms of contraction: we contract edges that have an endpoint that is not a solution vertex and has degree at most two until there is no such an edge left. Since L and O are feedback vertex sets of G, every cycle after the contraction must contain a vertex in L and a vertex in O. Since edges incident to vertices of $O \cap L$ are removed, there is no self-loop after this step.

Step 3 We keep the graph simple by removing all but one edge in each maximal set of parallel edges.

Let K be the resulting graph. Since K is a minor of G, it is H-minor-free and thus, divisible. It remains show that K satisfies three properties in Definition 2. Property (1) is obvious because we never delete a vertex in $L \cup O$ from K. To show property (3), let C be a cycle of G. If any edge of C is removed in Step 1, C must contain a vertex in $O \cap L$; implying (3a). Thus, we can assume that no edge of C is deleted after Step 1. Since contraction does not destroy cycles, after the contraction in Step 2, there is a cycle C' such that $V(C') \subseteq V(C)$. If $|V(C')| = 2$ (C' is a cycle of two parallel edges), then (3b) holds. Thus, we can assume that every edge of C' remains intact after removing parallel edges. But that implies (3c) since we never remove solution vertices from G. Thus, K satisfies property (3).

The most challenging part is showing property (2) in Definition 2, that is, $|V(K)| \le O(\sigma_H)(|L| + |O|)$. By Step 2, we have:

Observation 2. *Every Steiner vertex of K has degree at least 3.*

Since $O \cup L$ is a feedback vertex set of K, $K \setminus (O \cup L)$ is a forest F containing only Steiner vertices. For each tree T in F, we define the *degree* of T, denoted by $\deg_K(T)$, to be the number of edges in K between T and $O \cup L$.

Claim. $|V(T)| \leq \deg_K(T)$.

Proof. Let T' be obtained from T by adding every edge uv to T where $u \in V(T)$ and $v \in O \cup L$. Observe that no vertex in $(O \cup L) \setminus (O \cap L)$ can be adjacent to more than one vertex in T since otherwise, there would be a cycle that contains vertices from L or O only, contradicting that L and O are feedback vertex sets. Since vertices in $L \cap O$ are isolated in K, T' must be a tree. Let $\ell(T')$ be the number of leaves of T'. By Step 2, leaves of T' are vertices in $O \cup L$. Thus, $\deg_K(T) = \ell(T')$. Since every internal vertices of T' has degree at least three, $|V(T)| \leq \ell(T')$ which implies the claim.   □

We contract each tree T of F into a single Steiner vertex s_T. Let K' be the resulting graph. We observe that:

Observation 3. K' *is simple.*

Proof. Since every cycle of K must contain a vertex from L and a vertex from O, there cannot be any solution vertex in K that is adjacent to more than one vertex of a tree T of F. So there cannot be parallel edges in K'.   □

To bound the size of K', we need the following structural lemma. We remark that this lemma holds for general graphs.

Lemma 1. *For a graph G and any two disjoint nonempty vertex subsets A and B, let $D = V(G) \setminus (A \cup B)$. If (i) D is an independent set, (ii) every vertex in D has degree at least 3 in G and (iii) every cycle C contains at least one vertex in A and at least one vertex in B, then $|V(G)| \leq 2(|A| + |B|)$.*

Proof. We remove every edge that only has endpoints in $A \cup B$ and let the resulting graph be G'. Then G' is a bipartite graph with $A \cup B$ in one side and D in the other side since D is an independent set. Let D_A (D_B) be the subset of D containing every vertex that has at least two neighbors in A (B). Since every vertex of D has degree at least 3, we have $D_A \cup D_B = D$.

Let H_A be the subgraph of G' induced by $A \cup D_A$. Then H_A is acyclic since otherwise every cycle of H_A would correspond to a cycle in G that does not contain any vertex in B. We now construct a graph H_A^* on vertex set A. For each vertex $v \in D_A$, we choose any two neighbors x and y of v in A and add an edge between x and y in H_A^*. By construction, there is a one-to-one mapping between edges of H_A^* and vertices of D_A.

Since H_A is acyclic, H_A^* is also acyclic. Thus, $|E(H_A^*)| \leq V(H_A^*) = |A|$. That implies $|D_A| \leq |A|$. By a similar argument, we can show that $|D_B| \leq |B|$. Thus, $|D| = |D_A \cup D_B| \leq |A| + |B|$ which implies the lemma.   □

Let Z be an arbitrary component of K' that contains at least one Steiner vertex. Then two sets $V(Z) \cap O$ and $V(Z) \cap L$ must be disjoint since any vertex in $O \cap L$ is isolated in K'. If any of two sets $V(Z) \cap O$ and $V(Z) \cap O$, say $V(Z) \cap O$, is empty, then Z must be a tree. By Step 2, leaves of Z are in L. Thus, $|V(Z)| \leq |V(Z) \cap L|$ since internal vertices of Z have degree at least 3. Otherwise, both $V(Z) \cap O$ and

$V(Z) \cap O$ are non-empty. Let X be the set of Steiner vertices in Z. By the construction of K', X is an independent set of Z. By Observation 2, every vertex of X has degree at least 3. So we can apply Lemma 1 to X, $V(Z) \cap O$ and $V(Z) \cap L$, and obtain $|V(Z)| \leq 2(|V(Z) \cap O| + |V(Z) \cap L|) = 2(|V(Z) \cap O| + |V(Z) \cap (L \setminus O)|)$. Note that this bound holds trivially if Z does not contain any Steiner vertex. In both cases, $|V(Z)| \leq 2(|V(Z) \cap O| + |V(Z) \cap (L \setminus O)|)$. Summing over all components of K', we have $|V(K')| \leq 2(|V(K') \cap O| + |V(K') \cap (L \setminus O)|) \leq 2(|O| + |L|)$. Since K' is a minor of G, it is also H-minor-free. Thus, $|E(K')| = O(\sigma_H|V(K')|) = O(\sigma_H)(|O| + |L|)$. We now ready to bound the size of $V(K)$. We have:

$$
\begin{aligned}
|V(K) \setminus (O \cup L)| = \sum_{T \in F} |V(T)| &\leq \sum_{T \in F} \deg_K(T) \quad \text{(Claim 4)} \\
&= \sum_{T \in F} \deg_{K'}(s_T) \\
&\leq |E(K')| \quad (\text{ since } \{s_T | T \in F\} \text{ is an independent set}) \\
&= O(\sigma_H)(|O| + |L|)
\end{aligned}
\tag{2}
$$

That implies $V(K) \leq O(\sigma_H)(|O| + |L|)$. Thus K satisfies property (2) in Definition 2 with $c_{\text{ex}} = O(\sigma_H)$.

5   Negative Results

In this section, we show some negative results for the FVS problem and two closely related problems: odd cycle transversal and subset feedback vertex set. The odd cycle transversal (also called *bipartization*) problem asks for a minimum set of vertices in an undirected graph whose removal results in a bipartite graph. Given an undirected graph and a subset U of vertices, the subset feedback vertex set problem asks for a minimum set S of vertices such that after removing S the resulting graph contains no cycle that passes through any vertex of U.

We first show that the FVS problem is APX-hard in 1-planar graphs. A graph is *1-planar* if it can be drawn in the Euclidean plane such that every edge has at most one crossing.

Theorem 4. *Given a graph G, we can construct an 1-planar graph H in polynomial time, such that G has a feedback vertex set of size at most k if and only if H has a feedback vertex set of size at most k.*

Proof. Consider a drawing of G on the plane where each pair of edges can cross at most once. For each crossed edge e in G, we subdivide e into edges so that there is exactly one crossing per new edge. Let H be the resulting graph. By construction, H is 1-planar. Since we only subdivide edges, there is a one-to-one mapping between cycles of G and cycles of H. Observe that any feedback vertex set of G is also a feedback vertex set of H.

For any subdividing vertex v in the optimal feedback vertex set of H that is not the original vertex of v, we can replace v by one of the endpoint of the edge that v subdivides. Thus, if H has a feedback vertex set of size k, G has feedback vertex set of size at most k, which implies the lemma.   □

Since the FVS problem is APX-hard in general graphs (by an approximation preserving reduction [18] from vertex cover problem, which is APX-hard [11]), Theorem 4 implies that FVS is APX-hard in 1-planar graphs. In the full version of the paper, we show by examples that the simple local search with constant-size exchanges cannot give a constant approximation for the odd cycle transversal and the subset feedback vertex set problems in planar graphs.

Acknowledgement. We thank the anonymous reviewer who pointed out an error in our argument to bound the size of the exchange graph. This material is based upon work supported by the National Science Foundation under Grant No. CCF-1252833. This work was done while the first author was at Oregon State University.

References

1. Alon, N., Seymour, P.D., Thomas, R.: A separator theorem for nonplanar graphs. J. Am. Math. Soc. **3**(4), 801–808 (1990)
2. Antunes, D., Mathieu, C., Mustafa, N.H.: Combinatorics of local search: an optimal 4-local Hall's theorem for planar graphs. In: 25th Annual European Symposium on Algorithms (ESA 2017), vol. 87, pp. 8:1–8:13 (2017)
3. Bafna, V., Berman, P., Fujito, T.: A 2-approximation algorithm for the undirected feedback vertex set problem. SIAM J. Discrete Math. **12**(3), 289–297 (1999)
4. Becker, A., Geiger, D.: Optimization of Pearl's method of conditioning and greedy-like approximation algorithms for the vertex feedback set problem. Artif. Intell. **83**(1), 167–188 (1996)
5. Cabello, S., Gajser, D.: Simple PTAS's for families of graphs excluding a minor. Discrete Appl. Math. **189**(C), 41–48 (2015)
6. Chan, T.M., Har-Peled, S.: Approximation algorithms for maximum independent set of pseudo-disks. In: Proceedings of the Twenty-Fifth Annual Symposium on Computational Geometry, SocG 2009, pp. 333–340 (2009)
7. Cohen-Addad, V., Klein, P.N., Mathieu, C.: Local search yields approximation schemes for k-means and k-median in Euclidean and minor-free metrics. In: Proceedings of the 57th Annual IEEE Symposium on Foundations of Computer Science, FOCS 2016 (2016)
8. Cohen-Addad, V., de Verdière, É.C., Klein, P.N., Mathieu, C., Meierfrankenfeld, D.: Approximating connectivity domination in weighted bounded-genus graphs. In: Proceedings of the 48th Annual ACM SIGACT Symposium on Theory of Computing, pp. 584–597. ACM (2016)
9. Courcelle, B.: The monadic second-order logic of graphs. I. Recognizable sets of finite graphs. Inf. Comput. **85**(1), 12–75 (1990)
10. Demaine, E.D., Hajiaghayi, M.: Bidimensionality: new connections between FPT algorithms and PTASs. In: Proceedings of the Sixteenth Annual ACM-SIAM Symposium on Discrete Algorithms, SODA 2005, pp. 590–601 (2005)
11. Dinur, I., Safra, S.: On the hardness of approximating minimum vertex cover. Ann. Math. **162**(1), 439–485 (2005)

12. Even, G., Naor, J., Schieber, B., Zosin, L.: Approximating minimum subset feedback sets in undirected graphs with applications. SIAM J. Discrete Math. **13**(2), 255–267 (2000)
13. Fomin, F.V., Lokshtanov, D., Raman, V., Saurabh, S.: Bidimensionality and EPTAS. In: Proceedings of the Twenty-Second Annual ACM-SIAM Symposium on Discrete Algorithms, SODA 2011, pp. 748–759 (2011)
14. Frederickson, G.: Fast algorithms for shortest paths in planar graphs with applications. SIAM J. Comput. **16**, 1004–1022 (1987)
15. Gilbert, J.R., Hutchinson, J.P., Tarjan, R.E.: A separator theorem for graphs of bounded genus. J. Algorithms **5**(3), 391–407 (1984)
16. Har-Peled, S., Quanrud, K.: Approximation algorithms for polynomial-expansion and low-density graphs. In: Bansal, N., Finocchi, I. (eds.) ESA 2015. LNCS, vol. 9294, pp. 717–728. Springer, Heidelberg (2015). https://doi.org/10.1007/978-3-662-48350-3_60
17. Johnson, D.S.: The NP-completeness column: an ongoing guide (column 19). J. Algorithms **8**(3), 438–448 (1987)
18. Karp, R.M.: Reducibility among combinatorial problems. In: Miller, R.E., Thatcher, J.W., Bohlinger, J.D. (eds.) Complexity of Computer Computations. IRSS, pp. 85–103. Springer, Boston (1972). https://doi.org/10.1007/978-1-4684-2001-2_9
19. Kleinberg, J., Kumar, A.: Wavelength conversion in optical networks. J. Algorithms **38**, 25–50 (2001)
20. Kostochka, A.V.: The minimum hadwiger number for graphs with a given mean degree of vertices. Metody Diskretnogo Analiza **38**, 37–58 (1982). (in Russian)
21. Kostochka, A.V.: Lower bound of the hadwiger number of graphs by their average degree. Combinatorica **4**(4), 307–316 (1984)
22. Lipton, R., Tarjan, R.: A separator theorem for planar graphs. SIAM J. Appl. Math. **36**(2), 177–189 (1979)
23. Michiels, W., Aarts, E., Korst, J.: Theoretical Aspects of Local Search. Springer, Heidelberg (2010)
24. Mohar, B.: A linear time algorithm for embedding graphs in an arbitrary surface. SIAM J. Discrete Math. **12**(1), 6–26 (1999)
25. Mustafa, N.H., Ray, S.: Improved results on geometric hitting set problems. Discrete Comput. Geom. **44**(4), 883–895 (2010)
26. Bus, N., Garg, S., Mustafa, N.H., Ray, S.: Limits of local search: quality and efficiency. Discrete Comput. Geom. **57**(3), 607–624 (2017)
27. Qin, S., Zhou, H.: Solving the undirected feedback vertex set problem by local search. Eur. Phys. J. B **87**(11), 273 (2014)
28. Robertson, N., Seymour, P.D.: Graph minors. I. Excluding a forest. J. Comb. Theory Ser. B **35**(1), 39–61 (1983)
29. Robertson, N., Seymour, P.D.: Graph minors. XVI. Excluding a non-planar graph. J. Comb. Theory Ser. B **89**(1), 43–76 (2003)
30. Yannakakis, M.: Node-and edge-deletion NP-complete problems. In: Proceedings of the Tenth Annual ACM Symposium on Theory of Computing, pp. 253–264. ACM (1978)
31. Zhang, Z., Ye, A., Zhou, X., Shao, Z.: An efficient local search for the feedback vertex set problem. Algorithms **6**(4), 726–746 (2013)

The Seeding Algorithm for Functional k-Means Problem

Min Li[1], Yishui Wang[2(✉)], Dachuan Xu[3], and Dongmei Zhang[4]

[1] School of Mathematics and Statistics, Shandong Normal University,
Jinan 250014, People's Republic of China
liminEmily@sdnu.edu.cn
[2] Shenzhen Institutes of Advanced Technology, Chinese Academy of Sciences,
Shenzhen 518055, People's Republic of China
ys.wang1@siat.ac.cn
[3] Department of Operations Research and Scientific Computing,
Beijing University of Technology, Beijing 100124, People's Republic of China
xudc@bjut.edu.cn
[4] School of Computer Science and Technology, Shandong Jianzhu University,
Jinan 250101, People's Republic of China
zhangdongmei@sdjzu.edu.cn

Abstract. The functional k-means problem involves different data from k-means problem, where the functional data is a kind of dynamic data and is generated by continuous processes. By defining a new distance with derivative information, the functional k-means clustering algorithm can be used well for functional k-means problem. In this paper, we mainly investigate the seeding algorithm for functional k-means problem and show that the performance guarantee is obtained as $8(\ln k + 2)$. Moreover, we present the numerical experiment showing the validity of this algorithm, comparing to the functional k-means clustering algorithm.

Keywords: Functional k-means problem · k-means problem ·
Approximation algorithm

1 Introduction

In machine learning and computational geometry, the k-means problem is one of the most classic problems, which aims to separate the given data sets into k parts according to the minimization of the sum of squared distances. The Lloyd's algorithm [10,12], also known as k-means, is a very popular algorithm for k-means problem. Based on Lloyd's algorithm, there are several approximation algorithms designed, such as k-means++ [3], k-means|| [2].

Different from the characteristic in k-means problem where the given data are real vectors, the functional data in functional k-means problem are a kind of dynamic data, which are generated by continuous processes [6,13]. Based on this continuity property, there are several categories of clustering methods for

© Springer Nature Switzerland AG 2019
D.-Z. Du et al. (Eds.): COCOON 2019, LNCS 11653, pp. 387–396, 2019.
https://doi.org/10.1007/978-3-030-26176-4_32

functional k-means problem, such as two-stage methods [1,9,14], model-based clustering methods [5], non-parametric clustering methods [4,16], etc. For more information, one can refer to [8]. Recently, Meng et al. [11] define a new distance between two functional data involving the derivative information, and apply the k-means method to solve this functional k-means problem. However, there is no theoretical analysis to show how good this method can achieve. Our main contribution in this paper is to design one approximation algorithm for functional k-means problem, based on the distance defined in [11] and the seeding algorithm. And the approximation factor is obtained as $8(\ln k + 2)$.

The rest of this paper is organized as follows. In Sect. 2, we present the functional k-means problem and some basic notations. We introduce the seeding algorithm and the main result for functional k-means problem in Sect. 3. In Sect. 4, the proof to show the correctness of the algorithm is given. In Sect. 5, the numerical experiment about the seeding algorithm for functional k-means problem is presented. The final remarks are concluded in Sect. 6.

2 Preliminaries

In this section, the definition of functional k-means problem, some symbols and notations, as well as some important results are mainly introduced.

In general, given two real numbers T_1, T_2 such that $T_1 \leq T_2$, for any $t \in T = [T_1, T_2]$, the function $x(t) : T \to \mathbb{R}$ is defined as a functional curve, which is described using one real valued function of T. Then the d-dimensional functional data $X(t) = (x_1(t), x_2(t), \ldots, x_d(t))^T$ called functional sample can be given, where $x_1(t), x_2(t), \ldots, x_d(t)$ all are functional curves with the same ground set T. And we use $\mathfrak{F}^d(t)$ to denote all the d-dimensional functional samples with T as the ground set of their functional curves. Therefore, given two functional samples $X^i(t) = (x_1^i(t), x_2^i(t), \ldots, x_d^i(t))^T$ and $X^j(t) = (x_1^j(t), x_2^j(t), \ldots, x_d^j(t))^T$ in $\mathfrak{F}^d(t)$, one can define their similarity metric as follows,

$$d(X^i(t), X^j(t)) = \sqrt{\sum_{p=1}^{d} [\int_T (x_p^i(t) - x_p^j(t))^2 dt + \int_T (Dx_p^i(t) - Dx_p^j(t))^2 dt]}, \quad (1)$$

where $Dx_p^i(t)$ means the first order derivative of the p-th functional curve in the i-th functional sample. Moreover, the metric given in (1) can be proved to be a distance metric [11]. Then, we can define the distance from one functional sample $X(t) \in \mathfrak{F}^d(t)$ to some functional sample set $\Gamma(t) \subseteq \mathfrak{F}^d(t)$ in the following way,

$$d(X(t), \Gamma(t)) = \min_{Y(t) \in \Gamma(t)} d(X(t), Y(t)).$$

Moreover, we use $X(t)_{\Gamma(t)}$ to denote one closest functional sample in $\Gamma(t)$ to $X(t)$, i.e.,

$$X(t)_{\Gamma(t)} \in \arg \min_{Y(t) \in \Gamma(t)} d(X(t), Y(t)).$$

Given a set $\Gamma(t)$ in $\mathfrak{F}^d(t)$ with n functional samples and an integer k, as well as a k functional sample set $\Omega(t)$ in $\mathfrak{F}^d(t)$ (also called center set of functional samples or center set for short), one can define the loss or cost of $\Omega(t)$ over $\Gamma(t)$ as the sum of the squared similarity distances over each functional sample of $\Gamma(t)$ to the functional sample set $\Omega(t)$, i.e.,

$$\Phi(\Gamma(t), \Omega(t)) = \sum_{X(t)\in\Gamma(t)} d^2(X(t), \Omega(t)).$$

The *functional k-means problem* is to find one optimal center set $\Omega(t)^*_{\Gamma(t)}$ to minimize the loss function, i.e.,

$$\Omega(t)^*_{\Gamma(t)} \in \arg\min_{|\Omega(t)|=k, \Omega(t)\subseteq\mathfrak{F}^d(t)} \left(\sum_{X(t)\in\Gamma(t)} d^2(X(t), \Omega(t)) \right).$$

In the following part, we still apply $\Omega(t)^*_{\Gamma(t)}$ to denote one optimal clustering set of $\Gamma(t)$, and $\Phi^*(\Gamma(t))$ to denote its optimal cost, i.e., $\Phi^*(\Gamma(t)) = \Phi(\Gamma(t), \Omega(t)^*_{\Gamma(t)})$. Specially, when $k = 1$, the functional sample center of $\Gamma(t)$ denoted by $\mu(\Gamma(t))$ can be computed by $\mu(\Gamma(t)) = \frac{1}{n}\sum_{X(t)\in\Gamma(t)} X(t)$ in [11]. A proof in novel way will be given in Lemma 1.

Given any k-functional sample clustering $\Omega(t) = \{Q^1(t), Q^2(t), \ldots, Q^k(t)\}$, $\Gamma(t)$ can be separated into k parts depending on the closest distance to the functional sample in $\Omega(t)$, which consists of k functional sample clusters. In fact, for any $Q^i(t) \in \Omega(t)$, the cluster $\Gamma^i_{\Omega(t)}(t)$ is used to denote the functional samples clustered to this center in the following way,

$$\Gamma^i_{\Omega(t)}(t) = \{X(t) \in \Gamma(t) : d(X(t), Q^i(t)) \le d(X(t), Q^j(t)), \forall Q^j(t) \in \Omega(t), j \neq i\}.$$

In this way, one can present the loss function as follows,

$$\Phi(\Gamma(t), \Omega(t)) = \sum_{X(t)\in\cup_{i=1}^k \Gamma^i_{\Omega(t)}(t)} d^2(X(t), \Omega(t)).$$

Specially, for the optimal functional sample center $\Omega(t)^*_{\Gamma(t)}$, there also exist k clusters of $\Gamma(t)$. In the following part, we will denote these clusters by $\Gamma^1(t)^*, \Gamma^2(t)^*, \ldots, \Gamma^k(t)^*$.

Moreover, for any $\Delta(t) \subseteq \Gamma(t)$, $\Phi(\Delta(t), \Omega(t)) = \sum_{X(t)\in\Delta(t)} d^2(X(t), \Omega(t))$ is used to denote the contribution of $\Delta(t)$ to the cost. Specially, if $\Delta(t) = \{X(t)\}$ with only one functional sample, we can write it as $\Phi(X(t), \Omega(t))$.

Now, we show a novel proof of the case for functional 1-mean problem, which is different from the one given in [11].

Lemma 1. *Given any functional sample set* $\Gamma(t) = \{X^1(t), X^2(t), \ldots, X^n(t)\}$ *with n functional samples in $\mathfrak{F}^d(t)$, then for any $X(t) \in \mathfrak{F}^d(t)$, we have*

$$\sum_{i=1}^n d^2(X^i(t), X(t)) = \sum_{i=1}^n d^2(X^i(t), \mu(\Gamma(t))) + nd^2(\mu(\Gamma(t)), X(t)),$$

where $\mu(\Gamma(t)) = \sum\limits_{X(t) \in \Gamma(t)} X(t)/n$.

Proof. First, applying the definition of the metric between functional samples, one can obtain that

$$\sum_{i=1}^{n} d^2(X^i(t), X(t))$$

$$= \sum_{i=1}^{n} \sum_{l=1}^{d} \left[\int_T (x_l^i(t) - x_l(t))^2 dt + \int_T (Dx_l^i(t) - Dx_l(t))^2 dt \right]$$

$$= \sum_{i=1}^{n} \sum_{l=1}^{d} \left[\int_T (x_l^i(t) - \mu_l(\Gamma(t)) + \mu_l(\Gamma(t)) - x_l(t))^2 dt \right.$$

$$\left. + \int_T (Dx_l^i(t) - D\mu_l(\Gamma(t)) + D\mu_l(\Gamma(t)) - Dx_l(t))^2 dt \right]$$

$$= \sum_{i=1}^{n} \sum_{l=1}^{d} \left[\int_T (x_l^i(t) - \mu_l(\Gamma(t)))^2 dt + \int_T (Dx_l^i(t) - D\mu_l(\Gamma(t)))^2 dt \right]$$

$$+ \sum_{i=1}^{n} \sum_{l=1}^{d} \left[\int_T (\mu_l(\Gamma(t)) - x_l(t))^2 dt + \int_T (D\mu_l(\Gamma(t)) - Dx_l(t))^2 dt \right]$$

$$+ 2 \sum_{i=1}^{n} \sum_{l=1}^{d} \left[\int_T (x_l^i(t) - \mu_l(\Gamma(t)))(\mu_l(\Gamma(t)) - x_l(t)) dt \right.$$

$$\left. + \int_T (Dx_l^i(t) - D\mu_l(\Gamma(t)))(D\mu_l(\Gamma(t)) - Dx_l(t)) dt \right]$$

$$= \sum_{i=1}^{n} d^2(X^i(t), \mu(\Gamma(t))) + n d^2(\mu(\Gamma(t)), X(t))$$

$$+ 2 \sum_{i=1}^{n} \sum_{l=1}^{d} \left[\int_T (x_l^i(t) - \mu_l(\Gamma(t)))(\mu_l(\Gamma(t)) - x_l(t)) dt \right.$$

$$\left. + \int_T (Dx_l^i(t) - D\mu_l(\Gamma(t)))(D\mu_l(\Gamma(t)) - Dx_l(t)) dt \right]. \tag{2}$$

Thus, the proof can be elucidated if we prove that the last two terms of the right hand in (2) is zero, i.e.,

$$2 \sum_{i=1}^{n} \sum_{l=1}^{d} \left[\int_T (x_l^i(t) - \mu_l(\Gamma(t)))(\mu_l(\Gamma(t)) - x_l(t)) dt \right.$$

$$\left. + \int_T (Dx_l^i(t) - D\mu_l(\Gamma(t)))(D\mu_l(\Gamma(t)) - Dx_l(t)) dt \right] = 0. \tag{3}$$

In fact, by the definition of $\mu(\Gamma(t))$, for any $l \in \{1, 2, \ldots, d\}$, we obtain with not much effort that

$$\mu_l(\Gamma(t)) = \frac{1}{n} \sum_{j=1}^{n} x_l^j(t).$$

Then, we get the following results,

$$\sum_{i=1}^{n} (x_l^i(t) - \mu_l(\Gamma(t))) = \sum_{i=1}^{n} (x_l^i(t) - \frac{1}{n} \sum_{j=1}^{n} x_l^j(t)) = 0,$$

and

$$\sum_{i=1}^{n} D(x_l^i(t) - \mu_l(\Gamma(t))) = D \sum_{i=1}^{n} (x_l^i(t) - \mu_l(\Gamma(t))) = 0.$$

Therefore, (3) can be proved according to the commutation property of summation and integrals.   □

At last, it is easy to verify the following property.

Property 1. Given any functional sample set $\Gamma(t)$ in $\mathfrak{F}^d(t)$, and two clusterings $\Omega(t)$ and $\Omega(t)'$ in $\mathfrak{F}^d(t)$ such that $\Omega(t) \subseteq \Omega(t)'$, then

$$\Phi(\Gamma(t), \Omega(t)) \geq \Phi(\Gamma(t), \Omega(t)').$$

That is, when the elements of a clustering becomes more, the contribution of $\Gamma(t)$ to the new clustering cannot be increased.

3   The Seeding Algorithm and Our Main Result

In this section, we will mainly present the seeding algorithm for the functional k-means problem, which is generalized from both the Lloyd's method for functional k-means problem and k-means++ for k-means problem. From Step 1 to Step 5, one notices that the initial functional sample clustering $\Omega(t)$ are chosen from the functional sample set $\Gamma(t)$ with very specific probabilities, rather than being sampled at random without any rules like Lloyd's method. Then, the initial functional samples $\Gamma(t)$ can be separated into k clusters from Step 6 to Step 8, depending on the closest similarity metric distance. In the following processes from Step 9 to Step 11, we renew the functional sample clustering by applying the result to functional 1-mean problem. In fact, the updated clustering cannot increase the value of loss function. This can be explained by the Lloyd's method for k-means problem [7]. Therefore, this algorithm makes enough improvement in each cycle until the clustering is no longer changed. Moreover, if one can bound the cost function of the first clustering, the return clustering can be found with this property, too.

Then, we present our main result as follows.

Theorem 1. Suppose that $\Omega(t)$ is constructed in Algorithm 1 for $\Gamma(t) \subseteq \mathfrak{F}^d$, then the corresponding cost function satisfies

$$\mathrm{E}[\Phi(\Gamma(t), \Omega(t))] \leq 8(\ln k + 2)\Phi^*(\Gamma(t)).$$

Algorithm 1. The seeding algorithm for functional k-means problem

---

Input: A set of n functional samples $\Gamma(t) \subseteq \mathfrak{F}^d$, and $\Omega(t) := \emptyset$.
Output: An approximate functional k-means $\Omega(t)$ of $\Gamma(t)$.

1: Choose the first functional sample center $Q^1(t)$ uniformly at random from $\Gamma(t)$, then set $\Omega(t) := \Omega(t) \cup \{Q^1(t)\}$;
2: **for** i from 2 to k **do**
3:     Choose the functional sample center $Q^i(t)$ from $\Gamma(t)$ with probability $\frac{d^2(Q^i(t),\Omega(t))}{\sum_{X(t)\in\Gamma(t)} d^2(X(t),\Omega(t))}$;
4:     Set $\Omega(t) := \Omega(t) \cup \{Q^i(t)\}$;
5: **end for**
6: **for** i from 1 to k **do**
7:     Set the cluster $\Gamma^i_{\Omega(t)}(t) := \{X(t) \in \Gamma(t) : d^2(X(t), \Omega(t)) = d^2(X(t), Q^i(t))\}$;
8: **end for**
9: **for** i from 1 to k **do**
10:     Update the functional sample clustering $\Omega(t)$ by setting $Q^i(t) := \mu(\Gamma^i_{\Omega(t)}(t))$, which is the center functional sample of $\Gamma^i_{\Omega(t)}(t)$;
11: **end for**
12: Repeat Step 6 to Step 11 until $\Omega(t)$ no longer change;
13: Return $\Omega(t)$.

---

4 Proof of Correctness

In this section, we mainly discuss that the set $\Omega(t)$ returned by Algorithm 1 is an $8(\ln k + 2)$-approximate functional k clustering for functional k-means problem, and the main proof follows k-means++ for k-means problem. In fact, the returned set $\Omega(t)$ used in this part is just the one obtained in seeding part (from Step 1 to Step 5). From Algorithm 1, we know that the first center of functional sample is chosen uniformly from $\Gamma(t)$. Since the group of clusters $\Gamma^1(t)^*, \Gamma^2(t)^*, \ldots, \Gamma^k(t)^*$ is a division of $\Gamma(t)$, the chosen center obviously belongs to one of these groups. That is, there exists one joint cluster to $\Omega(t)$. If we cluster all the functional samples of this joint cluster to the chosen center, rather than its optimal functional 1-mean, we show that the value of loss function in the former case is exact twice as the one in the latter case in the following lemma, and its proof is presented in the supplementary material.

Lemma 2. *Let $\Delta(t)$ be an arbitrary functional sample cluster of $\Gamma(t)$ with respect to $\Omega^*(t)$, and suppose that $\Omega(t) = \{Q(t)\}$ is a functional sample clustering with only one center, which is chosen uniformly from $\Delta(t)$, then*

$$\mathrm{E}(\Phi(\Delta(t), \Omega(t))) = 2\Phi(\Delta(t), \Omega(t)^*_{\Gamma(t)}).$$

In the following lemma, we further give the bound when only one new functional sample center is added to the current center set.

Lemma 3. *Let $\Delta(t)$ is an arbitrary functional sample cluster of $\Gamma(t)$ with respect to $\Omega(t)^*_{\Gamma(t)}$. Suppose that $\Omega(t)$ is any clustering of $\Gamma(t)$ with less than k*

centers. If we choose one center $Q(t)$ from $\Delta(t)$ at random with probability

$$\frac{d^2(Q(t), \Omega(t))}{\sum_{X(t) \in \Gamma(t)} d^2(X(t), \Omega(t))},$$

and add $Q(t)$ to $\Omega(t)$, we have

$$\mathrm{E}[\Phi(\Delta(t), \Omega(t)')] \leq 8\Phi(\Delta(t), \Omega(t)^*_{\Gamma(t)}),$$

where $\Omega(t)' = \Omega(t) \cup \{Q(t)\}$.

By Lemma 3, one knows that if the k center functional samples are chosen from each cluster of $\Gamma(t)$ with respect to $\Omega(t)^*_{\Gamma(t)}$, the approximation rate can be obtained as a factor. The left question is what about the case that more than two center functional samples are chosen in the same cluster? We will solve this problem in the following lemma. Given any clustering of $\Gamma(t)$, for the convenience of discussing this special case, we will separate the functional sample clusters of $\Gamma(t)$ given by $\Omega(t)^*_{\Gamma(t)}$ into joint and disjoint clusters to $\Gamma(t)$. If $\Gamma^i(t)^* \cap \Omega(t) \neq \emptyset$ for some $i \in \{1, 2, \dots, k\}$, we call $\Gamma^i(t)^*$ a joint functional sample cluster of $\Omega(t)$. Otherwise, it is named as a disjoint cluster of $\Omega(t)$. And we use $n_{\Omega(t)}$ to denote the number of the disjoint functional sample clusters of $\Omega(t)$, $N_{\Omega(t)} = \{X(t) \in \Gamma^i(t)^* : \Gamma^i(t)^* \cap \Omega(t) \neq \emptyset\}$ to denote the functional samples in the disjoint clusters, and $D_{\Omega(t)} = \Gamma(t) - N_{\Omega(t)}$ to mean the functional samples in the joint clusters.

Lemma 4. *Let $\Omega(t)$ be any functional sample clustering of $\Gamma(t)$. Now we add $m\ (\leq n_{\Omega(t)})$ functional samples $Q^1(t), Q^2(t), \dots, Q^m(t)$ from $\Gamma(t)$ to $\Omega(t)$, where for each $Q^i(t), i = 1, 2, \dots, m$, its chosen probability is*

$$\frac{d^2(Q^i(t), \Omega(t))}{\sum_{X(t) \in \Gamma(t)} d^2(X(t), \Omega(t))}.$$

If we set $\Omega(t)' = \Omega(t) \cup \{Q^1(t), Q^2(t), \dots, Q^m(t)\}$, we have

$$\mathrm{E}(\Phi(\Gamma(t), \Omega(t)')) \leq [\Phi(D_{\Omega(t)}, \Omega(t)) + 8\Phi(N_{\Omega(t)}, \Omega(t)^*_{\Gamma(t)})] \cdot (1 + H_m)$$
$$+ \frac{n_{\Omega(t)} - m}{n_{\Omega(t)}} \Phi(N_{\Omega(t)}, \Omega(t)),$$

where $H_m = 1 + \frac{1}{2} + \cdots + \frac{1}{m}$.

Its proof is presented in the supplementary material.

Now, we finish the proof of Theorem 1 by using above lemmas.

Proof of Theorem 1. Let $\Omega(t)$ be the functional sample clustering obtained from Step 1 to Step 5 in Algorithm 1, and suppose that the current functional clustering $\Omega^1(t) = \{Q(t)\}$ just has one functional sample center. By Algorithm 1, we know this center should be chosen from some optimal cluster $\Delta(t)$. Then,

$D_{\Omega^1(t)} = \Delta(t)$, $N_{\Omega^1(t)} = \Gamma(t) \setminus \Delta(t)$, and there are $n_{\Omega^1(t)} = k - 1$ disjoint optimal clusters. Thus, by Lemmas 2 and 4, we obtain the following result,

$$E[\Phi(\Gamma(t), \Omega(t))]$$

$$\overset{\text{Lemma 4}}{\leq} \left[\Phi(\Delta(t), \Omega^1(t)) + 8\Phi(\Gamma(t), \Omega(t)^*_{\Gamma(t)}) - 8\Phi(\Delta(t), \Omega(t)^*_{\Gamma(t)}) \right]$$
$$\cdot (1 + H_{k-1})$$

$$\overset{\text{Lemma 2}}{=} \left[2\Phi(\Delta(t), \Omega(t)^*_{\Gamma(t)}) + 8\Phi(\Gamma(t), \Omega(t)^*_{\Gamma(t)}) - 8\Phi(\Delta(t), \Omega(t)^*_{\Gamma(t)}) \right]$$
$$\cdot (1 + H_{k-1})$$

$$\leq \qquad 8\Phi(\Gamma(t), \Omega(t)^*_{\Gamma(t)}) \cdot (1 + H_{k-1})$$

$$\leq \qquad 8\Phi(\Gamma(t), \Omega(t)^*_{\Gamma(t)}) \cdot (1 + \ln k).$$

$\square$

5   Numerical Experiments

In this section, we test the seeding algorithm (Algorithm 1) on the data sets Simudata [15] and Sdata [11], comparing to the functional k-means clustering algorithm provided in the reference [11]. Below we describe the two data sets.

- **Simudata**
  This data set has two clusters with respective to functions $X_1(t) = uh_1(t) + (1 - u)h_2(t) + \epsilon(t)$ and $X_2(t) = uh_1(t) + (1 - u)h_3(t) + \epsilon(t)$, where $h_1(t) = \max(6 - |t - 11|, 0)$, $h_2 = h_1(t - 4)$, $h_3 = h_1(t + 4)$, u is random number drawn uniformly from $[0, 1]$, and $\epsilon(t)$ is a white noise with expectation $E[\epsilon(t)] = 0$ and variance $\text{Var}[\epsilon(t)] = 1$. Each cluster includes 100 functional curves generated by observation points $t = 1, 1.2, 1.4, \ldots, 21$.
- **Sdata**
  This data set has three clusters with respective to functions $X_1(t) = \cos(1.5 \pi t) + \epsilon(t)$, $X_2(t) = \sin(1.5\pi t) + \epsilon(t)$, and $X_3(t) = \sin(\pi t) + \epsilon(t)$, where $\epsilon(t)$ is a white noise with expectation $E[\epsilon(t)] = 0$ and variance $\text{Var}[\epsilon(t)] = 1$. Each cluster includes 100 functional curves on $[0, 1]$.

In order to calculate the derivatives, we smooth the functional data by the three order polynomial fit technology. We use two measures, namely, adjusted rand index (ARI) and Davies Bouldin index (DBI) to evaluate the effectiveness of the two algorithms. ARI is an external clustering validation index, defined as follows.

$$\text{ARI} := \frac{\text{RI} - E(\text{RI})}{\max(\text{RI}) - E(\text{RI})}$$

where RI is the Rand index which is defined as the number of pairs of objects that are either in the same group or in different groups in both the partition returned by the algorithm and the real partition divided by the total number

Table 1. Comparison of Algorithm 1 and the functional k-means algorithm in [11]

Data set	Method	ARI	DBI	Initial cost	Returned cost	Time (s)
Simudata	SeedAlg	0.8642	0.8112	2949	1549	56
	FuncAlg	0.8642	0.8151	6403	1551	64
Sdata	SeedAlg	0.6601	1.2030	1513	701	184
	FuncAlg	0.6675	1.2052	2650	701	240

of pairs of objects. It is easy to see that ARI has a value between -1 and 1. A larger ARI indicates higher consistence between the clustering result and the real class labels, especially, the value 1 of ARI indicates the clustering result is same to the real class labels. DBI is an internal clustering validation index, defined as follows.

$$\text{DBI} := \frac{1}{n} \sum_{i=1}^{n} \max_{j \neq i} \left(\frac{S_i + S_j}{M_{ij}} \right),$$

where S_i is the average value of the Euclidean distances from all points in the cluster i to the centroid of this cluster, and M_{ij} is the Euclidean distance between the centroids of clusters i and j. The smaller DBI is, the less between-cluster similarity is. We also show the costs of the initial and returned solutions, as well as the running times of the algorithms. Since the two algorithms we compare are random, we run them for 10 times and compute the average values for all measures. Results are shown in Table 1 where the notations SeedAlg and FuncAlg represent Algorithm 1 and the functional k-means algorithm in [11] respectively.

From the results, we can observe that the clustering accuracies of the two algorithms are very close, since the relative differences of the ARIs, DBIs, and returned costs of these two algorithms are less than 1.2%. Nevertheless, the seeding algorithm we provide is competitive in efficiency, since it can product a better initial solution and consumes less time.

6 Conclusions

By defining a new distance with derivative information, the functional k-means clustering algorithm can be used well for functional k-means problem. In this paper, we first introduce a novel proof for functional 1-mean problem. Then we apply the seeding algorithm to functional k-means problem and obtain that the performance guarantee is $O(\log k)$. In future, we will focus on the parallel seeding algorithm and local search method for functional k-means problem.

Acknowledgments. The first author is supported by Higher Educational Science and Technology Program of Shandong Province (No. J17KA171). The second author is supported by National Natural Science Foundation of China (No. 61433012), Shenzhen research grant (KQJSCX20180330170311901, JCYJ20180305180840138 and GGFW2017073114031767). The third author is supported by National Natural Science Foundation of China (No. 11531014). The fourth author is supported by National Natural Science Foundation of China (No. 11871081).

References

1. Abraham, C., Cornillon, P.A., Matzner-Løber, E., Molinari, N.: Unsupervised curve clustering using B-splines. Scand. J. Stat. **30**(3), 581–595 (2003)
2. Aggarwal, A., Deshpande, A., Kannan, R.: Adaptive sampling for k-means clustering. In: Dinur, I., Jansen, K., Naor, J., Rolim, J. (eds.) APPROX/RANDOM -2009. LNCS, vol. 5687, pp. 15–28. Springer, Heidelberg (2009). https://doi.org/10.1007/978-3-642-03685-9_2
3. Arthur, D., Vassilvitskii, S.: k-means++: the advantages of careful seeding. In: Nikhil, B., Kirk, P., Clifford, S. (eds.) SODA 2007, Theory, pp. 1027–1035. SIAM, Philadelphia (2007). https://doi.org/10.1145/1283383.1283494
4. Boullé, M.: Functional data clustering via piecewise constant nonparametric density estimation. Pattern Recogn. **45**(12), 4389–4401 (2012)
5. Bouveyron, C., Brunet-Saumard, C.: Model-based clustering of high-dimensional data: a review. Comput. Stat. Data Anal. **71**, 52–78 (2014)
6. Gamasaee, R., Zarandi, M.: A new Dirichlet process for mining dynamic patterns in functional data. Inf. Sci. **405**, 55–80 (2017)
7. Har-Peled, S., Sadri, B.: How fast is the k-means method? Algorithmica **71**(3), 185–202 (2005)
8. Jacques, J., Preda, C.: Functional data clustering: a survey. Adv. Data Anal. Classif. **8**(3), 231–255 (2014)
9. Kayano, M., Dozono, K., Konishi, S.: Functional cluster analysis via orthonormalized Gaussian basis expansions and its application. J. Classif. **27**(2), 211–230 (2010)
10. Lloyd, S.: Least squares quantization in PCM. IEEE Trans. Inf. Theory **28**(2), 129–137 (1982)
11. Meng, Y., Liang, J., Cao, F., He, Y.: A new distance with derivative information for functional k-means clustering algorithm. Inf. Sci. **463–464**, 166–185 (2018)
12. Ostrovsky, R., Rabani, Y., Schulman, L., Swamy, C.: The effectiveness of Lloyd-type methods for the k-means problem. J. ACM **59**(6), 28:1–28:22 (2012)
13. Park, J., Ahn, J.: Clustering multivariate functional data with phase variation. Biometrics **73**(1), 324–333 (2017)
14. Peng, J., Müller, H.G.: Distance-based clustering of sparsely observed stochastic processes, with applications to online auctions. Ann. Appl. Stat. **2**(3), 1056–1077 (2008)
15. Preda, C., Saporta, G., Lévéder, C.: PLS classification of functional data. Comput. Statistics **22**(2), 223–235 (2007)
16. Tarpey, T., Kinateder, K.K.: Clustering functional data. J. Classif. **20**(1), 93–114 (2003)

More Efficient Algorithms for Stochastic Diameter and Some Unapproximated Problems in Metric Space

Daogao Liu$^{(\boxtimes)}$

Department of Physics, Tsinghua University, Beijing, China
liudg16@mails.tsinghua.edu.cn

Abstract. Dealing with data on uncertainty has appealed to many researchers as there may be many stochastic problems in a realistic situation. In this paper, we study two basic uncertainty models: Existential Uncertainty Model where the location of each node is fixed while it may be absent with some probability, and the Locational Uncertainty Model where each node must be present, but the situation is uncertain. We consider the problem of estimating the expectation and the tail bound distribution of the diameter, and obtain an improved FPRAS (Fully Polynomial Randomized Approximation Scheme) which requires much fewer samples. In the meanwhile, we prove some problems in the two uncertainty models can't be approximated within any factor unless NP $\subseteq$ BPP by simple reductions.

Keywords: FPRAS · Stochastic diameter · Hardness for approximation

1 Introduction

Models: As mentioned before, we focus on two stochastic geometry models, the existential uncertainty model and locational uncertainty model. We'll show the precise definition of these two models below:

Definition 1 *(Locational Uncertainty Model). We are given a metric space P. The location of each node $v \in V$ is a random point in the metric space P and the probability distribution is given as the input. Formally, we use the term nodes to refer to the vertices of the graph, points to describe the locations of the nodes in the metric space. We denote the set of nodes as $V = \{v_1, ..., v_n\}$ and the set of points as $P = \{s_1, ..., s_m\}$, where $n = |V|$ and $m = |P|$. A realization r can be represented by an n-dimensional vector $(r_1, ..., r_n) \in P_n$ where point r_i is the location of node v_i for $1 \le i \le n$. Let R denote the set of all possible realizations. We assume that the distributions of the locations of nodes in the metric space P are independent, thus r occurs with probability $Pr[r] = \prod_{i \in [n]} p_{v_i r_i}$, where p_{vs} represents the probability that the location of node v is point $s \in P$.*

© Springer Nature Switzerland AG 2019
D.-Z. Du et al. (Eds.): COCOON 2019, LNCS 11653, pp. 397–411, 2019.
https://doi.org/10.1007/978-3-030-26176-4_33

Definition 2 *(Existential Uncertainty Model). A closely related model is the existential uncertainty model where the location of a node is a fixed point in the given
metric space, but the existence of the node is probabilistic. In this model, we use p_i to
denote the probability that node v_i exists (if exists, its location is s_i). A realization
r can be represented by a subset $S \subset P$ and $Pr[r] = \prod_{s_i \in S} p_i \prod_{s_i \notin S}(1 - p_i)$.*

Problem Formulation. The natural problems in the above models are to estimate the expectation and the tail bound of distribution of certain combinatorial
objects, denoted by E(Obj) and P(Obj $\geq$ 1) (or the form P(Obj $\leq$ 1)). More
accurately, take the expectation of diameter (the longest distance between two
realized points) as an example. Note the expectation E(D), and D($\mathbf{r}$) be the
longest distance between two points in the realization $\mathbf{r}$. The precise definition
of the E(D) is:

$$E(D) = \sum_{r \in R} Pr[\mathbf{r}]D(\mathbf{r})$$

Similarly, note the probability that the diameter is no less than the given threshold, i.e. P(D $\leq$ 1). And what we estimate in this paper is E(D) and P(D $\leq$ 1).
The P(D $\geq$ 1) has been shown unapproximable [13].

Preliminaries. The most useful techniques in the estimation are the straightforward Monte Carlo strategy. We repeat the experiment and obtain the average
of the experiment results, and use the average as the estimation of the true value.
The number of samples required by this algorithm is suggested by the following
standard Chernoff bound.

Lemma 1 *(Chernoff bound). Let random variables $X_1, X_2, ..., X_N$ be independent random variables taking on values between 0 and U. Let $X = \frac{1}{N} \sum_{i=1}^{N} X_i$
and $\mu = E(X)$. Then for any $\epsilon > 0$, we have $P((1 - \epsilon)\mu \leq X \leq (1 + \epsilon)\mu) \geq
1 - 2e^{-N\frac{\mu}{U}\epsilon^2/4}$.*

Then if we want to get an $(1 \pm \epsilon)$ approximation with probability $1 - \frac{1}{poly(N)}$,
the number of samples needs to be $O(\frac{U}{\mu\epsilon^2} ln N)$.

Call one realization of all nodes in both models **one sample**. So the main
target of our algorithm in this paper is to bound the value $\frac{U}{\mu}$ and use as fewer
samples as possible.

Take the Locational Uncertainty Model as an example. To simplify the argument of the running time, we assume the running time of experimenting with
one node is nearly the same whatever its locational distribution is. However, it's
difficult to argue that how much time it will take to do an experiment with one
node. So we take one realization as one sample and use the necessary number of
samples as the evaluation criterion of our algorithms.

Our Contributions. Recall that the *fully polynomial randomized approximation scheme* (FPRAS) for a problem f is randomized algorithm A that
takes an input instance x, a real number $\epsilon > 0$, returns A(x) such that

$P[(1 - \epsilon)f(x) \leq A(x) \leq (1 + \epsilon)f(x)] \geq \frac{3}{4}$ and its running time is polynomial in both the size of the input n and $1/\epsilon$.

We designed the FPRAS for E(D) and $P(D \geq 1)$ in both models which are the best of our knowledge.

Huang et al. [13] gives the FPRAS for E(closet pair) and P(Closet pair ≤ 1) (denote by $P(C \leq 1)$ later). The FPRAS for $P(C \leq 1)$ can be used to estimate $P(D \geq 1)$ with some trivial operations. In the existential uncertainty model, suppose there are m nodes, we improve the FPRAS from needing $O(\frac{m^6}{\epsilon^4} lnm)$ independent samples to $O(\frac{m}{\epsilon^4} lnm)$. As for a locational uncertainty model with m nodes and n points, we improve the FPRAS from needing $O(\frac{m^6}{\epsilon^4} lnm)$ samples to $O(\frac{m^3}{\epsilon^4} lnm)$.

As for the E(D), note that we can get an FPRAS for E(D) by the FPRAS for $P(D \geq 1)$, but it will need $O(\frac{m^8}{\epsilon^4} lnm)$ independent samples. We give the first direct FPRAS for E(D) in this paper which only needs $O(\frac{m^2}{\epsilon^2} ln^2m)$ samples in the worst case and need $O(\frac{m^2}{\epsilon^2} lnm)$ samples in the best situation for both models. This direct FPRAS doesn't need to estimate $P(D \geq 1)$ anymore.

Moreover, we'll show some problems can't be approximated unless NP $\subseteq$ BPP, which answers one of the open problems given in [13]. The main results of unapproximation are shown in the below Table 1:

Table 1. Results for unapproximated problems

Unapproximable value	Model	NPC problem for reduction
P(k-th closest pair ≤ 1)	Loc	Max 2-SAT
P(k-th longest m-nearest neighbor ≤ 1)	Loc	Maximum clique
P(k-clustering ≥ 1)	Loc	3-coloring
P(Minimum cycle cover ≥ 1)	Loc	3-coloring
P(Minimum spanning tree ≥ 1)	Loc	3-coloring
E(k-th longest m-nearest neighbor)	Loc	Vertex cover
P(k-clustering ≥ 1)	Exis	Independent set

We will show the non-approximation of E(k-th longest m-nearest neighbor) in Locational model and P(k-clustering ≥ 1) in Existential model in Sect. 4. The exact definition and brief proof for other problems will be shown in the appendix due to space constraints.

Related Work. The uncertain or imprecise data has been studied extensively recently [7,9]. Consider the locational data collected by the Global-Positioning Systems (GPS), there are always some random measurement errors [27]. For another example, if we use a sensor network to monitor the living habits or migration of certain animals, there will also be some noise among the data we collected as the sensors won't be perfect [8,17,22]. Some people study the imprecise data in a model where each point may be in some region [4,23,26,29].

The existential uncertainty model and the locational uncertainty modes we mentioned before have been studied extensively in recent years (e.g. [1–3,14,17]). It's worth mentioning that when all the points follow the same distribution, it's a classic topic in stochastic geometry literature [5,6,27]. The asymptotics expectation for certain combinatorial problems (such as MST) is the main interest in that topic. The general locational uncertain model is also of fundamental interest in the area of wireless networks. There is a survey [12] and you can see more references about the stochastic model and wireless networks there.

There have been many works under the term *stochastic geometry* in the above uncertainty model and many other different stochastic models. For example, Huang et al. [14] initiate the study of constructing ϵ-kernel coresets for uncertain points in the above two models. The convex hull [11,20], minimum enclosing ball problem [24], shape fitting [19], MST [5] and many other problems have also been studied on the imprecise data.

The study of estimating the expectation of objects in the model is started by Kamousi, Chan and Suri [15,16]. They showed that the expectation of some values, such as nearest neighbor (NN) graph, the Gabriel graph (GG) and so on, can be solved in polynomial time. And they designed FPRAS for E(MST) and E(the closest pair) in the existential uncertainty model.

Huang et al. [13] gives the FPRAS for the expected values of closest pair, minimum spanning tree, k-clustering, minimum perfect matching, and minimum cycle cover in both models by several powerful techniques. And they also consider the problem of estimating the probability that the length of closest pairs at most, or at least, a given threshold.

Most recently, Li and Deshpande [18] observe that the expected value is inadequate in some problems and study the maximization of the expected utility of the solution for some given utility function. The initial motivation for the study is the stochastic shortest-path problem, which has been studied extensively [21,25,28].

2   The Expectation of Diameter

Existential Uncertainty Model: Let's show the FPRAS of E(D) in the Existential Uncertainty Model. First, let us show the meaning of the signals we use. Let U be the complete set of the points. And $S\langle \geq j \rangle$ means that there are at least j points realized in the set of points S. Suppose we have m points in total (we use m to describe the complexity of samples we need later). Then there are $l = \binom{m}{2}$ different pairs of points. W.l.o.g, suppose the lengths of the l pairs are distinct. And we sort them in ascending order of their length and index them. Let e_i represent the i-th pair, and d_i is its length. We have $d_1 < d_2 < ... < d_l$. And for a pair $e_i = (u, v)$, $P(e_i|\alpha)$ represent the probability that both u and v are realized conditioning on $event(\alpha)$.

What we want to estimate is indeed $E(D|U\langle \geq 2 \rangle)$, because the diameter doesn't make sense if there is only one or zero point realized. Now let us introduce the algorithm.

First, for pair $e_i = (u, v)$, we can calculate $P(e_i | U \langle \geq 2 \rangle) = \frac{P(e_i, U \langle \geq 2 \rangle)}{P(U \langle \geq 2 \rangle)} = \frac{P(e_i)}{P(U \langle \geq 2 \rangle)} = \frac{P_u P_v}{P(U \langle \geq 2 \rangle)}$, which can be calculated easily by the following lemma:

Lemma 2. *For a set of points C and $j \in \mathbf{Z}$, we can compute $P(C \langle \geq j \rangle)$ in polynomial time. Moreover, there exists a poly-time sampler to sample present points from C conditioning on $C \langle \geq j \rangle$ (Or $C \langle j \rangle$).*

Proof. The idea is essentially from [10]. W.l.o.g, we assume that the points in C are $x_1, x_2, ..., x_n$. We denote the event that among the first a points, at least b points are present by E(a, b) and denote the probability of E(a, b) by P(a, b). Note that our goal is to compute P(n, j), which can be solved by the following dynamic program:

1. If $a < b$, P(a, b) = 0. If $a = b$, $P(a, b) = \prod_{1 \leq l \leq a} P_l$. If b = 0, P(a, b) = 1.
2. For $a > b$ and $b > 0$, $P(a, b) = P_a P(a-1, b-1) + (1 - P_a) P(a-1, b)$.

We can also use this dynamic program to construct an efficient sampler. Consider the point x_n, with probability $P_n P(n-1, j-1) / P(n, j)$, we make it present and then recursively consider the point x_{n-1}, conditioning on the event $E(n-1, j-1)$. With probability $(1 - P_n) P(n-1, j) / P(n, j)$, we discard it and then recursively sample conditioning on the event $E(n-1, j)$.

The proof of $P(C \langle j \rangle)$ (i.e. there are exactly j points present in C) is similar and we skip it.

Now continue our algorithm.

There exists a set of pairs $S = \{ e_i | P(e_i | U \langle \geq 2 \rangle) \geq \frac{1}{m^2} \}$. S is non-empty, or $P(U \langle \geq 2 \rangle | U \langle \geq 2 \rangle) = 1 = P(\cup_i e_i | U \langle \geq 2 \rangle) \leq l P(e_i | U \langle \geq 2 \rangle) < 1$. Let Y be the largest index among all the pairs in S. Recall that the l-th pair is the longest one. If $d_Y \geq \frac{1}{lnm} d_l$, then $E(D | U \langle \geq 2 \rangle) \geq P(D \geq d_Y | U \langle \geq 2 \rangle) d_Y \geq \frac{1}{m^2 lnm} d_l$. By chernoff bound, we only need to take $O(\frac{m^2}{\epsilon^2} ln^2 m)$ independent samples. This is the worst case of our algorithm.

Now consider the other situation, i.e. $d_Y < \frac{1}{lnm} d_l$. Then we have a set of points $H = \{ u | \exists v \in U, (u, v) \in S \}$. It's obvious that $\forall u, v \in H, d(u, v) < \frac{3}{lnm} d_n$ due to the Triangle inequality. As if $d(u, v) \geq \frac{3}{lnm} d_l$, suppose $(u, u') \in S, (v, v') \in S$. Let $u'' = u$ if $P_u > P_{u'}$, or $u'' = u'$. And get v'' by the similar approach. Then $P((u'', v'') | U \langle \geq 2 \rangle) \geq \frac{1}{m^2}$ and $d(u'', v'') > \frac{1}{lnm} d_l$, which is impossible. (Remark: We can understand this property as those points with a relatively high probability of realization are surrounded by a small sphere.)

Suppose x is one of the points that have the largest probability of realization among U (if there are more than one points with largest probability, choose one arbitrarily), then we have the following property: $d(x, H) < \frac{2}{lnm} d_l$. The definition of $d(x, H)$ is $d(x, H) = \max_{u \in H} d(x, u)$. Let $H' = H \cup \{x\}$. And we can construct a set of points $H'' = \{ u | u = x \text{ or } d(u, x) < \frac{4}{lnm} d_n \}$. It's obvious that $H \subseteq H' \subseteq H''$. If $H'' = U$, we need only $O(\frac{m^2}{\epsilon^2} ln^2 m)$ independent samples. Else, we can use the following algorithm.

For any point t, let $P(\alpha | t)$ represent the probability of $event(\alpha)$ conditioning on that point t is realized, and $P(\alpha | \bar{t})$ correspond to the probability of $event(\alpha)$

Algorithm 1. construct event

1: $S_0 = U/H'', N_0 = \emptyset, i=0$
2: **while** S_i is not empty **do**
3:     $t_i = \arg\max\limits_{u \in S_i} d(u, H')$
4:     $S_{i+1} \leftarrow S_i/\{t_i\}$
5:     $N_{i+1} \leftarrow N_i \cup \{t_i\}$
6:     i←i+1
7: **Output:**S_i, t_i and N_i for all i

---

conditioning on that t is not realized. $P(t|\alpha)$ represents the probability that t is realized conditioning on $event(\alpha)$. Let $|S_0|$ denote the size of S_0. Then we have that $E(D|U\langle\geq 2\rangle) = \sum_{i=0}^{|S_0|-1} E(D|U\langle\geq 2\rangle, t_i, N_{i-1}\langle 0\rangle)P(t_i, N_{i-1}\langle 0\rangle|U\langle\geq 2\rangle) + E(D|N_{|S_0|-1}\langle 0\rangle, H''\langle\geq 2\rangle)P(H''\langle\geq 2\rangle, N_{|S_0|-1} < 0 > |U\langle\geq 2\rangle)$.

All of the probability can be calculated easily. What we need to get is the expected value in each part. $E(D|N_{|S_0|-1}\langle 0\rangle, H''\langle\geq 2\rangle)$ can be seen as we recurse our original problem into a smaller problem. Then for each i, we have the following lemma:

Lemma 3. *We only need $O(\frac{m}{\epsilon^2}logm)$ independent samples to estimate $E(D|U <\geq 2 >, t_i, N_{i-1} < 0 >)$.*

Proof. Let $U_i = U/N_i = U/(\{t_i\} \cup N_{i-1})$. We can rewrite $E(D|U\langle\geq 2\rangle, t_i, N_{i-1}\langle 0\rangle) = E(D|t_i, N_{i-1}\langle 0\rangle, U_i\langle\geq 1\rangle)$. We have a sampler condition on $event(t_i, N_{i-1} < 0 >, U_i <\geq 1 >)$ according to Lemma 1. And recall that x is the point that has the largest probability of realization, it's obvious that $x \in U_i$. Then $P(x|U_i\langle\geq 1\rangle) \geq 1/m$. Let D_i represent the maximum value of Diameter condition on $event(t_i, N_{i-1}\langle 0\rangle, U_i\langle\geq 1\rangle)$. We have $d(t_i, x) \geq \frac{2*d(t_i, H')}{6} \geq D_i/6$ condition on $event(t_i, N_{i-1}\langle 0\rangle, U_i\langle\geq 1\rangle)$. Then $E(D|t_i, N_{i-1}\langle 0\rangle, U_i\langle\geq 1\rangle) \geq d(t_i, x)P(x|U_i\langle\geq 1\rangle) \geq d(t_i, x)/m \geq D_i/(6m)$. And by Chernoff bound, we proved this lemma.

Let T(m) represent the independent samples we need to estimate $E(D|U <\geq 2 >)$ with $|U| = m$. We have the following recursive relation in the best case: $T(m) = T(m - |S_0|) + O(|S_0| * \frac{m}{\epsilon^2}lnm)$. Then we have $T(m) = O(\frac{m^2}{\epsilon^2}lnm)$.

Locational Uncertainty Model: Our algorithm is almost the same as the existential model with the assumption that at for each point, there is only one node that may be realized at this point. In principle, if more than one node may be realized at the same point, we can create multiple copies of the point co-located at the same place. We can't use the Monte Carlo method directly only when all points with high probability to be realized are 'wrapped in a small ball', we can use the similar algorithm like **Algorithm 1** to do the recursion and get the same required complexity as the existential model. For example, we can run the while loop only once and get a sub-problem with size $m - 1$.

Theorem 1. *There is an FPRAS for estimating the expected distance between the longest pair of nodes both existential and locational uncertainty models. It needs $O(\frac{m^2}{\epsilon^2} \ln^2 m)$ independent samples in the worst case and $O(\frac{m^2}{\epsilon^2} \ln m)$ in the best case in both models.*

3   The Tail Bound of Distribution

Existential Uncertainty Model: Now let us introduce the FPRAS of $P(D \geq 1)$. We can construct a set of points $H' = \{u|P_u \geq \frac{\epsilon}{m}\}$, and $H = U/H'$. We have that $P(D \geq 1) = \sum_{i=0}^{|H|} P(D \geq 1, H\langle i \rangle)$. We'll show that we only need to estimate $\sum_{i=0}^{2} P(D \geq 1, H\langle i \rangle)$ as the remaining is negligible.

Call a set of points S connected if $\forall u \in S, P_u \geq \frac{\epsilon}{m} \wedge \exists v \in S, d(u,v) \geq 1$. Call a point $u \in S$ is unique in S, if let $S' = S/\{u\}$, S' is not connected. Let $C = \{u|u \in H' \wedge \exists v \in H', d(u,v) \geq 1\}$. It's obvious that the C we constructed is connected.

Lemma 4. *For connected non-empty set S with all points unique, $P(D \geq 1|S\langle \geq 2\rangle) \geq \frac{\epsilon}{2m}$.*

Proof. It's obvious that S must have even points according to the definition. Call pair (u, v) a match if $d(u, v) \geq 1$. Suppose S has 2k points, then S has exactly k matches. Index the 2k points subject to that u_i and u_{k+i} is a match and $P_{u_i} \geq P_{u_{i+k}}$. Let $S_{a,b}$ denote the subset of points with index in $[a, b]$. Then we have $P(D \geq 1|S\langle \geq 2\rangle) = \sum_{i=1}^{k} P(D \geq 1|u_i, S_{1,i-1}\langle 0\rangle, S_{i+1,2k}\langle \geq 1\rangle)P(u_i, S_{1,i-1}\langle 0\rangle, S_{i+1,2k}\langle \geq 1\rangle|S\langle \geq 2\rangle)$.

When $i \leq k, P(D \geq 1|u_i, S_{1,i-1}\langle 0\rangle, S_{i+1,2k}\langle \geq 1\rangle) \geq \frac{\epsilon}{m}$, and we can get $P(u_i, S_{1,i-1}\langle 0\rangle, S_{i+1,2k}\langle \geq 1\rangle|S\langle \geq 2\rangle) \geq P(u_{i+k}, S_{1,i+k-1}\langle 0\rangle, S_{i+k+1,2k}\langle \geq 1\rangle)$. And notice $\sum_{i=1}^{2k-1} P(u_i, S_{1,i-1} < 0 >, S_{i+1,2k}\langle \geq 1\rangle|S\langle \geq 2\rangle) = 1$. Thus we proved this lemma.

Lemma 5. *We can estimate $P(D \geq 1, H\langle 0\rangle)$ with $O(\frac{m}{\epsilon^4} \ln m)$ independent samples.*

Proof. Then in order to prove this lemma, we only need to show that $P(D \geq 1|S\langle \geq 2\rangle) \geq \frac{\epsilon}{2m}$ for any non-empty connected set S by Mathematical induction.

Now prove Lemma 5. When $|S| = 2$, then $P(D \geq 1|S\langle \geq 2\rangle) = 1 \geq \frac{\epsilon}{2m}$. Suppose $P(D \geq 1|S\langle \geq 2\rangle) \geq \frac{\epsilon}{2m}$ when $|S| \leq n$ for any connected S and some integer n, then consider the situation when $|S| = n + 1$. When all points in C are unique, then we have $P(D \geq 1|S\langle \geq 2\rangle) \geq \frac{\epsilon}{2m}$ by Lemma 4. If there exists some point u that are not unique, we have $P(D \geq 1|S\langle \geq 2\rangle) = P(D \geq 1|u, S\langle \geq 1\rangle)P(u|S\langle \geq 2\rangle) + P(D \geq 1|\overline{u}, S\langle \geq 2\rangle)P(\overline{u}|S\langle \geq 2\rangle)$. Both $P(D \geq 1|u, S\langle \geq 1\rangle)$ and $P(D \geq 1|\overline{u}, S\langle \geq 2\rangle)$ are no less than $\frac{\epsilon}{2m}$. And $P(u|S\langle \geq 2\rangle) + P(\overline{u}|S\langle \geq 2\rangle) = 1$, thus we proved $P(D \geq 1|S\langle \geq 2\rangle) \geq \frac{\epsilon}{2m}$ when $|S| = n + 1$.

Then by Monte carlo directly, we proved this lemma.

Now let's show how to estimate $P(D \geq 1, H\langle 1 \rangle)$. Observe that $P(D \geq 1, H\langle 1 \rangle) = \sum_{u \in H} P(D \geq 1, u, H/\{u\}\langle 0 \rangle)$. For point u in H, denote $G_u = \{v | v \in H', d(u, v) \geq 1\}$. We can calculate $P(G_u \langle \geq 1 \rangle, u, H/\{u\}\langle 0 \rangle)$ exactly in linear time. We can use the value of $\sum_{u \in H} P(G_u \langle \geq 1 \rangle, u, H/\{u\}\langle 0 \rangle)$ as an estimation of $P(D \geq 1, H\langle 1 \rangle)$ because of the following claim.

Claim. $\sum_{u \in H} P(G_u \langle \geq 1 \rangle, u, H/\{u\}\langle 0 \rangle) \leq P(D \geq 1, H\langle 1 \rangle) \leq \sum_{u \in H} P(G_u \langle \geq 1 \rangle, u, H/\{u\}\langle 0 \rangle) + 2\epsilon P(D \geq 1, H\langle 0 \rangle)$.

Proof. Since we only miss the summation probability of these events: there are two points x, y in H'/G_u realized with d(x, y) ≥ 1 and there are no points present in G_u. Write down the expression: $\sum_{u \in H} P(D \geq 1, u, H/\{u\}\langle 0 \rangle, G_u\langle 0 \rangle)$. Denote the set of realization we may miss by M. Each realization **r** in M can be transferred to the $event(D \geq 1, H\langle 0 \rangle)$ by making the only present point in H absent. We denote the realization after the transform **r'**. We have $P(\mathbf{r'}) \geq \frac{m}{2\epsilon} P(\mathbf{r})$. And given **r'**, there are at most m different realizations can be transformed into it. We have $P(D \geq 1, H\langle 1 \rangle) = \sum_{u \in H} P(G_u\langle \geq 1 \rangle, u, H/\{u\}\langle 0 \rangle) + \sum_{\mathbf{r} \in M} P(\mathbf{r}) \leq \sum_{u \in H} P(G_u\langle \geq 1 \rangle, u, H/\{u\}\langle 0 \rangle) + 2\epsilon \sum_{\mathbf{r'}} P(\mathbf{r'}) \leq \sum_{u \in H} P(G_u\langle \geq 1 \rangle, u, H/\{u\}\langle 0 \rangle) + 2\epsilon P(D \geq 1, H\langle 0 \rangle)$.

Call the argument method of this claim **ARG**, which will be useful later.

As for $P(D \geq 1, H\langle 2 \rangle) = \sum_{u,v \in H} P(D \geq 1, u, v, H/\{u, v\}\langle 0 \rangle)$. Given u, v $\in H$. If d(u, v) ≥ 1, $P(D \geq 1, u, v, H/\{u, v\}\langle 0 \rangle) = P_u P_v P(H/\{u, v\}\langle 0 \rangle)$ which can be calculated directly. And $\sum_{u,v \in H \wedge d(u,v) < 1} P(D \geq 1, u, v, H/\{u, v\}\langle 0 \rangle) \leq 2\epsilon(P(D \geq 1, H\langle 0 \rangle) + P(D \geq 1, H\langle 1 \rangle))$ by the similar argument of **ARG**, which means it's negligible. So we can use the value of $\sum_{u,v \in H \wedge d(u,v) \geq 1} P(D \geq 1, u, v, H/\{u, v\}\langle 0 \rangle)$ as an estimation of $P(D \geq 1, H\langle 2 \rangle)$.

So far we have shown how to estimate $\sum_{i=0}^{2} P(D \geq 1, H\langle i \rangle)$. The last thing we have to do is to show $\sum_{i=3}^{|H|} P(D \geq 1, H\langle i \rangle)$ is negligible. In fact, we can prove $P(D \geq 1, H\langle 2 + i \rangle) \leq (2\epsilon)^i P(D \geq 1, H\langle 2 \rangle)$ with the similar method with **ARG**. So $\sum_{i \geq 3} P(D \geq 1, H\langle i \rangle) \leq$ is negligible compared with $\sum_{i=0}^{2} P(D \geq 1, H\langle i \rangle)$.

Theorem 2. *There is an FPRAS for estimating the probability of the distance between the furthest pair of nodes is at least 1 in the existential uncertainty model with only $O(\frac{m}{\epsilon^4} lnm)$ independent samples.*

Locational Uncertainty Model: Please pay attention that the *node* and *point* have different meaning in the locational model. And recall that we assume that at for each point, there is only one node that may be realized at this point. Suppose we have n nodes and m points. Huang et al. [13] has given a FPRAS for $P(D \geq 1)$ which needs $O(\frac{m^6}{\epsilon^4} lnm)$ independent samples. And we improved it and only need $O(\frac{m^3}{\epsilon^4} lnm)$ independent samples.

The thought of FPRAS for $P(D \geq 1)$ in locational uncertainty model is exactly the same as the existential model, while we need a little bit more samples because of the difference of the two models.

Call a point u *not-alone* in a point set H, if $\exists v \in H$, st. $d(u,v) \geq 1 \wedge$ u, v correspond to different nodes. And we call the set H single if H doesn't contain any not-alone points.

Let $H = \{u | P_u \geq \frac{\epsilon}{m^2}\}$. $F = V/H$. So similarly, $P(D \geq 1) = \sum_{i=0}^{|F|} P(D \geq 1, F\langle i \rangle)$. And we also only need to estimate $\sum_{i=0}^{2} P(D \geq 1, F\langle i \rangle)$ as $\sum_{i=3}^{|F|} P(D \geq 1, F\langle i \rangle)$ is negligible by the similar argument with **ARG**.

Lemma 6. *We can estimate $P(D \geq 1, F\langle 0 \rangle)$ by $O(\frac{m^3}{\epsilon^4} lnm)$ independent samples.*

Proof. 1. It's obvious that if H is single, then $P(D \geq 1, F\langle 0 \rangle) = 0$.
2. If H is not single, with the following Algorithm 2, we can estimate $P(D \geq 1, F\langle 0 \rangle)$ by $O(\frac{m^3}{\epsilon^4} lnm)$ independent samples.

Algorithm 2. Estimate $P(D \geq 1, F\langle 0 \rangle)$

1: $S_0 = H, N_0 = \emptyset, i=0$
2: **while** S_i not-single **do**
3:     find arbitrary not-alone point t_i
4:     $S_{i+1} \leftarrow S_i / \{t_i\}$
5:     $N_{i+1} \leftarrow N_i \cup \{t_i\}$
6:     i←i+1
7:     Estimate $P(D \geq 1 | t_i, N_i \langle 0 \rangle, F\langle 0 \rangle)$
8: **Output:**The summation of $P(D \geq 1 | t_i, N_i \langle 0 \rangle, F\langle 0 \rangle)$ for all i

---

Note that we can estimate $P(D \geq 1 | t_i, N_i \langle 0 \rangle, F\langle 0 \rangle)$ with $O(\frac{m^2}{\epsilon^4} lnm)$ for any given i. And $i \leq m$, thus we finish the proof.

As for the term $P(D \geq 1 | F\langle 1 \rangle)$. For point $u \in F$ corresponds to node n_i, then we can either estimate $P(D \geq 1 | u, F/\{u\} \langle 0 \rangle)$ with $O(\frac{m^2}{\epsilon^4} lnm)$ independent samples if there are $d(u,v) \geq 1$ for $v \in H \wedge v$ corresponds to a different node n_j, or this value can be neglected by the similar argument with **ARG**, as there will must be a point $u' \in H$ which also corresponds to n_i st. $P(u') \geq \frac{1}{m}$. Thus we can estimate $P(D \geq 1, F\langle 1 \rangle)$ with $O(\frac{m^3}{\epsilon^4} lnm)$ samples.

Similarly, we can estimate $P(D \geq 1 | F\langle 2 \rangle)$ by enumerating point pairs $(u,v) \in H$, and let $\sum_{u,v \in H, d(u,v) \geq 1} P(u,v)$ be the estimation of $P(D \geq 1 | F\langle 2 \rangle)$. There are at most $O(m^2)$ pairs.

Theorem 3. *There is an FPRAS for estimating the probability of the distance between the furthest pair of nodes is at least 1 in the Locational Uncertainty Model with only $O(\frac{m^3}{\epsilon^4} lnm)$ independent samples.*

4   Examples for Unapproximable Values

k-th Longest m-Nearest Neighbor: The precise description of this problem is under any realization, for each node, find the distance to its m-nearest neighbor, then compute the k-th longest one among these distances. Huang et al. [13] gives a FPRAS for this value in the existential model. And we'll show that this value can't be approximated in the locational uncertainty model unless $NP \subseteq BPP$.

Lemma 7. *Given the undirect graph G, we can construct a Locational Uncertainty Model G'. Then there is a vertex cover of size k iff E((n − k)th Longest (n + m − 1)Nearest Neighbor) > 0 in G'.*

Proof. Suppose for one point, there may be more than one node realized at it. Now let's show how to construct such an G' according to G. Suppose there are n vertices and m edges in G. Construct n points and n + m nodes in G'. Divide the n+m nodes into two disjoint sets S_1 and S_2, with $|S_1| = n, |S_2| = m$. The i-th node in S_1 can only be present at i-th point with probability equals 1. The n points in G' correspond to the n vertices in G each, the m nodes in S_2 correspond to the m edges in G. Then if vertex v_j is one of the end point of edge e_i in G, the corresponding node can be present at the corresponding point with probability $1/2$ in G'. As for the distance of point pairs in G', the distance of each pair is M, which can even be $+\infty$.

Under such a construction, it's obvious that E((n − k)th Longest (n + m − 1)Nearest Neighbor) = M * p with $p > 0$ strictly if and only if there exists a vertex cover with size k. Note that when $p > 0$, E((n − k)th Longest (n + m − 1)Nearest Neighbor) can be infinitely large.

Then if there exists a FPRAS or any other approximation algorithms for E(k-th Longest m-Nearest Neighbor) with finite approximation ratio and guaranteed accuracy, we can construct a Locational Uncertainty Model G' according to *Lemma 7*. Run the algorithm on G', and we can judge if there is a k-vertex cover in G by comparing the output of the algorithm with zero, and get the accurate result with guaranteed accuracy, which means $NP \subseteq BPP$.

Theorem 4. *E(k-th Longest m-Nearest Neighbor) in Locational Uncertainty Model is imapproximable within any finite ratio and guaranteed accuracy unless $NP \subseteq BPP$.*

k-clustering Problem: Not only in locational uncertainty model, the similar thought can also be used in the existential uncertainty model. In the deterministic kclustering problem, we want to partition all points into k disjoint subsets such that the spacing of the partition is maximized, where the spacing is defined to be the minimum of any d(u, v) with u, v in different subsets. We have the following lemma:

Lemma 8. *Given an undirect graph G, we can construct a Existential Uncertainty Model G', subject to there exists an independent set of k vertices G iff P(k − clustering ≥ 1) > 0 in G'.*

Proof. Suppose there are n vertices in G, then there will also be n nodes in G'. And there is a bijection between them. Each node will be present with probability $1/2$. As for the distance of nodes in G', for pair (n_i, n_j) in G', if there is an edge between the corresponding vertices in G, then $d(n_i, n_j) = 0.9$, else $d(n_i, n_j) = 1.8$.

Then if there is an independent set of size k in G, the output of the approximation algorithms for the $P(k - clustering \geq 1)$ in G' should be more than 0 strictly with guaranteed accuracy, or the approximation ratio will be ∞.

Theorem 5. $P(k - Clustering \geq 1)$ *in Existential Uncertainty Model is imapproximable within any finite ratio and guaranteed accuracy unless $NP \subseteq BPP$.*

5 Conclusion

In this paper, we studied the expectation and the tail bound of distribution of stochastic diameter, and prove some values can't be approximated. One remaining open problem is if there is FPRAS for k-Clustering problem and kth Closest Pair problem, or they are also imapproximable. And studying the threshold probabilities $P(Obj \geq 1)$ and $P(Obj \leq 1)$ for other values is also an interesting topic.

Acknowledgments. The author would like to thank Jian Li for several useful discussions and the help with polishing the paper. The research is supported in part by the National Basic Research Program of China Grant 2015CB358700, the National Natural Science Foundation of China Grant 61822203, 61772297, 61632016, 61761146003.

A Appendix

A.1 k-th Closest Pair

k-th closest pair means for all pairs of nodes, find the k-th closest one among them. We know that Max 2-SAT is a NP-Complete problem. Given a 2CNF with n clauses and a integer k < n, we would like to ask whether there is an assignment such that at least k clauses are satisfied. Let $P(kC \leq 1)$ represent $P(k\text{-th Closest Pair} \leq 1)$. We will show that

Lemma 9. *Given the 2CNF and the integer k, we can construct a Locational Uncertainty Model G. Then there is an assignment such that at least k clauses are satisfied iff $P(kC \leq 1) > 0$ in G.*

Proof. Suppose there are n clauses and m variables in the 2-CNF. And there is no clause containing both variable x_i and $\overline{x_i}$ for some i. Corresponding to each variable x_i, there are one node u_i and two possible points A_i and B_i for realization of u_i. We have $P_{u_i A_i} = P_{u_i B_i} = \frac{1}{2}$. Then $P_{u_i A_j} = P_{u_i B_j} = 0$ for $i \neq j$.

Then for each clause c_i, there will be one node v_i and two possible points C_i and D_i. We also have $P_{v_i C_i} = P_{v_i D_i} = \frac{1}{2}$.

We can set a bijection that $x_i = true$ iff u_i is realized to A_i. Then $\overline{x_i} = true$ iff u_i is realized to B_i.

Then we should give the distance of the pairs of points. We let the distance of any pairs of points be 1.8 for initialization. For the clause $c_i = [x_t \cup x_s]$. The distance of two pairs (C_i, A_t) and (D_i, A_s) should be changed to 0.9.

To see that even if both x_t and x_s are true, the clause $c_i = [x_t \cup x_s]$ can only contributes one pair with distance ≤ 1 in one possible realization. And for another example, if $c_i = [x_t \cup \overline{x_s}]$, we can let the distance of (C_i, A_t) and (D_i, B_s) to be 0.9.

And what we should pay attention is that even if we have two same clauses c_i and c_j, we still need to change the distance of four different pairs of points in G be 0.9, each clause corresponds to two pairs.

Then if there is an assignment such that at least k clauses are satisfied, there will be a realization that each node is realized in the corresponding point according to assignment and bijection. And there will be one possible realization that the k nodes corresponding to the k satisfied clauses are realized in the points whose closest pair $= 0.9$. Then $P(kC \leq 1) > 0$. And the reversal direction is similar.

Having proved this lemma, we can have the theorem below:

Theorem 6. $P(kC \leq 1)$ *in Locational Uncertainty Model is imapproximable within any finite ratio and guaranteed accuracy unless* $NP \subseteq BPP$.

A.2 kth Longest m-Nearest Neighbor

Lemma 10. *Given the undirect graph G, we can construct a Locational Uncertainty Model G'. Then there is a clique of size k iff P(longest k-1 nearest neighbor $\leq$ 1) > 0 in G'.*

Proof. Let k be the size of clique we want to find. And there are n vertices in G. We will have k nodes in G', denoted by $\{x_1, ..., x_k\}$. And we have k family of points $S_1, ..., S_k$. Each family S_i has n points. And the node x_i can be only realized at the n points in S_i with random probability, i.e. $P_{x_i A_j} = \frac{1}{n}$ for point $A_j \in S_i$.

As for the distance of pairs of nodes. For pair (u, v) when u and v are in the same family, let $d(u, v) = 0.9$. (In fact the distance of pair in the same family is not important, as there will be only one node realized in the same family). We want each vertex u in G corresponds to k points in G', and the k points are separated in the k disjoint family. Then there will be a bijection between the n points in one family and the n vertices in G. Then consider pair (u_i, v_j) with u_i in S_i, v_j in S_j and $i \neq j$. Denote the corresponding vertex u of u_i and v of v_j in G, if (u, v) is an edge in G, then let $d(u_i, v_j) = 0.9$, else $d(u_i, v_j) = 1.8$. (Remark: Even if u == v in G, $d(u_i, v_j) = 1.8$). Then all the pairs will have a distance and will meet the triangle inequality.

Theorem 7. $P(kmNN \leq 1)$ *in Locational Uncertainty Model is imapproximable within any finite ratio and guaranteed accuracy unless* $NP \subseteq BPP$.

A.3 K-clustering

We have shown that $P(k-clustering \geq 1)$ is hard to approximate in Existential Uncertainty Model, now we show it's also unapproximated in Locational Model. Note $P(k - clustering \geq 1)$ by $P(kCL \geq 1)$ later.

Lemma 11. *Given an undirect graph G, we can construct a Locational Uncertainty Model G', subject to G is 3-colorable iff $P(kCL \geq 1) > 0$ in G'.*

Proof. Suppose there a n vertices in G. We can construct G' with n nodes, and there is a bijection between these n vertices and n nodes. We have 3 family of points, noted by S_1, S_2, S_3. And each family contains n points, where there also is a bijection between n vertices in G and n points in S_i for all i $\in \{1, 2, 3\}$.

For each vertex x_i in G, it has bijection relationships with node u_i and three points A_i, B_i, C_i, where A_i, B_i, C_i are in the three different family. Then we let u_i can only be realized in A_i, B_i, C_i, with probability $\frac{1}{3}$ each.

As for the distance of pairs of points. For pair (u, v) with u and v are in different family, let $d(u, v) = 1.8$. For pair (u_i, u_t) in the same family, let x_i has the bijection relation with u_i and x_t for u_t. If there is an edge (x_i, x_t) in G, then $d(u_i, u_t) = 0.9$, else $d(u_i, u_t) = 1.8$.

Theorem 8. *$P(kCL \geq 1)$ in Locational Uncertainty Model is imapproximable within any finite ratio and guaranteed accuracy unless $NP \subseteq BPP$.*

A.4 Minimum Cycle Cover and MST Problem

In the deterministic version of the cycle cover problem, we are asked to find a collection of node-disjoint cycles such that each node is in one cycle and the total length is minimized. Here we assume that each cycle contains at least two nodes. If a cycle contains exactly two nodes, the length of the cycle is two times the distance between these two nodes. And we still starts from 3-coloring problem to show that P(Minimum Cycle Cover ≥ 1) is imapproximable. We denote Minimum Cycle Cover by MCC below.

With the same construction in A.3, we have following lemmas and theorems:

Lemma 12. *Given an undirect graph G, we can construct a Locational Uncertainty Model G', subject to G is 3-colorable iff $P(MCC \geq 1.8n) > 0$ in G'.*

Lemma 13. *Given an undirect graph G, we can construct a Locational Uncertainty Model G', subject to G is 3-colorable iff $P(MST \geq 1.8n) > 0$ in G'.*

With this lemma, we can have the following theorem:

Theorem 9. *$P(MCC \geq 1)$ and $P(MST \geq 1)$ in Locational Uncertainty Model are imapproximable within any finite ratio and guaranteed accuracy unless $NP \subseteq BPP$.*

References

1. Agarwal, P.K., Har-Peled, S., Suri, S., Yıldız, H., Zhang, W.: Convex hulls under uncertainty. In: Schulz, A.S., Wagner, D. (eds.) ESA 2014. LNCS, vol. 8737, pp. 37–48. Springer, Heidelberg (2014). https://doi.org/10.1007/978-3-662-44777-2_4
2. Agarwal, P.K., Cheng, S.-W., Yi, K.: Range searching on uncertain data. ACM Trans. Algorithms (TALG) 8(4), 43 (2012)
3. Atallah, M.J., Qi, Y., Yuan, H.: Asymptotically efficient algorithms for skyline probabilities of uncertain data. ACM Trans. Datab. Syst. 32(2), 12 (2011)
4. Bandyopadhyay, D., Snoeyink, J.: Almost-Delaunay simplices: nearest neighbor relations for imprecise points. In: Proceedings of the 15th ACM-SIAM Symposium on Discrete Algorithms, pp. 410–419 (2004)
5. Beardwood, J., Halton, J.H., Hammersley, J.M.: The shortest path through many points. Proc. Cambridge Philos. Soc. 55, 299–327 (1959)
6. Bertsimas, D.J., van Ryzin, G.: An asymptotic determination of the minimum spanning tree and minimum matching constants in geometrical probability. Oper. Res. Lett. 9(4), 223–231 (1990)
7. Cheng, R., Chen, J., Xie, X.: Cleaning uncertain data with quality guarantees. Proc. VLDB Endowment 1(1), 722–735 (2008)
8. Czumaj, A., et al.: Approximating the weight of the euclidean minimum spanning tree in sublinear time. SIAM J. Comput. 35(1), 91–109 (2005)
9. Dong, X., Halevy, A.Y., Yu, C.: Data integration with uncertainty. In: Proceedings of the 33rd International Conference on Very Large Data Bases, pp. 687–698. VLDB Endowment (2007)
10. Dyer, M.: Approximate counting by dynamic programming. In: ACM Symposium on Theory of Computing, pp. 693–699 (2003)
11. Evans, W., Sember, J.: The possible hull of imprecise points. In: Proceedings of the 23rd Canadian Conference on Computational Geometry (2011)
12. Haenggi, M., Andrews, J.G., Baccelli, F., Dousse, O., Franceschetti, M.: Stochastic geometry and random graphs for the analysis and design of wireless networks. IEEE J. Sel. Areas Commun. 27(7), 1029–1046 (2009)
13. Huang, L., Li, J.: Approximating the expected values for combinatorial optimization problems over stochastic points. In: Halldórsson, M.M., Iwama, K., Kobayashi, N., Speckmann, B. (eds.) ICALP 2015. LNCS, vol. 9134, pp. 910–921. Springer, Heidelberg (2015). https://doi.org/10.1007/978-3-662-47672-7_74
14. Huang, L., Li, J., Phillips, J.M., Wang, H.: ε-kernel coresets for stochastic points. arXiv preprint arXiv:1411.0194 (2014)
15. Kamousi, P., Chan, T.M., Suri, S.: Stochastic minimum spanning trees in Euclidean spaces. In: Proceedings of the 27th Annual ACM Symposium on Computational Geometry, pp. 65–74. ACM (2011)
16. Kamousi, P., Chan, T.M., Suri, S.: Closest pair and the post office problem for stochastic points. Comput. Geom. 47(2), 214–223 (2014)
17. Li, J., Deshpande, A.: Ranking continuous probabilistic datasets. Proc. VLDB Endowment 3(1–2), 638–649 (2010)
18. Li, J., Deshpande, A.: Maximizing expected utility for stochastic combinatorial optimization problems. Math. Oper. Res. (2018)
19. Löffler, M., Phillips, J.M.: Shape fitting on point sets with probability distributions. In: Fiat, A., Sanders, P. (eds.) ESA 2009. LNCS, vol. 5757, pp. 313–324. Springer, Heidelberg (2009). https://doi.org/10.1007/978-3-642-04128-0_29

20. Löffler, M., van Kreveld, M.: Approximating largest convex hulls for imprecise points. J. Discrete Algorithms **6**, 583–594 (2008)
21. Loui, R.P.: Optimal paths in graphs with stochastic or multidimensional weights. Commun. ACM **26**(9), 670–676 (1983)
22. Mainwaring, A., Culler, D., Polastre, J., Szewczyk, R., Anderson, J.: Wireless sensor networks for habitat monitoring. In: Proceedings of the 1st ACM International Workshop on Wireless Sensor Networks and Applications, pp. 88–97. ACM (2002)
23. Matoušek, J.: Computing the center of planar point sets. Discrete Comput. Geom. **6**, 221 (1991)
24. Munteanu, A., Sohler, C., Feldman, D.: Smallest enclosing ball for probabilistic data. In: Proceedings of the 30th Annual Symposium on Computational Geometry (2014)
25. Nikolova, E., Brand, M., Karger, D.R.: Optimal route planning under uncertainty. In: ICAPS, vol. 6, pp. 131–141 (2006)
26. Ostrovsky-Berman, Y., Joskowicz, L.: Uncertainty envelopes. In: Abstracts of the 21st European Workshop on Computational Geometry, pp. 175–178 (2005)
27. Pfoser, D., Jensen, C.S.: Capturing the uncertainty of moving-object representations. In: Güting, R.H., Papadias, D., Lochovsky, F. (eds.) SSD 1999. LNCS, vol. 1651, pp. 111–131. Springer, Heidelberg (1999). https://doi.org/10.1007/3-540-48482-5_9
28. Elliott Sigal, C., Pritsker, A.A.B., Solberg, J.J.: The stochastic shortest route problem. Oper. Res. **28**(5), 1122–1129 (1980)
29. van Kreveld, M., Löffler, M.: Largest bounding box, smallest diameter, and related problems on imprecise points. Comput. Geom. Theory Appl. **43**, 419–433 (2010)

Lower Bounds for Small Ramsey
Numbers on Hypergraphs

S. Cliff Liu[✉]

Princeton University, Princeton, NJ 08540, USA
sixuel@princeton.edu

Abstract. The Ramsey number $r_k(p,q)$ is the smallest integer N that satisfies for every red-blue coloring on k-subsets of $[N]$, there exist p integers such that any k-subset of them is red, or q integers such that any k-subset of them is blue. In this paper, we study the lower bounds for small Ramsey numbers on hypergraphs by constructing counter-examples and recurrence relations. We present a new algorithm to prove lower bounds for $r_k(k+1, k+1)$. In particular, our algorithm is able to prove $r_5(6,6) \geq 72$, where there is no lower bound on 5-hypergraphs before this work. We also provide several recurrence relations to calculate lower bounds based on lower bound values on smaller p and q. Combining both of them, we achieve new lower bounds for $r_k(p,q)$ on arbitrary p, q, and $k \geq 4$.

Keywords: Ramsey number · Lower bounds · Hypergraph

1 Introduction

At least how many guests you have to invite for a party to make sure there are either certain number of people know each other or certain number of people do not know each other? The answer is the classical Ramsey number. Ramsey theory generally concerns unavoidable structures in graphs, and has been extensively studied for a long time [4,7,14]. However, determining the exact Ramsey number is a notoriously difficult problem, even for small p and q. For example, it is only known that the value of $r_2(5,5)$ is between 43 to 48 inclusively, and for $r_2(10,10)$, people merely know a much rougher range from 798 to 23556 [10,13,15].

As for the hypergraph case of $k \geq 3$, our understanding of Ramsey number is even less. The only known exact value of Ramsey number is $r_3(4,4) = 13$, with only loose lower bounds for other values of p, q, and k [9,11]. Although some progresses have been made for $r_4(p,q)$, and particularly, lower bound for $r_4(5,5)$ has been continuously pushed forward in the past thirty years, the recurrence relations remain the same, i.e., one can immediately obtain better lower bounds for $p, q \geq 6$ by substituting into improved bound for $r_4(5,5)$, but there is no other way to push them further [12,16].

Another fruitful subject in Ramsey theory is the asymptotic order of Ramsey number. Using the so-called Stepping-up Lemma introduced by Erdős and

© Springer Nature Switzerland AG 2019
D.-Z. Du et al. (Eds.): COCOON 2019, LNCS 11653, pp. 412–424, 2019.
https://doi.org/10.1007/978-3-030-26176-4_34

Hajnal, the Ramsey number $r_k(p, n)$ is lower bounded by the tower function $t_k(c \cdot f(n))$ defined by $t_1(x) = x, t_{i+1}(x) = 2^{t_i(x)}$, where $f(n)$ is some function on n and c is a constant depending on p [3,6]. Recent research improves the orders of $r_4(5, n)$ and $r_4(6, n)$ and leads to similar bounds for $r_k(k + 1, n)$ and $r_k(k + 2, n)$ [2]. We point out that their lower bounds for r_k depends on r_{k-1}. In other words, to get a lower bound for $r_k(p, q)$, one must provide the lower bounds for some $r_{k-1}(p', q')$. More importantly, when focusing on Ramsey numbers on small p, q values, the Stepping-up Lemma cannot be applied directly. We refer readers to Chapter 4.7 in [6] for details.

It is well known that directly improving the lower bounds for Ramsey number is extremely hard, since it requires tremendous computing resources [5]. A possible method to attack this is to use recurrence relations based on the initial values. However, calculating a good initial value itself can be way beyond our reach. For instance, a simple attempt to push the current best lower bound $r_2(6, 6) \geq 102$ could be constructing a CNF (Conjunctive Normal Form) whose satisfying assignment is equivalent to a 6-clique free and 6-independent-set free graph on 102 vertices. This CNF has size (the number of literals in the formula) about 10^{10}, but state-of-the-art SAT solvers are only capable of solving CNF with size no more than 10^6, and is almost sure to not terminate in reasonable time [1,17].

Contributions. We prove several recurrence relations in the form of $r_k(p, q) \geq d \cdot (r_k(p - 1, q) - 1) + 1$, where d depends on p, q, and k. Two of them are for arbitrary integer $k \geq 4$. To the best of our knowledge, this is the first recurrence relation on $r_k(p, q)$ not depending on $r_{k-1}(p, q)$, but for arbitrary k. To build our proof, we introduce a method called *pasting*, which constructs a good coloring by combining colorings on smaller graphs. The recurrence relations are proven by inductions, where several base cases are proven by transforming to an equivalent CNF and solved by a SAT solver. Additionally, to obtain a good initial values of the recurrence relations, a new algorithm for constructing counter-example hypergraphs is proposed, which efficiently proves a series of lower bounds for Ramsey number on k-hypergraphs including $r_5(6, 6) \geq 72$: the first non-trivial result of lower bounds on 5-hypergraphs. The algorithm is based on local search and is easy to implement. Combining both techniques, we significantly improve the lower bounds for $r_4(p, q)$ and achieve new non-trivial lower bounds for $r_k(p, q)$ on arbitrary p, q, and $k \geq 5$.

Roadmap. The rest of this paper is organized as follows. In Sect. 2 we introduce fundamental definitions. The basic forms of recurrence relations are given in Sect. 3. In Sect. 4 we present proofs for the recurrence relations on several small values of k, followed by two recurrence relations on arbitrary k in Sect. 5. Finally, we summarize some of our new lower bounds in Sect. 6. The formal recurrence relations are given in Theorem 1, 2, 3, 4, and 5. Our algorithm for calculating lower bounds for $r_k(k + 1, k + 1)$ is deferred to the full version of this paper.

2 Preliminaries

In this section, basic notations in Ramsey theory are introduced, followed by a sketch of our proof procedure. Then we propose our key definitions and several useful conclusions.

2.1 Notations

A k-uniform hypergraph $G(V, E; k)$ is a tuple of vertex set V and a set E of hyperedges such that each hyperedge in E is a k-subset of V, where each $e \in E$ is called a k-hyperedge. If the context is clear, $G(V, E)$ or G is used instead. A complete k-uniform hypergraph consists of all possible k-subsets of V as its hyperedge set. Since we only deal with k-uniform hypergraphs, we may use k-graph (or graph) and edge for short. Given a vertex set V with $|V| \geq k$, we use $V^{(k)}$ to denote the complete k-uniform hypergraph.

A coloring is a mapping $\chi^{(k)} : E \to \{\text{red}, \text{blue}\}$ that maps all k-hyperedges in E to red or blue. We write $\chi^k(e) = \text{red}$ for coloring some edge $e \in E$ with red under χ^k. Given $G(V, E; k)$, we say $\chi^{(k)}$ is a $(p, q; k)$-coloring of G if there is neither red p-clique nor blue q-clique in G. We also use χ instead of $\chi^{(k)}$ if there is no ambiguity. A p-clique is a complete subgraph induced by p vertices, and a red (resp. blue) p-clique is a clique where all edges are red (resp. blue).

The Ramsey number $r_k(p, q)$ is the minimum integer N that satisfies there is no $(p, q; k)$-coloring for $G(V, E; k)$ on $|V| = N$ vertices. In other words, for any coloring on G, there is either a red p-clique or a blue q-clique.

2.2 A Proof Procedure

We prove recurrence relations in the form of $r_k(p, q) \geq d \cdot (r_k(p - 1, q) - 1) + 1$ by the following procedure Pasting:

1. Given integer d, for each $i \in [d]$, let $G_i(V_i, E_i)$ be a graph on $r_k(p - 1, q) - 1$ vertices with $(p - 1, q; k)$-coloring χ_i.
2. Add an edge for every k-subset of $\bigcup_{i \in [d]} V_i$ if there is no edge on it. Denote the set of added edges as $\mathfrak{E}$. Let the complete graph after adding all edges be $\mathbf{G}(\bigcup_{i \in [d]} V_i, (\bigcup_{i \in [d]} E_i) \bigcup \mathfrak{E})$.
3. Construct χ' on $\mathfrak{E}$ such that $\chi := (\bigcup_{i \in [d]} \chi_i) \bigcup \chi'$ on $\mathbf{G}$ satisfies that each p-clique of $\bigcup_{i \in [d]} V_i$ contains a blue edge and each q-clique of it contains a red edge.
4. It can be concluded that $r_k(p, q) \geq d \cdot (r_k(p - 1, q) - 1) + 1$ since χ is a $(p, q; k)$-coloring for $\mathbf{G}$ and $\left| \bigcup_{i \in [d]} V_i \right| = d \cdot (r_k(p - 1, q) - 1)$.

The non-trivial step in Pasting is Step 3 (*coloring construction*), which will be discussed in details in Sects. 4 and 5. Pasting(k, p, q, d) is *successful* if $\chi = (\bigcup_{i \in [d]} \chi_i) \bigcup \chi'$ can be found.

2.3   Primal Cardinality Vector

Observe that the coloring construction cannot depend on the order of G_i dues to symmetry, thus a primal order shall be fixed and our coloring depends only on the sequence of cardinalities of the intersections in non-increasing order. We introduce the following concepts concerning this.

Let $V_1, V_2, \ldots, V_d$ be d disjoint sets each with cardinality $r_k(p-1,q)-1$, and let V be $\bigcup_{i \in [d]} V_i$. For any σ-subset $X \subseteq V$, define cardinality vector $\hat{\mathbf{v}}(X) = (\hat{v}_1, \hat{v}_2, \ldots, \hat{v}_d)$ where $\hat{v}_i = |X \cap V_i|$. Let $\hat{v}_{(1)}, \hat{v}_{(2)}, \ldots, \hat{v}_{(d)}$ be the sequence after sorting the $\hat{v}_i$'s in a non-increasing order.

Definition 1. *Given V, X, and $\{\hat{v}_{(i)} \mid i \in [d]\}$ as above, define* primal cardinality vector *of X as* $\mathbf{v}(X) = (v_1, v_2, \ldots, v_{\pi(X)})$, *where* $v_i = \hat{v}_{(i)}$ *for all $i \in [\pi(X)]$, and $\pi(X)$ satisfies either (i) $\pi(X) = d$ or (ii) $\hat{v}_{(\pi(X))} > 0$ and $\hat{v}_{(\pi(X)+1)} = 0$.*

In a word, $\mathbf{v}(X)$ is a sequence of all positive coordinates of the cardinality vector $\hat{\mathbf{v}}(X)$ in a non-increasing order. Observe that when $\sigma = |X| = k$, X corresponds to some edge $e(X)$ in $\mathbf{G}$, and $\mathbf{v}(X) = (v_1, v_2, \ldots, v_{\pi(X)})$ essentially means that $e(X)$ has v_i endpoints in the i-th subgraph (in a non-increasing order of the cardinalities of intersections). Usually primal cardinality vector $\mathbf{v}$ shows up without indicating which set X it corresponds to, and we refer $\pi(\mathbf{v})$ to the length of $\mathbf{v}$.

The following remark captures the idea we proposed at the beginning of this subsection.

Remark 1. In Step 3 of Pasting, $\forall e_1, e_2 \in \mathfrak{E}$, $\chi'(e_1) = \chi'(e_2)$ if $\mathbf{v}(e_1) = \mathbf{v}(e_2)$.

We will write $\mathbf{v}(e)$ instead of $\mathbf{v}(X)$ when X corresponds to edge e. In this case, abusing the notation slightly, we write $\chi(\mathbf{v}(e))$ as the color under χ on edge e, since all edges with the same primal cardinality vector $\mathbf{v}$ are in same color. Furthermore, we write $\chi(\mathbf{v}) = c$ where c is red or blue for assigning all edges with primal cardinality vector $\mathbf{v}$ to color c. For any $i \in [\pi(X)]$, $v_i(X)$ is the i-th coordinate of $\mathbf{v}(X)$.

Remark 2. For any non-trivial σ-subset X, $\forall \tau$-subset $Y \subseteq X$, it must be that: (i) $\sum_{i \in [\pi(X)]} v_i(X) = \sigma > 0$, (ii) $\sum_{i \in [\pi(Y)]} v_i(Y) = \tau \leq \sigma$, (iii) $\pi(X) \geq \pi(Y)$, and (iv) $\forall i \in [\pi(Y)]$, $v_i(X) \geq v_i(Y)$.

Proof. The first three bullets are simple cardinality properties. To show that property (iv) holds, let j be the smallest index with $v_j(Y) > v_j(X)$. If $j = 1$, there is no way to fit the largest subset of Y into any subset of X. Else if $j > 1$, the only way to fit Y_j into a subset of X is to swap it with some Y_i ($i < j$), but $v_i(Y) \geq v_j(Y) > v_i(X)$, then Y_i cannot fit into the i-th subset of X. □

Definition 2. *Given two primal cardinality vectors $\mathbf{v_1}$ and $\mathbf{v_2}$, define* partial order *between them as: $\mathbf{v_1} \leq_c \mathbf{v_2}$ if and only if (i) $\pi(\mathbf{v_1}) \leq \pi(\mathbf{v_2})$ and (ii) $\forall i \in [\pi(\mathbf{v_1})]$, $\mathbf{v_1}_i \leq \mathbf{v_2}_i$. If equalities in (i) and (ii) do not hold at the same time, then $\mathbf{v_1} <_c \mathbf{v_2}$.*[1]

---

[1] $\mathbf{v_2} \geq_c \mathbf{v_1}$ reads "$\mathbf{v_2}$ contains v_1".

One can easily show that *reflexivity*, *antisymmetry* and *transitivity* for any partial order hold for $\leq_c$. Under this definition, with Remark 2 and subsets enumeration we can immediately conclude the following:

Corollary 1. *Given* $V = \bigcup_{i \in [d]} V_i$, $G = V^{(k)}$ *and* $X \subseteq V$, *we have* $\forall Y \subseteq X$, $\mathbf{v}(Y) \leq_c \mathbf{v}(X)$, *and* $\forall \mathbf{v}' \leq_c \mathbf{v}(X)$, $\exists Y \subseteq X$ *with* $\mathbf{v}(Y) = \mathbf{v}'$. *Specifically,* Y *corresponds to an edge* $e(Y)$ *of* G *when* $\sum_{i \in [\pi(\mathbf{v}')]} \mathbf{v}'_i = k$, *and* $e(Y)$ *is an edge of* $X^{(k)}$.

Given any subset of V, observe that there are at most d different subsets to be intersected with. As a result, we only concern subsets with primal cardinality vectors in the following set:

Definition 3. *Define* $\mathbf{V_s}(d)$ *as the set of all primal cardinality vectors* $\mathbf{v}$ *such that* $\pi(\mathbf{v}) \leq d$ *and* $\sum_{i \in [\pi(\mathbf{v})]} \mathbf{v}_i = s$.

Remark 3. $\forall d \geq s, \mathbf{V_s}(d) = \mathbf{V_s}(s)$.

Based on Corollary 1, we conclude this section with the following corollary:

Corollary 2. *Given integers* p, q, k, d, $V = \bigcup_{i \in [d]} V_i$, *and* $G = V^{(k)}$, *the following four statements are equivalent:*

1. $\exists \chi$ *such that* $\forall \mathbf{v} \in \mathbf{V_p}(d)$ *(resp.* $\forall \mathbf{v} \in \mathbf{V_q}(d)$*),* $\exists \mathbf{v}' \in \mathbf{V_k}(d)$ *such that* $\mathbf{v}' \leq_c \mathbf{v}$ *and* $\chi(\mathbf{v}') = $ *blue (resp. red).*
2. $\exists \chi$ *such that* $\forall p$*-subset (resp.* q*-subset)* $X \subseteq V$, $\exists k$*-hyperedge* e *of* $X^{(k)}$ *such that* $\chi(e) = $ *blue (resp. red).*
3. Pasting(k, p, q, d) *is successful.*
4. $r_k(p, q) \geq d \cdot (r_k(p-1, q) - 1) + 1$.

3   Forms of Recurrences

We prove $r_k(p, q) \geq d \cdot (r_k(p-1, q) - 1) + 1$ for three different forms of d under different conditions: (1) $d = 2$, (2) $d = p - 1$, and (3) $d = \lfloor \frac{q-1}{k-2} \rfloor$. Form (3) requires the strongest condition but its proof turns out to be simpler. For forms (1) and (2), we show that to prove recurrence relation on given k and arbitrary p, q, it is sufficient to prove the base case on p and q, i.e., prove the case on $p = p_0, q = q_0$ for some constants p_0, q_0.

Firstly we show that for a given integer d, if $r_k(p_0, q_0) \geq d \cdot (r_k(p_0-1, q_0) - 1) + 1$ is given by Pasting, then $r_k(p, q) \geq d \cdot (r_k(p-1, q) - 1) + 1$.

Lemma 1. *Given integer* d, *if* Pasting(k, p_0, q_0, d) *is successful, then* $\forall p \geq p_0, q \geq q_0$, Pasting$(k, p, q, d)$ *is successful, which is* $r_k(p, q) \geq d \cdot (r_k(p-1, q) - 1) + 1$.

Proof. The proof relies on Corollary 2. Let χ_0 be a $(p_0, q_0; k)$-coloring fed to Pasting. We have $\forall \mathbf{v} \in \mathbf{V_{p_0}}(d)$, $\exists \mathbf{v}' \in \mathbf{V_k}(d)$ such that $\mathbf{v}' \leq_c \mathbf{v}$ and $\chi_0(\mathbf{v}') = $ blue. Meanwhile, by Corollary 1 we know that $\forall p \geq p_0, \forall \mathbf{u} \in \mathbf{V_p}(d)$, $\exists \mathbf{v} \in$

$\mathbf{V_{p_0}}(d)$ such that $\mathbf{v} \leq_c \mathbf{u}$. By transitivity it must be that $\mathbf{v}' \leq_c \mathbf{u}$. This means $\forall p \geq p_0, \forall \mathbf{u} \in \mathbf{V_p}(d)$, $\exists \mathbf{v}' \in \mathbf{V_k}(d)$ such that $\mathbf{v}' \leq_c \mathbf{u}$ and $\chi_0(\mathbf{v}') =$ blue. Using the same reasoning we get $\forall q \geq q_0, \forall \mathbf{u} \in \mathbf{V_q}(d)$, $\exists \mathbf{v}' \in \mathbf{V_k}(d)$ such that $\mathbf{v}' \leq_c \mathbf{v}$ and $\chi_0(\mathbf{v}') =$ red, and the conclusion follows. □

Secondly we give the following lemma showing that the induction from the base case to arbitrary p, q also holds for form (2).

Lemma 2. *Given* $p_0 \geq q_0 + 1$, *if* $\mathsf{Pasting}(k, p_0, q_0, p_0 - 1)$ *is successful, then* $\forall p \geq p_0, q \geq q_0$, $\mathsf{Pasting}(k, p, q, p - 1)$ *is successful, which is* $r_k(p, q) \geq (p - 1) \cdot (r_k(p - 1, q) - 1) + 1$.

We give the sketch of the proof here, followed by two lemmas to integrate the formal proof.

Proof (Proof sketch of Lemma 2). The proof contains two parts. First, we need to show that $\mathsf{Pasting}(k, p_0, q_0, p_0 - 1)$ is successful implies that $\forall p \geq p_0$, $\mathsf{Pasting}(k, p, q_0, p - 1)$ is successful. Then we prove that for arbitrary fixed p, $\forall q \geq q_0$, $\mathsf{Pasting}(k, p, q, p - 1)$ is successful. Combining both of these we can conclude the proof. □

Lemma 3. *Given* $p_0 \geq q_0 + 1$, *if* $\mathsf{Pasting}(k, p_0, q_0, p_0 - 1)$ *is successful, then* $\forall p \geq p_0$, $\mathsf{Pasting}(k, p, q_0, p - 1)$ *is successful.*

Proof. By Corollary 2, if $\mathsf{Pasting}(k, p_0, q_0, p_0 - 1)$ is successful, we have that $\exists \chi_0$ such that the following two statements hold:

$$\forall \mathbf{v} \in \mathbf{V_{p_0}}(p_0 - 1) \; \exists \mathbf{v}' \in \mathbf{V_k}(p_0 - 1), \; \mathbf{v}' \leq_c \mathbf{v} \wedge \chi_0(\mathbf{v}') = \text{blue.} \tag{1}$$

$$\forall \mathbf{v} \in \mathbf{V_{q_0}}(p_0 - 1) \; \exists \mathbf{v}' \in \mathbf{V_k}(p_0 - 1), \; \mathbf{v}' \leq_c \mathbf{v} \wedge \chi_0(\mathbf{v}') = \text{red.} \tag{2}$$

By induction on p, it remains to prove the inductive step: $\mathsf{Pasting}(k, p_0 + 1, q_0, p_0)$ is successful, which is equivalent to that $\exists \chi_1$ such that the following two statements hold:

$$\forall \mathbf{v} \in \mathbf{V_{p_0+1}}(p_0) \; \exists \mathbf{v}' \in \mathbf{V_k}(p_0), \; \mathbf{v}' \leq_c \mathbf{v} \wedge \chi_1(\mathbf{v}') = \text{blue.} \tag{3}$$

$$\forall \mathbf{v} \in \mathbf{V_{q_0}}(p_0) \; \exists \mathbf{v}' \in \mathbf{V_k}(p_0), \; \mathbf{v}' \leq_c \mathbf{v} \wedge \chi_1(\mathbf{v}') = \text{red.} \tag{4}$$

We prove that any χ_0 satisfies (1) and (2) also satisfies (3) and (4). First we prove that (1) implies (3), then we prove that (2) implies (4). Noticing that $p_0, q_0 \geq k + 1$, otherwise the hypergraph is trivial. So by Remark 3 we know that $\mathbf{V_k}(p_0 - 1) = \mathbf{V_k}(p_0) = \mathbf{V_k}(k)$.

For the first implication, by Definition 3, $\mathbf{V_{p_0+1}}(p_0) = \mathbf{V_{p_0+1}}(p_0 - 1) \bigcup \mathbf{V}'$ where $\mathbf{V}' = \{\mathbf{v} \mid \pi(\mathbf{v}) = p_0, \sum_{i \in [\pi(\mathbf{v})]} v_i = p_0 + 1\} = \mathbf{1}^{p_0} + \mathbf{e_1}$.[2] Thus $\forall \mathbf{v} \in \mathbf{V_{p_0+1}}(p_0)$, there are two cases: (i) if $\mathbf{v} \in \mathbf{V_{p_0+1}}(p_0 - 1)$, then let $\mathbf{u} := \mathbf{v} - \mathbf{e_{\pi(\mathbf{v})}}$; (ii) else if $\mathbf{v} = \mathbf{1}^{p_0} + \mathbf{e_1}$, let $\mathbf{u} := \mathbf{1}^{p_0 - 1} + \mathbf{e_1}$. In either case we have $\mathbf{u} \in$

---

[2] Conventionally, $\mathbf{1}^n$ is a vector of length n with all coordinates being 1; $\mathbf{e_i}$ is a vector with the i-th coordinate being 1 and others being 0.

$\mathbf{V}_{\mathbf{p_0}}(p_0 - 1)$ and $\mathbf{u} \leq_c \mathbf{v}$. Also, by (1) we know that $\exists \mathbf{v}' \in \mathbf{V}_{\mathbf{k}}(k)$ such that $\mathbf{v}' \leq_c \mathbf{u}$ and $\chi_0(\mathbf{v}') = $ blue, so by transitivity $\mathbf{v}' \leq_c \mathbf{u} \leq_c \mathbf{v}$, we have that χ_0 satisfies (3). For the second implication, since $p_0 \geq q_0 + 1$, by Remark 3 we have $\mathbf{V}_{\mathbf{q_0}}(p_0) = \mathbf{V}_{\mathbf{q_0}}(p_0 - 1) = \mathbf{V}_{\mathbf{q_0}}(q_0)$, then (2) is equivalent to (4).

As a result, χ_1 satisfies (3) and (4), by which we finish the induction and conclude the proof.   □

Lemma 4. *Given integers p, q_0, if* Pasting$(k, p, q_0, p-1)$ *is successful, then $\forall q \geq q_0$,* Pasting$(k, p, q, p - 1)$ *is successful.*

Proof. Since p is fixed, by Lemma 1 with $d = p - 1$, we have that Pasting$(k, p', q_0, p-1)$ is successful for any $p' \geq p, q \geq q_0$. In particular, the conclusion holds for $p' = p$ and any $q \geq q_0$ □

By the proof sketch of Lemma 2, with Lemmas 3 and 4 we finish the proof of Lemma 2.

4   Recurrences for Small k

In this section, we give our main results on recurrence relations for small k, followed by their proofs and the relation to the satisfiability problem.

4.1   Main Results on Small k

Theorem 1. *For any integer $p \geq 6$ and $q \geq 5$, $r_4(p, q) \geq 2r_4(p-1, q) - 1$ holds. Furthermore, if $q \geq 7$ then $r_4(p, q) \geq (p - 1) \cdot (r_4(p - 1, q) - 1) + 1$ holds.*

Theorem 2. *There exists a constant $c \geq 25$, such that given integer $k \geq 5$ and $k \leq c$, for any integer $p \geq k+2$ and $q \geq k+2$, $r_k(p, q) \geq (p-1) \cdot (r_k(p-1, q)-1)+1$ holds.*

Theorem 3. *There exists a constant $c \geq 25$, such that given integer $k \neq 9$ and $8 \leq k \leq c$, for any integer $p \geq k + 2$ and $q \geq k + 1$, $r_k(p, q) \geq (p - 1) \cdot (r_k(p - 1, q) - 1) + 1$ holds.*

The difference between Theorems 2 and 3 is the base cases of q, which are $k + 2$ and $k + 1$ respectively. Note that the right-hand side of the recurrence relation in Theorem 3 on initial values is $r_k(k + 1, k + 1)$: the first non-trivial Ramsey number on k-hypergraphs.

4.2   Proof Sketch

Before proving the above theorems, we take a detour to revisit Corollary 2. We show that Statement 1 in Corollary 2 can be interpreted in a slightly different way.

Lemma 5. *Define* $\mathbf{P_p}(d) = \{\mathbf{v} \mid \sum_{i \in [\pi(v)]} v_i = p, \pi(\mathbf{v}) \leq d, v_1 \leq p - 2\}$. *Define* $\mathbf{Q_q}(d) = \{\mathbf{v} \mid \sum_{i \in [\pi(v)]} v_i = q, \pi(\mathbf{v}) \leq d, v_1 \leq q - 1\}$. *Given integers* p, q, k, d, $V = \bigcup_{i \in [d]} V_i$, *and* $\mathbf{G} = V^{(k)}$ *as before, the following two statements are equivalent:*

1. $\exists \chi$ *such that* $\forall \mathbf{v} \in \mathbf{P_p}(d)$, $\exists \mathbf{v'} \in \mathbf{V_k}(d)$ *such that* $\mathbf{v'} \leq_c \mathbf{v}$ *and* $\chi(\mathbf{v'}) = blue$. *Moreover,* χ *also satisfies that* $\forall \mathbf{v} \in \mathbf{Q_q}(d)$, $\exists \mathbf{v'} \in \mathbf{V_k}(d)$ *such that* $\mathbf{v'} \leq_c \mathbf{v}$ *and* $\chi(\mathbf{v'}) = red$.
2. $\exists \chi$ *such that* $\forall p$-subset (resp. q-subset) $X \subseteq V$, $\exists k$-hyperedge e of $X^{(k)}$ such that $\chi(e) = blue$ (resp. red).

Proof. Given p-subset $X \subseteq V$, if $\exists i \in [d]$ such that $|X \wedge V_i| \geq p - 1$, $X^{(k)}$ must contain a blue edge, because χ_i is a $(p - 1, q; k)$-coloring on $V_i^{(k)}$ and $X^{(k)}$ contains some $(p - 1)$-clique, which cannot be a red clique. Analogously, any q-subset Y intersecting with any V_i on more than $q-1$ vertices necessarily contains a red edge, because any $V_i^{(k)}$ has a $(p, q; k)$-coloring. □

This lemma enables us to consider only a proper subset of the previous primal cardinality vector set, leading to a simpler proof of our theorems. We give a simple proof of Theorem 1, and we prove Theorems 2 and 3 in the next subsection.

Proof (Proof of Theorem 1). Firstly we prove that for any integer $p \geq 6$ and $q \geq 5$, $r_4(p, q) \geq 2r_4(p - 1, q) - 1$. By Lemma 1, it is sufficient to prove that $r_4(6, 5) \geq 2(r_4(5, 5) - 1) + 1$. We give a $(6, 5; 4)$-coloring as follows:

$$\chi_1^{(4)} = \{\chi(3, 1) = red, \chi(2, 2) = blue\}.$$

To prove $\chi_1^{(4)}$ is a $(6, 5; 4)$-coloring, by Lemma 5, we need to check the following:

- $\forall \mathbf{v} \in \mathbf{P_6}(2)$, $\exists \mathbf{v'} \leq_c \mathbf{v}$, such that $\chi_1^{(4)}(\mathbf{v'}) = $ blue. This is true because $\mathbf{P_6}(2) = \{(4, 2), (3, 3)\}$, both $\geq_c (2, 2)$.
- $\forall \mathbf{v} \in \mathbf{Q_5}(2)$, $\exists \mathbf{v'} \leq_c \mathbf{v}$, such that $\chi_1^{(4)}(\mathbf{v'}) = $ red. Since $\mathbf{Q_5}(2) = \{(4, 1), (3, 2)\}$, each of them $\geq_c (3, 1)$.

Thus we proved $r_4(6, 5) \geq 2r_4(5, 5) - 1$.

Now we need to prove that for any integer $p \geq 6$ and $q \geq 7$, $r_4(p, q) \geq (p - 1)(r_4(p - 1, q) - 1) + 1$ by starting with proving the case of $p = 6, q = 7$. We give a $(6, 7; 4)$-coloring as following:

$$\chi_2^{(4)} = \{\chi(3, 1) = \chi(1, 1, 1, 1) = red\} \cup \{\chi(2, 2) = \chi(2, 1, 1) = blue\}.$$

The following needs to be checked:

- $\forall \mathbf{v} \in \mathbf{P_6}(5)$, we have $2 \leq v_1 \leq 4$, thus either $(2, 2) \leq_c \mathbf{v}$ or $(2, 1, 1) \leq_c \mathbf{v}$, which are blue.
- $\forall \mathbf{v} \in \mathbf{Q_7}(5)$, it must be that either $v_1 \geq 3$ and $2 \leq \pi(\mathbf{v}) \leq 3$ or $\pi(\mathbf{v}) \geq 4$. The first case $\geq_c (3, 1)$ and the second case $\geq_c (1, 1, 1, 1)$, which are both red.

By the same reasoning, one can show that $\chi_2^{(4)}$ is also a $(7, 7; 4)$-coloring and an $(8, 7; 4)$-coloring. Since now the recurrence relation holds for $p = 8, q = 7$, we can apply Lemma 2 to get $\forall p \geq 8, q \geq 7, r_4(p, q) \geq (p-1) \cdot (r_4(p-1, q) - 1) + 1$. Combining all these cases we proved the theorem.   □

4.3   Automated Theorem Proving

The "$\exists \forall$" structure of Statement 1 in Lemma 5 reminds us of Propositional Logic Satisfiablity (SAT). In fact, a $(p, q; k)$-coloring χ serves as a certificate of the proof for theorem $r_k(p, q) \geq d \cdot (r_k(p-1, q) - 1) + 1$. Thus it is nature to use automated theorem proving instead of proving it by hand. As we saw in the proof of Theorem 1, even the simplest case is time-consuming to verify, regardless of how to find that coloring.

Definition 4. *A Conjunctive Normal Form (CNF) is a conjunction of clauses, such that each clause is a disjunction of literals, where a literal can be positive of negative variable. A satisfying assignment of CNF is a mapping from all variables to true or false such that every clause has at least one true literal. A SAT solver takes a CNF as input and outputs a satisfying assignment or UNSAT if the CNF is unsatisfiable.*

We give the procedure to prove $r_k(p, q) \geq d \cdot (r_k(p-1, q) - 1) + 1$ for fixed p, q, then Lemmas 1 and 2 can be applied to prove it for arbitrary p, q:

1. For every $\mathbf{v} \in \mathbf{P_p}(d)$, construct a clause $C_p(\mathbf{v})$ as follow: For every $\mathbf{u} \in \mathbf{V_k}(d)$, if $\mathbf{u} \leq_c \mathbf{v}$, add a positive variable $x(\mathbf{u})$ in $C_p(\mathbf{v})$.
2. For every $\mathbf{v} \in \mathbf{Q_q}(d)$, construct a clause $C_q(\mathbf{v})$ as follow: For every $\mathbf{u} \in \mathbf{V_k}(d)$, if $\mathbf{u} \leq_c \mathbf{v}$, add a negative variable $\neg x(\mathbf{u})$ in $C_q(\mathbf{v})$.
3. Use SAT solver to solve the constructed CNF:

$$F = \left(\bigcup_{\mathbf{v} \in \mathbf{P_p}(d)} C_p(\mathbf{v}) \right) \bigcup \left(\bigcup_{\mathbf{v} \in \mathbf{Q_q}(d)} C_q(\mathbf{v}) \right).$$

4. If a satisfying assignment α is found, we construct a $(p, q; k)$-coloring χ as follows: if $\alpha(x(\mathbf{u})) =$ true, set $\chi(\mathbf{u}) :=$ blue; if $\alpha(x(\mathbf{u})) =$ false, set $\chi(\mathbf{u}) :=$ red.

It is easy to show that this procedure is a correct proof when SAT solver returns a satisfying assignment: $\forall \mathbf{v} \in \mathbf{P_p}(d), \exists \mathbf{u} \in \mathbf{V_k}(d)$ such that $\mathbf{u} \leq_c \mathbf{v}$ and $\chi(\mathbf{u}) =$ blue, because $\exists x(\mathbf{u}) \in C_p(\mathbf{v})$ such that $x(\mathbf{u}) =$ true; similarly, $\forall \mathbf{v} \in \mathbf{Q_q}(d), \exists \mathbf{u} \in \mathbf{V_k}(d)$ such that $\mathbf{u} \leq_c \mathbf{v}$ and $\chi(\mathbf{u}) =$ red, because $\exists x(\mathbf{u}) \in C_q(\mathbf{v})$ such that $x(\mathbf{u}) =$ false. So by Lemma 5 we proved the recurrence relation holds for p and q.

Proof (Proof of Theorem 2 and Theorem 3). We use the latest version of SAT solver from [8] to solve the following two kinds of CNFs:

$$F_1 = \left(\bigcup_{\mathbf{v} \in \mathbf{P_{k+2}}(k+1)} C_p(\mathbf{v}) \right) \bigcup \left(\bigcup_{\mathbf{v} \in \mathbf{Q_{k+2}}(k+1)} C_q(\mathbf{v}) \right).$$

$$F_2 = \left(\bigcup_{\mathbf{v} \in \mathbf{P_{k+3}}(k+2)} C_p(\mathbf{v}) \right) \bigcup \left(\bigcup_{\mathbf{v} \in \mathbf{Q_{k+2}}(k+2)} C_q(\mathbf{v}) \right).$$

Our SAT solver returns satisfying assignments on all $5 \le k \le 25$. The satisfying assignment of F_1 is a proof for the recurrence relation of case $p = k+2, q = k + 2$. While that of F_2 is a proof for the case $p = k + 3, q = k + 2$. Therefore, by Lemma 2 we proved Theorem 2.

We do the same for the CNF corresponding to $p = k + 2, q = k + 1$ on all $8 \le k \le 25$, and get satisfying assignments on all k except for $k = 9$ returning UNSAT, thus (with Lemma 2) proved Theorem 3.   □

Given more time on constructing more CNFs on larger k, it is almost sure that lower bound for c in Theorem 3 can be improved. As a result, we give the following conjecture as the c-unbounded version of Theorem 3.

Conjecture 1. Given integer $k \ge 10$, for any integer $p \ge k + 2$ and $q \ge k + 1$, $r_k(p, q) \ge (p - 1) \cdot (r_k(p - 1, q) - 1) + 1$.

5 Recurrences for Arbitrary k

In this section, we give two recurrence relations for arbitrary k. The recurrence forms align with forms (2) and (3) in Sect. 3.

Theorem 4. *Given even integer $k \ge 4$, for any integers $p \ge k + 2, q \ge k + 1$, $r_k(p, q) \ge 2 \cdot (r_k(p - 1, q) - 1) + 1$ holds. Given odd integer $k \ge 5$, for any integers $p \ge k + 2, q \ge k + 2$, the same recurrence relation holds.*

The proof of Theorem 4 can be found in the full version of this paper.

Theorem 5. *Given any integer $k \ge 4$, for any integers $p \ge k + 2, q \ge k + 1$, $r_k(p, q) \ge d \cdot (r_k(p - 1, q) - 1) + 1$ holds, where $d = \lfloor \frac{q-1}{k-2} \rfloor$.*

Proof. If $d \le 2$, this is implied by Theorem 4. Now assume $d \ge 3$. Define coloring as follows: $\chi^{(k)}(\mathbf{v}) = $ red if and only if $\mathbf{v} = (k - 1, 1)$; otherwise $\chi^{(k)}(\mathbf{v}) = $ blue. We show that under such $\chi^{(k)}$, $\forall \mathbf{v} \in \mathbf{P_p}(d)$, $\exists \mathbf{v'} \in \mathbf{V_k}(d)$ such that $\mathbf{v'} \le_c \mathbf{v}$ and $\chi^{(k)}(\mathbf{v'}) = $ blue; and $\forall \mathbf{v} \in \mathbf{Q_q}(d)$, $\exists \mathbf{v'} \in \mathbf{V_k}(d)$ such that $\mathbf{v'} \le_c \mathbf{v}$ and $\chi^{(k)}(\mathbf{v'}) = $ red.

Firstly, for any $\mathbf{v} \in \mathbf{P_p}(d)$, there are two cases. The first case is that if $\pi(\mathbf{v}) \ge 3$, then $\exists \mathbf{u} \in \mathbf{V_k}(d)$, such that $\pi(\mathbf{u}) = \pi(\mathbf{v})$ and $\mathbf{v} \ge_c \mathbf{u}$, so $\chi^{(k)} = $ blue. The existence of such $\mathbf{u}$ can be proven by induction on p: $\forall \mathbf{v} \in \mathbf{P_p}(d)$,

$\exists \mathbf{v}' \in \mathbf{P_{p-1}}(d)$ such that $\mathbf{v} \geq_c \mathbf{v}'$ and $\pi(\mathbf{v}') = \pi(\mathbf{v})$, because either $v_{\pi(\mathbf{v})} \geq 2$ or $\exists i \in [\pi(\mathbf{v}) - 1], v_i > v_{i+1}$ (since $\mathbf{v} \neq \mathbf{1}^k$). For the first one we set $\mathbf{v}' := \mathbf{v} - \mathbf{e}_{\pi(\mathbf{v})}$, and set $\mathbf{v}' := \mathbf{v} - \mathbf{e_i}$ for the second one. We do this until p reaches k, and the conclusion followed by transitivity. The second case is that $\pi(\mathbf{v}) = 2$. This is straightforward since $v_1 \leq p - 2$ (Lemma 5), it must be $v_2 \geq 2$. Just let $u_2 = \max(v_2 - \lceil \frac{p-k}{2} \rceil, 2)$ and $u_1 = k - u_2$, we have $u_1 \geq u_2 \geq 2$ and $\chi^{(k)}(\mathbf{u}) =$ blue by definition.

Secondly, $\forall \mathbf{v} \in \mathbf{Q_q}(d)$, by the Pigeonhole principle, it must be $v_1 \geq \lfloor \frac{q-1}{d} \rfloor + 1 = k - 1$. Additionally, by Lemma 5 we have $v_1 \leq q - 1$, thus $v_2 \geq 1$, so $(v_1, v_2) \geq_c (k - 1, 1)$. Since $\chi^{(k)}(k - 1, 1) =$ red by definition, we proved that $\forall \mathbf{v} \in \mathbf{Q_q}(d), \exists u \in \mathbf{V_k}(d)$, such that $\mathbf{v} \geq_c \mathbf{u}$ and $\chi^{(k)}(\mathbf{u}) =$ red.

Combining both we proved the theorem. $\square$

6 Improved Lower Bounds

We summarize some of our improved lower bounds for Ramsey numbers on hypergraphs in this section.

6.1 4-hypergraph

Previous best lower bounds for Ramsey number on 4-hypergraphs can be found in [16] and [11]. We point out that some of their values are based on [12] whose calculation of $r_4(7, 7)$ is wrong. The following lower bounds values in the "Previous" column are re-calculated in a corrected way using their methods. We also add a "Reference" column for the method we use to derive our results. Some representative results are displayed below.

	Previous	Our Result	Reference
$r_4(5,6) \geq$	37	67	Theorem 1 or 4
$r_4(6,6) \geq$	73	133	Theorem 4 or 5
$r_4(6,7) \geq$	361	661	Theorem 2
$r_4(6,13) \geq$	23041	50689	Theorem 5
$r_4(7,7) \geq$	2161	3961	Theorem 2
$r_4(8,8) \geq$	105841	194041	Theorem 2

Using our constructive algorithm (see the full version of this paper), a coloring for proving $r_4(5,5) \geq 34$ can be found. The subsequent lower bounds can be obtained using the corresponding recurrence relations.

6.2 5-hypergraph

Before this work, there is no constructive lower bounds for Ramsey numbers on 5-hypergraphs.

Using our constructive algorithm, a coloring for proving $r_5(6,6) \geq 72$ can be found, which serves as a certificate of the lower bound. Subsequently, lower bounds for $r_5(p,q)$ can be calculated using our Theorems 3 and 5.

6.3 $\geq$ 6-hypergraph

Previously, there is neither constructive nor recursive lower bounds for Ramsey number on $\geq$ 6-hypergraphs.

The base case of the recurrence relation is $r_k(k+1, k+1) \geq r_k(k+1, k) = k+1$. For any $k \geq 6$, lower bounds for $r_k(p,q)$ can be calculated using our Theorems 2, 3, 4 and 5.

Acknowledgments. The author wants to thank the anonymous reviewers for their valuable comments. Research at Princeton University partially supported by an innovation research grant from Princeton and a gift from Microsoft.

References

1. Biere, A., Heule, M., van Maaren, H., Walsh, T. (eds.): Handbook of Satisfiability, Frontiers in Artificial Intelligence and Applications, vol. 185. IOS Press, Amsterdam (2009)
2. Conlon, D., Fox, J., Sudakov, B.: Hypergraph ramsey numbers. J. Am. Math. Soc. **23**(1), 247–266 (2010)
3. Erdős, P., Hajnal, A., Rado, R.: Partition relations for cardinal numbers. Acta Mathematica Hungarica **16**(1–2), 93–196 (1965)
4. Erdös, P., Rado, R.: A partition calculus in set theory. Bull. Am. Math. Soc. **62**(5), 427–489 (1956)
5. Gaitan, F., Clark, L.: Ramsey numbers and adiabatic quantum computing. Phys. Rev. Lett. **108**(1), 010501 (2012)
6. Graham, R.L., Rothschild, B.L., Spencer, J.H.: Ramsey Theory, vol. 20. Wiley, New York (1990)
7. Heule, M.J.H., Kullmann, O., Marek, V.W.: Solving and verifying the boolean pythagorean triples problem via cube-and-conquer. In: Creignou, N., Le Berre, D. (eds.) SAT 2016. LNCS, vol. 9710, pp. 228–245. Springer, Cham (2016). https://doi.org/10.1007/978-3-319-40970-2_15
8. Liu, S., Papakonstantinou, P.A.: Local search for hard SAT formulas: the strength of the polynomial law. In: Proceedings of the Thirtieth AAAI Conference on Artificial Intelligence, pp. 732–738 (2016)
9. McKay, B.D., Radziszowski, S.P.: The first classical ramsey number for hypergraphs is computed. In: Proceedings of the Second Annual ACM/SIGACT-SIAM Symposium on Discrete Algorithms, vol. 1991, pp. 304–308 (1991)
10. McKay, B.D., Radziszowski, S.P.: Subgraph counting identities and ramsey numbers. J. Comb. Theory Ser. B **69**(2), 193–209 (1997)
11. Radziszowski, S.P., et al.: Small ramsey numbers. Electron. J. Combin. **1**(7) (1994)

12. Shastri, A.: Lower bounds for bi-colored quaternary ramsey numbers. Discrete Math. **84**(2), 213–216 (1990)
13. Shearer, J.B.: Lower bounds for small diagonal ramsey numbers. J. Comb. Theory Ser. A **42**(2), 302–304 (1986)
14. Shelah, S.: Primitive recursive bounds for van der waerden numbers. J. Am. Math. Soc. **1**(3), 683–697 (1988)
15. Shi, L.: Upper bounds for ramsey numbers. Discrete Math. **270**(1–3), 250–264 (2003)
16. Song, E., Ye, W., Liu, Y.: New lower bounds for ramsey number R (p, q; 4). Discrete Math. **145**(1–3), 343–346 (1995)
17. Tompkins, D.A.D., Hoos, H.H.: UBCSAT: an implementation and experimentation environment for SLS algorithms for SAT and MAX-SAT. In: Hoos, H.H., Mitchell, D.G. (eds.) SAT 2004. LNCS, vol. 3542, pp. 306–320. Springer, Heidelberg (2005). https://doi.org/10.1007/11527695_24

APTER: Aggregated Prognosis Through Exponential Re-weighting

Yang Liu[1](✉) and Kristiaan Pelckmans[2]

[1] Teesside University, Middlesbrough, UK
sjtuly@gmail.com
[2] Uppsala University, Uppsala, UK
http://user.it.uu.se/~liuya610/index.html

Abstract. This paper considers the task of learning how to make a prognosis of a patient based on his/her micro-array expression levels. The method is an application of the aggregation method as recently proposed in the literature on theoretical machine learning, and excels in its computational convenience and capability to deal with high-dimensional data. This paper gives a formal analysis of the method, yielding rates of convergence similar to what traditional techniques obtain, while it is shown to cope well with an exponentially large set of features. Those results are supported by numerical simulations on a range of publicly available survival-micro-array data sets. It is empirically found that the proposed technique combined with a recently proposed pre-processing technique gives excellent performances. All used software files and data sets are available on the authors' website http://user.it.uu.se/~liuya610/index.html.

Keywords: Survival analysis · Bioinformatics · Machine learning · Data mining

1 Introduction

Learning how to make a prognosis of a patient is an important ingredient to the task of building an automatic system for personalised medical treatment. A prognosis here is understood as a useful characterisation of the (future) time of an event of interest. In cancer research, a typical event is the relapse of a patient after receiving treatment. The traditional approach to process observed event times is addressed in the analysis of survival data, see e.g. [12] for an excellent review of this mature field in statistics. Most of those techniques are based on parametric or semi-parametric assumptions on how the data was generated.

Probably the most prevalent technique is Cox' Proportional Hazard (PH) approach, where inference is made by maximising a suitable partial likelihood function. This approach has proven to be very powerful in many applications of survival analysis, but it is not clear that the basic assumption underlying this technique holds in the analysis of micro-array data sets. Specifically, the proportional hazard assumption is hard to verify and might not even be valid.

© Springer Nature Switzerland AG 2019
D.-Z. Du et al. (Eds.): COCOON 2019, LNCS 11653, pp. 425–436, 2019.
https://doi.org/10.1007/978-3-030-26176-4_35

This in turn jeopardises the interpretation of the results. This is especially so since the data has typically a high dimensionality while typically a few (complete) cases are available, incurring problems of ill-conditioning. Many authors suggested fixes to this problem. Some of such work proposed in the early 2000, was studied numerically and compared in [3]. In applied work, one often resorts to a proper form of pre-processing in order to use Cox' PH model, see e.g. [18].

Since prognosis involves essentially a form of prediction, it is naturally to phrase this problem in a context of modern machine learning. This insight allowed a few authors to come up with algorithms which are deviating from a likelihood-based approach.

This work takes this route even further. It studies the question *how can new insights in machine learning help to build a more powerful algorithm?* As dictated by the application, we are especially interested in dealing with high-dimensional data. That is, cases where many ($O(10^4)$) co-variates might potentially be relevant, while only relatively few cases ($O(10^2)$) are available. Furthermore, we are not so much interested in *recovering* the mechanisms underlying the data since that is probably too ambitious a goal. Instead, we merely aim at making a good *prognosis*. It is this rationale that makes the present technique essentially different from likelihood-based, or penalised likelihood-based approaches as e.g. the PH-L$_1$ [8,16] or the Danzig Selector for survival analysis [1], and points us resolutely to methods of machine learning and empirical risk minimisation.

The contribution of this work is threefold. Firstly, discussion of the application of prognosis leads us to formulate a criterion which does not resort to a standard approach of classification, function approximation or maximum (partial) likelihood inference. Secondly, we point to the use of aggregation methods in a context of bioinformatics, give a subsequent algorithm (APTER) and derive a competitive performance guarantee. Thirdly, we present empirical evidence which supports the theoretical insights, and affirms its use for the analysis of micro-array data for survival analysis. The experiments can be reproduced using the software made public at http://user.it.uu.se/~liuya610/index.html.

1.1   Organization and Notation

This paper is organized as follows. The next section discusses the setting of survival analyses and the aim of prognosis. Section 3 describes and analyses the proposed algorithm. Section 4 gives empirical results of this algorithms on artificial and micro-array data sets. Section 4 concludes with a number of open questions.

This paper follows the notational convention to represent deterministic single quantities as lower-case letters, vectors are denoted in bold-face, and random quantities are represented as upper-case letters. Expectation with respect to any random variable in the expression is denoted as $\mathbb{E}$. The shorthand notation $\mathbb{E}_n[\cdot]$ denotes expectation with respect to all n samples seen thus far, while $\mathbb{E}_{n-1}[\cdot]$ denotes expectation with respect to the first $n-1$ samples. $\mathbb{E}^n[\cdot]$ denotes expectation with respect to the nth sample only, such that the rules of probability imply that $\mathbb{E}_n[\cdot] = \mathbb{E}_{n-1}\mathbb{E}^n[\cdot]$.

The data is represented as a set of size n of tuples $\{(\mathbf{x}_i, Y_i, \delta_i)\}_{i=1}^{n}$. Let $0 < Y_1 \leq Y_2 \leq \cdots \leq Y_n$ be an ordered sequence of observed event times associated to n subjects. An event can be either a failure with time T_i, or a (right) censoring time C_i, expressed as the time elapse from t_0. In this paper we assume that all n subjects share the same time of origin t_0. It will be convenient to assume that each subject has a failure and right censoring time with values T_i and C_i respectively. Then only the minimum time can be observed, or $Y_i = \min(T_i, C_i)$. It will be convenient to define the *past event set* $P(t) \subset \{1, \ldots, n\}$ at time t. That is, $P(t)$ denotes the set of all subjects which have experienced an *event* strictly before time t. Let for $i = 1, \ldots, n$ the indicator $\delta_i \in \{0, 1\}$ denote whether the event (failure) is directly observed ($\delta_i = 1$), or if the subject i is censored ($\delta_i = 0$), or $\delta_i = I(Y_i < C_i)$. Then $P(t) = \{i : Y_i < t, \delta_i = 1\}$. Furthermore, associate to each subject $i = 1, \ldots, n$ a co-variate $\mathbf{x}_i \in \mathbb{R}^d$ of dimension d. In the present setting, $d = O(1000)$, while $n = O(100)$ at best.

2   Prognosis in Survival Analysis

In this section we formalize the task of learning how to make a prognosis, based on observed cases. The general task of prognosis in survival analysis can be phrased as follows:

Definition 1 (Prognosis). *Given a subject with co-variate $\mathbf{x}_* \in \mathbb{R}^d$, what can we say about the value of its associated T_*?*

Motivated by the popular essay by S.J. Gould[1], we like to make statements as 'my co-variates indicate that with high probability I will outlive 50% of the subjects suffering the same disease', or stated more humanely as 'my co-variates indicate that I belong to the *good* half of the people having this disease'. The rationale is that this problem statement appears easier to infer than estimating the full conditional hazard or conditional survival functions, while it is more informative than single median survival rates.

Specifically, we look for an *expert* $f : \mathbb{R}^d \to \mathbb{R}$ which can decide for any 2 different subjects $0 < i, j \leq n$ which one of them will *fail* first. In other words, we look for an f such that for as many couples (i, j) as possible, one has $(T_i - T_j)(f(\mathbf{x}_i) - f(\mathbf{x}_j)) \geq 0$. Since T_k is not observed in general due to censoring, the following (re-scaled) proxy is used instead

$$\sum_{i=1}^{n} \frac{1}{|P(Y_i)|} \sum_{j \in P(Y_i)} I(f(\mathbf{x}_i) < f(\mathbf{x}_j)), \tag{1}$$

where $I(z) = 1$ if z holds true, and equals zero otherwise. In case $|P(Y_i)| = 0$, the ith summand in the sum is omitted. This is standard practice in all subsequent formulae. Note that this quantity is similar to the so called Concordance Index

---

[1] 'The Median Isn't the Message' as in http://www.prognosis.org/what_does_it_mean.php.

(C_n) as proposed by Harell [9]. The purpose of this paper is to propose and analyze an algorithm for finding such f from a large set $\{f\}$, based on observations and under the requirements imposed by the specific setup.

If given one expert $f : \mathbb{R}^d \rightarrow \mathbb{R}$, its 'loss' of a prognosis of a subject with co-variate $\mathbf{x}_* \in \mathbb{R}^d$ and time of event Y_* would be

$$\ell_*(f) = \frac{1}{|P(Y_*)|} \sum_{k \in P(Y_*)} I\left(f(\mathbf{x}_*) \leq f(\mathbf{x}_k)\right). \tag{2}$$

That is, $\ell_*(f)$ is the fraction of samples which experience an event before the time Y_* associated to the subject with the co-variate $\mathbf{x}_*$, although they were prognosis with a higher score by expert f. Now we consider having m experts $\{f_i\}_{i=1}^m$, and we will learn which of them performs best. We represent this using a vector $\mathbf{p} \in \mathbb{R}^m$ with $\mathbf{p}_i \geq 0$ for all $i = 1, \ldots, m$, and with $1_m^T \mathbf{p} = 1$. Then, we will use this *weighting* of the experts to make an informed prognosis of the event to occur at T_*, of a subject with co-variate $\mathbf{x}_* \in \mathbb{R}^d$. Its associated loss is given as

$$\ell_*(\mathbf{p}) = \sum_{i=1}^m \mathbf{p}_i \left(\frac{1}{|P(T_*)|} \sum_{k \in P(T_*)} I\left(f_i(\mathbf{x}_*) \leq f_i(\mathbf{x}_k)\right) \right). \tag{3}$$

This represents basically which expert is assigned most value to for making a prognosis. For example, in lung-cancer we may expect that an expert based on smoking behaviour of a patient has a high weight. Note that we include the $' ='$ case in (3) in order to avoid the trivial cases where f is constant. So, we have formalised the setting as learning such $\mathbf{p}$ in a way that the smallest possible loss $\ell_*(\mathbf{p})$ will be (or can be expected to be) made.

3   The APTER Algorithm

When using a *fixed* vector $\hat{\mathbf{p}}$, we are interested in the *expected* loss of the rule. The expected loss of the nth sample $(\mathbf{x}_n, T_n)$ becomes $\mathbb{L}(\hat{\mathbf{p}}) = \mathbb{E}^n \ell_n(\hat{\mathbf{p}}) =$

$$\mathbb{E}^n \left[\sum_{i=1}^m \hat{\mathbf{p}}_i \frac{1}{|P(T_n)|} \sum_{k \in P(T_n)} I\left(f_i(\mathbf{x}_n) \leq f_i(\mathbf{x}_k)\right) \right]. \tag{4}$$

Note that bounds will be given for this quantity which are valid for any $\mathbf{x}_n \in \mathbb{R}^d$ which may be provided. In order to device a method which guarantees properties of this quantity, we use the mirror averaging algorithm as studied in Tsybakov, Rigollet, Juditsky in [11]. This algorithm is based on ideas set out in [14]. It is a highly interesting result of those authors that the resulting estimate has better properties in terms of oracle inequalities compared to techniques based on sample averages. Presently, such fast rate is not obtained since the involved loss function is not exponentially concave as in [11], Definition 4.1. Instead of this property, we resort to use of Hoeffding's inequality which gives us a result with rate $O(\sqrt{\frac{\ln m}{n}})$. In order to give a formal guarantee of the algorithm, the following property is needed:

Algorithm 1. APTER: Aggregate Prognosis Through Exponential Re-weighting

(0) Let $\mathbf{p}_i^0 = \frac{1}{m}$ for $i = 1, \ldots, m$.

for all $k = 1, \ldots, n$ **do**

(1) The prognosis associated to the m experts $\{f_i\}_{i=1}^m$ are scored whenever *any* new event (censored or not) is recorded for a subject $k \in \{1, \ldots, n\}$ at time Y_k as

$$\ell_k(f_i) = \frac{1}{|P(Y_k)|} \sum_{l \in P(Y_k)} I\left(f_i(\mathbf{x}_k) \leq f_i(\mathbf{x}_l)\right) \tag{5}$$

and the cumulative loss is $L_k(f_i) = \sum_{s=1}^k \ell_s(f_i)$.

(2) The vector $\mathbf{p}^k$ is computed for $i = 1, \ldots, m$ as follows

$$\mathbf{p}_i^k = \frac{\exp(-\nu L_k(f_i))}{\sum_{j=1}^m \exp(-\nu L_k(f_j))}. \tag{6}$$

end for

(3) Aggregate the hypothesis $\{\mathbf{p}^k\}_k$ into $\hat{\mathbf{p}}$ as follows:

$$\hat{\mathbf{p}} = \frac{1}{n} \sum_{k=0}^{n-1} \mathbf{p}^k. \tag{7}$$

Definition 2. *For any $t = 1, \ldots, n$ and $i = 1, \ldots, m$ we have that*

$$\mathbb{E}_n\left[g\left(\frac{L_n(f_i)}{n}\right)\right] = \mathbb{E}_n[g(\ell_t(f_i))]. \tag{8}$$

for any regular function $g : \mathbb{R} \rightarrow \mathbb{R}$.

This essentially means that we do not expect the loss to be different when it is measured at different points in time (different subjects).

Theorem 1 (APTER). *Given m experts $\{f_i\}_{i=1}^m$, and the loss function ℓ as defined in Eq. (3). Then run the APTER algorithm with $\nu = \sqrt{\frac{2\ln m}{n}}$ resulting in $\hat{\mathbf{p}}$. Then*

$$\mathbb{E}_{n-1}\left[\mathbb{L}(\hat{\mathbf{p}}) - \min_{i=1,\ldots,m} \mathbb{L}(f_i)\right] \leq \sqrt{\frac{2\ln m}{n}}. \tag{9}$$

This result is in some way surprising. It says that we can get competitive performance guarantees without a need for explicitly (numerically) optimizing the performance over a set of hypothesis. Note that an optimization formulation lies on the basis of a maximum (partial) likelihood method or a risk minimization technique as commonly employed in a machine learning setting. There is an implicit link with optimization and aggregation through the method of mirror descent, see e.g. [10] and [2]. The lack of an explicit optimization stage results in the considerable computational speedups. Note further that the performance guarantee degrades only as $\sqrt{\log(m)}$ in terms of the number of experts m.

3.1   Choice of Experts and APTER$_p$

The following experts are used in the application in micro-array case studies. Here, we use simple uni-variate rules. That is, the experts are based on individual features (gene expression levels) of the data set. The rationale is that a single gene expression might well be indicative for the observed behaviours.

Let $\mathbf{e}_i$ be the ith unit vector, and let $\pm$ denote both the positive as well as the negated version. Then, the experts $\{f_i\}$ are computed as $f_i(\mathbf{x}) = \pm\mathbf{e}_i^T\mathbf{x}$, so that $m = 2d$, and every gene expression level can both be used for *over-expression* or *under-expression*.

In practice however, evidence is found that the following set of experts result in better performance: $f_i(\mathbf{x}) = s_i\mathbf{e}_i^T\mathbf{x}$, where the sign $s_i \in \{-1, 1\}$ is given by whether the ith expression has a concordance index with the observed outcome larger or equal to 0.5, as estimated on the set used for training. This means that $m = d$. This technique is referred to as APTER$_p$. Note that this subtlety needs also to be addressed in the application of Boosting methods.

3.2   Pre-processing Using SIS and ISIS

It is found empirically that pre-processing using the Iterative Sure Independence Screening ISIS as described in [5] improves the numerical results. However, the rational for this technique comes from an entirely different angle. That is, it is conceived as a screening technique for PH-L$_1$-type of algorithms.

Let $\mathbf{m} = (\mathbf{m}_1, \ldots \mathbf{m}_d)^T \in \mathbb{R}^d$ be defined as $\mathbf{m} = \sum_{i=1}^{n} Y_i\mathbf{x}_i$. For any given $\gamma \in (0, 1)$, define the set M_γ as [5]: $M_\gamma = \{1 \leq i \leq d :\}$, where $|\mathbf{m}_i|$ is among the first $[\gamma n]$ largest entries of $\mathbf{m}$.

Here, $[\gamma n]$ denotes the integer part of γn. This set then gives the indices of the features which are retained in the further analysis. It is referred to as Sure Independence Screening (SIS) [5]. In the second step, APTER is applied using only the retained features. Note that in the paper [5], one suggests instead using a Cox partial Likelihood approach with a SCAD penalty (for numerical comparison with such scheme, see the next section).

An extension of SIS is Iterative SIS (ISIS), see [5]. The idea is to pick up important features, missed by SIS. This goes as follows: rather than having a single pre-processing (SIS) step, the procedure is repeated as follows. At the end of a SIS-APTER step, a new (semi-) response vector Y' can be computed by application of the found regression coefficients. This new response variables can then be reused in a SIS step, resulting in fresh $[\gamma n]$ features. This procedure is repeated until one has enough *distinct* features.

Since $[\gamma n]$ features are then given as input to the actual training procedure, we will refer to this value as m in the experiments, making this connection between screening and training more explicit.

4   Empirical Results

This section present empirical results supporting the claim of efficiency. First, we describe the setup of the experiments.

4.1 Setup

The following measure of quality (the Concordance index or C_n, see e.g [15]) of a prognostic index scored by the function $f : \mathbb{R}^d \to \mathbb{R}$ is used. Let again the data be denoted as $\{(\mathbf{x}_i, Y_i, \delta_i)\}_{i=1}^n$, where $\mathbf{x}_i$ are the co-variates, Y_i contains the survival- or censoring time, and δ_i is the censoring indicator as before. Consider any $f : \mathbb{R}^d \to \mathbb{R}$, then C_n is defined as

$$C_n(f) = \frac{\sum_{i:\delta_i=0} \sum_{Y_j > Y_i} I(f(\mathbf{x}_i) < f(\mathbf{x}_j))}{|\varepsilon|}. \tag{10}$$

Here $|\varepsilon|$ denotes the number of the pairs which have $Y_i < Y_j$ when Y_i is not censored. The indicator function $I(\pi) = 1$ if π holds, and equals 0 otherwise. That is, if $C_n(f) = 1$, one has that f scores a higher prognostic index to the subject with will experience the event later ('good'). A $C_n(f) = 0.5$ says that the prognostic index given by f is arbitrary with respect to event times ('bad'). Observe that this measure is not quite the same as $\ell_n(f)$ or $L_n(f)$ as were used in the design of the APTER algorithm. Note that this function goes along the lines of the Area under the ROC curve or the Mann-Whitney statistic, adapted to handling censored data.

The data is assigned randomly to training data of size $n_t = \lfloor 2n/3 \rfloor$ and test data of size $n - n_t$. The training data is used to follow the training procedures, resulting in $\hat{f}$. The test data is used to compute the performance expressed as $C_n(\hat{f})$. The results are randomised 50 times (i.e. a random assignments to training and test set), and we report the median value as well as $\pm$ the variance. The parameter $\nu > 0$ is tuned in the experiments using cross-validation on the data set which is used for training. It was found that proper tuning of this parameter is crucial for achieving good performance.

The following ten algorithms are run on each of these data sets:

(a) APTER: The approach as given in Algorithm 1 where experts $\{f_i, f_i'\}$ are taken as $f_i(\mathbf{x}) = \mathbf{e}_i^T \mathbf{x}$ and $f_i'(\mathbf{x}) = -\mathbf{e}_i^T \mathbf{x}$. In this way we can incorporate positive effects due to over-expression and under-expression of a gene. This means that $m = 2d$.

(b) APTER$_p$: The approach as given in 1 where experts $\{f_i\}$ are given as $f_i(\mathbf{x}) = s_i \mathbf{e}_i^T \mathbf{x}$ where the sign $s_i \in \{-1, 1\}$ is given by the C_n of the ith expression with the observed effect, estimated on the set used for training. This means that $m = d$.

(c) MINLIP$_p$: The approach based on ERM and s_i as discussed in [17].

(d) MODEL2: Another approach based on ERM as discussed in [17].

(g) PLS: An approach based on pre-processing the data using PLS and application of Cox regression, as described in [3].

(f) PH-L$_1$: An approach based on a L_1 penalized version of Cox regression, as described in [8].

(g) PH-L$_2$: An approach based on a L_2 penalized version of Cox regression, as described in [7].

(h) ISIS-APTER$_p$: An approach which uses ISIS as pre-processing, and applies APTER$_p$ on the resulting features [5].
(i) ISIS-SCAD: An approach which uses ISIS as pre-processing, and applies SCAD on the resulting features [5].
(j) Rankboost: An approach based on boosting the c-index [6].

Those algorithms are applied to an artificial data set (as described below) as well as on a host of real-world data sets (as can be found on the website). Those data sets are publicly available, and all experiments can be reproduced using the code available at[2].

4.2   Artificial Data

The technique is tested on artificial data which was generated as follows. A disjunct training set and test set, both of size 100 'patients' was generated. For each 'patient', d features are sampled randomly from a standard distribution, so that $\mathbf{x}_i \in \mathbb{R}^d$. We say that we have only k *informative* features when an event occurs at time T_i computed for $i = 1, \ldots, n$ as $T_i = \dfrac{-\log Z_i}{10 \exp\left(\sum_{j=1}^{k} \mathbf{x}_{i,j}\right)}$, where Z_i is a random value generated from a uniform distribution on the unit interval $[0, 1]$, and $\mathbf{x}_{i,j}$ is the jth co-variate for the ith patient. The right-censoring time is randomly generated from the exponential distribution with rate 0.1. After application of the censoring rule, we arrive at the *survival* time Y_i.

In a first experiment, d is fixed as 100, but only the first $k \leq d$ features have an effect on the outcome ('informative'). Figure 1a shows the evolution of the performance $(C_n(\hat{f}))$ for increasing values of k. In a second experiment we fix $k = 10$, and record the performance for increasing values of d, investigating the effect of a growing number of *ambient* dimension on the performance of APTER. Results are displayed in Fig. 1b.

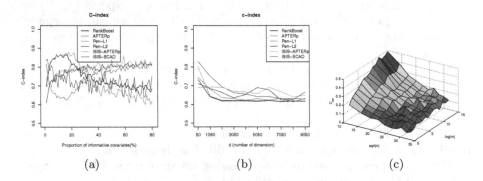

(a)   (b)   (c)

Fig. 1. Comparison of the numerical results obtained on the artificial data sets (a) when keeping $d = 100$ fixed, and (b) when keeping $k = 10$ fixed. (c) The evolution of the 'C-index error' C_{err} obtained by APTER$_p$ for different values of (n, m).

---

[2] The software is available at http://user.it.uu.se/~liuya610/index.html.

Thirdly, we investigate how well the numerical results align with the result of Theorem 1. We take results of APTER using uni-variate experts, so that $m = 2d$. The "c-index error" (C_{err}) is given for different values of d and n. C_{err} is computed as the difference between the C_n obtained by APTER - denoted as $\hat{f}$ - and the C_n of the single "best" expert $f_j(\mathbf{x}_i) = \mathbf{x}_{i,j}$:

$$C_{err} = \max_j C_n(f_j) - C_n\left(\hat{f}\right). \tag{11}$$

This formula is similar to equation (9). The numerical performances are displayed in Fig. 1c. This figure indicates that C_{err} increases logarithmically in d, and in terms of $\frac{1}{\sqrt{n}}$. This supports the result of Theorem 1.

4.3 Real Datasets

In order to benchmark APTER and its variations against state-of-the-art approaches, we run the algorithms on a wide range of large-dimensional real data sets. The data set are collected in a context of bioinformatics, and a full description of this data can be found on the website. The experiments are divided into three categories: (i) The algorithms are run on 8 micro-array data sets, in order to asses performance on typical sizes for those data sets. Here we see that there is no clear overall winner amongst the algorithms, but the proposed algorithm (ISIS-APTER$_p$) does do repeatedly very well, and performs best on most (3) data sets. Results are given in Table 1. (ii) In order to see whether the positive performance is not due to irregularities of the data, we consider the following *null* experiment. Consider the AML data set, but lets shuffle the observed phenotype (the observed Y) between different subjects. So any relation between the expression level and the random phenotype must be due to plain chance (by construction). We see in Fig. 2 that indeed the distribution of

Table 1. Numerical results of the experiments of 10 different methods on 8 micro-array datasets.

	NSBCD	DBCD	DLBCD	Veer	Vijver	Beer	AML	FL
APTER	0.73	0.69	0.58	0.65	0.44	0.60	0.58	0.70
APTER$_p$	0.77	0.74	0.59	**0.68**	0.62	0.73	0.60	0.73
MINLIP$_p$	0.74	0.71	0.59	0.65	0.61	0.69	0.55	0.70
MODEL2	0.75	0.74	0.62	**0.67**	0.61	**0.74**	0.56	0.72
PLS	**0.78**	0.74	0.53	0.58	0.62	0.66	0.57	0.66
PH-L2	0.69	0.73	**0.65**	0.64	0.61	0.73	0.54	0.69
PH-L1	0.69	0.74	0.60	0.60	**0.65**	0.69	0.61	0.67
Rankboost	0.75	0.72	0.62	0.62	**0.65**	0.71	0.53	0.67
ISIS-SCAD	0.69	0.72	**0.65**	**0.68**	0.62	0.72	**0.63**	0.71
ISIS-APTER$_p$	**0.78**	**0.76**	0.62	0.66	0.62	**0.75**	0.59	**0.74**

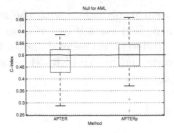

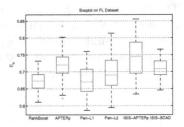

Fig. 2. Performances of APTER and APTER$_p$ on the AML dataset when the reposes are randomly shuffled.

Fig. 3. Boxplots of the numerical results obtained on the FL data set. Results are expressed in terms of the $C_n(f)$.

the methods based on this *shuffled* data nears a neutral C_n on the test set of 0.50. This means that the 10% improvement as found in the real experiment (see table) is substantial with respect to the randomization, and are not due to chance alone. (iii) The results of the algorithm is compared on the micro-array data set as reported in [4], and analysed further in [13]. Here we found that the obtained performance is significantly larger than what was reported earlier, while we do not have to resort to the *clustering* pre-processing as advocated in [4,13]. This data has a very high dimensionality ($d = 44.928$) and has only a few cases ($n = 191$). Results are given in the lat row of Table 1 and the box plots of the performances due to the 50 randomization, are given in Fig. 3.

Finally, we discuss the application of the method on the same high dimensional ($d = 44.928$) data set as before, but we study the impact of the parameter m given to ISIS, which returns in turn the data to be processed by APTER$_p$. The performances for different values of m are given in Fig. 4a. The best performance is achieved for $m = 800$, which is the value which was used in the earlier experiment reported in Fig. 3. Here we compare only to a few other approaches, namely the PH-L$_1$, MINLIP$_p$ and MODEL2 approach which make all use of an explicit optimisation scheme. Panel Fig. 4b reports the time needed to perform training/ tuning and randomisation corresponding to a fixed value of m. Panel Fig. 4c reports the size of the memory used up for the same procedure. Here it is clearly seen that APTER$_p$ results in surprisingly good performance, given that it uses up less computations and memory. It is even so that the optimisation-based techniques cannot finish for large m in reasonable time or without problems of the memory management, despite the fact that a very efficient optimisation solver (Yalmip) was used to implement those.

4.4   Discussion of the Results

These results uncover some interesting properties of the application of the proposed algorithms in this bio-informatics setting.

First of all, the APTER and APTER$_p$ methods are orders of magnitudes faster (computationally) compared to the bulk of methods based on optimization

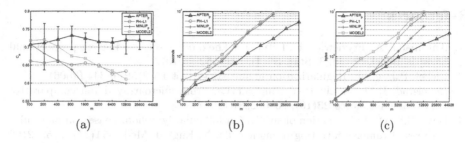

(a)   (b)   (c)

Fig. 4. results of the choice of m in ISIS, based on the Follicular Lymphoma data set [4,13]. (a) Performance expressed as $C_n(\hat{f})$ on the test sets (medium of 50 randomization). (b) Computation time for running tuning, training and randomisation for a fixed value of m. (c) Usage of memory of the same procedure.

formulations (either using Maximum (penalized) Partial Likelihood, Empirical Risk Minimization or multivariate pre-processing techniques). This does not affect the performance in any way, contrary to what intuition would suggest. In fact, the performance on typical micro-array data of the *vanilla* APTER or APTER$_p$ (without ISIS) is often amongst the better.

Secondly, inclusion of pre-processing with ISIS - also very attractive from a computational perspective - is boosting up significantly the performance of APTER. We have no theoretical explanation for this, since ISIS was designed to complement L_1 or Danzig-selector approaches. While the authors of ISIS advocate the used of a SCAD norm, we find that APTER$_p$ is overall a better choice for the mentioned data sets.

Furthermore, the empirical results indicate that the statistical performance is preserved by using APTER$_p$ combined with ISIS, and may even improve over performances obtained using existing approaches. This is remarkable since the computational power is orders of magnitude smaller than most existing approaches based on (penalised) PL of ERM. We find also that empirical results align quite closely the theoretical findings as illustrated with an experiment on artificial data.

5   Conclusions

This paper presents statistically and computationally compelling arguments for a method based on aggregation can be used for analysis of survival data in high dimensions. Theoretical findings are complemented with empirical results on micro-array data sets. We feel that this result is surprising not only in that it outperforms methods in ERM or (penalised) PL, but provides as well a tool with much lower computational complexity as the former ones since no direct optimization is involved. We present empirical, reproducible results which support this claim of efficiency. This analysis presents many new opportunities, both applied (towards Genome Wide Analysis, or GWAs) as well as theoretical (can we improve the rates of convergence by choosing other loss functions?).

References

1. Antoniadis, A., Fryzlewicz, P., Letué, F.: The dantzig selector in cox's proportional hazards model. Scand. J. Stat. **37**(4), 531–552 (2010)
2. Bickel, P.J., et al.: Regularization in statistics. Test **15**(2), 271–344 (2006)
3. Bøvelstad, H.M., et al.: Predicting survival from microarray dataa comparative study. Bioinformatics **23**(16), 2080–2087 (2007)
4. Dave, S.S., et al.: Prediction of survival in follicular lymphoma based on molecular features of tumor-infiltrating immune cells. N. Engl. J. Med. **351**(21), 2159–2169 (2004)
5. Fan, J., Lv, J.: Sure independence screening for ultrahigh dimensional feature space. J. Roy. Stat. Soc.: Ser. B (Stat. Methodol.) **70**(5), 849–911 (2008)
6. Freund, Y., Iyer, R., Schapire, R.E., Singer, Y.: An efficient boosting algorithm for combining preferences. J. Mach. Learn. Res. **4**, 933–969 (2003)
7. Goeman, J.: Penalized: L1 (lasso) and l2 (ridge) penalized estimation in GLMs and in the Cox model. R package version 09–21 2008 (2008)
8. Goeman, J.J.: L1 penalized estimation in the Cox proportional hazards model. Biometrical J. **52**(1), 70–84 (2010)
9. Gönen, M., Heller, G.: Concordance probability and discriminatory power in proportional hazards regression. Biometrika **92**(4), 965–970 (2005)
10. Jakobson, C., Feinsod, M., Nemirovsky, Y.: Low frequency noise and drift in ion sensitive field effect transistors. Sens. Actuators B: Chemical **68**(1), 134–139 (2000)
11. Juditsky, A., Rigollet, P., Tsybakov, A.B., et al.: Learning by mirror averaging. Ann. Stat. **36**(5), 2183–2206 (2008)
12. Kalbfleisch, J.D., Prentice, R.L.: The Statistical Analysis of Failure Time Data. Wiley, New York (2011)
13. Lin, R.S.: Re-analysis of molecular features in predicting survival in follicular lymphoma. report (2006)
14. Nemirovsky, A., Yudin, D.: Problem Complexity and Method Efficiency in Optimization. Wiley, Chichester (1983)
15. Raykar, V.C., Steck, H., Krishnapuram, B., Dehing-Oberije, C., Lambin, P.: On ranking in survival analysis: bounds on the concordance index. In: NIPS (2007)
16. Simon, N., Friedman, J., Hastie, T., Tibshirani, R.: Regularization paths for Coxs proportional hazards model via coordinate descent. J. Stat. Softw. **39**(5), 1–13 (2011)
17. Van Belle, V., Pelckmans, K., Van Huffel, S., Suykens, J.A.: Improved performance on high-dimensional survival data by application of survival-SVM. Bioinformatics **27**(1), 87–94 (2011)
18. van't Veer, L.J., et al.: Gene expression profiling predicts clinical outcome of breast cancer. Nature **415**(6871), 530–536 (2002)

An Erdős–Pósa Theorem on Neighborhoods and Domination Number

Jayakrishnan Madathil[1]([✉]), Pranabendu Misra[2], and Saket Saurabh[1,3]

[1] The Institute of Mathematical Sciences, HBNI, Chennai, India
{jayakrishnanm,saket}@imsc.res.in
[2] Max Planck Institute for Informatics, Saarbrucken, Germany
pmisra@mpi-inf.mpg.de
[3] Department of Informatics, University of Bergen, Bergen, Norway

Abstract. The neighborhood packing number of a graph is the maximum number of pairwise vertex disjoint closed neighborhoods in the graph. This number is a lower bound on the domination number of the graph. We show that the domination number of a graph of girth at least 7 is bounded from above by a (quadratic) function of its closed neighborhood packing number, and further that no such bound exists for graphs of girth at most 6. We then show that as girth of the graph increases, the upper bound on the domination number drops as a function of girth.

Keywords: Erdős–Pósa property · Domination number · Closed neighborhood packing number

1 Packing and Covering: The Erdős–Pósa Framework

A celebrated result by Paul Erdős and Lajos Pósa [2] says that there exists a function $f : \mathbb{N} \to \mathbb{R}$ such that given any $k \in \mathbb{N}$, every graph contains either k disjoint cycles or a set of vertices of size at most $f(k)$ whose deletion will render G acyclic. There is nothing sacrosanct about cycles in this result. Indeed, we may consider any class $\mathcal{G}$ of graphs, and given an arbitrary graph G, we may consider the largest number of disjoint copies of graphs in $\mathcal{G}$ that can be packed in G against the least number of vertices in G that cover all such graphs in $\mathcal{G}$. Formally, a class of graphs $\mathcal{G}$ is said to have the *Erdős–Pósa property* if there exists a function $f : \mathbb{N} \to \mathbb{R}$ such that given any $k \in \mathbb{N}$ and any graph G, either G contains k disjoint subgraphs, each isomorphic to a graph in $\mathcal{G}$, or there exists a subset $V' \subseteq V(G)$ of size at most $f(k)$ such that $G - V'$ has no subgraph isomorphic to any graph in $\mathcal{G}$. Note that f depends only on $\mathcal{G}$.

Many well known combinatorial results fit into the Erdős–Pósa mold. For instance, by taking $\mathcal{G}$ to be a singleton set containing an edge and $f(k) = k - 1$, König's theorem (which says that the sizes of maximum matching and minimum

This work is supported by the European Research Council (ERC) via grant LOPPRE, reference no. 819416.

D.-Z. Du et al. (Eds.): COCOON 2019, LNCS 11653, pp. 437–444, 2019.
https://doi.org/10.1007/978-3-030-26176-4_36

vertex cover are equal in bipartite graphs) can be rephrased as "given any $k \in \mathbb{N}$, every *bipartite* graph G contains either k disjoint subgraphs from $\mathcal{G}$ or a set of $f(k)$ vertices that meets all subgraphs in G that are isomorphic to any graph in $\mathcal{G}$." Note that this result holds for all graphs (not necessarily bipartite) with the same $\mathcal{G}$ and $f(k) = 2(k-1)$. Such Erdős–Pósa results have been obtained for several graph classes. See the survey by Raymond and Thilikos [4] for a summary of similar results. In this paper, we take $\mathcal{G}$ to be the collection of all closed neighborhoods.

2   Preliminaries

For $n \in \mathbb{N}$, $[n]$ denotes the set $\{1, 2, \ldots, n\}$. All graphs in this paper are simple and undirected. For a graph G, $V(G)$ and $E(G)$ denote the vertex set and edge set of G, respectively. For a vertex v of G, the (open) neighborhood of v, denoted by $N(u)$, is the set of all vertices in V that are adjacent to v; the elements of $N(v)$ are called the neighbors of v; and we denote by $N[v]$, the *closed neighborhood* of v, defined as $N[v] = N(u) \cup \{v\}$. For a subset $V' \subseteq V(G)$, $N(V') = (\cup_{v \in V'} N(v)) \setminus V'$, and $N[V'] = \cup_{v \in V'} N[v']$.

For a graph G, a set $S \subseteq V(G)$ is said to be a *closed neighborhood packing* (or *packing*, for short), in G if for every distinct $x, y \in S$, we have $N[x] \cap N[y] = \emptyset$. The *closed neighborhood packing number* $\rho(G)$ of G is the size of a largest closed neighborhood packing in G.

The dual of the notion of a closed neighborhood packing is that of a dominating set, a set of vertices that "covers" or "hits" all closed neighborhoods in a graph. We say that a vertex "dominates" itself and all its neighbors, i.e, $v \in V(G)$ dominates all vertices in $N[v]$, and similarly, a subset $V' \subseteq V(G)$ of vertices dominates all vertices in $N[V']$. A subset $D \subseteq V(G)$ is said to be a *dominating set* of G if $N[D] = V(G)$, i.e., if D dominates the entire vertex set $V(G)$. Or equivalently, $D \subseteq V(G)$ is a dominating set of G if for every $v \in V(G) \setminus D$, at least one of the neighbors of v is in D. The *domination number* $\gamma(G)$ of G is the size of a smallest dominating set.

The *girth* of G is the length of a shortest cycle in G. For a vertex v of G, we denote by $d_G(v)$, the degree of v in G and by $\Delta(G)$, the maximum degree of a vertex in G. The following lemma is very well-known and straightforward to prove. For the sake of completeness, we provide a proof below.

Lemma 1. *For a graph* G,

(i) $\gamma(G) \geq \rho(G)$, *and*
(ii) $\gamma(G) \geq \frac{|V(G)|}{\Delta(G)+1}$.

Proof. (i) If S is a closed neighborhood packing in G of size $\rho(G)$, then *every* dominating set of G must contain either v or one of its neighbors for every $v \in S$. Since $N[v]$s are pairwise disjoint for $v \in S$, we get $\gamma(G) \geq |S| = \rho(G)$.

(ii) Let D be a dominating set of G of size $\gamma(G)$. Since each vertex in D dominates at most $\Delta(G)+1$ vertices, and since every vertex in $V(G)$ is dominated by D, we have $|V(G)| \leq |D|(\Delta(G) + 1)$. $\square$

The first part of the above lemma says that the domination number $\gamma(G)$ is bounded from below by the closed neighborhood packing number $\rho(G)$. Our results show that $\gamma(G)$ is bounded from above by a function of $\rho(G)$ for graphs of girth at least 7, and that no such bound exists for graphs of girth less than 7.

3   Main Result

Theorem 1. *Given any $k \in \mathbb{N}$, every graph of girth at least 7 has a closed neighborhood packing of size $k + 1$ or a dominating set of size at most $k^2 + k$.*

Proof. Let G be a graph of girth at least 7 and assume that every closed neighborhood packing in G has size at most k, i.e., $\rho(G) \leq k$. Let S be a maximal closed neighborhood packing in G. Then $|S| \leq \rho(G)$. First, observe that $N[S]$ is a dominating set of G, because if $v \in V(G) \setminus N[S]$ has no neighbor in $N[S]$, then $S \cup \{v\}$ will be a closed neighborhood packing in G, which contradicts the maximality of S. Note that S dominates $N[S]$ and $N[S] \setminus S = N(S)$ dominates $V \setminus N[S]$. Therefore, we can remove from $N(S)$ the vertices that do not dominate any vertex in $V(G) \setminus N[S]$, and we will still be left with a set of vertices that together with S dominates the entire vertex set $V(G)$. For every $v \in S$, let $N^*(v) = \{u \in N(v) \mid N(u) \cap V \setminus N[S] \neq \emptyset\}$. That is, every vertex in $N^*(v)$ dominates at least one vertex outside of $N[S]$, and on the other hand, every vertex in $V \setminus N[S]$ is dominated by $N^*(v)$ for some $v \in S$. Hence, $D = S \cup N^*(S)$ is a dominating set of G, where $N^*(S) = \cup_{v \in S} N^*(v)$. We claim that $|N^*(v)| \leq k$ for every $v \in S$ so that $|D| = |S| + \sum_{v \in S} |N^*(v)| \leq k + k^2$, which proves the theorem.

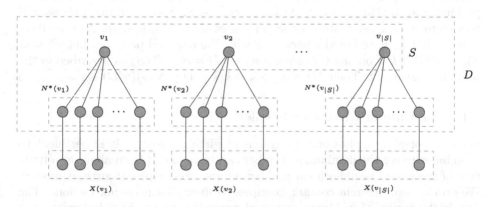

Fig. 1. S is a maximum closed neighborhood packing; D is a dominating set; $|N^*(v_i)| = |X(v_i)|$ for every $v_i \in S$.

Consider $v \in S$. Every vertex in $N^*(v)$ has at least one neighbor in $V \setminus N[S]$; corresponding to every $x \in N^*(v)$, fix one of its neighbors, say x' in $V \setminus N[S]$. Let $X(v)$ be the set of all such vertices (See Fig. 1). That is, every vertex in

$N^*(v)$ has exactly one representative in $X(v)$. Since no two distinct vertices $x_1, x_2 \in N^*(v)$ can have a common neighbor apart from v (because such a common neighbor along with x_1, v, x_2 would form a 4-cycle), the x' vertices that we fix for each vertex in $N^*(v)$ are all distinct. That is, we have a distinct $x' \in X(v)$ corresponding to each $x \in N^*(v)$ and hence $|X(v)| = |N^*(v)|$. We shall show that $|X(v)| \leq k$. No two distinct $x_1', x_2' \in X(v)$ are adjacent, for otherwise v, x_1, x_1', x_2', x_2 would form a cycle of length 5; nor can x_1' and x_2' have a common neighbor, say y, for otherwise $v, x_1, x_1', y, x_2', x_2$ would form a cycle of length 6. Therefore, $N[x_1'] \cap N[x_2'] = \emptyset$ for every distinct pair of vertices $x_1', x_2' \in X(v)$. In other words, $X(v)$ is a closed neighborhood packing and hence $|X(v)| \leq k$. Hence, $|N^*(v)| \leq k$ and the theorem follows.   □

The proof of Theorem 1 can be easily converted into an approximation algorithm (with an approximation ratio of opt+1) for the DOMINATING SET problem on graphs of girth at least 7. Here, opt denotes the value of an optimum solution for the DOMINATING SET problem. That is, for a graph G, opt $= \gamma(G)$.

Corollary 1. *There is a polynomial time algorithm that takes as input a graph G of girth at least 7, and returns a dominating set of G of size $\gamma(G)(\gamma(G) + 1)$.*

Proof. The algorithm works as follows. Given a graph G on n vertices with girth at least 7, find a maximal closed neighborhood packing S in G such that $|X(v)| \leq |S|$ for every $v \in S$, where $X(v)$ is as defined in the proof of Theorem 1. To do this, greedily find a maximal closed neighborhood packing S in G. For $v \in S$, $X(v)$ is a closed neighborhood packing. So, if $|X(v)| > |S|$ for some $v \in S$, then set $S = X(v)$ and grow S into a maximal closed neighborhood packing and start over. Keep updating S this way until we get a maximal closed neighborhood packing S such that $|X(v)| \leq |S|$ for every $v \in S$. Note that this process will terminate in at most n steps as $|S|$ increases with each update. Assume from now on that we have found a packing S with the required property. Let $|S| = k$. Then, $\gamma(G) \geq k$. Construct a dominating set $D = S \cup N^*(S)$ as described in the proof of Theorem 1. Then, $|D| \leq k^2 + k = k(k + 1) \leq \gamma(G)(\gamma(G) + 1)$.   □

3.1   The Case of Girth 6 or Smaller

Since Theorem 1 applies only to graphs of girth at least 7, it is pertinent to examine whether the domination number can be bounded from above as a function of the closed neighborhood packing number for graphs of girth at most 6. We provide two different counter examples to show that it cannot be done. The first is the family of incidence graphs of projective planes of finite order. (See Chapter 5 of [3] for a detailed introduction to projective planes and incidence graphs.) A projective plane of order n consists of a set P of $n^2 + n + 1$ points and a set L of $n^2 + n + 1$ lines with the following properties:

(i) For any two distinct points, there is exactly one line incident with both of them.

(ii) For any two distinct lines, there is exactly one point incident with both of them.

(iii) There are four points such that no line is incident with more than two of them.

(iv) Every point is incident with exactly $n + 1$ lines.

(v) Every line is incident with exactly $n + 1$ points.

It can be shown that projective planes of order n exists if n is the power of a prime, i.e., if $n = q^m$ for some prime number q and $m \in \mathbb{N}$.

The incidence graph of a projective plane $\mathcal{P}_n$ of order n with point set P and line set L is the bipartite graph $G(\mathcal{P}_n)$, where $V(G(\mathcal{P}_n)) = P \cup L$ and vertex $p \in P$ is adjacent to vertex $\ell \in L$ if and only if point p is incident with line ℓ in $\mathcal{P}_n$. By property (i), any two distinct vertices $p, p' \in P$ have a unique common neighbor in L, and hence $N(p) \cap N(p') \neq \emptyset$; and by property (ii), any two distinct vertices $\ell, \ell' \in L$ have a (unique) common neighbor in P, and hence $N(\ell) \cap N(\ell') \neq \emptyset$. Therefore, any closed neighborhood packing in $G(\mathcal{P}_n)$ contains at most one $p \in P$ and at most one $\ell \in L$, and thus $\rho(G(\mathcal{P})) \leq 2$. Properties (i) and (ii) also imply that $G(\mathcal{P}_n)$ has no 4-cycle, for existence of a 4-cycle would mean that $G(\mathcal{P}_n)$ has a pair of vertices with two common neighbors; and that $G(\mathcal{P}_n)$ has a 6-cycle—formed by three distinct vertices in P and the three common neighbors of pairs of those vertices. This shows that the girth of $G(\mathcal{P}_n)$ is 6. By properties (iv) and (v), every vertex in $G(\mathcal{P}_n)$ has degree $n + 1$. Therefore, by Lemma 1, $\gamma(G(\mathcal{P}_n)) \geq 2(n^2 + n + 1)/(n + 1 + 1) \geq 2n - 1$.

Our next counter example is somewhat similar to the above one, but of smaller girth. For $n \in \mathbb{N}$, the bipartite graph G_n is defined as follows. We have the bipartition $V(G_n) = X \cup Y$, such that $X = \cup_{i \in [n]} X_i$ and $Y = \cup_{i \in [n]} Y_i$, where $X_i = \{x_{i1}, x_{i2}, \ldots, x_{in}\}$ and $Y = \{y_{i1}, y_{i2}, \ldots, y_{in}\}$ for every $i \in [n]$. Thus $|X| = |Y| = n^2$. For each $i, j, j' \in [n]$, $x_{ij} y_{ij'}$ is an edge. That is, for each i, the subgraph of G_n induced by $X_i \cup Y_i$ is isomorphic to the complete bipartite graph $K_{n,n}$. In addition, for each $i, i', j \in [n]$, $x_{ij} y_{i'j}$ is an edge. It can be seen that the girth of G_n is 4, each vertex has degree $2n - 1$ and $|V(G_n)| = 2n^2$. Thus, by the second part of Lemma 1, $\gamma(G_n) \geq n$. Now, observe that for $(i, j) \neq (i', j')$, both $y_{ij'}$ and $y_{i'j}$ are common neighbors of x_{ij} and $x_{i'j'}$. Therefore, any closed neighborhood packing in G_n can contain at most one vertex from X and similarly, at most one vertex from Y, and hence $\rho(G_n) \leq 2$.

These families of graphs demonstrate the following result.

Theorem 2. *There exists no function $f : \mathbb{N} \to \mathbb{R}$ such that $\gamma(G) \leq f(\rho(G))$ for all graphs G of girth at most 6.*

4   Further Results

Observe from the proof of Theorem 1 that the quadratic term in the upper bound on $|D|$ comes from the upper bound we obtained for $|N^*(S)|$. If it can be shown that $|N^*(S)|$ is asymptotically smaller than k^2, consequently we will obtain a better bound for $|D|$. We shall show that this can be done, but with a stronger premise. We will have to exclude cycles of length greater than 7 to show the existence of a $o(k^2)$ sized dominating set. Specifically, we prove the following theorem.

Theorem 3. *Given any $k \in \mathbb{N}$, every graph of girth at least 17 has a closed neighborhood packing of size $k+1$ or a dominating set of size at most $(k/2)(1 + \sqrt{4k - 3}) + 2k$.*

The proof of this theorem relies on the following lemma.

Lemma 2 (Reiman [5]). *If a graph H on n vertices has no 4-cycles, then $|E(H)| \leq (n/4)(1 + \sqrt{4n - 3})$.*

Proof (Proof of Theorem 3). Let G be a graph of girth at least 17, and assume that $\rho(G) \leq k$. Let S be a closed neighborhood packing in G of size $\rho(G)$, and let $N^*(v), X(v)$ for every $v \in S$ and $N^*(S)$ be as defined in the proof of Theorem 1. Consider $v \in S$. Let x_1 and x_2 be two distinct vertices in $X(v)$. As argued earlier, x_1 and x_2 are not adjacent and they do not have a common neighbor. Now suppose that neither x_1 nor x_2 is adjacent to any vertex in $N(v')$ for every $v' \in S, v \neq v'$. Then, $(S \setminus \{v\}) \cup \{x_1, x_2\}$ will be a closed neighborhood packing of size $\rho(G) + 1$, which is not possible. Therefore, for every $v \in S$, $X(v)$ contains at most one vertex which has no neighbor in $N(v')$ for every $v' \in S, v \neq v'$. In other words, for every $v \in S$, there are at least $|X(v)| - 1$ edges between $X(v)$ and $\cup_{v' \in S: v' \neq v} N^*(v')$. On the other hand, since G is of girth 17, for every $v, v' \in S, v \neq v'$ there can be at most one edge between $X(v)$ and $N^*(v')$. Notice that more than one such edge would constitute a cycle of length at most 8.

Construct an auxiliary graph H on $|S|$ vertices as follows. Corresponding to every $v \in S$, introduce a vertex h_v in $V(H)$, and corresponding to every $v, v' \in S, v \neq v'$, introduce an edge $h_v h_{v'}$ in $E(G)$ if there is an edge in G between either $X(v)$ and $N^*(v')$ or $X(v')$ and $N^*(v)$. Since there are least $|X(v)| - 1$ edges in G between $X(v)$ and $\cup_{v' \in S: v' \neq v} N^*(v')$, each h_v has degree at least $|X(v)| - 1$. Now observe that if H has a cycle of length 4, then G will have a cycle of length at most 16. Since the girth of G is at least 17, we can conclude that H has no 4-cycle. Therefore, by Lemma 2, $|E(H)| \leq (|S|/4)(1 + \sqrt{4|S| - 3})$. Hence,

$$\sum_{v \in S}(|X(v)| - 1) \leq \sum_{v \in S} d_H(h_v) = 2|E(H)| \leq 2(|S|/4)(1 + \sqrt{4|S| - 3}).$$

Now, since $|S| = \rho(G) \leq k$, and as argued in the proof of Theorem 1, $|X(v)| = |N^*(v)|$ for every $v \in S$, and hence $\sum_{v \in S} |X(v)| = |N^*(S)|$, we get $|N^*(S)| - k \leq (k/2)(1 + \sqrt{4k - 3})$. Therefore, $D = S \cup N^*(S)$ is a dominating set of G of size at most $(k/2)(1 + \sqrt{4k - 3}) + 2k$. $\qquad \square$

4.1   Generalizing Theorem 3

As is clear from the proof of Theorem 3, the bound on $|D|$ hinges on the number of edges of the auxiliary graph H. Also, note that if H has a cycle of length g, then G will have a cycle of length at most $4g$. Therefore, if it can be shown that $|E(H)| \leq f(k)$ for some function $f : \mathbb{N} \to \mathbb{R}$, provided that H is g-cycle free for some g (or, under a stronger assumption that the girth of H is at least $g+1$), then, as shown in the proof of Theorem 3, we will have $|N^*(S)| - k \leq 2|E(H)| \leq 2f(k)$. This will immediately imply that $\gamma(G) \leq |S \cup N^*(S)| \leq 2(f(k) + k)$, provided that G is of girth at least $4g + 1$.

The problem now boils down to answering the following question: what is the maximum number of edges in a graph on k vertices and girth $g + 1$? This number is often denoted by $\mathbf{ex}(k, \{C_3, C_4, \ldots, C_g\})$. Observe that if a graph on k vertices has e edges and average degree d, then $d = 2e/k$. In light of this observation, the above question can be considered in a dual form: what is the least number of vertices in a graph of girth $g + 1$ and average degree at least d? The following result due to Alon, Hoory and Linial answers this question.

Lemma 3 (Alon, Hoory, Linial [1]). *The number of vertices k in a graph of girth $g + 1$ and average degree at least $d \geq 2$ satisfies $k \geq n_0(d, g + 1)$, where*

$$n_0(d, 2r + 1) = 1 + d \sum_{i=0}^{r-1} (d-1)^i \tag{1}$$

and

$$n_0(d, 2r + 2) = 2 \sum_{i=0}^{r} (d-1)^i. \tag{2}$$

In the above lemma, the right side of (1) is greater than $(d-1)^r$. To see this, note that

$$n_0(d, 2r + 1) = 1 + d \sum_{i=0}^{r-1} (d-1)^i$$

$$= 1 + \frac{d[(d-1)^r - 1]}{(d-1) - 1}$$

$$> 1 + \frac{d[(d-1)^r - 1]}{d}$$

$$= (d-1)^r.$$

That is, when the girth is $2r + 1$, the number of vertices $k \geq n_0(d, 2r + 1) > (d-1)^r$. Equivalently, $2e/k = d < k^{1/r} + 1$, which implies $e < (k/2)(k^{1/r} + 1)$. In other words,

$$\mathbf{ex}(k, \{C_3, C_4, \ldots, C_{2r}\}) < \frac{1}{2} k^{1+(1/r)} + \frac{1}{2} k. \tag{3}$$

Similarly, from (2), we get $d < (k/2)^{1/r} + 1$, or in other words,

$$\mathbf{ex}(k, \{C_3, C_4, \ldots, C_{2r+1}\}) < \frac{1}{2^{1+(1/r)}} k^{1+(1/r)} + \frac{1}{2} k. \tag{4}$$

Now, getting back to the original problem, if a graph G has girth at least $4g+1$ and $\rho(G) \leq k$, then the corresponding auxiliary graph H (on $\rho(G)$ vertices) has girth at least $g + 1$, and hence

$$|E(H)| \leq \mathbf{ex}(k, \{C_3, C_4, \ldots, C_g\}) < f(k) = \begin{cases} \frac{1}{2}(k^{2/g} + k), & \text{if } g \text{ is even, and} \\ (\frac{k}{2})^{1+(2/(g-1))} + \frac{1}{2}k, & \text{if } g \text{ is odd.} \end{cases}$$

Then, as argued earlier, $\gamma(G) \leq 2f(k) + 2k$. We thus have the following result.

Theorem 4. *Given any $k, g \in \mathbb{N}$, $g \geq 3$, every graph of girth at least $4g + 1$ has a closed neighborhood packing of size $k + 1$ or a dominating set of size at most $k^{1+(2/g)} + 3k$, if g is even, and $2(k/2)^{1+(2/(g-1))} + 3k$, if g is odd.*

References

1. Alon, N., Hoory, S., Linial, N.: The Moore bound for irregular graphs. Graphs and Combinatorics **18**(1), 53–57 (2002)
2. Erdős, P., Pósa, L.: On independent circuits contained in a graph. Canad. J. Math. **17**, 347–352 (1965)
3. Godsil, C.D., Royle, G.F.: Algebraic Graph Theory. Graduate texts in mathematics, 1st edn. Springer, New York (2001). https://doi.org/10.1007/978-1-4613-0163-9
4. Raymond, J., Thilikos, D.M.: Recent techniques and results on the erdős-pósa property. CoRR abs/1603.04615 arxiv:1603.04615 (2016)
5. Reiman, I.: Über ein Problem von K. Zarankiewicz. Acta Mathematica Academiae Scientiarum Hungaricae **9**, 269–279 (1958)

On the Hardness of Reachability Reduction

Dongjing Miao[1] and Zhipeng Cai[2]($\boxtimes$)

[1] Harbin Institute of Technology, Harbin 150001, Heilongjiang, China
miaodongjing@hit.edu.cn
[2] Georgia State University, Atlanta, GA 30303, USA
zcai@gsu.edu

Abstract. A class of reachability reduction problem is raised in the area of computer network security and software engineering. We revisit the reachability reduction problem on vertex labeled graphs in which labels are representing multiple roles. In this paper, reachability reduction is modeled based on a kind of graph cut problem aiming to disconnect those reachable label pairs specified in advance by edge deletion while minimizing the number of pairs have to be disconnected, and followed by some potential applications where our model can be applied. Based on our unified model, we provide a comprehensive complexity analysis of this problem under different conditions, to provide the hardness hierarchy of reachability preserved cut problem. Result in this paper implies that reachability reduction is typically at least harder in directed graph than its undirected counterpart, even beyond NP under generalized inputs.

Keywords: Reachability reduction · Graph cut · Complexity

1 Introduction

Reducing reachability of a graph is to delete some edges from it so that any source node cannot reach the rest of the graph. A simple understanding of graph reachability reduction could be performing a partial graph cut. In the setting of classic graph cut, the goal is to find some edges such that deleting them will disconnect the graph into several components. The traditional concern on the cut is to minimize the sum of edge costs. However, as stated in [1], it is difficult to quantify the edge costs which usually indicate un-parametrizable preferences in practice. Instead, we consider more on the effect of applying some cut, that is, to prevent the reachability of several pairs and preserve that of other pairs as much as possible.

In fact, It has a broad application in many fields of science and engineering, especially computer science. For example, in the area of network security, an effective way to prevent network security attack (e.g., *identity snowball attack* [1,2]) is to reduce the reachability of the network. Similarly, to achieve the denial of malicious ddos attacks, it is also a alternative way to reduce some

© Springer Nature Switzerland AG 2019
D.-Z. Du et al. (Eds.): COCOON 2019, LNCS 11653, pp. 445–455, 2019.
https://doi.org/10.1007/978-3-030-26176-4_37

reachability by link breaking (e.g., bad flow removing [3]). The major concern here as argued in [3], the hard constraint is to remove the undesired flow by edge deletion, meanwhile, the other good flow should be preserved as much as possible. In software engineering, a huge body of literature studies discovering logical causality or sufficient condition of finite-state models (such as non-deterministic automata network [4,5], etc.), which is to find a set of transitions such that disabling them could prevent the reachability to some specific states exactly. Another application would be in the area of database, side effect minimizing problem is a fundamental problem in the topic of data provenancing [6]. In such context, deleting any result tuple of a database query can be understood as prevent the reachability of its generating tuple pairs by deleting tuples in those corresponding join paths. The goal of minimizing the side-effect is just to preserve several necessary reachable tuple pairs as possible.

To understand this class of reachability reduction problem, we propose a unified model to depict graph reachability reduction in this paper. We provide a comprehensive analysis on the complexity of its different fragments so that this series of problems can be better understood.

1.1   Model

Vertex Labeled Graph. A vertex labeled graph G is a quintuple (V, E, L, ρ), in which V is the vertex set $\{v_1, v_2, \ldots, v_n\}$, E is the edge set $\{e_1, e_2, \ldots, e_m\}$, L is the given label set $\{b_1, \ldots, b_l\}$, and label mapping $\rho : V \to L$ assigns each vertex a label from L. Especially, when ρ is a *bijection*, then the input graph could be regarded as a classic graph without labels. Given a path P of G, any label b is said to be contained in P if path P contains at least a vertex labeled by b.

Label Reachable Set. Given a vertex labeled graph $G = (V, E, L, \rho)$, a natural pair-wise label reachable set $S \subseteq L^2$ could be found in G, such that $(b_i, b_j) \in S$ if there is at least a path in G which contains both b_i and b_j. We can also generalize such a pair-wise reachable set in a different granularity to r-ary label reachable set $S \subseteq L^r$ such that $\bar{b} = (b_1, \ldots, b_r) \in S$ if there is at least a path in G which contains all the labels from $\bar{b}$. Note that, labels in G could follow any order. Obviously, the label reachable set is fixed once given a vertex labeled graph G, which is denoted as $S(G)$. If G is a connected graph with n vertices and ρ is a *bijection*, then $|S(G)| = n^r$ for each $r \geq 2$. In fact, the input of BFR problem [3] is just a non-uniform case of label reachable set when ρ is a bijection.

To study complexity and algorithm of this problem, we next define the decision and optimization versions of its cut problem.

Decision Version: exact reachability preserved cut. Given a vertex labeled graph $G = (V, E, L, \rho)$ where $S(G)$ is the r-ary label reachable set of G, a negative set $S^- \subseteq S(G)$ and a positive set $S^+ \subseteq S(G) \setminus S^-$, an **exact reachability preserved cut** is an edge set $C \subseteq E$ such that for $G' = (V, E \setminus C)$ obtained by removing the edges in C from G where

$$(a)\ S^- \cap S(G') = \emptyset, \text{ and } (b)\ S^+ \subseteq S(G').$$

It is easy to see for any graph G, a negative set and a positive set, it is not possible that there always exists a corresponding exact reachability preserved cut. Therefore, we usually want to find a maximum reachability preserved cut, which is the optimization version of the cut problem defined as follows.

Optimization Version: maximum reachability preserved cut. Given a vertex labeled graph $G = (V, E, L, \rho)$ where $S(G)$ is the r-ary label reachable set of G, a negative set $S^- \subseteq S(G)$ and a positive set $S^+ \subseteq S(G) \setminus S^-$, a *maximum reachability preserved cut* is an edge set $C \subseteq E$ such that for $G' = (V, E \setminus C)$ obtained by removing the edges in C from G, we have

$$(a)\ S^- \cap S(G') = \emptyset, \text{ and } (b)\ |S^+ \cap S(G')| \text{ is maximum.}$$

Obviously, condition (b) is the only difference between *exact reachability preserved cut* and *maximum reachability preserved cut*. And we denote the reachability preserving cut problem as *r-RP-cut* when the label reachable set is defined as an r-ary one, e.g., *2-RP-cut* and *2-MRP-cut* when r is 2.

1.2 Related Work

Reachability preserved cut is a new problem in the realm of cut problems. Since Ford and Fulkerson [7] proposed the famous max-flow min-cut theorem for network flow, the cut problem has been extensively investigated for decades. The general objective of the cut problem is to find a set of edges with optimal cost whose deletion disconnects the given graph. By the different definition of *cost*, related work could be mainly categorized as following.

Minimum cut is to find a cut to partition the given graph into two or more disjoint components by deleting as few edges as possible [8]. *Minimum k-cut* [9] is to partition the given graph into at least k connected components. *Minimum multi-cut problem* is to find the multi-cut with the minimum total cost for a graph when each edge in the graph is associated with a positive cost. This problem can be reduced to the traditional minimum (S, T)-cut problem in undirected graphs and has been proved to be NP-hard [10] as well.

Maximum cut is to find a cut with the maximum size in a graph [11], its weighted version is one of the 21 NP-complete problems in Karp's famous work [12]. Same as the *minimum cut, maximum cut* is to disconnect a graph with maximum cost rather than minimizing the side effect.

Sparsest cut is to bipartition the vertices of a graph into S and $V - S$ while minimizing the sparsity ratio, *i.e.*, the number of edges across S and $V - S$ to the number of vertices in the smaller half of the partition [13]. *Balanced cut* is an important variant [14], but still not related to our problem.

All the aforementioned cut problems focus on disconnecting a given graph, but do not take the reachability preservation into account, since our goal is to cut off specific vertex pairs while minimizing the side-effect on other reachable pairs. It also follows in the other related work, such as reachability cuts [15] for the vehicle routing problem, local cut [16], minimum feedback arc set [17], maximum acyclic subgraph [18] and so on.

2   Results for Undirected Graphs

For undirected graphs, we show the polynomially intractable and tractable cases at the domination position, then the theory bounds of other cases follow immediately. We here provide the proof sketches.

Theorem 1. *If the input vertex labeled graph G is undirected and ρ is a bijection, then*

(a) *when the underlying graph of G is a forest, r-RP-cut ($r \geq 2$) is* NP-complete *if $|S^-| = 1$ and $S^+ \cup S^- = S(G)$;*

(b) *when the underlying graph of G is a connected graph, 2-MRP-cut, the optimal version, is* NP-complete *even if S^- is complete bipartite and $S^+ \cup S^- = S(G)$; (That is, given $r > 0$, decide if there is a cut C such that all pairs in S^- can be cut off while preserving at least r pairs from S^+, i.e., $|S^+ \cap S(G(V, E \setminus C))| \geq r$)*

(c) *when the underlying graph of G is a forest (with no common label in two disjoint connected components), 2-MRP-cut is* PTime *if $|S^-|$ is of constant size.*

Proof. (i) Reachability checking of multiple vertices in an undirected graph could be easily carried out by a depth-first search in polynomial time, then r-RP-cut is trivially in NP.

We build a simple reduction from 3sat to the **exact reachability preserved cut** problem, which is well known in NP-complete, then the hardness of the **exact reachability preserved cut** problem follows immediately. Given a 3sat instance ϕ which is a 3dnf with n existential variables and m clauses, we build a vertex labeled graph G_ϕ as follows.

For each variable x_i in ϕ, build a tree rooted by label x, and fork to two sub-trees rooted at '$-$' and '$+$' respectively. The leaves in the sub-tree rooted at '$+$' correspond to the clauses including x_i, and the leaves in the sub-tree rooted at '$-$' correspond to the clauses including $\bar{x}_i$.

Initially, $S(G_\phi)$ has the following pairs

$$(i, j), (\mathsf{x}/i, +/-), (\mathsf{x}, i), (+, -).$$

The idea is that we intent to cut off label pair $(+, -)$ to guarantee a valid assignment on variables, while preserving all the label pairs (x, i), so that every clause is satisfied by the corresponding assignment of variables. However, the problem is that the destroy of edges connecting 0 and 1 in each variable tree will impact the reachability from x to c_i. To make up the pairs potentially dead in the solution cut, for each variable tree, duplicate the two sub-trees rooted at $+$ and $-$, while swapping labels $+$ and $-$ in the duplication (Fig. 1). Then, build another tree and distribute all the labels j belonging to the same variable gadget to all the tree vertices. Moreover, add two edges e_1 and e_2 into G_ϕ, with label pairs $(\mathsf{x}, -)$ and $(\mathsf{x}, +)$, so that they will survive from the solution cut. At last, let

$$S^- = \{(+, -)\} \quad \text{and} \quad S^+ = S(G_\phi) \setminus S^-.$$

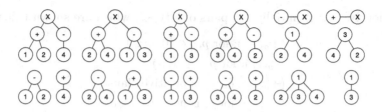

Fig. 1. Undirected graph G_ϕ for 3sat instance $\phi = (x_1 + \bar{x}_2 + x_3)(x_1 + x_2 + \bar{x}_4)(\bar{x}_2 + \bar{x}_3 + x_4)(\bar{x}_1 + x_2 + x_4)$.

One can easily verify that a 3sat instance ϕ is satisfiable if and only if there is a 2-*RP-cut* in G_ϕ such that $(+, -)$ is never reachable while every (x, i) is still in S^+.

(*ii*) We build a reduction from monotone-3sat to 2-*MRP-cut*, which is in NP-complete, then the hardness of the 2-*MRP-cut* problem with arbitrary S^+ and S^- follows immediately.

Roughly speaking, the instance of monotone-3sat is also a 3cnf formula. The only difference is that each clause contains either 3 negative variables or 3 positive ones, *e.g.*, instance $\phi = (x_1 + x_2 + x_3)(\bar{x}_1 + \bar{x}_2 + \bar{x}_4)(x_2 + x_3 + x_4)(\bar{x}_1 + \bar{x}_2 + \bar{x}_4)$ is a monotone 3cnf formula.

Given an instance ϕ of monotone 3sat, for its variables $x_1, \ldots, x_n$, create n vertices $v_1, \ldots, v_n$ in G_ϕ, then add another two vertices $+_0$ and $-_0$, and connect every v_i to $+_0$ and $-_0$.

Additionally, take an integer $z > \max\{m, n\}$, append to $+$ and $-$ by mz^4 distinct vertices, respectively, say $+_1, \ldots, +_{mz^4-1}$ and $-_1, \ldots, -_{mz^4-1}$.

The idea is that we expect to guarantee the consistency of assignment on variables by cutting all the possible paths from $+$ to $-$.

For each clause $c_i = (x_j + x_k + x_l)$, add edges

$$(i_0, v_j), (i_0, v_k), (i_0, v_l).$$

After this, we append $z^4 - 1$ distinct vertices, $i_1, \ldots, i_{z^4-1}$, to each vertex i_0 (Fig. 2).

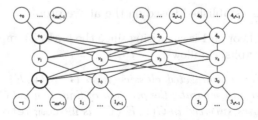

Fig. 2. The undirected graph G_ϕ built for monotone-3sat instance $(x_1 + x_2 + x_3)(\bar{x}_1 + \bar{x}_2 + \bar{x}_4)(x_2 + x_3 + x_4)(\bar{x}_1 + \bar{x}_2 + \bar{x}_4)$.

Obviously, $S(G_\phi)$ initially has pairs of 5 types, which are shown below

$$(+_p, -_q): \text{ for } p, q \in [0, mz^4)$$
$$(v_p, v_q): \text{ for } p, q \in [1, n)$$
$$(+_p, i_q): \text{ for } c_i \text{ is positive clause}$$
$$(-_p, i_q): \text{ for } c_i \text{ is negative clause}$$
$$(i_p, j_q): \text{ for any } c_i \text{ and } c_j$$

Then define S^- (since $S^+ = S(G_\phi) \setminus S^-$ is required). Let

$$D = \{(i_p, j_q)| \text{ exactly one of } c_i \text{ and } c_j \text{ is positive}\},$$
$$D_+ = \{(+_p, i_q)| c_i \text{ is negative}\},$$
$$D_- = \{(-_p, i_q)| c_i \text{ is positive}\},$$
$$D_x = \{(+_p, -_q)| \forall p, q\}.$$

Then let

$$S^- = D \cup D_+ \cup D_- \cup D_x.$$

Note that S^- is complete bipartite, *i.e.*, exactly two groups of vertices. At last, let

$$r = (mz^4 + m_+ z^4)^2 + (mz^4 + (m - m_+)z^4)^2$$

i.e.,

$$r = \left[(m + m_+)^2 + (m + m_-)^2 \right] z^8.$$

One can verify that the monotone-3-sat instance ϕ is satisfiable if and only if there is a cut for G_ϕ which will cut off all the pairs in S^- while preserving at least k pairs in S^+. The key idea of this reduction is that *"if isolate at least one i_0, then there won't be a makeup enough to n^8 by all the vertices corresponding to variables (at most $n^2 \cdot 2mz^4 \ll z^8$)"*. Therefore, for the '$+_0$' part and '$-_0$' part, any i_p should not be cut off. The correctness of assignment on the variables of ϕ will be guaranteed by cutting off all the reachable pairs from S^-.

(*iii*) We will provide an optimal algorithm with time complexity $O(|V| + |E|)$ in the next section.

The Theorem is established based on the above proofs.

Note that the Theorem 1 is surprising since the part (*ii*) implies an important result for simple graphs as follows,

Corollary 1. *Given a connected classic graph $G = (V, E)$ without label and reachable vertex pairs to separate, the problem of finding a cut $C \subseteq E$ maximizing reachable vertex pairs survived in $G(V, E \setminus C)$ is* NP-complete.

Moreover, part (*iii*) of the theorem identifies a class of cases that can be solved optimally in polynomial time, which will be explained in the next section and an efficient algorithm with a linear time complexity is proposed.

3   Results for Digraphs

Now we show that directed graphs make this problem harder than undirected graphs. This is caused by the fact that the number of possible maximal connected components could be exponential with respect to the number of directed edges (think about a "chain"), and reachability checking is somehow harder in directed graphs if input ρ is not a *bijection*.

Theorem 2. *In a weakly connected DAG, for each $r \geq 2$, the r-RP-cut problem is* NP-*hard even if $|S^-| = 1$ and $S^+ \cup S^- = S(G)$.*

Proof. We modify the reduction in part (ii) of Theorem 1 to directed version as shown in Fig. 3.

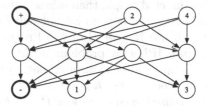

Fig. 3. The DAG G_ϕ built for monotone-3sat instance $(x_1+x_2+x_3)(\bar{x}_1+\bar{x}_2+\bar{x}_4)(x_2+x_3+x_4)(\bar{x}_1+\bar{x}_2+\bar{x}_4)$.

Given an instance ϕ of 3sat, for variables $x_1, \ldots, x_n$, create n vertices $v_1, \ldots, v_n$ without label and two another vertices with label $+$ and $-$, then connect $+$ to every v_i, connected each v_i to $-$. The idea is that we expect to guarantee the consistency of assignment on variables by cut all the possible paths $+$ to $-$.

For negative clause $c_i = (x_j + x_k + x_l)$, add paths

$$i \to v_j, \quad i \to v_k, \quad i \to v_l.$$

For positive clause $c_i = (\bar{x}_j + \bar{x}_k + \bar{x}_l)$, add paths

$$v_j \to i, \quad v_k \to i, \quad v_l \to i.$$

Obviously, $S(G_\phi)$ initially has pairs of another 3 types, besides pair $(+, -)$, as follows

$(+, i)$: c_i is a positive clause;
$(i, -)$: c_i is a negative clause;
(i, j): negative c_i shares a variable in positive c_j.

At last, set $S^- = \{(+, -)\}$ and $S^+ = S(G_\phi) \setminus S^-$.
The idea of this reduction is that the destroy of paths from $+$ to $-$ will impact the pair-wise reachability of that 3 types above, and it is to mimic x_i

taking assignment of 1 that edge $+, v_i$ survived from cut, otherwise taking 0. And if there is a path '$+ \to i$' or '$i \to -$' survived after cutting G, then c_i is true since at least one of its 3 variables is satisfied.

Note that, all the pairs of form (i, j) is preserved by monotone property, correctness follows immediately. Intuitively, the proper survival makes all the clauses true. So, if there is a 2-RP-cut preserving S^+, then we can find a valid corresponding assignment making the monotone-3sat instance formula true.

We claim that, this reduction runs in time of $O(m)$, where m is the number of clauses.

One can verify that a monotone-3sat instance ϕ is satisfiable if and only if there is a 2-*RP-cut* in G_ϕ which exactly cuts the only pair $(+, -)$ of S^- while preserving pairs of S^+.

For the exact upper bound of Reachability preserved cut problem, we show that if the input graph is a directed graph, then it is so hard inside the *Polynomial Hierarchy* for the general case, even beyond NP but never exceeds NP^{NP}. That means we cannot expect to use Integer (Linear) Programming to model and solve it approximately, since IPL is a problem in NP. Moreover, to the best of our knowledge, there is no sub-linear approximation ratio of any complete problem in NP^{NP}. We may not expect a good approximation of this problem in the future for such generalized version, unless $P = NP$.

Theorem 3 (Lower bound for the general *exact reachability preserved cut* problem). *If the vertex labeled graph G is a weakly connected DAG, both S^+ and S^- are of polynomial size, then the general **exact reachability preserved cut** problem is NP^{NP}-complete.*

Proof. We show that condition 'r-ary' makes the **exact reachability preserved cut** problem much harder. However, it is easy to verify that r-*RP-cut* is still bounded in NP^{NP}. For the lower bound, we use the $\exists\forall$-3sat problem belonging to NP^{NP}-*complete*. Given an instance $\exists\bar{x}\forall\bar{x}' \ \phi(\bar{x}, \bar{x}')$ where $\phi(\bar{x}, \bar{x}')$ is a 3dnf formula, the question is whether there is an assignment of the existential variables in $\bar{x}$ which is a mapping from $\bar{x}$ to $\{0, 1\}^n$ such that for all the possible assignments of universal variables in $\bar{x}'$, the 3dnf formula $\phi(\bar{x}, \bar{x}')$ is always true, *i.e.*, it is a *tautology*.

Given gadget G_i for each variable x_i, suppose x_i occurs in clauses $c_{i'}, c_{j'}, \ldots, c_{l'}$ and $\bar{x}_i$ occurs in clauses $c_i, c_j, \ldots, c_l$. Then G_i contains two disjoint paths, both starting with label A and ending with label B. One connects labels $-i', -j', \ldots, -l'$, and the other connects labels $-i, -j, \ldots, -l$, as shown in Fig. 4.

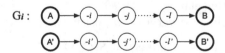

Fig. 4. Gadget G_i.

For any variable x_i, connect its corresponding gadget G_i with its subsequent G_{i+1} with 4 disjoint paths as shown in Fig. 5.

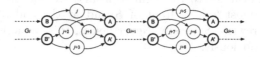

Fig. 5. Connecting gadget G_i with its subsequent G_{i+1}.

$$G_i.B \rightarrow j \rightarrow G_{i+1}.A, \qquad G_i.B \rightarrow j+1 \rightarrow G_{i+1}.A'$$
$$G_i.B' \rightarrow j+2 \rightarrow G_{i+1}.A, \quad G_i.B' \rightarrow j+3 \rightarrow G_{i+1}.A'$$

we then require that (For simplicity, we define S^+ by using non-uniform notations. At the end of the proof, we will show a simple uniform modification.)

- for each variable, S^+ should contain such 4 paths;
- for each *universal* variable, e.g., x_{i+1}, S^+ should include 16 paths containing all the combinations of $\{j, j+1, j+2, j+3\}$ and $\{j+5, j+6, j+7, j+8\}$;

The idea here is to restrict the two paths inside G_{i+1} which should not be cut, so as to mimic arbitrary assignment of both 0 and 1 of universal variable.

After connecting all the gadgets one by one, append a starting label S at the front of G_1 by two disjoint paths $s \rightarrow a \rightarrow G_1.A$ and $s \rightarrow b \rightarrow G_1.A'$, then append an ending label T at the end of G_m also by two disjoint paths $G_m.B \rightarrow c \rightarrow T$ and $G_m.B' \rightarrow d \rightarrow T$. We also require S^+ includes the 4 disjoint paths and another pair (S, T).

Then, append S by another redundant path

$$-1 \rightarrow -2 \cdots \rightarrow -m$$

to make sure such a path is in the initial set $S(G_\phi)$.

The entire constructed directed graph G_ϕ is shown in Fig. 6.

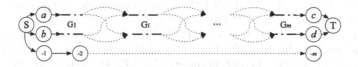

Fig. 6. Directed graph G_ϕ.

Finally, let $S^- = \{(-1, -2, \ldots, -m)\}$, note that it is valid since $S^- \subseteq S(G_\phi)$.

The overall idea is that reachability of (S, T) makes sure that one and only one path of $A \rightarrow B$ and $A' \rightarrow B'$ could be cut for each *existential* variable, and cut $(-1, -2, \ldots, -m)$ is to guarantee at least one clause of the 3dnf formula is satisfied under any assignment on universal variables. If there is no assignment on the existential variables making the 3dnf formula ϕ to be a *tautology*, then every possible cut in G_ϕ preserving S^+ is not able to destroy $(-1, -2, \ldots, -m)$ of S^-.

At last, to uniform S^+ to the form of m-ary, we can add enough distinct dummy labels, at most $2m$, in front of S and at the end of T. We can use these dummies to rewrite S^+ easily. Nevertheless, this reduction could be done in $O(n + m)$ time obviously.

One can verify that an $\exists\forall$-3sat instance ϕ is satisfiable if and only if there is an *r-RP-cut* in G_ϕ to cut off all the pairs of S^- while preserving S^+ entirely. $\square$

So we know that under a very generalized input, the case of directed graphs is much harder than that of undirected graphs.

4   Conclusions

The reachability reduction problem can be modeled as reachability preserved cut on vertex labeled graphs. Based on the unified model, we show a theoretical complexity analysis and find that it is polynomially intractable in several cases from NP-complete to Σ_2^P-complete. And for the cases in directed graphs (even DAG), it is harder than the undirected graph cases. Moreover, a surprising finding is for classic graphs, on which reachability reducing is still NP-complete.

Acknowledgments. This work is partly supported by the National Natural Science Foundation of China (NSFC) under grant NOs. 61832003, U1811461, 61732003 and the National Science Foundation (NSF) under grant NOs. 1252292, 1741277, 1829674 and 1704287.

References

1. Zheng, A.X., Dunagan, J., Kapoor, A.: Active graph reachability reduction for network security and software engineering. In: Twenty-Second International Joint Conference on Artificial Intelligence (2011)
2. Dunagan, J., Zheng, A.X., Simon, D.R.: Heat-ray: combating identity snowball attacks using machine learning, combinatorial optimization and attack graphs. In: Proceedings of the ACM SIGOPS 22nd Symposium on Operating Systems Principles, pp. 305–320. ACM (2009)
3. Polevoy, G., Trajanovski, S., Grosso, P., de Laat, C.: Removing undesirable flows by edge deletion. In: Kim, D., Uma, R.N., Zelikovsky, A. (eds.) COCOA 2018. LNCS, vol. 11346, pp. 217–232. Springer, Cham (2018). https://doi.org/10.1007/978-3-030-04651-4_15
4. Folschette, M., Paulevé, L., Magnin, M., Roux, O.: Sufficient conditions for reachability in automata networks with priorities. Theor. Comput. Sci. **608**, 66–83 (2015)
5. Paulevé, L., Andrieux, G., Koeppl, H.: Under-approximating cut sets for reachability in large scale automata networks. In: Sharygina, N., Veith, H. (eds.) CAV 2013. LNCS, vol. 8044, pp. 69–84. Springer, Heidelberg (2013). https://doi.org/10.1007/978-3-642-39799-8_4

6. Buneman, P., Khanna, S., Tan, W.C.: On the propagation of deletions and annotations through views. In: Proceedings of 21st ACM Symposium on Principles of Database Systems, Madison, Wisconsin, pp. 150–158 (2002)
7. Ford, L.R., Fulkerson, D.R.: Maximal flow through a network. In: Gessel, I., Rota, G.C. (eds.) Classic papers in Combinatorics, 243–248. Birkhäuser, Boston (2009)
8. Smith, J.M.: Optimization theory in evolution. Ann. Rev. Ecol. Syst. **9**, 31–56 (1978)
9. Goldschmidt, O., Hochbaum, D.S.: Polynomial algorithm for the k-cut problem. In: 29th Annual Symposium on Foundations of Computer Science, pp. 444–451 (1988)
10. Dahlhaus, E., Johnson, D.S., Papadimitriou, C.H., Seymour, P.D., Yannakakis, M.: The complexity of multiway cuts. In: Proceedings of the Twenty-Fourth Annual ACM Symposium on Theory of Computing, pp. 241–251. ACM (1992)
11. Ausiello, G., Cristiano, F., Laura, L.: Syntactic isomorphism of CNF Boolean formulas is graph isomorphism complete. In: Electronic Colloquium on Computational Complexity, vol. 19, p. 122 (2012)
12. Karp, R.M.: Reducibility among combinatorial problems. In: Miller, R.E., Thatcher, J.W., Bohlinger, J.D. (eds.) Complexity of Computer Computations. The IBM Research Symposia Series, pp. 85–103. Springer, Boston (1972). https://doi.org/10.1007/978-1-4684-2001-2_9
13. Shmoys, D.B.: Cut problems and their application to divide-and-conquer. In: Approximation Algorithms for NP-Hard Problems, pp. 192–235 (1997)
14. Andreev, K., Racke, H.: Balanced graph partitioning. Theory Comput. Syst. **39**, 929–939 (2006)
15. Lysgaard, J.: Reachability cuts for the vehicle routing problem with time windows. Eur. J. Oper. Res. **175**, 210–223 (2006)
16. Chung, F.: Random walks and local cuts in graphs. Linear Algebra Appl. **423**, 22–32 (2007)
17. Younger, D.: Minimum feedback arc sets for a directed graph. IEEE Trans. Circuit Theory **10**, 238–245 (1963)
18. Guruswami, V., Manokaran, R., Raghavendra, P.: Beating the random ordering is hard: inapproximability of maximum acyclic subgraph. In: 49th Annual IEEE Symposium on Foundations of Computer Science, pp. 573–582. IEEE (2008)

Give and Take: Adaptive Balanced Allocation for Peer Assessments

Hideaki Ohashi[✉], Yasuhito Asano, Toshiyuki Shimizu,
and Masatoshi Yoshikawa

Graduate School of Informatics, Kyoto University,
Yoshida Honmachi, Sakyo-ku, Kyoto 606-8501, Japan
ohashi@db.soc.i.kyoto-u.ac.jp,
{asano,tshimizu,yoshikawa}@i.kyoto-u.ac.jp

Abstract. Peer assessments, in which people review the works of peers and have their own works reviewed by peers, are useful for assessing homework, reviewing academic papers and so on. In conventional peer assessment systems, works are usually allocated to people before the assessment begins; therefore, if people drop out (abandoning reviews) during an assessment period, an imbalance occurs between the number of works a person reviews and that of peers who have reviewed the work. When the total imbalance increases, some people who diligently complete reviews may suffer from a lack of reviews and be discouraged to participate in future peer assessments. Therefore, in this study, we adopt a new adaptive allocation approach in which people are allocated review works only when requested and propose an algorithm for allocating works to people, which reduces the total imbalance. To show the effectiveness of the proposed algorithm, we provide an upper bound of the total imbalance that the proposed algorithm yields. In addition, we experimentally compare the proposed adaptive allocation to existing nonadaptive allocation methods.

Keywords: Peer assessment · Task allocation · Allocation algorithm

1 Introduction

Peer assessments, in which people review the works of peers and have their own works reviewed by peers, are useful for reviewing homework and academic papers. Particularly, peer assessment is effective when the number of participants is large, such as in a massive open online course (MOOC), in which people can attend various lectures on the Internet. Lecturers and teaching assistants (TAs) alone are unable to review large volumes of works [10,13,14].

However, some reports indicate that people are not willing to participate in peer assessments; one reason is that people are disheartened by the lack of reviews [2,12]. Therefore, we need to develop methods of peer assessment that allow people to receive sufficient feedback based on the number of reviews to increase the number of people who participate in peer assessments.

© Springer Nature Switzerland AG 2019
D.-Z. Du et al. (Eds.): COCOON 2019, LNCS 11653, pp. 456–468, 2019.
https://doi.org/10.1007/978-3-030-26176-4_38

A major reason for the existence of insufficient review numbers is that peers dropout without reviewing allocated works [2,8]. In existing peer assessment systems, each person is usually asked to review a predefined number of works, and works are allocated to people before the peer assessments start. If a certain number of people drop out of the review process, an imbalance occurs between the number of works a person reviews (termed the "reviewing number") and the number of peers who review the work of the same person (termed the "reviewed number"). When the total imbalance increases, people who diligently finish reviews may suffer from a lack of reviews and be discouraged to participate in future peer assessments.

To address this problem, we develop a new adaptive allocation approach in which people are allocated works only when requested. People can request one work to review at any time; they can request second and subsequent works to review only after they have finished the review of the previously requested work. This rule is more suitable for a realistic situation in which some people drop out during peer assessments.

Under the above approach, our goal is to reduce the sum of the absolute values of the differences between the reviewing number and reviewed number of each person, termed RR imbalance (reviewing-reviewed imbalance).

We propose an allocation algorithm called the RRB (reviewing-reviewed balanced) allocation algorithm, which reduces the RR imbalance, which means that it is highly possible that the work of one person will be reviewed (Taken) as many times as that same person reviews the works of others (Given). It can be expected that this algorithm resolves dissatisfaction about the lack of reviews and incentivizes people to review the works of their peers.

To demonstrate the usefulness of the RRB algorithm, we theoretically prove that the RRB algorithm guarantees an upper bound of the RR imbalance, which does not depend on the number of people; instead, it depends on the maximum reviewing number among people. In practical situations, the maximum reviewing number usually does not increase, even if the number of people grows. Therefore, our results show that the average difference between the reviewing number of each person and the reviewed number decreases as the number of people increases. This property is desirable in MOOC settings from the viewpoint of fairness among people.

However, unfairness still remains up to the amount of the upper bound. To reduce the RR imbalance, extra effort is required. For instance, in MOOC settings, lecturers and TAs could perform extra reviews for people whose reviewing number is above their reviewed number at the end of the peer assessment. In this case, the obtained upper bound can be used to estimate the number of reviews the lecturers and TAs need to perform.

To show the effectiveness of the RRB algorithm, we experimentally compare its performance with that of the existing nonadaptive allocation.

The remainder of this paper is organized as follows. In Sect. 2, we introduce the related works. We describe the problem definitions in this research in Sect. 3. In Sect. 4, we describe the RRB algorithm. In Sect. 5, we prove the upper bound of the RR imbalance by the RRB algorithm. We present the experimental results in Sect. 6, and finally, we conclude this work and suggest future work in Sect. 7.

2   Related Work

Crowdsourcing has attracted much attention, and studies on crowdsourcing and peer assessment are closely related [18]. Many task allocation methods have been proposed for crowdsourcing [1,6,11,19]; however, there have been few proposals for task allocation methods in peer assessments. The difference between task allocations for crowdsourcing and those for peer assessment is the strength of the incentive provided; crowdsourcing can use clear incentives, such as money, that are unavailable in peer assessment situations. Consequently, dropout is more likely to occur in peer assessments; thus, peer assessment research must consider the effect of dropout.

Estévez-Ayres et al. [9] proposed an allocation mechanism to avoid lack of reviews due to dropout and confirmed its usefulness through a simulation. They assumed that some people were willing to review other works even when their reviewing number exceeded their reviewed number. We do not assume such optimistic person characteristics in this study.

There are some methods for improving the quality of reviews in peer assessments. A method of automatically assessing review content (automated meta-reviewing) that prompts the reviewer to correct and improve review content has been proposed [15]. In addition, another method was proposed in which the reviewee scores the reviewer on his or her review content [3]. Increasing the quality of the rubrics (reviewing standards) used in peer assessments leads directly to improved review quality; therefore, some studies have verified the effect of rubrics [4,10]. In addition, many studies exist that aggregate reviewer scores in peer assessments; these studies apply quality control research in the context of crowdsourcing [5,7,14,16,17]. The above studies are orthogonal to our study; hence, we can combine their methodologies and results with ours.

3   Problem Setting

Initially, we explain our problem setting intuitively through Fig. 1. In this research, to deal with realistic situations in which some people drop out during the peer assessment process, we propose an allocation algorithm that uses an adaptive allocation approach. Under this approach, people are allocated a new work only when they request one, and they can request an additional work to review only after they have finished the review of the previous work. In addition, we assume that people always complete the requested review. This assumption is considered to be valid because people who are not willing to review do not request a work in the first place.

In Fig. 1, we assume that there are five people, a, b, c, d, and e, and each vertex represents a person. First, a requests a work; then, the work of d is allocated to a. This allocation is denoted by the directed edge from a to d. We assume that no person can review his or her own work and that each person can review a given work only once. After the first allocation, the next allocation occurs when another person requests a work, and then, a directed edge is drawn. These steps are repeated under an adaptive allocation approach.

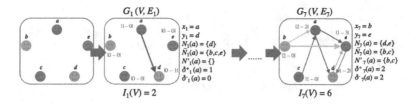

Fig. 1. Example of adaptive allocation behavior.

Let V be a set of people, E_i be the edge set and G_i be the graph created up to the i-th allocation. Note that $E_0 = \emptyset$. The RR imbalance (reviewing-reviewed imbalance) in graph G_1, which consists of single edge, is the sum of all the absolute values of the differences between the reviewing number (outdegree) and the reviewed number (indegree) as follows: $|1 - 0| + |0 - 0| + |0 - 0| + |0 - 1| + |0 - 0| = 2$. Now, let us assume that there are seven allocations during this peer assessment. The final RR imbalance in graph G_7 is $|2 - 2| + |2 - 0| + |1 - 0| + |1 - 2| + |1 - 3| = 6$. In this study, we propose an allocation algorithm that reduces the RR imbalance at the end of a peer assessment.

Some definitions are provided below. Let a person doing the i-th request under the adaptive allocation approach be $x_i \in V$. A work by a person $y_i(\neq x_i) \in V$ is allocated to x_i before a person x_{i+1} can request a work. This allocation is represented by a directed edge from x_i to y_i. In the graph G_i, let the set of people whose works are allocated to person $v \in V$ be $N_i(v)$ and $\bar{N}_i(v) = V \setminus \{N_i(v) \cup \{v\}\}$; then, $y_{i+1} \in \bar{N}_i(x_{i+1})$. Moreover, use $N_i'(v)$ to denote the set of people who review the work of person $v \in V$. The reviewing number (outdegree) of person v in graph G_i is defined as $\delta_i^+(v)(= |N_i(v)|)$, and the reviewed number (indegree) is defined as $\delta_i^-(v)(= |N_i'(v)|)$.

We explain the above definitions using Fig. 1. In Fig. 1, we assume five people, a, b, c, d, and e; thus, $V = \{a, b, c, d, e\}$. Initially, person a requests a work, and the work of person d is allocated to a; therefore, x_1 and y_1 are a and d, respectively. The edge set E_1 of the graph $G_1(V, E_1)$ contains only one directed edge from a toward d. In addition, $N_1(a) = \{d\}$, $\bar{N}_1(a) = \{b, c, e\}$ and $N_1'(a) = \{\}$, and the node a has an outdegree of 1 and an indegree of 0; consequently, $\delta_1^+(a) = 1$ and $\delta_1^-(a) = 0$.

Our goal is to reduce the RR imbalance when the last allocation is done during the peer assessment period. The RR imbalance is defined as the sum of the absolute values of the difference between the reviewing number and the reviewed number for all people. That is, when the t-th allocation is finished, RR imbalance $I_t(V)$ can be calculated by the following equation:

$$I_t(V) = \sum_{v \in V} |\delta_t^+(v) - \delta_t^-(v)|$$

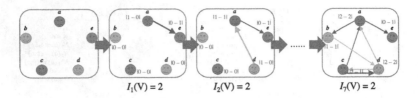

$$I_1(V) = 2 \qquad I_2(V) = 2 \qquad I_7(V) = 2$$

Fig. 2. Example of RRB behavior.

4   Algorithm

In this section, we propose an allocation algorithm to reduce the RR imbalance, termed the RRB algorithm. A theoretical analysis of the RRB algorithm is given in Sect. 5, and experiments to evaluate the performance of the RRB algorithm are presented in Sect. 6.

The RRB algorithm adopts a greedy approach to reduce the RR imbalance. We propose an algorithm which selects a person whose difference between reviewing number and reviewed number is the maximum at each request.

Algorithm 1. The RRB algorithm
___
INPUT: V ▷ a set of people
INPUT: $\langle x_1, \ldots, x_t \rangle$ ▷ the sequence of people who request a work
OUTPUT: $\langle y_1, \ldots, y_t \rangle$ ▷ the sequence of people whose works are reviewed
 1: **for** $i \leftarrow 0$ to $t - 1$ **do**
 2:     $y_{i+1} \in \arg\max\limits_{v \in \tilde{N}_i(x_{i+1})} (\delta_i^+(v) - \delta_i^-(v))$
 3: **end for**

Note that y_{i+1} is selected randomly when multiple candidates exist.

We provide an intuitive explanation of the above algorithm using Fig. 2. In this figure, it is assumed that there are five people, a, b, c, d, and e, whose requesting order is $\langle a, d, b, a, c, d, c \rangle$. First, the difference between the reviewing number and the reviewed number of every person is 0; therefore, the work is randomly allocated to a. Let us assume that the work of e is randomly selected. Next, because the difference between the reviewing number and reviewed number of a is the maximum, the work of a is allocated to d. Subsequently, a's difference between reviewing number and reviewed number becomes 0, while for d, the difference becomes 1. Therefore, the work of d is preferentially allocated in the next step. In Fig. 2, the above allocation is repeated, showing intuitively how the RRB algorithm aims to reduce the RR imbalance.

5   Theoretical Analysis for the RRB Algorithm

In this section, we show that when the maximum outdegree of graph G_i is k and the number of people exceeds $k^2 + k + 1$, the RRB algorithm ensures that the upper bound of the RR imbalance in the graph G_i is $\mathcal{O}(k^2)$. The upper bound does not depend on the total number of people n; it depends only on the maximum number of reviews performed by any one reviewer. When an enormous number of people exist, such as in an MOOC, k is expected to be considerably smaller than n because one person cannot review works by everyone. In other words, the proposed algorithm should be extremely effective on MOOCs. Although we assume that the number of people is larger than $k^2 + k + 1$, this is equivalent to the assumption that the total number of people is larger than the square of the reviewing number of any one person. It is natural to use this assumption when many people are participating. In the following section, after presenting two lemmas, we prove our assertion of the upper bound.

Lemma 1. *For a vertex subset $V' \subseteq V$ of graph G_i, suppose that the following inequality holds for all vertices $v \in V'$:*

$$\delta_i^+(v) - \delta_i^-(v) \leq 0$$

We define the set of edges from $V \setminus V'$ to V' as $E_I \subseteq E_i$ and the set of edges from V' to $V \setminus V'$ as $E_O \subseteq E_i$. Then, the following equation is satisfied:

$$I_i(V') = |E_I| - |E_O|$$

Proof. From the assumption, $|\delta_i^+(v) - \delta_i^-(v)| = \delta_i^-(v) - \delta_i^+(v) \geq 0$ is satisfied for any $v \in V'$. Therefore, the RR imbalance on V' is as follows:

$$I_i(V') = \sum_{v \in V'} \delta_i^-(v) - \delta_i^+(v) = \sum_{v \in V'} \delta_i^-(v) - \sum_{v \in V'} \delta_i^+(v)$$

Here, we define the edge set in V' as $E' \subseteq E_i$, and the following two equations are satisfied:

$$\sum_{v \in V'} \delta_i^-(v) = |E'| + |E_I|$$

$$\sum_{v \in V'} \delta_i^+(v) = |E'| + |E_O|$$

Hence, $I_i(V') = (|E'| + |E_I|) - (|E'| + |E_O|) = |E_I| - |E_O|$ □

Lemma 2. *The maximum outdegree $\max_{v \in V}\{\delta_i^+(v)\}$ in G_i is defined as k_i. Assuming that $n > k_i^2 + k_i + 1$, the case that the RR imbalance increases with the $i+1$-th allocation, or $I_{i+1}(V) > I_i(V)$, is limited to the following case, and the increment is 2.*

$$\delta_i^+(x_{i+1}) - \delta_i^-(x_{i+1}) \geq 0 \quad \text{and} \quad \delta_i^+(y_{i+1}) - \delta_i^-(y_{i+1}) = 0$$

Proof. We separate the cases as follows:

0. $\delta_i^+(x_{i+1}) - \delta_i^-(x_{i+1}) < 0$ & $\delta_i^+(y_{i+1}) - \delta_i^-(y_{i+1}) > 0$
1. $\delta_i^+(x_{i+1}) - \delta_i^-(x_{i+1}) \geq 0$ & $\delta_i^+(y_{i+1}) - \delta_i^-(y_{i+1}) > 0$
2. $\delta_i^+(x_{i+1}) - \delta_i^-(x_{i+1}) < 0$ & $\delta_i^+(y_{i+1}) - \delta_i^-(y_{i+1}) \leq 0$
3. $\delta_i^+(x_{i+1}) - \delta_i^-(x_{i+1}) \geq 0$ & $\delta_i^+(y_{i+1}) - \delta_i^-(y_{i+1}) < 0$
4. $\delta_i^+(x_{i+1}) - \delta_i^-(x_{i+1}) \geq 0$ & $\delta_i^+(y_{i+1}) - \delta_i^-(y_{i+1}) = 0$

Adding the edges (x_{i+1}, y_{i+1}) means that $\delta_i^+(x_{i+1})$ and $\delta_i^-(y_{i+1})$ are incremented by 1. That is, $\delta_i^+(x_{i+1}) - \delta_i^-(x_{i+1})$ increases by 1 and $\delta_i^+(y_{i+1}) - \delta_i^-(y_{i+1})$ decreases by 1. Therefore, it is obvious that the RR imbalance decreases for case 0. Next, in cases 1 and 2, the RR imbalance does not change because either $|\delta_i^+(x_{i+1}) - \delta_i^-(x_{i+1})|$ or $|\delta_i^+(y_{i+1}) - \delta_i^-(y_{i+1})|$ increases by 1, but the other decreases by 1. In case 3, because the RRB algorithm chooses a y_{i+1} that meets $\delta_i^+(y_{i+1}) - \delta_i^-(y_{i+1}) < 0$, we require the condition that $\delta_i^+(v) - \delta_i^-(v) \leq \delta_i^+(y_{i+1}) - \delta_i^-(y_{i+1}) < 0$ for any $v \in \bar{N}_i(x_{i+1})$. That is, $|\delta_i^+(v) - \delta_i^-(v)| \geq 1$ for any $v \in \bar{N}_i(x_{i+1})$. Here, because $|N_i(x_{i+1})| \leq k_i$, $|\bar{N}_i(x_{i+1})| \geq n - k_i - 1$, the RR imbalance on $\bar{N}_i(x_{i+1})$ satisfies the following inequality:

$$I_i(\bar{N}_i(x_{i+1})) \geq n - k_i - 1 \tag{1}$$

In contrast, the number of edges from $N_i(x_{i+1})$ to $\bar{N}_i(x_{i+1})$ is at most k_i^2 because $|N_i(x_{i+1})| \leq k_i$; therefore, the following inequality holds by Lemma 1:

$$I_i(\bar{N}_i(x_{i+1})) \leq k_i^2 \tag{2}$$

From the above two inequalities (1 and 2), $n - k_i - 1 \leq k_i^2$. However, this contradicts the assumption of Lemma 2 $n > k_i^2 + k_i + 1$. Therefore, case 3 cannot occur.

In addition, the RR imbalance increases by two in case 4. Thus, we complete the proof of Lemma 2. □

Theorem 1. *We assume that $n > k_i^2 + k_i + 1$. After the i-th allocation based on the RRB algorithm is completed, the RR imbalance in graph G_i satisfies the following condition:*

$$I_i(V) \leq 4k_i^2 - 4k_i + 2$$

Proof. We provide an outline of the proof and prove Theorem 1 using mathematical induction. First, using Lemma 2, we show two conditions where the RR imbalance increases during the $i + 1$-th allocation. Then, we divide the person sets into $\{x_{i+1}\}$, $N_i(x_{i+1})$ and $\bar{N}_i(x_{i+1})$ and consider the number of edges between sets and in each set to derive the upper bound of the RR imbalance.

We begin our proof of Theorem 1 by mathematical induction on the number of allocations i. The proposition clearly holds when $i = 1$. We assume that the proposition holds in the case of $i = l(\geq 2)$. $1 \leq k_l \leq k_{l+1}$; thus, the condition when $4k_l^2 - 4k_l + 2 \leq 4k_{l+1}^2 - 4k_{l+1} + 2$ is satisfied. Then, when the RR imbalance does not increase in the $l + 1$-th allocation—that is, when $I_{l+1}(V) \leq I_l(V)$ is satisfied—the following condition is met:

$$I_{l+1}(V) \leq I_l(V) \leq 4k_l^2 - 4k_l + 2 \leq 4k_{l+1}^2 - 4k_{l+1} + 2$$

Therefore, from Lemma 2, we should consider only the following equation:

$$\delta_l^+(x_{l+1}) - \delta_l^-(x_{l+1}) \geq 0 \ \& \ \delta_l^+(y_{l+1}) - \delta_l^-(y_{l+1}) = 0 \tag{3}$$

In addition, if $\delta_l^+(x_{l+1}) = k_l$, then $k_{l+1} = k_l + 1$ holds. From Lemma 2, the RR imbalance increment is at most 2. Consequently, the following holds:

$$I_{l+1}(V) \leq (4k_l^2 - 4k_l + 2) + 2 \leq 4(k_l + 1)^2 - 4(k_l + 1) + 2 = 4k_{l+1}^2 - 4k_{l+1} + 2$$

Therefore, we need to consider only the following case:

$$\delta_l^+(x_{l+1}) \leq k_l - 1 \tag{4}$$

Since the vertex set of graph G_l is $\{x_{l+1}\} \oplus \bar{N}_l(x_{l+1}) \oplus N_l(x_{l+1})$ (see Fig. 3), $I_l(V) = I_l(\{x_{l+1}\}) + I_l(\bar{N}_l(x_{l+1})) + I_l(N_l(x_{l+1}))$. Subsequently, the values on the right side of the expression can be calculated individually.

1. $I_l(\{x_{l+1}\})$: We consider the edge sets $E1, E2$, and $E3$ in Fig. 3. From conditions (3) and (4), the following condition holds:

$$I_l(\{x_{l+1}\}) = |\delta_l^+(x_{l+1}) - \delta_l^-(x_{l+1})| = \delta_l^+(x_{l+1}) - \delta_l^-(x_{l+1}) \leq \delta_l^+(x_{l+1}) \leq k_l - 1$$

2. $I_l(\bar{N}_l(x_{l+1}))$: We consider the edge sets $E2, E4$, and $E5$ and the edges in $\bar{N}_l(x_{l+1})$ in Fig. 3. From condition (3), the RRB algorithm selects a y_{+1} that meets $\delta_l^+(y_{l+1}) - \delta_l^-(y_{l+1}) = 0$. Then, because the RRB algorithm chooses a $v \in \bar{N}_l(x_{l+1})$ with the maximum $\delta_l^+(v) - \delta_l^-(v)$, the following condition holds:

$$\forall v \in \bar{N}_l(x_{l+1}), \ \delta_l^+(v) - \delta_l^-(v) \leq 0 \tag{5}$$

Therefore, from Lemma 1, the RR imbalance on $\bar{N}_l(x_{l+1})$ is less than $|E4|$ (the number of edges from $N_l(x_{l+1})$ to $\bar{N}_l(x_{l+1})$). From condition (4), $|N_l(x_{l+1})| \leq k_l - 1$ holds. Then, because the maximum outdegree is k_l, the following is satisfied:

$$I_l(\bar{N}_l(x_{l+1})) \leq |E4| \leq k_l(k_l - 1) \tag{6}$$

3. $I_l(N_l(x_{l+1}))$: We consider the edge sets $E1, E3, E4$, and $E5$ and the edges in $N_l(x_{l+1})$ in Fig. 3. We utilize the fact that the RR imbalance on $N_l(x_{l+1})$ is less than the sum of the outdegree and indegree in $N_l(x_{l+1})$—which can be written as follows:

$$I_l(N_l(x_{l+1})) = \sum_{v \in V} |\delta_l^+(v) - \delta_l^-(v)| \leq \sum_{v \in V} (\delta_l^+(v) + \delta_l^-(v))$$

From condition (4), because $|N_l(x_{l+1})| \leq k_l - 1$, the outdegree is less than $k_l(k_l - 1)$, and the indegree is the sum of the edges from $\{x_{l+1}\}$, $N_l(x_{l+1})$ and $\bar{N}_l(x_{l+1})$.

(a) Edges from $\{x_{l+1}\}$ (E1): From condition (4), the number of edges is less than $k_l - 1$.
(b) Edges between $N_l(x_{l+1})$: From condition (4), $|N_l(x_{l+1})| \leq k_l - 1$. Then, the number of edges is less than $(k_l - 1)(k_l - 2)$ because no self-loop occurs.
(c) Edges from $\bar{N}_l(x_{l+1})$ (E5): From condition (5) and Lemma 1, the following is satisfied:

$$I_l(\bar{N}_l(x_{l+1})) = |E4| - (|E2| + |E5|) \geq 0$$

Therefore, from condition (6), $|E5| \leq |E2| + |E5| \leq |E4| \leq k_l(k_l - 1)$ holds.

Hence, the sum of the indegree is less than $(k_l-1)+(k_l-1)(k_l-2)+k_l(k_l-1) = 2k_l^2 - 3k_l + 1$. Then, the sum of the outdegree and indegree is less than $k_l(k_l - 1) + 2k_l^2 - 3k_l + 1 = 3k_l^2 - 4k_l + 1$, and $I_l(N_l(x_{l+1})) \leq 3k_l^2 - 4k_l + 1$.

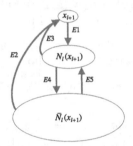

Fig. 3. Grouping for proof of Theorem 1.

Therefore, after the l-th allocation, the following condition holds:

$$I_l(V) \leq k_l - 1 + k_l(k_l - 1) + 3k_l^2 - 4k_l + 1 = 4k_l^2 - 4k_l$$

The RR imbalance increment is 2 from Lemma 2, and $k_l = k_{l+1}$ because of condition (4); thus, the following condition is satisfied after the $l + 1$-th allocation:

$$I_{l+1}(V) \leq 4k_l^2 - 4k_l + 2 = 4k_{l+1}^2 - 4k_{l+1} + 2$$

which concludes the proof of Theorem 1.   □

Based on the above proof, when using the RRB algorithm, the upper bound of the RR imbalance in the graph G_i is $\mathcal{O}(k^2)$, when the maximum outdegree of graph G_i is k and the number of people exceeds $k^2 + k + 1$. By Theorem 1, even if the number of people is large, when $k = 5$, we can know beforehand that the upper bound becomes $4 \cdot 5^2 - 4 \cdot 5 + 2 = 82$.

6   Experiments

We experimentally compare the proposed algorithm under the adaptive allocation approach to an algorithm under the existing nonadaptive allocation approach using two types of data. First, we describe the data characteristics, and
then, we describe a baseline and present the experimental results.

6.1   Experimental Data

We create experimental data based on the data published by Canvas Network[1].
These data are comprised of de-identified data from March 2014 - September
2015 of Canvas Network open courses. In our experiments, we refer to those data
whose class ID is 770000832960949 and whose assignment ID is 770000832930436
(denoted as data 1) and those data whose class ID is 770000832945340 and
assignment ID is 770000832960431 (denoted as data 2). Figure 4 shows a plot of
the number of reviewers for each number of reviews from the datasets.

Then, we generate the requesting order because we cannot read it from the
Canvas Network data. The requesting order is generated as follows. We set the
probability that reviewer x_{i+1} is the same as the previous reviewer x_i to P and
arrange the reviewers according to this probability. Note that when the previous
reviewer x_i cannot review another work, the reviewer x_{i+1} is randomly selected
regardless of x_i. For example, when $P = 0$, reviewer x_{i+1} is randomly chosen
regardless of the previous reviewer x_i, and when $P = 1$, reviewer x_{i+1} is selected
to be the previous reviewer x_i.

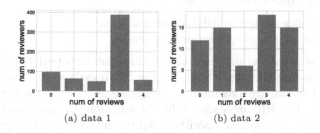

(a) data 1   (b) data 2

Fig. 4. The number of reviewers for each number of reviews.

6.2   Baseline

As a baseline, we utilize random allocations so that both the reviewing number
and the reviewed number for all people are the same before the assessment starts.
We denote this algorithm as Random. Most of the existing peer assessment

---

[1] https://dataverse.harvard.edu/dataset.xhtml?persistentId=doi:10.7910/DVN/
XB2TLU.

systems adopt this allocation approach[2,3]. Note that we allow for new works to be allocated to people who wish to review more than the predefined number of works, similar to the study in [9]. In this experiment, we set the number of works allocated to one person to 3.

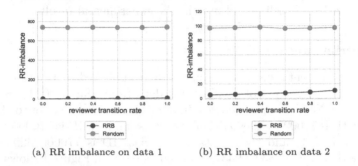

(a) RR imbalance on data 1   (b) RR imbalance on data 2

Fig. 5. Experimental results (Color figure online)

6.3 Experimental Results

For each method, we generate 100 data points and apply the algorithms to these data; then, we obtain the average value of the RR imbalance. The results are shown in Fig. 5. The vertical axis represents the RR imbalance, and the horizontal axis represents the probability value P. The two types of lines plotted in each figure represent the following.

- *RRB:* Imbalance using the RRB algorithm (blue)
- *Random:* Imbalance using Random (orange)

The result shows that the proposed algorithm greatly outperforms the baseline. We can confirm that the upper bound of the RR imbalance described in Sect. 5 is established because the maximum reviewing number is 4 in the both data, and the upper bound of RR imbalance is 50. In addition, we can also see that the performances of the proposed algorithm slightly deteriorate when probability P is high; that is, the same reviewers continue reviewing.

7 Conclusion

In this study, we propose an allocation algorithm to achieve fair peer assessments with respect to the number of reviews using an adaptive allocation approach and considering a situation where dropout can occur during peer assessment. We analyze the RRB algorithm theoretically and show its robustness. In future work, we plan to propose a framework that could be useful throughout the peer assessment process, including aggregating reviewer scores.

---

[2] https://www.coursera.org.
[3] https://www.edx.org.

References

1. Abraham, I., Alonso, O., Kandylas, V., Slivkins, A.: Adaptive crowdsourcing algorithms for the bandit survey problem. In: Conference on Learning Theory, pp. 882–910 (2013)
2. Acosta, E.S., Otero, J.J.E., Toletti, G.C.: Peer review experiences for MOOC. Development and testing of a peer review system for a massive online course. New Educ. Rev. **37**(3), 66–79 (2014)
3. de Alfaro, L., Shavlovsky, M.: CrowdGrader: a tool for crowdsourcing the evaluation of homework assignments. In: Proceedings of the 45th ACM Technical Symposium on Computer Science Education, pp. 415–420 (2014)
4. Babik, D., Gehringer, E.F., Kidd, J., Pramudianto, F., Tinapple, D.: Probing the landscape: toward a systematic taxonomy of online peer assessment systems in education. In: Educational Data Mining (Workshops) (2016)
5. Chan, H.P., King, I.: Leveraging social connections to improve peer assessment in MOOCs. In: Proceedings of the 26th International Conference on World Wide Web Companion, pp. 341–349 (2017)
6. Chen, X., Lin, Q., Zhou, D.: Optimistic knowledge gradient policy for optimal budget allocation in crowdsourcing. In: International Conference on Machine Learning, pp. 64–72 (2013)
7. Díez Peláez, J., Luaces Rodríguez, Ó., Alonso Betanzos, A., Troncoso, A., Bahamonde Rionda, A.: Peer assessment in MOOCs using preference learning via matrix factorization. In: NIPS Workshop on Data Driven Education (2013)
8. Er, E., Bote-Lorenzo, M.L., Gómez-Sánchez, E., Dimitriadis, Y., Asensio-Pérez, J.I.: Predicting student participation in peer reviews in moocs. In: Proceedings of the Second European MOOCs Stakeholder Summit 2017 (2017)
9. Estévez-Ayres, I., García, R.M.C., Fisteus, J.A., Kloos, C.D.: An algorithm for peer review matching in massive courses for minimising students' frustration. J. UCS **19**(15), 2173–2197 (2013)
10. Gehringer, E.F.: A survey of methods for improving review quality. In: Cao, Y., Väljataga, T., Tang, J.K.T., Leung, H., Laanpere, M. (eds.) ICWL 2014. LNCS, vol. 8699, pp. 92–97. Springer, Cham (2014). https://doi.org/10.1007/978-3-319-13296-9_10
11. Karger, D.R., Oh, S., Shah, D.: Budget-optimal task allocation for reliable crowdsourcing systems. Oper. Res. **62**(1), 1–24 (2014)
12. Onah, D.F., Sinclair, J., Boyatt, R.: Dropout rates of massive open online courses: behavioural patterns. In: International Conference on Education and New Learning Technologies, pp. 5825–5834 (2014)
13. Pappano, L.: The year of the MOOC. New York Times **2**(12), 2012 (2012)
14. Piech, C., Huang, J., Chen, Z., Do, C., Ng, A., Koller, D.: Tuned models of peer assessment in MOOCs. In: Educational Data Mining 2013 (2013)
15. Ramachandran, L.: Automated Assessment of Reviews. North Carolina State University (2013)
16. Raman, K., Joachims, T.: Methods for ordinal peer grading. In: Proceedings of the 20th ACM SIGKDD International Conference on Knowledge Discovery and Data Mining, pp. 1037–1046 (2014)

17. Shah, N.B., Bradley, J.K., Parekh, A., Wainwright, M., Ramchandran, K.: A case for ordinal peer-evaluation in MOOCs. In: NIPS Workshop on Data Driven Education (2013)
18. Weld, D.S., et al.: Personalized online education-a crowdsourcing challenge. In: Workshops at the Twenty-Sixth AAAI Conference on Artificial Intelligence, pp. 1–31 (2012)
19. Yan, Y., Rosales, R., Fung, G., Dy, J.G.: Active learning from crowds. In: International Conference on Machine Learning, vol. 11, pp. 1161–1168 (2011)

LIKE Patterns and Complexity

Holger Petersen[(✉)]

Reinsburgstr. 75, 70197 Stuttgart, Germany
dr.holger.petersen@googlemail.com

Abstract. We investigate the expressive power and complexity questions for the LIKE operator in SQL. The languages definable by a single
LIKE pattern and generalizations are related to a well-known hierarchy
of classes of formal languages, namely the dot-depth hierarchy introduced
by Cohen and Brzozowski. Then we turn to natural decision problems
and show that membership is likely easier for LIKE patterns than for
more powerful regular expressions. Equivalence is provably harder for
general regular expressions. More complex conditions based on LIKE
patterns are also considered.

1 Introduction

Regular expressions conveniently support the analysis of software defects involving strings stored in a data base and the subsequent selection of test data for
checking the effectiveness of data cleansing. As an example take a list of values
separated by a special symbol. When manipulating strings of this form, it might
happen that separator symbols are stored consecutively or strings start with a
separator. This data corruption possibly leads to problems when displaying the
data or generating export files.

The full power of regular expressions is not required in many situations
encountered in practice. A very restricted variant of regular expressions we will
consider here are patterns for the LIKE operator available in SQL (Structured
Query Language) [1]. It admits defining patterns including constants and wildcard symbols representing single letters or arbitrary strings. Since our investiga-
tions are motivated by defect analysis and test data selection, which by definition
may not modify data, we assume that new auxiliary columns for holding intermediate values cannot be defined.

Continuing the example given above, we can select corrupt strings in an
obvious way even by LIKE patterns. After data cleansing, the same selections
can verify the correctness of the resulting data. It is known that LIKE pattern
matching can define star-free languages only [4, Section 4.2]. In Sect. 3 we will
explore what classes of languages known from the literature are characterized
by LIKE patterns and their boolean combinations.

A more extensive set of operations than those available with the LIKE operator (including concatenation and closure) is employed in classical regular expressions studied in Theoretical Computer Science. An even more powerful set of

D.-Z. Du et al. (Eds.): COCOON 2019, LNCS 11653, pp. 469–477, 2019.
https://doi.org/10.1007/978-3-030-26176-4_39

operations is offered by practical regular expressions, which may contain back references [2]. The additional power of classical or practical regular expressions is not for free. There decision problems appear to be more complex, with equivalence of practical regular expressions even undecidable. These theoretical results are supported by the observation that selections based on LIKE patterns can be significantly more efficient than those utilizing regular expressions.

2 Preliminaries

For basic definitions related to formal languages, finite automata, and computational complexity we refer to [13].

The *star-free languages* are those regular languages obtained by replacing the star-operator with complement in regular expressions. Cohen and Brzozowski [5] defined a hierarchy of star-free languages according to the notion of *dot-depth*. For an alphabet $\Sigma = \{a_1, \ldots, a_k\}$ the family E_0 consists of the *basic languages* $\{a_1\}, \ldots, \{a_k\}, \{\varepsilon\}$ (where ε denotes the empty string). If X is a family of languages, then we denote by $B(X)$ the boolean closure of X and by $M(X)$ the closure of X under concatenation. Define the following hierarchy of language families:

$$B_1 = B(E_0)$$
$$M_n = M(B_n) \text{ for } n \geq 1$$
$$B_n = B(M_{n-1}) \text{ for } n \geq 2$$

Obviously, these families form a hierarchy:

$$E_0 \subseteq B_1 \subseteq M_1 \subseteq B_2 \subseteq M_2 \cdots$$

The *dot-depth* $d(R)$ for a language R is defined to be n if $R \in B_{n+1} \setminus B_n$. In [5] its is shown that the hierarchy is strict up to dot-depth 2 (B_3), leaving open whether the upper levels can be separated. This open problem was resolved in [3] by showing that the hierarchy is strict.

The LIKE operator of SQL admits defining patterns in WHERE clauses which can be matched against string valued columns. Each symbol represents itself except for certain meta-characters, among which the most important is % as a wildcard matching zero or more characters. Symbol _ is a substitute for an arbitrary single character. Similarly, as further syntactic enhancements (character sets and complements of such sets), the _ wildcard can be seen as a (very convenient) shorthand for an enumeration of patterns for every symbol in the alphabet. If wildcard symbols are required in a pattern, an escape symbol can be declared that enforces a literal interpretation of % and _.

The more powerful operator SIMILAR TO or the Oracle® function REG-EXP_LIKE implement general regular expression matching in SQL (the latter even for extended regular expressions).

The following table compares different notations of the variants Practical Regular Expressions (PRE), Classical Regular Expressions (CRE) [13], Star-Free Expressions (SFE), and LIKE Patterns[1]:

	PRE	CRE	SFEs	LIKE Patterns
letter x	x	x	x	x
empty string	impl.	ε	ε	impl.
union	|	$\cup, +$	$\cup, +$	n/a
concatenation	impl.	impl., $\circ$, $\cdot$	impl., $\circ$, $\cdot$	impl.
closure	*	*	n/a	n/a
any character	.	Σ	Σ	-
any string	.*	Σ^*	$\overline{\emptyset}$	%

CRE and SFE include a notation for the empty set, which is not relevant for practical purposes and thus does not have a counterpart in PRE or LIKE patterns. PRE may include as "syntactic sugar" the notations $[\alpha_1\alpha_2 \ldots \alpha_n]$ for the set of characters $\{\alpha_1, \alpha_2, \ldots, \alpha_n\}$ and $[\alpha_1 - \alpha_n]$ for the range of consecutive characters α_1 to α_n (this assumes some specific encoding). Notation $[\hat{}\,\alpha_1\alpha_2 \ldots \alpha_n]$ and $[\hat{}\,\alpha_1 - \alpha_n]$ denote the complements of these sets of characters. Other extensions are the notation e? that denotes zero or one occurrence of expression e and $e\{n\}$ that denotes exactly n occurrences. None of these operators increases the expressive power of regular expressions, but they may lead to significantly shorter expressions than possible with CRE.

One extension of PRE that goes beyond regular languages is the use of back references. The k-th subexpression put into parentheses can be referenced by $\backslash k$, which matches the same string as matched by the subexpression.

For CRE the membership problem asks whether the *entire* input text matches a given pattern. In practice we are more interested on one or even all substrings within the input text matching the pattern. From the latter set of substrings the answer to the decision problem can easily be derived and lower bounds above polynomial time carry over (notice that the number of substrings of a text of length n is $\binom{n+1}{2} = \Theta(n^2)$). We can enforce a match of a PRE α with the entire input text by enclosing it into "anchors" and matching with $\hat{}\,\alpha\$$. Conversely, the CRE $\Sigma^*\alpha\Sigma^*$ simulates the PRE α. We conclude that upper and lower bounds for CRE membership and PRE matching coincide.

Since LIKE patterns are rather restricted (see Sect. 3) we also consider boolean formulas containing LIKE patterns (LIKE expressions) and boolean formulas without negations (monotone LIKE expressions).

---

[1] By 'impl.' we denote the implicit notation of the empty string or concatenation by juxtaposition of neighboring symbols. Σ is not part of the syntax of CRE or SFE but a common abbreviation.

Definition 1. *A language $L \subseteq \Sigma^*$ is LIKE-characterizable if it is a set of strings satisfying a boolean combination of LIKE pattern matching conditions.*

We summarize known complexity results for some decision problems related to regular expressions:

	PRE	CRE	SFE
matching, membership	NP-complete [2, Thm. 6.2]	NL-complete [9, Thm. 2.2]	P-complete [12, Thm. 1]
equivalence	undecidable [6, Thm. 9]	PSPACE-compl. [11, Lem. 2.3]	$\mathrm{NSPACE}\left(2^{2^{\cdot^{\cdot^{\cdot^2}}}}\right) \Big\} g(n)$ $g(n) = n$ [14] (u. b.) $g(n) = \frac{c \cdot n}{(\log^* n)^2}$ [7] (l. b.)
non-emptiness	$\in$ ALOGTIME (see CRE)	$\in$ ALOGTIME [12, Intr.]	see equivalence

3   Expressive Power

In this section we briefly discuss the power of LIKE patterns and LIKE expressions in comparison to the dot-depth hierarchy as defined in [5].

It is clear that the languages of family E_0 can be characterized by LIKE patterns of the form a_i. Family B_1 is incomparable to the languages characterized by LIKE patterns: For an alphabet Σ with $|\Sigma| \geq 2$ the language $L_1 = \{a_1, \varepsilon\}$ is clearly in B_1 (a boolean combination of basic languages), but a LIKE pattern characterizing a finite language can contain different words via _ only, which does not allow for words of the different lengths. Thus L_1 cannot be characterized by a LIKE pattern. Conversely, the LIKE pattern 00 defines the language $L_2 = \{00\}$, which cannot be expressed as a boolean combination of basic languages. Monotone LIKE expressions can describe all finite languages, but also all co-finite languages. Therefore, B_1 is properly contained in the languages characterized by monotone LIKE expressions (separation by L_2).

Every language in family B_2 can be denoted in the form

$$\bigcup_{k=1}^{\ell} \left(\left[\bigcap_{i=1}^{m(k)} \overline{w_0^{k,i} \Sigma^* w_1^{k,i} \Sigma^* \cdots \Sigma^* w_{s(k,i)}^{k,i}} \right] \cap \left[\bigcap_{j=1}^{n(k)} u_0^{k,j} \Sigma^* u_1^{k,j} \Sigma^* \cdots \Sigma^* u_{t(k,j)}^{k,j} \right] \right)$$

with $w_p^{k,i}$, $u_q^{k,j}$ words and $m(k)$, $n(k)$, ℓ, $s(k,i)$, $t(k,j)$ non-negative integers [5, Lemma 2.8]. This representation translates directly to a LIKE expression. Given a LIKE expression, every pattern containing wildcards _ can be replaced by an enumeration of patterns substituting the alphabet symbols for wildcards. All negations can be moved to the LIKE operators applying De Morgan's laws. The resulting expression characterizes a set in B_2.

We are thus led to the following observation:

Observation 1. *The class of LIKE-characterizable languages coincides with the class of languages of dot-depth 1.*

An example of a star-free language shown to be of dot-depth 2 (and therefore not LIKE-characterizable) is $(0 + 1 + 2)^*02^*$ from [5, Lemma 2.9].

Finally, we sketch why monotone LIKE expressions are weaker than general LIKE expressions. We claim that monotone LIKE expressions cannot express that strings are formed over a proper subset Σ' of the underlying alphabet Σ (which we assume to contain at least two symbols). Suppose a monotone LIKE expression e can express this restriction. Choose a string w over Σ' which is longer than e. Then w matches e and at least one symbol of w matches wildcards only. This symbol can be substituted by a symbol from $\Sigma \setminus \Sigma'$. The resulting string still matches e, contradicting the assumption.

4   Computational Complexity

We first introduce a syntactical transformation of patterns that will simplify the subsequent algorithms.

Definition 2. *A LIKE pattern is called* normalized, *if it contains none of the substrings %_ and %%.*

Consider an arbitrary string $w \in \{\%, \_\}^*$ consisting of wildcards. If w matches a string over the base alphabet, then a string w' containing the same number of the symbol _ and a trailing % if and only if w contains % matches as well. Since w' is normalized we obtain:

Proposition 1. *For every LIKE pattern there is an equivalent normalized LIKE pattern.*

Normalization cannot in general identify equivalent patterns. As an example take the patterns %01% and %0%1% over the binary alphabet $\{0, 1\}$. Obviously, any string matching the first pattern matches the second. But the converse is also true, because there is a left-most 1 between the two constants of the pattern (including the 1) and it is preceded by a 0. Over the alphabet $\{0, 1, 2\}$, the patterns are separated by 021.

Lemma 1. *LIKE patterns can be normalized in deterministic logarithmic space.*

Proof. If the pattern contains to wildcard, it is normalized. A pattern containing at least one wildcard can be written as $x_0 w_0 \cdots x_n w_n x_{n+1}$ where $w_0, \ldots, w_n \in \{\%, \_\}^+$ and $x_0 \cdots x_{n+1} \in \Sigma^*$ for the underlying alphabet Σ.

A deterministic Turing machine M scans the input and directly outputs any symbol from Σ. For every string w_i of consecutive wildcards, the number m of occurrences of _ is counted and a flag is maintained indicating the presence of %. At the end of w_i, machine M outputs m symbols _ and an optional % if the flag is set.

Since M has to store counters bounded by the input length, it can do so in logarithmic space if the counters are encoded in binary notation.   □

Theorem 1. *Matching with a LIKE pattern can be done in deterministic logarithmic space.*

Proof. If the pattern contains no %, in a single scan the constant symbols in the pattern are compared and for every _ in the pattern a symbol in the text is skipped.

By Lemma 1 we can assume that any LIKE pattern containing % has the form $p = a_1 \% a_2 \% \cdots \% a_n$ where $a_i \in \Sigma^*\_^*$. We first argue that a greedy matching strategy suffices for checking whether p matches a text t. Suppose in a given matching i is minimal with the property that a_i could be matched further to the start of the text (but after a_{i-1}). Then a new match can be obtained by moving a_i to the first occurrence. Carrying out this operation for all a_i leads to a greedy matching.

For every a_i a left-most match can be determined by comparing the constant part and shifting the position in the text if a mis-match occurs. Once an a_i has been matched, it is not necessary to reconsider it by the argument above.

In logarithmic space pointers into pattern and text can be stored and by scanning p and t in parallel a greedy matching can be determined.   □

We have the following (weaker) lower bound for the membership problem:

Theorem 2. *Matching with a LIKE pattern cannot be done by constant-depth, polynomial-size, unbounded fan-in circuits (it is not in AC^0).*

Proof. Recall from [8] that the majority predicate on n binary variables is 1 if and only if more than half of the input values are 1. We map a given input x for the majority predicate to the pattern $\%(1\%)^{\lceil(|x|+1)/2\rceil}$. String x matches the pattern only if x contains at least $\lceil(|x|+1)/2\rceil > |x|/2$ symbols 1, which is majority. By the result [8, Theorem 4.3] this predicate is not in AC^0.   □

Since the evaluation of boolean formulas is possible in logarithmic space [10], we obtain from Theorem 1:

Corollary 1. *Matching with a LIKE expression can be done in deterministic logarithmic space.*

Considering equivalence of LIKE patterns, a test using syntactical properties alone seems to be impossible because of the example given above.

Based on Theorem 1 we can obtain the following upper bound:

Corollary 2. *Equivalence of LIKE patterns is in nondeterministic logarithmic space.*

Proof. Guess a separating text symbol by symbol and match with the given patterns in logarithmic space. By the closure under complement of NL [13, Theorem 8.27] the result follows.   □

Theorem 3. *Nonemptiness of monotone LIKE expressions is complete in NP.*

Proof. For membership in NP consider a string w matching a given expression e. We claim that there is no loss of generality in assuming $|w| \leq |e|$. We fix a matching of w by e. For every OR in expression e there has to be at least one sub-expression matching w. We delete the other sub-expression and continue this process until there is no OR left obtaining e'. Clearly $|e'| \leq |e|$. Now we mark every symbol of w matched by a constant or _. At most $|e'|$ symbols of w will thus be marked and the others have to be matched by %. Deleting these symbols yields a string w' matching e with $|w'| \leq |e'| \leq |e|$. The NP algorithm simply consists in guessing a string w with $|w| \leq |e|$, writing it onto the work tape, and checking membership according to Corollary 1.

For hardness we reduce the satisfiability problem of boolean formulas in 3-CNF (3SAT) to the nonemptiness problem. It is well-known that 3SAT is complete in NP [13]. Let

$$F = (\alpha_1 \vee \beta_1 \vee \gamma_1) \wedge \cdots \wedge (\alpha_m \vee \beta_m \vee \gamma_m)$$

be a formula in CNF over variables $x_1, \ldots, x_n$. The idea is to enumerate all satisfied literals in a string that matches a monotone LIKE expression. We form a set of LIKE patterns over the alphabet $\{x_1, \ldots, x_n, \bar{x}_1, \ldots, \bar{x}_n\}$ that are joined by AND:

- $\_^n$ (there are exactly n literals).
- For $1 \leq i \leq n$ an OR of the patterns x_i and $\bar{x}_i$ (for every variable at least one literal is true).
- For every clause $\alpha_k \vee \beta_k \vee \gamma_k$ an OR of the patterns α_k, β_k, and γ_k (at least one literal is true in every clause).

Suppose that F is satisfied by some assignment of boolean values to $x_1, \ldots, x_n$. Concatenate the satisfied literal for each variable to form a string to be matched. This string clearly matches all patterns defined above. Conversely, if a string matches all patterns it contains at least one literal per variable by the second item. The length restriction to n symbols implies that exactly one literal per variable is included. These literals define a truth assignment in the obvious way and by the third item every clause is satisfied by this assignment. □

Lemma 2. *For a deterministic Turing machine M with input w and space bound $s(|w|)$, a LIKE expression e with the following properties can be constructed:*

1. *All LIKE conditions are negative.*
2. *The LIKE expression e is of size $O(s^2(|w|))$.*
3. *If M accepts w within space $s(|w|)$, there is a single string matching e.*
4. *If M does not accept w within space $s(|w|)$, the language described by e is empty.*

Proof. Without loss of generality we assume that M accepts with a blank tape and the tape head on the left-most tape cell. We denote the input length by $n = |w|$.

In order to simplify the presentation we first use arbitrary LIKE conditions. We encode a computation of M as a sequence of *configurations* over the alphabet $\Gamma \cup Q$ (tape alphabet and set of states). A configuration uqv encodes the tape inscription uv, current state q and head position on the first symbol of v. A computation consisting of k steps is encoded as $\#c_0\#c_1\# \cdots \#c_k\#$. Configuration c_0 is q_0w followed by $s(n) - n$ blanks and for $i \geq 1$ configuration c_{i-1} yields c_i by M's transition function. We therefore identify the following patterns:

1. $\#c_0\#\%$ (start configuration).
2. $\%\#c_{\text{accept}}\#$ (accepting configuration).
3. For every $\delta(q_i, b) = (q_j, c, L)$ negative patterns $aq_ib\_{}^{s(n)}def$ with $def \neq q_jac$.
4. For every $\delta(q_i, b) = (q_j, c, R)$ negative patterns $aq_ib\_{}^{s(n)}def$ with $def \neq acq_j$.
5. Negative patterns $abc\_{}^{s(n)}d$ with $a, b, c \in \Gamma \cup \{\#\}$ and $b \neq d$ (portions of the tape not affected by the computation).

For each of the patterns in item 1 and 2 we can substitute $(s(n) + 2)(|\Gamma| + |Q|)$ equivalent negative patterns that exclude all but one symbol from $\Gamma \cup Q \cup \{\#\}$ at position i with $1 \leq i \leq s(n) + 2$ from the start resp. end of the string.   □

Lemma 3. *Inequivalence of LIKE expressions can be decided nondeterministically in linear space.*

Proof. For two given expressions guess a string symbol by symbol and mark in every pattern the positions reachable by matching the guessed string. When a separating string has been found, both expressions are evaluated and it is checked that exactly one of the expressions matches.   □

The previous lemmas can be summarized in the following way:

Theorem 4. *Equivalence of monotone as well as of arbitrary LIKE expressions is complete in PSPACE.*

Proof. By Savitch's theorem [13, Theorem 8.5] the upper bound from Lemma 3 yields deterministic algorithm in quadratic space. For the lower bound we observe that complementing the expression from Lemma 2 for a polynomial space Turing machine M results in a monotone expression that describes all strings if and only if M does not accept its input.   □

5   Discussion

We investigated the expressive power and computational complexity of the LIKE operator. For the more powerful monotone and general LIKE expressions we classified the complexity of nonemptiness and equivalence in terms of well-known complexity classes. In case of membership we could establish the upper bound L (deterministic logarithmic space). This is believed to be of lower complexity than the general membership problem for CRE, which is complete in NL [9]. Membership for a single LIKE pattern is not decidable by the highly parallel AC^0 circuits. It remains open, what the exact complexity of the latter problem and inequivalence is.

Acknowledgement. Many thanks to Manfred Kufleitner for information about star-free languages.

References

1. Oracle® Database SQL Reference 10g Release 1. https://docs.oracle.com/cd/B13789_01/server.101/b10759/conditions016.htm
2. Aho, A.V.: Algorithms for finding patterns in strings. In: van Leeuwen, J. (ed.) Handbook of Theoretical Computer Science. Algorithms and Complexity, vol. A, pp. 255–300. MIT Press, Cambridge (1990)
3. Brzozowski, J.A., Knast, R.: The dot-depth hierarchy of star-free languages is infinite. JCSS **16**, 37–55 (1978)
4. Benedikt, M., Libkin, L., Schwentick, T., Segoufin, L.: String operations in query languages. In: Buneman, P. (ed.) Proceedings of the Twentieth ACM SIGACT-SIGMOD-SIGART Symposium on Principles of Database Systems, Santa Barbara, California, USA, 21–23 May 2001, pp. 183–194 (2001)
5. Cohen, R.S., Brzozowski, J.A.: Dot-depth of star-free events. JCSS **5**, 1–16 (1971)
6. Freydenberger, D.D.: Extended regular expressions: succinctness and decidability. In: Schwentick, T., Dürr, C. (eds.) Proceedings of the 28th Annual Symposium on Theoretical Aspects of Computer Science (STACS 2011), Leibniz International Proceedings in Informatics, pp. 507–518. Schloss Dagstuhl: Leibniz-Zentrum für Informatik. Dagstuhl Publishing, Germany (2011)
7. Fürer, M.: Nicht-elementare untere Schranken in der Automaten-Theorie. Ph.D. thesis, ETH Zürich (1978)
8. Furst, M., Saxe, J.B., Sipser, M.: Parity, circuits, and the polynomial-time hierarchy. Math. Syst. Theory **17**, 13–27 (1984)
9. Jiang, T., Ravikumar, B.: A note on the space complexity of some decision problems for finite automata. Inf. Process. Lett. **40**, 25–31 (1991)
10. Lynch, N.: Log space recognition and translation of parenthesis languages. J. Assoc. Comput. Mach. **24**, 583–590 (1977)
11. Meyer, A.R., Stockmeyer, L.J.: The equivalence problem for regular expressions with squaring requires exponential space. In: Proceedings of the 13th Annual IEEE Symposium on Switching and Automata Theory, College Park (Maryland), pp. 125–129 (1972)
12. Petersen, H.: Decision problems for generalized regular expressions. In: Proceedings of the 2nd International Workshop on Descriptional Complexity of Automata, Grammars and Related Structures, London (Ontario), pp. 22–29 (2000)
13. Sipser, M.: Introduction to the Theory of Computation, 2nd edn. Thomson, Boston (2006)
14. Stockmeyer, L.J., Meyer, A.R.: Word problems requiring exponential time. In: Proceedings of the 5th ACM Symposium on Theory of Computing (STOC 1973), Austin (Texas), pp. 1–9 (1973)

Data Structures for Incremental Interval Coloring

J. Girish Raguvir, Manas Jyoti Kashyop$^{(\boxtimes)}$, and N. S. Narayanaswamy

Department of Computer Science and Engineering, Indian Institute of Technology
Madras, Chennai, India
girishraguvir@gmail.com, {manasjk,swamy}@cse.iitm.ac.in

Abstract. We consider the dynamic graph coloring problem restricted to the class of interval graphs. At each update step the algorithm is presented with an interval to be colored, or a previously colored interval to delete. The goal of the algorithm is to efficiently maintain a proper coloring of the intervals with as few colors as possible by an online algorithm. In the incremental model, each update step presents the algorithm with an interval to be colored. The problem is closely connected to the online vertex coloring problem of interval graphs for which the Kierstead-Trotter (KT) algorithm achieves the best possible competitive ratio. We first show that a sub-quadratic time direct implementation of the KT-algorithm is unlikely to exist conditioned on the correctness of the Online Boolean Matrix Vector multiplication conjecture due to Henzinger et al. [9]. We then design an incremental algorithm that is subtly different from the KT-algorithm and uses at most $3\omega - 2$ colors, where ω is the maximum clique in the interval graph associated with the set of intervals. Our incremental data structure maintains a proper coloring in amortized $O(\log n + \Delta)$ update time where n is the total number of intervals inserted and Δ is the maximum degree of a vertex in the interval graph.

1 Introduction

The graph coloring problem is an extensively studied problem. Similarly, maintenance of data structures for dynamic graphs has been extensively studied. The dynamic graph coloring problem is as follows: There is an online update sequence of insertion or deletion of edges or vertices and the goal is to maintain proper coloring after every update. Several works [1,13,14] and [5] propose heuristic and experimental results on the dynamic graph coloring problem. To the best of our knowledge, the only formal analysis of data structures for dynamic graph coloring are [3,4,15], and [2]. Bhattacharya et al. give the current best fully dynamic randomized algorithm which maintains $\Delta + 1$ vertex coloring in $O(\log \Delta)$ expected amortized update time [4]. They also give the current best deterministic algorithm which maintains $\Delta + o(\Delta)$ vertex coloring in $O(polylog\Delta)$ amortized update time [4].

D.-Z. Du et al. (Eds.): COCOON 2019, LNCS 11653, pp. 478–489, 2019.
https://doi.org/10.1007/978-3-030-26176-4_40

In this work we study dynamic data structures for coloring of interval graphs when the input is an online update sequence of intervals. The goal is to efficiently maintain a proper coloring with as few colors as possible by an online algorithm. In our dynamic model, at each update step an interval is inserted or deleted. Thus, a single update may insert or delete many edges in the underlying interval graph. This is different from the commonly studied case in the area of dynamic graph algorithms where on each update an edge is inserted or deleted. In the incremental model intervals are inserted one after the other and we aim to efficiently maintain a proper coloring using as few colors as possible after every update. Our approach is to consider efficient implementations of well-studied online algorithms for interval coloring. Note that an online algorithm is not allowed to re-color a vertex during the execution of the algorithm. On the other hand, an incremental algorithm is not restricted in anyway during an update step except that we desire that the updates be done as efficiently as possible. Naturally, an online algorithm is a good candidate for an incremental algorithm as it only assigns a color to the current interval, and does not change the colour of any of the other intervals. Online algorithms for interval coloring and its variants is a rich area with many results. Epstein et al. studied online graph coloring for interval graphs [7]. They studied four variants of the problem: online interval coloring with bandwidth, online interval coloring without bandwidth, lazy online interval coloring with bandwidth, and lazy online interval coloring without bandwidth. For online interval coloring with bandwidth, Narayanaswamy presented an algorithm with competitive ratio 10 [12] and Epstein et al. showed a lower bound of 3.2609 [7]. For lazy online interval coloring with bandwidth and lazy online interval coloring without bandwidth, Epstein et al. proved that competitive ratio can be arbitrarily bad for any online algorithm [7]. For the online interval coloring problem, Kierstead and Trotter presented a 3 competitive algorithm and they also proved that their result is tight [10]. In other words, The online algorithm (KT-algorithm) due to Kierstead and Trotter [10] is known to have the optimum competitive ratio. The tightness is by showing the existence of an adaptive adversary that forces an online algorithm to use $3\omega - 2$ colors where ω is the maximum clique size in the interval graph formed by the given set of n intervals. On the other hand, the KT-algorithm uses at most $3\omega - 2$ colors. The KT-algorithm computes a proper coloring by assigning to each vertex a color which is 2-tuple denoted by (level,offset). The value of level is in the range $[0, \omega - 1]$ and the value of offset is from the set $\{1, 2, 3\}$. Further, all the vertices whose level value is 0 form an independent set. Therefore, an efficient implementation of KT-algorithm gives us an incremental algorithm that maintains a proper coloring with at most $3\omega - 2$ colors.

Our Work: In Sect. 2.2, we show that we are unlikely to be able to design a sub-quadratic implementation of the KT-algorithm. The reason for this is that the level value assigned to an interval depends on the size of the maximum clique in the graph induced by the intervals intersecting with it. We show a reduction from the Online Boolean Matrix-Vector Multiplication (OMv) problem [9] to the problem of computing the induced subgraph of the neighbours of a vertex.

By the conjecture on the OMv problem, it is unlikely to have a sub-quadratic algorithm for the Induced Subgraph Computation problem. Thus, we believe that any algorithm that depends on computing an induced subgraph is unlikely to have a sub-quadratic dynamic algorithm, even when the graph is an interval graph. We design an incremental algorithm which avoids this limitation by using a different approach to compute level value for an interval. Thus, we differ from KT-algorithm in computing the level value for an interval. However, our algorithm uses the same number of colors as the KT-algorithm. Our incremental algorithm (Theorem 1) supports insertion of a sequence of intervals in amortized $O(\log n + \Delta)$ update time. We have also considered the fully dynamic framework in which an interval that has already been colored can be deleted, apart from the insertions. At the end of each update, our aim is to maintain a $3\omega - 2$ coloring of the remaining set of intervals, where ω is the maximum clique in the interval graph associated with the remaining set of intervals. Our fully dynamic algorithm supports insertion of an interval in $O(\log n + \Delta \log \omega)$ worst case update time and deletion of an interval in $O(\Delta^2 \log n)$ worst case update time, and this will be included in the journal version of the work. Finally, the question of significant interest to us is whether the dependence on Δ can be sub-linear in the incremental case and whether it can be sub-quadratic in the fully dynamic case. Another interesting direction is the nature of the trade-off between the number of colors used and the update time if we allow a change of color assigned to an interval.

2   Preliminaries

$\mathcal{I}$ denotes a set of intervals and the associated interval graph [8] is denoted by $G(\mathcal{I})$. For an undirected graph G, $\omega(G)$ denotes the size of the maximum cardinality clique in G, $\Delta(G)$ denotes the maximum degree of a vertex in G and $\chi(G)$ denotes the chromatic number of G. It is well-known that for interval graphs $\omega(G) = \chi(G)$. When the context is clear we denote $\omega(G) = \chi(G)$ as ω and $\Delta(G)$ as Δ.

2.1   Kierstead-Trotter Algorithm - Overview

Let $\sigma = v_1, v_2, v_3, \ldots, v_n$ be the ordering of vertices of an interval graph $G = G(\mathcal{I})$. Each vertex in σ is presented to an online algorithm as the corresponding interval. Let $\mathcal{I}$ denote the set of intervals corresponding to the vertices and vertex v_j corresponds to the interval $I_j \in \mathcal{I}$. We also refer to each vertex and interval along with a subscript whose range is from 1 to n. For example, when we use v_j or I_j we mean the j-th vertex in σ. Throughout the paper, v_j will be the vertex corresponding to the interval I_j. For a given vertex v_i from σ the algorithm computes a color based on the color given to the vertices $v_1, \ldots, v_{i-1}$. The color assigned to a vertex v is a tuple of two values and is denoted as $(p(v), o(v))$. In Step I, the first value called the level or position of the presented vertex denoted by $p(v)$, is computed, and in Step II the second value, called the offset denoted by $o(v)$, is computed from the set $\{1, 2, 3\}$. The key property is that for each edge $\{u, v\}$, the tuple $(p(u), o(u))$ is different from $(p(v), o(v))$.

Step I: For $r \geq 0$, let $G_r(v_i)$ denote the induced subgraph of G on the vertex set $\{v_j | v_j \in V(G), j < i, p(v_j) \leq r, (v_i, v_j) \in E(G)\}$. Define $p(v_i) = \min\{r | \omega(G_r(v_i)) \leq r\}$.

Key Properties maintained by Step I [10]:

- For each vertex v_i, $p(v_i) \leq \omega - 1$.
- **Property P:** The set $\{v | p(v) = 0\}$ is an independent set. For each i, $1 \leq i \leq \omega - 1$, the subgraph of G induced on $\{v_j \mid p(v_j) = i\}$ has maximum degree at most 2.

Step II: Since there are at most two neighbours of v_i such that their level is $p(v_i)$, $o(v_i)$ is chosen to be the smallest value from the set $\{1, 2, 3\}$ different from the offset of these neighbours whose level is $p(v_i)$.

Analysis: All the vertices in level 0 form an independent set, and for all these vertices the offset value is 1. For the vertices in levels 1 to $\omega - 1$, maximum degree is 2. Therefore, the algorithm uses at most 3 colors to color all the vertices belonging to a particular level l where $1 \leq l \leq \omega - 1$. Hence, total colors used by the algorithm is $1 + 3(\omega - 1) = 3\omega - 2$.

Crucial Step in the Implementation of KT-Algorithm: Given a vertex v_i, the subgraph $G_r(v_i)$ is the induced subgraph among the neighbours of v_i which have level value at most r. Computing $G_r(v_i)$ is a very crucial step in KT-algorithm for different values of r starting from $r = 0$ until the value of r for which $\omega(G_r(v_i)) \leq r$ is true. In Sect. 3 we show that the search for such an r can be done without computing the induced subgraphs $G_r(v_i), r \geq 0$. The motivation for this line of research is the result in Sect. 2.2.

2.2   KT-Algorithm Must Avoid Computing Induced Subgraphs

In this Section we show that the problem of computing the induced subgraph of the closed neighbourhood of a set of vertices is unlikely to have a sub-quadratic time algorithm. The input to the *Induced Neighbourhood Subgraph Computation* problem consists of the adjacency matrix M of a directed graph and a set S of vertices. The aim is to compute the graph induced by $N_{out}(S) \cup S$ and output the subgraph as adjacency lists. Here $N_{out}(S)$ is the set of those vertices which have a directed edge from some vertex in S. In other words, there is a directed edge from v_j to v_k iff the entry $M[k][j]$ is 1. Next we show that Induced Neighbourhood Subgraph Computation problem is at least as hard as the following problem.

Online Boolean Matrix-Vector Multiplication [9]: The input for this online problem consists of an $n \times n$ matrix M, and a sequence of n boolean column vectors $v_1, \ldots, v_n$, presented one after another to the algorithm. For each $1 \leq i \leq n - 1$, the online algorithm should output $M \cdot v_i$ before v_{i+1} is presented to the algorithm. Note that in this product, a multiplication is an AND operation and the addition is an OR operation. According to the OMv conjecture, due to Henzinger et al. [9], the Online Boolean Matrix-Vector Multiplication problem does not have a $O(n^{3-\epsilon})$ algorithm for any $\epsilon > 0$. The current best algorithm for the Online Boolean Matrix-Vector Multiplication problems has an expected running time of $O(\frac{n^3}{2\sqrt{\log n}})$ [11].

We now show that an algorithm to solve the Induced Neighbourhood Subgraph Computation problem can be used to solve the Online Boolean Matrix-Vector Multiplication problem. Let $\mathcal{A}$ be an algorithm for the Induced Neighbourhood Subgraph Computation problem with a running time of $O(n^{2-\epsilon})$, for some $\epsilon > 0$. Then, we use algorithm $\mathcal{A}$ to solve the Online Boolean Matrix-Vector Multiplication problem in $O(n^{3-\epsilon})$ time as follows: Let M be the input matrix for the Online Boolean Matrix-Vector Multiplication problem and let $V_1, \ldots V_n$ be the column vectors presented to the algorithm one after another. For the column vector V_i, let set $S_i = \{v_j | V_i[j] = 1, 0 \leq j \leq n-1\}$. To compute $M \cdot V_i$, we invoke $\mathcal{A}$ on input $\{M, S_i\}$. Let G_{S_i} denote the induced subgraph on $N_{out}(S_i) \cup S_i \subseteq V$ computed by the algorithm $\mathcal{A}$. Note that G_{S_i} is an induced subgraph of the directed graph whose adjacency matrix is M. To output the column vector $M \cdot V_i$, we observe that the j-th row in the output column vector is 1 if and only if $v_j \in G_{S_i}$ and there is an edge (u, v_j) in G_{S_i} such that $u \in S_i$. Given that G_{S_i} has been computed in $O(n^{2-\epsilon})$ time, it follows that the number of edges in G_{S_i} is $O(n^{2-\epsilon})$ and consequently the column vector $M \cdot V_i$ can be computed in $O(n^{2-\epsilon})$ time. Therefore, using the $O(n^{2-\epsilon})$ algorithm $\mathcal{A}$ we can solve Boolean Matrix-Vector Multiplication problem in $O(n^{3-\epsilon})$ time. If we believe that the OMv conjecture is indeed true, then it follows that the Induced Neighbourhood Subgraph Computation problem cannot have a $n^{2-\epsilon}$ algorithm for any $\epsilon > 0$. This conditional lower bound on the Induced Neighbourhood Subgraph Computation problem deters us from coming up with a direct implementation of the KT algorithm for the incremental setting of the online interval coloring problem. In the following section we design an online interval coloring algorithm that avoids an explicit computation of the induced subgraph on the neighbourhood of an input interval.

3   An Incremental Data Structure for Interval Coloring

In this section we present an incremental algorithm which is essentially an implementation of the KT-algorithm [10]. The subtle difference is that our algorithm has a different definition of the level value of an interval. The level value that we assign is at most the level value assigned by the KT-algorithm, and thus our algorithm uses the same number of colors as the KT-algorithm. Let $\mathcal{W} = \{0, 1, \ldots, \omega - 1\}$. We use $L(v_i) \in \mathcal{W}$ to denote the level value computed by our algorithm and $p(v_i)$ to denote the level value computed by the KT-algorithm. To compute the offset value $o(v_i)$ we use the same approach as KT-algorithm as described in Sect. 2.1. We design appropriate data structures to compute the level $L(v)$ and the offset $o(v)$ associated with a vertex v.

For a set of intervals $\mathcal{J}$, define the set $levels(\mathcal{J}) = \{L(v_j) \in \mathcal{W} | I_j \in \mathcal{J}\}$ to be set of levels assigned to intervals in $\mathcal{J}$. Recall from Sect. 2 that v_j is the j-th vertex in σ. Let t be a non-negative real number and $\mathcal{I}_t$ be the set of all intervals in $\mathcal{I}$ which contain the point t. Define $h_t = \min(\{y \in \mathcal{W} | y \notin levels(\mathcal{I}_t)\})$ i.e. h_t is the smallest non-negative integer which is not the level value for any interval containing t. If $h_t \geq 1$, then the *Supporting Line Segment* (SLS) at t is defined to be the set $e_t = \{(t, 0) \ldots (t, h_t - 1)\}$, and h_t is called the *height* of SLS e_t.

3.1 Insertion

We show in Lemma 2 that the level $L(v_i)$ of interval I_i, which is given by the maximum height of the SLS at any point contained in I_i, is at most $p(v_i)$. We prove in Lemma 3 that level values computed by our algorithm satisfy Property **P** (described in Sect. 2.1). We also prove using Lemma 1 that the $L(v_i)$ can be computed based on the height of the SLS at finite set of points in interval I_i. This proof is one of our key contributions. In particular, we show that the finite set of points that we need to consider in interval I_i is the set of endpoints of the intervals intersecting with I_i. For n intervals we have at most $2n$ distinct endpoints and we denote this set of endpoints by $\mathcal{E}$. Therefore, on insertion of interval I_i, we first compute the set S of endpoints which are in I_i. We then query the height of the SLS at each of the points in S and take the SLS with maximum height to compute $L(v_i)$.

Insertion Algorithm: Let $I_i = [l_i, r_i]$ be the interval that is inserted in the current update step. Our algorithm computes the color $(L(v_i), o(v_i))$ by implementing Step 1 and Step 2 as described below:

1. **Computing $L(v_i)$:**
   Step 1: Insert l_i and r_i into the set $\mathcal{E}$.
   Step 2: Compute $S = \mathcal{E} \cap I_i$. For each $t \in S$, compute h_t, the height of the SLS e_t at t. Let $L(v_i) = \max_{t \in S} h_t$.
   Step 3: Update the SLS e_t for each point $t \in S$.
2. **Computing $o(v_i)$:** We prove in Lemma 3 that level values computed by our algorithm in Step I satisfy Property **P**. From Property **P**, v_i has at most two neighbours whose levels are $L(v_i)$. We compute $o(v_i)$ to be that value from the set $\{1, 2, 3\}$ that is different from the offset values of the neighbours of v_i which have the level $L(v_i)$.

Proof of Correctness: As mentioned before the description of the algorithm, the correctness of the insertion algorithm follows from the Lemmas 1, 2, and 3. In the following Lemma, we show that it is sufficient to use the endpoints of the intervals to compute the level of an interval.

Lemma 1. *Let t be a non-negative real number and let $\mathcal{I}_t$ be the set of intervals that contain t. Then there exists at least one endpoint in the set $\mathcal{E}$ which is contained in each interval in $\mathcal{I}_t$. Further, the height of the SLS at this endpoint is at least the height of the SLS at t.*

Proof. If t is an endpoint of an interval, then $t \in \mathcal{E}$ and hence the Lemma is proved. Suppose t is not an endpoint. Let l_t denote the largest left endpoint among all the intervals in $\mathcal{I}_t$ and r_t denotes the smallest right endpoint among all the intervals in $\mathcal{I}_t$. By definition, $l_t \in \mathcal{E}$ and $r_t \in \mathcal{E}$. Since $\mathcal{I}_t$ is a set of intervals, it follows that l_t and r_t are present in all the intervals in $\mathcal{I}_t$. Further, since both l_t and r_t are present in each interval in $\mathcal{I}_t$, it follows that the set of intervals that contain them is a superset of $\mathcal{I}_t$. Therefore, the height of the SLS at l_t and r_t is at least the height of the SLS at t. Hence the Lemma. □

From the description in Sect. 2.1 we know that the level of v_i computed by the KT-algorithm is given by $p(v_i) = \min\{r | \omega(G_r(v_i)) \leq r\}$. In our algorithm, we define $L(v_i)$ to be the maximum height of the SLS at any point contained in v_i. In the following Lemma we prove $L(v_i) \leq p(v_i)$.

Lemma 2. *For each $i \in [n]$, $p(v_i)$ is at least the maximum height of the SLS at any point in the interval I_i.*

Proof. By definition, $L(v_i)$ is the maximum height of SLS at any point contained in v_i. By the definition of the height of an SLS at a point t, we know that for each $0 \leq r \leq h_t - 1$ there is an interval $v \in \mathcal{I}_t$ such that $L(v) = r$, and all these intervals form a clique of size h_t. Therefore, it follows that $G_{L(v_i)}(v_i)$ has a clique of size at least $L(v_i)$. Therefore, it follows that $p(v_i) \geq L(v_i)$. Hence the Lemma. □

Lemma 3. *Our algorithm satisfies Property $\mathbf{P}$ and thus uses at most $3\omega - 2$ colors.*

Proof. We first prove that the set $\{v_j | L(v_j) = 0\}$ is an independent set. Suppose not, and let two vertices v_i and v_j be adjacent and both have level values 0. Without loss of generality, let us assume that v_i appeared before v_j in σ. Therefore, at the time when v_j is presented to the algorithm, there is an endpoint of v_i contained in v_j where the height of the SLS is non-zero. Therefore, the insertion algorithm does not assign $L(v_j)$ to be 0. Therefore, $\{v | L(v) = 0\}$ is an independent set. Further, for each $1 \leq l \leq \omega - 1$, the same argument is used to show that for each pair of intervals I_i and I_k corresponding to two distinct vertices in $\{v_j \mid L(v_j) = l\}$, $I_i \not\subseteq I_k$. We prove that for each l, $1 \leq l \leq \omega - 1$, the subgraph of G induced on $\{v_j | L(v_j) = l\}$ has maximum degree at most 2. Suppose not, and if some vertex is of degree 3 in level l. We know that for two vertices with the same level value, the corresponding intervals cannot have a containment relationship between them. Therefore, it follows that the vertex of degree 3 is in a clique of 3 vertices in level l. Let v_i, v_j, and v_k be the clique of 3 vertices in level l. The intervals corresponding to the three vertices are such that one of the intervals is contained in the union of the other two. Therefore, one of the 3 intervals contains a point t for which the SLS has height $l+1$. Consequently, all the 3 intervals cannot be assigned the same level value. Therefore, for each level l, the maximum vertex degree in the graph induced on $\{v_j | L(v_j) = l\}$ is at most 2.

Since we have proved that the level value computed by our algorithm satisfies Property **P**, we use the same procedure as KT-algorithm (described in Sect. 2.1) to compute the offset value. From Lemma 2, we know that for any interval v_i, we have $L(v_i) \leq p(v_i)$. Therefore, maximum level value of any vertex is $\omega - 1$. For level 0 we use one color and for every other level we use 3 colors. Therefore, total colors used by our algorithm is $3(\omega - 1) + 1 = 3\omega - 2$. Hence the Lemma.

$\square$

3.2   Implementation of the Incremental Algorithm

We use the following data structures in our incremental algorithm.

1. $\mathcal{I}$ **of type Interval Tree:** Every interval I_i is of the form $[l_i, r_i]$ where l_i denotes the left endpoint of the interval and r_i denotes the right endpoint of the interval. We maintain all the vertices corresponding to the intervals in $\mathcal{I}$ in an interval tree $\mathcal{I}$. For every v_i, along with maintaining the endpoints $\{l_i, r_i\}$, we also maintain the level $L(v_i)$ and offset $o(v_i)$.

2. V_t **of type dynamic array,** V_t' **of type dynamic array, and** Z_t **of type doubly linked list:** We maintain the supporting line segment e_t at point t and the height of supporting line segment h_t using dynamic arrays V_t, V_t' and doubly linked list Z_t. If the supporting line segment e_t intersects with an interval whose level value is i then we set $V_t[i]$ to 1, otherwise we set $V_t[i]$ to 0. We define the height h_t of the supporting line segment e_t as the index of the first 0 in V_t. To maintain h_t, we define a doubly linked list Z_t which stores every index i in V_t where $V_t[i]$ is 0 in the increasing order of the value of i. Note that the value stored at the head node of Z_t is h_t. We augment Z_t with another dynamic array V_t'. For an index i, if $V_t[i]$ is 0 then $V_t'[i]$ stores a pointer to the node in Z_t which stores the index i. If $V_t[i]$ is 1 then $V_t'[i]$ stores NULL. Therefore, using the dynamic array V_t', insert, delete, and search operation in Z_t can be performed in constant time. Since our algorithm is incremental, the dynamic arrays only expand. Insertion into a dynamic array takes amortized constant time [6]. The size of the array V_t is the length of V_t. During insertion, whenever size of V_t is increased, size of V_t' is also increased and appropriate nodes are inserted in Z_t. A Query for the value of h_t can be answered in constant time by returning the value stored in head node of Z_t. Due to an insert if h_t changes, then we need to change the head node of Z_t to the next node in the list and delete the previous head node of Z_t. This operation also takes constant time.

3. $\mathcal{E}$ **of type Interval Tree:** We maintain the set of endpoints ($\mathcal{E}$) as an interval tree. For every interval $I_i = [l_i, r_i]$, we maintain the left endpoint and the right endpoint as interval $[l_i, l_i]$ and $[r_i, r_i]$ respectively in $\mathcal{E}$.

4. T **of type Map with domain as integers and range as Interval Tree:** We use the interval tree $T[h]$ for maintaining the intervals at level h to enable allocation of the offset values to the intervals whose level is h.

3.3 Analysis of the Incremental Algorithm

Let $I_i = [l_i, r_i]$ be the interval inserted in the current update step. We analyse our algorithm by computing the time required in every step.

1. **Computing $L(v_i)$:**
   Step 1: $I_i = [l_i, r_i]$ is inserted into $\mathcal{I}$. Let $I_i^l = [l_i, l_i]$. We check if I_i^l is present in $\mathcal{E}$ by an intersection query. This query takes $O(\log n)$ time in the worst case.
   - If I_i^l is in $\mathcal{E}$. Then proceed to Step 2.
   - If I_i^l is not in $\mathcal{E}$. We use the procedure $GET\text{-}SLS(\mathcal{I}, l_i)$ to create the SLS e_{l_i} at endpoint l_i. The procedure $GET\text{-}SLS$ works as follows :
   It performs an intersection query on $\mathcal{I}$ with I_i^l. The query returns all the intervals in $\mathcal{I}$ which contains I_i^l. Let $\mathcal{I}_{I_i^l}$ denote the set returned by the intersection query. The worst case time required for this query is $O(\log |\mathcal{I}| + |\mathcal{I}_{I_i^l}|)$. We create dynamic arrays V_{l_i} and V'_{l_i}, each of size $\max(levels(\mathcal{I}_{I_i^l}))$. For every i in the range $[0, \max(levels(\mathcal{I}_{I_i^l}))]$, we set $V_{l_i}[i] = 1$ if $i \in levels(\mathcal{I}_{I_i^l})$ and $V_{l_i}[i] = 0$ otherwise. For every $V_{l_i}[i] = 0$, we insert a node to the doubly linked list Z_{l_i} storing index i and store the pointer to that node in $V'_{l_i}[i]$. For every $V_{l_i}[i] = 1$, we store a NULL in $V'_{l_i}[i]$. Time taken for creating the dynamic arrays and the associated linked list is $O(\max(levels(\mathcal{I}_{I_i^l})))$. Since $\max(levels(\mathcal{I}_{I_i^l})) \leq \omega$, time taken for the operations on dynamic arrays and associated linked list is $O(\omega)$. Total time taken by procedure $GET\text{-}SLS(\mathcal{I}, l_i)$ is $O(\log |\mathcal{I}| + |\mathcal{I}_{I_i^l}|) + O(\omega) = O(\log |\mathcal{I}| + |\mathcal{I}_{I_i^l}| + \omega)$. At any level, e_{l_i} intersects with at most 2 intervals and we have ω many levels. Hence, $|\mathcal{I}_{I_i^l}| = O(\omega)$. Again, $|\mathcal{I}| \leq 2n$. Therefore, time taken by procedure $GET\text{-}SLS$ in the worst case is $O(\log n + \omega)$.

Same processing is repeated for $I_i^r = [r_i, r_i]$. Therefore, we have the following Lemma.

Lemma 4. *Worst case time taken by Step 1 in computing $L(v_i)$ of interval v_i is $O(\log n + \omega)$.*

Step 2: We use procedure $MAX\text{-}HEIGHT\text{-}OF\text{-}SLS\text{-}IN\text{-}INTERVAL(\mathcal{E}, v_i)$ for this step. The procedure works as follows: It performs an intersection query of $I_i = [l_i, r_i]$ on $\mathcal{E}$. This query returns the set S of all the endpoints that intersect with I_i. The worst case time taken by intersection query is $O(\log |\mathcal{E}| + |S|)$. Further, finding the height of the SLS at one endpoint in S takes constant time. Therefore, finding the maximum height of the SLS at any point in S takes time $O(|S|)$. Let h denote the maximum height for any endpoint $t \in S$. Therefore, the level value of I_i, $L(v_i)$ is set to h. Since Δ is the maximum degree in the graph, any interval I_i can intersect with at most Δ intervals. Therefore, $|S| = O(\Delta)$. Again, $|\mathcal{E}| \leq 2|\mathcal{I}| \leq 2n$. Thus the worst case time taken by Step 2 is $O(\log n + \Delta)$. Therefore, we have the following Lemma.

Lemma 5. *Worst case time taken by Step 2 in computing $L(v_i)$ of vertex v_i is $O(\log n + \Delta)$.*

Step 3: We use procedure *UPDATE-END-POINTS*$(S, L(v_i))$ to perform this step. The procedure works as follows:

We use the set S and the level value $L(v_i)$ of v_i computed in Step 2 to update the endpoints. Let $l = L(v_i)$. For every endpoint $t \in S$ we do the following: We check the length of V_t which is the size of the array V_t in constant time.

(a) **Case A:** If $l < len(V_t)$. In this case, we set $V_t[l]$ to 1. We use the pointer in $V_t'[l]$ to delete the node in Z_t storing the value l and set $V_t'[l]$ to NULL. If deleted node in Z_t was the head node, then we update the head node to the next node in Z_t and thus the value of h_t also gets updated. All these operations take constant time in the worst case.

(b) **Case B:** If $l \geq len(V_t)$. In this case, we use the standard doubling technique for expansion of dynamic arrays [6] until $len(V_t)$ becomes strictly greater than l. We also expand V_t' along with V_t and insert appropriate nodes to Z_t. Following the standard analysis for dynamic array expansion as shown in [6], one can easily show that all these steps take amortized constant time. Once $len(V_t) > l$, the remaining operations are same as in the case **A**.

To analyse the time required in Step 3, we observe that every update must perform the operations as described in case **A**. We refer to these operations as *task M*(M stands for mandatory). Some updates have to perform additional operations as described in case **B**. We refer to these operations as *task A*(A stands for additional). The time taken by each update to perform *task M* is $|S| \times O(1) = O(|S|)$. Since Δ is the maximum degree, hence $|S| \leq \Delta$. Therefore, every update takes $O(\Delta)$ time to perform *task M* in the worst case. To analyse the time required to perform *task A*, we crucially use the fact that our algorithm is incremental and hence only expansions of the dynamic arrays take place. Since ω is the size of the maximum clique, it follows that the maximum size of a dynamic array throughout the entire execution of the algorithm is upper bounded by 2ω. Over a sequence of n insertions, the total number of endpoints is upper bounded by $2n$. Therefore, we maintain at most $4n$ dynamic arrays. For every such array, total number of inserts in the array and the associated doubly linked list is at most 2ω in the entire run of the algorithm. An insertion into the dynamic array takes constant amortized time and insertion into doubly linked list takes constant worst case time. Therefore, during the entire run of the algorithm total time required to perform *task A* on one dynamic array and its associated doubly linked list is $O(\omega)$. This implies that during the entire run of the algorithm total time spent on *task A* over all the updates is $\leq 4n \times O(\omega)$. Let T be total time spent on Step 3 at the end of n insertions. This is the sum of the total time for *task A* and the total time in *task M*. Therefore,

$\mathsf{T} \leq 4n \times O(\omega) + n \times O(\Delta)$

$\mathsf{T} \leq 4n \times O(\Delta) + n \times O(\Delta)$ [since $\omega \leq \Delta + 1$]

Therefore, $\mathsf{T} = O(n\Delta)$ and we have the following Lemma.

Lemma 6. *The Amortized time taken by Step 3 in computing* $L(v_i)$ *of vertex* v_i *is* $O(\Delta)$.

2. **Computing** $o(v_i)$:

We have assigned level $L(v_i)$ to interval I_i. We use the map T to assign color to interval I_i. If $T[L(v_i)]$ is NULL, then we create an interval tree $T[L(v_i)]$ with v_i as the first node. This takes constant time. Otherwise, $T[L(v_i)]$ gives us the interval tree which stores all the intervals at level $L(v_i)$. We perform an intersection query on $T[L(v_i)]$ with I_i to obtain all the intervals that intersect with I_i. From Property **P**, the maximum intervals returned by the above query is 2. Therefore, worst case time taken by the intersection query is $O(\log |\mathcal{I}| + 2) = O(\log n + 2) = O(\log n)$. $o(v_i)$ is the smallest color from $\{1, 2, 3\}$ not assigned to any of the at most two neighbours of v_i in level $L(v_i)$.

Thus we have the following Lemma.

Lemma 7. *Computing the offset value of the vertex v_i with level value $L(v_i)$ takes $O(\log n)$ time in the worst case.*

The amortized update time of our incremental algorithm is given by the following theorem.

Theorem 1. *There exist an incremental algorithm which supports insertion of a sequence of n intervals in amortized $O(\log n + \Delta)$ time per update.*

Proof. For interval graphs, it is well known that $\omega = \chi(G) \le \Delta + 1$. Therefore, using Lemmas 4, 5, 6 and 7, we conclude that total time taken by our incremental algorithm for insertion of n intervals is:

$\mathcal{T}$ = Total time for Step 1 + Total Time for Step 2 + Total Time for Step 3 + Total time for computing offset

$$\mathcal{T} = n \times O(\log n + \omega) + n \times O(\log n + \Delta) + n \times O(\Delta) + n \times O(\log n)$$
$$\mathcal{T} = O(n \log n + n\Delta)$$

Therefore, the amortized update time over a sequence of n interval insertions is $O(\log n + \Delta)$. □

References

1. Dutot, A., Olivier, D., Guinand, F., Pign, Y.: On the decentralized dynamic graph coloring problem. In: Complex Systems and Self Organization Modelling, pp. 259–261 (2007)
2. Barba, L., et al.: Dynamic graph coloring. In: Ellen, F., Kolokolova, A., Sack, J.R. (eds.) Algorithms and Data Structures. LNCS, vol. 10389, pp. 97–108. Springer, Cham (2017). https://doi.org/10.1007/978-3-319-62127-2_9
3. Barenboim, L., Maimon, T.: Fully-dynamic graph algorithms with sublinear time inspired by distributed computing. In: International Conference on Computational Science, ICCS 2017, Zurich, Switzerland, 12–14 June 2017, pp. 89–98 (2017)
4. Bhattacharya, S., Chakrabarty, D., Henzinger, M., Nanongkai, D.: Dynamic algorithms for graph coloring. In: Proceedings of the Twenty-Ninth Annual ACM-SIAM Symposium on Discrete Algorithms, SODA 2018, New Orleans, LA, USA, 7–10 January 2018, pp. 1–20 (2018)

5. Hardy, B., Lewis, R., Thompson, J.: Tackling the edge dynamic graph coloring problem with and without future adjacency information. J. Heuristics **24**, 1–23 (2017)
6. Cormen, T.H., Leiserson, C.E., Rivest, R.L., Stein, C.: Introduction to Algorithms, 3rd edn, pp. 463–467. The MIT Press, Cambridge (2009). Chapter 17
7. Epstein, L., Levy, M.: Online interval coloring and variants. In: Caires, L., Italiano, G.F., Monteiro, L., Palamidessi, C., Yung, M. (eds.) ICALP 2005. LNCS, vol. 3580, pp. 602–613. Springer, Heidelberg (2005). https://doi.org/10.1007/11523468_49
8. Golumbic, M.C.: Algorithmic Graph Theory and Perfect Graphs. Academic Press, New York (1980)
9. Henzinger, M., Krinninger, S., Nanongkai, D., Saranurak, T.: Unifying and strengthening hardness for dynamic problems via the online matrix-vector multiplication conjecture. In: Proceedings of the Forty-Seventh Annual ACM on Symposium on Theory of Computing, STOC 2015, Portland, OR, USA, 14–17 June 2015, pp. 21–30 (2015)
10. Kierstead, H.A., Trotter, W.T.: An extremal problem in recursive combinatorics. Congressus Numerantium **33**(143–153), 98 (1981)
11. Larsen, K.G., Williams, R.: Faster online matrix-vector multiplication. In: Proceedings of the Twenty-Eighth Annual ACM-SIAM Symposium on Discrete Algorithms, SODA 2017, Barcelona, Spain, Hotel Porta Fira, 16–19 January 2017, pp. 2182–2189 (2017)
12. Narayanaswamy, N.S.: Dynamic storage allocation and on-line colouring interval graphs. In: Chwa, K.-Y., Munro, J.I.J. (eds.) COCOON 2004. LNCS, vol. 3106, pp. 329–338. Springer, Heidelberg (2004). https://doi.org/10.1007/978-3-540-27798-9_36
13. Ouerfelli, L., Bouziri, H.: Greedy algorithm for dynamic graph coloring. In: Communications, Computing and Control Applications, pp. 1–5 (2011)
14. Poudel, S., Gokhale, M., Ripeanu, M., Sallinen, S., Iwabuchi, K., Pearce, R.A.: Graph coloring as a challenge problem for dynamic graph processing on distributed systems. In: International Conference for High Performance Computing, Networking, Storage and Analysis, pp. 347–358 (2016)
15. Solomon, S., Wein, N.: Improved dynamic graph coloring. In: 26th Annual European Symposium on Algorithms, ESA 2018, 20–22 August 2018, Helsinki, Finland, pp. 72:1–72:16 (2018)

Lower Bounds for the Happy Coloring Problems

Ivan Bliznets[1,2] and Danil Sagunov[1(✉)]

[1] St. Petersburg Department of Steklov Institute of Mathematics of the Russian Academy of Sciences, Saint Petersburg, Russia
iabliznets@gmail.com, danilka.pro@gmail.com
[2] National Research University Higher School of Economics, Saint Petersburg, Russia

Abstract. In this paper, we study the MAXIMUM HAPPY VERTICES and the MAXIMUM HAPPY EDGES problems (MHV and MHE for short). Very recently, the problems attracted a lot of attention and were studied in Agrawal '17, Aravind et al. '16, Choudhari and Reddy '18, Misra and Reddy '17. Main focus of our work is lower bounds on the computational complexity of these problems. Established lower bounds can be divided into the following groups: NP-hardness of the above guarantee parameterization, kernelization lower bounds (answering questions of Misra and Reddy '17), exponential lower bounds under the SET COVER CONJECTURE and the EXPONENTIAL TIME HYPOTHESIS, and inapproximability results. Moreover, we present an $\mathcal{O}^*(\ell^k)$ randomized algorithm for MHV and an $\mathcal{O}^*(2^k)$ algorithm for MHE, where ℓ is the number of colors used and k is the number of required happy vertices or edges. These algorithms cannot be improved to subexponential taking proved lower bounds into account.

1 Introduction

In this paper, we study MAXIMUM HAPPY VERTICES and MAXIMUM HAPPY EDGES. The problems are motivated by a study of algorithmic aspects of homophyly law in large networks and were introduced by Zhang and Li in 2015 [21]. The law states that in social networks people are more likely to connect with people they like. Social network is represented by a graph, where each vertex corresponds to a person of the network, and an edge between two vertices denotes that the corresponding persons are connected within the network. Furthermore, we let vertices have a color assigned. The color of a vertex indicates type, character or affiliation of the corresponding person in the network. An edge is called *happy* if its endpoints are colored with the same color. A vertex is called *happy* if all its neighbours are colored with the same color as the vertex itself. Equivalently, a vertex is happy if all edges incident to it are happy. Formal definition of MAXIMUM HAPPY VERTICES and MAXIMUM HAPPY EDGES is the following.

This research was supported by the Russian Science Foundation (project 16-11-10123).

D.-Z. Du et al. (Eds.): COCOON 2019, LNCS 11653, pp. 490–502, 2019.
https://doi.org/10.1007/978-3-030-26176-4_41

MAXIMUM HAPPY VERTICES (MHV)
Input: A graph G, a partial coloring of vertices $p : S \to [\ell]$ for some $S \subseteq V(G)$ and an integer k.   **Question:** Is there a coloring $c : V(G) \to [\ell]$ extending partial coloring p such that the number of happy vertices with respect to c is at least k?

MAXIMUM HAPPY EDGES (MHE)
Input: A graph G, a partial coloring of vertices $p : S \to [\ell]$ for some $S \subseteq V(G)$ and an integer k.   **Question:** Is there a coloring $c : V(G) \to [\ell]$ extending partial coloring p such that the number of happy edges with respect to c is at least k?

Recently, MHV and MHE have attracted a lot of attention and were studied from parameterized [1–4,18] and approximation [19–22] points of view as well as from experimental perspective [17].

NP-hardness of MHV and MHE was proved by Zhang and Li even in case when only three colors are used. Later, Misra and Reddy [18] proved NP-hardness of both MHV and MHE on split and on bipartite graphs. However, MHV is polynomially time solvable on cographs and trees [2,18]. Approximation results for MHV are presented in Zhang et al. [22]. They showed that MHV can be approximated within $\frac{1}{\Delta+1}$, where Δ is the maximum degree of the input graph, and MHE can be approximated within $\frac{1}{2} + \frac{\sqrt{2}}{4} f(\ell)$, where $f(\ell) = \frac{(1-1/\ell)\sqrt{\ell(\ell-1)}+1/\sqrt{2}}{\ell-1+1/2\ell}$. From parameterized point of view the following parameters were studied: pathwidth [1,3], treewidth [1,3], neighbourhood diversity [3], vertex cover [18], distance to clique [18], distance to threshold graphs [4]. Kernelization questions were studied in works [1,13]. Agrawal [1] provided a $\mathcal{O}(k^2\ell^2)$ kernel for MHV where ℓ is the number of colors used and k is the number of desired happy vertices. Independently, Gao and Gao [13] present $2^{k\ell+k} + k\ell + k + \ell$ kernel for general case and $7(k\ell + k) + \ell - 10$ in case of planar graphs.

Short summary of our results can be found below.

No polynomial kernels: If NP $\not\subseteq$ coNP/poly then there are no polynomial kernels for MHV parameterized by vertex cover, and no polynomial kernels for MHE under the following parameterizations: number of uncolored vertices, number of happy edges, and distance to almost any reasonable graph class. Moreover, under NP $\not\subseteq$ coNP/poly, there is no $\mathcal{O}((k^d\ell)^{2-\epsilon})$ and no $\mathcal{O}((k^d h)^{2-\epsilon})$ bitsize kernel for MHV. Note that these results answer question from [18]: "Do the MAXIMUM HAPPY VERTICES and MAXIMUM HAPPY EDGES problems admit polynomial kernels when parameterized by either the vertex cover or the distance to clique parameters?"

Above guarantee: Above-greedy versions of MHV and MHE are NP-complete even for budget equal 1.

Exponential lower bounds: Assuming the Set Cover Conjecture, MHV and MHE do not admit $\mathcal{O}^*((2 - \epsilon)^{n'})$ algorithms, where n' is the number of uncolored vertices in the input graph. Even with $\ell = 3$, there is no $2^{o(n+m)}$ algorithm for MHV and MHE, unless ETH fails.

Innaproximability: Unless P = NP, MHV does not admit approximation algorithm with factors $\mathcal{O}(n^{\frac{1}{2}-\epsilon})$, $\mathcal{O}(m^{\frac{1}{2}-\epsilon})$, $\mathcal{O}(h^{1-\epsilon})$, $\mathcal{O}(\ell^{1-\epsilon})$, for any $\epsilon > 0$.

Algorithms: We present $\mathcal{O}^*(\ell^k)$ randomized algorithm for MHV and $\mathcal{O}^*(2^k)$ algorithm for MHE. Running time of this algorithms match with the corresponding lower bounds. We should note that an algorithm with the running time of $\mathcal{O}^*(2^k)$ for MHE was also presented by Aravind et al. in [3].

2   Preliminaries

Basic Notation. We denote the set of positive integer numbers by $\mathbb{N}$. For each positive integer k, by $[k]$ we denote the set of all positive integers not exceeding k, $\{1, 2, \ldots, k\}$. We use $\sqcup$ for the disjoint union operator, i.e. $A \sqcup B$ equals $A \cup B$, with an additional constraint that A and B are disjoint.

We use traditional $\mathcal{O}$-notation for asymptotical upper bounds. We additionally use $\mathcal{O}^*$-notation that hides polynomial factors. Many of our results concern the parameterized complexity of the problems, including fixed-parameter tractable algorithms, kernelization algorithms, and some hardness results for certain parameters. For detailed survey in parameterized algorithms we refer to the book of Cygan et al. [7].

Throughout the paper, we use standard graph notation and terminology, following the book of Diestel [11]. All graphs in our work are undirected simple graphs. We may refer to the distance to $\mathcal{G}$ parameter, where $\mathcal{G}$ is an arbitrary graph class. For a graph G, we say that a vertex subset $S \subseteq V(G)$ is a $\mathcal{G}$ *modulator* of G, if G becomes a member of $\mathcal{G}$ after deletion of S, i.e. $G \setminus S \in \mathcal{G}$. Then, the *distance to* $\mathcal{G}$ parameter of G is defined as the size of its smallest $\mathcal{G}$ modulator.

Graph Colorings. When dealing with instances of MAXIMUM HAPPY VERTICES or MAXIMUM HAPPY EDGES, we use a notion of colorings. A *coloring* of a graph G is a function that maps vertices of the graph to the set of colors. If this function is partial, we call such coloring *partial*. If not stated otherwise, we use ℓ for the number of distinct colors, and assume that colors are integers in $[\ell]$. A partial coloring p is always given as a part of the input for both problems, along with graph G. We also call p a *precoloring* of the graph G, and use (G, p) to denote the graph along with the precoloring. The goal of both problems is to extend this partial coloring to a specific coloring c that maps each vertex to a color. We call c a *full coloring* (or simply, a coloring) of G that extends p. We may also say that c is a coloring of (G, p). For convenience, introduce the notion of potentially happy vertices, both for full and partial colorings.

Definition 1. We call a vertex v of (G, p) *potentially happy*, if there exists a coloring c of (G, p) such that v is happy with respect to c. In other words, if u and w are precolored neighbours of v, then $p(u) = p(w)$. We denote the set of all potentially happy vertices in (G, p) by $\mathcal{H}(G, p)$.

By $\mathcal{H}_i(G, p)$ we denote the set of all potentially happy vertices in (G, p) such that they are either precolored with color i or have a neighbour precolored with color i:

$$\mathcal{H}_i(G, p) = \{v \in \mathcal{H}(G, p) \mid N[v] \cap p^{-1}(i) \neq \emptyset\}.$$

In other words, if a vertex $v \in \mathcal{H}_i(G, p)$ is happy with respect to some coloring c of (G, p), then necessarily $c(v) = i$.

For a graph with precoloring (G, p), by $h = |\mathcal{H}(G, p)|$ we denote the number of potentially happy vertices in (G, p). Note that if c is a full coloring of a graph G, then $|\mathcal{H}(G, c)|$ is equal to the number of vertices in G that are happy with respect to c.

Due to lack of space, we omit proofs of some theorems and lemmata. We mark such theorems and lemmata with the '⋆' sign. Missing proofs can be found in the full version of the paper.

3 Polynomial Kernels for Structural Graph Parameters

In this section, we study existence of polynomial kernels for MHV or MHE under several parameterizations. We start with proving lower bounds for structural graph parameters. We provide reductions to both MHV and MHE from the following problem.

BOUNDED RANK DISJOINT SETS [4]

Input: A set family $\mathcal{F}$ over a universe U with every set $S \in \mathcal{F}$ having size at most d, and a positive integer k.

Question: Is there a subfamily $\mathcal{F}'$ of $\mathcal{F}$ of size at most k such that every pair of sets S_1, S_2 in $\mathcal{F}'$ we have $S_1 \cap S_2 = \emptyset$?

Theorem 1 ([12]). BOUNDED RANK DISJOINT SETS *parameterized by kd does not admit a polynomial compression even if every set $S \in \mathcal{F}$ consists of exactly d elements and $|U| = kd$, unless $NP \subseteq coNP/poly$.*

The following two theorems answer open questions posed in [18].

Theorem 2. MAXIMUM HAPPY VERTICES *parameterized by the vertex cover number does not admit a polynomial compression, unless NP $\subseteq$ coNP/poly.*

Proof. We give a polynomial reduction from the SET PACKING problem, such that the vertex cover number of the constructed instance of MHV is at most the size of the universe of the initial instance of SET PACKING plus one. Since BOUNDED RANK DISJOINT SETS is a special case of SET PACKING, from Theorem 1 the theorem statement will then follow. The reduction is as follows.

Given an instance $(U = [n], \mathcal{F} = \{S_1, S_2, \ldots, S_m\}, k)$ of SET PACKING, construct an instance (G, p, k) of MHV. For each $i \in U$, introduce vertex u_i in G and left it uncolored. For each set $S_j \in \mathcal{F}$, introduce a vertex s_j in G and precolor it with color j, i.e. $p(s_j) = j$. Thus, the set of colors used in precoloring p is exactly $[m]$. Then, for each $i \in [n]$ and $j \in [m]$ such that $i \in S_j$, introduce an edge between u_i and s_j in G. Additionally, introduce two vertices t_1 and t_2 to G and precolor them with colors 1 and 2 respectively. Then, introduce an edge (t_1, t_2) to G and for every $i \in [n]$ and $j \in [2]$, introduce an edge (u_i, t_j) in G. Thus, vertices t_1 and t_2 never become happy and ensure that u_i never become happy for any $i \in [n]$. Finally, set the number of required happy vertices to k. Observe that $\{u_1, \ldots, u_n\} \cup \{t_1\}$ forms a vertex cover of G, hence the vertex cover number of G is at most $n + 1$.

We now claim that $(U, \mathcal{F}, k)$ is a yes-instance of SET PACKING if and only if (G, p, k) is a yes-instance of MHV. Let $S_{i_1}, S_{i_2}, \ldots, S_{i_k}$ be the answer to $(U, \mathcal{F}, k)$, i.e. $S_{i_p} \cap S_{i_q} = \emptyset$ for every distinct $p, q \in [k]$. Since $S_{i_1}, S_{i_2}, \ldots, S_{i_k}$ are disjoint, $s_{i_1}, s_{i_2}, \ldots, s_{i_k}$ do not have any common neighbours in G. Hence, we can extend coloring p to coloring c in a way that $s_{i_1}, s_{i_2}, \ldots, s_{i_k}$ are happy with respect to c ($c(u_i)$ is then, in fact, the index of the set containing u_i, i.e. $u_i \in S_{c(u_i)}$). At least k vertices become happy in G, hence (G, p, k) is a yes-instance of MHV.

In the other direction, let c be a coloring of G extending p so that at least k vertices in G are happy with respect to c. Only vertices that can be happy in (G, p) are vertices of type s_i, hence there are vertices $s_{i_1}, s_{i_2}, s_{i_3}, \ldots, s_{i_k}$ that are happy in G with respect to c. Since these vertices are precolored with pairwise distinct colors and are simultaneously happy, they may have no common neighbours in G. This implies that the corresponding sets of the initial instance $S_{i_1}, S_{i_2}, \ldots, S_{i_k}$ are pairwise disjoint. Hence, they form an answer to the initial instance $(U, \mathcal{F}, k)$ of SET PACKING. This completes the proof. $\square$

Theorem 3. MAXIMUM HAPPY EDGES *parameterized by the number of uncolored vertices or by the number of happy edges does not admit a polynomial compression, unless* $NP \subseteq coNP/poly$.

Proof. As in the proof of Theorem 2, we again provide a polynomial reduction from BOUNDED RANK DISJOINT SETS and then use Theorem 1. In this proof though, we will use the restricted version BOUNDED RANK DISJOINTS SETS problem itself (and not the SET PACKING problem), formulated in Theorem 1. That is, we will use the constraint that all sets in the given instance are of the same size d, and the size of the universe $|U|$ is equal to kd. We note that the following reduction has very much in common with the reduction described in the proof of Theorem 2. $\square$

Given an instance $([n], \mathcal{F} = \{S_1, S_2, \dots, S_m\}, k)$ of BOUNDED RANK DIS-JOINT SETS with $n = kd$ and $|S_i| = d$ for every $i \in [m]$, we construct an instance (G, p, k') of MHE. We assume that each element of the universe $[n]$ is contained in at least one set, otherwise the given instance is a no-instance. Firstly, as in the proof of Theorem 2, for each element of the universe $i \in [n]$, introduce a cor-responding vertex u_i in G. For each set S_j, $j \in [m]$, introduce not just one, but n corresponding vertices $s_{j,1}, s_{j,2}, \dots, s_{j,n}$. Then again, similarly to the proof of Theorem 1, for each i, j such that $i \in S_j$, introduce edges between u_i and *each* vertex $s_{j,t}$ corresponding to the set S_j, i.e. n edges in total. To finish the construction of G, introduce every possible edge (u_i, u_j) in G.

Thus, $V(G) = \{u_i \mid i \in [n]\} \cup \{s_{j,t} \mid j \in [m], t \in [n]\}$ and $E(G) = \{(u_i, s_{j,t}) \mid i \in S_j, t \in [n]\} \cup \{(u_i, u_j) \mid i, j \in [n], i \neq j\}$. Then, precolor the vertices of G in the usual way, i.e. set $p(s_{j,t}) = j$ for every $j \in [m]$ and $t \in [n]$, and leave each vertex u_i uncolored. Finally, we set the number of required happy edges to $k' = n^2 + k\binom{d}{2} = (kd)^2 + k\binom{d}{2}$. Construction of (G, p, k') is done in polynomial time. Observe that the number of uncolored vertices in (G, p, k) equals the size of the universe n, and the number of required happy edges is polynomial of n. Hence, existence of a polynomial kernel respectively to any of these two parameters for MHE contradicts the statement of Theorem 1. We argue that the initial instance is a yes-instance if and only if (G, p, k') is a yes-instance of MHE.

We prove first that if $([n], \mathcal{F}, k)$ is a yes-instance, then (G, p, k') is a yes-instance. Let $([n], \mathcal{F}, k)$ be a yes-instance of the restricted version of BOUNDED RANK DISJOINT SETS, and let $S_{i_1}, S_{i_2}, \dots, S_{i_k}$ be the instance solution. As usual, extend p to a coloring c of G by setting $c(u_i)$ to the index of the set in the solution containing u_i, i.e. $c(u_i) = i_t$ for some $t \in k$ and $i \in S_{c(u_i)}$. Since $S_{i_1}, S_{i_2}, \dots, S_{i_k}$ are disjoint, and their total size equals the size of the universe, such coloring c always exists uniquely for a fixed solution of $([n], \mathcal{F}, k)$. We claim that there are exactly k' happy edges in G with respect to c.

All edges in G are either of type $(u_i, s_{j,t})$ or of type (u_i, u_j). Consider edges of type $(u_i, s_{j,t})$ for a fixed $i \in [n]$. Happy edges among them are those with $c(s_{j,t}) = c(u_i)$. Since $c(s_{j,t}) = p(s_{j,t}) = j$ and $i \in S_{c(u_i)}$, these edges are exactly $(u_i, s_{c(u_i),t})$. Hence, there are n happy edges of this type for a fixed $i \in [n]$ and n^2 happy edges of this type in total. It is left to count the number of happy edges of the clique, i.e. edges of type (u_i, u_j). Observe that each u_i is colored with a color corresponding to a containing set of the answer. Since each set is of size d, the vertices u_i are split by color into k groups of size d. Each group contributes exactly $\binom{d}{2}$ happy edges, and no edge connecting vertices from different groups is happy. Thus, there are exactly $k\binom{d}{2}$ happy edges of type (u_i, u_j) in G with respect to c. We get that exactly $n^2 + k\binom{d}{2}$ edges of G are happy with respect to c, hence (G, p, k') is a yes-instance of MHE.

In the other direction, let (G, p, k') be a yes-instance of MHE, and let c be an optimal coloring of G extending p. At least k' edges are happy in G with respect to c. Let us show that *exactly* k' edges are happy in G with respect to c.

Claim 1. In any optimal coloring c of G extending p, $i \in S_{c(u_i)}$ for each $i \in [n]$.

Proof of Claim 1. Suppose it is not true, and c is an optimal coloring of (G, p) and $i \notin S_{c(u_i)}$ for some $i \in [n]$. For each j with $i \in S_j$, u_i is adjacent to n vertices $s_{j,t}$, which are precolored with color j. None of edges $(u_i, s_{j,t})$ are happy with respect to c, since $j \neq c(u_i)$. The only other edges incident to u_i are $n - 1$ edges of the clique. Thus, u_i is incident to at most $n - 1$ happy edges.

Choose arbitrary j with $i \in S_j$, and put $c(u_i) = j$. u_i becomes incident to at least n happy edges. Happiness of edges not incident with u_i has not changed. Thus, the change yields at least one more happy edge. A contradiction with the optimality of c. ∎

Claim 2. In any optimal coloring c of G extending p, there are at most $k\binom{d}{2}$ happy edges of type (u_i, u_j) in G with respect to c.

Proof of Claim 2. The vertices u_i are split into groups containing vertices of the same color by c, so the happy edges of type (u_i, u_j) are exactly the edges inside the groups. By Claim 1, each u_i is colored with a color corresponding to a set containing i in c. Hence, each group contains vertices corresponding to elements of the same set, and thus contains at most d vertices. So each u_i is incident to at most $d - 1$ happy edges of type (u_i, u_j), and in total there are at most $n \cdot (d - 1)/2 = k\binom{d}{2}$ such happy edges in G with respect to c. ∎

From Claims 1 and 2 follows that at most $n^2 + k\binom{d}{2} = k'$ edges are happy in G with respect to c. And as seen in the proof of Claim 2, the only way that yields exactly k' happy edges is when u_i are split by color into disjoint groups of size d, each containing vertices corresponding to a set of the initial instance. Hence, if c yields k' happy edges in G, $\{S_{c(u_i)} \mid i \in [n]\}$ is a solution to $([n], \mathcal{F}, k)$. Thus, $([n], \mathcal{F}, k)$ is a yes-instance of BOUNDED RANK DISJOINT SETS. This finishes the whole proof. □

Definition 2. We call a graph family $\mathcal{G}$ *uniformly polynomially instantiable*, if there is an algorithm that, given positive integer n as input, outputs a graph G, such that $|V(G)| \geq n$ and $G \in \mathcal{G}$, in poly(n) time.

Corollary 1 ($\star$). *For any uniformly polynomially instantiable graph family $\mathcal{G}$, MAXIMUM HAPPY EDGES, parameterized by the distance to graphs in $\mathcal{G}$, does not admit a polynomial compression, unless NP $\subseteq$ coNP/poly.*

In the rest of the section we study kernel bitsize lower bounds for MHV, parameterized by either $k + \ell$ or $k + h$, where h is the number of potentially happy vertices. This relates to the result of Agrawal in [1], where the author showed that MHV admits a polynomial kernel with $\mathcal{O}(k^2 \ell^2)$ vertices. We show that, for any $d > 0$ and any $\epsilon > 0$, there is no kernel of bitsize $\mathcal{O}(k^d \cdot \ell^{2-\epsilon})$ for MHV. Similarly, we show that there is no kernel of bitsize $\mathcal{O}(k^d \cdot h^{2-\epsilon})$ for MHV. To prove these lower bounds, we refer to the framework of weak cross-compositions, that originates from works of Dell and van Mekelbeek [10], Dell and Marx [9] and Hermelin and Wu [15]. These results are finely summarized by Cygan et al. in the chapter on lower bounds for kernelization [6]. We recall the notion of weak cross-compositions.

Definition 3 ([6,9,15]). Let $L \subseteq \Sigma^*$ be a language and $Q \subseteq \Sigma^* \times \mathbb{N}$ be a parameterized language. We say that L *weakly-cross-composes* into Q if there exists a real constant $d \geq 1$, called the dimension, a polynomial equivalence relation $\mathcal{R}$, and an algorithm $\mathcal{A}$, called the *weak cross-composition*, satisfying the following conditions. The algorithm $\mathcal{A}$ takes as input a sequence of $x_1, x_2, \ldots, x_t \in \Sigma^*$ that are equivalent with respect to $\mathcal{R}$, runs in time polynomial in $\sum_{i=1}^{t} |x_i|$, and outputs one instance $(y, k) \in \Sigma^* \times \mathbb{N}$ such that:

(a) for every $\delta > 0$ there exists a polynomial $p(\cdot)$ such that for every choice of t
and input strings $x_1, x_2, \ldots, x_t$ it holds that $k \leq p(\max_{i=1}^{t} |x_i|) \cdot t^{\frac{1}{d} + \delta}$, and
(b) $(y, k) \in Q$ if and only if there exists at least one index i such that $x_i \in L$.

The framework of weak cross-compositions is used for proving conditional lower bounds on polynomial compression bitsize. This is formulated in the following theorem.

Theorem 4 ([6,9,15]). *If an NP-hard language L admits a weak cross-composition of dimension d into a parameterized language Q. Then for any $\epsilon > 0$, Q does not admit a polynomial compression with bitsize $\mathcal{O}(k^{d-\epsilon})$, unless* NP $\subseteq$ coNP/poly.

Dell and Marx [9] use this framework to show that the VERTEX COVER problem parameterized by the solution size does not admit a kernel with subquadratic bitsize. Their result is the following.

Lemma 1 ([6,9]). *There exists a weak cross composition of dimension 2 from an NP-hard problem* MULTICOLORED BICLIQUE *into the* VERTEX COVER *problem parameterized by the solution size. In fact, this weak cross-composition $\mathcal{A}$, given instances $x_1, x_2, \ldots, x_t$ of* MULTICOLORED BICLIQUE *as input, outputs an instance (G, k') of* VERTEX COVER *satisfying*

$-\ |V(G)| \leq p(\max_{i=1}^{t} |x_i|) \cdot \sqrt{t}$, *and*
$-\ |V(G)| - k' \leq q(\max_{i=1}^{t} |x_i|)$,

for some polynomials p and q.

The bound for $|V(G)| - k'$ is given because one can look at an instance (G, k') of VERTEX COVER as at an instance $(G, |V(G)| - k')$ of INDEPENDENT SET. Then, the solution parameter of INDEPENDENT SET is bounded with polynomial of the maximum input size, independently of the number of instances t. We are ready to prove the theorem.

Theorem 5 ($\star$). *For any fixed constant d and any $\epsilon > 0$,* MAXIMUM HAPPY VERTICES *does not admit polynomial compressions with bitsizes $\mathcal{O}((k^d \cdot \ell)^{2-\epsilon})$ and $\mathcal{O}((k^d \cdot h)^{2-\epsilon})$, where h is the number of potentially happy vertices, unless* NP $\subseteq$ coNP/poly.

4   Parameterization Above Guarantee

This section concerns the above guarantee parameter for MHV and MHE. By *guarantee* we mean the number of happy vertices or edges that can be obtained with a trivial extension of the precoloring given in input. The definition of trivial extensions follows.

Definition 4. For a graph with precoloring (G, p), we call a full coloring c a *trivial extension* of p, if p can be extended to c by choosing a single color i and assigning color i to every uncolored vertex. In other words, $p(v) = c(v)$ for every $v \in p^{-1}([\ell])$, and $c(u) = c(v)$ for every $u, v \notin p^{-1}([\ell])$.

We formulate the version of MHV where the above guarantee parameter equals one.

ABOVE GUARANTEE HAPPY VERTICES
Input: A graph G, a partial coloring $p : S \to [\ell]$ for some $S \subseteq V(G)$ and integer k, such that there is a trivial extension of p that yields exactly k happy vertices in G.
Question: Is $(G, p, k + 1)$ a yes-instance of MHV?

The ABOVE GUARANTEE HAPPY EDGES is formulated analogously. We show that both these problems cannot be solved in polynomial time, unless P = NP. We start with ABOVE GUARANTEE HAPPY VERTICES. To prove that it is computationally hard, we provide a chain of polynomial reductions. An intermediate problem in this chain is the WEIGHTED MAX-2-SAT problem.

WEIGHTED MAX-2-SAT
Input: A boolean formula in 2-CNF with integer weights assigned to its clauses, an integer w.
Question: Is there an assignment of the variables of ϕ satisfying clauses of total weight at least w in ϕ?

Lemma 2 (⋆). WEIGHTED MAX-2-SAT *is NP-complete even when the inputs* ϕ *and* w *satisfy*

1. *The total weight of all positive clauses (i.e., clauses containing at least one positive literal) of* ϕ *equals* $w - 1$;
2. *Each clause of* ϕ *is assigned either weight* 1 *or weight* 13;
3. *Each variable appears exactly three times in* ϕ, *at least once positively in a clause containing also a negative literal, and at least once negatively in a clause containing also a positive literal.*

The chain continues with the following version of the INDEPENDENT SET problem.

INDEPENDENT SET ABOVE COLORING
Input: A graph G, properly colored with ℓ colors: $V(G) = V_1 \sqcup V_2 \sqcup \ldots \sqcup V_\ell$.
Question: Is there an independent set of size at least $\max\limits_{i=1}^{\ell} |V_i| + 1$ in G?

Lemma 3 $(\star)$. INDEPENDENT SET ABOVE COLORING *is NP-complete for* $\ell = 3$.

Theorem 6 $(\star)$. ABOVE GUARANTEE HAPPY VERTICES *is NP-complete even when* $\ell = 3$.

We now turn onto ABOVE GUARANTEE HAPPY EDGES. We provide a reduction from the following well-known NP-complete problem.

EXACT 3-COVER (X3C) [14, 16]

Input: An integer n, a collection $\mathcal{S} = \{S_1, S_2, \ldots, S_m\}$ of three-element subsets of $[3n]$.

Question: Is there an exact cover of $[3n]$ with elements of $\mathcal{S}$, i.e. is there a sequence $i_1, i_2, \ldots, i_n$, such that $S_{i_1} \cup S_{i_2} \cup \ldots \cup S_{i_n} = [3n]$?

Theorem 7 $(\star)$. ABOVE GUARANTEE HAPPY EDGES *is NP-complete.*

5 ETH and Set Cover Conjecture Based Lower Bounds

In this section, we show lower bounds for exact algorithms for MHV and MHE, based on the popular Exponential Time Hypothesis and the Set Cover Conjecture. We start with the Set Cover Conjecture and the following problem.

SET PARTITIONING

Input: An integer n, a set family $\mathcal{F} = \{S_1, S_2, \ldots, S_m\}$ over a universe U with $|U| = n$.

Question: Is there a sequence of pairwise disjoint sets $S_{i_1}, S_{i_2}, \ldots, S_{i_k}$ in $\mathcal{F}$, such that $\bigsqcup_{j=1}^{k} S_{i_j} = U$?

Theorem 8 ([5]). *For any* $\epsilon > 0$, SET PARTITIONING *cannot be solved in time* $\mathcal{O}^*((2 - \epsilon)^n)$, *unless the Set Cover Conjecture fails.*

Theorem 9 $(\star)$. *For any* $\varepsilon > 0$, MAXIMUM HAPPY VERTICES *cannot be solved in time* $\mathcal{O}^*((2 - \varepsilon)^{n'})$, *where* n' *is the number of uncolored vertices, unless the Set Cover Conjecture fails.*

Theorem 10 $(\star)$. *For any* $\varepsilon > 0$, MAXIMUM HAPPY EDGES *cannot be solved in time* $\mathcal{O}^*((2 - \varepsilon)^{n'})$, *where* n' *is the number of uncolored vertices, unless the Set Cover Conjecture fails.*

We now turn onto ETH-based lower bounds.

Theorem 11 $(\star)$. MAXIMUM HAPPY VERTICES *with* $\ell = 3$ *cannot be solved in time* $2^{o(n+m)}$, *unless ETH fails.*

We now prove another computational lower bound for MHV that is based on the reduction from INDEPENDENT SET to MHV discussed above in the proofs of Theorems 5 and 6. This reduction also implies some approximation lower bounds.

Theorem 12 ($\star$). MAXIMUM HAPPY VERTICES *cannot be solved in* $\mathcal{O}(n^{o(k)})$ *time, unless ETH fails. Also, for any* $\epsilon > 0$, MAXIMUM HAPPY VERTICES *cannot be approximated within* $\mathcal{O}(n^{\frac{1}{2}-\epsilon})$, $\mathcal{O}(m^{\frac{1}{2}-\epsilon})$, $\mathcal{O}(h^{1-\epsilon})$ *or* $\mathcal{O}(\ell^{1-\epsilon})$ *in polynomial time, unless* P = NP.

Theorem 13 ($\star$). MAXIMUM HAPPY EDGES *with* $\ell = 3$ *cannot be solved in time* $2^{o(n+m)}$, *unless ETH fails.*

6   Algorithms

In this section, we present two algorithms solving MHV or MHE. We start with a randomized algorithm for MHV that runs in $\mathcal{O}^*(\ell^k)$ time and recognizes a yes-instance and finds the required coloring with a constant probability. The algorithm is based on the following lemma.

Lemma 4 ($\star$). *Let* (G, p) *be a graph with precoloring, and* $P = \bigcup\limits_{i=1}^{\ell} \mathcal{H}_i(G, p)$. *Let* c *be a coloring that yields the maximum possible number of happy vertices in* (G, p), *and let* $H = \mathcal{H}(G, c)$ *be the set of these vertices. Then* $|H \cap P| \geq \frac{1}{\ell} \cdot |P|$.

Theorem 14 ($\star$). *There is a* $\mathcal{O}^*(\ell^k)$ *running time randomized algorithm for* MAXIMUM HAPPY VERTICES.

Note that by Theorem 12, no algorithm with running time $\mathcal{O}(\ell^{o(k)})$ exists for MHV, unless ETH fails. Similarly, no $\mathcal{O}(\ell^{o(k)})$ running time *randomized* algorithm exists for MHV under the *randomized* ETH [8]. The algorithm given above is optimal in that sence.

We now turn onto MHE and give an exact algorithm with $\mathcal{O}^*(2^k)$ running time for this problem. In its turn, this algorithm optimal in a sence that no $2^{o(k)}$ running time algorithm exists for MHE under ETH (see Theorem 13). The algorithm relies on the following kernelization result. We note that this kernelization result and an algorithm with the running time of $\mathcal{O}^*(2^k)$ was already presented by Aravind et al. in [3]. We believe that our kernelization algorithm is short and somewhat simpler, since it relies on a single reduction rule.

Theorem 15 ([3], $\star$). MAXIMUM HAPPY EDGES *admits a kernel with at most* k *uncolored vertices.*

Theorem 16 ([3], $\star$). *There is a* $\mathcal{O}^*(2^k)$ *running time algorithm for* MAXIMUM HAPPY EDGES.

References

1. Agrawal, A.: On the parameterized complexity of happy vertex coloring. In: Brankovic, L., Ryan, J., Smyth, W.F. (eds.) IWOCA 2017. LNCS, vol. 10765, pp. 103–115. Springer, Cham (2018). https://doi.org/10.1007/978-3-319-78825-8_9
2. Aravind, N.R., Kalyanasundaram, S., Kare, A.S.: Linear time algorithms for happy vertex coloring problems for trees. In: Mäkinen, V., Puglisi, S.J., Salmela, L. (eds.) IWOCA 2016. LNCS, vol. 9843, pp. 281–292. Springer, Cham (2016). https://doi.org/10.1007/978-3-319-44543-4_22
3. Aravind, N., Kalyanasundaram, S., Kare, A.S., Lauri, J.: Algorithms and hardness results for happy coloring problems. arXiv preprint arXiv:1705.08282 (2017)
4. Choudhari, J., Reddy, I.V.: On structural parameterizations of happy coloring, empire coloring and boxicity. In: Rahman, M.S., Sung, W.-K., Uehara, R. (eds.) WALCOM 2018. LNCS, vol. 10755, pp. 228–239. Springer, Cham (2018). https://doi.org/10.1007/978-3-319-75172-6_20
5. Cygan, M., et al.: On problems as hard as CNF-SAT. ACM Trans. Algorithms 12(3), 1–24 (2016)
6. Cygan, M., et al.: Lower bounds for kernelization. Parameterized Algorithms, pp. 523–555. Springer, Cham (2015). https://doi.org/10.1007/978-3-319-21275-3_15
7. Cygan, M., et al.: Parameterized Algorithms, vol. 3. Springer, Cham (2015). https://doi.org/10.1007/978-3-319-21275-3
8. Dell, H., Husfeldt, T., Marx, D., Taslaman, N., Wahlén, M.: Exponential time complexity of the permanent and the Tutte polynomial. ACM Trans. Algorithms 10(4), 1–32 (2014)
9. Dell, H., Marx, D.: Kernelization of packing problems. In: Proceedings of the Twenty-Third Annual ACM-SIAM Symposium on Discrete Algorithms. Society for Industrial and Applied Mathematics (2012)
10. Dell, H., Melkebeek, D.V.: Satisfiability allows no nontrivial sparsification unless the polynomial-time hierarchy collapses. J. ACM 61(4), 1–27 (2014)
11. Diestel, R.: Graph Theory. Springer, Heidelberg (2018)
12. Dom, M., Lokshtanov, D., Saurabh, S.: Kernelization lower bounds through colors and IDs. ACM Trans. Algorithms 11(2), 1–20 (2014)
13. Gao, H., Gao, W.: Kernelization for maximum happy vertices problem. In: Bender, M.A., Farach-Colton, M., Mosteiro, M.A. (eds.) LATIN 2018. LNCS, vol. 10807, pp. 504–514. Springer, Cham (2018). https://doi.org/10.1007/978-3-319-77404-6_37
14. Garey, M., Johnson, D., Stockmeyer, L.: Some simplified NP-complete graph problems. Theoret. Comput. Sci. 1(3), 237–267 (1976)
15. Hermelin, D., Wu, X.: Weak compositions and their applications to polynomial lower bounds for kernelization. In: Proceedings of the Twenty-Third Annual ACM-SIAM Symposium on Discrete Algorithms. Society for Industrial and Applied Mathematics (2012)
16. Karp, R.M.: Reducibility among combinatorial problems. In: Miller, R.E., Thatcher, J.W., Bohlinger, J.D. (eds.) Complexity of Computer Computations, pp. 85–103. Springer, Boston (1972). https://doi.org/10.1007/978-1-4684-2001-2_9
17. Lewis, R., Thiruvady, D., Morgan, K.: Finding happiness: an analysis of the maximum happy vertices problem. Comput. Oper. Res. 103, 265–276 (2019)
18. Misra, N., Reddy, I.V.: The parameterized complexity of happy colorings. In: Brankovic, L., Ryan, J., Smyth, W.F. (eds.) IWOCA 2017. LNCS, vol. 10765, pp. 142–153. Springer, Cham (2018). https://doi.org/10.1007/978-3-319-78825-8_12

19. Xu, Y., Goebel, R., Lin, G.: Submodular and supermodular multi-labeling, and vertex happiness. CoRR (2016)
20. Zhang, P., Jiang, T., Li, A.: Improved approximation algorithms for the maximum happy vertices and edges problems. In: Xu, D., Du, D., Du, D. (eds.) COCOON 2015. LNCS, vol. 9198, pp. 159–170. Springer, Cham (2015). https://doi.org/10.1007/978-3-319-21398-9_13
21. Zhang, P., Li, A.: Algorithmic aspects of homophyly of networks. Theoret. Comput. Sci. **593**, 117–131 (2015)
22. Zhang, P., Xu, Y., Jiang, T., Li, A., Lin, G., Miyano, E.: Improved approximation algorithms for the maximum happy vertices and edges problems. Algorithmica **80**(5), 1412–1438 (2018)

An Efficient Decision Procedure for Propositional Projection Temporal Logic

Xinfeng Shu[1]([⊠]) and Nan Zhang[2]

[1] School of Computer Science and Technology,
Xi'an University of Posts and Communications, Xi'an 710061, China
shuxf@xupt.edu.cn
[2] Institute of Computing Theory and Technology, Xidian University,
Xi'an 710071, China
nanzhang@xidian.edu.cn

Abstract. The decision problem for Propositional Projection Temporal Logic (PPTL) has been solved successfully, however time complexity of the procedure is increased exponentially to the length of the formula. To solve the problem, a Labeled Unified Complete Normal Form is introduced as an intermediate form to rewrite a PPTL formula into its equivalent Labeled Normal Form, based on which the Labeled Normal Form Graph is constructed and an efficient decision procedure for PPTL is formalized with the time complexity linear to the length of the formula and the size of the power set of the atomic propositions in the formula.

Keywords: Projection Temporal Logic · Decision procedure ·
Labeled Unified Complete Normal Form · Labeled Normal Form Graph

1 Introduction

Projection Temporal Logic (PTL) [1–3] is an extension of Interval Temporal Logic (ITL) [4] by introducing a new projection construct, $(P_1, \ldots, P_m)\, prj\, Q$, and supporting both finite and infinite time. Within the PTL framework, a unified model checking approach [5] is advocated, which employs an executable subset of PTL with a framing technique, named Modeling, Simulation and Verification Language (MSVL) [3], to model systems, and uses Propositional Projection Temporal Logic (PPTL), the propositional subset of PTL, formulas to specify desired properties. PPTL has the expressive power of the full regular language [6], and hence enables us to verify more properties of the computer systems [7–9] compared to available methods [10,11].

The decision problem for Propositional Projection Temporal Logic (PPTL) has been solved successfully based on the techniques of Normal Form (NF) and Normal Form Graph (NFG) in recently years [13–15], however, time complexities

This research is supported by the Industrial Research Project of Shaanxi Province No. 2017GY-076, and NSFC Grant No. 61572386 and No. 61672403.

D.-Z. Du et al. (Eds.): COCOON 2019, LNCS 11653, pp. 503–515, 2019.
https://doi.org/10.1007/978-3-030-26176-4_42

of the available decision procedures are very high. For example, let the NF of formula P be $(p_e \wedge \varepsilon) \vee \bigvee_{i=1}^{n}(p_i \wedge \bigcirc P_i')$, then the NF of $\neg P$ is $\neg P \equiv \neg p_e \wedge \varepsilon \vee \bigvee_{\Delta \subseteq \Psi}(true \wedge \bigwedge_{j \in \Delta} p_j \wedge \bigwedge_{k \in \Psi - \Delta} \neg p_k \wedge \bigcirc(true \wedge \bigwedge_{j \in \Delta} \neg P_j'))$, where $\Psi = \{1, .., n\}$. Intuitively, the NF of $\neg P$ is computed by enumerating the composition of each p_j, $\neg p_k$ and $\bigcirc \neg P_j'$ with the time complexity of $O(2^n)$, where $1 \leq j \leq n$, $1 \leq k \leq n$, $j \neq k$, and $p_j \wedge \bigcirc P_j'$ is a future product in the NF of P. However, if negation operators are nested in many layers, e.g., $\underbrace{\neg ... \neg}_{L} P$, the size of the future products in the NF increases exponentially to the nested layers of negation operators, i.e., $O(2^{\cdot^{\cdot^{\cdot^{n}}}}\}L)$, which greatly impacts the efficiency for verifying the software and hardware systems with PPTL.

To solve the problem, in this paper, we are motivated to formalize an efficient decision procedure for PPTL. To this end, a Labeled Unified Complete Normal Form (LCCNF) is introduced as an intermediate form while rewriting a PPTL formula into its equivalent Labeled Normal Form (LNF). With the new approach, the time complexity for transforming a PPTL P into its LNF is $O(L*2^{|\Phi|})$, where L is the length of P, and Φ is the set of atomic propositions in P. Based on the LNF, the Labeled Normal Form Graph (LNFG) for describing models of a PPTL formula is constructed, and the efficiency of the decision procedure for PPTL is greatly improved.

The rest of paper is organized as follows. In the next section, the syntax and semantics of PPTL are briefly introduced. In Sect. 3, the unified complete normal form and normal form graph are introduced. In Sect. 4, the techniques of LCCNF, LNF and LNFG are presented, and the improved decision procedure for PPTL is given. Finally, conclusions are drawn in Sect. 5.

2 Propositional Projection Temporal Logic

Propositional Projection Temporal Logic (PPTL) is an extension of Propositional Interval Temporal Logic (PITL) [4] with infinite models and a new projection construct prj. In this section, the syntax and semantics of PPTL are briefly introduced. More details can be found in literature [1].

Syntax. Let Φ be a finite set of atomic propositions, and $B = \{true, false\}$ the boolean domain. The formulas P of PPTL are inductively defined as follows:

$$P ::= p \mid \neg P \mid P_1 \wedge P_2 \mid \bigcirc P \mid P^+ \mid (P_1, \ldots, P_m) \, prj \, P$$

where $p \in \Phi$ is an atomic proposition; $\bigcirc$ (next), $+$ (chop-plus) and prj (projection) are temporal operators, and $\neg$ and $\wedge$ are identical to those in the classical propositional logic. A formula is called a $state$ formula if it contains no temporal operators. The conventional constructs $true, false, \wedge, \rightarrow$ as well as $\leftrightarrow$ are defined as usual. Furthermore, we use the following abbreviations:

$$\varepsilon \stackrel{\text{def}}{=} \neg \bigcirc true \qquad\qquad \bar{\varepsilon} \stackrel{\text{def}}{=} \neg \varepsilon$$

$$P^* \stackrel{\text{def}}{=} \varepsilon \vee P^+ \qquad\qquad P;Q \stackrel{\text{def}}{=} (P, Q) \, prj \, \varepsilon$$

$$\Diamond P \stackrel{\text{def}}{=} true; P \qquad\qquad \Box P \stackrel{\text{def}}{=} \neg \Diamond \neg P$$

Semantics. A state s over Φ is a mapping from Φ to B, i.e., $s : \Phi \rightarrow B$. We use notation $s[p]$ to denote the valuation of p at state s. An interval (i.e., model) σ is a non-empty sequence of states $\sigma = \langle s_0, \ldots, s_{|\sigma|} \rangle$, which $|\sigma|$ denotes the length of σ and is ω if σ is infinite, or the number of states minus one if σ is finite. Let N_0 be the set of non-negative integers and $N_\omega = N_0 \cup \{\omega\}$, we extend the comparison operators, $=, <, \leq$, to N_ω by considering $\omega = \omega$, and for all $i \in N_0$, $i < \omega$. Moreover we define $\preceq$ as $\leq -\{(\omega, \omega)\}$. We use notation $\sigma_{(i..j)}$ to mean that a subinterval $\langle s_i, \ldots, s_j \rangle$ of σ with $0 \leq i \preceq j \leq |\sigma|$. The *concatenation* of a finite interval $\sigma = \langle s_0, \ldots, s_{|\sigma|} \rangle$ with another interval $\sigma' = \langle s'_0, \ldots, s'_{|\sigma'|} \rangle$ (may be infinite) is denoted by $\sigma \bullet \sigma'$ and $\sigma \bullet \sigma' = \langle s_0, \ldots, s_{|\sigma|}, s'_0, \ldots, s'_{|\sigma'|} \rangle$. Further, let $\sigma = \langle s_0, \ldots, s_{|\sigma|} \rangle$ be an interval and $r_1, \ldots, r_h$ be integers ($h \geq 1$) such that $0 \leq r_1 \leq \ldots \leq r_h \preceq |\sigma|$, the *projection* of σ onto $r_1, \ldots, r_h$ is the interval (called projected interval) $\sigma \downarrow (r_1, \ldots, r_h) = \langle s_{t_1}, \ldots, s_{t_l} \rangle$, $(t_1 < t_2 < \ldots < t_l)$, where $t_1, \ldots, t_l$ is obtained from $r_1, \ldots, r_h$ by deleting all duplicates. For example, $\langle s_0, s_1, s_2, s_3, s_4, s_5 \rangle \downarrow (0, 2, 2, 2, 4, 4, 5) = \langle s_0, s_2, s_4, s_5 \rangle$.

An interpretation for PPTL is a triple $\mathcal{I} = (\sigma, i, j)$, where σ is an interval, $i \in N_0$ and $j \in N_\omega$, and $0 \leq i \preceq j \leq |\sigma|$. We use notation (σ, i, j) to mean that a formula is interpreted over a subinterval $\langle s_i, \ldots, s_j \rangle$ of σ with the current state being s_i. The satisfaction relation ($\models$) for PPTL formulas is defined as follows:

$\mathcal{I} \models p$ iff $s_i[p] = true$, for any given atomic proposition p.

$\mathcal{I} \models \neg P$ iff $\mathcal{I} \nvDash P$.

$\mathcal{I} \models P \wedge Q$ iff $\mathcal{I} \models P$ and $\mathcal{I} \models Q$.

$\mathcal{I} \models \bigcirc P$ iff $i < j$ and $(\sigma, i+1, j) \models P$.

$\mathcal{I} \models P^+$ iff there exist finite many integers $i = r_0 \leq \ldots \leq r_{m-1} \preceq r_m = j$
    such that $(\sigma, r_{l-1}, r_l) \models P$ for all $1 \leq l \leq m$, or there exist infinite
    many integers $i = r_0 \leq r_1 \leq r_2 \leq \ldots$ such that $\lim_{l \to \infty} r_l = \infty$ and for
    all $1 \leq l$, $(\sigma, r_{l-1}, r_l) \models P$.

$\mathcal{I} \models (P_1, ..., P_m) \, prj \, Q$ iff there exist integers $i = r_0 \leq \ldots \leq r_{m-1} \leq r_m \preceq j$
    such that $(\sigma, r_{l-1}, r_l) \models P_l$ for all $1 \leq l \leq m$, and $(\sigma', 0, |\sigma'|) \models Q$ for
    one of the following σ':
    (1) $r_m < j$ and $\sigma' = \sigma \downarrow (r_0, \ldots, r_m) \bullet \sigma_{(r_m+1..j)}$.
    (2) $r_m = j$ and $\sigma' = \sigma \downarrow (r_0, \ldots, r_h)$ for some $0 \leq h \leq m$.

A formula P is satisfied by an interval σ, denoted by $\sigma \models P$, if $(\sigma, 0, |\sigma|) \models P$. A formula P is called *satisfiable* if $\sigma \models P$ for some σ. A formula P is *valid*, denoted by $\models P$, if $\sigma \models P$ for all σ. Usually, we denote $\models \Box(P \leftrightarrow Q)$ by $P \equiv Q$ and $\models \Box(P \rightarrow Q)$ by $P \sqsupset Q$. The following are some useful logic laws, where $m > 1$ and p_s is a state formula, and the related proofs can be found in [12].

L1 $\bigcirc(P \wedge Q) \equiv \bigcirc P \wedge \bigcirc Q$ L2 $\bigcirc(P \vee Q) \equiv \bigcirc P \vee \bigcirc Q$

L3 $\bigcirc P \equiv \overline{\varepsilon} \wedge \bigcirc P$ L4 $\varepsilon \wedge \bigcirc P \equiv false$

L5 $\bigcirc(P;Q) \equiv \bigcirc P;Q$ L6 $\neg \bigcirc P \equiv \varepsilon \vee \bigcirc \neg P$

L7 $\varepsilon \, prj \, Q \equiv Q$ L8 $P \, prj \, \varepsilon \equiv P$

L9 $\varepsilon ; P \equiv P$ L10 $P; \varepsilon \equiv P \wedge \Diamond \varepsilon$

L11 $(P_1 \vee P_2);Q \equiv (P_1;Q) \vee (P_2;Q)$ L12 $P;(Q_1 \vee Q_2) \equiv (P;Q_1) \vee (P;Q_2)$

L13 $(p_s \wedge P;Q) \equiv p_s \wedge (P;Q)$ L14 $(P_1, \ldots, P_m) \, prj \, \varepsilon \equiv (P_1; \ldots ;P_m)$

L15 $(P \wedge \overline{\varepsilon} \, prj \, \bigcirc Q) \equiv (P \wedge \overline{\varepsilon};Q)$ L16 $P^+ \equiv P \vee (P \wedge \overline{\varepsilon};P^+)$

L17 $(p_s \wedge P_1, \ldots, P_m) \, prj \, Q \equiv p_s \wedge (P_1, \ldots, P_m) \, prj \, Q$

L18 $(P_1, \ldots, P_m) \, prj \, p_s \wedge Q \equiv p_s \wedge (P_1, \ldots, P_m) \, prj \, Q$

L19 $(P_1 \wedge \overline{\varepsilon}, P_2 \ldots, P_m) \, prj \, \bigcirc Q \equiv (P_1 \wedge \overline{\varepsilon};(P_2, \ldots, P_m) \, prj \, Q)$

L20 $(P_1, \ldots, P_m) \, prj \, (Q_1 \vee Q_2) \equiv ((P_1, \ldots, P_m) \, prj \, Q_1) \vee ((P_1, \ldots, P_m) \, prj \, Q_2)$

L21 $(P_1, \ldots, (P_i \vee P_i'), \ldots, P_m) \, prj \, Q \equiv ((P_1, \ldots, P_i, \ldots, P_m) \, prj \, Q)$
$$\vee ((P_1, \ldots, P_i', \ldots, P_m) \, prj \, Q)$$

3 Improved Method for Constructing NFG

The techniques of Normal Form (NF) and Normal Form Graph (NFG) [14] are the basis of the decision procedure of PPTL formulas. In this section, we present an efficient method for computing NFs and constructing NFGs of PPTL formulas.

3.1 Unified Complete Normal Form

The Complete Normal Form (CNF) was introduced in [16] to compute the normal form (NF) for negation formulas. In this subsection, we put forward a special kind of CNF, named Unified CNF (UCNF), to accelerate computing NF for all PPTL formulas. In the following, we first give the definition of UCNF, and then prove that any PPTL formula can be rewritten into its equivalent UCNF.

Definition 1. Let Γ_1 and Γ_2 be any two sets of PPTL formulas, the conjunction of Γ_1 and Γ_2, denoted by $\Gamma_1 \wedge \Gamma_2$, is defined as:

$$\Gamma_1 \wedge \Gamma_2 = \begin{cases} \{P \wedge Q \mid P \in \Gamma_1, Q \in \Gamma_2\}, & \text{if } \Gamma_1 \neq \varnothing \text{ and } \Gamma_2 \neq \varnothing \\ \Gamma_1, & \text{if } \Gamma_2 = \varnothing \\ \Gamma_2, & \text{if } \Gamma_1 = \varnothing \end{cases}$$

Definition 2 (Unified Complete Normal Form). Let Φ be the finite set of atomic propositions of PPTL and Ψ the set of min-products over Φ, i.e., $\Psi = \bigwedge_{p \in \Phi} \{p, \neg p\}$. For any PPTL formula P, the *unified complete normal form* of P can be defined as follows:

$$P \equiv \bigvee_{j=1}^{|\Psi^1|} (\omega_j \wedge \varepsilon) \vee \bigvee_{i=1}^{|\Psi|} (\omega_i \wedge \bigcirc P_i'),$$

where $\Psi^1 \subseteq \Psi$, $\omega_j \in \Psi^1$, $\omega_i \in \Psi$ and $P_i' (1 \leq i \leq |\Psi|)$ is a general PPTL formula. We call the product of the form $\omega_j \wedge \varepsilon$ *terminal product*, whereas the product of the form $\omega_i \wedge \bigcirc P_i'$ *future product*. Besides, we call P_i' the *successor formula* of P.

It is readily to prove that $|\Psi| = 2^{|\Phi|}$, and for any $\omega_i, \omega_j \in \Psi$, $\bigvee_{i \neq j} \omega_i \wedge \omega_j \equiv false$, and $\bigvee_{i=1}^{|\Psi|} \omega_i \equiv true$. In the following, we employ the specific symbols Φ and Ψ to represent the finite set of atomic propositions of PPTL and the set of min-products over Φ respectively in default.

We claim that any PPTL formula P can be rewritten into its equivalent UCNF with the time complexity $O(L * 2^{|\Phi|})$, where L is the length of formula P. The fact concludes in the following theorem.

Theorem 1. Any PPTL formula P can be rewritten into its equivalent UCNF $P \equiv \bigvee_{j=1}^{|\Psi^1|}(\omega_j \wedge \varepsilon) \vee \bigvee_{i=1}^{|\Psi|}(\omega_i \wedge \bigcirc P_i')$ with the time complexity $O(L * 2^{|\Phi|})$.

Proof. We first prove the existence of the UCNF, and then analyze the time complexity of the rewriting process. Make an induction on the structure of P:

Base Case: P is an atomic proposition p.

$$p \equiv \bigvee_{j=1}^{|\Psi^1|} \omega_j \wedge \varepsilon \vee \bigvee_{i=1}^{|\Psi^1|} \omega_i \wedge \bigcirc true \vee \bigvee_{i=1}^{|\Psi^2|} \omega_i \wedge \bigcirc false$$

where $\Psi^1 = \{\omega | \omega \in \Psi, \text{ and } \omega \rightarrow p\}$, $\Psi^2 = \Psi - \Psi^1$.

Induction Step: Suppose the theorem holds for formulas $P_k (1 \leq k \leq m)$ and Q. Let the UCNFs of P_k be $P_k \equiv \bigvee_{j_k=1}^{|\Psi_k^1|} \omega_{j_k} \wedge \varepsilon \vee \bigvee_{i=1}^{|\Psi|}(\omega_i \wedge \bigcirc P_i^{k'})$, we have

- P is next formula $\bigcirc P_1$: $\bigcirc P \equiv \bigvee_{i=1}^{|\Psi|} \omega_i \wedge \bigcirc P$.
- P is conjunction formula $P_1 \wedge P_2$:

$$P_1 \wedge P_2 \equiv \bigvee_{j=1}^{|\Psi^1|} \omega_j \wedge \varepsilon \vee \bigvee_{i=1}^{|\Psi|} \omega_i \wedge \bigcirc(P_i^{1'} \wedge P_i^{2'})$$

  where $\Psi^1 = \Psi_1^1 \cap \Psi_2^1$.
- P is chop formula $P_1 ; P_2$:

$$\begin{aligned} P_1 ; P_2 &\equiv (\bigvee_{j_1=1}^{|\Psi_1^1|} \omega_{j_1} \wedge \varepsilon \vee \bigvee_{i=1}^{|\Psi|}(\omega_i \wedge \bigcirc P_i^{1'})) ; P_2 \\ &\equiv \bigvee_{j=1}^{|\Psi^1|} \omega_j \wedge \varepsilon \vee \bigvee_{i=1}^{|\Psi^1|} \omega_i \wedge \bigcirc(P_i^{2'} \vee (P_i^{1'} ; P_2)) \\ &\quad \vee \bigvee_{i=1}^{|\Psi^2|} \omega_i \wedge \bigcirc(P_i^{1'} ; P_2) \end{aligned}$$

  where $\Psi^1 = \Psi_1^1 \cap \Psi_2^1$, $\Psi^2 = \Psi - \Psi_1^1$.
- P is chop-plus formula $P_1 +$:

$$\begin{aligned} P_1 + &\equiv P_1 \vee (P_1 \wedge \overline{\varepsilon} ; P_1^+) \\ &\equiv \bigvee_{j=1}^{|\Psi_1^1|} \omega_j \wedge \varepsilon \vee \bigvee_{i=1}^{|\Psi|} \omega_i \wedge \bigcirc(P_i' \vee (P_i^{1'} ; P_1^+)) \end{aligned}$$

- P is negation formula $\neg P_1$: $\neg P_1 \equiv \bigvee_{j=1}^{|\Psi^{1'}|} \omega_j \wedge \varepsilon \vee \bigvee_{i=1}^{|\Psi|} \omega_i \wedge \bigcirc \neg P_i'$, where $\Psi^{1'} = \Psi - \Psi^1$.
- P is projection formula $(P_1, \ldots, P_m) \, prj \, Q$: Similarly to the proof of Lemma 1 in literature [13].

Now we review the transformation process. Obviously, atomic proposition p and next formula $\bigcirc P_1$ can be written into their UCNFs directly with the time complexity $O(|\Psi|)$. For other composite formulas P, we first need to transform the subformulas of P into their UCNFs, and then traverse each ω in Ψ and check whether the corresponding terminal products and future products of its sub-formulas should be properly composed and added to the result UCNF according to the structure of P. In consideration of computing the UCNFs of the subformulas, the total time consumed in computing the UCNF of P is $O(L * |\Psi|) = O(L * 2^{|\Phi|})$.

3.2 Normal Form Graph

For any formula P, let the UCNF of P be $\bigvee_{j=1}^{|\Psi^1|}(\omega_j \wedge \varepsilon) \vee \bigvee_{i=1}^{n}(\omega_i \wedge \bigcirc P_i')$. Intuitively, the UCNF of P characterizes under what circumstances P can be satisfied, that is, P can be satisfied if and only if either there exists a ω_j ($1 \leq j \leq |\Psi^1|$) holding over a single-state interval, or there exists a $\omega_i \wedge \bigcirc P_i'$ ($1 \leq i \leq n$) holding over an interval with length greater than zero such that p_i must hold at the first state and its associated P_i' must hold over the remainder of the interval. Thus, if we repeatedly rewrite P, P_i' and their successor formulas into UCNFs, a normal form graph (NFG) showing the decomposition relationships of PPTL formulas and their UCNFs can be constructed. For more details of the NFG, please refer to literature [13].

Three examples of NFGs of PPTL formulas are shown in Fig. 1. In an NFG, root note is denoted by a double circle, ϵ node by a circle with a black dot in it, and each of other nodes by a single circle. Each edge is denoted by a directed arc connecting two nodes. Intuitively, the NFG of a formula P describes all models of P, i.e., each finite path from root node to node ϵ corresponds to a finite model of P; each infinite path emanating from root node may (not definitely) correspond to an infinite model of P. Along a path, each edge describes a state in the corresponding model of the path if any.

For instance, as we can see in Fig. 1(a), formula $p \wedge (\varepsilon \vee \square \bigcirc p)$ can be satisfied by a single state finite interval with the atomic proposition p holding at the only state, or by an infinite model with the atomic proposition p holding at every state. For simplicity, while drawing the NFG, we usually use the simplified state formula labeled on an edge instead of enumerating each min-product, e.g., the formula *true* labeled on the edge $\langle \square \bigcirc p, p \wedge \square \bigcirc p \rangle$ stands for $p, \neg p$ labeled on the edge, i.e., there exist two edges from node $\square \bigcirc p$ to node $p \wedge \square \bigcirc p$ labeled with p and $\neg p$ respectively.

However, the NFG of a PPTL formula P describing models of P is not always true for formulas containing chop construct. For instance, formula $\square \bigcirc p; q$ is equivalent to *false*, but there exists an infinite path in its NFG as shown in Fig. 1(c). Obviously, the NFG of formula $\square \bigcirc p; q$ is isomorphic to that of formula $\square \bigcirc p$ as shown in Fig. 1(b), and along the infinite path the decomposition of $\square \bigcirc p; q$ occurs always on sub-formula $\square \bigcirc p$ and never on q, which in fact describes an infinite model of $\square \bigcirc p$ but not $\square \bigcirc p; q$, and hence must be removed. To solve

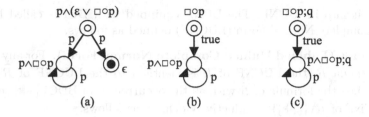

Fig. 1. NFGs of formulas $p \wedge (\varepsilon \vee \Box \bigcirc p), \Box \bigcirc p$ and $\Box \bigcirc p; q$

the problem, literature [15] introduces the techniques of Labeled Normal Form (LNF) and Labeled Normal Form Graph (LNFG) to trace the decomposition of chop formulas, and identify and remove the infinite paths describing the wrong models of PPTL formulas.

4 Improved Decision Procedure for PPTL

The decision procedure for PPTL is based on the technique of LNFG, which in turn is constructed with the LNFs of PPTL formulas [15]. Following the idea of computing UCNF, in this section, we employ the labeled UCNF to accelerate computing the LNFs to improve the efficiency of the decision procedure.

4.1 Labeled Unified Complete Normal Form

For each strongly connected component (SCC) Ω_P in the NFG of PPTL formula P, it is not hard to prove that there exists a corresponding SCC $\Omega_{P;Q}$ in the NFG of formula $P; Q$, within which the decomposition of $P; Q$ always occurs on sub-formula P and never on Q. In the NFG of a chop formula, such a SCC is called a *chop decomposition cycle* (CDC). Each CDC is named with a unique *decomposition cycle identifier* (DCI) '$[\Bbbk]$' ($\Bbbk \in N_0$), and all the nodes in a CDC share a same DCI. Further, when a node R directly comes out from a CDC $[\Bbbk]$, we bound it with a special mark $[-\Bbbk]$, i.e., $(R)[-\Bbbk]$, which is called *boundary decomposition cycle identifier* (BDCI).

The PPTL formula possibly with some subformulas bounded with DCI $[\Bbbk]$ or $[-\Bbbk]$ ($\Bbbk \in N_0$) is called *labeled PPTL formula*. For any PPTL formula R, we use symbol $\widetilde{R}$ to denote any labeled formula of R with some subformulas R' possibly bounded with $[\Bbbk]$ or $[-\Bbbk]$, i.e., $(R')[\Bbbk]$ or $(R')[-\Bbbk]$. Note that, DCI and BDCI do not belong to the logical system, and do not participate into logic calculus, they just help to construct LNFGs for PPTL formulas. The equivalence between two labeled PPTL formulas relies on the equivalence of the corresponding PPTL formulas obtained by removing all DCIs and BDCIs.

To ensure each node in a CDC shares a same DCI $[\Bbbk]$ as well as to compute the label of each node correctly during constructing the LNFG for a PPTL formula, we introduce the DCI to trace the decomposition of each chop formula when rewriting it into its UCNF. The UCNF equipped with DCI is called Labeled Unified Complete Normal Form (LUCNF) defined as follows.

Definition 3 (Labeled Unified Complete Normal Form). For any labeled PPTL formula $\widetilde{R}$, the LUCNF of $\widetilde{R}$ is identical to the LUCNF of $\widetilde{R}\backslash\{[-\Bbbk]\}$ ($\Bbbk \in N_0$), i.e., the formula of $\widetilde{R}$ with all the occurrences of BDCI $[-\Bbbk]$ removed. The LUCNF of $\widetilde{R}\backslash\{[-\Bbbk]\}$ is inductively defined as follows:

- $\widetilde{R}\backslash\{[-\Bbbk]\}$ is p, $\bigcirc P$, $(P_1, \ldots, P_m)\, prj\, Q$ or P^+: the LUCNF of $\widetilde{R}\backslash\{[-\Bbbk]\}$ is identical to the UCNF of R.
- $\widetilde{R}$ is a chop formula $(P;Q)$ or $(P;Q)[\Bbbk]$: let the UCNFs of P and Q be

$$P \equiv \bigvee_{j_1=1}^{|\Psi_1^1|} \omega_{j_1} \wedge \varepsilon \vee \bigvee_{i_1=1}^{|\Psi_{1\_1}|}(\omega_{i_1} \wedge \bigcirc P'_{i_1}) \vee \bigvee_{k_1=1}^{|\Psi_{1\_2}|}(\omega_{k_1} \wedge \bigcirc P'_{k_1}),$$
$$Q \equiv \bigvee_{j_2=1}^{|\Psi_2^1|} \omega_{j_2} \wedge \varepsilon \vee \bigvee_{i_2=1}^{|\Psi|}(\omega_{i_2} \wedge \bigcirc Q'_{i_2}),$$

where $\Psi_{1\_1} \cup \Psi_{1\_2} = \Psi$, nodes $P_{i_1}^{1'}$ ($1 \leq i_1 \leq |\Psi_{1\_1}|$) locate in the SCCs of the NFG of P. The LUCNF of formulas $P;Q$ is defined as

$$P;Q \equiv (\bigvee_{j_1=1}^{|\Psi_1^1|} \omega_{j_1} \wedge \varepsilon \vee \bigvee_{i_1=1}^{|\Psi_{1\_1}|}(\omega_{i_1} \wedge \bigcirc P'_{i_1}) \vee \bigvee_{k_1=1}^{|\Psi_{1\_2}|}(\omega_{k_1} \wedge \bigcirc P'_{k_1}));Q$$
$$\equiv \bigvee_{j=1}^{|\Psi^1|} \omega_j \wedge \varepsilon \vee \bigvee_{i_1=1}^{|\Psi_{1\_1}^1|} \omega_{i_1} \wedge \bigcirc((P'_{i_1};Q)[\Bbbk_{i_1}] \vee Q'_{i_1})$$
$$\vee \bigvee_{k_1=1}^{|\Psi_{1\_2}^1|} \omega_{k_1} \wedge \bigcirc((P'_{k_1};Q) \vee Q'_{i_1}) \vee \bigvee_{i_1=1}^{|\Psi_{1\_1}^2|} \omega_{i_1} \wedge \bigcirc(P'_{i_1};Q)[\Bbbk_{i_1}]$$
$$\vee \bigvee_{k_1=1}^{|\Psi_{1\_2}^2|} \omega_{k_1} \wedge \bigcirc(P'_{k_1};Q)$$

where $\Psi^1 = \Psi_1^1 \cap \Psi_2^1$, $\Psi_{1\_1}^1 = \Psi_{1\_1} \cap \Psi_1^1$, $\Psi_{1\_1}^2 = \Psi_{1\_1} - \Psi_{1\_1}^1$, $\Psi_{1\_2}^1 = \Psi_{1\_2} \cap \Psi_1^1$, $\Psi_{1\_2}^2 = \Psi_{1\_2} - \Psi_{1\_2}^1$; each $[\Bbbk_{i_1}]$ ($\Bbbk_{i_1} \in N_0, \omega_{i_1} \in \Psi_{1\_1}$) is the DCI of the CDC of which $(P'_{i_1};Q)$ locates in. The LUCNF of formula $(P;Q)[\Bbbk]$ is defined as

$$(P;Q)[\Bbbk] \equiv ((\bigvee_{j_1=1}^{|\Psi_1^1|} \omega_{j_1} \wedge \varepsilon \vee \bigvee_{i_1=1}^{|\Psi_{1\_10}|}(\omega_{i_1} \wedge \bigcirc P'_{i_1})$$
$$\vee \bigvee_{k_1=1}^{|\Psi_{1\_2+}|}(\omega_{k_1} \wedge \bigcirc P'_{k_1}));Q)[\Bbbk]$$
$$\equiv \bigvee_{j=1}^{|\Psi^1|} \omega_j \wedge \varepsilon \vee \bigvee_{i_1=1}^{|\Psi_{1\_10}^1|} \omega_{i_1} \wedge \bigcirc((P'_{i_1};Q)[\Bbbk] \vee Q'_{i_1}[-\Bbbk])$$
$$\vee \bigvee_{k_1=1}^{|\Psi_{1\_2+}^1|} \omega_{k_1} \wedge \bigcirc((P'_{k_1};Q)[-\Bbbk] \vee Q'_{i_1}[-\Bbbk])$$
$$\vee \bigvee_{i_1=1}^{|\Psi_{1\_10}^2|} \omega_{i_1} \wedge \bigcirc(P'_{i_1};Q)[\Bbbk] \vee \bigvee_{k_1=1}^{|\Psi_{1\_2+}^2|} \omega_{k_1} \wedge \bigcirc(P'_{k_1};Q)[-\Bbbk]$$

where Ψ_{1_10} is the set of mini-products to which the corresponding successor formula $(P'_{i_1};Q)$ locates in a same CDC with $(P;Q)$; $\Psi^1 = \Psi_1^1 \cap \Psi_2^1$, $\Psi_{1\_2+} = \Psi_{1\_2} \cup \Psi_{1\_1} - \Psi_{1\_10}$, $\Psi_{1\_10}^1 = \Psi_{1\_10} \cap \Psi_1^1$, $\Psi_{1\_10}^2 = \Psi_{1\_10} - \Psi_{1\_10}^1$, $\Psi_{1\_2+}^1 = \Psi_{1\_2+} \cap \Psi_1^1$, $\Psi_{1\_2+}^2 = \Psi_{1\_2+} - \Psi_{1\_2+}^1$.
- $\widetilde{R}\backslash\{[-\Bbbk]\}$ is a conjunction formula $\widetilde{P} \wedge \widetilde{Q}$: the LUCNF of $\widetilde{R}\backslash\{[-\Bbbk]\}$ is the conjunction of the LCCNFs of $\widetilde{P}$ and $\widetilde{Q}$ by applying Theorem 1.
- $\widetilde{R}\backslash\{[-\Bbbk]\}$ is a negation formula $\neg\widetilde{P}$: the LCNF of $\widetilde{R}\backslash\{[-\Bbbk]\}$ is the negation of the LUCNF of $\widetilde{P}$ by applying Theorem 1.

Theorem 2. Any labeled PPTL formula $\widetilde{P}$ can be rewritten into its equivalent LUCNF.

Proof. Direct consequence of Theorem 1 and definition of LUCNF.

4.2   Labeled Normal Form and Labeled Normal Form Graph

In case of a labeled PPTL formula having been rewritten into its equivalent LUCNF, we need to further transform the LUCNF into the LNF to construct the LNFG for the decision procedure of PPTL. In the following, we first give the definition of LNF and argue that the LUCNF of any labeled PPTL formula can be rewritten into its equivalent LNF, then present the formal definition of LNFG.

Definition 4. A labeled formula $\overline{R}$ is called a *min-conjunctive* labeled PPTL formula if it is generated by the following grammar:

$$L ::= p \mid P;Q \mid (P;Q)[\Bbbk] \mid (P)[-\Bbbk] \mid P^+ \mid \bigcirc P \mid (P_1,\ldots,P_m)\,prj\,Q$$
$$\overline{R} ::= L \mid \neg L \mid \overline{R}_1 \wedge \overline{R}_2$$

where p is an atomic proposition; $P, Q, P_1, \ldots, P_m$ are PPTL formulas.

Definition 5 (Labeled Normal Form). Let Φ be the finite set of atomic propositions of PPTL and Ψ the set of min-products over Φ. For any labeled PPTL formula $\widetilde{P}$, the *labeled normal form* of $\widetilde{P}$ can be defined as follows:

$$\widetilde{P} \equiv \bigvee_{j=1}^{|\Psi^1|}(\omega_j \wedge \varepsilon)\vee \bigvee_{i=1}^{n}(\omega_i \wedge \bigcirc\overline{P}'_i),$$

where $\Psi^1 \subseteq \Psi$, $\omega_i \in \Psi$ and $\overline{P}'_i(1 \leq i \leq n)$ is a min-conjunctive labeled PPTL formula. We call the product of the form $\omega_j \wedge \varepsilon$ *terminal product*, whereas the product of the form $\omega_i \wedge \bigcirc\overline{P}'_i$ *future product*.

While rewriting a LUCNF into its equivalent LNF, the terminal products $\bigvee_{j=1}^{|\Psi^1|}(\omega_j\wedge \varepsilon)$ keep unchanged. For each future product $\omega_i\wedge\bigcirc\widetilde{P}'_i$ $(1 \leq i \leq |\Psi|)$, by classical propositional calculus, $\widetilde{P}'_i$ can be written into an equivalently disjunctive normal form $\bigvee_{k=1}^{m}\overline{P}'_{i_k}$, by logical Law L2, $\omega_i \wedge \bigcirc\widetilde{P}'_i\equiv\bigvee_{k=1}^{m}(\omega_i \wedge \bigcirc\overline{P}'_{i_k})$, it is replaced with the future products $\bigvee_{k=1}^{m}(\omega_i \wedge \bigcirc\overline{P}'_{i_k})$ in the LNF of P.

Definition 6 (Labeled Normal Form Graph). For any labeled PPTL formula $\widetilde{P}$, the *labeled normal form graph* of $\widetilde{P}$ is a directed graph, $G = (\widetilde{CL}(\widetilde{P}), \widetilde{EL}(\widetilde{P}), v_0, V_f)$, where $\widetilde{CL}(\widetilde{P})$ denotes the set of nodes and $\widetilde{EL}(\widetilde{P})$ denotes the set of directed edges among $\widetilde{CL}(\widetilde{P})$, $v_0 \in \widetilde{CL}(\widetilde{P})$ is the root (or initial) node named by $\widetilde{P}$, and V_f denotes the set of pairs of nodes and their finite labels. Each node is specified by a labeled PPTL formula; each edge is a directed arc labeled with a state formula $\omega \in \Psi$ from node $\widetilde{Q}$ to node $\widetilde{R}$ and identified by a triple $\langle\widetilde{Q},\omega,\widetilde{R}\rangle$; each node $\widetilde{Q}$ and its label $\{[\Bbbk_1],\ldots,[\Bbbk_m]\}$ $(m \geq 1)$, i.e., the set of DCIs of $\widetilde{Q}$, is depicted by a pair $(\widetilde{R}, \{[\Bbbk_1],\ldots,[\Bbbk_m]\})$ in V_f. The sets $\widetilde{CL}(\widetilde{P})$, $\widetilde{EL}(\widetilde{P})$ and V_f can be defined inductively as follows:

(1) $\widetilde{P} \in \widetilde{CL}(\widetilde{P})$.

(2) For every node $\widetilde{Q} \in \widetilde{CL}(\widetilde{P}) \backslash \{\epsilon, false\}$, if node $\widetilde{Q}$ needs finite label, say $\{[\Bbbk_1], \ldots, [\Bbbk_m]\}$, according to Definition 7, $(\widetilde{Q}, \{[\Bbbk_1], \ldots, [\Bbbk_m]\}) \in V_f$. Further, let the LNF of $\widetilde{Q}$ be $\widetilde{Q} \equiv \bigvee_{j=1}^{|\Psi^1|}(\omega_j \wedge \epsilon) \vee \bigvee_{i=1}^{n}(\omega_i \wedge \bigcirc \widetilde{Q}'_i)$, then $\epsilon \in \widetilde{CL}(\widetilde{P})$, and $\langle \widetilde{Q}, \omega_j, \epsilon \rangle \in \widetilde{EL}(\widetilde{P})$ for each $j (1 \le j \le |\Psi^1|)$; $\overline{Q}'_i \in \widetilde{CL}(\widetilde{P})$, and $\langle \widetilde{Q}, \omega_i, \overline{Q}'_i \rangle \in \widetilde{EL}(\widetilde{P})$ for all $i (1 \le i \le n)$.

Definition 7. The finite label of node $\widetilde{R}$ is computed inductively according to the structure of $\widetilde{R}$ as follows:

- If $\widetilde{R}$ is $(P;Q)[\Bbbk]$, the finite label of $\widetilde{R}$ is $\{[\Bbbk]\}$.
- If $\widetilde{R}$ is $(P)[-\Bbbk]$, the finite label of $\widetilde{R}$ is $\{[-\Bbbk]\}$.
- If $\widetilde{R}$ is $\widetilde{P} \wedge \widetilde{Q}$, let the finite labels of $\widetilde{P}$ and $\widetilde{Q}$ be $\Gamma_{\widetilde{P}}$ and $\Gamma_{\widetilde{Q}}$ respectively, the finite label Γ of $\widetilde{R}$ is $\Gamma = (\Gamma_{\widetilde{P}} \cup \Gamma_{\widetilde{Q}}) \backslash \{[\Bbbk] | [\Bbbk] \in \Gamma_{\widetilde{P}} \cup \Gamma_{\widetilde{Q}}, [-\Bbbk] \in \Gamma_{\widetilde{P}} \cup \Gamma_{\widetilde{Q}}\}$, i.e., Γ is the union of $\Gamma_{\widetilde{P}}$ and $\Gamma_{\widetilde{Q}}$ under the constraint that for any $\Bbbk \in N_0$, if $[\Bbbk]$ and $[-\Bbbk]$ both occur in $\Gamma_{\widetilde{P}} \cup \Gamma_{\widetilde{Q}}$, then only $[-\Bbbk]$ is remained.
- If $\widetilde{R}$ is none of the above cases, $\widetilde{R}$ does not need finite label.

Based on the Definition 6, algorithm LNFG for constructing the LNFG of a given labeled PTL formula $\widetilde{P}$ is presented in Table 1. In the algorithm, three global sets Γ_N, Γ_C and Γ_S, which help to rewrite a labeled chop formula into its LNF, are initialized to empty. Then, global variable $DCISeed$, which works as a seed to generate the unique DCI for each CDC, is set to 0. Subsequently, function COMATOMPROD is employed to compute the set of min-products over the set atom propositions in $\widetilde{P}$. Further, functions LUCNF and LNF are used in sequence to rewrite a node $\widetilde{R}$ into its LNF Moreover, functions GETFL is employed to compute the finite label for the given node. These functions can be easily formalized according to Definitions 2–7 and Theorem 1, so their code is omitted here.

4.3 Decision Procedure for PPTL

In the LNFG of a labeled PPTL formula $\widetilde{P}$, a cycle $\Pi_C = \langle \widetilde{P}_1, \ldots, \widetilde{P}_h, \widetilde{P}_1 \rangle$ $(h \ge 1)$ is called an F *cycle* if there exists a DCI $[\Bbbk]$ such that all the nodes $\widetilde{P}_i$ $(1 \le i \le h)$ are labeled with the same DCI $[\Bbbk]$, otherwise, the cycle Π_C is called an *acceptable cycle*. Further, a *finite acceptable path*, $\Pi = \langle P, p_0, \widetilde{P}_0, \ldots, \widetilde{P}_{m-1}, p_m, \epsilon \rangle$, is an alternate sequence of nodes and edges from the root P to node ϵ; while an *infinite acceptable path*, $\Pi_\omega = \langle P, p_0, \widetilde{P}_0, \ldots, \widetilde{P}_h, p_{h+1}, \ldots \rangle$, is an infinite alternate sequence of nodes and edges emanating from the root node and cannot finally enter into infinite circles among nodes all labeled with a same DCI $[\Bbbk]$.

It has been proved in [15] that the LNFG of a PPTL formula P enjoys the following properties: (1) Each F cycle in the LNFG corresponds to a CDC; (2) For any node $\widetilde{R}$ in the LNFG, if $\widetilde{R}$ is reachable to neither node ϵ nor an acceptable cycle in the LNFG of P, then $\widetilde{R} \equiv false$; (3) Each finite (infinite) acceptable

Table 1. Algorithm for constructing LNFG of a labeled PPTL formula

---

function LNFG($\widetilde{P}$)
/*precondition: $\widetilde{P}$ is any labeled PPTL formula*/
/*postcondition: LNFG($\widetilde{P}$) computes LNFG of $\widetilde{P}$, $G = (\widetilde{CL}(\widetilde{P}), \widetilde{EL}(\widetilde{P}), v_0, V_f)$*/

---

begin function

    $\widetilde{CL}(\widetilde{P}) = \{\widetilde{P}\}$; $\widetilde{EL}(\widetilde{P}) = \phi$; $v_0 = \widetilde{P}$; $V_f = \phi$; $mark[\widetilde{P}] = 0$;

    $\Gamma_N = \Gamma_C = \Gamma_S = \Phi$;   /*The three sets help to compute LUCNF*/

    $DCISeed = 0$;   /*The variable helps to compute LUCNF*/

    $\Psi = \text{COMATOMPROD}(\widetilde{P})$;   /*The set of min-products of atom propositions in $\widetilde{P}$*/

    while there exists $\widetilde{R} \in \widetilde{CL}(\widetilde{P}) \backslash \{\epsilon, false\}$ and $mark[\widetilde{R}] == 0$ **do**

        $mark[\widetilde{R}] = 1$;

        if GETFL($\widetilde{R}$) $\neq \Phi$ **then**   /*Add finite label to $\widetilde{R}$*/

            $V_f = V_f \cup \{(\widetilde{R}, \text{GETFL}(\widetilde{R}))\}$;

        $\widetilde{Q} = \text{LUCNF}(\Psi, \widetilde{R})$;   /*Rewrite $\widetilde{R}$ into LUCNF*/

        $\widetilde{Q} = \text{LNF}(\widetilde{Q})$;   /*Rewrite the LUCNF of $\widetilde{R}$ into LNF*/

        $AddE = AddN = 0$;

        case

            $\widetilde{Q}$ is $\bigvee_{j=1}^{|\Psi_1|}(\omega_j \wedge \varepsilon)$: AddE=1;

            $\widetilde{Q}$ is $\bigvee_{i=1}^{n}(\omega_i \wedge \bigcirc\overline{Q}'_i)$: AddN=1;

            $\widetilde{Q}$ is $\bigvee_{j=1}^{|\Psi_1|}(\omega_j \wedge \varepsilon) \vee \bigvee_{i=1}^{n}(\omega_i \wedge \bigcirc\overline{Q}'_i)$: AddE=AddN=1;

        end case

        if $AddE == 1$ **then**

            $\widetilde{CL}(\widetilde{P}) = \widetilde{CL}(\widetilde{P}) \cup \{\epsilon\}$;

            $\widetilde{EL}(P) = \widetilde{EL}(P) \cup \bigcup_{j=1}^{\Psi_1}\{\langle\widetilde{R}, \omega_j, \epsilon\rangle\}$

        if $AddN == 1$ **then**

            for $i = 1$ to n **do**

                if $\widetilde{Q}'_i \notin CL(P)$ **then**

                    $\widetilde{CL}(\widetilde{P}) = \widetilde{CL}(\widetilde{P}) \cup \{\overline{Q}'_i\}$;

                    $mark[\overline{Q}'_i] = 0$;   /*$\overline{Q}'_i$ needs decomposed*/

            end for

            $\widetilde{EL}(\widetilde{P}) = \widetilde{EL}(\widetilde{P}) \cup \bigcup_{i=1}^{n}\{\langle\widetilde{R}, \omega_i, \overline{Q}'_i\rangle\}$

    end while

    return G;

end function

Table 2. Algorithm for checking the satisfiability of a PPTL formula

---

function CHECKPPTL(P)
/*precondition: P is any PPTL formula*/
/*postcondition: CHECKPPTL(P) returns *true* if P is satisfiable, otherwise *false**/

---

begin function

    $G = \text{LNFG}(P)$;

    $G' = \text{SIMPLIFY}(G)$;

    if G' is not empty **then**   **return** *true*;

    else   **return** *false*;

end function

---

path corresponds to a finite (infinite) model of P. With the properties of LNFG, a decision procedure for PPTL is formalized in Table 2, where algorithm SIMPLIFY is used to remove the redundant nodes by property (2) of LNFG and its details can be found in [15].

5 Conclusion

In this paper, we present a novel way to compute the normal form and labeled normal form as well as construct the LNFG. Compared with the existing method, the time complexity of the new way for rewriting a PPTL formula into its equivalent LNF improves from $O(2^{\cdot^{\cdot^{n}}}\}L)$ to $O(L * 2^{|\Phi|})$. Accordingly, the efficiency of the decision procedure for PPTL, which is based on the LNFG, is greatly enhanced. In the near future, we will apply the improved decision procedure to refine the model checking tool MSV to verify more complicate software and hardware systems.

References

1. Duan, Z.: An extended interval temporal logic and a framing technique for interval temporal logic programming. Ph.D. thesis, University of Newcastle Upon Tyne (1996)
2. Duan, Z., Koutny, M.: A framed temporal logic programming language. J. Comput. Sci. Technol. **19**, 333–344 (2004)
3. Duan, Z., Yang, X., Koutny, M.: Framed temporal logic programming. Sci. Comput. Program. **70**(1), 31–61 (2008)
4. Moszkowski, B.: Executing Temporal Logic Programs. Cambridge University Press, Cambridge (1986)
5. Duan, Z., Tian, C.: A unified model checking approach with projection temporal logic. In: Liu, S., Maibaum, T., Araki, K. (eds.) ICFEM 2008. LNCS, vol. 5256, pp. 167–186. Springer, Heidelberg (2008). https://doi.org/10.1007/978-3-540-88194-0_12
6. Tian, C., Duan, Z.: Expressiveness of propositional projection temporal logic with star. Theor. Comput. Sci. **412**(18), 1729–1744 (2011)
7. Wang, M., Duan, Z., Tian, C.: Simulation and verification of the virtual memory management system with MSVL. In: CSCWD, pp. 360–365 (2014)
8. Yu, Y., Duan, Z., Tian, C., Yang, M.: Model checking C programs with MSVL. In: Liu, S. (ed.) SOFL 2012. LNCS, vol. 7787, pp. 87–103. Springer, Heidelberg (2013). https://doi.org/10.1007/978-3-642-39277-1_7
9. Ma, Q., Duan, Z., Zhang, N., Wang, X.: Verification of distributed systems with the axiomatic system of MSVL. Formal Asp. Comput. **27**(1), 103–131 (2015)
10. Holzmann, G.J.: The model checker spin. IEEE Trans. Softw. Eng. **23**(5), 279–295 (1997)
11. McMillan, K.: Symbolic Model Checking: An Approach to the State Explosion Problem. Kluwer Academic Publisher, Dordrecht (1993)
12. Duan, Z.: Temporal Logic and Temporal Logic Programming Language. Science Press, Beijing (2006)

13. Duan, Z., Tian, C., Zhang, L.: A decision procedure for propositional projection temporal logic with infinite models. Acta Inf. **45**(1), 43–78 (2008)
14. Duan, Z., Tian, C.: A practical decision procedure for propositional projection temporal logic with infinite models. Theor. Comput. Sci. **554**, 169–190 (2014)
15. Shu, X., Duan, Z., Du, H.: A decision procedure and complete axiomatization for projection temporal logic Theor. Comput. Sci. (2017). https://doi.org/10.1016/j.tcs.2017.09.026
16. Bowman, H., Thompson, S.: A decision procedure and complete axiomatization of finite interval temporal logic with projection. J. Log. Comput. **13**(2), 195–239 (2003)

On the Relationship Between Energy Complexity and Other Boolean Function Measures

Xiaoming Sun[1,2], Yuan Sun[1,2(✉)], Kewen Wu[3], and Zhiyu Xia[1,2]

[1] CAS Key Lab of Network Data Science and Technology, Institute of Computing Technology, Chinese Academy of Sciences, Beijing, China
{sunxiaoming,sunyuan2016,xiazhiyu}@ict.ac.cn
[2] University of Chinese Academy of Sciences, Beijing, China
[3] School of Electronics Engineering and Computer Science, Peking University, Beijing, China
shlw_kevin@pku.edu.cn

Abstract. In this work we investigate *energy complexity*, a Boolean function measure related to circuit complexity. Given a circuit $\mathcal{C}$ over the standard basis $\{\vee_2, \wedge_2, \neg\}$, the energy complexity of $\mathcal{C}$, denoted by $\mathrm{EC}(\mathcal{C})$, is the maximum number of its activated inner gates over all inputs. The energy complexity of a Boolean function f, denoted by $\mathrm{EC}(f)$, is the minimum of $\mathrm{EC}(\mathcal{C})$ over all circuits $\mathcal{C}$ computing f.

Recently, Dinesh et al. [3] gave $\mathrm{EC}(f)$ an upper bound in terms of the decision tree complexity, $\mathrm{EC}(f) = O(\mathrm{D}(f)^3)$. They also showed that $\mathrm{EC}(f) \leq 3n-1$, where n is the input size. For the lower bound, they show that $\mathrm{EC}(f) \geq \frac{1}{3}\,\mathrm{psens}(f)$, where $\mathrm{psens}(f)$ is the *positive sensitivity*. They asked whether $\mathrm{EC}(f)$ can be lower bounded by a polynomial of $\mathrm{D}(f)$. We improve both the upper and lower bounds in this paper. For upper bounds, We show that $\mathrm{EC}(f) \leq \min\{\frac{1}{2}\mathrm{D}(f)^2 + O(\mathrm{D}(f)), n + 2\mathrm{D}(f) - 2\}$. For the lower bound, we answer Dinesh et al.'s question by proving that $\mathrm{EC}(f) = \Omega(\sqrt{\mathrm{D}(f)})$. For non-degenerated functions, we also give another lower bound $\mathrm{EC}(f) = \Omega(\log n)$ where n is the input size. These two lower bounds are incomparable to each other. Besides, we examine the energy complexity of OR functions and ADDRESS functions, which implies the tightness of our two lower bounds respectively. In addition, the former one answers another open question in [3] asking for non-trivial lower bound for energy complexity of OR functions.

Keywords: Energy complexity · Decision tree · Boolean function · Circuit complexity

This work was supported in part by the National Natural Science Foundation of China Grants No. 61433014, 61832003, 61761136014, 61872334, 61502449, 61602440, 61801459, the 973 Program of China Grant No. 2016YFB1000201, K.C. Wong Education Foundation.

1   Introduction

Given a gate basis $\mathcal{B}$ and a circuit $\mathcal{C}$ over $\mathcal{B}$, the *energy complexity* of $\mathcal{C}$, defined as $\mathrm{EC}_{\mathcal{B}}(\mathcal{C})$, is the maximum number of activated gates (except input gates) in $\mathcal{C}$ over all possible inputs. Spontaneously, the *energy complexity* of a Boolean function $f : \{0,1\}^n \to \{0,1\}$ over gate basis $\mathcal{B}$ is defined as $\mathrm{EC}_{\mathcal{B}}(f) := \min_{\mathcal{C}} \mathrm{EC}_{\mathcal{B}}(\mathcal{C})$, where $\mathcal{C}$ is a circuit over $\mathcal{B}$ computing f.

When $\mathcal{B}$ is composed of *threshold* gates, this model simulates the neuron activity [10,11], as the transmission of a 'spike' in neural network is similar with an activated threshold gate in the circuit. A natural question readily comes up: can Boolean functions be computed with rather few activated gates over threshold gate basis? In order to answer this question, or more precisely, to give lower and upper bounds, plenty of studies were motivated [1,4,5,9,10].

Despite motivated by neurobiology in modern world, tracing back into history, this concept is not brand-new. Let $\mathrm{EC}_{\mathcal{B}}(n)$ be the maximum energy complexity among all Boolean functions on n variables over basis $\mathcal{B}$, i.e., the maximum $\mathrm{EC}_{\mathcal{B}}(f)$ among all possible $f : \{0,1\}^n \to \{0,1\}$. Vaintsvaig [8] proved that asymptotically, if $\mathcal{B}$ is finite, the lower and upper bounds of $\mathrm{EC}_{\mathcal{B}}(n)$ are n and $2^n/n$ respectively. Then this result was further refined by the outstanding work from Kasim-Zade [6], which states that $\mathrm{EC}_{\mathcal{B}}(n)$ could be $\Theta(2^n/n)$, between $\Omega(2^{n/2})$ and $O(\sqrt{n}2^{n/2})$, or between $\Omega(n)$ and $O(n^2)$.

When it comes to a specific gate basis, a natural thought is to discuss the energy complexity over the standard Boolean basis $\mathcal{B} = \{\vee_2, \wedge_2, \neg\}$. (From now on we use $\mathrm{EC}(f)$ to represent $\mathrm{EC}_{\mathcal{B}}(f)$ for the standard basis.) Towards this, Kasim-zade [6] showed that $\mathrm{EC}(f) = O(n^2)$ for any n variable Boolean function f by constructing an explicit circuit, which was further improved by Lozhkin and Shupletsov [7] to $4n$ and then $(3 + \epsilon(n))n$ where $\lim_{n\to\infty} \epsilon(n) = 0$.

Recently, Dinesh, Otiv, Sarma [3] discovered a new upper bound which relates energy complexity to decision tree complexity, a well-studied Boolean function complexity measure. In fact, they proved that for any Boolean function $f : \{0,1\}^n \to \{0,1\}$, $\frac{\mathrm{psens}(f)}{3} \leq \mathrm{EC}(f) \leq \min\{O(\mathrm{D}(f)^3), 3n-1\}$ holds, where the function $\mathrm{psens}(f)$ is defined as the positive sensitivity of f, i.e., the maximum of the number of sensitive bits $i \in \{1, 2, \ldots, n\}$ with $x_i = 1$ over all possible inputs x [3]. However, positive sensitivity may give weak energy complexity lower bounds for some rather fundamental functions. For example, the positive sensitivity of OR function $f(x_1, \ldots, x_n) = x_1 \vee \ldots \vee x_n$ is only 1. Therefore, Dinesh et al. asked 2 open problems on this issue:

1. Does the inequality $\mathrm{D}(f) \leq \mathrm{poly}(\mathrm{EC}(f))$ always hold?
2. Give a non-trivial lower bound of the energy complexity for OR function.

Throughout this paper, we use completely different method to achieve better bounds from both sides, which explores a polynomial relationship between energy complexity and decision tree complexity and answers two open problems asked by Dinesh et al. Furthermore, we also construct an explicit circuit computing OR function to match this lower bound.

First, in Sect. 3 we show the upper and lower bounds of energy complexity by decision tree complexity:

Theorem 1. *For any Boolean function* $f : \{0,1\}^n \rightarrow \{0,1\}$,

$$\mathrm{EC}(f) \leq \frac{1}{2}D(f)^2 + O(D(f)).$$

Theorem 2. *For any Boolean function* $f : \{0,1\}^n \rightarrow \{0,1\}$,

$$\mathrm{EC}(f) = \Omega(\sqrt{D(f)}).$$

Second, in Sect. 4 we also show upper and lower bounds of energy complexity with respect to the number of variables.

Theorem 3. *For any Boolean function* $f : \{0,1\}^n \rightarrow \{0,1\}$,

$$\mathrm{EC}(f) \leq n - 2 + 2D(f) \leq 3n - 2.$$

Theorem 4. *For any non-degenerated Boolean function* $f : \{0,1\}^n \rightarrow \{0,1\}$,

$$\mathrm{EC}(f) = \Omega(\log_2 n).$$

Note that these lower bounds are incomparable with each other, since for any non-degenerated Boolean function $f : \{0,1\}^n \rightarrow \{0,1\}$, we have $\Omega(\log_2 n) \leq D(f) \leq O(n)$ and this result is essentially tight from both sides.

Finally, in order to show the tightness of lower bounds, we examine the energy complexity of two specific function classes: OR functions and EXTENDED ADDRESS functions (see the definition of EXTENDED ADDRESS in Sect. 2).

Proposition 1. *For any positive integer* n, $\mathrm{EC}(\mathrm{OR}_n) = \Theta(\sqrt{n})$.

Proposition 2. *For any positive integer* n *and an arbitrary Boolean function* $g : \{0,1\}^n \rightarrow \{0,1\}$, $\mathrm{EC}(\mathrm{EADDR}_{n,g}) = \Theta(n)$.

Note that $D(\mathrm{OR}_n) = n$ and the number of variables in $\mathrm{EADDR}_{n,g}$ is $n + 2^n$, which shows that the lower bounds in Theorems 2 and 4 are tight.

2   Preliminaries

In the following context, we denote $(\underbrace{0,\ldots,0}_{i-1},1,\underbrace{0,\ldots,0}_{n-i})$ as e_i, $\{1,2,\ldots,n\}$ as $[n]$, and the cardinality of set S as $|S|$ or $\#S$.

A Boolean function f is a function mapping $\{0,1\}^n$ to $\{0,1\}$, where n is a positive integer. We say a Boolean function $f : \{0,1\}^n \rightarrow \{0,1\}$ *depends on* m *variables* if there exists $S \subseteq [n], |S| = m$ and for any $i \in S$, there exists $x \in \{0,1\}^n$ such that $f(x) \neq f(x \oplus e_i)$; and when f depends on all n variables, we say f is *non-degenerated*.

We define a new Boolean function class called EXTENDED ADDRESS function, which is an extension of the well-known ADDRESS function.

Definition 1. *Given integer n, the address function* $\text{ADDR}_n : \{0,1\}^{n+2^n} \rightarrow \{0,1\}$ *is defined as* $\text{ADDR}_n(x_1,..,x_n,y_0,\ldots,y_{2^n-1}) = y_{x_1 x_2 \ldots x_n}$.

Definition 2. *Given integer n and an arbitrary Boolean function* $g : \{0,1\}^n \rightarrow \{0,1\}$, *we define the extended address function* $\text{EADDR}_{n,g} : \{0,1\}^{n+2^n} \rightarrow \{0,1\}$ *as*

$$\text{EADDR}_{n,g}(x_1,..,x_n,y_0,\ldots,y_{2^n-1}) = \begin{cases} y_{x_1 x_2 \ldots x_n}, & g(x_1,\ldots,x_n) = 1 \\ \bar{y}_{x_1 x_2 \ldots x_n}, & g(x_1,\ldots,x_n) = 0. \end{cases}$$

Following from [2], a (deterministic) decision tree is a rooted ordered binary tree, where each internal node is labeled with a variable x_i and each leaf is labeled with value 0 or 1. Given an input x, the tree is evaluated as follows. First, we start at the root. Then continue the process until we stop:

- if we reach a leaf v, then stop;
- otherwise query x_i, which is the labelled variable on current node,
  - if $x_i = 0$, then move to its left child,
  - if $x_i = 1$, then move to its right child.

The output of the evaluation is the value on the final position. A decision tree is said to compute f, if for any input x the output after evaluation is $f(x)$. The complexity of a decision tree T, denoted by $D(T)$, is its depth, i.e., the number of queries, made on the worst-case input. The decision tree complexity of a Boolean function f, denoted by $D(f)$, is the minimum $D(T)$ among all decision trees T computing f.

A Boolean circuit $\mathcal{C}$ over a basis $\mathcal{B}$ is a directed acyclic graph which has an output gate, input gates with in-degree 0 representing variables, and other gates among the circuit basis $\mathcal{B}$. For convenience, we have several definitions related with circuit gates here. For two gates u, v in a Boolean circuit, we say

- u is an *inner gate* if and only if u is not an input gate.
- u is *activated* under input x if and only if u outputs 1 when the input of the circuit is x.
- u is *deactivated* under input x if and only if u outputs 0 when the input of the circuit is x.
- u is an *incoming gate* of v if and only if u is an input of v.
- u *covers* v if and only if there exists a directed path in circuit from v to u.

The circuit basis we mainly discuss is the standard basis $\mathcal{B} = \{\vee_2, \wedge_2, \neg\}$, which means $\vee$-gate with fan-in 2, $\wedge$-gate with fan-in 2 and $\neg$-gate with fan-in 1. The fan-out of all kinds of gates is unlimited. Particularly, a circuit over standard basis is called *monotone* if it does not contain any $\neg$-gate. For convenience, from now on, Boolean circuits and energy complexity are over the standard basis if not specified. In addition, if a circuit $\mathcal{C}$ computes $g : \{0,1\}^n \rightarrow \{0,1\}$ which depends on m variables, we say $\mathcal{C}$ *depends on m input gates*.

Next we give energy complexity a mathematical definition.

Definition 3. *For a Boolean circuit C and an input x, the energy complexity of C under x (denoted by $\mathrm{EC}(C,x)$) is defined as the number of activated inner gates in C under the input x. Define the energy complexity of C as $\mathrm{EC}(C) = \max_{x \in \{0,1\}^n}\{\mathrm{EC}(C,x)\}$ and the energy complexity of a Boolean function f as*

$$\mathrm{EC}(f) = \min_{\substack{C|C(x)=f(x) \\ \forall x \in \{0,1\}^n}} \mathrm{EC}(C).$$

Remark 1. W.l.o.g, the first and third structures in Fig. 1 are forbidden in the circuit, since we can replace them by the second or fourth one without increasing the energy complexity. So we can assume that any ¬-gate in the circuit has a non-¬ incoming gate and any two ¬-gates do not share a same incoming gate.

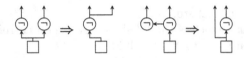

Fig. 1. Substructures related to ¬-gates

3   Upper Bounds of Energy Complexity

In this section, we will show upper bounds of energy complexity with respect to decision tree complexity, which improves the result in [3]. We first prove Theorem 3, then prove Theorem 1.

Proof (Theorem 3). Since $\mathrm{D}(f) \leq n$, $n + 2(\mathrm{D}(f) - 1) \leq 3n - 2$ holds naturally. Now suppose $\mathcal{T}$ is a decision tree of f with depth $\mathrm{D}(f)$. Denote the node set of $\mathcal{T}$ (including leaves) as S, where $v_{root} \in S$ is the root, $v_{left}, v_{right} \in S$ are the left and right children of v_{root} respectively. We also define $F : S \backslash \{v_{root}\} \to S$, where $F(v)$ is the father of node v in $\mathcal{T}$. Furthermore, define $vbs : S \to \{x_1, \ldots, x_n\} \cup \{0,1\}$, where $vbs(v)$ indicates the label on node v, i.e., $vbs(v) = x_i$ means v is labelled with x_i and $vbs(v) = 0$ (or 1) means v is a leaf with value 0 (or 1). Define $S_0 = \{v \in S \mid vbs(v) = 0\}$ and $\widetilde{S} = S \backslash (\{v_{root}\} \cup S_0)$.

Based on $\mathcal{T}$, a circuit C can be constructed such that $\mathrm{EC}(C) \leq n + 2(\mathrm{D}(f) - 1)$ as follows. First, define all gates in C: the input gates are $g_{x_1}, \ldots, g_{x_n}$; the ¬-gates are $g_{x_1}^{\neg}, \ldots, g_{x_n}^{\neg}$ and the ∧-gates are $g_v^{\wedge}, v \in \widetilde{S}$; furthermore, C contains a unique ∨-gate $g^{\vee}$ as output gate with fan-in size $\#\{v \in S \mid vbs(v) = 1\}$. Actually, $g^{\vee}$ is a sub-circuit formed by $\#\{v \in S \mid vbs(v) = 1\} - 1$ ∨-gates. These gates are connected in following way:

1. For all $i \in [n]$, the input of $g_{x_i}^{\neg}$ is g_{x_i}.
2. For all $v \in \widetilde{S} \backslash \{v_{left}, v_{right}\}$, if v is the right child of $F(v)$, the input of $g_v^{\wedge}$ is $g_{F(v)}^{\wedge}$ and $g_{vbs(F(v))}$; otherwise, the input of $g_v^{\wedge}$ is $g_{F(v)}^{\wedge}$ and $g_{vbs(F(v))}^{\neg}$.

3. Merge $g^\wedge_{v_{left}}$ with $g^\neg_{vbs(v_{root})}$; and merge $g^\wedge_{v_{right}}$ with $g_{vbs(v_{root})}$.
4. The input of $g^\vee$ is all the gates in $\{g^\wedge_v \mid vbs(v) = 1\}$.

See Fig. 2 as an example, where $vbs(v_{root}) = x_1$, $vbs(v_{left}) = x_2$, $vbs(v_{right}) = x_3$, $vbs(v_1) = vbs(v_2) = x_4$, $vbs(v_3) = vbs(v_5) = vbs(v_8) = 1$, $vbs(v_4) = vbs(v_6) = vbs(v_7) = 0$ and $S_0 = \{v_4, v_6, v_7\}$, $\widetilde{S} = \{v_{left}, v_{right}, v_1, v_2, v_3, v_5, v_8\}$.

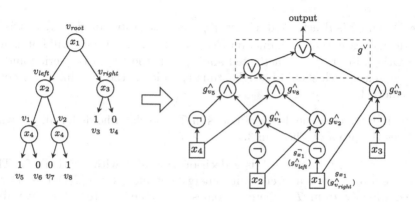

Fig. 2. Decision tree $\mathcal{T}$ and circuit $\mathcal{C}$

The construction of $\mathcal{C}$ implies several facts:

- Under any input, $g^\wedge_{u_1}, \ldots, g^\wedge_{u_{k-1}}$ is activated if $g^\wedge_{u_k}$ is activated, where $u_i = F(u_{i+1})$.
- For sibling nodes $u, w \in S$, that $g^\wedge_u$ is activated implies $g^\wedge_w$ is deactivated, since one of $g^\wedge_u, g^\wedge_w$ receives $g_{vbs(F(u))}$ as input and the other uses $g^\neg_{vbs(F(u))}$, which means they can not output 1 simultaneously.

These facts imply there are at most $D(f) - 1$ activated $\wedge$-gate under any input. Furthermore, at most one gate in $\{g^\wedge_v \mid vbs(v) = 1\}$ is activated. It is easy to construct a circuit computing OR_m for $g^\vee$, whose energy complexity is no more than $\lceil \log m \rceil$ when promised that the input bits include at most one 1. Thus, the contribution from $g^\vee$ is no more than

$$\lceil \log \left(\#\{v \in S \mid vbs(v) = 1\} \right) \rceil \leq D(f) - 1.$$

Also, the $\neg$-gates in $\mathcal{C}$ contribute at most n to the whole energy complexity under any input. Thus, $EC(\mathcal{C}) \leq n + 2(D(f) - 1)$.

To justify that circuit $\mathcal{C}$ actually computes f, it suffices to show $g^\wedge_v$ outputs 1 if and only if v is queried in $\mathcal{T}$ during the evaluation process under some input. The proof goes as follows:

- First, for v_{left} and v_{right}, the claim holds immediately.
- Then assume that for any node whose depth is less than k in $\mathcal{T}$, the claim holds. Consider any $v \in \mathcal{T}$ of depth k. Without loss of generality, assume v is the left child of $F(v)$; then the input of $g_v^\wedge$ is $g_{F(v)}^\wedge$ and $g_{vbs(F(v))}^\neg$. Let $x_i = vbs(F(v))$.
  - When $g_v^\wedge$ is activated, $g_{F(v)}^\wedge$ is activated and $x_i = 0$. By induction, $F(v)$ is queried in $\mathcal{T}$ and the chosen branch after querying is left, which is exactly v.
  - When $g_v^\wedge$ is deactivated, either $g_{F(v)}^\wedge$ is deactivated or $x_i = 1$. If it is the former case, then by induction $F(v)$ is not queried; thus v will not as well. Otherwise if $g_{F(v)}^\wedge$ is activated and $x_i = 1$, then $F(v)$ is queried and the chosen branch should be right; thus v, which is the left child, will not be queried.

Thus by induction on the depth of nodes in $\mathcal{T}$, the claim holds for all $g_v^\wedge$, which completes the proof of Theorem 3.

Proof (Theorem 1). Suppose $\mathcal{T}$ is a decision tree of f with depth $D(f)$. Then by Theorem 3, there is a circuit with energy complexity $n + 2(D(f) - 1)$ constructed directly from $\mathcal{T}$, where n comes from the $\neg$-gates of all variables. In order to reduce the number of $\neg$-gates, we introduce $D(f)$ additional variables $y_1, y_2, \ldots, y_{D(f)}$ in each level of $\mathcal{T}$ as a record log of the evaluation process on the tree, where $y_i = 0$ means in the i-th level of $\mathcal{T}$, it chooses the left branch, and $y_i = 1$ means to choose the right branch. For example, in Fig. 3 these additional variables are computed by $y_1 = x_1$, $y_2 = \bar{y}_1 x_2 + y_1 x_3$, $y_3 = \bar{y}_1 \bar{y}_2 x_4 + \bar{y}_1 y_2 x_5 + y_1 \bar{y}_2 x_6 + y_1 y_2 x_7$, etc. Given the value of all y_i's, the output of f can be determined by reconstruct the evaluation path in $\mathcal{T}$; thus f can be viewed as a function on y_i's. Therefore, define

$$y_{D(f)+1} = \sum_{z \in \prod\{y_i, \bar{y}_i\}} f(z) \prod_{i=1}^{D(f)} z_i.$$

Then given any input x, after determine all y_i's, $y_{D(f)+1} = f(x)$. Now construct a circuit using these temporary variables. (See Fig. 4 as an example of the gates for second level of the decision tree.) Notice that for any $1 \le k \le D(f) - 1$, to compose y_{k+1}, an OR_{2^k} gadget is required in the k-th level sub-circuit, which induces a k-level of $\vee$-gates. After computing y_{k+1}, we also need two additional levels of gates to compute $\bar{y}_{k+1}$ and $\prod_{i=1}^{k+1} z_i, z_i \in \{y_i, \bar{y}_i\}$. In order to compute $y_{D(f)+1}$, an $OR_{2^{D(f)}}$ gadget is required, which brings a $D(f)$-level sub-circuit of $\vee$-gates. Thus summing up all sub-circuits, the circuit depth is $\sum_{i=1}^{D(f)-1}(i+2) + D(f) = \frac{1}{2}D(f)^2 + O(D(f))$.

For any fixed $k, 1 \le k \le D(f)$, only one of all 2^k cases in $\prod_{i=1}^{k} z_i, z_i \in \{y_i, \bar{y}_i\}$ is true, thus each level of the circuit provides at most one activated gate. Then the whole energy complexity is $\frac{1}{2}D(f)^2 + O(D(f))$.

4   Lower Bounds of Energy Complexity

In this section we will give two theorems on lower bounds of energy complexity. The first one relates decision tree complexity to energy complexity by an intricately constructed decision tree with respect to a given circuit. The second one provides a lower bound depending on the number of variables. In the meantime, we will offer cases where these bounds are tight. We will use the following two lemmas (since their proofs are trivial, we omit them):

Lemma 1. *If $\mathcal{C}$ is a monotone circuit depending on m inputs, $\mathrm{EC}(\mathcal{C}) \geq m - 1$.*

Lemma 2. *If $\mathcal{C}$ is a circuit with k $\neg$-gates, then $\mathrm{EC}(\mathcal{C}) \geq k$.*

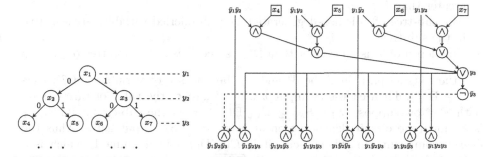

Fig. 3. A decision tree with temporary variables

Fig. 4. Sub-circuit representing the second level of the decision tree

Now we can give the proofs of the main results in this section.

Proof (Theorem 2). For any Boolean function $f : \{0,1\}^n \to \{0,1\}$ and any circuit $\mathcal{C}$ computing f, suppose $\mathrm{EC}(\mathcal{C}) = o(\sqrt{\mathrm{D}(f)})$ and let m be the number of $\neg$-gates in $\mathcal{C}$, then $m = o(\sqrt{\mathrm{D}(f)})$ by Lemma 2. List all the $\neg$-gates with topological order $\neg_1, \neg_2, \ldots, \neg_m$ such that for any $1 \leq i < j \leq m$, $\neg_i$ does not cover $\neg_j$. Suppose the set of all variables (input gates) covered by $\neg_i$ is $\widetilde{S}_i$, then S_i is defined as $\widetilde{S}_i \backslash \left(\bigcup_{j=1}^{i-1} \widetilde{S}_j \right)$. (See also the left side of Fig. 5.) Define $S_{m+1} = [n] \backslash \left(\bigcup_{j=1}^{m} \widetilde{S}_j \right)$. Also let k_i be the number of elements of S_i; thus $S_i = \{x_{i,j} \mid j \in [k_i]\}$. Notice that the set collection $S_1, S_2, \ldots, S_{m+1}$ is a division of all variables.

Consider the query algorithm by the order $x_{i,j}$, where $x_{i,j}$ precedes $x_{i',j'}$ if and only if $(i < i') \vee (i = i' \wedge j < j')$. This algorithm induces a decision tree $\mathcal{T}'$ with depth n immediately. (See also the middle part of Fig. 5.)

Since $\mathcal{T}'$ may be redundant, the simplification process goes as follows: From the root to leaves, check each node whether its left sub-tree and right sub-tree are identical. If so, this node must be inconsequential when queried upon. Thus delete this node and its right sub-tree, and connect its parent to its left child. (See also the right side of Fig. 5.)

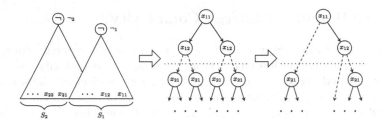

Fig. 5. Circuit $\mathcal{C}$, its induced decision tree $\mathcal{T}'$, and the simplified decision tree $\mathcal{T}$

After this process, the new decision tree $\mathcal{T}$ satisfies:

- In any path from the root to a leaf, if $x_{i,j}$ is queried before $x_{i',j'}$, i is not greater than i'.
- Any sub-tree of $\mathcal{T}$ is non-degenerated, i.e., all queried variables are sensitive in the sub-tree.
- The depth of $\mathcal{T}$ is no smaller than $D(f)$ since $\mathcal{T}$ is a decision tree of f.

Let the longest path in $\mathcal{T}$ be $\mathcal{P}$ and $S_{\mathcal{P}}$ be the set of variables on $\mathcal{P}$; thus $|S_{\mathcal{P}}| \geq D(f)$. Then choose an input $\hat{x}$ which matches the value of variables on path $\mathcal{P}$. Now suppose $|S_1 \cap S_{\mathcal{P}}| \geq \Omega(\sqrt{D(f)})$, then the sub-circuit under $\neg_1$ is a monotone circuit depending on at least $|S_1 \cap S_{\mathcal{P}}|$ input gates. Thus the energy complexity in this sub-circuit is $\Omega(\sqrt{D(f)})$ by Lemma 1, which is a contradiction. Therefore $|S_1 \cap S_{\mathcal{P}}| = o(\sqrt{D(f)})$. Then set variables in S_1 to the same value in $\hat{x}$. Thus the restricted circuit has fewer $\neg$-gates and computes a restricted f function whose decision tree is a sub-tree of $\mathcal{T}$ with depth at least $|S_{\mathcal{P}}| - o(\sqrt{D(f)})$. Now consider $|S_2 \cap S_{\mathcal{P}}|$ in the restricted circuit and the same analysis follows. Continue this restriction process until the value of all $\neg$-gates are determined.

By then, the depth of the decision tree is still at least $|S_{\mathcal{P}}| - m \times o(\sqrt{D(f)}) \geq D(f) - o(D(f))$. Thus the remaining monotone circuit depends on at least $D(f) - o(D(f))$ input gates. By Lemma 1, the energy complexity is at least $D(f) - o(D(f)) = \Omega(\sqrt{D(f)})$, which is a contradiction.

The tightness of this lower bound is shown in $EC(OR_n) = \Theta(\sqrt{n})$ in Proposition 1 as $D(OR_n) = n$.

Proof (Theorem 4). Assume $\mathcal{C}$ is an arbitrary circuit computing f. It suffices to show $EC(\mathcal{C}) \geq \frac{1}{2} \log_2 n$. Since f is non-degenerated, the output gate must cover all input gates. Since the fan-in of $\wedge_2, \vee_2, \neg$ gate is no more than 2, a k-depth circuit can cover at most 2^k input gates. Thus removing all gates, of which the shortest path to output gate is less than $\log n$ in $\mathcal{C}$, some input gate x_i will be disconnected with the output gate.

Choose an input $\hat{x}$ satisfying $f(\hat{x}) = 0$, $f(\hat{x} \oplus e_i) = 1$. Note that when inputted $\hat{x}$, the output of $\mathcal{C}$ is different from that when inputted $\hat{x} \oplus e_i$. Therefore, there exists a path $\mathcal{P}$ from input gate x_i to output gate under $\hat{x}$, such that the value of any gate on $\mathcal{P}$ changes after flipping x_i. Let ℓ be the length of path $\mathcal{P}$, then $\ell \geq \log n$. Also, when inputted $\hat{x}$ and $\hat{x} \oplus e_i$, the total number of activated inner gates in $\mathcal{P}$ is ℓ. It follows immediately

$$\mathrm{EC}(\mathcal{C}) \geq \max\left\{\mathrm{EC}(\mathcal{C}, \hat{x}), \mathrm{EC}(\mathcal{C}, \hat{x} \oplus e_i)\right\}$$
$$\geq \frac{\left(\mathrm{EC}(\mathcal{C}, \hat{x}) + \mathrm{EC}(\mathcal{C}, \hat{x} \oplus e_i)\right)}{2} \geq \frac{\ell}{2} \geq \frac{\log n}{2}.$$

The tightness of this lower bound is shown by $\mathrm{EC}(\mathrm{EADDR}_{n,g}) = \Theta(n)$ in Proposition 2 as $\mathrm{EADDR}_{n,g} : \{0,1\}^{n+2^n} \to \{0,1\}$ is non-degenerated.

5   Tight Bounds of Energy Complexity on Specific Functions

In this section, we discuss two specific function classes, OR_n and $\mathrm{EXTENDED}$ $\mathrm{ADDRESS}$, to obtain the tightness of lower bounds of energy complexity. Precisely, OR_n function shows the lower bound in Theorem 2 is tight, and $\mathrm{EXTENDED}$ $\mathrm{ADDRESS}$ function corresponds with the lower bound in Theorem 4.

First, we discuss the energy complexity of the OR_n function and prove Proposition 1. Given integer n, $\mathrm{OR}_n : \{0,1\}^n \to \{0,1\}$ is defined as

$$\mathrm{OR}_n(x_1, x_2, \ldots, x_n) = x_1 \vee x_2 \vee \cdots \vee x_n.$$

Proof (Proposition 1). The lower bound follows from $\mathrm{D}(\mathrm{OR}_n) = n$ and Theorem 2. To prove $\mathrm{EC}(\mathrm{OR}_n) = O(\sqrt{n})$, a circuit is constructed as follows (see Fig. 6):

1. Divide all n variables into $\sqrt{n}$ blocks, each block contains $\sqrt{n}$ variables. For variables in the first block, use $\sqrt{n} - 1$ $\vee$-gates to connect them as an $\mathrm{OR}_{\sqrt{n}}$ function and mark the output gate of the sub-circuit as g_1.
2. Add a $\neg$-gate h_1 linked from g_1; and for each variable in the second block, feed it into a $\wedge$-gate together with h_1. Then use $\sqrt{n} - 1$ $\vee$-gates to connect these $\sqrt{n}$ $\wedge$-gates and mark the output gate of the sub-circuit as g_2'.
3. Add a $\vee$-gate which has incoming gates g_1 and g_2'; and insert a $\neg$-gate h_2 linked from g_2. For each variable in the second block, connect it with h_2 by a $\wedge$-gate. Then use $\sqrt{n} - 1$ $\vee$-gates to connect these $\sqrt{n}$ $\wedge$-gates.
4. Repeat this process until all blocks are constructed. Then $g_{\sqrt{n}}$ shall be the output gate of the whole circuit.

The main idea is to view each block as a switch so that if it has an activated gate then it can "switch of" all blocks behind it with low cost. Consider a specific input x. If $x = 0^n$, then the activated gates are h_i's, whose number is $\sqrt{n}$. Otherwise if $x \neq 0^n$, then at least one bit is 1. Suppose all variables in the first $k-1$ blocks are 0 and in the k-th block there exists a value-1 input bit. Then in the first $k-1$ blocks, only $\neg$-gates $h_1, \ldots, h_{k-1}$ are activated. And in the k-th $\mathrm{OR}_{\sqrt{n}}$ sub-circuit, at most $\sqrt{n} - 1$ gates are activated. Thus $g_i, i \geq k$ is activated, indicating $h_i, i \geq k$ is deactivated. Therefore, all variables in blocks after k-th block are "switched off". To sum up, all the activated gates are among g_i's, g_i''s, h_i's, and k-th $\mathrm{OR}_{\sqrt{n}}$ gadget. So the energy complexity is $\Theta(\sqrt{n})$.

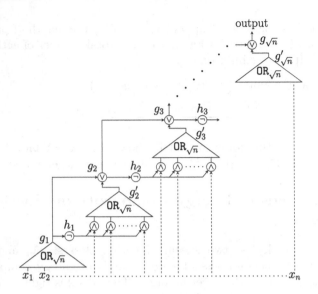

Fig. 6. OR_n circuit

Second, we discuss the **EXTENDED ADDRESS** function, which is defined in Definition 2, thus complete the proof of Theorem 4. Note that although **ADDRESS** function in itself verifies Theorem 4, the low-energy circuit for **ADDR** actually gives rise to tight bounds of the more generalized $\text{EADDR}_{n,g}$ function.

Lemma 3. *For any positive integer n, $\text{EC}(\text{ADDR}_n) = \Theta(n)$.*

Proof. The lower bound can be deduced from Theorem 4. It suffices to prove the upper bound by construction. Let $\mathcal{T}$ be the natural decision tree of ADDR_n, where all the nodes in the i-th ($i \leq n$) level of $\mathcal{T}$ are labelled with x_i, and y_j's are queried in $(n+1)$-th level. Thus, $\mathcal{T}$ is a full binary tree with depth $n+1$. Now consider the circuit $\mathcal{C}$ constructed in Theorem 3 based on $\mathcal{T}$. Note that the output of $g_{y_i}^{\neg}$ are not received by any gate as input. Thus these redundant gates can be safely removed; and the remaining circuit $\mathcal{C}'$ still computes ADDR_n. Therefore, $\text{EC}(\text{ADDR}_n) \leq \text{EC}(\mathcal{C}') \leq 2\big(\text{D}(\text{ADDR}_n) - 1\big) + \#\{\neg\text{-gates in } \mathcal{C}'\} = 3n$. $\qquad\square$

Proof (Proposition 2). Apply the construction in Lemma 3, and prepare two copies of the circuit computing ADDR_n and denote them as $\mathcal{C}_0$ and $\mathcal{C}_1$. Then modify them into a new circuit $\mathcal{C}'$ for $\text{EADDR}_{n,g}$ as follows:

1. $\forall x \in \{0,1\}^n$, if $g(x) = 0$, change y_x's input gate in $\mathcal{C}_1$ into constant input gate 0; otherwise, change y_x's input gate in $\mathcal{C}_0$ into constant input gate 1.
2. $\forall i \in [n]$, merge x_i's input gate in $\mathcal{C}_0, \mathcal{C}_1$ together as x_i's new input gate.
3. Add a $\neg$-gate $\widetilde{g}$ linked from the output gate of $\mathcal{C}_0$.
4. Add a $\vee$-gate h as the new output gate, which takes $\widetilde{g}$ and the output gate of $\mathcal{C}_1$ as input.

Thus $\mathcal{C}'$ has exactly $n + 2^n$ input gates. To show $\mathcal{C}'$ actually computes $\text{EADDR}_{n,g}$, it suffices to consider an arbitrary $x \in \{0,1\}^n$. If $g(x) = 0$, sub-circuit $\mathcal{C}_1$ outputs y_x which becomes 0 after modification, and $\mathcal{C}_0$ still outputs y_x; thus after $\tilde{g}$ and h, $\mathcal{C}'$ gives $\bar{y}_x$ correctly. Similar argument holds when $g(x) = 1$.

It is also easy to verify that $\text{EC}(\mathcal{C}')$ is bounded by $\text{EC}(\mathcal{C}_0)$ and $\text{EC}(\mathcal{C}_1)$:

$$\text{EC}(\text{EADDR}_{n,g}) \leq \text{EC}(\mathcal{C}') \leq \max_x \big(\text{EC}(\mathcal{C}_0, x) + \text{EC}(\mathcal{C}_1, x) \big) + 2$$

$$\leq \text{EC}(\mathcal{C}_0) + \text{EC}(\mathcal{C}_1) + 2 = O(n).$$

6   Conclusion and Open Problems

Throughout this paper, we build polynomial relationship between energy complexity and other well-known measures of Boolean functions. Precisely, we prove that $\text{EC}(f) \leq \min\{\frac{1}{2}\text{D}(f)^2 + O(\text{D}(f)), n + 2\text{D}(f) - 2\}$ and $\text{EC}(f) = \Omega(\sqrt{\text{D}(f)})$, as well as a logarithmic lower bound in terms of the input size for non-degenerated functions. We also show the tightness of lower bounds by examining OR functions and EADDR functions. However, some fascinating problems still remain open.

1. Two Boolean functions f, $g : \{0,1\}^n \rightarrow \{0,1\}$ are called *co-isomorphic* if there exists a subset $S \subseteq [n]$ such that $\forall\, x \in \{0,1\}^n$, $f(x) = 1 - g(x \bigoplus_{i \in S} e_i)$. For example, AND_n and OR_n are co-isomorphic, with a quadratic separation between their energy complexity: $\text{EC}(\text{AND}_n) = \Theta(n)$ and $\text{EC}(\text{OR}_n) = \Theta(\sqrt{n})$. What is the largest gap between two co-isomorphic Boolean functions?
2. Is the upper bound $\text{EC}(f) = O(\text{D}(f)^2)$ tight?

Acknowledgement. The authors want to thank Krishnamoorthy Dinesh for answering some questions with [3].

References

1. Amano, K., Maruoka, A.: On the complexity of depth-2 circuits with threshold gates. In: Jędrzejowicz, J., Szepietowski, A. (eds.) MFCS 2005. LNCS, vol. 3618, pp. 107–118. Springer, Heidelberg (2005). https://doi.org/10.1007/11549345_11
2. Buhrman, H., De Wolf, R.: Complexity measures and decision tree complexity: a survey. Theoret. Comput. Sci. **288**(1), 21–43 (2002)
3. Dinesh, K., Otiv, S., Sarma, J.: New bounds for energy complexity of Boolean functions. In: Wang, L., Zhu, D. (eds.) COCOON 2018. LNCS, vol. 10976, pp. 738–750. Springer, Cham (2018). https://doi.org/10.1007/978-3-319-94776-1_61
4. Hajnal, A., Maass, W., Pudlák, P., Szegedy, M., Turán, G.: Threshold circuits of bounded depth. J. Comput. Syst. Sci. **46**(2), 129–154 (1993)
5. Håstad, J., Goldmann, M.: On the power of small-depth threshold circuits. Comput. Complex. **1**(2), 113–129 (1991)
6. Kasim-Zade, O.M.: On a measure of active circuits of functional elements. Math. Prob. Cybern. **4**, 218–228 (1992)
7. Lozhkin, S., Shupletsov, M.: Switching activity of Boolean circuits and synthesis of Boolean circuits with asymptotically optimal complexity and linear switching activity. Lobachevskii Journal of Mathematics **36**(4), 450–460 (2015)

8. Vaintsvaig, M.N.: On the power of networks of functional elements. In: Proceedings of the USSR Academy of Sciences, vol. 139, pp. 320–323. Russian Academy of Sciences (1961)
9. Razborov, A., Wigderson, A.: $n^{\Omega(\log n)}$ lower bounds on the size of depth-3 threshold cicuits with AND gates at the bottom. Inf. Process. Lett. **45**(6), 303–307 (1993)
10. Uchizawa, K., Douglas, R., Maass, W.: On the computational power of threshold circuits with sparse activity. Neural Comput. **18**(12), 2994–3008 (2006)
11. Uchizawa, K., Takimoto, E.: Exponential lower bounds on the size of constant-depth threshold circuits with small energy complexity. Theoret. Comput. Sci. **407**(1–3), 474–487 (2008)

The One-Round Multi-player Discrete Voronoi Game on Grids and Trees

Xiaoming Sun[1,2], Yuan Sun[1,2], Zhiyu Xia[1,2(✉)], and Jialin Zhang[1,2]

[1] CAS Key Lab of Network Data Science and Technology, Institute of Computing Technology, Chinese Academy of Sciences, Beijing, China
{sunxiaoming,sunyuan2016,xiazhiyu,zhangjialin}@ict.ac.cn
[2] University of Chinese Academy of Sciences, Beijing, China

Abstract. Basing on the two-player Voronoi game introduced by Ahn et al. [1] and the multi-player diffusion game introduced by Alon et al. [2] on grids, we investigate the following *one-round multi-player discrete Voronoi game* on grids and trees. There are n players playing this game on a graph $G = (V, E)$. Each player chooses an initial vertex from the vertex set of the graph and tries to maximize the size of the nearest vertex set. As the main result, we give sufficient conditions for the existence/non-existence of pure Nash equilibrium in 4-player Voronoi game on grids and only a constant gap leaves unknown. We further consider this game with more than 4 players and construct a family of strategy profiles, which are pure Nash equilibrium on sufficiently narrow graphs. Besides, we investigate the game with 3 players on trees and design a linear time/space algorithm to decide the existence of a pure Nash equilibrium.

Keywords: Game theory · Nash equilibrium · Location game · Graph theory

1 Introduction

1.1 Model Description

Consider the following scene: Several investors plan to set up laundries in a city and each of them is permitted to manage only one. There are some residents in the city whose addresses have been obtained by the investors. Residents in the city would only choose the nearest laundry and if the nearest ones of some residence are not unique, he/she will choose one of them randomly. In this game, the investors try to attract more customers by locating their laundry at an advantageous position and the payoff of each investor is the number of residents who are certain to choose his/her laundry. Our goal is to determine whether

This work was supported in part by the National Natural Science Foundation of China Grants No. 61433014, 61832003, 61761136014, 61872334, 61502449, the 973 Program of China Grant No. 2016YFB1000201, and K.C. Wong Education Foundation.

there exists a stable profile in which no investor would change his/her choice to improve the payoff.

In previous works, there are two ways to model this problem:

Voronoi Game. The competitive facility location game is a well-studied topic in game theory. Ahn et al. introduced the Voronoi diagram to characterize players' payoff and proposed the Voronoi game [1]. In the Voronoi game, two players alternately locate their facilities. After locating, each player will control the area closer to one of the facilities. The aim of the two players is to maximize the area controlled by them, respectively. Cheong et al. then modified this model and introduced the *one-round Voronoi game* [10]. In their version, the first player locates n facilities first. Then, the second player locates n facilities. Their locating is only allowed to perform in one round. Dürr et al. introduced an interesting multi-player version on graphs, where each player can only choice one facility [11]. A small modification in the model of Dürr et al. is that a "tie" vertex, who has k nearest facilities, contributes $1/k$ to the payoff of the k players.

Diffusion Game. A similar model is raised by Alon et al. [2] which describes the following competitive process with k players on graph $G = (V, E)$. Let player-i's influenced set $I_i = \emptyset$ for all $i \in [k]$, at the beginning. The game contains several rounds. In round-0, each player-i chooses a vertex $v_i \in V$ as the initial vertex simultaneously, and v_i is gathered by I_i. In round-$(t + 1)$ for $t \geq 0$, if some vertex v is not gathered by any I_j for all $j \in [k]$ until the round-t but has a neighbor gathered in I_i, then v is gathered by I_i in this round. However, if some v is gathered by more than one set in any round, including round-0, v will be deleted from G and all influenced set after this round. The process iterates until each I_i is invariant after some round and the payoff of each player-i is $|I_i|$.

We investigate a one-round discrete version of Voronoi games with k players on a given graph G denoted by $\Gamma(G, k)$, which is also a simplified version of diffusion games. Note that in diffusion games, the duplicate vertices, i.e., the vertices gathered by more than one player, are deleted immediately after each round. In our model, the duplicate vertices will be deleted at the end of the game. The game is also in the Voronoi style, i.e., the owner of each vertex could be determined by which initial vertex is the nearest one from it. We describe the one-round discrete Voronoi game in an equivalent but more natural way. First, each player chooses a vertex simultaneously in a given graph as the *initial vertex*. The payoff of each player is the number of vertices, which have smaller distance to this player's initial vertex than to all the others. Note that the only difference between this model with the model of Dürr et al. is that "tie" vertices will contribute nothing to the payoff of any player.

Remark. It is easy to see that a diffusion game and a one-round Voronoi game are equivalent on paths, circles and trees. But in the general case, they could be totally different. Figure 1 is an example.

In game theory, the concept Nash equilibrium, named after John Forbes Nash Jr., takes a central position. In a Nash equilibrium, no player can improve the payoff by change the choice unilaterally. John Nash proved every game with

Fig. 1. The results of these games are different in a grid graph with 2 players. The left one is the result of the diffusion game, while the right one is the result of our model.

finite number of players and a finite strategy space has a mixed-strategy Nash equilibrium in his famous paper [13]. If we just allow pure strategies, in which each player can only make a deterministic choice, the existence of a Nash equilibrium cannot be ensured. In this paper, we investigate this existence in one-round discrete Voronoi games on grids and trees.

1.2 Our Results

In this paper, we consider the one-round Voronoi game on girds or trees. First, we show that there always exists a pure Nash profile for sufficiently narrow grids when $k \neq 3$ by a family of constructions.

Theorem 1. *For any $k \in \mathbb{Z}^+$ ($k \neq 3$) and sufficiently large integers m, n, there exists a pure Nash equilibrium in $\Gamma(\text{Grid}_{n \times m}, k)$ if $m \leq n/\lceil k/2 \rceil$.*

Furthermore, we attempt to characterize the existence of a pure Nash for small k. The results are enumerated as following:

1. In $\Gamma(\text{Grid}_{n \times m}, k)$ with $k \leq 2$, there always exists a pure Nash equilibrium;
2. In $\Gamma(\text{Grid}_{n \times m}, 3)$, there exists no pure Nash equilibrium;
3. For $\Gamma(\text{Grid}_{n \times m}, 4)$, we prove the following theorem.

Theorem 2. *In $\Gamma(\text{Grid}_{n \times m}, 4)$ ($n \geq m$) where n and m are sufficiently large, if $4 \mid n$ and m is odd, there exists a pure Nash profile if $m \leq n/2 + 2\lfloor \sqrt{n} \rfloor + 1$, and does not exist if $m \geq n/2 + 2\lfloor \sqrt{n} \rfloor + 5$. Otherwise, there exists a pure Nash profile if $m \leq n/2$, and does not exist if $m > n/2 + 6$.*

Then, we consider the game on a tree. The case with two players on a tree is solved in [14]. In this paper we solve the case with 3 players:

Theorem 3. *In $\Gamma(T, 3)$ where T is a tree with size n, there exists a pure Nash equilibrium if and only if there exists a vertex v such that $n - st(i_1, v) - st(i_2, v) \geq \max\{st(j_1, v), st(j_2, v)\}$ and $st(i_2, v) \geq st(i_1, v)$, where i_1 and i_2 are the children of v with maximum and second maximum $st(\cdot, v)$, and j_k is the child of i_k with maximum $st(\cdot, v)$ ($k = 1, 2$). Here, $st(v_1, v_2)$ represents the number of vertices in the subtree with root v_1, when the tree is rebuilt as a rooted tree with root v_2.*

Furthermore, if such a vertex v exists, then (i_1, v, i_2) is a Nash equilibrium, and we can design an algorithm to determine whether a pure Nash equilibrium exists in $O(n)$ time/space.

1.3   Related Works

Ahn et al. proved that the first player has a winning strategy in Voronoi games on a circle or a line segment [1]. Teramoto et al. showed it is NP-hard to decide whether the second player can win in a 2-player discrete Voronoi game on graphs [15]. Durr and Thang studied a one-round, multi-player version and proved it is also NP-hard to decide the existence of a pure Nash equilibrium in a discrete Voronoi game on a given graph [11]. Besides, several works focus on a one-round 2-player discrete version and try to compute a winning strategy for each player efficiently [3–8].

For diffusion games, it is also proved as a NP-hard problem to decide whether there exists a pure Nash equilibrium with multi-player on general networks [12]. Roshanbin et al. studied this game with 2 players [14] and showed the existence or nonexistence of a pure Nash equilibrium in paths, cycles, trees and grids. Especially, on a tree with 2 players, there is always a Nash equilibrium under the diffusion game model. Bulteau et al. proved a pure Nash equilibrium does not exist on $\text{Grid}_{n \times m}$ where $n, m \leq 5$ with 3 players and got some extra conclusions about the existence of pure Nash equilibrium in several special graph classes [9].

1.4   Organization

In Sect. 2, we define some notations and concepts used in the latter sections. We discuss about this game on grids and trees in Sect. 3 and Sect. 4 respectively. In Sect. 5, we summarize all results and raise some ideas about the future work on this topic. Due to the space limitation, we just show our proof ideas for each important lemma and theorem. The complete proofs are available in the full version of this paper.

2   Preliminaries

Denote $[n, m]$ as the set $\{n, n+1, \ldots, m\}$, and $[n]$ as $[1, n]$ for integers n, m ($n \leq m$). An (undirected) *graph* is an ordered pair $G = (V, E)$ where V is a set of *vertices* and E is a set of *edges*. We say $v_1, \ldots, v_\ell$ is a *path* if $\forall i \ (v_i, v_{i+1}) \in E(G)$. We say vertices $v, w \in V(G)$ are connected if there exists a path between v and w in G. For convenience of discussions, we denote that $v(G) = |V|$ and $e(G) = |E|$ for a graph $G = (V, E)$.

We indicate a *grid* with a set of integer pairs. Define $\text{Grid}_{n \times m} := (V, E)$ with the vertex set $V = [n] \times [m]$ and an edge set E. For vertices $v_1, v_2 \in V$, L_1 norm distance between them is defined as $\|v_1 - v_2\|_1 := |x_1 - x_2| + |y_1 - y_2|$, where $v_1 = (x_1, y_1)$ and $v_2 = (x_2, y_2)$. Then the edge set E of $\text{Grid}_{n \times m}$ is

$$\{(v_1, v_2) \mid \|v_1 - v_2\|_1 = 1\}.$$

A graph $T = (V, E)$ is called a *tree* if $v(T) = e(T) + 1$ and every 2 vertices $v, w \in V(T)$ are connected. We also define a function $st(w, v)$ $(w, v \in V(T))$, as the number of vertices in the subtree rooted at vertex w when T is rebuilt as a rooted tree with root v.

A one-round multi-player Voronoi game, denoted by $\Gamma(G, k)$, processes on the undirected graph G with k players. Each player, player-i for example, can choose one vertex v_i from G. v_i is called *initial vertex* of player-i and we also say player-i *takes* v_i. Player-i will influence the vertices in

$$I_i := \{v \in V(G) \mid \forall j \in [k]\backslash\{i\}(\|v - v_i\|_1 < \|v - v_j\|_1)\},$$

where v_j is the player-j's initial vertex. In this game the payoff of player-i is $U_i := |I_i|$. The aim of each player is to maximize the payoff in the game, respectively. Obviously, the vertex set influenced by a specific player totally depends on every player's initial vertex. So, a strategy profile is defined as a tuple of vertices $p := (v_1, v_2, \ldots, v_k)$ where v_i is player-i's initial vertex. The influenced vertex set and payoff of player-i in some strategy profile p are donated as $I_i(p)$ and $U_i(p)$. When a strategy profile p is implied in context, we use tuple (v_i, v_{-i}) to emphasize the player-i's choice where v_{-i} represents the others' choices. We say a strategy profile v is pure Nash equilibrium if

$$\forall i \in [k] \ \forall v_i' \in V(G) \ (U_i(v_i, v_{-i}) \geq U_i(v_i', v_{-i})).$$

Namely, no player can improve the payoff by moving the initial vertex in non-cooperative cases. In this paper, we always use the corresponding initial vertex v_i to represent player-i, and (x_i, y_i) to represent the position of v_i.

3 Voronoi Games on Grids

3.1 Voronoi Game on the Narrow Grids

In this section, we discuss the case in which k players $(k \neq 3)$ take part in a Voronoi game on $\text{Grid}_{n \times m}$. As the main result, a family of pure Nash equilibrium constructions will be given for narrow grids.

Bulteau et al. study multi-player diffusion games on graphs, which is equivalent to Voronoi games on some special kinds of graph classes, such as paths, circles and trees. Bulteau et al. construct a profile on paths and prove that a pure Nash equilibrium always exists in a game on Path_n, a path with n vertices, with k players, except the case where $k = 3$ and $n \geq 6$. When the number of players k is even, they set the initial vertices $\{v_i\}_{1 \leq i \leq k}$ as:

$$v_i := \begin{cases} \lfloor \frac{n}{k} \rfloor \cdot i + \min\{i, n \bmod k\} & \text{if } i \text{ is odd}, \\ v_{i-1} + 1, & \text{if } i \text{ is even}. \end{cases}$$

For the case with odd number players, Bulteau et al. reduce it to the even case. By constructing the pure Nash-equilibria $(v_1', \ldots, v_{k+1}')$ for P_{n+1}, they get a pure Nash-equilibria, $(v_1, \ldots, v_k) := (v_1', \ldots, v_{k-2}', v_k' - 1, v_{k+1}' - 1)$, on Path_n.

We find that such construction on path sometimes works on grids. Note we might treat a grid as a path when the ratio of the width to the height is sufficiently large (Fig. 2).

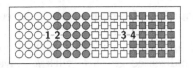

Fig. 2. Basing on the construction of Bulteau et al., construct a Nash in $\Gamma(\mathrm{Grid}_{17\times5}, 4)$.

For the odd m, we just embed the construction by Bulteau et al. into the middle row (i.e., the $(m+1)/2$-th row). Whereas, for the even m we would also embed it into one of the two middle rows (i.e., the m/2-th or $(m+2)/2$-th row). But in the latter case, we can only promise the pure Nash-property when $(n+k \bmod 2) \bmod (k+k \bmod 2) = 0$ holds. To fix it, we STRETCH or COMPRESS a pure Nash equilibrium profile. More precisely, we show the process to construct a pure Nash equilibrium profile v as following, where the routines COMPRESS and STRETCH are given in Algorithm 1:

Case 1: If k is even, consider the parity of m.
    Case 2.1: If m is odd, construct v as:

$$
v_i := \begin{cases} (\lfloor \frac{n}{k} \rfloor \cdot i + \min\{i, n \bmod k\}, \frac{m+1}{2}), & \text{if } i \text{ is odd;} \\ (x_{i-1}+1, \frac{m+1}{2}), & \text{if } i \text{ is even.} \end{cases}
$$

    Case 2.2: If m is even, define $r := n \bmod k$.
        Case 2.2.1: In the case where $r = 0$, construct a pure Nash equilibrium profile v for $\Gamma(\mathrm{Grid}_{n \times (m+1)}, k)$.
        Case 2.2.2: Otherwise, stretch or compress the profile in the following way:
            Case 2.2.2.1: If $r \leq k/2$ holds, construct a pure Nash equilibrium v' for $\Gamma(\mathrm{Grid}_{(n-r)\times m}, k)$ and construct v by STRETCH($n - r, m, k, v', r$).
            Case 2.2.2.2: Otherwise, construct a pure Nash profile v' for $\Gamma(\mathrm{Grid}_{(n+k-r)\times m}, k)$. Then, construct v by COMPRESS($n + k - r, m, k, v', k - r$).
Case 2: If k is odd but $k \neq 3$, construct a pure Nash equilibrium profile v' for $\Gamma(\mathrm{Grid}_{(n+1)\times m}, k+1)$. Then, construct $\{v_i\}_{1\leq i < k}$ as

$$
v_i := \begin{cases} v_i' & \text{if } i \leq k - 2 \\ (x_{i-1} - 1, y_{i-1}) & \text{otherwise} \end{cases}
$$

For convenience, we use a superscript to represent the profile. Namely, $v_i = (x_i^v, y_i^v)$ means that player-i's initial vertex is located at (x_i^v, y_i^v) in profile v.

Algorithm 1. Stretch or compress a profile

```
function STRETCH(n, m, k, v', i)
    return STRETCH-OR-COMPRESS(n, m, k, v', i, 1)

function COMPRESS(n, m, k, v', i)
    return STRETCH-OR-COMPRESS(n, m, k, v', i, −1)

function STRETCH-OR-COMPRESS(n, m, k, v', i, soc)
    for j ← 1 to k/2 do
        a ← soc · min{i, j − 1}
        v_{2j−1} ← (x^{v'}_{2j−1} + a, y^{v'}_{2j−1} + [a is odd])
        v_{2j} ← (x^{v'}_{2j} + a, y^{v'}_{2j} + [a is odd])
    return v
```

For our constructed strategy profile, the positions of the initial vertices are specific. Thus, the payoff of each players can be calculated, and the payoff of their changes can be predicted, wherever the changes are located. Actually, only constant number of initial vertices need to be checked due to the symmetry of our construction. For each of them, we claim most changes of them cannot be a pure Nash, which can be asserted by some lemmas given in the following section. Thus, the number of changes that need to be checked is quite limited and it is an tedious but feasible task to check all of them.

3.2 Non-existence Conditions

In this section, we give 3 sufficient conditions for the non-existence of pure Nash profiles, Lemmas 1, 2 and 3. They will be used as tools to find an boost and certify a profile is not a pure Nash.

Intuitively, we can perceive if an initial vertex v_1 moves closer to another initial vertex v_2, it will snatch some vertices from I_2. If some change makes an initial vertex closer to all the other initial vertex, it may be an boost. The first lemma is a formal statement for such idea.

Lemma 1. *For some player occupying $v_i = (x_i, y_i)$, $v'_i = (x_i + 1, y_i + 1)$ is an boost of v_i if $x_j + y_j > x'_i + y'_i$ holds for all $j \in [k] \backslash \{i\}$.*

It can be generalized due to the symmetry of grids. Namely, after reflecting coordinates along a dimension or rotating the grid by 90°, it also hold. The statement about such generalization will be omitted in this paper.

The second lemma shows the profile in which all the initial vertices are bounded by a small block cannot be pure Nash equilibrium.

Lemma 2. *In $\Gamma(\text{Grid}_{n \times m}, k)$ where $k \geq 3$ and n, m are sufficiently large, the profile is not pure Nash equilibrium if for some constant c there exists no distinct $i, j \in [k]$ such that $x_i - x_j > c$ or $y_i - y_j > c$.*

Without loss of generality, assume v_1 is the minimal-payoff one among all initial vertex, which implies $U_1 \leq nm/k$. We "merge" the other initial vertices and treat roughly it as a 2-player Voronoi game. Note that such merging is reasonable since the initial vertices are close to each other. Thus, v_1 can always move to some position and get the payoff of almost a half of the total vertex number, which is a significant boost for v_1 when $k \geq 3$ holds.

The third lemma describes another non-Nash case, where the 4 players are divided into 2 pairs, initial vertices in each pair are constantly close and all initial vertex is constantly close to a slash.

Lemma 3. *In* $\Gamma(\mathrm{Grid}_{n \times m}, 4)$ *where* n, m *are sufficiently large, a profile* v *is not a pure Nash where* $\|v_1 - v_2\|_1, \|v_3 - v_4\|_1$ *and* $|(x_1 + y_1) - (x_3 + y_3)|$ *(or* $|(y_1 - x_1) - (y_3 - x_3)|)$ *are upper bounded by a constant.*

The basic idea to prove Lemma 3 is as following. Suppose that $\{v_1, v_2\}$ contains the leftmost and the topmost initial vertices. Thus, there exists a position $v = (x, y)$ such that $x + y > x_i + y_i$ and $y - x > y_i - x_i$ hold for all $i \in [4]$, and $\|v - v_1\|$ can be bounded by a constant. If vertex pair $\{v_1, v_2\}$ is $(1/4 + \epsilon) \cdot m$-far from the upper bound, position v can provide payoff of at least $(1/4 + \epsilon/2) \cdot nm$, which is sufficient to improve the minimal-payoff initial vertex. Similarly, vertex pair $\{v_1, v_2\}$ cannot be too far from the left boundary, as well as pair $\{v_3, v_4\}$ cannot be too far from the bottom or the right boundary. In this case, we show there exists a position providing the payoff of at least $3nm/8 - o(nm)$.

3.3   Voronoi Game on Grids Within 4 Players

Naturally, it is a good start point to consider the cases with a few players. In this section, we answer the question that if there exists a pure Nash in $\Gamma(\mathrm{Grid}_{n \times m}, k)$ when $k \leq 4$.

In the first non-trivial case $\Gamma(\mathrm{Grid}_{n \times m}, 2)$, there is always a pure Nash equilibrium. If the grid has more than one centroid, v_1 and v_2 take a pair of adjacent centroids; Otherwise, v_1 takes the centroid and v_2 an adjacent vertex of v_1. It is easy to verify the pure Nash-property. For the case $\Gamma(\mathrm{Grid}_{n \times m}, 3)$, we prove the following conclusion.

Theorem 4. *For any sufficiently large integer* n, m, *there exists no pure Nash equilibrium in* $\Gamma(\mathrm{Grid}_{n \times m}, 3)$.

Roughly said, either there exists a player who can get closer to the others, or the 3 players are close to each others when $k = 3$. Thus, the profile in neither the 2 cases can be a pure-Nash due to Lemmas 1 and 2.

Next, we discuss a much more challenging case $\Gamma(\mathrm{Grid}_{n \times m}, 4)$. Intuitively, to construct a pure Nash profile, a "isolated" initial vertex, i.e. an initial vertex with no neighbor one, should be avoid, otherwise some initial vertex could get closer to the others. Unfortunately, the intuition is not so precise. We find that in the following special case every initial is isolated, but neither of them could get closer to all the other initial vertex simultaneously in the style of Lemma 1:

the case where the topmost, leftmost, rightmost and bottommost initial vertices are distinct. We say an initial vertex is a *controller* if such initial vertex is both the leftmost and the topmost one. This concept can also generalized due to the symmetry of grids. The condition of such special case is equivalent to there exists no controller. As the first step, we deal with this special case:

Lemma 4. *In $\Gamma(\text{Grid}_{n \times m}, 4)$ where n, m are sufficiently large, a profile is not pure Nash equilibrium if there is no controller.*

The related position of the leftmost initial vertex $v_3 = (x_3, y_3)$ and the rightmost one $v_4 = (x_4, y_4)$ is discussed first. As the result, an boost can be easily found, unless $x_3 + y_3 < x_4 + y_4$ and $y_3 - x_3 > y_4 - x_4$ hold. Next, we consider the positions the topmost initial vertex v_2 and the bottommost one v_1, and analyze the following two cases: (1) the case where $y_2 - x_2 \geq y_3 - x_3$ and $y_1 + x_1 \leq y_3 + x_3$ hold; (2) the case where $y_2 - x_2 > y_3 - x_3$, $y_3 + x_3 < y_1 + x_1$, $y_4 - x_4 > y_1 - x_1$ and $y_4 + x_4 > y_2 + x_2$ holds. For the first case, the key point is that v_3 can snatch significant payoff improvement from either v_1 or v_2, since $(y_2 - x_2) - (y_3 - x_3) \leq 2$ and $(y_3 + x_3) - (y_1 + x_1) \leq 2$ must hold (otherwise Lemma 1 will show an boost for v_1 or v_2). In the second case, there always exists a player who can improve the payoff, except all the players are close to each other, which cannot be a pure Nash due to Lemma 2. The result that profiles in these 2 cases cannot be a pure Nash can be generalized due to the symmetry of grids, which finishes the proof.

After excluding the special case, we prove Lemmas 5 and 6 to show profiles with some initial vertex isolated cannot be a pure Nash.

Lemma 5. *In $\Gamma(\text{Grid}_{n \times m}, 4)$ where n, m are sufficiently large, the profile is not pure Nash equilibrium, if there exists a controller and the distance between any 2 players' initial vertices is not shorter than 2.*

Lemma 6. *In $\Gamma(\text{Grid}_{n \times m}, 4)$ where n, m are sufficiently large, the profile is not pure Nash equilibrium if $\|v_1 - v_2\|_1 = 1$ and $\|v_3 - v_4\|_1 > 1$.*

By utilizing Lemma 1 and investigating potential boosts of a controller, it not a hard task to prove Lemma 5. To prove Lemma 6, assume $x_2 = x_1 + 1$ and $y_1 = y_2$ hold without loss of generality. We divide a grid in the following way (Fig. 3):

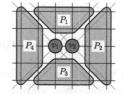

$$P_1 = \{(x, y) \mid x + y > x_1 + y_1 \wedge y - x > y_2 - x_2\}$$
$$P_2 = \{(x, y) \mid x + y \geq x_2 + y_2 \wedge y - x \leq y_2 - x_2\} \setminus \{v_2\}$$
$$P_3 = \{(x, y) \mid x + y < x_2 + y_2 \wedge y - x < y_1 - x_1\}$$
$$P_4 = \{(x, y) \mid x + y \leq x_1 + y_1 \wedge y - x \geq y_1 - x_1\} \setminus \{v_1\}.$$

Fig. 3. The grid is divided into P_1, P_2, P_3 and P_4.

Due to Lemma 1, we find the following facts: (1) If v_3 and v_4 are located in the same part, the profile is not a pure-Nash unless $\|v_3 - v_4\|_1 = 1$; (2) If v_3

and v_4 are located in 2 opposite parts, i.e. P_1, P_3 or P_2, P_4, the profile is not a pure-Nash unless $\|v_3 - v_4\|_1 \leq 5$; (3) Otherwise, the profile is not a pure-Nash unless $\|v_3 - v_4\|_1 \leq 5$. Furthermore, condition $\|v_3 - v_4\|_1 > 1$, Lemmas 2 and 3 imply respectively profiles in all the 3 cases cannot be pure Nash profiles.

Then, we prove Lemma 7 and give a non-existence condition of pure Nash profiles for this game.

Lemma 7. *In* $\Gamma(\mathrm{Grid}_{n \times m}, 4)$ *$(n \geq m)$ where n and m are sufficiently large and* $\|v_1 - v_2\|_1 = 1$ *and* $\|v_3 - v_4\|_1 = 1$ *hold, in the case where $4 \mid n$ and m is odd, there does not exist a pure Nash profile if $m \geq n/2 + 2\lfloor\sqrt{n}\rfloor + 5$ holds. In the other cases, there does not exist a pure Nash profile if $m > n/2 + 6$.*

To inherit notations $P_1, \ldots, P_4$, we abandon restriction $n \geq m$ temporarily. First, assume $x_2 = x_1 + 1$ and $y_1 = y_2$ hold. If both v_3 and v_4 are located in P_1, v_1 and v_2 cannot be $(1/4 + \epsilon)m$-far from the bottom boundary in a pure Nash profile, due to a similar argument to the one mentioned in the proof sketch of Lemma 2. We consider $v_1' = (x_1 - 1, y_1 + 1)$ and $v_2' = (x_2 + 1, y_2 + 1)$. Note that $U_1(v_1', v_{-1}) = U_1 + (x_1 - 1 - y_1)$ and $U_2(v_2', v_{-2}) = U_2 + (n - x_1 - y_2)$, which implies v_1' and v_2' are both not boosts only if $y_2 \geq (n - 1)/2$ holds. Due to the symmetry of girds, such argument works for most of the cases to give a $n > m/2 + 6$ bound, except the one where both v_3 and v_4 are located in P_2 (or P_4) and $y_3 = y_4$ holds. Second, suppose $y_3 = y_4$ and $v_3, v_4 \in P_2$ hold. We consider $v_2' = (x_2 + 1, y_2 + 1)$ and $v_2'' = (x_2 + 1, y_2 - 1)$. Note that $U_2(v_2', v_{-2}) = U_2 + (m - 2y_2)$ and $U_2(v_2'', v_{-2}) = U_2 + (2y_2 - m + y_2 - 2)$, which implies v_1, v_2 must be located roughly on the middle row in a pure Nash profile, as well as v_3, v_4 for the same reason. Up to now, the structure is simple enough such that it becomes feasible to show Lemma 7 by more detailed analysis.

Meanwhile, we can construct a pure Nash-profile if the grid is sufficiently narrow according to Theorem 1. Besides, when $4 \mid n$ and m is odd, our construction can be ensured as a pure Nash in a loosened condition, as the following lemma stated.

Lemma 8. *In* $\Gamma(\mathrm{Grid}_{n \times m}, 4)$ *$(n \geq m)$ where n, m are sufficiently large, $4 \mid n$ and m is odd, there exists a pure Nash profile if $m \leq n/2 + 2\lfloor\sqrt{n}\rfloor + 1$ holds.*

Combining all the results in this section and Lemma 8, our main result Theorem 2 can be shown.

4 Voronoi Game on Trees

In this section, we discuss the Voronoi game on trees among 3 players. For the limit of pages, we only show the algorithm induced by Theorem 3. The complete proof of Theorem 3 is in the full version of this paper.

The main idea towards Theorem 3 is that if there exists a Nash equilibrium on a tree with 3 players, then their positions must induce a Path$_3$ on the tree. Based on this conclusion, it is easy to verify the group of conditions in Theorem 3 is sufficient and necessary.

This theorem can induce an efficient algorithm as a judgment. See details in Algorithm 2, where each $val_{v,i}(v \in V(T), i = 1, 2)$ represents the value of $st(w, v)$, where w is a neighbor of v with the i-th maximum $st(\cdot, v)$, and each $lab_{v,i}(v \in V(T), i = 1, 2)$ records the vertex w. If there is a Nash equilibrium, the algorithm will return a triple of three different vertices as a solution, otherwise it will return -1.

The correctness of this algorithm is promised by Theorem 3. Now let us analyze its performance. Let $n = v(T)$. The main space cost of this algorithm is the tree structure and the arrays $\{val_{v,i}\}$ and $\{lab_{v,i}\}$, so it is $O(n)$. For the time cost, it is easy to see that the two arrays $\{val_{v,i}\}$ and $\{lab_{v,i}\}$ can be computed by a dynamic programming on the tree T within $O(n)$ time. Combine the $O(n)$ time cost by Algorithm 2, the total time cost is $O(n)$ too.

Algorithm 2. Nash equilibrium on a tree T with 3 players

$\quad$ **for** $v \in V(T)$ **do**
$\quad\quad$ $sti1 \leftarrow val_{v,1}$
$\quad\quad$ $sti2 \leftarrow val_{v,2}$
$\quad\quad$ **if** $lab_{lab_{v,1},1} = v$ **then**
$\quad\quad\quad$ $stj1 \leftarrow val_{lab_{v,1},2}$
$\quad\quad$ **else**
$\quad\quad\quad$ $stj1 \leftarrow val_{lab_{v,1},1}$
$\quad\quad$ **if** $lab_{lab_{v,2},1} = v$ **then**
$\quad\quad\quad$ $stj2 \leftarrow val_{lab_{v,2},2}$
$\quad\quad$ **else**
$\quad\quad\quad$ $stj2 \leftarrow val_{lab_{v,2},1}$
$\quad\quad$ **if** $(v(T) - sti1 - sti2 \geq \max\{stj1, stj2\}) \wedge (sti2 \geq stj1)$ **then**
$\quad\quad\quad$ **return** $(lab_{v,1}, v, lab_{v,2})$
$\quad$ **return** -1

---

5 Conclusion

In this paper, we consider one-round multi-player Voronoi games on grids and trees. First, we answer such a question: is there a pure Nash equilibrium in a one-round k-player Voronoi game processing on $\text{Grid}_{n \times m}$ $(n \geq m)$? For the game on sufficiently large grids within 4 players, we provide almost complete characterization of the existence of a pure Nash equilibrium. Besides, we raise a method to construct a pure Nash equilibrium profile for the game with k ($k \neq 3$) players on sufficiently narrow grids. Second, for the game on a tree with 3 players, we give a sufficient-necessary condition for the existence of a pure Nash equilibrium, as well as a linear time/space algorithm to check the condition.

The non-existence condition of a pure Nash in $\Gamma(\text{Grid}_{n \times m}, k)$ where $k > 4$ is unknown. We conjecture our construction is optimal, i.e., there exists no pure Nash if $n \geq m > n/\lceil k/2 \rceil + o(n)$ holds. For the case of tree, it is worth thinking how to determine whether there exists a pure Nash equilibrium with 4 players. It is also interesting to determine whether there is a Nash equilibrium with multi-players on other graph classes.

References

1. Ahn, H., Cheng, S., Cheong, O., Golin, M., Van Oostrum, R.: Competitive facility location: the Voronoi game. Theoret. Comput. Sci. **310**(1–3), 457–467 (2004)
2. Alon, N., Feldman, M., Procaccia, A.D., Tennenholtz, M.: A note on competitive diffusion through social networks. Inf. Process. Lett. **110**(6), 221–225 (2010)
3. Banik, A., Bhattacharya, B.B., Das, S.: Optimal strategies for the one-round discrete Voronoi game on a line. J. Comb. Optim. **26**(4), 655–669 (2013)
4. Banik, A., Bhattacharya, B.B., Das, S., Das, S.: Two-round discrete voronoi game along a line. In: Fellows, M., Tan, X., Zhu, B. (eds.) AAIM/FAW -2013. LNCS, vol. 7924, pp. 210–220. Springer, Heidelberg (2013). https://doi.org/10.1007/978-3-642-38756-2_22
5. Banik, A., Bhattacharya, B.B., Das, S., Mukherjee, S.: The discrete Voronoi game in R2. Comput. Geom. **63**, 53–62 (2017)
6. Banik, A., De Carufel, J.L., Maheshwari, A., Smid, M.: Discrete Voronoi games and ε-nets, in two and three dimensions. Comput. Geom. **55**, 41–58 (2016)
7. de Berg, M., Kisfaludi-Bak, S., Mehr, M.: On one-round discrete Voronoi games. arXiv preprint arXiv:1902.09234 (2019)
8. Berg, M.D., Gudmundsson, J., Mehr, M.: Faster algorithms for computing plurality points. ACM Trans. Algorithms (TALG) **14**(3), 36 (2018)
9. Bulteau, L., Froese, V., Talmon, N.: Multi-player diffusion games on graph classes. In: Jain, R., Jain, S., Stephan, F. (eds.) TAMC 2015. LNCS, vol. 9076, pp. 200–211. Springer, Cham (2015). https://doi.org/10.1007/978-3-319-17142-5_18
10. Cheong, O., Har-Peled, S., Linial, N., Matousek, J.: The one-round Voronoi game. Discrete Comput. Geom. **31**(1), 125–138 (2004)
11. Dürr, C., Thang, N.K.: Nash equilibria in Voronoi games on graphs. In: Arge, L., Hoffmann, M., Welzl, E. (eds.) ESA 2007. LNCS, vol. 4698, pp. 17–28. Springer, Heidelberg (2007). https://doi.org/10.1007/978-3-540-75520-3_4
12. Etesami, S.R., Basar, T.: Complexity of equilibrium in diffusion games on social networks. In: American Control Conference (ACC), pp. 2065–2070. IEEE (2014)
13. Nash, J.: Non-cooperative games. Ann. Math. 286–295 (1951)
14. Roshanbin, E.: The competitive diffusion game in classes of graphs. In: Gu, Q., Hell, P., Yang, B. (eds.) AAIM 2014. LNCS, vol. 8546, pp. 275–287. Springer, Cham (2014). https://doi.org/10.1007/978-3-319-07956-1_25
15. Teramoto, S., Demaine, E. D., Uehara, R.: Voronoi game on graphs and its complexity. In: IEEE Symposium on Computational Intelligence and Games, pp. 265–271. IEEE (2006)

Competitive Auctions and Envy-Freeness for Group of Agents

Taiki Todo[1,3](✉), Atsushi Iwasaki[2,3], and Makoto Yokoo[1,3]

[1] Kyushu University, Fukuoka, Japan
{todo, yokoo}@inf.kyushu-u.ac.jp
[2] The University of Electro-Communications, Chofu, Japan
iwasaki@is.uec.ac.jp
[3] RIKEN Center for Advanced Intelligence Project (AIP), Tokyo, Japan

Abstract. In mechanism design, fairness is one of the central criteria for analyzing mechanisms. Recently, a new fairness concept called envy-freeness of a group toward a group (GtG-EFness) has received attention, which requires that no group of agents envies any other group. In this paper, we consider GtG-EFness in more general combinatorial auctions, including several subclasses of the multi-unit auction domain (unit-demand, diminishing marginal values, and all-or-nothing), and reveal the tight bound of the competitive ratios. In particular, we prove that the tight bound of the competitive ratio is $1/k$ (where k is the number of items) for the general combinatorial auction domain. We also clarify the relationship with Walrasian equilibria and conclude that no group envies any other group in any Walrasian equilibrium.

Keywords: Combinatorial auctions · Competitive analysis · Envy-freeness

1 Introduction

Since *fairness* is one of the central criteria for evaluating mechanisms, several concepts concerned with it have been studied so far. In particular, a concept called *envy-freeness* has been scrutinized in economics [5]. A mechanism is envy-free if no individual (agent) envies any other individual. This concept has also attracted considerable attention from computer scientists, who are interested in designing and analyzing fair resource allocations, e.g., [6,12].

Todo et al. [14] proposed an extension of traditional envy-freeness called *envy-freeness of a group toward a group* (GtG-EFness), which requires that no group of agents envies any other group. In domains with complementarities, such as combinatorial auctions or task scheduling, an agent or a group of agents might desire a set of items/tasks that are assigned to other agents. Assuming monetary transfers are possible, it would be *fair* to allocate a set of items/tasks to a set of agents that is willing to pay more. Thus, GtG-EFness seems to be

© Springer Nature Switzerland AG 2019
D.-Z. Du et al. (Eds.): COCOON 2019, LNCS 11653, pp. 541–553, 2019.
https://doi.org/10.1007/978-3-030-26176-4_45

a natural extension of traditional envy-freeness and can open up interesting research directions in mechanism design with monetary transfers.

Many open problems remain on GtG-EF mechanisms. In particular, almost all the results presented in [14] are on a very restricted case called the *single-minded* combinatorial auction domain. This restriction is useful for simplifying theoretical analysis, but agents' preferences are much more complicated in real application fields. When agents are single-minded, we can safely assume a winner never has envy. On the other hand, if an agent is not single-minded, even if she wins something, she still might envy another agent who obtained more desirable items, or equally desirable items by paying less.

In this paper, we analyze the worst-case performances of GtG-EF mechanisms using a competitive ratio, which is the ratio of the welfare obtained by the mechanism to the Pareto efficient welfare in the worst-case. We address several domains beyond the single-minded combinatorial auction domain, in particular, the multi-unit auction domain, in which multiple identical items are going to be sold. In this domain, agents' preferences can be more concisely represented than in the general combinatorial auction domain (i.e., k vs. 2^k, where k indicates the number of items). The multi-unit auction domain is crucial in practice because it can represent many applications, such as telecommunication spectrum auctions in which identical channels are auctioned, job scheduling on multiple machines, and the assignment of the rights to use finite/infinite identical resources.

We consider several subclasses of the multi-unit auction domain, i.e., unit-demand, diminishing marginal values, and all-or-nothing domains, as well as the general combinatorial auction domain. We obtain the tight bounds of the competitive ratio of GtG-EF mechanisms in those domains, considering two incentive concepts of strategy-proofness and false-name-proofness [19]. In particular, we prove that the tight bound of the competitive ratio is $1/k$ for the general combinatorial auction domain. We believe this work is an important first step toward a fair mechanism design for real market environments.

We then shed light on the relationship between GtG-EFness and a notion on market-clearing called *Walrasian equilibrium*, which has been discussed in both economics and computer science literatures. Informally, a Walrasian equilibrium is a pair of an assignment of items and a vector of item prices in which all agents' utilities are maximized. We prove that in any Walrasian equilibrium, no group of agents envies any other group in the general combinatorial auction domain.

2   Related Works

Many works have addressed envy-freeness in computer science and economics. Foley [5] proposed envy-freeness in resource allocations. Haake et al. [8] characterized the allocation rules of envy-free mechanisms by a property called *local efficiency*. For unlimited-supply multi-unit auctions, Goldberg and Hartline [6] discussed strategy-proof and envy-free auctions and analyzed their competitive ratios. Cohen et al. [2] showed that incentive conditions and envy-freeness cannot coexist in combinatorial auctions with *capacitated valuations*, where agents have a limit on the number of items they may receive. Tsuruta et al. [16] studied the relation with false-name-proofness in the allocation of a divisible object.

Vind [18] and Varian [17] proposed an extension of envy-freeness called *coalition fairness*, which requires that no group envies any other group *of the same size*. Coalition fairness, which is a conventional concept in economic theory, has been attracting less attention than traditional envy-freeness. One reason might be that coalition fairness is too specific since it restricts the scope of envy-freeness within same-sized groups. On the other hand, GtG-EFness puts no restriction on the size of groups. In this sense, GtG-EFness can be considered an extension of coalition fairness. GtG-EFness also resembles *Walrasian equilibrium*. Gul and Stacchetti [7] discussed Walrasian equilibrium in economies that satisfy the *gross substitutes* condition. Conen and Sandholm [3] showed that Walrasian equilibrium always exists, even in the general combinatorial auction domain, if *nonlinear* pricing is allowed. Quite recently, such group fairness has also been studied in (indivisible) resource allocation without monetary transfers [1,4,11,13].

3   Model

Consider a set of (identical or heterogeneous) items $K = \{g_1, \ldots, g_k\}$ for sale and a set of agents (bidders) $N = \{1, \ldots, n\}$. Each agent $i \in N$ has her value function $v_i \in V$ that maps a set of items into $\Re$. Here V is a set of possible value functions (*value domain*). We assume a *quasi-linear, private value* model with *no allocative externality*; the utility of agent i, who obtains a set of items (bundle) $B_i \subseteq K$ and pays t_i, is represented as $v_i(B_i) - t_i$. We also assume v_i is normalized so that $v_i(\emptyset) = 0$ holds.

Let $v = (v_1, \ldots, v_n) \in V^n$ be a value profile. An (direct revelation, deterministic) auction mechanism $M(f, t)$ consists of an *allocation rule* and a *transfer (payment) rule*. An allocation rule is defined as $f : V^n \to X$, where X is a set of the possible assignments of items over N. For an assignment $x \in X$, let x_i indicate the bundle allocated to agent i. Note that an assignment $x \in X$ must satisfy *allocation feasibility*; $\bigcup_{i \in N} x_i \subseteq K \wedge \forall i, j (\neq i), x_i \cap x_j = \emptyset$. A transfer rule is defined as $t : V^n \to \Re$. For a mechanism $M(f, t)$ and a value profile v, let $f_i(v)$ and $t_i(v)$ respectively denote the bundle allocated to agent i and the amount agent i must pay. We use notations $f(v_i, v_{-i})$ and $t(v_i, v_{-i})$ to represent the assignment and transfer when agent i declares value function v_i and the other agents declare value profile v_{-i}.

We restrict our attention to mechanisms that satisfy *individual rationality* (IR) with *non-positive transfer*. IR means that no agent obtains negative utility by reporting her true value. Formally, $\forall i \in N, \forall v, v_i(f_i(v)) - t_i(v) \geq 0$. We also assume a mechanism is *almost anonymous* across agents; the results obtained from a mechanism are invariant under the permutation of the identifiers of agents except for the case of ties. This paper also discusses two incentive constraints: *strategy-proofness* and *false-name-proofness* (for a detailed discussion, see Yokoo et al. [19] for example). A mechanism is strategy-proof if for every agent, reporting her true value is a weakly dominant strategy.

Definition 1. *A mechanism* $M(f,t)$ *is* strategy-proof *if* $\forall i$, $\forall v_{-i}$, $\forall v_i$, $\forall v_i'$, $v_i(f_i(v_i, v_{-i})) - t_i(v_i, v_{-i}) \geq v_i(f_i(v_i', v_{-i})) - t_i(v_i', v_{-i})$.

Definition 2. *A mechanism* $M(f,t)$ *is* false-name-proof *if for all agents, truth-telling only using one identifier is a weakly dominant strategy.*

Notice that false-name-proofness implies strategy-proofness by definition.

Here we define the five value domains. The first two are for combinatorial auctions with k heterogeneous items, and the last three are for multi-unit auctions with k identical items.

General Combinatorial Auction (GCA)[1] Domain: the value domain V is a set of all value functions that satisfy free-disposal and are normalized with no allocative externalities. We say a value function v_i satisfies *free-disposal* if $\forall B_i \subseteq B_i' \subseteq K$, $v_i(B_i) \leq v_i(B_i')$ holds.

Single-Minded Combinatorial Auction (SMCA) Domain: the value domain V is a set of all single-minded (SM) value functions. A value function v_i is SM if there exists a bundle $B_i \subseteq K$ s.t., $v_i(B_i) > 0$, $v_i(B_i') = v_i(B_i)$ for all $B_i' \supseteq B_i$, and $v_i(B_i'') = 0$ for all $B_i'' \not\supseteq B_i$.

Unit-Demand (UD) Domain: the value domain V is a set of all UD value functions. A value function v_i is UD if there exists a non-negative value w_i such that for all $B_i \subseteq K$, $v_i(B_i) = w_i$ if $|B_i| \geq 1$ and 0 otherwise.

Diminishing Marginal Values (DMV) Domain: the value domain V is a set of all value functions that have DMV. A value function v_i has DMV if it has an associated non-increasing sequence of k non-negative values (*marginal values*) $w_{i,1} \geq \ldots \geq w_{i,k}$ such that for all $B_i \subseteq K$, $v_i(B_i) = \sum_{j=1}^{|B_i|} w_{i,j}$.

All-or-Nothing Values (ANV) Domain: the value domain V is a set of all the value functions that have ANV. A value function v_i has ANV if it has an associated positive number $k_i \in \{1, \ldots, k\}$ and a non-negative value w_i such that for every B_i, $v_i(B_i) = w_i$ if $|B_i| \geq k_i$ and 0 otherwise.

Next we introduce canonical *envy-freeness* and its extension. We refer to the canonical one as the *envy-freeness of an individual toward an individual* (ItI-EFness), and the extended concept as the *envy-freeness of a group toward another group* (GtG-EFness). A mechanism $M(f,t)$ is ItI-EF if $\forall i$, $\forall j \neq i$, $\forall v$, $v_i(f_i(v)) - t_i(v) \geq v_i(f_j(v)) - t_j(v)$. Todo et al. [14] proposed the following extension.

Definition 3. *A mechanism* $M(f,t)$ *is* GtG-EF *if* $\forall v$, $\forall C, S \subseteq N$,

$$\sum_{i \in C} [v_i(f_i(v)) - t_i(v)] \geq V^*(C, \bigcup_{j \in S} f_j(v)) - \sum_{j \in S} t_j(v),$$

where $V^*(C, B)$ *is the surplus for a set of agents* C *when a set of items* $B \subseteq K$ *is optimally allocated to* C.

---

[1] We abbreviate the GCA domain as GCA interchangeably, and the others as well.

Intuitively, this concept implies that no group of agents envies any other group. Todo et al. [14] completely characterized GtG-EF mechanisms in SMCA by introducing a condition called *locally efficient bundle assignment with split/merger* (LEBA-SM). Before introducing it, let us first define *permutation with split/merger*; an assignment x' is a permutation with split/merger of another assignment x if $\bigcup_{i \in N} x'_i = \bigcup_{i \in N} x_i$.

Definition 4. *An assignment x is a LEBA-SM w.r.t. v if $\sum_{i \in N} v_i(x_i) \geq \sum_{i \in N} v_i(x'_i)$ for any x', where x' is a permutation with split/merger of x.*

In other words, an LEBA-SM x maximizes the social welfare among all possible assignments that allocate the same set $\bigcup_{i \in N} x_i \subseteq K$ of items to N.

Theorem 1 (Todo et al., [14]). *In SMCA, there is a transfer rule t s.t., $M(f,t)$ is GtG-EF if and only if $f(v)$ is an LEBA-SM w.r.t. v for every v.*

For the proofs in Section 4, the following corollary is useful, which can be easily derived from Theorem 1.

Corollary 1. *In any domain V, if there exists a transfer rule t s.t. a mechanism $M(f,t)$ is GtG-EF, then $f(v)$ is an LEBA-SM w.r.t. v for all value profile $v \in V$.*

We evaluate mechanisms based on *competitive analysis*, which is commonly used in recent algorithmic mechanism design literature.

Definition 5. *A mechanism $M(f,t)$ has a competitive ratio of c if*

$$\inf_{v \in V} \frac{\sum_{i \in N} v_i(f_i(v))}{\max_{x \in X} \sum_{i \in N} v_i(x_i)} \geq c.$$

The denominator is the social welfare by a *Pareto efficient* assignment, which maximizes the social welfare when agents' utilities are assumed to be transferrable. When a mechanism always produces a Pareto efficient assignment, it achieves the optimal competitive ratio of one. For example, both the Vickrey-Clarke-Groves (VCG) and the first-price combinatorial auction (FPCA) mechanisms have the optimal competitive ratio.

However, these well-known mechanisms are not always strategy-proof and GtG-EF simultaneously. Precisely, even in SMCA, VCG is not GtG-EF but strategy-proof, and FPCA is not strategy-proof but GtG-EF.

Example 1. Consider two items, $\{g_1, g_2\}$, and three agents, $\{1, 2, 3\}$, whose values are drawn from the SMCA domain: agent 1 values 8 on g_1, agent 2 values 7 on g_2, and agent 3 values 10 on $\{g_1, g_2\}$.

In VCG, agent 1 wins g_1 and pays 3, and agent 2 wins g_2 and pays 2. Thus, agent 3 envies the set of winners $\{1, 2\}$, since her value on $\{g_1, g_2\}$ is strictly greater than the sum of the transfers $3 + 2 = 5$.

In FPCA, the assignment of items is identical to VCG. Each winner pays her reported value on the bundle she receives. In the above example, since agent 1 pays 8 and agent 2 pays 7, agent 3 does not envy the winners. However, agent 1 can increase her utility by reporting $3 + \epsilon$ rather than her true value of 8.

Thus, FPCA is an optimal GtG-EF mechanisms without incentive requirements in SMCA. However, the tight bound of GtG-EF and strategy-proof (or false-name-proof) mechanisms remains open in SMCA, although a non-trivial mechanism called *average-max minimal-bundle* (AM-MB, Ito et al. [10]) is GtG-EF and strategy-proof, which has a competitive ratio of $1/k$ in GCA, including SMCA. In the rest of this paper, we first examine the upper bounds of the three subclasses of the multi-unit auction domains (UD, DMV, and ANV). Then, instead of considering the general multi-unit auction domain, we focus on GCA. Actually, the bounds of the general multi-unit auction domain are identical to those of GCA. The obtained results are summarized in Table 1.

Table 1. Tight bounds of competitive ratios achieved by GtG-EF mechanisms. SP and FNP denote strategy-proofness and false-name-proofness.

Domain	w/o SP	w/ SP	w/ FNP
SMCA [14]	1	Open	Open
UD	1	1	1
DMV (Theorems 2 and 4)	1	$1/k$	$1/k$
ANV (Theorems 3 and 4)	$1/k \, 1/k$	$1/k$	$1/k$
GCA (Theorem 5)	$1/k$	$1/k$	$1/k$

4   Multi-unit Auctions

This section analyzes GtG-EF mechanisms in the three multi-unit auction domains defined above, where k identical items are going to be sold.

First, as an exercise, we examine the tight bounds of the competitive ratio for UD. The analysis for UD is quite easy, since VCG is simultaneously GtG-EF, strategy-proof, and false-name-proof and has a competitive ratio of 1. More precisely, GtG-EFness is equivalent to ItI-EFness in UD. Thus, all entries in the UD row of Table 1 have the tight bound of 1.

4.1   Diminishing Marginal Values

Omitting the incentive issues, the $(k + 1)$-st price auction mechanism, which is also known as the uniform-price auction, satisfies GtG-EFness because every winner pays the same amount of transfer per unit, a.k.a., the market-clearing price. Since the mechanism always produces a Pareto efficient assignment, it achieves the optimal competitive ratio. However, after taking either strategy-proofness or false-name-proofness into account, no mechanism can achieve the optimal ratio, i.e., the following theorem holds.

Theorem 2. *In DMV, any mechanism that is simultaneously GtG-EF and strategy-proof has a competitive ratio of at most $1/k$.*

Proof. We assume for a contradiction that a GtG-EF and strategy-proof mechanism $M(f,t)$ has a competitive ratio of $(1/k) + \epsilon$ $(\epsilon > 0)$. Consider three agents, $\{1, 2, 3\}$, and the following k value profiles defined by parameter s $(1 \le s \le k)$. We call each profile *Profile s*.

- Agent 1's value is the same for all profiles. She has a marginal value of $\alpha + \gamma$ on the first item and zero on additional items with arbitrary constant $\alpha > 0$, where γ satisfies $\frac{\alpha+\gamma}{k\alpha+\gamma} < \frac{1}{k} + \epsilon$. Here, γ is chosen so that agent 1 wins one item in all of these profiles.
- Agent 2 has a marginal value $w_{2,j}^s$ for the j-th item s.t., $w_{2,j}^s = \alpha + 2^{(s(s+1)/2)-j}\delta$, where constant δ (> 0) is chosen s.t., $\gamma > 2^{(k(k+1)/2)-1}\delta$ holds. In other words, for each profile, the additional term $2^{(s(s+1)/2)-j}\delta$ for the j-th item is $1/2$ of the one for the $(j-1)$-th item. By increasing parameter s, these marginal values also increase.
- Agent 3's value is the same for all the profiles; she has a marginal value of α on each item.

Figure 1 corresponds to the case where $k = 3, \gamma = 1$, and $\delta = 2^{-6}$. Assuming that the competitive ratio exceeds $1/k$, in each of these profiles, allocating either no item or just one is not allowed.

First, we show that for Profiles $s \ge 2$, i.e., except for Profile 1, allocating two items is impossible. For example, consider the situation described in Fig. 1. In Profile 3, when two items are allocated, both agents 1 and 2 win one item (Fig. 1(a)) from Corollary 1. Then agent 1 pays at least $\alpha + 2^{-2}$ to avoid envy from agent 2 toward a set of agents $\{1, 2\}$. Also from ItI-EFness, agents 1 and 2 pay the same amount. Thus, the utility of agent 2 is at most $2^{-1} - 2^{-2}$. On the other hand, in Profile 1, agent 2 wins at least one item. When she wins one item in Profile 1 (Fig. 1(e)), she pays at most $\alpha + 2^{-6}$ from IR. Thus, in Profile 3, if agent 2 underbids her values and creates a situation identical to Profile 1, her utility increases from $2^{-1} - 2^{-2}$ to $2^{-1} - 2^{-6}$. This contradicts the condition of strategy-proofness. Also, when agent 2 wins two items in Profile 1 (Fig. 1(f)), she pays at most $2\alpha + 2^{-6} + 2^{-7}$ from IR. Thus, in Profile 3, if agent 2 underbids her values and creates a situation identical to Profile 1, her utility increases from $2^{-1} - 2^{-2}$ to $2^{-1} + 2^{-2} - 2^{-6} - 2^{-7}$. This contradicts the condition of strategy-proofness. Thus, in Profile 3, allocating two items is impossible. From a similar argument, allocating two items is impossible for Profiles $s \ge 2$. In Fig. 1, (a) and (d) are impossible.

Next, we show that for Profiles $s \ge 3$, i.e., except for Profile 1 and 2, allocating exactly three items is not allowed. Again, consider the situation described in Fig. 1. In Profile 3, when three items are allocated, agent 1 wins one item and agent 2 wins two items (Fig. 1(b)) from Corollary 1. Here, agent 1 must pay at least $\alpha + 2^{-3}$ to avoid envy from agent 2 toward a group $\{1, 2\}$. Then, agent 2 must pay at least $2\alpha + 2^{-3}$ to avoid envy from a set of agents $\{1, 3\}$ toward agent 2. Thus, the utility of agent 2 is at most $2^{-1} + 2^{-2} - 2^{-3}$. On the other hand, in Profile 2, agent 2 must win two items (i.e., the situation must be (c) in Fig. 1), since we already proved that (d) is impossible. Here, agent 2 pays at most $2\alpha + 2^{-4} + 2^{-5}$ from IR. Thus, in Profile 3, if agent 2 underbids her values and creates a situation identical to Profile 2, her utility increases from $2^{-1} + 2^{-2} - 2^{-3}$ to $2^{-1} + 2^{-2} - 2^{-4} - 2^{-5}$. This contradicts the condition of strategy-proofness. Thus, in Fig. 1, (b) is impossible. From a similar argument, allocating exactly three items is impossible for Profiles $s \geq 3$.

Using a similar argument, we can prove that for each $s \geq s'$, allocating exactly s' items is impossible. Then, for Profile k, no allocation is possible. For example, in Fig. 1, no allocation is possible for Profile 3. This contradicts the assumption that the competitive ratio exceeds $1/k$.   □

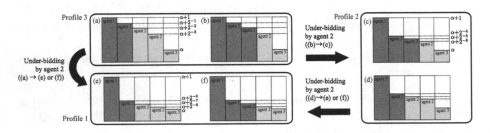

Fig. 1. Proof of Theorem 2 for $k = 3$, $\gamma = 1$, and $\delta = 2^{-6}$.

4.2 All-or-Nothing Values

In ANV, the value of an agent may have complementarity across items. For an auction mechanism, complementarity often causes difficulties. This is also true for a GtG-EF mechanism; even without any incentive issues, the competitive ratio of any GtG-EF mechanism is at most $1/k$.

Theorem 3. *In ANV, any GtG-EF mechanism has a competitive ratio of at most $1/k$.*

Proof. Consider a GtG-EF mechanism that is going to sell k items to $2k - 2$ agents. Based on a numerical sequence $(c_i)_{i=2,\dots,k}$ defined by a recurrence relation $c_i = 2c_{i-1} + 1$ and $c_2 = 1$, the agents' values are defined as follows:

- Agent 1 values $\alpha + c_k + 1$ on one item with a sufficiently large constant α.
- Agent $i \in \{2, \dots, k\}$ values $i\alpha + c_i$ on i items.
- Agent $j \in \{k+1, \dots, 2k-2\}$ values α on one item.

Figure 2 corresponds to $k = 4$. Note that the optimal social welfare is $k\alpha + c_k + c_{k-1} + 1$, which occurs when agents 1 and $k - 1$ win.

From Corollary 1, an assignment must be an LEBA-SM. Thus, only the following k cases are possible. Case 1: only agent 1 wins one item. Case 2: agent 1 wins one item and an agent in $\{k+1, \dots, 2k-2\}$ (wlog., agent $k+1$) wins one item. Case s $(3 \leq s \leq k)$: agent 1 wins one item and agent $s-1$ wins $s-1$ items. We then show that Cases $2, \ldots, k$ do not occur in the GtG-EF mechanism.

In Case 2, both agents 1 and $k+1$ must pay the same amount from ItI-EFness. Then, the sum of the transfers is at most 2α from IR. Thus, agent 2 with value $2\alpha + 1$ on two items envies the set of winners, which violates GtG-EFness.

In Case s $(3 \leq s \leq k)$, agent $s - 1$ pays at most $(s - 1)\alpha + c_{s-1}$ from IR. Furthermore, to avoid envy from agent s toward the winners (i.e., 1 and $s - 1$), the sum of the transfers is at least $s\alpha + c_s$. Therefore, agent 1 must pay at least $\alpha + c_s - c_{s-1}$ and then her utility is at most $c_k - c_s + c_{s-1} + 1 = c_k - c_{s-1}$. On the other hand, the total value on the $s - 1$ items by a set of agents $\{1, k + 1, \ldots, k + s - 2\}$ is $(s - 1)\alpha + c_k + 1$, and the transfer of agent $s - 1$ is at most $(s - 1)\alpha + c_{s-1}$. Thus, if this set of agents obtains $s - 1$ items that are currently assigned to agent $s - 1$, their total utility becomes at least $c_k - c_{s-1} + 1$, which is strictly larger than their current utility $c_k - c_{s-1}$ (here, the utilities of agents $k + 1, \ldots, k + s - 2$ are 0). This implies that the set of agents $\{1, k + 1, \ldots, k + s - 2\}$ envies agent $s - 1$.

As a result, for the above value profile, Case 1 occurs; only agent 1 wins. The ratio $(\alpha + c_k)/(k\alpha + c_k + c_{k-1} + 1)$ converges to $1/k$ for sufficiently large α. $\square$

In contrast to the result on DMV, this result seems quite negative because an optimal GtG-EF mechanism achieves only $1/k$ of the Pareto efficient welfare in the worst case in ANV, even without any incentive issues. This also implies that LEBA-SM does not characterize the allocation rules of GtG-EF mechanisms, since a Pareto efficient allocation satisfies LEBA-SM. A characterization of GtG-EF mechanisms in ANV remains open.

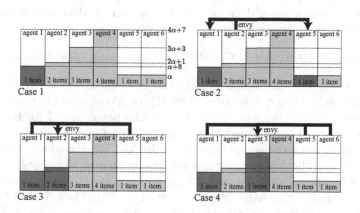

Fig. 2. Proof of Theorem 3 for $k = 4$.

5 General Combinatorial Auctions

In the previous section, we identified the upper bounds of $1/k$ for DMV (with strategy-proofness/false-name-proofness) and ANV, but we have not yet proved that they are tight. Nor have we obtained an upper bound for the general multi-unit auction domain. In this section, by considering a more general domain, i.e., the GCA domain, we answer the above questions.

We show that the following *average-max (AM)* mechanism is GtG-EF, strategy-proof and false-name-proof, and has a competitive ratio of $1/k$. Since GCA includes ANV, the ratio $1/k$ remains valid as an upper bound. Thus, AM gives a tight lower bound for GCA. Since AM can be applied for the multi-unit auction domain, it also gives a tight lower bounds for DMV (with strategy-proofness/false-name-proofness) and ANV. Furthermore, the general multi-unit auction domain includes ANV, and GCA includes the general multi-unit auction domain. Since the tight bounds of ANV and GCA are identical, they are also applicable to the general multi-unit auction domain.

Definition 6 (Average-Max). *For each agent $i \in N$ and each bundle $B_i \subseteq K$, compute average value $av_i(B_i) = v_i(B_i)/|B_i|$ for every agent i and every bundle B_i. An agent i with the highest average value is the winner, and obtains a bundle x_i that solves the following maximization problem:*

$$x_i = \arg\max_{B_i \subseteq K} \{v_i(B_i) - |B_i| \cdot \bar{av}_{-i}\}, \tag{1}$$

where $\bar{av}_{-i} = \max_{j \neq i, B_j \subseteq K} av_j(B_j)$. She pays $|x_i| \cdot \bar{av}_{-i}$.

First, note that AM is strategy-proof and false-name-proof in GCA. AM is obviously strategy-proof, since its allocation rule is monotone and the winner's transfer does not depend on her value. It is also straightforward to show that AM is false-name-proof. In AM, there is always only one winner. Furthermore, the winner's transfer never decreases as the number of identifiers grows. Thus, agents have no incentive to use multiple identifiers.

Theorem 4. *In GCA, AM is GtG-EF and has a competitive ratio of $1/k$.*

Proof. First, we show that AM is GtG-EF. Assume group C envies another group S. The only winner, denoted as $S := \{l\}$, in AM obtains her most desirable bundle under the unit price $\bar{av}_{-i}$, i.e., $l \notin C$. The payment of l for bundle x_l is at least as large as $V^*(C, x_l)$, which guarantees that C do not envy l.

Next, we examine the competitive ratio. In AM, there is always exactly one winner, who has the highest average valuations among all agents. When she obtains the bundle on which her average value is maximum, the ratio is obviously at least $1/k$. On the other hand, if the winner wins another bundle, say B, it must contain at least two items to solve the maximization problem in Eq. 1. Thus, the ratio is never smaller than $1/k$. Furthermore, there exists a value profile in GCA where the ratio of AM becomes essentially $1/k$. The following is one example of such a value profile: agent 1 has a value of $1 + \epsilon$ for item g_1, and agent 2 has a value of k for a bundle $\{g_1, \ldots, g_k\}$. AM allocates g_1 to agent 1, and the ratio is $(1 + \epsilon)/k$.   □

Since AM is also false-name-proof, the following holds.

Theorem 5. *In GCA, the tight bound of the competitive ratio achieved by GtG-EF and false-name-proof mechanisms is $1/k$.*

AM is a modified version of AM-MB, which allows us to allocate items to more than one agent, while AM always chooses only one winner. As we stated in Section 2, AM-MB is also GtG-EF and strategy-proof and has a competitive ratio of $1/k$. However, it is known that AM-MB is not false-name-proof. To the best of our knowledge, AM is the first mechanism that is both GtG-EF and false-name-proof with a non-zero competitive ratio.

6   Relation with Walrasian Equilibrium

In this section, we clarify the relationship between GtG-EFness and a notion on market clearing called *Walrasian equilibrium*, which has been widely discussed in economics literature. A Walrasian equilibrium is a pair of an assignment of items and a vector of item prices that maximizes all agents' utilities. In this sense, the idea of Walrasian equilibrium closely resembles GtG-EFness.

Definition 7. *A pair of an assignment and a price vector (x, p) is a Walrasian equilibrium for $v \in V^n$ if $\forall g \in K \setminus \bigcup_{i \in N} x_i, p_g = 0$ and $\forall i \in N, \forall B_i \subseteq K, v_i(x_i) - \sum_{g \in x_i} p_g \geq v_i(B_i) - \sum_{g' \in B_i} p_{g'}$.*

Theorem 6. *In GCA, no group of agents envies any other group of agents in any Walrasian equilibrium.*

Proof. We assume for a contradiction that a group C envies another group S in a Walrasian equilibrium (x, p). It does not matter in the argument below whether C and S intersect. Let x_i^* be the optimal assignment of items $\bigcup_{j \in S} x_j$ to the set of agents C, i.e., $V^*(C, \bigcup_{j \in S} x_j) = \sum_{i \in C} v_i(x_i^*)$. Note that a strict subset of $\bigcup_{j \in S} x_j$ may maximize the total welfare of the set of agents C, i.e., $\bigcup_{i \in C} x_i^* \subseteq \bigcup_{j \in S} x_j$. Thus, $\sum_{i \in C} \sum_{g \in x_i^*} p_g \leq \sum_{j \in S} \sum_{g \in x_j} p_g$ holds. Since (x, p) is a Walrasian equilibrium, $\forall i \in C, v_i(x_i) - t_i \geq v_i(x_i^*) - \sum_{g \in x_i^*} p_g$, where $t_i = \sum_{g \in x_i} p_g$. Summing this inequality over all the agents in C, we have

$$
\begin{aligned}
\sum_{i \in C} \left[v_i(x_i) - t_i \right] &\geq \sum_{i \in C} \left[v_i(x_i^*) - \sum_{g \in x_i^*} p_g \right] \\
&\geq V^*(C, \bigcup_{j \in S} x_j) - \sum_{j \in S} \sum_{g \in x_j} p_g \\
&= V^*(C, \bigcup_{j \in S} x_j) - \sum_{j \in S} t_j,
\end{aligned}
$$

which contradicts the assumption that C envies S.   □

There is a value profile s.t., no Walrasian equilibrium exists (e.g., see Gul and Stacchetti [7]). However, when the market size is sufficiently small (precisely, $|K| = 1$), we can easily see that a Walrasian equilibrium always exists.

From this observation, by withholding some item, we can guarantee the existence of a Walrasian equilibrium *on the set of allocated items and their prices.* Let $\mathcal{M}_{K'}$ be a modified environment in which only the set of items $K' \subseteq K$

is going to be sold and the remaining items $K \setminus K'$ are never sold. Let $v^{K'}$ be the value profile for environment $\mathcal{M}_{K'}$ derived from the original value profile v: $\forall i \in N, \forall B_i \subseteq K', v_i^{K'}(B_i) = v_i(B_i)$. We then construct a class of mechanisms that satisfies GtG-EFness in GCA.

Definition 8. *Given v, choose a set of items $K' \subseteq K$ s.t. there is a Walrasian equilibrium in the modified environment $\mathcal{M}_{K'}$. Set the price of each item g as follows; the equilibrium price of g in $\mathcal{M}_{K'}$ if $g \in K'$, and ∞ otherwise. For each agent, allocate the bundle that maximizes her utility under the price vector. An agent's transfer equals the sum of the prices of the items that she obtains.*

As a corollary of Theorem 6, the following statement holds.

Corollary 2. *In GCA, any mechanism described in Definition 8 is GtG-EF.*

Interestingly, in GCA, AM and AM-MB can be described in Definition 8. For example, AM can be described as follows. Given value profile v, choose $l = \arg\max_{i \in N} a v_i(B_i)$ and $K' = x_i$, where x_i is the solution of Eq. 1. Then allocate K' to agent l and set p s.t., $p_g = \bar{a}v_{-l}$ if $g \in K'$.

7 Conclusions

One interesting future direction of GtG-EF mechanisms is to completely characterize them in several domains, including GCA. Also, to achieve better competitive ratios, considering randomized mechanisms is a natural direction. Furthermore, we have not obtained the tight bound of the competitive ratio of GtG-EF and strategy-proof (or false-name-proof) mechanisms in SMCA. There remains a large gap between the lower and upper bounds; the AM gives the lower bound $1/k$, while the upper bound obtained in Todo et al. [14] is 2/3. Considering envy-freeness in dynamic auctions [9,15] would also be an interesting extension.

Acknowledgments. This work is partially supported by JSPS KAKENHI Grants JP17H00761 and JP17H04695. The authors thank Takayuki Mouri for his helpful comments and discussions. All errors are our own.

References

1. Bade, S., Segal-Halevi, E.: Fair and efficient division among families. arXiv preprint arXiv:1811.06684 (2018)
2. Cohen, E., Feldman, M., Fiat, A., Kaplan, H., Olonetsky, S.: Truth, envy, and truthful market clearing bundle pricing. In: Chen, N., Elkind, E., Koutsoupias, E. (eds.) WINE 2011. LNCS, vol. 7090, pp. 97–108. Springer, Heidelberg (2011). https://doi.org/10.1007/978-3-642-25510-6_9
3. Conen, W., Sandholm, T.: Anonymous pricing of efficient allocations in combinatorial economies. In: Proceedings of the AAMAS 2004, pp. 254–260 (2004)
4. Conitzer, V., Freeman, R., Shah, N., Vaughan, J.W.: Group fairness for indivisible good allocation. In: Proceedings of the AAAI 2019 (2019, to appear)

5. Foley, D.: Resource allocation and the public sector. Yale Econ. Essays **7**, 45–98 (1967)
6. Goldberg, A.V., Hartline, J.D.: Envy-free auctions for digital goods. In: Proceedings of the EC 2003, pp. 29–35 (2003)
7. Gul, F., Stacchetti, E.: Walrasian equilibrium with gross substitutes. J. Econ. Theory **87**(1), 95–124 (1999)
8. Haake, C.J., Raith, M.G., Su, F.E.: Bidding for envy-freeness: a procedural approach to n-player fair-division problems. Soc. Choice Welf. **19**, 723–749 (2002)
9. Hajiaghayi, M.T., Kleinberg, R.D., Parkes, D.C.: Adaptive limited-supply online auctions. In: Proceedings of the EC 2004, pp. 71–80 (2004)
10. Ito, T., Yokoo, M., Iwasaki, A., Matsubara, S.: A new strategy-proof greedy-allocation combinatorial auction protocol and its extension to open ascending auction protocol. In: Proceedings of the AAAI 2005, pp. 261–268 (2005)
11. Manurangsi, P., Suksompong, W.: Asymptotic existence of fair divisions for groups. Math. Soc. Sci. **89**, 100–108 (2017)
12. Mu'alem, A.: On multi-dimensional envy-free mechanisms. In: Rossi, F., Tsoukias, A. (eds.) ADT 2009. LNCS (LNAI), vol. 5783, pp. 120–131. Springer, Heidelberg (2009). https://doi.org/10.1007/978-3-642-04428-1_11
13. Suksompong, W.: Approximate maximin shares for groups of agents. Math. Soc. Sci. **92**, 40–47 (2018)
14. Todo, T., Li, R., Hu, X., Mouri, T., Iwasaki, A., Yokoo, M.: Generalizing envy-freeness toward group of agents. In: Proceedings of the IJCAI 2011, pp. 386–392 (2011)
15. Todo, T., Mouri, T., Iwasaki, A., Yokoo, M.: False-name-proofness in online mechanisms. In: Proceedings of the AAMAS 2012, pp. 753–762 (2012)
16. Tsuruta, S., Oka, M., Todo, T., Sakurai, Y., Yokoo, M.: Fairness and false-name manipulations in randomized cake cutting. In: Proceedings of the AAMAS 2015, pp. 909–917 (2015)
17. Varian, H.R.: Equity, envy, and efficiency. J. Econ. Theory **9**, 63–91 (1974)
18. Vind, K.: Special topics in mathematical economics. Stanford Lecture Notes (1971)
19. Yokoo, M., Sakurai, Y., Matsubara, S.: The effect of false-name bids in combinatorial auctions: new fraud in internet auctions. Games Econ. Behav. **46**(1), 174–188 (2004)

Upper and Lower Bounds on Approximating Weighted Mixed Domination

Mingyu Xiao$^{(\boxtimes)}$

University of Electronic Science and Technology of China, Chengdu, China
myxiao@gmail.com

Abstract. A mixed dominating set of a graph $G = (V, E)$ is a mixed set D of vertices and edges, such that for every edge or vertex, if it is not in D, then it is adjacent or incident to at least one vertex or edge in D. The mixed domination problem is to find a mixed dominating set with a minimum cardinality. It has applications in system control and some other scenarios and it is NP-hard to compute an optimal solution. This paper studies approximation algorithms and hardness of the weighted mixed dominating set problem. The weighted version is a generalization of the unweighted version, where all vertices are assigned the same nonnegative weight w_v and all edges are assigned the same nonnegative weight w_e, and the question is to find a mixed dominating set with a minimum total weight. Although the mixed dominating set problem has a simple 2-approximation algorithm, few approximation results for the weighted version are known. The main contributions of this paper include:

1. for $w_e \geq w_v$, a 2-approximation algorithm;
2. for $w_e \geq 2w_v$, inapproximability within ratio 1.3606 unless $P = NP$ and within ratio 2 under UGC;
3. for $2w_v > w_e \geq w_v$, inapproximability within ratio 1.1803 unless $P = NP$ and within ratio 1.5 under UGC;
4. for $w_e < w_v$, inapproximability within ratio $(1 - \epsilon) \ln |V|$ unless $P = NP$ for any $\epsilon > 0$.

Keywords: Approximation algorithms · Inapproximability · Domination

1 Introduction

Domination is an important concept in graph theory. In a graph, a vertex *dominates* itself and all neighbors of it, and an edge *dominates* itself and all edges sharing an endpoint with it. The VERTEX DOMINATING SET problem [10] (resp., EDGE DOMINATING SET problem [22]) is to find a minimum set of vertices to dominate all vertices (resp., a minimum set of edges to dominate all edges) in a graph. These two domination problems have many applications in different fields. For example, in a network, structures like dominating sets play an important role in global flooding to alleviate the so-called broadcast storm problem.

© Springer Nature Switzerland AG 2019
D.-Z. Du et al. (Eds.): COCOON 2019, LNCS 11653, pp. 554–566, 2019.
https://doi.org/10.1007/978-3-030-26176-4_46

A message broadcast only in the dominating set is an efficient way to ensure that it is received by all transmitters in the network, both in terms of energy and interference [17]. More applications and introduction to domination problems can be found in the literature [9].

Domination problems are rich problems in the field of algorithms. Both VERTEX DOMINATING SET and EDGE DOMINATING SET are NP-hard [7,22]. There are several interesting algorithmic results about the polynomial solvability on special graph [14,23], approximation algorithms [5,6,12], parameterized algorithms [20,21] and so on.

In this paper, we consider a related domination problem, called the MIXED DOMINATION problem. Mixed domination is a mixture concept of vertex domination and edge domination, and MIXED DOMINATION requires to find a set of edges and vertices with the minimum cardinality to dominate other edges and vertices in a graph. MIXED DOMINATION was first proposed by Alavi et al. based on some specific application scenarios and it was named as the TOTAL COVERING problem initially [2]. Although we prefer to call this problem a "domination problem" at present, it has some properties of "covering problems" and can also be treated as a kind of covering problems. For applications of MIXED DOMINATION, a direct application in system control was introduced by Zhao et al. [23]. They used it to minimize the number of phase measurement units (PMUs) needed to be placed and maintain the ability of monitoring the entire system. We can see that MIXED DOMINATION has drawn certain attention since its introduction [3,11,14,15,23].

MIXED DOMINATION is NP-hard even on bipartite and chordal graphs and planar bipartite graphs of maximum degree 4 [15]. Most of known algorithmic results of MIXED DOMINATION are about the polynomial-time solvable cases on special graphs. Zhao et al. [23] showed that this problem in trees can be solved in polynomial time. Lan et al. [14] provided a linear-time algorithm for MIXED DOMINATION in cacti, and introduced a labeling algorithm based on the primal-dual approach for MIXED DOMINATION in trees. Recently, MIXED DOMINATION was studied from the parameterized perspective [11]. Several parameterized complexity results under different parameters have been proved.

In terms of approximation algorithms, domination problems have also been extensively studied. It is easy to observe that a maximum matching in a graph is a 2-approximation solution to EDGE DOMINATING SET. But for VERTEX DOMINATING SET, the best known approximation ratio is $\log |V| + 1$ [12]. As a combination of EDGE DOMINATING SET and VERTEX DOMINATING SET, MIXED DOMINATION has a simple 2-approximation algorithm [8].

We will study approximation algorithms for weighted mixed domination problems. A mixed dominating set contains both edges and vertices. MIXED DOMINATION does not distinguish them in the solution set, and only considers the cardinality. However, edge and vertex are two different elements and they may have different contributions or prices in practice. In the application example in [23], we select vertices and edges to place phase measurement units (PMUs) on them to monitor their mixed neighbors' state variables in an electric power system. The price to place PMUs on edges and vertices may be different due to the different physical structures. It is reasonable to distinguish edge and vertex by setting different weights to them. So we introduce the following weighted version problem.

WEIGHTED MIXED DOMINATION (WMD)
Instance: A single undirected graph $G = (V, E)$, and two nonnegative values w_v and w_e.
Question: To find a vertex subset $V_D \subseteq V$ and an edge subset $E_D \subseteq E$ such that
(i) any vertex in $V \setminus V_D$ is either an endpoint of an edge in E_D or adjacent to a vertex in V_D;
(ii) any edge in $E \setminus E_D$ has at least one endpoint that is either an endpoint of an edge in E_D or a vertex in V_D;
(iii) the value $w_v|V_D| + w_e|E_D|$ is minimized under the above constraints.

In WEIGHTED MIXED DOMINATION, all vertices (resp., edges) receive the same weight. Although the weight function may not be very general, the hardness of the problem increases dramatically, especially in approximation algorithms. It is easy to see that the 2-approximation algorithm for the unweighted version in [8] cannot be extended to the weighted version. In fact, for most domination problems, the weight version may become much harder. For example, it is trivial to obtain a 2-approximation algorithm for EDGE DOMINATING SET. But for the weighted version of EDGE DOMINATING SET, it took years to achieve the same approximation ratio [6]. In order to obtain more tractability results for WEIGHTED MIXED DOMINATION, we consider two cases: VERTEX-FAVORABLE MIXED DOMINATION (VFMD) and EDGE-FAVORABLE MIXED DOMINATION (EFMD). If we add one more requirement $w_v \leq w_e$ in WEIGHTED MIXED DOMINATION, then it becomes VERTEX-FAVORABLE MIXED DOMINATION. EDGE-FAVORABLE MIXED DOMINATION is defined in a similar way by adding a requirement $w_e \leq w_v$. In fact, we will further distinguish two cases of VERTEX-FAVORABLE MIXED DOMINATION to study its complexity. We summarize our main algorithmic and complexity results for WEIGHTED MIXED DOMINATION in Table 1, where ε is any value > 0.

This paper is organized as follows. Sections 2 and 3 introduce some basic notations and properties. Section 4 deals with VERTEX-FAVORABLE MIXED DOMINATION. The results for the case that $2w_v \leq w_e$ are obtained by proving its equivalence to the VERTEX COVER problem. The case that $w_v \leq w_e < 2w_v$ is harder. Our 2-approximation algorithm is based on a linear programming for VERTEX COVER. The lower bounds are obtained by a nontrivial reduction from VERTEX COVER. The lower bounds for EDGE-FAVORABLE MIXED DOMINATION and the proofs of some lemmas are omitted due to the space limitation, which can be found in the full version of this paper.

Table 1. Upper and lower bounds on approximating WMD

Problems		Approximation ratio		
		Upper bounds	Lower bounds	
VFMD	$2w_v \leq w_e$	2 (Theorems 2 and 5)	$10\sqrt{5} - 21 - \varepsilon$	if $P \neq NP$ (Theorem 3)
			$2 - \varepsilon$	under UGC (Theorem 3)
	$w_v \leq w_e < 2w_v$		$5\sqrt{5} - 10 - \varepsilon$	if $P \neq NP$ (Theorem 6)
			$1.5 - \varepsilon$	under UGC (Theorem 6)
EFMD	$w_v > w_e$	–	$(1 - \varepsilon)\ln n$	if $P \neq NP$ [19]

2 Preliminaries

In this paper, a graph $G = (V, E)$ stands for an undirected simple graph with a vertex set V and an edge set E. We use $n = |V|$ and $m = |E|$ to denote the sizes of the vertex set and edge set, respectively. Let X be a subset of V. We use $G - X$ to denote the graph obtained from G by removing vertices in X together with all edges incident to vertices in X. Let $G[X]$ denote the graph induced by X, i.e., $G[X] = G - (V \setminus X)$. For a subgraph or an edge set G', we use $V(G')$ to denote the set of vertices in G'.

In a graph, a vertex *dominates* itself, all of its neighbors and all edges taking it as one endpoint; an edge *dominates* itself, the two endpoints of it and all other edges having a common endpoint. A mixed set of vertices and edges $D \subseteq V \cup E$ is called a *mixed dominating set*, if any vertex and edge are dominated by at least one element in D. For a mixed set D of vertices and edges, a vertex (resp., edge) in D is called a *vertex element* (resp., *edge element*) of D, and the set of vertex elements (resp., edge elements) may be denoted by V_D (resp., E_D). Thus $V_D = V(G) \cap D$. The set of vertices that appear in any form in D is denoted by $V(D)$, i.e., $V(D) = \{v \in V(G) | v \in D$ or v is adjacent to an edge in $D\}$. It holds that $V_D \subseteq V(D)$. MIXED DOMINATION is to find a mixed dominating set of the minimum cardinality, and WEIGHTED MIXED DOMINATION is to find a mixed dominating set D such that $w_v|V_D| + w_e|E_D|$ is minimized. A *weighted instance* is a graph with each vertex assigned the same nonnegative weight w_v and each edge assigned the same nonnegative weight w_e. In a weighted instance, for a mixed set D of vertices and edges (it may only contain vertices or edges), we define $w(D) = w_v|D \cap V| + w_e|D \cap E|$.

A vertex set in a graph is called a *vertex cover* if any edge has at least one endpoint in this set and a vertex set is called an *independent set* if any pair of vertices in it are not adjacent in the graph. The VERTEX COVER problem is to find a vertex cover of the minimum cardinality. We may use S_{md}, S_{wmd} and

S_{vc} to denote an optimal solution to MIXED DOMINATION, WEIGHTED MIXED DOMINATION and VERTEX COVER, respectively.

3 Properties

We introduce some basic properties of MIXED DOMINATION and WEIGHTED MIXED DOMINATION in this section.

Lemma 1. *Any mixed dominating set of a graph contains all isolating vertices (i.e. the vertices of degree 0) as vertex elements.*

This lemma follows from the definition of mixed dominating sets directly. Based on this lemma, we can simply include all isolating vertices in the graph to the solution set and assume the graph has no isolating vertices. We have said that MIXED DOMINATION is also related to covering problems. Next, we reveal some relations between MIXED DOMINATION and VERTEX COVER. By the definitions of vertex covers and mixed dominated sets, we get

Lemma 2. *In a graph without isolating vertices, any vertex cover is a mixed dominating set.*

Recall that for a mixed dominating set D, we use $V(D)$ to denote the set of vertices appearing in D. On the other hand, we have that

Lemma 3. *For any mixed dominating set D, the vertex set $V(D)$ is a vertex cover.*

Recall that S_{wmd} and S_{vc} denote an optimal solution to WEIGHTED MIXED DOMINATION and VERTEX COVER respectively. It is easy to get the following results from above lemmas.

Corollary 1. *For any mixed dominating set D, it holds that*

$$2|D| \geq |V_D| + 2|E_D| \geq |S_{vc}|.$$

Lemma 4. *Let G be an instance of VERTEX-FAVORABLE MIXED DOMINATION having no isolating vertices. For any mixed dominating set D and vertex cover C in G, it holds that*

$$w(S_{wmd}) \leq w(C) \text{ and } w(S_{vc}) \leq 2w(D).$$

Corollary 2. *Let G be an instance of VERTEX-FAVORABLE MIXED DOMINATION having no isolating vertices. It holds that*

$$w(S_{wmd}) \leq w(S_{vc}) \leq 2w(S_{wmd}).$$

Lemma 4 and Corollary 2 imply the following result.

Theorem 1. *For any $\alpha \geq 1$, given an α-approximation solution to VERTEX COVER, a 2α-approximation solution to VERTEX-FAVORABLE MIXED DOMINATION on the same graph can be constructed in linear time.*

VERTEX COVER allows 2-approximation algorithms and then we have that

Corollary 3. VERTEX-FAVORABLE MIXED DOMINATION *allows polynomial-time 4-approximation algorithms.*

4   Vertex-Favorable Mixed Domination

We have obtained a simple 4-approximation algorithm for VERTEX-FAVORABLE MIXED DOMINATION. In this section, we improve the ratio to 2 and also show some lower bounds. We will distinguish two cases to study it: $2w_v \leq w_e$; $w_v \leq w_e < 2w_v$.

4.1   The Case that $2w_v \leq w_e$

This is the easier case. In fact, we will reduce this case to VERTEX COVER and also reduce VERTEX COVER to it, keeping the approximation ratio. Thus, for this case we will get the same approximation upper and lower bounds as that of VERTEX COVER.

Lemma 5. *Let G be a graph having no isolating vertices. Any minimum vertex cover S_{vc} in G is also an optimal solution to WEIGHTED MIXED DOMINATION with $w_e \geq 2w_v$ in G.*

Lemma 6. *For a weighted instance G having no isolating vertices, if it holds that $w_e \geq 2w_v$, then any α-approximation solution to VERTEX COVER is also an α-approximation solution to WEIGHTED MIXED DOMINATION in G.*

The best known approximation ratio for VERTEX COVER is 2. Theorem 6 implies that

Theorem 2. WEIGHTED MIXED DOMINATION *with* $2w_v \leq w_e$ *allows polynomial-time 2-approximation algorithms.*

For lower bounds, we show a reduction from another direction.

Lemma 7. *Let G be an instance having no isolating vertices, where $w_e \geq 2w_v$. For any α-approximation solution D to WEIGHTED MIXED DOMINATION in G, the vertex set $V(D)$ is an α-approximation solution to VERTEX COVER in G.*

Proof. Let S_{wmd} and S_{vc} be an optimal solution to WEIGHTED MIXED DOMINATION and VERTEX COVER, respectively. By Lemma 5, we have that $w(S_{wmd}) = w(S_{vc})$. Then $w(D) \leq \alpha w(S_{wmd}) = \alpha w(S_{vc}) = \alpha w_v |S_{vc}|$. Note that $w(D) = w_v |D_v| + w_e |D_e| \geq w_v |D_v| + 2w_v |D_e|$ and $|V(D)| \leq |D_v| + 2|D_e|$. Thus, $|V(D)| \leq \alpha |S_{vc}|$. Furthermore, $V(D)$ is a vertex cover by Lemma 3. We know that $V(D)$ is an α-approximation solution to VERTEX COVER.   □

Dinur and Safra [4] proved that it is NP-hard to approximate VERTEX COVER within any factor smaller than $10\sqrt{5} - 21$. Khot and Regev [13] also prove that VERTEX COVER cannot be approximated to within $2 - \varepsilon$ for any $\varepsilon > 0$ under UGC. Those results and Lemma 7 imply

Theorem 3. *For any $\varepsilon > 0$, WEIGHTED MIXED DOMINATION with $2w_v \leq w_e$ is not $(10\sqrt{5} - 21 - \varepsilon)$-approximable in polynomial time unless $P = NP$, and not $(2 - \varepsilon)$-approximable in polynomial time under UGC.*

4.2 The Case that $w_v \leq w_e < 2w_v$

To simplify the arguments, in this section, we always assume the initial graph
has no degree-0 vertices. Note that we can include all degree-0 vertices to the
solution set directly according to Lemma 1, which will not affect our upper and
lower bounds.

Upper Bounds. We show that this case also allows polynomial-time 2-
approximation algorithms. Our algorithm is based on a linear programming
model for VERTEX COVER. Note that we are not going to build a linear programming for our problem WEIGHTED MIXED DOMINATION directly. Instead,
we use a linear programming for VERTEX COVER.

Linear programming is a powerful tool to design approximation algorithms for VERTEX COVER and many other problems. Lemma 4 and Theorem 1
reveal some connections between WEIGHTED MIXED DOMINATION and VER-
TEX COVER. Inspired by these, we investigate approximation algorithms for
WEIGHTED MIXED DOMINATION starting from a linear programming model for
VERTEX COVER. For a graph $G = (V, E)$, we assign a variable $x_v \in \{0, 1\}$ for
each vertex $v \in V$ to denote whether it is in the solution set. We can use the
following integer programming model (IPVC) to solve VERTEX COVER:

$$\min \sum_{v \in V} x_v$$
$$\text{s.t.}\quad x_u + x_v \geq 1, \forall uv \in E$$
$$x_v \in \{0, 1\}, \forall v \in V.$$

If relax the binary variable x_v to $0 \leq x_v \leq 1$, we get a linear relaxation for
VERTEX COVER, called LPVC. We will use $\mathcal{X}' = \{x_v'|v \in V\}$ to denote a feasible
solution to LPVC and $w(\mathcal{X}')$ to denote the objective value under $\mathcal{X}'$ on the graph
G. LPVC can be solved in polynomial time. However, a feasible solution $\mathcal{X}'$ to
LPVC may not be corresponding to a feasible solution to VERTEX COVER since
the values in $\mathcal{X}'$ may not be integers. A feasible solution $\mathcal{X}'$ to LPVC is *half
integral* if $x_v' \in \{0, \frac{1}{2}, 1\}$ for all $x_v' \in \mathcal{X}'$. Nemhauser and Trotter [16] proved
some important properties for LPVC.

Theorem 4 [16]. *Any basic feasible solution $\mathcal{X}'$ to LPVC is half integral. A
half-integral optimal solution to LPVC can be computed in polynomial time.*

We use $\mathcal{X}^* = \{x_v^*|v \in V\}$ to denote a half-integral optimal solution to LPVC.
We partition the vertex set V into three parts V_1, $V_{\frac{1}{2}}$ and V_0 according to $\mathcal{X}^*$,
which are the sets of vertices with the corresponding value x_v^* being 1, $\frac{1}{2}$ and 0,
respectively. There are several properties for the half-integral optimal solution.

Lemma 8 [16]. *For a half-integral optimal solution, all neighbors of a vertex in
V_0 are in V_1, and there is a matching of size $|V_1|$ between V_0 and V_1.*

Lemma 8 implies that $(V_0, V_1, V_{\frac{1}{2}})$ is a *crown decomposition* (see [1] for the
definition) and a half-integral optimal solution can be used to construct a 2-
approximation solution and a $2k$-vertex kernel for VERTEX COVER.

Lemma 9. *For a half-integral optimal solution $\mathcal{X}$ to LPVC, we use $G_{\frac{1}{2}}$ to denote the subgraph induced by The size of a minimum vertex cover in $G_{\frac{1}{2}}$ is at least $|V_{\frac{1}{2}}| - m$, where m is the size of a maximum matching in $G_{\frac{1}{2}}$.*

We are ready to describe our algorithm now. Our algorithm is based on a half-integral optimal solution $\mathcal{X}^*$ to LPVC. We first include all vertices in V_1 to the solution set as vertex elements, which will dominate all vertices in $V_0 \cup V_1$ and all edges incident on vertices in V_1. Next, we consider the subgraph $G[V_{\frac{1}{2}}]$ induced by $V_{\frac{1}{2}}$. We find a maximum matching M in $G[V_{\frac{1}{2}}]$ and include all edges in M to the solution set as edge elements. Last, for all remaining vertices in $V_{\frac{1}{2}}$ not appearing in M, include them to the solution set as vertex elements. The main steps of the whole algorithm are listed in Algorithm 2.

---

1. Compute a half-integral optimal solution $\mathcal{X}^*$ for the input graph G and let $\{V_1, V_{\frac{1}{2}}, V_0\}$ be the vertex partition corresponding to $\mathcal{X}^*$.
2. Include all vertices in V_1 to the solution set as vertex elements and delete $V_0 \cup V_1$ from the graph (the remaining graph is the induced graph $G[V_{\frac{1}{2}}]$).
3. Find a maximum matching M in $G = G[V_{\frac{1}{2}}]$ and include all edges in M to the solution set as edge elements.
4. Add all remaining vertices in $V_{\frac{1}{2}} \setminus V(M)$ to the solution set as vertex elements.

Algorithm 2. The main steps of the 2-approximation algorithm

We prove the correctness of this algorithm. First, the algorithm can stop in polynomial time, because Step 1 uses polynomial time by Theorem 4 and all other steps can be executed in polynomial time. Second, we prove that the solution set returned by the algorithm is a mixed dominating set.

All vertices in $V_0 \cup V_1$ and all edges incident on vertices in $V_0 \cup V_1$ are dominated by vertices in V_1 because the graph has no degree-0 vertices and $\mathcal{X}^*$ is a feasible solution to LPVC. All vertices and edges in $G[V_{\frac{1}{2}}]$ are dominated because all vertices in $V_{\frac{1}{2}}$ are included to the solution set either as vertex elements or as the endpoints of edge elements. We get the following lemma.

Lemma 10. *Algorithm 2 runs in polynomial time and returns a mixed dominating set.*

Last, we consider the approximation ratio. Lemma 8 implies that the size of a minimum vertex cover in the induced subgraph $G[V_0 \cup V_1]$ is at least $|V_1|$. By Lemma 9, we know that the size of a minimum vertex cover in the induced subgraph $G[V_{\frac{1}{2}}]$ is at least $|V_{\frac{1}{2}}| - m$, where m is the size of a maximum matching in $G_{\frac{1}{2}}$. So the size of a minimum vertex cover of G is at least $|V_1| + |V_{\frac{1}{2}}| - m$, i.e.,

$$|S_{vc}| \geq |V_1| + |V_{\frac{1}{2}}| - m. \tag{1}$$

Let D denote an optimal mixed dominating set in G. By Corollary 1, we have that $|V_D| + 2|E_D| \geq |S_{vc}|$. By this and $2w_v > w_e$, we have that

$$w(D) = |V_D|w_v + |E_D|w_e > \frac{w_e}{2}|V_D| + w_e|E_D| \geq \frac{w_e}{2}|S_{vc}|. \qquad (2)$$

Let D' denote a mixed dominating set returned by Algorithm 2. We have that

$$
\begin{aligned}
w(D') &= |V_1|w_v + mw_e + (|V_{\frac{1}{2}}| - 2m)w_v & \\
&\leq (|V_1| + |V_{\frac{1}{2}}| - m)w_e & \text{by } w_v \leq w_e \\
&\leq |S_{vc}|w_e & \text{by (1)} \\
&\leq 2w(D). & \text{by (2)}
\end{aligned}
$$

Theorem 5. WEIGHTED MIXED DOMINATION *with* $w_v \leq w_e < 2w_v$ *allows polynomial-time 2-approximation algorithms.*

Lower Bounds. In this section, we give lower bounds for WEIGHTED MIXED DOMINATION with $w_v \leq w_e < 2w_v$. These hardness results are also obtained by a reduction preserving approximation from VERTEX COVER. Lemma 1 shows that an α-approximation algorithm for VERTEX COVER implies a 2α-approximation algorithm for VERTEX-FAVORABLE MIXED DOMINATION. For WEIGHTED MIXED DOMINATION with $w_e \geq 2w_v$, we have improved the expansion from 2α to α in Lemma 7. For WEIGHTED MIXED DOMINATION with $w_v \leq w_e < 2w_v$, it becomes harder. We will improve the expansion from 2α to $2\alpha - 1$.

Lemma 11. *For any $\alpha \geq 1$, if there is a polynomial-time α-approximation algorithm for* WEIGHTED MIXED DOMINATION *with* $w_v \leq w_e < 2w_v$, *then there exists a polynomial-time $(2\alpha - 1)$-approximation algorithm for* VERTEX COVER.

Proof. For each instance $G = (V, E)$ of VERTEX COVER, we construct $|V|$ instances $G_i = (V_i, E_i)$ of WEIGHTED MIXED DOMINATION with $w_v \leq w_e < 2w_v$ such that a $(2\alpha - 1)$-approximation solution to G can be found in polynomial time based on an α-approximation solution to each G_i.

For each positive integer $1 \leq i \leq |V|$, the graph $G_i = (V_i, E_i)$ is constructed in the same way. Informally, G_i contains a star T of $2n + 1$ vertices and an auxiliary graph G'_i such that the center vertex c_0 of the star T is connected to all vertices in G'_i, where G'_i contains a copy of G, an induced matching M_i with size $|M_i| = i$, and a complete bipartite graph between the vertices of G and the left part of the induced matching M_i. This is to say, $V_i = V \cup \{a_j\}_{j=1}^i \cup \{b_j\}_{j=1}^i \cup \{c_j\}_{j=0}^{2n}$ and $E_i = E \cup M_i \cup H_i \cup F_i$, where $M_i = \{a_j b_j\}_{j=1}^i$, $H_i = \{va_j | v \in V, j \in \{1, \ldots, i\}\}$, and $F_i = \{c_0 u | u \in V_i \setminus \{c_0\}\}$. We give an illustration of the construction of G_i for $i = 3$ in Fig. 1. In the graphs G_i, the values of w_v and w_e can be any values satisfying $w_v \leq w_e < 2w_v$.

Let τ be the size of a minimum vertex cover of G. We first show that we can get a $(2\alpha - 1)$-approximation solution to G in polynomial time based on an α-approximation solution to G_τ.

We define a function $w^*(G')$ on subgraphs G' of G as follows. For a subgraph G' of G,

$$w^*(G') = \min_{D \in \mathcal{D}} \{ w_v |V(G') \cap V_D| + \frac{1}{2} w_e |V(G') \cap V(E_D)|\}.$$

It is easy to see that

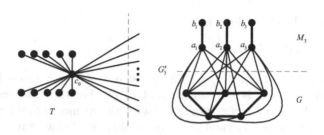

Fig. 1. An illustration of the construction of G_3

Lemma 12. *Let S_{wmd} be an optimal solution to* WEIGHTED MIXED DOMINA-TION *on G. It holds that*

$$w(S_{wmd}) \geq w^*(G),$$

and for any subgraph G' of G and any subgraph G_1 of G', it holds that

$$w^*(G') \geq w^*(G_1) + w^*(G' - V(G_1)).$$

Let D_τ be an optimal solution to G_τ and S_{vc} be a minimum vertex cover of G. By Lemma 12 and the definition of the function $w^*()$, we know that

$$w(D_\tau) \geq w^*(G_\tau) \geq w^*(T) + w^*(G'_\tau).$$

Note that T is a star and then $w^*(T) = w_v$. For G'_τ, we know that the size of a minimum vertex cover of it is at least 2τ because M_τ is an induced matching of size τ that needs at least τ vertices to cover all edges and the size of a minimum vertex cover of G is τ. By Lemma 3 and $w_e < 2w_v$, we know that $w^*(G'_\tau) \geq \tau w_e$. Thus, $w(D_\tau) \geq w_v + \tau w_e$.

On the other hand, $D'_\tau = \{c_0\} \cup M'$ is a mixed dominating set with $w(D'_\tau) = w_v + \tau w_e$, where M' is a perfect matching between S_{vc} and $\{a_j\}_{j=1}^\tau$ with size $|M'| = \tau$. So we have

$$w(D_\tau) = w_v + \tau w_e.$$

Let D^*_τ be an α-approximation solution to G_τ. We consider two cases. Case 1: the vertex c_0 is not a vertex element in D^*_τ. We will show that the whole vertex set V of G is of size at most $(2\alpha - 1)\tau$, which implies that the whole vertex set is a $(2\alpha - 1)$-approximation solution to G. For all the degree-1 vertices $\{c_j\}_{j=1}^{2n}$

in G_τ, Since all the degree-1 vertices $\{c_j\}_{j=1}^{2n}$ in G_τ should be dominated and their only neighbor c_0 is not a vertex element in the mixed dominating set, we know that $\{c_j\}_{j=1}^{2n} \subseteq V(D_\tau^*) \cap V(T)$. For G_τ', an induced subgraph of G_τ, the size of a minimum vertex cover of it is at least 2τ. Let $D_\tau'' \subseteq D_\tau^*$ be the set of vertices and edges in G_τ'. By $w_e < 2w_v$, we know that $w(D_\tau'') \geq \tau w_e$. Thus,

$$w(D_\tau^*) \geq 2nw_v + \tau w_e > (n + \tau)w_e.$$

On the other hand, we have that

$$w(D_\tau^*) \leq \alpha w(D_\tau) = \alpha(w_v + \tau w_e) \leq \alpha(1 + \tau)w_e.$$

Therefore, $(n + \tau)w_e < \alpha(1 + \tau)w_e$. Thus, $n < \alpha + \alpha\tau - \tau \leq (2\alpha - 1)\tau$.

Case 2: the vertex c_0 is a vertex element in D_τ^*. For this case, we show that $U_\tau = V(D_\tau^*) \cap V(G)$ is a vertex cover of G with size at most $(2\alpha-1)\tau+(2\alpha-1)$. Since $w(D_\tau^*) \leq \alpha(w_v + \tau w_e)$ and $w_v \leq w_e < 2w_v$, we know that $|V(D_\tau^*)|$ is at most $\alpha(2 + 2\tau)$. Since M_τ is an induced matching and T is a star, we know that $V(D_\tau^*)$ contains at least τ vertices in M_τ and at least one vertex in T. Therefore,

$$|U_\tau| \leq \alpha(2 + 2\tau) - \tau - 1 = (2\alpha - 1)\tau + 2\alpha - 1.$$

We know that U_τ is a $(2\alpha-1+\epsilon)$-approximation algorithm for G, where $\epsilon = \frac{2\alpha-1}{\tau}$. In fact, we can also get rid of ϵ in the above ratio by using one more trick. We let G' be $2\lceil\alpha\rceil$ copies of G, and construct G_i in the same way by taking G' as G. The size of the minimum vertex cover of G' is $2\lceil\alpha\rceil\tau$ now. For this case, we will get $|U_\tau| \leq (2\alpha - 1)2\lceil\alpha\rceil\tau + 2\alpha - 1$. Due to the similarity of each copy of G in G', we know that for each copy of G the number of vertices in $U_\tau \cap V(G)$ is at most $(2\alpha - 1)\tau + \frac{2\alpha-1}{2\lceil\alpha\rceil}$. The number of vertices is an integer. So we know that $U_\tau \cap V(G)$ is a vertex cover of G with size at most $(2\alpha - 1)\tau$.

However, it is NP-hard to compute the size τ of the minimum vertex cover of G. we cannot construct G_τ in polynomial time directly. Our idea is to compute U_i for each G_i with $i \in \{1, \cdots, |V(G)|\}$ and return the minimum one U_{i^*}. Therefore, U_{i^*} is a vertex cover of G with size $|U_{i^*}| \leq |U_\tau|$. □

VERTEX COVER cannot be approximated within any factor smaller than $10\sqrt{5} - 21$ in polynomial time unless $P = NP$ [4] and cannot be approximated within any factor smaller than 2 in polynomial time under UGC [13]. These results and Lemma 11 imply that

Theorem 6. *For any $\varepsilon > 0$, WEIGHTED MIXED DOMINATION with $w_v \leq w_e < 2w_v$ is not $(5\sqrt{5} - 10 - \varepsilon)$-approximable in polynomial time unless $P = NP$, and not $(\frac{3}{2} - \varepsilon)$-approximable in polynomial time under UGC.*

5   Concluding Remarks

Domination problems are important problems in graph theory and graph algorithms. In this paper, we give several approximation upper and lower bounds on WEIGHTED MIXED DOMINATION, where all vertices have the same weight and all edges have the same weight. For the general weighted version of MIXED DOMINATION such that each vertex and edge may receive a different weight, the hardness results in this paper show that it will be even harder and we may not be easy to get significant upper bounds. For further study, it will be interesting to reduce the gap between the upper and lower bounds in this paper.

Acknowledgements. This work was supported by the National Natural Science Foundation of China, under grants 61772115 and 61370071.

References

1. Abu-Khzam, F.N., Fellows, M.R., Langston, M.A., Suters, W.H.: Crown structures for vertex cover kernelization. Theory Comput. Syst. **41**(3), 411–430 (2007)
2. Alavi, Y., Behzad, M., Lesniak-Foster, L.M., Nordhaus, E.A.: Total matchings and total coverings of graphs. J. Graph Theory **1**(2), 135–140 (1977)
3. Alavi, Y., Liu, J., Wang, J., Zhang, Z.: On total covers of graphs. Discrete Math. **100**(1–3), 229–233 (1992)
4. Dinur, I., Safra, M.: The importance of being biased. In: Proceedings of STOC 2002, pp. 33–42 (2002)
5. Escoffier, B., Monnot, J., Paschos, V.T., Xiao, M.: New results on polynomial inapproximability and fixed parameter approximability of edge dominating set. Theory Comput. Syst. **56**(2), 330–346 (2015)
6. Fujito, T., Nagamochi, H.: A 2-approximation algorithm for the minimum weight edge dominating set problem. Discrete Appl. Math. **118**(3), 199–207 (2002)
7. Garey, M.R., Johnson, D.S.: Computers and Intractability: A Guide to the Theory of NP-Completeness. W.H. Freeman and Company (1979)
8. Hatami, P.: An approximation algorithm for the total covering problem. Discuss. Math. Graph Theory **27**(3), 553–558 (2007)
9. Haynes, T.W., Hedetniemi, S., Slater, P.: Fundamentals of Domination in Graphs. CRC Press, Boca Raton (1998)
10. Hedetniemi, S.T., Laskar, R.C.: Bibliography on domination in graphs and some basic definitions of domination parameters. Discrete Math. **86**(1), 257–277 (1991)
11. Jain, P., Jayakrishnan, M., Panolan, F., Sahu, A.: MIXED DOMINATING SET: a parameterized perspective. In: Bodlaender, H.L., Woeginger, G.J. (eds.) WG 2017. LNCS, vol. 10520, pp. 330–343. Springer, Cham (2017). https://doi.org/10.1007/978-3-319-68705-6_25
12. Johnson, D.S.: Approximation algorithms for combinatorial problems. J. Comput. System Sci. **9**(3), 256–278 (1973)
13. Khot, S., Regev, O.: Vertex cover might be hard to approximate to within $2 - \varepsilon$. J. Comput. Syst. Sci. **74**(3), 335–349 (2008)
14. Lan, J.K., Chang, G.J.: On the mixed domination problem in graphs. Theor. Comput. Sci. **476**, 84–93 (2013)
15. Manlove, D.F.: On the algorithmic complexity of twelve covering and independence parameters of graphs. Discrete Appl. Math. **91**(1–3), 155–175 (1999)
16. Nemhauser, G.L., Trotter, L.E.: Properties of vertex packing and independence system polyhedra. Math. Program. **6**(1), 48–61 (1974)

17. Nieberg, T., Hurink, J.: A PTAS for the minimum dominating set problem in unit disk graphs. In: Erlebach, T., Persinao, G. (eds.) WAOA 2005. LNCS, vol. 3879, pp. 296–306. Springer, Heidelberg (2006). https://doi.org/10.1007/11671411_23
18. Raz, R., Safra, S.: A sub-constant error-probability low-degree test, and a sub-constant error-probability PCP characterization of NP. In: Twenty-Ninth ACM Symposium on Theory of Computing, pp. 475–484 (1997)
19. Xiao, M.: Upper and Lower Bounds on Approximating Weighted Mixed Domination. arXiv:1906.10801 (2019)
20. Xiao, M., Kloks, T., Poon, S.H.: New parameterized algorithms for edge dominating set. Theor. Comput. Sci. **511**, 147–158 (2013)
21. Xiao, M., Nagamochi, H.: Parameterized edge dominating set in graphs with degree bounded by 3. Theor. Comput. Sci. **508**, 2–15 (2013)
22. Yannakakis, M., Gavril, F.: Edge dominating sets in graphs. SIAM J. Appl. Math. **38**(3), 364–372 (1980)
23. Zhao, Y., Kang, L., Sohn, M.Y.: The algorithmic complexity of mixed domination in graphs. Theor. Comput. Sci. **412**(22), 2387–2392 (2011)

Parameterized Algorithms
for the Traveling Purchaser Problem
with Additional Constraints

Mingyu Xiao$^{(\boxtimes)}$ (iD), Jianan Zhang, and Weibo Lin

University of Electronic Science and Technology of China, Chengdu, China
myxiao@gmail.com, willjnzhang@tencent.com, lweb1688@gmail.com

Abstract. The traveling purchaser problem (TPP), a generalization of the traveling salesman problem, is to determine a tour of suppliers and purchase needed products from suppliers, while minimizing the traveling and purchasing cost. This problem finds applications in the routing and scheduling contexts and its variants with different constraints have been widely studied. Motivated by the phenomenon that most real-world instances of TPP have a small parameter (such as the number of suppliers, the number of products to purchase and others), we study TPP and its variants from the view of parameterized complexity. We show that TPP and some variants are fixed-parameter tractable by taking the number k of products or the number m of suppliers as the parameter, and W[2]-hard by taking the number q of visited suppliers as the parameter. Furthermore, we implement some of our fixed-parameter tractable algorithms to show that they are practically effective when the parameters are not very large.

1 Introduction

The traveling purchaser problem (TPP) is a single vehicle routing problem that has been widely studied. In this problem, we need to buy several products from some suppliers and the objective is to minimize the total amount of traveling and purchasing costs. Let s_0 denote the home, which is the starting and ending point of the tour. We use $\mathcal{M} = \{s_0, s_1, s_2, \ldots, s_{m-1}\}$ to denote the set of suppliers together with the home and $\mathcal{K} = \{g_1, g_2, \ldots, g_k\}$ to denote the set of products to purchase, where $|\mathcal{M}| = m$ and $|\mathcal{K}| = k$. The input of the problem consists of an $m \times k$ matrix $\mathcal{P} = \{p_{ij}\}$ to indicate the price of product g_j at supplier s_i, where we may let $p_{i_0 j_0}$ be ∞ or empty if a supplier s_{i_0} does not provide product g_{j_0}, and an $m \times m$ matrix $\mathcal{D} = \{d_{ij}\}$ to indicate the traveling costs (distances) from site s_i to site s_j. The goal is to find a tour (cycle) starting and ending at home s_0, visiting a subset of the suppliers in $\mathcal{M}$ to buy all products in $\mathcal{K}$, while minimizing the composed cost of traveling and purchasing. For the distances between sites, our problem does not require that $d_{ij} = d_{ji}$ (the symmetry assumption). However, we assume that the distances satisfy the triangle inequality, i.e., $d_{ij} \leq d_{il} + d_{lj}$ holds for any i, j, l. We also assume that each supplier has enough amount of

© Springer Nature Switzerland AG 2019
D.-Z. Du et al. (Eds.): COCOON 2019, LNCS 11653, pp. 567–579, 2019.
https://doi.org/10.1007/978-3-030-26176-4_47

each provided product and then we do not buy a product from two different suppliers.

TPP is NP-hard, since it contains the well-known traveling salesman problem (TSP) as a special case, where each supplier provides only one different product. TPP combines the optimization of routing decisions and supplier selections together and fits well in many contexts, such as routing and scheduling problems. It can be straightforwardly interpreted to machine scheduling problems [9]. An application of the telecommunication network design was proposed in [17]. Many problems in location based services can be formulated as a traveling purchaser problem [12]. More applications of TPP can be found in [16] and [18].

Problem Variants. Due to the importance of TPP, many variants of it have been widely studied. Motivated by a scheduling problem (to assign some jobs to some machines), Gouveia *et al.* [9] considered TPP with two constraints: (I) the maximum number of suppliers to be visited is limited to q, where we can simply assume that $q \leq k$ since the distances satisfy the triangle inequality; (II) the maximum number of products can be bought from each supplier is limited to u. The two constraints are also called *side-constraints*. We will use TPP-S1 to denote TPP with only constraint (I) and TPP-S2 to denote TPP with both two constraints (I) and (II). To model a problem in telecommunication network designs, Ravi and Salman [17] introduced the traveling purchaser problem with *budget-constraint* (TPP-B). In TPP-B, a budget B on the purchasing cost is given, and the goal is to minimize the traveling cost such that we can buy all the products within the budget B. Two heuristic algorithms for this problem were studied in [14]. TPP with time windows can also be found in several real contexts [6,11]. In this problem each supplier has a time window and it only serves in this time window. Recently, a multi-vehicle variant of TPP, called MVTPP, was introduced by Choi and Lee [5]. In MVTPP, the optimization has to be done over a fleet of homogeneous vehicles instead of a single vehicle, each vehicle in the fleet has the same capacity (the amount of product can be carried on). Several constraints on MVTPP have also been studied, such as the constraint on the traveling distance of each vehicle [4], the incompatibility constraint under which some products are not allowed to be loaded on the same vehicle [13], and so on. MVTPP is related to the vehicle routing problem [19].

In real-world instances of TPP, some values are small. Here are examples: In the model of stocking products from suppliers by a supermarket, the number of different products may be large while the number of suppliers may be small; In the model where a person wants to purchase something in a weekend, the number of potential shops in the city may be large while the number of things to purchase may be small; In some problems, the number of sites to be visited may also be small due to the time limitation and some other reasons; The model from some scheduling problems in [9] also assumes that the number of products is small compared to the number of suppliers. Motivated by the phenomenon of small values of parameters in these problems, we study TPP and its variants in parameterized complexity.

Parameterized Complexity. Parameterized complexity has attracted much attention in both theory and practice since the first introduction of it by Downey and Fellows [7]. An instance of a parameterized problem consists of an instance I of the original (NP-hard) problem and a parameter l. We want to design an algorithm for the problem with running time in the form of $f(l)poly(|I|)$, where $f(l)$ is a computable function on l only, and $poly(|I|)$ is a polynomial function on the input size. These kinds of algorithms are called *fixed-parameter tractable* (FPT) algorithms. A parameterized problem is *fixed-parameter tractable* (FPT) if and only if it has FPT algorithms. Under some reasonable assumptions, some parameterized problems do not allow FPT algorithms, which are called *W[1]-hard*. For FPT algorithms, the running time bound is exponential only on the parameter l and not related to the whole instance size. When the parameter l is small or a constant, FPT algorithms may run fast and solve practical problems exactly in a short time.

In this paper, we will study TPP, TPP with side-constraints (TPP-S1 and TPP-S2) and TPP with the budget-constraint (TPP-B) under three parameters: the number k of products, the number m of suppliers and the maximum number q of suppliers to be visited. For each parameterized problem, we will either design fast FPT algorithms for it or prove the W[1]-hardness or W[2]-hardness, where W[2]-hard problems may be harder than W[1]-hard problems under some reasonable assumptions.

Our Contributions. To the best of our knowledge, this is the first paper that contributes to the parameterized complexity of TPP and its variants. Our results are summarized in Table 1.

Table 1. Our results in parameterized complexity

Problems	Parameters		
	$k = |\mathcal{K}|$	$m = |\mathcal{M}|$	q
TPP	FPT (Theorem 1)	FPT (Theorem 4)	W[2]-hard (Theorem 7)
TPP-S1	FPT (Theorem 2)	FPT (Theorem 5)	W[2]-hard (Theorem 7)
TPP-S2	FPT (Theorem 3)	FPT (Theorem 6)	W[2]-hard (Theorem 7)
TPP-B	W[1]-hard (Theorem 8)	FPT (Theorem 5)	W[2]-hard (Theorem 7)

Our hardness results are obtained by reductions from two known hard problems: the set cover problem and the multi-subset sum problem. The main techniques used to design our FPT algorithms are dynamic programming and color coding. In fact, we will design two practical dynamic programming algorithms for TPP, which run in time $O(2^k(m^2 + mk))$ and $O(2^m(m^2 + mk))$ respectively and can be modified for TPP with several additional constraints without exponentially increasing the running time bound. For TPP-S2 parameterized by k, we need to use the color coding technique to design an FPT algorithm.

Our algorithms imply that TPP is polynomial-time solvable when $k = O(\log m)$ or $m = O(\log k)$. This is the reason why we can solve TPP quickly when one of k and m is small. Furthermore, the polynomial part of the running time of most algorithms is small, which is linear on the input size of the problem, because the input size of TPP is $O(m^2 + mk)$.

In practice, our dynamic programming algorithms are effective and easy to implement. Compared with previous algorithms for TPP with additional constraints, our algorithms can quickly solve instances with one of k and m being a small value. To show the advantage of our FPT algorithms in practice, we also implement some of our FPT algorithms to test their experimental performances. However, the experimental part is omitted due to the space limitation and it can be found in the full version of this paper.

2   Algorithms for Small Number of Products

In this section, we will design an $O(2^k(m^2 + mk))$-time algorithm for TPP and then modify it to an $O(2^k q(m^2 + mk))$-time algorithm for TPP-S1. Then we give an $O(u^2 m^2 2^{k+q} e^q q^{O(\log q)} \log m)$-time algorithm for TPP-S2, where $e = 2.71828\ldots$ is a constant and it holds that $q \leq k$. When we take k as the parameter, the three problems are FPT.

2.1   TPP with Parameter k

First we consider TPP. For each subset $K \subseteq \mathcal{K}$ of products and each supplier $s_i \in \mathcal{M}$, we consider the subproblem Sub-TPP(K, s_i): buy all the products in K from suppliers in a tour starting from home s_0 and ending at s_i, while minimizing the traveling and purchasing cost. Note that in this subproblem, we require that we finally arrive at s_i even we may not buy any product from s_i. If we let $K = \mathcal{K}$ and $s_i = s_0$, then this problem becomes the original TPP problem. Our idea is to solve Sub-TPP(K, s_i) for each $K \subseteq \mathcal{K}$ and $s_i \in \mathcal{M}$ in a dynamic programming method. To solve Sub-TPP(K, s_i) efficiently, we will also solve a variant of Sub-TPP(K, s_i), called Sub-TPP$'(K, s_i)$: buy all the products in K from suppliers in a tour starting from home s_0 and ending at s_i such that at least one product in K is bought from s_i, while minimizing the traveling and purchasing cost. In Sub-TPP$'(K, s_i)$, we have one more constraint that is to buy some product g_j from s_i. We can also assume that the product g_j was bought when visit s_i for the last time, i.e., we buy g_j after arriving the final site s_i of our tour.

Let $SOL[K, s_i]$ denote an optimal solution to Sub-TPP(K, s_i) (the information of the tour and where to buy each product) and $OPT[K, s_i]$ denote the total traveling and purchasing cost of $SOL[K, s_i]$. Let $SOL'[K, s_i]$ and $OPT'[K, s_i]$ denote an optimal solution and the optimal cost to Sub-TPP$'(K, s_i)$, respectively. We have that $OPT[K, s_i] \leq OPT'[K, s_i]$.

There are two cases for an optimal solution $SOL[K, s_i]$: after arriving the final site s_i we buy at least one product $g_{i_0} \in K$ from s_i, and before arriving the final site s_i we have already bought all the products in K. For the former case, we have that $OPT[K, s_i] = OPT'[K, s_i]$ and we can solve Sub-TPP(K, s_i) by solving Sub-TPP$'(K, s_i)$. For the latter case, we can see that $OPT[K, s_i] = OPT[K, s_j] + d_{ji} \geq OPT[K, s_j]$ for some j, where s_j is the last but one site of the optimal tour in $SOL[K, s_i]$, and we can solve Sub-TPP(K, s_i) by solving Sub-TPP(K, s_j). Thus, we have

$$OPT[K, s_i] = \min\{OPT'[K, s_i], \min_{j \neq i}\{OPT[K, s_j] + d_{ji}\}\}. \qquad (1)$$

We will solve Sub-TPP(K, s_i) in order of increasing cardinality of K. For each fixed K and all $s_i \in \mathcal{M}$, we first solve Sub-TPP$'(K, s_i)$ by using the following recurrence relation

$$OPT'[K, s_i] = \min_{g_j \in K}\{OPT[K\backslash\{g_j\}, s_i] + p_{ij}\}, \qquad (2)$$

and then compute $OPT[K, s_i]$ based on $OPT'[K, s_i]$ in a greedy method similar to Dijkstra's shortest path algorithm. Note that Eq. (2) allows many products in K to be bought in supplier s_i (not just g_j).

Assume that we have computed $OPT'[K, s_i]$ for a fixed K and all $s_i \in \mathcal{M}$. We are going to compute $OPT[K, s_i]$ for all $s_i \in \mathcal{M}$. Our algorithm will maintain two subsets $M_1, M_2 \subseteq \mathcal{M}$ such that $M_2 = \mathcal{M} \setminus M_1$, where for each $s_{i_0} \in M_1$ we have computed $OPT[K, s_{i_0}]$, and for each $s_{i_0} \in M_2$ we have not. Initially, we have $M_1 = \emptyset$. The algorithm iteratively selects an element $s_j \in M_2$, compute $OPT[K, s_j]$ and move it from M_2 to M_1 until M_2 becomes $\emptyset$. We select $s_j \in M_2$ such that

$$OPT'[K, s_j] \leq OPT'[K, s_r] \quad \text{for any } s_r \in M_2, \qquad (3)$$

and compute $OPT[K, s_j]$ by

$$OPT[K, s_j] = \min\{OPT'[K, s_j], \min_{s_r \in M_1}\{OPT[K, s_r] + d_{rj}\}\}. \qquad (4)$$

Next we prove the correctness of (4). Consider a supplier $s_{i_0} \in M_2$. If $OPT[K, s_{i_0}] \geq OPT'[K, s_j]$, then we get that $OPT[K, s_{i_0}] + d_{i_0 j} \geq OPT'[K, s_j]$. Otherwise we assume that $OPT[K, s_{i_0}] < OPT'[K, s_j]$ and then $OPT[K, s_{i_0}] < OPT'[K, s_{i_0}]$. By (3) and (1), it holds that $OPT[K, s_{i_0}] = OPT[K, s_{i_l}] + d_{i_l i_{l-1}} + \cdots + d_{i_2 i_1} + d_{i_1 i_0}$ for some $s_{i_l} \in M_1$ and $\{s_{i_{l-1}}, s_{i_{l-2}}, \ldots, s_{i_0}\} \subseteq M_2$. By the triangle inequality, we get $OPT[K, s_{i_0}] \geq OPT[K, s_{i_l}] + d_{i_l i_0}$ and $OPT[K, s_{i_0}] + d_{i_0 j} \geq OPT[K, s_{i_l}] + d_{i_1 j}$. By (1) again, we get (4). After computing $OPT[K, s_j]$ according to (4), we move s_j from M_2 to M_1.

We use A-k to denote the above algorithm for TPP. In this algorithm, we first let $OPT[\emptyset, s_j] = d_{0j}$ for each s_j since the length of the shortest path from s_0 to s_j is d_{0j} by the triangle inequality, and then compute $OPT[K, s_j]$ for $K \neq \emptyset$ in order of nondecreasing size by using the above method.

In this algorithm, we need to solve $O(2^k m)$ subproblems Sub-TPP(K, s_i). For each subproblem, we use $|K| \leq k$ basic computations to solve Sub-TPP$'(K, s_i)$ and use $|\mathcal{M}| = m$ basic computations to compute $OPT[K, s_i]$ based on $OPT'[K, s_i]$. We have the following Theorem 1.

Theorem 1. *TPP can be solved in $O(2^k m(m+k))$ time and it is FPT by taking the number k of products as the parameter.*

2.2 TPP-S1 with Parameter k

Now we consider TPP-S1. In fact, the algorithm for TPP-S1 is modified from the above algorithm for TPP. In our algorithm for TPP-S1, the subproblem has one more input parameter (one more dimension) and the running time of it also increases. For each subset $K \subseteq \mathcal{K}$ of products, each supplier $s_i \in \mathcal{M}$ and each nonnegative integer $q^* \leq q$, we define the subproblem Sub-TPPS1(K, s_i, q^*): buy all the products in K from exactly q^* suppliers in a tour starting from home s_0 and ending at s_i, while minimizing the traveling and purchasing cost. The value $OPT[K, s_i, q^*]$ of an optimal solution to Sub-TPPS1(K, s_i, q^*), which is defined to be ∞ if no solution exists, can be computed by the following recurrence relation

$$OPT[K, s_i, q^*] = \min\{\min_{j \neq i}\{OPT[K, s_j, q^* - 1] + d_{ji}\}, \atop \min_{g_i \in K}\{OPT[K \setminus \{g_j\}, s_i, q^*] + p_{ij}\}\}. \tag{5}$$

We can compute $OPT[K, s_i, q^*]$ in an order of increasing q^* and the size of K. The detailed steps of this algorithm are omitted since they are similar to these of the algorithm for TPP. We need to compute $O(2^k mq)$ subproblems and each subproblem takes $O(m + k)$ basic computations.

Theorem 2. *TPP-S1 can be solved in $O(2^k qm(m + k))$ time and it is FPT by taking the number k of products as the parameter.*

2.3 TPP-S2 with Parameter k

Now we consider TPP-S2. Compared with TPP-S1, TPP-S2 has one more restriction, which requires that at most u pieces of products can be bought from each supplier. The algorithm for TPP-S1 parameterized by the number k of products, can not be directly modified to an algorithm for TPP-S2, since it is hard to control the number of products purchased from each supplier. To get an FPT algorithm for TPP-S2 parameterized by k, we need to use the color coding technique [2] together with dynamic programming.

For a graph G, we say that G is q-*colored* if each vertex of G is colored by one of q different colors. For a q-colored graph G, if there is a path (resp., circle) such that all the vertices of the path are colored with pairwise distinct colors, we call it a *colorful path* (resp., *colorful circle*). We first solve a special case of TPP-S2, called Colored-TPP-S2, in which the input graph is q-colored, and we are asked to solve the TPP-S2 under the constraint that the traveling path (circle) is colorful.

We use χ to denote the set of q colors used in the graph G and $\chi(s_i)$ denote the color of supplier s_i. For each subset $K \subseteq \mathcal{K}$ of products, each supplier

$s_i \in \mathcal{M}$, each subset $X \subseteq \chi$ and each nonnegative integer $u^* \leq u$, we define the subproblem Sub-TPPS2(K, s_i, X, u^*): buy all the products in K by traveling from a colorful path starting from home s_0 and ending at s_i, the set of colors used in the colorful path is X, at most u products are bought from each supplier, and exactly u^* products are bought from supplier s_i, while minimizing the traveling and purchasing cost. The value of an optimal solution to Sub-TPPS2(K, s_i, X, u^*), which is defined to be ∞ if no solution exists, is denoted by $OPT[K, s_i, X, u^*]$.

It is not hard to see the result of Colored-TPP-S2 is equal to

$$\min_{s_i \in \mathcal{M}, X \subseteq \chi, u^* \leq u} \{OPT[K, s_i, X, u^*] + d_{i0}\}. \tag{6}$$

Our idea is to compute all $OPT[K, s_i, X, u^*]$ in a dynamic programming way, in order of nondecreasing values of $|K|, |X|$ and u^* by using the following two state transition process.

When $u^* > 0$, it holds the *purchasing recurrence relation*

$$OPT[K, s_i, X, u^*] = \min_{g_j \in K} \{OPT[K \setminus \{g_j\}, s_i, X, u^* - 1] + p_{ij}\}. \tag{7}$$

When $u^* = 0$, it holds the *traveling recurrence relation*

$$OPT[K, s_i, X, 0] = \begin{cases} \infty, \text{ if } \chi(s_i) \notin X, \\ \min_{s_j \in \mathcal{M} \setminus \{s_i\} \& u^* \leq u} \{OPT[K, s_j, X \setminus \{\chi(s_i)\}, u^*] + d_{ji}\}, \text{ otherwise.} \end{cases} \tag{8}$$

Note that the number of sets K is 2^k, the number of sets X is 2^q, the number of possible values for u^* is $u + 1$, and s_i can be any candidate in $\mathcal{M}$. Thus, the number of subproblems Sub-TPPS2(K, s_i, X, u^*) is $O(um2^{k+q})$. When $u^* \neq 0$, we may use $|K| \leq k$ basic computations to compute $OPT[K, s_i, X, u^*]$ by (7). When $u^* = 0$, we may use $|\mathcal{M}|(u + 1) = m(u + 1)$ basic computations to compute $OPT[K, s_i, X, u^*]$ by (8). Note that $um \geq k$ otherwise the problem has no solution. So the total running time of the dynamic programming algorithm for Colored-TPP-S2 is $O(u^2m^2 2^{k+q})$.

Next we solve TPP-S2 by reducing it to Colored-TPP-S2. Consider an instance of TPP-S2. We use M^* to denote the set of suppliers visited in an optimal solution to it. We randomly color vertices of the instance graph by using q different colors with the same probability for each color. The probability such that all vertices in M^* get different colors is

$$\frac{|M^*|! \binom{q}{|M^*|}}{q^{|M^*|}}. \tag{9}$$

Since $|M^*| \leq q$, we know that

$$\frac{|M^*|! \binom{q}{|M^*|}}{q^{|M^*|}} \geq \frac{q!}{q^q}. \tag{10}$$

By using the well-known inequality $q! > (q/e)^q$, where e is the base of natural logs, we know that the probability of all vertices in M^* being colored with different colors is at least e^{-q}.

Thus, we can get a randomized algorithm of TPP-S2 by applying the random coloring operation and then solving each Colored-TPP-S2. We repeat the random coloring operation for e^q times and then get a colored instance such that vertices in M^* are colored with different colors with a constant probability.

There is also a technique to derandomize the above coloring operation with an additional running time factor of $q^{O(\log q)} \log m$ [15]. We have that

Theorem 3. *TPP-S2 can be solved in $O(u^2 m^2 2^{k+q} e^q q^{O(\log q)} \log m)$ time and it is FPT by taking the number k of products as the parameter.*

Note that we always assume $q \leq k$ due to the triangle inequality, and thus this algorithm is an FPT algorithm for the problem with parameter k.

3   Algorithms for Small Number of Suppliers

In this section, we will design an $O(2^m(m^2+mk))$-time algorithm for TPP, TPP-S1 and TPP-B, and then modify it to an $O(2^m(m^2 + mk\sqrt{k}))$-time algorithm for TPP-S2. When we take m as the parameter, the four problems are FPT.

3.1   TPP with Parameter m

We still consider TPP first. In TPP, we have to decide two things, a tour of some suppliers and a purchasing plan of products. A tour of suppliers is a permutation of suppliers in the visiting order, and a purchasing plan of products is a decision that decides to buy each product from which supplier. This problem is hard because we need to optimize the traveling cost and purchasing cost at the same time. However, if we have decided the suppliers where we should visit (the set of which is M), then the problem can be reduced to the normal TSP problem.

In our algorithm, for each subset $M \subseteq \mathcal{M}$ of suppliers, we solve the subproblem: find a tour (circle) starting and ending at home s_0, visiting all suppliers in M to buy all products in $\mathcal{K}$, while minimizing the total cost of purchasing and traveling. We use $c(M)$ to denote the cost of an optimal solution to the above subproblem, $d(M)$ to denote the cost of a minimum tour starting and ending at home s_0 and visiting all suppliers in M, and $p(M)$ to denote the optimal cost to buy all products in $\mathcal{K}$ from suppliers only in M, where $p(M)$ may be ∞ if some product is not sold in any suppliers in M. Then for each $M \subseteq \mathcal{M}$,

$$c(M) = d(M) + p(M). \tag{11}$$

We compute $c(M)$ for all $M \subseteq \mathcal{M}$ using (11).

The next target is to compute $d(M)$ and $p(M)$. For each $M \subseteq \mathcal{M}$ and $s_i \in M$, we use $OPT[M, s_i]$ to denote the minimum distance of a tour which starts from home s_0, visits all suppliers in M and ends at s_i. Then $OPT[M, s_i]$

can be computed in a dynamic programming method by the following recurrence relation (see [3, 10])

$$OPT[M, s_i] = \min_{s_j \in M \setminus \{s_i\}} \{OPT[M \setminus \{s_i\}, s_j] + d_{ji}\}. \tag{12}$$

We get that

$$d(M) = \min_{s_j \in M} \{OPT[M, s_j] + d_{j0}\}. \tag{13}$$

For a subset $M \subseteq \mathcal{M}$, we use $p_M(g_i)$ to denote the minimum price of product g_i in all suppliers in M, where $p_M(g_i) = \infty$ if no supplier in M provides g_i. Then

$$p(M) = \sum_{g_i \in \mathcal{K}} p_M(g_i). \tag{14}$$

The whole algorithm is denoted by A-m for TPP. In this algorithm, to compute $OPT[M, s_i]$ we use $|M| \leq m$ basic computations in (12). It takes $O(2^m m^2)$ time to compute all values of $OPT[M, s_i]$ for $M \subseteq \mathcal{M}$ and $s_i \in M$ in a dynamic programming method. For each fixed M, it takes at most m and mk basic computations to compute $d(M)$ in (13) and $p(M)$ in (14), respectively. The values of $c(M) = d(M) + p(M)$ for all M can be computed in $O(2^m mk)$ time. In total, this algorithm uses $O(2^m m^2 + 2^m mk)$ time.

Theorem 4. *TPP can be solved in $O(2^m m(m+k))$ time and it is FPT by taking the number m of suppliers as the parameter.*

3.2   TPP-S1 and TPP-B with Parameter m

The above algorithm can be easily modified for TPP-S1. We only need to compute $c(M)$ for $|M| \leq q$. Then TPP-S1 can be solved in $O(2^m m(m + k))$ time. For TPP-B, the goal is to find an M such that $d(M)$ is minimized under the budget constraint $p(M) \leq B$. We can also use the above algorithm to compute $d(M)$ and $p(M)$ and solve TPP-B in the same time.

Theorem 5. *TPP-S1 and TPP-B can be solved in $O(2^m m(m + k))$ time and they are FPT by taking the number m of suppliers as the parameter.*

3.3   TPP-S2 with Parameter m

We can also modify the algorithm for TPP to an algorithm for TPP-S2. However, we need one more technique to find a minimum cost matching on a bipartite graph. In TPP-S2, we have a part of input q to indicate the maximum number of suppliers can be visited and $\{u_i\}_{i=1}^{m}$ to indicate that at most u_i products can be bought from supplier s_i. For each $M \subseteq \mathcal{M}$ and $|M| \leq q$, we still use $d(M)$ to denote the cost of a minimum tour starting and ending at home s_0 and visiting all suppliers in M, and $p(M)$ to denote the optimal cost to buy all products in $\mathcal{K}$ from suppliers only in M under the constraints in TPP-S2, where $p(M) = \infty$ if we can not buy all the products from M under the constraints. To solve TPP-S2, we only need to find an M with $|M| \leq q$ such that the cost $c(M) = d(M) + p(M)$ is minimized. The above method to compute $d(M)$ is still suitable for TPP-S2. The hard part is to compute $p(M)$.

We construct a bipartite graph $H = (V_K \cup V_M, E)$ and compute $p(M)$ by finding a minimum cost matching in H. For each product $g_i \in \mathcal{K}$ we generate a vertex a_i in V_K. For each supplier $s_i \in M$ we generate u_i different vertices in V_M, each of which is adjacent to each $a_j \in V_K$ with the edge cost being the price p_{ij} of product g_j at supplier s_i. We can see that the minimum cost of a matching of size $|V_K| = k$ in H is equal to $p(M)$. By using the algorithm developed in [1], the minimum cost matching can be found in $O(k\sqrt{k}m)$ time.

Theorem 6. *TPP-S2 can be solved in $O(2^m m(m + k\sqrt{k}))$ time and it is FPT by taking the number m of suppliers as the parameter.*

4   Parameterized by the Number q of Suppliers to be Visited

In some real-world problems, usually the number q of suppliers to be visited is small. It is natural to consider q as the parameter. Different from the above two sections, this section will show that it is unlikely to have FPT algorithms by proving the W[2]-hardness of TPP parameterized by q. We will reduce from the well-known W[2]-hard problem: the set cover problem parameterized by the solution size.

An instance of the set cover problem is given by $(U, \mathcal{C})$, where U is the universe of elements and $\mathcal{C}$ is a collection of subsets of U. The target of the problem is to find a subset $\mathcal{A} \subseteq \mathcal{C}$ of minimum size such that $\cup_{A \in \mathcal{A}} A = U$.

For an instance $I = (U, \mathcal{C})$ of the set cover problem, we construct an instance I' of TPP. In I', each product g is corresponding to an element g in U, and each supplier s is corresponding to a set s in $\mathcal{C}$. The price of a product g in a supplier s is 0 if the corresponding element g is contained in the corresponding set s and is ∞ otherwise. The distance between any two sites (home and suppliers) is 1. We can see that I has a set cover of size at most q if and only if I' has a solution with cost $q + 1$ (the purchasing cost is 0 and the traveling cost is $q + 1$).

Note that TPP-S1, TPP-S2 and TPP-B are general cases of TPP with more constraints. The above reduction also implies the hardness of TPP-S1, TPP-S2 and TPP-B.

Theorem 7. *TPP, TPP-S1, TPP-S2 and TPP-B are W[2]-hard by taking the number q of suppliers to be visited as the parameter.*

5   A Hardness Result for TPP-B

We have shown that TPP-B parameterized by m is FPT. The algorithm A-k for TPP can not be modified to TPP-B. In this section, we will show that TPP-B parameterized by k is indeed W[1]-hard. Our proof is based on a reduction from the multi-subset sum problem.

In the subset sum problem, we are given a set of integers $U = \{x_1, x_2, \ldots, x_{|U|}\}$ and two integers w and k, and the task is to find a subset $S \subseteq U$ such that $|S| = k$ and the sum of the elements in S is equal to w. For the multi-subset sum problem, the input is the same and the task is to find a multi-subset S of U with k elements such that the sum of the elements in S is equal to w (i.e., an integer in U can appear more than one time in S). It is known that the subset sum problem with parameter k is W[1]-hard [8]. The proof in [8], without any modification, can also prove the W[1]-hardness of the multi-subset sum problem with parameter k.

For an instance $I = (U = \{x_1, x_2, \ldots, x_{|U|}\}, w, k)$ of the multi-subset sum problem, we construct an instance I' of TPP-B. In I', we have k different products to be bought and $k|U|$ suppliers. We partition the suppliers into k groups $G_1, G_2, \ldots, G_k$, each group G_i has exactly $|U|$ suppliers.

We use $s_{i,j}$ $(j = 1, 2, \ldots, |U|)$ to denote the jth supplier in group G_i. Any supplier in the same group G_i only sales one product g_i. However, the price of g_i in supplier $s_{i,j}$ is x_j. Next we define the distance between each pair of sites. Let $X = \sum_{x \in U} x$. The distance from home s_0 to each supplier $s_{1,j}$ in group 1 is $X - x_j/2$. For any $i = 1, 2, \ldots, k - 1$, the distance from each supplier s_{i,j_1} in group i to each supplier s_{i+1,j_2} in group $i+1$ is $X - (x_{j_1} + x_{j_2})/2$. The distance from each supplier $s_{k,j}$ in group k to home s_0 is $X - x_j/2$. Any other distance between two sites not defined above is a very large number such that any optimal solution will not choose the path. The budget b is w. See Fig. 1 for an illustration for the construction.

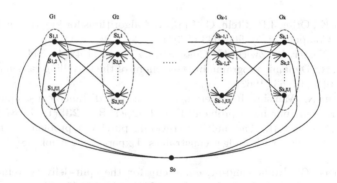

Fig. 1. Reduction from the multi-subset sum problem

In I', an optimal solution will visit exact one supplier in each group. Assume that in an optimal solution, we buy product g_i with price y_i in group i ($i = 1, 2, \ldots, k$). Then the traveling cost is $c_1 = (X - y_1/2) + (X - (y_1 + y_2)/2) + (X - (y_2 + y_3)/2) + \cdots + (X - y_k/2) = (k + 1)X - \sum_{i=1}^{k} y_i$ and the purchasing cost is $c_2 = \sum_{i=1}^{k} y_i$. We can see that $c_1 + c_2 = (k + 1)X$ is a constant. When the purchasing cost reaches the budget w, the traveling cost reaches the optimal value of $(k+1)X - w$. So the instance I of the multi-subset sum problem has a solution of size k if and only if the instance I' of TPP-B has a solution with traveling cost $(k+1)X - w$.

Theorem 8. *TPP-B is W[1]-hard by taking the number k of products as the parameter.*

6   Conclusion

To deal with NP-hard problems, approximation algorithms relax the accuracy, heuristic methods loss the certainty (or accuracy), and parameterized algorithms find certain tractable ranges. Parameterized algorithms restrict the exponential part of the running time to parameters only. When the parameter is small, parameterized algorithms can optimally solve problems in a short time. Many real-world instances of TPP and its variants have the property of small parameters. In this paper, we establish the parameterized complexity of TPP with additional constraints and different parameters. In practice, the experimental results show the advantages of parameterized algorithms on instances with small parameters.

Acknowledgements. This work was supported by the National Natural Science Foundation of China, under grants 61772115 and 61370071.

References

1. Ahuja, R.K., Orlin, J.B., Stein, C.: Improved algorithms for bipartite network flow. SIAM J. Comput. **23**(5), 906–933 (1994)
2. Alon, N., Yuster, R., Zwick, U.: Color-coding. J. ACM **42**(4), 844–856 (1995)
3. Bellman, R.: Dynamic programming treatment of the travelling salesman problem. J. ACM **9**(1), 61–63 (1962)
4. Bianchessi, N., Mansini, R., Speranza, M.G.: The distance constrained multiple vehicle traveling purchaser problem. Eur. J. Oper. Res. **235**(1), 73–87 (2014)
5. Choi, M.J., Lee, S.H.: The multiple traveling purchaser problem for maximizing system's reliability with budget constraints. Expert Syst. Appl. **38**(8), 9848–9853 (2011)
6. Desaulniers, G.: Branch-and-price-and-cut for the split-delivery vehicle routing problem with time windows. Oper. Res. **58**(1), 179–192 (2010)
7. Downey, R.G., Fellows, M.R.: Fixed-parameter tractability and completeness. Cornell University, Mathematical Sciences Institute (1992)
8. Downey, R.G., Fellows, M.R.: Fixed-parameter tractability and completeness II: on completeness for W[1]. Theor. Comput. Sci. **141**(1), 109–131 (1995)
9. Gouveia, L., Paias, A., Voß, S.: Models for a traveling purchaser problem with additional side-constraints. Comput. Oper. Res. **38**(2), 550–558 (2011)
10. Held, M., Karp, R.M.: A dynamic programming approach to sequencing problems. J. Soc. Ind. Appl. Math. **10**, 196–210 (1962)

11. Ho, S.C., Haugland, D.: A tabu search heuristic for the vehicle routing problem with time windows and split deliveries. Comput. Oper. Res. **31**(12), 1947–1964 (2004)

12. Kang, S., Ouyang, Y.: The traveling purchaser problem with stochastic prices: exact and approximate algorithms. Eur. J. Oper. Res. **209**(3), 265–272 (2011)

13. Manerba, D., Gendreau, M., Mansini, R.: The multi-vehicle traveling purchaser problem with pairwise incompatibility constraints and unitary demands: a branch-and-price approach. Eur. J. Oper. Res. **148**, 59–71 (2016)

14. Mansini, R., Tocchella, B.: The traveling purchaser problem with budget constraint. Comput. Oper. Res. **36**(7), 2263–2274 (2009)

15. Narayanaswamy, N., Raman, V., Ramanujan, M., Saurabh, S.: LP can be a cure for parameterized problems. In: 29th Symposium on Theoretical Aspects of Computer Science, STACS 2012, vol. 14, pp. 338–349. LIPIcs (2012)

16. Ramesh, T.: Traveling purchaser problem. Opsearch **18**(1–3), 78–91 (1981)

17. Ravi, R., Salman, F.S.: Approximation algorithms for the traveling purchaser problem and its variants in network design. In: Algorithms-ESA 1999, pp. 29–40 (1999)

18. Singh, K.N., van Oudheusden, D.L.: A branch and bound algorithm for the traveling purchaser problem. Eur. J. Oper. Res. **97**(3), 571–579 (1997)

19. Zhang, Z., He, H., Luo, Z., Qin, H., Guo, S.: An efficient forest-based tabu search algorithm for the split-delivery vehicle routing problem. In: Twenty-Ninth AAAI Conference on Artificial Intelligence, pp. 3432–3438 (2015)

An Approximation Algorithm for Sorting by Bounded Singleton Moves

Shengjun Xie, Haodi Feng, Haitao Jiang[(✉)], Junfeng Luan, and Daming Zhu

School of Computer Science and Technology, Shandong University,
Qingdao, People's Republic of China
1640893020@qq.com, {fenghaodi,htjiang,jfluan,dmzhu}@sdu.edu.cn

Abstract. Sorting permutations by block moves is a fundamental combinatorial problem in genome rearrangements. The classic block move operation is called transposition, which switches two consecutive blocks, or equivalently, moves a block to some other position. But large blocks movement rarely occurs during real evolutionary events. A natural restriction of transposition is to bound the length of the blocks to be switched. In this paper, we investigate the problem called sorting by bounded singleton moves, where one block is exactly a singleton while the other is of length at most c. This problem generalizes the sorting by short block moves problem proposed by Heath and Vergara [10], which requires the total length of blocks switched bounded by 3. By exploring some properties of this problem, we devise a $\frac{9}{5}$-approximation algorithm for $c = 3$. Our algorithm can be extended to any constant $c \geq 3$, guaranteeing an approximation factor of $\frac{3c}{5}$.

1 Introduction

In the 1980s, some evidence was found that some species have essentially the same set of genes, but their gene order differs [1,2]. Sorting permutations with the fewest number of operations has gained much interest in the area of computational biology during the last thirty years, because it helps to reconstruct the sequence of evolutionary events. Sankoff was probably the first researcher who proposed the three basic operations of genome rearrangement, i.e., reversals, transpositions and translocations [3]. A transposition, which is also called a block-move, is a rearrangement operation that cuts a segment out of the permutation and pastes it in a different location, i.e., it swaps two adjacent sub-permutations. Of interest to biologists is how to transform one permutation to another by the minimum number of transpositions. The problem was first studied by Bafna and Pevzner, who devised a 1.5-approximation algorithm, which runs in quadratic time [4]. Elias and Hartman improved the approximation factor to 1.375 [5]. Feng and Zhu proposed the "permutation tree" to store the permutation, and improved the time complexity of the algorithms from $O(n^2)$ to $O(n\log n)$ [6]. Eriksson et al. showed that the transposition diameter is bounded by $\lceil (2*n-2)/3 \rceil$ for $n \geq 9$, and gave an exact expression for sorting the reverse

D.-Z. Du et al. (Eds.): COCOON 2019, LNCS 11653, pp. 580–590, 2019.
https://doi.org/10.1007/978-3-030-26176-4_48

permutation [7]. Recently, Bulteau et al. proved that sorting permutations by transpositions is NP-complete [8], answering the long lasting open problem.

Actually, in the process of the genomes evolution, a segment is rarely moved far away from its original position. Naturally, Heath and Vergara proposed the problem of sorting by bounded block-moves [9], where the blocks must be moved within a bounded distance. Among which, the problem of sorting by short block moves is well studied. A short block move is a transposition on a permutation such that the total length of the two segments swapped is at most three. Heath and Vergara presented a 4/3-approximation algorithm for this problem, as well as polynomial algorithms for some special permutations [10]. Short block move is also called 3-bounded transposition in [11], where Mahajan et al. simplified Heath and Vergara's approximation algorithm, and described a linear-time algorithm to optimally sort the correcting-hop-free permutations [11]. Jiang et al. devised an $O(n^2)$ algorithm to sort a special permutation with a structure called "umbrella", by use of the umbrellas. They proposed a $(1 + \varepsilon)$-approximation algorithm for sorting permutations with many inversions [12]. Then, they devised a 14/11-approximation algorithm for sorting general permutations by short block moves [13], later, the approximation factor was improved to 5/4 [14].

In light of the limitation of short block move, in this paper, we investigate the problem of sorting by bounded singleton moves, which requires a single element to move at most c positions away from its original position. We observe that this problem shows distinct properties from the problem of sorting by short block moves, and we present a polynomial $\frac{9}{5}$-approximation algorithm for $c = 3$. Our algorithm can be extended to any constant $c \geq 3$, guaranteeing an approximation factor of $\frac{3c}{5}$.

2 Preliminaries

In the context of genome rearrangements, generally, genomes are presented by permutations, where each element stands for a gene. For example, $\pi = [7, 1, 3, 5, 4, 2, 6, 8]$ is a permutation of eight elements. Let $\iota_n = [1,2,...,n-1,n]$ be the identity permutation of n elements. A block is a segment of contiguous elements or just one element, a single element is called a *singleton*. A *block move (transposition)* switches two adjacent blocks, it is called *c-bounded singleton move* when one block is a singleton while the other block has at most c elements. Equivalently, a *c-bounded singleton move* asks a singleton to move at most c positions away from its original position in the permutation. Observe that a 2-bounded singleton move is actually a short block move.

Let $\pi = [g_1, g_2, ..., g_n]$ be a permutation. Note that a singleton can be moved either to its left or to its right, then there are three types of c-bounded singleton moves.

1. *k-right-move*: $\rho(g_i, [g_{i+1} \cdots g_{i+k}])$, which moves g_i to the immediate right of g_{i+k}, $i \geq 1$, $i + k \leq n$, $1 < k \leq c$.
2. *k-left-move*: $\rho([g_{i-k} \cdots g_{i-1}], g_i)$, which moves g_i to the immediate left of g_{i-k}, $1 < k < i \leq n$, $k \leq c$.

3. skip: $\rho(g_i, g_{i+1})$, which switches the two singletons g_i and g_{i+1}.

Either a k-right-move or a k-left-move is also called a k-move. Applying a bounded singleton move ρ to π yields $\pi' = \pi \cdot \rho$. We now formally define the problem of sorting by bounded singleton moves.

Definition 1. *Sorting by Bounded Singleton Moves, abbreviated as SBSM*
  Input: $\pi = [g_1, g_2, ..., g_n]$, *a constant* c.
  Question: *Is there a sequence of* c-*bounded singleton moves:* $\rho_1, \rho_2, ..., \rho_t$, *such that* $\pi \cdot \rho_1 \cdot \rho_2 ... \cdot \rho_t = \iota_n = [1, 2, ..., n]$, *and* t *is minimized?*

The minimum integer t is the c-bounded singleton move distance of π, denoted by $BSM_c(\pi)$.

An *inversion* in a permutation is a pair of elements $\{g_i, g_j\}$ that are not in their correct order (i.e., $i < j$ and $g_i > g_j$, or vice versa). There is no inversion in the identity permutation. A *correcting* c-bounded singleton move corrects the relative order of the moved elements and does not bring any new inversions, i.e., a correcting skip erases a single inversion, a correcting k-move erases k inversions. It is interesting that the problem of sorting by short block moves, which is called the 2-bounded singleton move problem in this paper, has the following property.

Theorem 1. *For a permutation* π, *there exists an optimal sequence of short block moves* $\rho_1, \rho_2, ..., \rho_t$ *that sorts* π *such that each short block move is a correcting short block-move* [10].

Then the subsequent researches on this problem could ignore the non-correcting short block moves. Unfortunately, when it comes to the sorting by c-bounded singleton moves problem for $c \geq 3$, the above property does no longer hold. Sorting the permutation $\pi = [2, 4, 5, 7, 1, 10, 3, 12, 6, 8, 9, 11]$ by 3-bounded singleton moves becomes a counter example to the property. An optimal sorting sequence should be: $\rho(7, [1\,10\,3])$, $\rho([2\,4\,5], 1)$, $\rho([4\,5\,10], 3)$, $\rho([10\,7\,12], 6)$, $\rho(12, [8\,9\,11])$, $\rho(10, [7\,8\,9])$, which contains 6 3-bounded singleton moves, among which $\rho(7, [1\,10\,3])$ is not a correcting 3-bounded singleton move. One could check that it needs at least 7 correcting 3-bounded singleton moves to sort π. So the sorting by bounded singleton moves problem seems more complicated than sorting by short block moves, and all the previous algorithms can not be extended to solve this problem straightforwardly.

The following graph representation of a permutation serves as a fundamental tool for solving the sorting by c-bounded singleton moves problem. The permutation graph of π is a graph $G(\pi) = (V, E)$, where $V = \{g_1, g_2, ..., g_n\}$, $E = \{<g_i, g_j> \mid i < j$ and $g_i > g_j\}$(See Fig. 1 for an example). All the arcs of E direct from left to right, for sake of simplicity, we ignore their directions. Each arc of $G(\pi)$ represents an inversion in π. An arc $<g_i, g_j>$ is short if $j = i+1$. Two arcs in the permutation graph are *compatible* if they share an identical starting element or ending element. A lone arc in the permutation graph is an arc that is not compatible with any other arc.

As mentioned above, from what a c-bounded singleton move does, we have

Lemma 1. *A correcting c-bounded singleton move can remove at most c inversions, which correspond to c compatible arcs.*

Without causing any confusion, in the rest of this paper, we will not distinguish between inversions and arcs, as well as between a permutation and its permutation graph. So sorting a permutation is equal to removing all the arcs from its permutation graph.

Though the optimal sorting sequence may contain non-correcting c-bounded singleton moves, we still use the following lower bound which is not so tight for the c-bounded singleton move distance.

Lemma 2. *The c-bounded singleton move distance of a permutation π satisfies that*

$$\frac{\mid E(G(\pi)) \mid}{c} \leq BSM_c(\pi) \leq \mid E(G(\pi)) \mid$$

3   A $\frac{9}{5}$ - Approximation Algorithm for 3-Bounded Singleton Moves

In this section, we will present an algorithm for 3-bounded singleton moves that guarantees, for any permutation, a sorting sequence whose length is within $\frac{9}{5}$ of the optimal. The algorithm's main idea is trying to perform as many high efficiency moves (correcting 3-moves and 2-moves) as possible and avoiding low efficiency moves (skips). Now, we introduce some special subpermutations.

Definition 2. *A dome in a permutation $\pi = [g_1, g_2, ..., g_n]$ is a subpermutation of three consecutive elements $D = [g_i, g_{i+1}, g_{i+2}]$ that satisfies $g_i > g_{i+1} > g_{i+2}$. The dome $D = [g_i, g_{i+1}, g_{i+2}]$ is lone if it also satisfies $g_j < g_{i+2}$ for all $1 \leq j < i$ and $g_k > g_i$ for all $i + 2 < k \leq n$.*

Definition 3. *A rainbow in a permutation $\pi = [g_1, g_2, ..., g_n]$ is a subpermutation of four consecutive elements $R = [g_i, g_{i+1}, g_{i+2}, g_{i+3}]$ that satisfies $g_{i+1} > g_{i+3} > g_i > g_{i+2}$. The rainbow $R = [g_i, g_{i+1}, g_{i+2}, g_{i+3}]$ is lone if it also satisfies $g_j < g_{i+2}$ for all $1 \leq j < i$ and $g_k > g_{i+1}$ for all $i + 3 < k \leq n$.*

Definition 4. *A mushroom in a permutation $\pi = [g_1, g_2, ..., g_n]$ is a subpermutation of four consecutive elements $M = [g_i, g_{i+1}, g_{i+2}, g_{i+3}]$ that satisfies $g_{i+2} > g_i > g_{i+3} > g_{i+1}$. The rainbow $M = [g_i, g_{i+1}, g_{i+2}, g_{i+3}]$ is lone if it also satisfies $g_j < g_{i+1}$ for all $1 \leq j < i$ and $g_k > g_{i+2}$ for all $i + 3 < k \leq n$.*

One could find that lone domes, lone mushrooms as well as lone rainbows can not be sorted only by correcting 3-moves or 2-moves, so we call them *barriers*.

Lemma 3. *Removing each lone arc needs one correcting skip.*

Proof. Removing a lone arc $<g_i, g_{i+1}>$ requires moving g_i backward or g_{i+1} forward. Since all the elements ahead g_i are smaller than g_{i+1} and all the elements followed g_{i+1} are greater than g_i, then moving g_i or g_{i+1} too far away (not by a skip) would bring new arcs. Thus, it just needs a correcting skip.   □

Lemma 4. *Removing each lone barrier needs two correcting moves.*

Proof. A correcting k-left move can remove k compatible arcs with a common ending element, while a correcting k-right move can remove k compatible arcs with a common starting element. Each lone barrier has three arcs, which could be partitioned into at least two group of compatible arcs. Moving some elements beyond the range of the lone barrier would bring new arcs. Thus, it needs two correcting moves to remove the three arcs.   □

Lemma 5. *Given a permutation π, let a, b be the number of lone arcs and lone barriers respectively in $G(\pi)$. Then,*

$$BSM_3(\pi) \geq \frac{\mid E(G(\pi)) \mid -a - 3b}{3} + a + 2b = \frac{\mid E(G(\pi)) \mid +2a + 3b}{3}.$$

Our strategy is exhausting correcting 3-moves and then correcting 2-moves, after that, the resulting permutation would have a fixed structure, which could be sorted by a polynomial algorithm.

Lemma 6. *Let α and β be the number of lone arcs and barriers introduced by applying a correcting 3-move respectively, then $\alpha \leq 3$, $\beta \leq 2$, and $\alpha + \beta \leq 3$.*

Proof. W.L.O.G., assume that we apply a correcting 3-right move $\rho(g_i, [g_{i+1} \ g_{i+2} \ g_{i+3}])$ on the permutation $\pi = [g_1, g_2, \ldots, g_n]$, which yields π'. Note that a lone arc must appear on two adjacent elements, and a barrier contains at least three elements. $\rho(g_i, [g_{i+1} \ g_{i+2} \ g_{i+3}])$ generates two new neighborhoods on π': (g_{i-1}, g_{i+1}) and (g_i, g_{i+4}). π' can be partitioned into three parts: $[g_1, \ldots, g_{i-1}, g_{i+1}]$, $[g_{i+2}, g_{i+3}]$, and $[g_i, g_{i+4}, \ldots, g_n]$. There could be a newly introduced lone arc or a lone barrier in part $[g_1, \ldots, g_{i-1}, g_{i+1}]$, as well as in part $[g_i, g_{i+4}, \ldots, g_n]$. However, there could not exist a newly introduced lone barrier but a lone arc in $[g_{i+2}, g_{i+3}]$. By symmetry, it is similar while applying a correcting 3-left move. See Fig. 3 for an example.   □

A correcting 3-move is *acceptable* if it fulfills one of the following two conditions: (1) $\alpha \leq 1$ and $\beta \leq 2$; (2) $\alpha \leq 2$ and $\beta = 0$. From Lemma 6, there are two types of correcting 3-moves: (I) $\alpha = 3$ and (II) $\alpha = 2$ and $\beta = 1$, which are not acceptable. We will replace it according to the following two lemmas.

Lemma 7. *If there exists a correcting 3-move which would introduce three lone arcs, then there also exist three correcting 2-moves to remove the six arcs.*

Proof. W.L.O.G., as what are shown in Fig. 3, assume that there exists a correcting 3-right move $\rho(g_i, [g_{i+1} \quad g_{i+2} \quad g_{i+3}])$, applying which would introduce three long arcs: $<g_{i-1}, g_{i+1}>$, $<g_{i+2}, g_{i+3}>$, $<g_i, g_{i+4}>$. Thus, it needs four steps to remove the total six arcs. As a twist, applying the correcting 2-move $\rho([g_{i-1} \quad g_i], g_{i+1})$ followed by the correcting 2-move $\rho([g_i \quad g_{i+2}], g_{i+3})$ and then the correcting 2-move $\rho(g_i, [g_{i+2} \quad g_{i+4}])$ will, of course, remove the same six arcs, but consume only three steps. It is similar when applying a correcting 3-left move. $\square$

Lemma 8. *If there exists a correcting 3-move which would introduce two lone arcs and a barrier, then there also exist four correcting 2-moves to remove the eight arcs.*

Proof. W.L.O.G., assume that there exists a correcting 3-right move $\rho(g_i, [g_{i+1} \quad g_{i+2} \quad g_{i+3}])$, applying which would introduce two lone arcs and a barrier. As shown in Fig. 5, there are 6 cases:
(*a*) $<g_{i+2}, g_{i+3}>$, $<g_i, g_{i+4}>$ become lone arcs and $[g_{i-2}, g_{i-1}, g_{i+1}]$ forms a lone dome. Apply four correcting 2-moves: $\rho([g_{i-1} \quad g_i], g_{i+1})$, $\rho(g_{i-2}, [g_{i+1} \quad g_{i-1}])$, $\rho([g_i \quad g_{i+2}], g_{i+3})$, $\rho(g_i, [g_{i+2} \quad g_{i+4}])$. (*b*) $<g_{i-1}, g_{i+1}>$, $<g_{i+2}, g_{i+3}>$ become lone arcs and $[g_i, \ g_{i+4}, \ g_{i+5}]$ forms a lone dome. Apply four correcting 2-moves: $\rho([g_{i-1} \quad g_i], g_{i+1})$, $\rho([g_i \quad g_{i+2}], g_{i+3})$, $\rho(g_i, [g_{i+2} \quad g_{i+4}])$, $\rho([g_{i+4} \quad g_i], g_{i+5})$. (*c*) $<g_{i+2}, g_{i+3}>$, $<g_i, g_{i+4}>$ become lone arcs and $[g_{i-3}, \ g_{i-2}, \ g_{i-1}, \ g_{i+1}]$ forms a lone rainbow. Apply four correcting 2-moves: $\rho([g_{i-3} \quad g_{i-2}], g_{i-1})$, $\rho([g_{i-2} \quad g_i], g_{i+1})$, $\rho([g_i \quad g_{i+2}], g_{i+3})$, $\rho(g_i, [g_{i+2} \quad g_{i+4}])$. (*d*) $<g_{i-1}, g_{i+1}>$, $<g_{i+2}, g_{i+3}>$ become lone arcs and $[g_i, \ g_{i+4}, \ g_{i+5}, \ g_{i+6}]$ forms a lone rainbow. Apply four correcting 2-moves: $\rho([g_{i-1} \quad g_i], g_{i+1})$, $\rho([g_i \quad g_{i+2}], g_{i+3})$, $\rho(g_{i+4}, [g_{i+5} \quad g_{i+6}])$, $\rho(g_i, [g_{i+2} \quad g_{i+5}])$. (*e*) $<g_{i+2}, g_{i+3}>$, $<g_i, g_{i+4}>$ become lone arcs and $[g_{i-3}, \ g_{i-2}, \ g_{i-1}, \ g_{i+1}]$ forms a lone mushroom. Apply four correcting 2-moves: $\rho([g_{i-1} \quad g_i], g_{i+1})$, $\rho(g_{i-3}, [g_{i-2} \quad g_{i+1}])$, $\rho([g_i \quad g_{i+2}], g_{i+3})$, $\rho(g_i, [g_{i+2} \quad g_{i+4}])$. (*f*) $<g_{i-1}, g_{i+1}>$, $<g_{i+2}, g_{i+3}>$ become lone arcs and $[g_i, \ g_{i+4}, \ g_{i+5}, \ g_{i+6}]$ forms a lone mushroom. Apply four correcting 2-moves: $\rho([g_{i-1} \quad g_i], g_{i+1})$, $\rho([g_i \quad g_{i+2}], g_{i+3})$, $\rho(g_i, [g_{i+2} \quad g_{i+4}])$, $\rho([g_i \quad g_{i+5}], g_{i+6})$.
It is similar for the case of applying a correcting 3-left move. $\square$

Lemma 9. *Let α and β be the number of lone arcs and barriers introduced by applying a correcting 2-move respectively, then $\alpha \leq 2$, $\beta \leq 2$, and $\alpha + \beta \leq 2$.*

Proof. W.L.O.G., assume that we apply a correcting 2-right move $\rho(g_i, [g_{i+1} \quad g_{i+2}])$ on the permutation $\pi = [g_1, g_2, \ldots, g_n]$, which yields π'. Note that a lone arc must appear on two adjacent elements, and a barrier contains at least three elements. $\rho(g_i, [g_{i+1} \quad g_{i+2}])$ generates two new neighborhoods on π': (g_{i-1}, g_{i+1}) and (g_i, g_{i+3}). π' can be partitioned into two parts: $[g_1, \ldots, g_{i-1}, g_{i+1}, g_{i+2}]$ and $[g_i, g_{i+3}, \ldots, g_n]$. In each part, there could be at most one newly formed lone arc or barrier. Then the lemma follows. $\square$

Lemma 10. *If there exists a correcting 2-move which would introduce two lone arcs, then there must be a correcting 3-move.*

Proof. W.L.O.G., assume that there exists a correcting 2-right move $\rho(g_i, [g_{i+1} \quad g_{i+2}])$, applying which would introduce three possible lone arcs: $<g_{i-1}, g_{i+1}>$, $<g_{i+1}, g_{i+2}>$, $<g_i, g_{i+3}>$. But $<g_{i-1}, g_{i+1}>$ and $<g_{i+1}, g_{i+2}>$ can not be lone arcs meanwhile. So, $<g_i, g_{i+3}>$ must be a lone arc. Thus, $\rho(g_i, [g_{i+1} \quad g_{i+2}])$ introduces two lone arcs, which implies $\rho(g_i, [g_{i+1} \quad g_{i+2} \quad g_{i+3}])$ is a correcting 3-move. $\qquad\square$

Lemma 10 can also be stated in the converse-negative form: if there is no correcting 3-move, then every correcting 2-move, if exists, would not introduce two lone arcs.

Lemma 11. *Supposing that there is no barrier and correcting 3-move, then there must be a correcting 2-move which would not introduce any lone arc unless there is no correcting 2-move.*

Proof. Assume on the contrary that every correcting 2-move introduces lone arcs. Since there is no correcting 3-move, from Lemma 10, every correcting 2-move introduces exactly one lone arc. W.L.O.G., let $\rho(g_i, [g_{i+1} \quad g_{i+2}])$ be such a 2-move. There can not exist the arc $<g_i, g_{i+3}>$, since otherwise there would be a correcting 3-move. Thus, the newly introduced lone arc may be either $<g_{i-1}, g_{i+1}>$ or $<g_{i+1}, g_{i+2}>$. For the former case, $\rho([g_{i-1} \quad g_i], g_{i+1})$ becomes a correcting 2-move, applying which would not introduce lone arc $<g_{i-2}, g_{i+1}>$. That is because there is no correcting 3-move, as well as lone arc $<g_i, g_{i+2}>$, and $[g_{i-1}, g_i, g_{i+1}, g_{i+2}]$ is not a lone rainbow. For the latter case, $\rho([g_i \quad g_{i+1}], g_{i+2})$ becomes a correcting 2-move, applying which would not introduce lone arc $<g_{i-1}, g_{i+2}>$. That is because there is no correcting 3-move, as well as lone arc $<g_i, g_{i+1}>$, and $[g_i, g_{i+1}, g_{i+2}]$ is not a lone rainbow. It is similar for the case of applying a correcting 2-left move. $\qquad\square$

Lemma 12. *Supposing that there is no barrier and correcting 3-move, if there is a correcting 2-move applying which would introduce two barriers, then there exist four correcting 2-moves to remove the eight arcs.*

Proof. W.L.O.G., let $\rho(g_i, [g_{i+1} \quad g_{i+2}])$ be a correcting 2-move applying which would introduce two barriers. Since there is no correcting 3-move, one could check that a correcting 2-right move can not introduce any dome, and also a mushroom on the rightside. Thus, we have the following three cases,

1. $[g_{i-3}, g_{i-2}, g_{i-1}, g_{i+1}]$ forms a rainbow and $[g_i, g_{i+3}, g_{i+4}, g_{i+5}]$ forms a rainbow. Apply four correcting 2-moves: $\rho([g_{i-3} \quad g_{i-2}], g_{i-1})$, $\rho([g_{i-2} \quad g_i], g_{i+1})$, $\rho(g_{i+3}, [g_{i+4} \quad g_{i+5}])$, $\rho(g_i, [g_{i+2} \quad g_{i+4}])$.
2. $[g_{i-3}, g_{i-2}, g_{i-1}, g_{i+1}]$ forms a mushroom and $[g_i, g_{i+3}, g_{i+4}, g_{i+5}]$ forms a rainbow. Apply four correcting 2-moves: $\rho([g_{i-1} \quad g_i], g_{i+1})$, $\rho(g_{i-3}, [g_{i-2} \quad g_{i+1}])$, $\rho(g_{i+3}, [g_{i+4} \quad g_{i+5}])$, $\rho(g_i, [g_{i+2} \quad g_{i+4}])$.
3. $[g_{i-2}, g_{i-1}, g_{i+1}, g_{i+2}]$ forms a mushroom and $[g_i, g_{i+3}, g_{i+4}, g_{i+5}]$ forms a rainbow. Apply four correcting 2-moves: $\rho([g_i \quad g_{i+1}], g_{i+2})$, $\rho(g_{i-2}, [g_{i-1} \quad g_{i+2}])$, $\rho(g_{i+3}, [g_{i+4} \quad g_{i+5}])$, $\rho(g_i, [g_{i+1} \quad g_{i+4}])$. $\qquad\square$

A correcting 2-move is *acceptable* if applying it would not introduce either lone arcs or two barriers.

Definition 5. *A multi-mushroom in a permutation* $\pi = [g_1, g_2, ..., g_n]$ *is a sub-permutation of* $2k + 2$ *consecutive elements* $[g_i, g_{i+1}, \ldots, g_{i+2k}, g_{i+2k+1}](k \geq 2, 1 \leq i \leq n-5)$ *that satisfies* $g_{i+2j+1} < g_{i+2j+3} < g_{i+2j} < g_{i+2j+2}$, *equivalently,* $[g_{i+2j}, g_{i+2j+1}, g_{i+2j+2}, g_{i+2j+3}]$ *forms a mushroom, for every* $0 \leq j \leq k - 1$.

Lemma 13. *Supposing that there are no correcting 2-moves in an unsorted permutation* π, *then each connected component in the permutation graph of* π *is a lone arc or a lone mushrooms or a multi-mushroom.*

Proof. Assume that there is a connected component $CC = [g_i, g_{i+1}, \ldots, g_{i+m-1}]$ of m elements. It is trivial that $m \geq 2$, since otherwise CC is sorted.

Claim 1. $g_{i+j} < g_{i+j+2}$, *for all* $0 \leq j \leq m - 3$ *and* $m \geq 3$.

Proof. Assume to the contrary that $g_{i+j} > g_{i+j+2}$, then g_{i+j+1} is either greater than g_{i+j+2} or smaller than g_{i+j+2}. In the former case $\rho([g_{i+j} \ g_{i+j+1}], g_{i+j+2})$ is a correcting 2-move, and in the latter case $\rho(g_{i+j}, [g_{i+j+1} \ g_{i+j+2}])$ is a correcting 2-move. Thus the claim follows. □

Claim 2. $g_{i+2j} > g_{i+2j+1}$, *for all* $0 \leq j \leq \lfloor (m - 1)/2 \rfloor$.

Proof. If $m = 2$, $CC = [g_i, g_{i+1}]$. Since CC is a connected component, $g_i > g_{i+1}$. If $m \geq 2$, the proof is by induction on j. Initially, when $j = 0$, assume to the contrary that $g_i < g_{i+1}$. From Claim 1, $g_i < g_{i+2}$. Since g_i is in this connected component, there must be an element which is smaller than g_i and also appears on the right of g_{i+1}. Let g_{i+r} $(r > 1)$ be the fist such element while searching CC from g_{i+1} to its right. Then $g_i, g_{i+1}, \ldots, g_{i+r-1}$ are all greater than g_{i+r}, as a result, $\rho([g_{i+r-2} \ g_{i+r-1}], g_{i+r})$ is a correcting 2-move, which is a contradiction. Then we have $g_i > g_{i+1}$.

For the inductive step, assume that $g_{i+2j} > g_{i+2j+1}$, we must prove that $g_{i+2j+2} > g_{i+2j+3}$, $2j + 3 \leq m - 1$. If $2j + 3 < m - 1$, assume to the contrary that $g_{i+2j+2} < g_{i+2j+3}$. From Claim 1, $g_{i+2j} < g_{i+2j+2}$. Since $[g_i, g_{i+1}, \ldots, g_{i+2j+1}]$ and $[g_{i+2j+2}, g_{i+2j+3}, \ldots, g_{i+m-1}]$ must be a single connected component, from Claim 1 and inductive basis, g_{i+2j} is the greatest element in $[g_i, g_{i+1}, \ldots, g_{i+2j+1}]$, then there must be arcs going from g_{i+2j} to some elements of $[g_{i+2j+2}, g_{i+2j+3}, \ldots, g_{i+m-1}]$. Let $g_{i+2j+3+r}$ $(r > 1)$ be the first such element to the right of g_{i+2j+3}. Then $g_{i+2j+2}, g_{i+2j+3}, \ldots, g_{i+2j+2+r}$ are all greater than $g_{i+2j+3+r}$. As a result, $\rho([g_{i+2j+1+r} \ g_{i+2j+2+r}], g_{i+2j+3+r})$ is a correcting 2-move, which is a contradiction. If $2j + 3 = m - 1$, to keep CC a single connected component, $g_{i+2j} > g_{i+2j+3}$, from Claim 1, $g_{i+2j+2} > g_{i+2j}$, then we have $g_{i+2j+2} > g_{i+2j+3}$. □

From Claim 2, we can also conclude that m is even, since otherwise g_{i+m-1} is the greatest element but appears on the rightmost of the subpermutation, thus it could not be in this connected component.

Claim 3. $g_{i+2j} > g_{i+2j+3}$, *for all* $0 \leq j \leq (m - 4)/2$ *and* $m \geq 4$.

Algorithm 1. sorting a multi-mushroom

Input: A multi-mushroom $MM = [g_i, g_{i+1}, \ldots, g_{i+2*k}, g_{i+2*k+1}]$
Output: A collection of correcting moves to sort MM

1: Apply the skip $\rho(g_{i+2k}, g_{i+2k+1})$.
2: **for** $j = k - 1$ to 0 **do**
3:     Apply 2-right moves and 3-right moves to move g_{i+2j} to its right position.
4: **end for**

Proof. From Claims 1 and 2, g_{i+2j} is the greatest element of $[g_i, g_{i+1}, \ldots, g_{i+2j+1}]$ and g_{i+2j+3} is the smallest element of $[g_{i+2j+2}, g_{i+2j+3}, \ldots, g_{i+m-1}]$. To make a single connected component, there must an arc between g_{i+2j} and g_{i+2j+3}.   □

From the above three claims, $[g_i, g_{i+1}]$ forms a lone arc while $m = 2$, and $[g_{i+2j}, g_{i+2j+1}, g_{i+2j+2}, g_{i+2j+3}]$ forms a mushroom, for every $0 \leq j \leq (m-4)/2$ and $m \geq 4$, thus the proof of lemma 13 is done.   □

Definition 6. *A k-claw (k $\geq$ 2) in a permutation $\pi = [g_1, g_2, \ldots, g_n]$ is a subpermutation of $k + 1$ consecutive elements $C = [g_i, g_{i+1}, g_{i+k}]$ that satisfies $g_{i+1} < g_{i+2} < \cdots < g_{i+k} < g_i$.*

From its definition, A k-claw can be sorted by $\lfloor \frac{k}{2} \rfloor$ 2-moves and 3-moves. From Claims 1, 2 and 3, each round of the FORLOOP in Algorithm 1 sorts a k-claw.

Now, we present our algorithm for sorting an arbitrary permutation by bounded singleton moves. In Algorithm 2: **sorting by bounded singleton moves**, line 1 handles the original lone arcs and barriers; line 2–8 handle correcting 3-moves if exist; line 9–15 handle correcting 2-moves if exist; line 16 handles multi-mushrooms. As shown in line 4, an acceptable correcting 3-move will be applied straightforwardly, on the other side, a not acceptable correcting 3-move will be handled in line 6. From Lemma 11, there always exists a correcting 2-move without introducing any lone arc unless there is no correcting 2-move. If such a correcting 2-move brings two barriers, the algorithm handles them in line 11, otherwise it is acceptable and will be applied directly in line 13. Finally, the remained multi-mushrooms are handled in line 16.

Theorem 2. *Algorithm 2 approximates the 3-bounded singleton move distance within a factor of 9/5.*

Proof. For the lone arcs and barriers, from Lemmas 3 and 4, Algorithm 2 adopts the same steps as the optimal solution. It is sufficient to show that each move in Algorithm 2 removes at least 5/3 arcs averagely.

(I) As in line 4, an acceptable correcting 3-move, may introduce one lone arc as well as at most two barriers, Algorithm 2 adopts six moves to remove ten arcs, on average 10/6, also an acceptable correcting 3-move may introduce at most two lone arcs but no barriers, Algorithm 2 adopts three moves to remove five arcs, on average 5/3. (II) From Lemma 7, Algorithm 2 adopts three correcting 2-moves to remove six arcs, on average 6/3. (III) From Lemma 8, Algorithm 2 adopts

---

Algorithm 2. sorting by bounded singleton moves

Input: A permutation π
Output: A collection of correcting moves that sort π

1: Remove the lone arcs and barriers following Lemma 3 and Lemma 4.
2: **while** there exists a correcting 3-move **do**
3:     **if** it is acceptable **then**
4:       apply this correcting 3-move and remove the lone arcs and barriers introduced by this correcting 3-move.
5:     **else**
6:       handle it according to Lemma 7 or Lemma 8.
7:     **end if**
8: **end while**
9: **while** there exists a correcting 2-move (choose a correcting 2-move following Lemma 11) **do**
10:     **if** it introduces two barriers **then**
11:       handle it according to Lemma 12.
12:     **else**
13:       apply this correcting 2-move and remove the barrier introduced by this correcting 2-move.
14:     **end if**
15: **end while**
16: Sort each connected component by **Algorithm 1: sorting a multi-mushroom**.

---

four correcting 2-moves to remove eight arcs, on average 8/4. (IV) As in line 13, an acceptable correcting 2-move introduces at most one barrier, so totally, Algorithm 2 adopts three moves to remove five arcs, on average 5/3. (V) From Lemma 12, Algorithm 2 adopts four correcting 2-moves to remove eight arcs, on average 8/4. (VI) For sorting a multi-mushroom $[g_i, g_{i+1}, \cdots, g_{i+2*k}, g_{i+2*k+1}]$ $(k \geq 2)$, Algorithm 1 adopts one skip and at least k correcting 2-moves, since $k \geq 2$, $k+1$ moves remove at least $2k+1$ arcs, averagely $\frac{2k+1}{k+1} \geq \frac{5}{3}$. Above all, each move in the optimal solution removes at most three arcs, thus, the approximation factor reaches to $\frac{3}{5/3} = \frac{9}{5}$. $\qquad\qquad\qquad\qquad\qquad\qquad\qquad\qquad\qquad\qquad$ $\square$

For the c-bounded singleton move distance, since each move in its optimal solution can remove at most c arcs, then we have,

Theorem 3. *Algorithm 2: **sorting by bounded singleton moves** approximates the c-bounded singleton move distance within a factor of $\frac{3c}{5}$.*

4 Concluding Remarks

This paper investigates the problem of sorting by bounded singleton moves, which requires a single element to move at most c positions away from its original position. We present a polynomial approximation algorithm with a factor of $\frac{9}{5}$ for $c = 3$. The complexity of this problem is still open. We think better approximations with good structure analysis are preferable.

Acknowledgments. This research is supported by NSF of China under grant 61872427, 61732009 and 61628207, by NSF of Shandong Provence under grant ZR201702190130. Haitao Jiang is also supported by Young Scholars Program of Shandong University.

References

1. Hoot, S.B., Palmer, J.D.: Structural rearrangements, including parallel inversions within the choroplast genome of anemone and related genera. J. Mol. Evol. **38**, 274–281 (1994)
2. Palmer, J.D., Herbon, L.A.: Tricicular mitochondrial genomes of Brassica and Raphanus: reversal of repeat configurations by inversion. Nuc. Acids Res. **14**, 9755–9764 (1986)
3. Sankoff, D., Leduc, G., Antoine, N., Paquin, B., Lang, B.F., Cedergran, R.: Gene order comparisons for phylogenetic interferce: evolution of the mitochondrial genome. Proc. Nat. Acad. Sci. U.S.A. **89**, 6575–6579 (1992)
4. Bafna, V., Pevzner, P.: Sorting by transposition. SIAM J. Discrete Math. **11**(2), 224–240 (1998)
5. Elias, I., Hartman, T.: A 1.375-approximation algorithm for sorting by transpositions. IEEE/ACM Trans. Comput. Biol. Bioinform. **3**(4), 369–379 (2006)
6. Feng, J., Zhu, D.: Faster algorithms for sorting by transpositions and sorting by block-interchanges. ACM Trans. Algorithms **3**(3), 25 (2007)
7. Eriksson, H., Eriksson, K., Karlander, J., Svensson, L.: Watlund: sorting a bridge hand. J. Discrete Appl. Math. **241**(1–3), 289–300 (2001)
8. Bulteau, L., Fertin, G., Rusu, I.: Sorting by transpositions is difficult. In: Proceeding of 38th International Colloquium on Automata, Languages and Programming (ICALP), vol. 1, pp. 654–665 (2011)
9. Heath, L.S., Vergara, J.P.C.: Sorting by bounded blockmoves. Discrete Appl. Math. **88**, 181–206 (1998)
10. Heath, L.S., Vergara, J.P.C.: Sorting by short blockmoves. Algorithmica **28**(3), 323–354 (2000)
11. Mahajan, M., Rama, R., Vijayakumar, S.: On sorting by 3-bounded transpositions. Discrete Math. **306**(14), 1569–1585 (2006)
12. Jiang, H., Zhu, D., Zhu, B.: A $(1+\varepsilon)$-approximation algorithm for sorting by short block-moves. Theor. Comput. Sci. **54**(2), 279–292 (2011)
13. Jiang, H., Zhu, D.: A 14/11-approximation algorithm for sorting by short block-moves. Sci. China Inf. Sci. **54**(2), 279–292 (2011)
14. Jiang, H., Feng, H., Zhu, D.: An 5/4-approximation algorithm for sorting permutations by short block moves. In: Ahn, H.-K., Shin, C.-S. (eds.) ISAAC 2014. LNCS, vol. 8889, pp. 491–503. Springer, Cham (2014). https://doi.org/10.1007/978-3-319-13075-0_39

Universal Facility Location in Generalized Metric Space

Yicheng Xu[1], Dachuan Xu[2], Yong Zhang[1($\boxtimes$)], and Juan Zou[3]

[1] Chinese Academy of Sciences, Shenzhen Institutes of Advanced Technology,
Shenzhen 518055, People's Republic of China
{yc.xu,zhangyong}@siat.ac.cn
[2] Department of Operations Research and Scientific Computing,
Beijing University of Technology, Beijing 100124, People's Republic of China
xudc@bjut.edu.cn
[3] School of Mathematical Sciences, Qufu Normal University, Qufu 273165,
Shandong, People's Republic of China
zoujuanjn@163.com

Abstract. We consider the universal facility location that extends several classical facility location problems like the incremental-cost facility location, concave-cost facility location, hard-capacitated facility location, soft-capacitated facility location, and of course, uncapacitated facility location. In this problem we are given a set of facilities $\mathcal{F}$ and clients $\mathcal{C}$, as well as the distances between any pair of facility and client. Each facility i has its specific cost function $f_i(\cdot)$ depending on the amount of clients assigned to that facility. The goal is to assign the clients to facilities such that the sum of facility and service costs is minimized. In metric facility location, the service cost is proportional to the distance between the client and its assigned facility. We study a cost measure known as l_2^2 considered by Jain and Vazirani [J. ACM'01] and Fernandes et al. [Math. Program.'15] where the service cost is proportional to the squared distance. We extend their work to include the aforementioned variants of facility location. As our main contribution, a local search based $(11.18 + \varepsilon)$-approximation algorithm is proposed.

Keywords: Universal facility location · Capacitated facility location · Approximation algorithm · Squared metric

1 Introduction

Facility location is no doubt one of the most classical and fundamental NP-hard problems. In the location problem, we are given a set of clients as well as possible locations of facilities. Each client has its demand and each facility has an open cost. Any unit of demand should be severed by an open facility and has to pay a unit of assignment/service cost. The objective is to search a subset of possible locations/facilities to open and assign all demands to the opened facilities such that the total cost of opening facilities and the assignment is minimized.

© Springer Nature Switzerland AG 2019
D.-Z. Du et al. (Eds.): COCOON 2019, LNCS 11653, pp. 591–602, 2019.
https://doi.org/10.1007/978-3-030-26176-4_49

Many variants of facility locations in metric space are studied. Linear programming based technique has been successfully applied to metric uncapacitated facility location to obtain good approximations. The state-of-the-art algorithm for metric uncapacitated facility location is a 1.488-approximation by Li [12], which is very close to the lower bound of 1.463 proved by Guha and Khuller [8] under the assumption of NP $\nsubseteq$ DTIME$[n^{O(\log \log n)}]$ and strengthened by Sviridenko that it holds as long as P $\neq$ NP (Personal Communication). Linear programming based technique yields a 2-approximation for soft-capacitated facility location presented by Mahdian et al. [14] that achieves the integrality gap of the natural LP relaxation of the problem. However, this technique seems to be inefficient for solving hard-capacitated facility locations. The only known result is a 288-approximation proposed by An et al. [2]. Most results are based on local search heuristic including the state-of-the-art 5-approximation algorithm proposed by Bansal et al. [4]. It is also the case for universal facility location which is first proposed by Mahdian and Pál [13] who present the first constant approximation for this problem. The proposed approximation ratio is then improved to 6.702 and 5 by Vygen [18], and Bansal et al. [5] respectively. Local search also achieves series of results in variants of universal facility location, see [19,20].

Facility location in Euclidean space is somehow well-studied. As its name implies, Euclidean facility location considers the service cost proportional to the Euclidean distance. Note Euclidean space is metric thus any approximation results in metric space are also valid for that of Euclidean space. The first PTAS result [3] for 2-dimensional uncapacitated Euclidean facility location appears in STOC'98. This work is then improved by Kolliopoulos and Rao [11] who propose an efficient PTAS for it in any fixed dimensional Euclidean space. Another particular case of facility location considered in literature is so-called squared Euclidean facility location where the service cost is proportional to the squared Euclidean distance. This variant is first considered by Jain and Vazirani [9] as a cost measure named l_2^2. Their work implies a 9-approximation for it. This result is improved afterwards by Fernandes et al. [7] who reanalyze the primal-dual algorithms of uncapacitated metric facility locations and obtain a best possible 2.04-approximation for squared metric uncapacitated facility location.

To the best of our knowledge, seldom work on capacitated or universal facility location in squared metric space are known. We first consider this problem and propose a constant $(11.18 + \varepsilon)$-approximation algorithm based on local search. This work from one side extends the study of the aforementioned universal facility location as well as the squared metric facility location, and from other side allows the inputs to locate in a wider space. In addition, we will extend this work to allow any positive distance inputs and an input-related performance ratio distribution for the proposed algorithm will be presented. Due to space constraint, we mainly propose our work in squared metric space and the extension work will further appear in the full version of this paper. The remainder of this paper is organized as: The notations and models are presented in definition section. Follows by the main body of algorithm and analysis section. In the last section we conclude our results and possible applications. Several missing proofs refer to the full version of this paper.

2 Definitions

The inputs of the universal facility location include: a set of facilities $\mathcal{F}$, a set of clients $\mathcal{C}$, distances between any facility-client pair $c_{ij} \in c_{\mathcal{F} \times \mathcal{C}}$ and a well-defined facility cost function $f_i(\cdot)$ for each $i \in \mathcal{F}$. Let m represent the number of facilities and n represent the number of clients. For the sake of simplicity, we force the demand of each client to be one. Due to Pál et al. [16] and Mahdian et al. [13], arbitrary demand can be handled easily by making slight modifications.

We abuse the definition of distance a little for representing the unit service/assignment cost as most facility location works do, as there is no need to introduce a constant multiplier. In classical facility location, the distances $c_{\mathcal{F} \times \mathcal{C}}$ form a *metric*, which means the distances are nonnegative, symmetric, and obeying the triangle inequality. In this paper, we assume the distances form a *squared metric* where the only difference is that the distances are not necessary obeying the triangle inequality but the squared triangle inequality, a relaxed property. More formally, we call $c_{\mathcal{F} \times \mathcal{C}}$ obeys squared triangle inequality if for any facilities i and i', clients j and j', we have $\sqrt{c_{ij}} \leq \sqrt{c_{ij'}} + \sqrt{c_{i'j'}} + \sqrt{c_{i'j}}$. Note a *metric* must be a *squared metric* (but not vice versa). Therefore any approximation algorithm for squared metric facility location is also that for metric facility location, and the inapproximability bound for metric facility location is also valid for squared metric facility location. And the facility location dose not differ from its cost measure in the mathematical programming point of view.

The opening cost function of facilities is so well-defined that the universal facility location can be a general model of a variety of facility locations. We assume each $f_i(\cdot)$ is a non-decreasing and left-continuous mapping from non-negative reals to non-negative reals with infinity. And it makes sense by assuming $f_i(0) = 0$ since we do not need to pay for what we do not use. Due to [13], these assumptions guarantee the existence of a globally optimal solution to the following mathematical programming, and we simply ignore the proof.

$$\min \sum_{i \in \mathcal{F}} f_i(u_i) + \sum_{i \in \mathcal{F}, j \in \mathcal{C}} c_{ij} x_{ij} \tag{1}$$

$$\text{s.t. } \sum_{i \in \mathcal{F}} x_{ij} = 1, \qquad \forall \, j \in \mathcal{C}, \tag{2}$$

$$\sum_{j \in \mathcal{C}} x_{ij} \leq u_i, \qquad \forall \, i \in \mathcal{F}, \tag{3}$$

$$x_{ij} \in \{0, 1\}, \qquad \forall \, i \in \mathcal{F}, j \in \mathcal{C}, \tag{4}$$

$$u_i \geq 0, \qquad \forall \, i \in \mathcal{F}, j \in \mathcal{C}. \tag{5}$$

For notational convenience, u_i is introduced to represent the allocation of facility i. Since $f_i(\cdot)$ is nondecreasing, we assume w.l.o.g. that for any facility

i, it must be the case that $u_i = |\{j : x_{ij} = 1\}|$ (thus integral), otherwise one can reduce u_i without increasing the opening cost of facility i. Therefore, any solution to programming (1–5) can be specified by pair (u, x) in which u is a m-dimensional vector representing the allocation and x is a mn-dimensional vector representing the assignment. Also note, the above programming is actually tractable in polynomial time when given allocation vector u. In fact, as long as (1) the coefficient matrix is totally unimodular and (2) u is an integral vector, the above programming is equivalent to its linear relaxation and thus polynomial time tractable. Obviously, (1) and (2) are satisfied in (1–5). Detail see [17] as a reference. Therefore either u or x contains the complete information of a solution to the above programming.

The universal facility location is general enough to include several important variants of facility location in the literature as special cases. For instance, by setting $f_i(u_i) = F_i \cdot 1_{\{u_i > 0\}}$ we obtain uncapacitated facility location, where F_i is the constant opening cost of facility i and the indicator function $1_{\{u_i > 0\}}$ equals 1 when $u_i > 0$ and 0 otherwise. By setting $f_i(u_i) = F_i \cdot \lceil u_i/U_i \rceil$ where F_i is the constant opening cost for one copy of facility i and U_i is its capacity, we obtain the soft-capacitated variant. Any arbitrary concave function $f_i(u_i)$ end up in the concave-cost facility location variant. Note uncapacitated facility location and incremental-cost (linear-cost) facility location are also concave facility locations. See figures below the opening cost functions in the above three facility location variants, all of which are non-decreasing and left-continuous (Fig. 1).

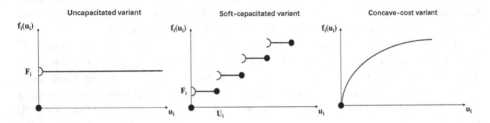

Fig. 1. Opening cost functions in uncapacitated, soft-capacitated and concave-cost facility locations

3   Local Search Based Algorithm

We borrow the main idea of local search heuristic from literature as it is the only known efficient technique in universal facility location study. When given current feasible solution $S = (u, x)$, as stated in last section, all we have to cope with is the allocation vector u. Let σ be the mapping from clients to facilities that reflects the assignment. That is, $\sigma(j)$ represents the facility with $x_{ij} = 1$ and $\sigma^{-1}(i)$ represents all the clients with $x_{ij} = 1$.

3.1   Moves and Moves Finding

The following defined moves are to reduce the current cost and by iteratively taking these moves we obtain the locally optimal solution to our problem.

- $add(s, \delta)$: Increase the allocation of facility s by an amount $\delta > 0$ and solve the optimal assignment under the adjusted allocation. This move is also considered by many previous work to reduce the service cost.
- $pivot(s, \Delta)$: Adjust the allocation of all facilities except s by a vector Δ (the allocation of each facility i is adjusted by an amount of Δ_i). We find the pivot facility s to connect the source facilities (with $\Delta_i < 0$) with sink ones (with $\Delta_i > 0$). That means we must ship $|\Delta_i|$ clients from i to s for facilities with $\Delta_i < 0$, and ship $|\Delta_i|$ clients from s to those with $\Delta_i > 0$. This move only makes sense when $\sum_{i \in \mathcal{F}} \Delta_i \geq 0$ as the total number of clients must be conserved.

Note for *pivot* move, $\sum_{i \in \mathcal{F}} \Delta_i = 0$ is satisfied after reducing the allocation from $u_i + \Delta_i$ to the real amount of clients assigned to i for those facilities that these two amount are not equal. To obtain a locally optimal solution, any move with positive cost save is acceptable. We define the cost function (a negative value implies cost save) as estimate cost for the two types of moves below respectively. And we claim that both moves with minimum cost function values can be computed in polynomial time. Note the estimate cost of *pivot* move is tighter than previous work [5,13,18] for metric facility locations in order to bound the opening cost efficiently in this generalized metric space.

$$c_{add(s,\delta)} = f_s(u_s + \delta) - f_s(u_s) + c_s(S') - c_s(S)$$

The cost function $c_{add(s,\delta)}$ consists of two parts, opening cost increase and service cost increase. We assume S is the initial solution and S' is the optimal solution under the adjusted allocation after one $add(s, \delta)$ move. $c_s(S)$ denotes the service cost of solution S, and later we use $c_f(S)$ denoting the opening cost of S. The cost increase for $pivot(s, \Delta)$ can be formulated as follows.

$$c_{pivot(s,\Delta)} = \sum_{i \in \mathcal{F}} (f_i(u_i + \Delta_i) - f_i(u_i)) + c_s(S') - c_s(S)$$

Also, $c_{pivot(s,\Delta)}$ consists of opening cost increase and service cost increase. And we abuse the notations $c_s(S)$ and $c_s(S')$ a little to represent the service cost of initial solution S and one *pivot* move solution S' respectively. More accurately, it is equivalent to the following expression.

$$c_{pivot(s,\Delta)} = \sum_{i \in \mathcal{F}} (f_i(u_i + \Delta_i) - f_i(u_i)) + \sum_{i \in \mathcal{F}} \sum_{j \in \sigma^{-1}(i)_{|\Delta_i|}} (c_{sj} - c_{ij})(-1)^{1\{\Delta_i > 0\}}$$

We introduce the indicator for event $\Delta_i > 0$ to separate the cost increase of s to i from those i to s. And $\sigma^{-1}(i)_{|\Delta_i|}$ represents the $|\Delta_i|$-element subset of $\sigma^{-1}(i)$ with minimum values of $(c_{sj} - c_{ij})(-1)^{1\{\Delta_i > 0\}}$. In other words, one can order the clients in $\sigma^{-1}(i)$ according to the nondecreasing rule of $(c_{sj} - c_{ij})(-1)^{1\{\Delta_i > 0\}}$, and the first $|\Delta_i|$ clients at the top of the list are subset $\sigma^{-1}(i)_{|\Delta_i|}$.

Observe that once we are given pair (s, δ) or (s, Δ), we are able to compute $c_{add(s,\delta)}$ or $c_{pivot(s,\Delta)}$ immediately. Now we claim that both moves with minimum cost function increase can also be found in polynomial time. The proof for add move is quite straightforward. Since we have only mn choices for (s, δ) pair, we can simply try all of them. But it is not the case for $pivot$ move.

Lemma 1. *Given current solution $S = (u, x)$, we can in polynomial time find s and Δ such that $c_{pivot(s,\Delta)}$ is minimized.*

Proof. Refer to the full version of this paper.   □

We fulfill the above proof (in the full version) mainly to see that we do not require $c_{\mathcal{F} \times \mathcal{C}}$ be *metric* as previous work ([5, 13, 18]) do.

3.2   Significant Moves and Polynomial-Time Proof

Even though we have one optimal move executed in polynomial time, we cannot guarantee a polynomial-time algorithm. This because of the unbounded number of moves. For example, a sequence of moves with cost save $1/k^2$ in kth step cannot end up in a locally optimum within finite steps. Because $\sum_{k=1}^{\infty} 1/k^2$ converges to a finite real $\pi^2/6$, the iteration will never terminate before it hits a minimum. Thus we introduce a standard idea of significant move that requires a lower bound of step length. With such requirement we guarantee that our algorithm only takes polynomial many moves. Specifically, let the step length be $p(n, \varepsilon)$ times the current cost of solution, where $p(n, \varepsilon)$ is a suitable chosen polynomial w.r.t. n and small ε. Note in this way, we find the approximate locally optimal solution instead of the real locally optimal one, with only a factor of $p(n, \varepsilon)$ difference.

Lemma 2. *If one only takes any of the two moves whenever he finds one with cost save more than $p(n, \varepsilon)$ times the current, then after at most $O(p^{-1}(n, \varepsilon) \log \frac{c(S_0)}{c(S_k)})$ moves, the algorithm terminates, where S_0 denotes the initial solution and S_k denotes the approximate locally optimal solution.*

Proof. Refer to the full version of this paper.   □

4   Analysis

In this section, we mainly aim to compare the performance of the locally optimal solution with the global one, and we use $S := (u, x)$ and $S^* := (u^*, x^*)$ representing them respectively. Remember $c(S)$ is the total cost of a solution S; $c_s(S)$ and $c_f(S)$ are the service cost and opening cost. We separate this section into two parts, one of which is to bound the service cost using add move and the other is to bound the opening cost using $pivot$ move.

4.1 Bounding the Service Cost

In the facility location literature, the *add* move is widely used to efficiently reduce the high service cost of a locally optimal solution. The lemma first proposed by Guha, Khuller [8] and Korupolu et al. [10] for the capacitated facility location is also valid for universal facility location.

Lemma 3. *Any locally optimal solution w.r.t. the add move has service cost no larger than the total cost of the global optimal solution, i.e., $c_s(S) \leq c(S^*)$.*

Proof. For locally optimal solution S together with its allocation and assignment (u, x) and global optimal solution S^* together with (u^*, x^*), consider the facility i with $u_i < u_i^*$. If none, x^* is a feasible assignment and we are done. Otherwise take moves $add(i, u_i^* - u_i)$ for all such facilities. Observe that x^* must be feasible to the new solution. Thus the cost increase is no more than $c_f(S^*) + c_s(S^*) - c_s(S)$. From the local optimality of S we have $c_f(S^*) + c_s(S^*) - c_s(S) \geq 0$, implying the lemma. $\qquad\square$

This lemma is simple but powerful. We sketch the proof mainly in order to find out that it does not require any *metric* property.

4.2 Bounding the Opening Cost

We show in this subsection that whenever a feasible solution has high opening cost, an efficient *pivot* move exists. In other words, a locally optimal solution w.r.t. the *pivot* move will never have a high opening cost. We begin with a transhipment problem that will help to build the connection between S and S^*. We shall show that by taking several *pivot* moves in a well-defined graph we can switch S to S^* and afterwards an analysis on the cost increase will result in the bound for opening cost.

Let $\mathcal{F}_{out} = \{i \in \mathcal{F} : u_i > u_i^*\}$ and $\mathcal{F}_{in} = \{i \in \mathcal{F} : u_i^* > u_i\}$. We set up a transhipment that move a number of clients from $\mathcal{F}_{out}$ to $\mathcal{F}_{in}$, where each $s \in \mathcal{F}_{out}$ has $u_s - u_s^*$ clients shipping out and each $t \in \mathcal{F}_{in}$ has $u_t^* - u_t$ clients shipping in. Note this is feasible as $\sum_{s \in \mathcal{F}_{out}} (u_s - u_s^*) = \sum_{t \in \mathcal{F}_{in}} (u_t^* - u_t)$ which is derived from the fact that the total number of assigned clients are equal in S and that in S^*. Note we do not specify the reassignment. Specifically, we model this transhipment using the following programming.

$$\min \quad \sum_{s \in \mathcal{F}_{out}, t \in \mathcal{F}_{in}} c_{st} r(s, t) \tag{6}$$

$$\text{s.t.} \quad \sum_{t \in \mathcal{F}_{in}} r(s, t) = u_s - u_s^*, \qquad \forall\, s \in \mathcal{F}_{out}, \tag{7}$$

$$\sum_{s \in \mathcal{F}_{out}} r(s, t) = u_t^* - u_t, \qquad \forall\, t \in \mathcal{F}_{in}, \tag{8}$$

$$r(s, t) \geq 0, \qquad \forall\, s \in \mathcal{F}_{out}, t \in \mathcal{F}_{in}. \tag{9}$$

Any feasible solution to the above programming implies an feasible assignment of clients that switch the allocation from S to S^*. But remember we do not have the distances between facilities as inputs. Unlike metric facility location algorithms do, we first define a fictitious distance $c'_{st} := \min_{j \in \mathcal{C}}(\sqrt{c_{sj}} + \sqrt{c_{tj}})^2$ for each facility pairs $s, t \in \mathcal{F}$, and update that distance to be the shortest path between facilities if we can reduce the distance by doing this update. Particularly, define $c_{st} := \min\{c'_{st}, \min_{p \in P(s,t)} length(p)\}$, where $P(s,t)$ denotes the set of all paths consisting of facilities that start in s and end in t, and $length(p)$ is the sum of distances of all passing edges in path p. Note the above definition meet the squared metric triangle inequality, i.e., for any $s, t \in \mathcal{F}$ and $j \in \mathcal{C}$, $\sqrt{c_{st}} \leq \sqrt{c_{sj}} + \sqrt{c_{tj}}$. Under this definition, we are able to find a feasible way to transfer the allocation of S to that of S^* with an acceptable cost.

Lemma 4. *There exists a feasible transhipment at cost no more than $2(c_s(S) + c_s(S^*))$.*

Proof. Refer to the full version of this paper. □

To simplify the analysis, we start with the optimal transhipment w.r.t. the programming (6–9) denoting by y, of which the cost is at most $2(c_s(S) + c_s(S^*))$. We interpret the element of y in each edge a nonnegative arc flow. From flow decomposition theorem [1], we can represent the arc flow as path and cycle flow such that each directed path flow connects a source facility to a sink one. Augmenting the flow along the cycle must not increase the cost (since y is optimal) and by iteratively doing this we can remove all the cycles without increase the cost. Thus, the modified y forms a forest.

We focus on this forest and partition it into simple subtrees that have depth at most 2 and rooted at facilities in $\mathcal{F}_{in}$. Let $\mathcal{T}_t$ be such subtree rooted at t. For any node s in the tree, let $C(s)$ be its children and $p(s)$ be its parent. We sometimes abuse this notation by replacing s with a set S if no confusion. Let $y(s,t)$ be the directed flow value on edge (s,t), $y(s,\cdot)$ be the total flow value going out of s and $y(\cdot,t)$ be the total flow value going in t. We sometimes abuse this notation by replace a facility by a set of facilities, for example $y(s,S) := \sum_{i \in S} y(s,i)$. In fact, $y(s,\cdot) = u_s - u_s^*$ for any $s \in \mathcal{F}_{out}$ and $y(\cdot,t) = u_t^* - u_t$ for any $t \in \mathcal{F}_{in}$. By closing a facility $s \in \mathcal{F}_{out}$ we mean to reassign $y(s,\cdot)$ out of s, and open a $t \in \mathcal{F}_{in}$ means we reassign $y(\cdot,t)$ clients to t. Note we may open a facility t several times, which means an excess use of $y(\cdot,t)$. Now we build a set of *pivot* moves to reassign the excess allocations from $\mathcal{F}_{out}$ to $\mathcal{F}_{in}$.

Let us begin with a simple particular case that $\mathcal{T}_t$ has no grandchildren, that is to say, $\mathcal{T}_t$ is a subtree of depth one. In this case we consider the move $pivot(t, \Delta)$ in which Δ can be defined to close all children of t and reassign all their allocations to t. This move is obviously feasible since we open the only neighbor of every nodes in $C(t)$.

For nontrivial subtree $\mathcal{T}_t$ of depth 2, we partition $C(t)$ into two subsets that will be separately cope with. Particulary, for arbitrary $s \in \mathcal{F}_{out}$, we say s is an up-facility if $y(s, p(s)) \geq y(s, C(s))$, otherwise it is a down-facility. Note an up

s also means $y(s, p(s)) \geq \frac{1}{2} y(s, \cdot)$. We denote all up-facilities by $\mathcal{F}_u$ and down-facilities by $\mathcal{F}_d$. For the nodes in $C(t)$ that are up-facilities, which for notation convenience we denote by $C_u(t) := C(t) \cap \mathcal{F}_u$, consider a single $pivot(t, \Delta)$ move that closes all facilities in $C_u(t)$ and open t as well as all the children of $C_u(t)$. This move ships the allocations of $C_u(t)$ up to pivot t and then reassigns to the children of $C_u(t)$.

For the rest of the facilities $C_d(t)$, we cannot close all these facilities through a single $pivot(t, \Delta)$ move. Because it is infeasible to ship all the allocations from $C_d(t)$ to the pivot t as the flow downwards may be arbitrarily larger than upwards. We consider these facilities in an nondecreasing order of the value $y(s, t)$. That is, assume $C_d(t) = \{s_1, s_2, \ldots, s_k\}$ is so labeled that $y(s_1, t) \leq y(s_2, t) \leq \ldots \leq y(s_k, t)$. For an arbitrary $s_v \in C_d(t)$ with $v \in \{1, 2, \ldots, k-1\}$, consider the $pivot(s_v, \Delta)$ move that closes s_v and open all facilities in $C(s_v) \cup C(S_{v+1})$. Note this move is pivot at s_v, so we can directly ship the $y(s_v, u)$ amount of clients from s_v to u for an arbitrary $u \in C(s_v)$. And the rest of the allocation should be reassigned along the edges (s_v, t), (t, s_{v+1}), and set of edges $(s_{v+1}, C(s_{v+1}))$. For the special facility s_k, we consider such $pivot(s_k, \Delta)$ that closes s_k and open all its neighbors including the root t. Later we will prove all these moves are feasible and economical. First of all, let us officially define the aforementioned $pivot$ moves as follows.

Trivial $\mathcal{T}_t$: Consider $pivot(t, \Delta)$ move with Δ defined as:

$$\Delta_i = \begin{cases} -(u_i - u_i^*), & i \in C(t); \\ u_t^* - u_t, & i = t; \\ 0, & otherwise. \end{cases}$$

Nontrivial $\mathcal{T}_t$:

– $C_u(t)$: Consider $pivot(t, \Delta)$ move with Δ defined as:

$$\Delta_i = \begin{cases} -(u_i - u_i^*), & i \in C_u(t); \\ u_i^* - u_i, & i \in C(C_u(t)); \\ 0, & otherwise. \end{cases}$$

– $C_d(t) \setminus \{s_k\}$: Consider a set of $pivot(s_v, \Delta)$ moves for all $s_v \in C_d(t) \setminus \{s_k\}$ with the corresponding Δ defined as:

$$\Delta_i = \begin{cases} -(u_{s_v} - u_{s_v}^*), & i = s_v; \\ u_i^* - u_i, & i \in C(s_v) \cup C(s_{v+1}); \\ 0, & otherwise. \end{cases}$$

– s_k: Consider $pivot(s_k, \Delta)$ move with Δ defined as:

$$\Delta_i = \begin{cases} -(u_{s_k} - u_{s_k}^*), & i = s_k; \\ u_i^* - u_i, & i \in C(s_k) \cup \{t\}; \\ 0, & otherwise. \end{cases}$$

Note in the above definitions, the negative sign is a reminder that the corresponding value is negative. We analyze the feasibility case by case in the following lemma. Remember a $pivot(s, \Delta)$ move is feasible only if $\sum_{i \in \mathcal{F}} \Delta_i \geq 0$.

Lemma 5. *All the pivot moves considered above are feasible.*

Proof. Refer to the full version of this paper.   □

Observing the above, we are able to close all the facilities in $\mathcal{F}_{out}$ once, at the same time open as few times as possible the facilities in $\mathcal{F}_{in}$ and obey the transhipment flow along the edges. Particularly, we conclude the following.

Lemma 6. *Combining all the pivot moves above, we close all the facilities in $\mathcal{F}_{out}$ exactly once, at the same time open an arbitrary $t \in \mathcal{F}_{in}$ at most 4 times, and use at most 3 times the flow capacity along the edges.*

Proof. The conclusion about closing $\mathcal{F}_{out}$ is obvious. For the open times of $\mathcal{F}_{in}$, do not forget we are considering the whole forest which are partitioned into above subtrees. We only consider the worst case when a facility $t \in \mathcal{F}_{in}$ is the root of a nontrivial subtree $\mathcal{T}_t$ and a leaf in another subtree $\mathcal{T}_{t'}$ where it happen to be a child node of an facility s_k (same property as s_k described in previous analysis). Such t is opened 2 times as a root and 2 times as a leaf, implies the conclusion.

Also, we only focus on the worst case for the edge capacity and the rest cases can be similarly checked. In the nontrivial case, consider the move closing all $s \in C_u(t)$. We first ship $y(s, \cdot)$ clients to t, leave $y(s, t)$ clients at t and then ship the rest clients to $C(s)$. This ends in using (s, t) totally $2y(s, \cdot) - y(s, t)$ flow of clients. Remember the capacity is $y(s, t)$ and s is an up-facility which means $y(s, \cdot) \leq 2y(s, t)$. So $2y(s, \cdot) - y(s, t) \leq 3y(s, t)$, completing the proof.   □

The feasibility and economy of the defined *pivot* moves end in the following main result of a locally optimal solution.

Lemma 7. $c_f(S) \leq 10c_f(S^*) + 12c_s(S^*)$

Proof. This upper bound for opening cost can be directly derived from Lemma 6. Since we reassign all excess allocations of facilities in S to that of S^*, we have $c_f(S) \leq 4c_f(S^*) + 3c(y)$. And from Lemma 4 we know $c(y) \leq 2(c_s(S) + c_s(S^*))$, combing with $c_s(S) \leq c(S^*)$ obtain, $c_f(S) \leq 10c_f(S^*) + 12c_s(S^*)$.   □

4.3   Scaling and Polynomial-Time Establishing

Combine Lemma 7 with aforementioned upper bound for service cost, we have a straightforward performance guarantee of 13. But we can do slightly better than this by employing a standard scaling technique. That is, we scale the unit service/opening cost by a factor of λ as a data preprocessor. Then applying the algorithm on the modified inputs, we conclude that by doing this we obtain a slightly better return from the algorithm. Formally speaking, the modified upper bounds for service and opening cost are $\lambda c_s(S) \leq c_f(S^*) + \lambda c_s(S^*)$ and $c_f(S) \leq 10c_f(S^*) + 12\lambda c_s(S^*)$ respectively. By setting $\lambda \approx 0.8482$ we obtain a performance guarantee of 11.18.

Recall we restrict a polynomial step length of the proposed local search algorithm by $p(n, \varepsilon)$. In the above we analyze the performance of a locally optimal solution, not the approximate locally optimal one. There is only ϵ difference of these two solutions by setting $p(n, \varepsilon) = \varepsilon/6n$ and substitute into the proof of previous lemmas. Putting all things together, we have the following theorem.

Theorem 1. *The proposed local search algorithm w.r.t. add and pivot moves outputs an approximate locally optimal solution with performance ratio* $11.18 + \varepsilon$.

5   Conclusion

Metric property plays an important role in many combinatorial problems. For example, we all know that under the assumption P $\neq$ NP, there does not exist an α-approximation algorithm (for any constant $\alpha > 1$) for the traveling salesman problem. Even an $O(2^n)$ approximation ratio implies P = NP, where n is the number of cities. But under the *metric* assumption, one can easily obtain a 2-approximation or 1.5-approximation algorithm [6]. And the *metric* traveling salesman problem has a much lower inapproximability bound of $220/219 \approx 1.0045$ [15], provided by P $\neq$ NP.

However, it is time-consuming to check whether a set of distance inputs are *metric* or not if we are not informed beforehand. Our algorithm allows distance inputs to locate in a more generalized space. And for the worst case in this generalized space, we have a constant approximation ratio of $11.18 + \varepsilon$.

Acknowledgement. The first author is supported by China Postdoctoral Science Foundation funded project (No. 2018M643233), Hong Kong GRF 17210017 and Shenzhen Discipline Construction Project for Urban Computing and Data Intelligence. The second author is supported by Natural Science Foundation of China (Nos. 11531014, 11871081). The third author is supported by Natural Science Foundation of China (No. 61433012), Shenzhen research grant (KQJSCX20180330170311901, JCYJ20180305180840138 and GGFW2017073114031767). The fourth author is supported by Natural Science Foundation of China (No. 11801310).

References

1. Ahuja, R.K., Magnanti, T.L., Orlin, J.B.: Network Flows: Theory, Algorithms, and Applications. Prentice Hall Englewood Cliffs (1993)
2. An, H.C., Singh, M., Svensson, O.: LP-based algorithms for capacitated facility location. SIAM J. Comput. **46**(1), 272–306 (2017)
3. Arora, S., Raghavan, P., Rao, S.: Approximation schemes for Euclidean k-medians and related problems. In: Proceedings of the 30th Annual ACM Symposium on the Theory of Computing, pp. 106–113 (1998)

4. Bansal, M., Garg, N., Gupta, N.: A 5-approximation for capacitated facility location. In: Proceedings of the 20th Annual European Symposium on Algorithms, pp. 133–144 (2012)

5. Bansal, M., Garg, N., Gupta, N.: A 5-approximation for universal facility location. In: Proceedings of the 38th Annual Conference on Foundations of Software Technology and Theoretical Computer Science (2018)

6. Christofides, N.: Worst-case analysis of a new heuristic for the travelling salesman problem. No. RR-388. Carnegie-Mellon University Pittsburgh Pa Management Sciences Research Group (1976)

7. Fernandes, C.G., Meira, L.A.A., Miyazawa, F.K., Pedrosa, L.L.C.: A systematic approach to bound factor-revealing LPs and its application to the metric and squared metric facility location problems. Math. Program. **153**(2), 655–685 (2015)

8. Guha, S., Khuller, S.: Greedy strikes back: improved facility location algorithms. J. Algorithms **31**(1), 228–248 (1999)

9. Jain, K., Vazirani, V.V.: Approximation algorithms for metric facility location and k-median problems using the primal-dual schema and Lagrangian relaxation. J. ACM **48**(2), 274–296 (2001)

10. Korupolu, M.R., Plaxton, C.G., Rajaraman, R.: Analysis of a local search heuristic for facility location problems. J. Algorithms **37**(1), 146–188 (2000)

11. Kolliopoulos, S.G., Rao, S.: A nearly linear-time approximation scheme for the Euclidean k-median problem. SIAM J. Comput. **37**(3), 757–782 (2007)

12. Li, S.: A 1.488 approximation algorithm for the uncapacitated facility location problem. Inf. Comput. **222**, 45–58 (2013)

13. Mahdian, M., Pál, M.: Universal facility location. In: Proceedings of the 11th Annual European Symposium on Algorithms, pp. 409–421 (2003)

14. Mahdian, M., Ye, Y., Zhang, J.: A 2-approximation algorithm for the soft-capacitated facility location problem. In: Proceedings of the Approximation, Randomization, and Combinatorial Optimization, Algorithms and Techniques, pp. 129–140 (2003)

15. Papadimitriou, C.H., Vempala, S.: On the approximability of the traveling salesman problem. Combinatorica **26**(1), 101–120 (2006)

16. Pál, M., Tardos, É., Wexler, T.: Facility location with nonuniform hard capacities. In: Proceedings of the 42nd IEEE Symposium on Foundations of Computer Science, pp. 329–338 (2001)

17. Schrijver, A.: Combinatorial Optimization: Polyhedra and Efficiency. Springer, Heidelberg (2003)

18. Vygen, J.: From stars to comets: improved local search for universal facility location. Oper. Res. Lett. **35**(4), 427–433 (2007)

19. Xu, Y., Xu, D., Du, D., Wu, C.: Local search algorithm for universal facility location problem with linear penalties. J. Global Optim. **67**(1–2), 367–378 (2017)

20. Xu, Y., Xu, D., Du, D., Wu, C.: Improved approximation algorithm for universal facility location problem with linear penalties. Theoret. Comput. Sci. **774**, 143–151 (2019)

Activation Probability Maximization for Target Users Under Influence Decay Model

Ruidong Yan[1], Yi Li[2], Deying Li[1(✉)], Yuqing Zhu[3], Yongcai Wang[1], and Hongwei Du[4]

[1] School of Information, Renmin University of China, Beijing, China
{yanruidong,deyingli,ycw}@ruc.edu.cn
[2] Department of Computer Science, University of Texas at Dallas, Richardson, USA
yi.li@utdallas.edu
[3] Department of Computer Science, California State University at Los Angeles, Los Angeles, USA
yuqing.zhu@calstatela.edu
[4] Department of Computer Science and Technology, Harbin Institute of Technology, Shenzhen, China
hongwei.du@ieee.org

Abstract. In this paper, we study how to activate a specific set of targeting users $\mathcal{T}$, e.g., selling a product to a specific target group, is a practical problem for using the limited budget efficiently. To address this problem, we first propose the *Activation Probability Maximization* (APM) problem, i.e., to select a seed set S such that the activation probability of the target users in $\mathcal{T}$ is maximized. Considering that the influence will decay during information propagation, we propose a novel and practical *Influence Decay Model* (IDM) as the information diffusion model in the APM problem. Based on the IDM, we show that the APM problem is NP-hard and the objective function is monotone non-decreasing and submodular. We provide a $(1 - 1/e)$-approximation *Basic Greedy Algorithm* (BGA). Furthermore, a speed-up *Scalable Algorithm* (SA) is proposed for online large social networks. Finally, we run our algorithms by simulations on synthetic and real-life social networks to evaluate the effectiveness and efficiency of the proposed algorithms. Experimental results validate our algorithms are superior to the comparison algorithms.

Keywords: Social network · Influence decay model · Seed selection · Sub-modularity · Target user

This work is partly supported by National Natural Science Foundation of China under grant 11671400, 61672524.

D.-Z. Du et al. (Eds.): COCOON 2019, LNCS 11653, pp. 603–614, 2019.
https://doi.org/10.1007/978-3-030-26176-4_50

1   Introduction

Currently, online social networks such as Facebook, Twitter, Wechat and Google+ have become major social tools for users to share and to disseminate information. Information quickly and widely spread on social networks with "Word-of-mouth" effects compared to traditional ways such as TV, radio, etc. According to statistics, there are 2.34 billion people around the world frequently visiting social networks.

Online marketing is the most representative application of social networks. In viral marketing, the *Influence Maximization* (IM) problem has been extensively studied. Specifically, a company launches a kind of novel product and wants to market it by social networks. Due to the limited budget, it can only choose a small number of initial users (seeds) to provide them free or discounted samples. The company hopes that these initial users would like the products and recommend them to their friends on the social networks. Similarly, their friends will influence more friends in the same way. Finally, the company wants to maximize the number of users who would like to adopt the product.

However, in some practical scenarios, one may consider to maximize the activation probability of a set of target users for efficiently using the limited budget. More specifically, assume that each user in a social network has a potential value[1] for a company. The company may pay more attention to the users who have higher potential values. These higher potential value users are called the *target users* and denoted by a set T. These target users may be potential adopters, such as Make-up artists for a specific kind of cosmetics, or the highly influential and authoritative users who have high probability of activating other users. Intuitively, the company will benefit from maximizing the activation probability of these target users. In most cases, the company cannot directly reach these target users, so the company aims to find the optimal seed set such that the sum of activation probability of the target users is maximized. We call this problem *Activation Probability Maximization* (APM).

It's obvious that the IM problem is different from the APM problem. The former selects a seed set from all nodes in network within a budget such that the expected number of nodes influenced by the seed set through information diffusion is maximized. However, the latter selects a seed set from all nodes except the target set such that the sum of activation probabilities of the target users is maximized.

To the best of our knowledge, only a few studies explored the APM problem even though it plays an essential role in viral marketing. The similar studies have been done in some researches such as [5,19]. Guo et al. [5] propose a problem to find the top-k most influential nodes to a given user. They develop a simple greedy algorithm to solve the problem based on the sub-modularity. We expand their work and solve the APM problem from different perspectives. In [19], Yang et al. advocate recommendation support for active friending, where

---

[1] The potential value can be obtained through statistical or machine learning based methods. And this is beyond the scope of this paper.

a user actively specifies a friending target. In other words, they want to maximize the probability that the friending target would accept an invitation from the source user. The differences between our APM and the previous works are: (1) APM chooses seeds from all nodes except the target nodes instead of all nodes on social network. (2) APM maximizes the activation probability of the target users instead of the expected number of influenced nodes.

Another key issue is about propagation probability. According to [14], a user's influence in a social network decays with time, i.e., the influence propagation probability is dynamic. For example, a famous football player Bailey said, "Brazil is the biggest favorite in the 2018 World Cup. Neymar can take advantage of the 2018 World Cup shame, optimistic about Brazil's finals." This news is quickly reported by major media websites and newspapers in February 2018 such as Baidu, Sina, etc. As time goes by, to the end of the World Cup, the news is significantly less affected by the Brazilian fans than before. Indeed, the influence to users after the World Cup would not bring any profit to the marketers. In the previous literature, well-studied influence propagation models such as *Independent Cascade* (IC) and *Linear Threshold* (LT) models overlook the fact that users' influence on others will decrease with time, which can lead to an over-estimation of the activation probabilities. In this paper, we consider a more realistic *Influence Decay Model* (IDM) that gives a more accurate estimation of the activation probability of the target users. We summarize main contributions as following:

- We propose the *Activation Probability Maximization* (APM) problem for the a specific set of target users.
- We propose a new influence diffusion model named *Influence Decay Model* (IDM) based on the IC model.
- We show that APM is NP-hard and objective function is monotone non-decreasing as well as submodular. Furthermore, we show that computing the activation probability of the target users is $\#P - hard$.
- We propose a *Basic Greedy Algorithm* (BGA) that has a $(1 - 1/e)$ approximation ratio. In addition, a speed-up *Scalable Algorithm* (SA) is proposed for online large social networks.
- To evaluate proposed algorithms, we use a synthetic network and three real-life social networks in experiments. Experimental results validate the proposed algorithms are superior to the comparison methods.

The rest of this paper is organized as follows. In Sect. 2, we begin by recalling some existing related work. In Sect. 3, the influence diffusion model is presented. In Sect. 4, we introduce the problem description and show the properties of the objective function. Algorithms are designed for solving the APM problem in Sect. 5. The experiment results are shown in Sect. 6. We draw conclusions in Sect. 7.

2   Related Work

Kempe et al. [7] model viral marketing as a discrete optimization problem, which is named *Influence Maximization* (IM). They propose a greedy algorithm with

$(1 - 1/e)$-approximation ratio since the objective function is submodular under the *Independent Cascade* (IC) or *Linear Threshold* (LT) model.

Note that the previous researches [8,16,18] study social influence problem without the target users. Therefore they cannot be directly convert to the APM problem. Specifically, Tong et al. [16] present a *Dynamic Independent Cascade* (DIC) model to study the strategies selecting seed users in an adaptive manner. Yan et al. [18] investigate the problem of group-level influence maximization with budget constraint. They introduce a statistical method to reveal the influence relationship among the groups. Furthermore, theoretical analysis shows that their algorithm can guarantee an approximation ratio of at least $1 - \sqrt{e}$. In [8], Kuhnle et al. consider *Threshold Activation Problem* (TAP) which finds a minimum size set triggering expected activation at a certain threshold. They exploit the bicriteria nature of solutions to TAP and control the running time by a parameter.

The related works involve the target users such as [13,20]. In [20], Zhou et al. study a new problem: Give an activatable set A and a targeted set T, finding the k nodes in A with the maximal influence in T. They give a greedy algorithm with a $(1 - 1/e)$-approximation ratio. In [13], Song et al. formalize the problem targeted influence maximization in social networks. They adopt a login model where each user is associated with a login probability and he can be influenced by his neighbors only when he is online. Moreover, they develop a sampling based algorithm that returns a $(1 - 1/e - \varepsilon)$-approximate solution.

The other category involve the decay models such as [3,9,10]. In [3], Feng et al. study the influence maximization from the impact of *novelty decay* on influence propagation, i.e., repeated exposures will have diminishing influence on users. Furthermore, they propose the *IC model with Novelty Decay* (ICND) as their diffusion model. Liu et al. [9] propose algorithms for the time constrained influence maximization problem. In [10], Mohammadi et al. propose the *Time-Sensitive Influence Maximization* (TSIM) problem, which takes into account the time dependence of the information value. They develop two diffusion models based on the time delay, namely the *Delayed Independent Cascade Model* (DICM) and the *Delayed Linear Threshold Model* (DLTM).

In this paper, we simultaneously consider maximizing the activation probability of the target users and dynamic influence diffusion model. And we believe that it can provide a good seed selection for the company in social marketing.

3   Influence Diffusion Model

Independent Cascade Model: A directed social network is denoted as $G = (V, E, p)$, where V is the node set (users) and $E \subseteq V \times V$ is the edge set (the relationships between users). In the IC model, $e_{vu} \in E$ denotes a directed edge from v to u and p_{vu} of edge e_{vu} denotes the probability that node v can successfully activate node u. We call a node *active* if it accepts the product or the information from other nodes, *inactive* otherwise. Influence propagation process unfolds in discrete time steps. The initial seed set is S_0. Let S_t denote the *active*

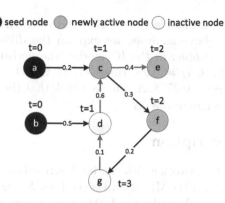

Fig. 1. An example for the influence decay model.

nodes in time t, and each node v in S_t has single chance to activate each *inactive* neighbor u through its out-edge with probability p_{vu} at time $t + 1$. Then repeat this process until no more new nodes can be activated. Note that a node can only switch from *inactive* to *active*, but not in reverse direction.

Influence Decay Model: We propose the *Influence Decay Model* (IDM) based on the IC model. The biggest difference between our model and the IC model is that the propagation probability of each edge on the network decays with time. More specifically, at any time step t $(t = 0, 1, ...)$, let v be an *active* node in time step $t - 1$. From the work in [17], we have learned that the propagation probability of each edge will decay in a logarithmic manner over the time. In this paper, we give the modified propagation probability from v to u at time t in following equation

$$p_{vu}^{act}(t) = \frac{p_{vu}}{\log(10 + t)}, \tag{1}$$

where p_{vu} is the initial propagation probability from v to u at time step $t = 0$. It can be seen clearly from (1) that the propagation probability decreases as time increases. Based on the above discussion, the influence propagation process is similar to the IC model.

Figure 1 shows an example of the influence decay model in a social network. In the figure, the number embedded on each edge indicates the propagation probability at time step $t = 0$. Influence propagation process unfolds as follows. At time step $t = 0$, without loss of generality, we assume that nodes a and b are *active* (seed nodes) and other nodes are *inactive*. a and b attempt to activate c and d with probability 0.2 and 0.5, respectively. For ease of exposition, we suppose that c becomes *active* and d is still *inactive*[2]. At time step $t = 1$, c tries to activate e with probability $\frac{0.4}{\log(10+1)}$ and c tries to activate f with probability $\frac{0.3}{\log(10+1)}$. As a result, the nodes e and f become *active*. The influence continues to spread. At time stamp $t = 2$, only f attempts to activate g. Finally, g is still

---

[2] In fact, the seeds activating the inactive nodes is a stochastic process.

inactive. At this time, no new nodes become *active* so the propagation process stops. Based on the above analysis, we explain the differences between IDM and IC model. For example, in the IC model, the activation probability of e by a is $1 - (1 - 0.2 \times 0.4) = 0.08$. However, in the DTM, this probability is $1 - (1 - 0.2 \times \frac{0.4}{\log(10+1)}) = 0.0768$. Again, we think that the IC model is a special case of the IDM if we ignore the decay.

4   Problem Description

Given a directed social network which can be described as a graph $G = (V, E, p)$, an influence decay model (IDM) $\mathcal{M}$ and a seed set S, we define the activation probability of a node $u \in V$ under the IDM at time step t as follow.

$$
Pr_{\mathcal{M}}(u, S) = \begin{cases} 1, & \text{if } u \in S \\ 0, & \text{if } N^{in}(u) = \emptyset \\ 1 - \displaystyle\prod_{v \in N^{in}(u)} (1 - Pr_{\mathcal{M}}(v, S) p_{vu}^{act}(t)). \end{cases} \tag{2}
$$

Where $N^{in}(u)$ is the set of in-neighbors of node u. And $Pr_{\mathcal{M}}(v, S) p_{vu}^{act}(t)$ shows the probability v successfully activates u. We define the activation probability of the target user set $\mathcal{T} = \{\mathcal{T}_1, \mathcal{T}_2, ..., \mathcal{T}_q\}$ as following equation

$$
Pr_{\mathcal{M}}(\mathcal{T}, S) = \sum_{u \in \mathcal{T}} Pr_{\mathcal{M}}(u, S). \tag{3}
$$

Now, we can formally define the *Acceptance Probability Maximization* (APM) problem. Given a directed social network $G = (V, E, p)$, an influence decay model $\mathcal{M}$ and a non-negative integer budget b, our APM aims to find a seed set S^* such that

$$
S^* = \arg \max_{S \subseteq V \setminus \mathcal{T}, |S| \leq b} Pr_{\mathcal{M}}(\mathcal{T}, S). \tag{4}
$$

We show the hardness result of APM as following theorem.

Theorem 1. *The Activation Probability Maximization (APM) problem is NP-hard under the Influence Decay Model (IDM).*

It's easy to prove this theorem with reduction from the set cover problem [6]. In this paper, we omit the proof due to space constraints.

From the Theorem 1, we can clearly know the APM problem is NP-hard. However, there is still a question what's the hardness of calculating the activation probability of the target users with respect to a given seed set S? We answer this question by the following theorem.

Theorem 2. *Given a seed set S and a target set $\mathcal{T}$, computing the activation probability from the seed set S to the target set $\mathcal{T}$ is $\#P - hard$ under the IDM.*

Theorem 3. *The objective function (4) is monotone non-decreasing and sub-modular under the IDM.*

5 Algorithms

Basic Greedy Algorithm: Submodular functions have a variety of nice tractability properties [11] which provide us a good way to obtain an approximation algorithm. From previous Theorem 3, we know the objective function (4) is submodular. Thus we propose a *Basic Greedy Algorithm* (BGA) to solve APM. The details can be found in Algorithm 1.

Algorithm 1. Basic Greedy Algorithm (BGA)

Input: $G = (V, E, p)$, b, $\mathcal{T}$ and $\mathcal{M}$.
Output: seed set S.
$S \leftarrow \emptyset$;
for $|S| < b$ **do**
    Select $v \leftarrow \arg\max_{v \in V \setminus \mathcal{T}} Pr(v|S)$;
    $S \leftarrow S \cup \{v\}$;
end
return S.

In Algorithm 1, the key step is to select a node with the largest marginal activation probability $Pr(v|S) = Pr_{\mathcal{M}}(\mathcal{T}, S \cup \{v\}) - Pr_{\mathcal{M}}(\mathcal{T}, S)$ based on current seed set S in each iteration. According to [11], a non-decreasing and non-negative submodular function can provide a $(1 - 1/e)$-approximation ratio by the basic greedy algorithm. We use the following theory to give the approximation ratio analysis.

Theorem 4. *Let S_{BGA} and S^* denote the solution returned by the Algorithm 1 and the optimal solution, respectively. Then we have $Pr_{\mathcal{M}}(\mathcal{M}, S_{BGA}) \geq (1 - 1/e) \cdot Pr_{\mathcal{M}}(\mathcal{T}, S^*)$.*

Scalable Algorithm: In the basic greedy algorithm, computing activation probability is time consuming. Intuitively, computing activation probability of the target users in a local way such as tree structure is an effective method [1]. Furthermore, it's relatively easy to approximate local influence from the seed set S to the target user $T_j \in \mathcal{T}$ $(j = 1, 2, ...q)$ through the directed tree structure.

For a path $Path(v, u) = < v = x_1, x_2, ..., x_l = u >$ from v to u in social network $G = (V, E, p)$, we define the probability of this path as $\mathcal{P}(v, u) = \prod_{k=1}^{k=l-1} p_{x_k x_{k+1}}$. Since v attempts to activate u through the $Path(v, u)$, which leads to v activating all the nodes along the path. Let $Path_G(v, u)$ denote the set of all paths from v to u in G. In particular, we focus on the path whose probability is maximum. And we define the *Maximum Influence Path* (MIP) as follow.

Definition 1 *(Maximum Influence Path (MIP)).* *Given a social network $G = (V, E, p)$, we define maximum influence path $MIP_G(v, u)$ from v to u in G as $MIP_G(v, u) = \arg\max\{\mathcal{P}(v, u) | Path(v, u) \in Path_G(v, u)\}$.*

Note that if we convert the propagation probability p_{vu} to $1/p_{uv}$ for each edge (v, u) in G, $MIP_G(v, u)$ is equivalent to the shortest path from v to u in G. As we all know, the shortest path problem can be solved by polynomial time algorithms, e.g., Floyd-Warshall and Dijkstra algorithms. For a target user $T_j \in T$, we create a directed tree structure that is the union of maximum influence paths to estimate the activation probability of the target user T_j from the seed set. Moreover, we use a threshold θ to delete MIPs which have small activation probabilities.

Definition 2 *(Maximum Activation Probability Tree (MAPT)). For a threshold θ $(0 < \theta \leq 1)$, the maximum activation probability tree rooted a target user $T_j \in T$ in G is $MAPT(T_j, \theta) = \bigcup_{MIP_G(u, T_j) \geq \theta} MIP_G(u, T_j)$.*

Algorithm 2. Scalable Algorithm (SA)

Input: $G = (V, E, p)$, b, T, M and θ.
Output: seed set S.
$S \leftarrow \emptyset$;
for *each $T_j \in T$* **do**
   if $|S| < b$ **then**
      Create a $MAPT(T_j, \theta)$ for T_j;
      for *each node $v \in MAPT(T_j, \theta) \backslash T_j$* **do**
         Calculate $Pr_M(T_j, v)$;
      end
      Select $v \leftarrow \arg \max\limits_{v \in MAPT(T_j, \theta) \backslash T_j} Pr_M(T_j, v)$;
      $S \leftarrow S \cup \{v\}$;
   end
end
return S.

Based on these directed trees, we propose a *Scalable Algorithm* (SA) in Algorithm 2. Instead of computing APM in the entire network, the scalable algorithm constructs local tree structures consisting of only the shortest paths between the seed nodes and the target nodes with the parameter θ. Then we restrict computations within the shortest paths of tree structures.

6   Experiments

Experiment Setup: We use one synthetic network and three real-life networks with various scale from (SNAP)[3] and [15]. Note that Amazon and Youtube are undirected networks. Therefore we convert these two undirected graphs to directed graphs. Specifically, for each undirected edge (v, u) in Amazon and

---

[3] http://snap.stanford.edu/data.

Youtube networks, we randomly generate a directed edge (v, u) or (u, v) with a probability of 0.5. Since all networks are without probability, we first assign a uniform probability $p = 0.5$ for each edge on Synthetic and Email networks. Second, we employ a trivalency model [2] to uniformly select a value from $\{0.1, 0.3, 0.5\}$ at random for each edge on Amazon and Youtube networks (e.g., $p = TRI$).

Synthetic. We randomly generate a relatively small directed network to evaluate the proposed algorithms and compare the results with the real-life datasets. This synthetic network includes 2K nodes and 5.2K edges.

E-mail is generated using email data from a large European research institution. Each node represents a researcher and each directed edge (u, v) means that u sent at least one email to v. It includes 1K nodes and 25.9K edges.

Amazon is based on *Customers Who Bought This Item Also Bought* feature of the Amazon website. Each node is a product. If a product u is frequently co-purchased with product v, thus there is an edge between u and v. It includes 334.8K nodes and 925.8K edges.

Youtube is a video-sharing social network. Each node is a user on network. Users can form friendship if they share same videos. It includes 1134.8K nodes and 2987.6K edges.

Comparison Methods: We use the following three *Heuristic Algorithms* (HA): Degree Discount (DD) [2], Local Structural Centrality (LSC) [4] and PageRank (PR) [12] to discover influential nodes on the network and to select these influential nodes as seed nodes. Furthermore, we adopt the algorithm [5] as a benchmark method.

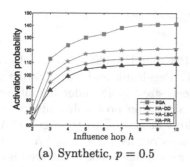
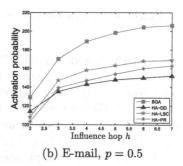

(a) Synthetic, $p = 0.5$ (b) E-mail, $p = 0.5$

Fig. 2. Influence hop h study: target set size $|T| = 500$, seed set size $b = 45$, $p = 0.5$ for Synthetic and E-mail networks.

Results: Influence Hop h Study: We study the effect of influence hop on the activation probability of the target users. We evaluate the performance of *Basic Greedy Algorithm* (BGA) and *Heuristic Algorithm* (HA) on Synthetic and E-mail networks by varying h from $h = 2$ to $h = 10$. Figure 2 shows the results. In Fig. 2, the vertical and horizontal axes represent the total activation probability of the target users $Pr_{\mathcal{M}}(T, S)$ and the influence hop h, respectively. Specifically, we first randomly select 500 nodes (i.e., $|T| = 500$) as target users.

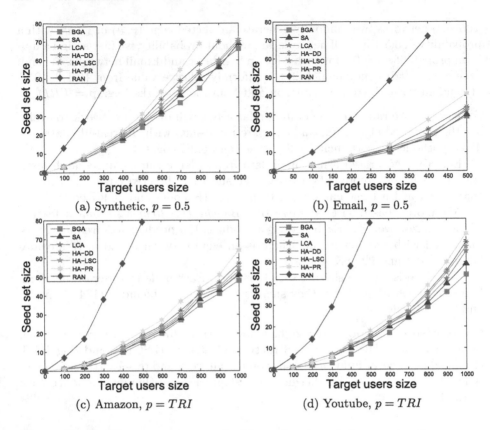

(a) Synthetic, $p = 0.5$ (b) Email, $p = 0.5$

(c) Amazon, $p = TRI$ (d) Youtube, $p = TRI$

Fig. 3. The target users set size Vs. seed nodes set size on networks

Then we run BGA and HA (HA-DD, HA-PR and HA-LSC denote selecting global influential nodes by Degree Discount, PageRank and Local Structural Centrality respectively.) with restriction of seed size $b = 45$ under the influence decay model $\mathcal{M}$. We observe that the total activation probability increases as the influence hop h increases in all algorithms according to the figures. And the total activation probability sharply increases between $h = 2$ and $h = 5$ but goes slowly after $h = 6$. This phenomenon is reasonable because the seed nodes may not successfully activate the target nodes when the number of influence hop is small. In the subsequent experiments, we set the influence hop $h = 6$. In addition, we can see that the total activation probability of BGA is greater than that of HA (BGA>HA-LSC>HA-PR>HA-DD). This is because BGA has performance guarantee as we analyzed before. On the other hand, the heuristic algorithm is not as good as we expected. The reason is that heuristic rules select the global influential nodes as seed nodes and these nodes may be far away from the target users, which leads to the less activation probability.

Target Users Size vs. Seed Size: In order to investigate the relationship between seed size b and target users size $|\mathcal{T}|$ with different methods, we let influence propagation probability $p = 0.5$ for Synthetic as well as E-mail networks or $p = TRI$ for Amazon as well as Youtube networks. We let $\theta = 0.02$ and influence hop $h = 6$. Experiments are conducted by varying the target users size $|\mathcal{T}|$ from 0 to 1000 while $|\mathcal{T}|$ from 0 to 500 on E-mail network. The average the number of seeds are calculated with different methods (BGA, SA, LCA, HA-DD, HA-LSC, HA-PR and RAN where RAN means that it randomly selects nodes as seed nodes). Note that all experimental results are averaged from 10 independent randomized experiments. The results are plotted in Fig. 3.

In the figure, the horizontal and vertical axis indicate the target users size and seed nodes size, respectively. On each network, the seed nodes size increases as the target users size increases. And we find that seed size steadily increases from 0 and 300 while sharply increases from 400 and 1000. With the same target users size, our BGA method needs the minimum seed nodes while the RAN method needs the maximum seed nodes. BGA is better than HA since it has performance guarantee as we analyzed in previous section. In all heuristic algorithms, HA-DD, HA-LSC and HA-PR are close. Moreover, we observe that LCA is better than HA. And SA is better than LCA since SA needs less seed nodes to activate same target nodes than LCA. On the other hand, instead of utilizing Monte-Carlo simulation, it indicates MAPT is an effective approximation to calculate the activation probability.

7   Conclusion

In this paper, we study the *Acceptance Probability Maximization* (APM) problem that finds a seed set such that the activation probability of the target users is maximized. And we propose an influence propagation model *Influence Decay Model* (IDM). Based on the IDM, we show APM is NP-hard and computing APM is $\#p - hard$. We also prove objective functions satisfy monotonicity and submodularity. We develop a *Basic Greedy Algorithm* (BGA) which has a $(1 - 1/e)$-approximation ratio when $|S| \leq b$. In addition, a speed-up *Scalable Algorithm* (SA) is developed for online large social networks. Finally, in order to evaluate our proposed algorithms, extensive experiments have been conducted on synthetic and real-life networks. And the experiment results show that our proposed algorithms outperform other comparison methods.

References

1. Chen, W., Wang, C., Wang, Y.: Scalable influence maximization for prevalent viral marketing in large-scale social networks. In: Proceedings of the 16th ACM SIGKDD International Conference on Knowledge Discovery and Data Mining, pp. 1029–1038. ACM (2010)
2. Chen, W., Wang, Y., Yang, S.: Efficient influence maximization in social networks. In: Proceedings of the 15th ACM SIGKDD International Conference on Knowledge Discovery and Data Mining, pp. 199–208. ACM (2009)
3. Feng, S., Chen, X., Cong, G., Zeng, Y., Chee, Y.M., Xiang, Y.: Influence maximization with novelty decay in social networks. In: AAAI, pp. 37–43 (2014)

4. Gao, S., Ma, J., Chen, Z., Wang, G., Xing, C.: Ranking the spreading ability of nodes in complex networks based on local structure. Phys. A Stat. Mech. Appl. **403**, 130–147 (2014)
5. Guo, J., Zhang, P., Zhou, C., Cao, Y., Guo, L.: Personalized influence maximization on social networks. In: Proceedings of the 22nd ACM International Conference on Information & Knowledge Management, pp. 199–208. ACM (2013)
6. Karp, R.M.: Reducibility among combinatorial problems. In: Miller, R.E., Thatcher, J.W., Bohlinger, J.D. (eds.) Complexity of Computer Computations. The IBM Research Symposia Series, pp. 85–103. Springer, Boston (1972). https://doi.org/10.1007/978-1-4684-2001-2_9
7. Kempe, D., Kleinberg, J., Tardos, É.: Maximizing the spread of influence through a social network. In: Proceedings of the Ninth ACM SIGKDD International Conference on Knowledge Discovery and Data Mining, pp. 137–146. ACM (2003)
8. Kuhnle, A., Pan, T., Alim, M.A., Thai, M.T.: Scalable bicriteria algorithms for the threshold activation problem in online social networks. In: INFOCOM 2017-IEEE Conference on Computer Communications, pp. 1–9. IEEE (2017)
9. Liu, B., Cong, G., Xu, D., Zeng, Y.: Time constrained influence maximization in social networks. In: 2012 IEEE 12th International Conference on Data Mining (ICDM), pp. 439–448. IEEE (2012)
10. Mohammadi, A., Saraee, M., Mirzaei, A.: Time-sensitive influence maximization in social networks. J. Inf. Sci. **41**(6), 765–778 (2015)
11. Nemhauser, G.L., Wolsey, L.A., Fisher, M.L.: An analysis of approximations for maximizing submodular set functions-I. Math. Program. **14**(1), 265–294 (1978)
12. Page, L., Brin, S., Motwani, R., Winograd, T.: The PageRank citation ranking: bringing order to the web. Technical report, Stanford InfoLab (1999)
13. Song, C., Hsu, W., Lee, M.L.: Targeted influence maximization in social networks. In: Proceedings of the 25th ACM International on Conference on Information and Knowledge Management, pp. 1683–1692. ACM (2016)
14. Sun, H., Cheng, R., Xiao, X., Yan, J., Zheng, Y., Qian, Y.: Maximizing social influence for the awareness threshold model. In: Pei, J., Manolopoulos, Y., Sadiq, S., Li, J. (eds.) DASFAA 2018, Part I. LNCS, vol. 10827, pp. 491–510. Springer, Cham (2018). https://doi.org/10.1007/978-3-319-91452-7_32
15. Tang, J., Lou, T., Kleinberg, J.: Inferring social ties across heterogenous networks. In: Proceedings of the Fifth ACM International Conference on Web Search and Data Mining, pp. 743–752. ACM (2012)
16. Tong, G., Wu, W., Tang, S., Du, D.Z.: Adaptive influence maximization in dynamic social networks. IEEE/ACM Trans. Netw. (TON) **25**(1), 112–125 (2017)
17. Wang, B., Chen, G., Fu, L., Song, L., Wang, X., Liu, X.: DRIMUX: dynamic rumor influence minimization with user experience in social networks. In: AAAI, vol. 16, pp. 791–797 (2016)
18. Yan, Q., Huang, H., Gao, Y., Lu, W., He, Q.: Group-level influence maximization with budget constraint. In: Candan, S., Chen, L., Pedersen, T.B., Chang, L., Hua, W. (eds.) DASFAA 2017, Part I. LNCS, vol. 10177, pp. 625–641. Springer, Cham (2017). https://doi.org/10.1007/978-3-319-55753-3_39
19. Yang, D.N., Hung, H.J., Lee, W.C., Chen, W.: Maximizing acceptance probability for active friending in online social networks. In: Proceedings of the 19th ACM SIGKDD International Conference on Knowledge Discovery and Data Mining, pp. 713–721. ACM (2013)
20. Zhou, C., Guo, L.: A note on influence maximization in social networks from local to global and beyond. Procedia Comput. Sci. **30**, 81–87 (2014)

Maximization of Constrained Non-submodular Functions

Ruiqi Yang[1], Dachuan Xu[1], Donglei Du[2], Yicheng Xu[3(⊠)], and Xihong Yan[4]

[1] Department of Operations Research and Scientific Computing,
Beijing University of Technology, Beijing 100124, People's Republic of China
yangruiqi@emails.bjut.edu.cn, xudc@bjut.edu.cn
[2] Faculty of Business, University of New Brunswick,
Fredericton, NB E3B 5A3, Canada
ddu@unb.ca
[3] Shenzhen Institutes of Advanced Technology, Chinese Academy of Sciences,
Shenzhen 518055, People's Republic of China
yc.xu@siat.ac.cn
[4] Department of Mathematics, Taiyuan Normal University,
Jinzhong 030619, Shanxi, People's Republic of China
xihong1@e.ntu.edu.sg

Abstract. We investigate a non-submodular maximization problem subject to a p-independence system constraint, where the non-submodularity of the utility function is characterized by a series of parameters, such as submodularity (supmodularity) ratio, generalized curvature, and zero order approximate submodularity coefficient, etc. Inspired by Feldman et al. [15] who consider a non-monotone submodular maximization with a p-independence system constraint, we extend their Repeat-Greedy algorithm to non-submodular setting. While there is no general reduction to convert algorithms for submodular optimization problems to non-submodular optimization problems, we are able to show the extended Repeat-Greedy algorithm has an almost constant approximation ratio for non-monotone non-submodular maximization.

Keywords: Approximation algorithms ·
Non-submodular maximization · Independence system constraint

1 Introduction

Submodular optimization is widely studied in optimization, computer science, and economics, etc. Submodularity is a very powerful tool in many optimization applications such as viral marketing [9,18], recommendation system [12,15,21], nonparametric learning [1,16], and document summarization [20], etc.

The greedy algorithm introduced by Nemhauser et al. [22] gave the first $(1 - e^{-1})$-approximation for monotone submodular maximization with a cardinality constraint (SMC). Feige [13] considered a maximal k-cover problem, which is a special case of SMC, and showed that there is no algorithms with approximation

© Springer Nature Switzerland AG 2019
D.-Z. Du et al. (Eds.): COCOON 2019, LNCS 11653, pp. 615–626, 2019.
https://doi.org/10.1007/978-3-030-26176-4_51

ratio greater than $(1 - e^{-1} + \epsilon)$ for any $\epsilon > 0$, under the assumption $P \neq NP$. Sviridenko [23] considered a monotone submodular maximization with a knapsack constraint, and provided a tight $(1 - e^{-1})$-approximation algorithm with time complexity $O(n^5)$. Calinescu et al. [8] provided a $(1 - e^{-1})$-approximation algorithm for monotone submodular maximization with a matroid constraint. All extant results of constrained submodular maximization assume monotonicity of the objective functions. In this paper, we consider a non-monotonic and non–submodular maximization problem subject to a more general p-independence system constraint.

1.1   Our Contributions

In this work, we consider a non-submodular maximization with an independence system constraint. Specifically, all feasible solutions associated with this model generate a p-independence system and the objective function is characterized by a series of parameters, such as submodularity (supmodularity) ratio, generalized curvature, and zero order approximate submodularity coefficient, etc. Our main results can be summarized as follows.

- We first investigate the efficiency of a greedy algorithm with two scenarios. Firstly, the objective function is non-submodularity and non-monotonic. Secondly, the feasible solution belongs to a p-independence system. We show that some good properties are still retained in the non-submodular setting (Theorem 1).
- Second we study the non-monotone non-submodular maximization problem without any constraint. Based on a simple approximate local search, for any $\epsilon > 0$ we show that there exists a polynomial time $(3/c^2 + \epsilon)$-approximation algorithm, where c is the zero order approximate submodularity coefficient of objective function (Theorem 2).
- Finally, we apply the first two algorithms as the subroutines to solve a non-monotone non-submodular maximization problem with p-independence system constraint. Our algorithm is an extension of the Repeat-Greedy introduced in [15]. Based on a multiple times rounding of the above subroutine algorithms, we derived a nearly constant approximation ratio algorithm (Theorem 3).

1.2   The Organization

We give a brief summary of related work in Sect. 2. The necessary preliminaries and definitions are presented in Sect. 3. The main algorithms and analyses are provided in Sect. 4. We present a greedy algorithm in Sect. 4.1, an approximate local search in Sect. 4.2, and the core algorithm is provided in Sect. 4.3. In Sect. 5, we offer a conclusion for our work.

2   Related Works

Non-monotone Submodular Optimization. Unlike the monotone submodular optimization, there exists a natural obstacle in the study of non-monotonic case. For example, direct application of the greedy algorithm introduced by Nemhauser et al. [22] to the non-monotonic case does not yield a constant approximation guarantee. Some previous work is summarized below. For the non-monotone submodular maximization problem without any constraint (USM), Feige et al. [14] presented a series of algorithms. They first showed a uniform random algorithm has a 1/4-approximation ratio. Second, they gave a deterministic local search 1/3-approximation algorithm and a random 2/5-approximation algorithm. For symmetric submodular functions, they derived a 1/2-approximation algorithm and showed that any $(1/2 + \epsilon)$-approximation for symmetric submodular functions must need an exponential number of queries for any fixed $\epsilon > 0$. Based on local search technique, Buchbinder et al. [7] provided a random linear time 1/2-approximation algorithm.

Optimization of non-monotone submodular with complex constraints are also considered previously. Buchbinder and Feldman [5] gave a deterministic 1/2-approximation algorithm for USM with time complexity $O(n^2)$. For non-monotone submodular maximization with cardinality constraint, they derived a deterministic 1/e-approximation algorithm, which has a slightly better approximation ration than the random $(1/e + 0.004)$-approximation ratio by [6]. Buchbinder and Feldman [4] considered a more general non-monotone submodular maximization problem with matroid constraint and presented the currently best random 0.385-approximation algorithm. Lee et al. [19] derived a $1/(p + 1 + 1/(p - 1) + \epsilon)$-approximation algorithm as well as non-monotone submodular maximization algorithm with a constraint of the intersection of p matroids. For a more general p-independence system constraint, Gupta et al. [17] derived a $1/3p$-approximation, which needs $O(np\ell)$ function value oracles, where ℓ is the maximum size of feasible solutions. Mirzasoleiman et al. [21] improved the approximation ratio to $1/2k$, while the time complexity was still bounded by $O(np\ell)$. Recently, with improved time complexity of $O(n\ell\sqrt{p})$, the approximation ratio was improved to $1/(p + \sqrt{p})$ by Feldman et al. [15].

Non-submodular Maximization. There are also many problems in optimization and machine learning whose utility functions do not possess submodularity. Das and Kempe [11] introduced a definition of submodularity ratio γ to measure the magnitude of submodularity of the utility function. For the maximization of monotone non-submodular function with cardinality constraint (NSMC), they showed the greedy algorithm can achieve a $(1 - e^{-\gamma})$-approximation ratio. Conforti and Cornuéjols [10] studied the efficiency of the greedy algorithm by defining curvature κ of submodular objective functions for SMC and showed the approximation could be improved it to $1/\kappa(1 - e^{-\kappa})$. Bian et al. [2] introduced a more expressive formulation by providing a definition of the generalized curvature α of any non-negative set function. Combining the submodularity ratio with the generalized curvature, they derived the tight $1/\alpha(1 - e^{1/(\alpha\gamma)})$-approximation ratio of the greedy algorithm for NSMC. Inspired by these work, Bogunovic et al. [3] introduced further parameters, such as supmodularity ratio, inverse generalized curvature, etc., to characterize the utility function. They derived the first constant approximation algorithm for monotone robust non-submodular maximization problem with cardinality constraint.

3 Preliminaries

In this section, we present some necessary notations. We are given a ground set $\mathcal{V} = \{u_1, ..., u_n\}$, and a utility function $f : 2^{\mathcal{V}} \to R_+$. The function f may not be submodular; namely the following zero order condition of submodular may not hold

$$f(A) + f(B) \geq f(A \cup B) + f(A \cap B), \forall A, B \subseteq \mathcal{V}.$$

For our purpose, we define a parameter c to approximate the submodularity of our utility function.

Definition 1. *Given an integer k, let $\{A_i\}_{i=1}^k$ be a collection of subsets of $\mathcal{V}$. The zero order approximate submodularity coefficient is the largest $c_k \in [0, 1]$ such that*

$$\sum_{i=1}^{k} f(A_i) \geq c_k \cdot [f(\cup_i A_i) + f(\cap_i A_i)].$$

For any $A, B \subseteq \mathcal{V}$, we set $f_A(B) = f(A \cup \{B\}) - f(A)$ as the amount of change by adding B to A. For the sake of brevity and readability, we set $f_A(u) = f(A + u) - f(A)$ for any singleton element $u \in \mathcal{V}$. Then we restate the submodularity ratio γ in the following definition. The submodularity ratio measures how close of f being submodular.

Definition 2 *([2, 3, 11]). Given an integer k, the submodularity ratio of a non-negative set function f with respect to $\mathcal{V}$ is*

$$\gamma_{\mathcal{V}, k}(f) = \min_{A \subseteq \mathcal{V}, B: |B| \leq k, A \cap B = \emptyset} \frac{\sum_{u \in B} f_A(u)}{f_A(B)}.$$

Let k be the maximum size of any feasible solution, and omit signs $k, \mathcal{V}$ and f for clarity. Bian et al. [2] introduced an equivalent formulation of submodularity ratio γ by the largest γ such that

$$\sum_{u \in B} f_A(u) \geq \gamma \cdot f_A(B), \forall A, B \subseteq \mathcal{V}, A \cap B = \emptyset.$$

Bogunovic et al. [3] defined *supmodularity ratio* $\check{\gamma}$ that measures how close a utility function is supermodular.

Definition 3 *([3]). The supmodularity ratio of a non-negative set function f is the largest $\check{\gamma}$ such that*

$$f_A(B) \geq \check{\gamma} \cdot \sum_{u \in B} f_A(u), \forall A, B \subseteq \mathcal{V}, A \cap B = \emptyset.$$

For a monotone submodular function, Conforti and Cornuéjols [10] introduced the definition of *total curvature* κ_f and *curvature* $\kappa_f(S)$ w.r.t. a set $S \subseteq V$ as follows. Denote $\kappa_f = 1 - \min_{u \in V} f_{V \setminus \{u\}}(u)/f(u)$, and $\kappa_f(S) = 1 - \min_{u \in S} f_{S \setminus \{u\}}(u)/f(u)$. Sviridenko et al. [24] provided a definition of curvature from submodular to non-submodular functions. The expanded *curvature* is defined as $\kappa^o = 1 - \min_{u \in V} \min_{A,B \subseteq V \setminus \{u\}} f_A(u)/f_B(u)$. Bian et al. [2] presented a more expressive formulation of curvature, which measures how close a set function is to being supmodular.

Definition 4 *([2]). The generalized curvature of a non-negative function f is the smallest scalar α such that*

$$f_{A \setminus \{u\} \cup B}(u) \geq (1 - \alpha) \cdot f_{A \setminus \{u\}}(u), \forall A, B \subseteq V, u \in A \setminus B.$$

Recently, Bogunovic et al. [3] introduced the concept of *inverse generalized curvature* $\check{\alpha}$, which can be described as follows:

Definition 5 *([3]). The inverse generalized curvature of a non-negative function f is the smallest scalar $\check{\alpha}$ such that*

$$f_{A \setminus \{u\}}(u) \geq (1 - \check{\alpha}) f_{A \setminus \{u\} \cup B}(u), \forall A, B \subseteq V, u \in A \setminus B.$$

The above parameters are used to characterize a non-negative set function from different points of view. We provide a lower bound of zero order approximate submodularity coefficient c by inverse generalized curvature $\check{\alpha}$. I.e., $c \geq 1 - \check{\alpha}$. The proof is referred to the full version. We omit the relation of the other parameters as they can be found in [2,3]. In the rest of this part, we restate the concept of the p-independence system.

Let $\mathcal{I} = \{A_i\}_i$ be a finite collection of subsets chosen from V. We say the tuple $(V, \mathcal{I})$ is an *independence system* if for any $A \in \mathcal{I}$, $A' \subseteq A$ implies that $A' \in \mathcal{I}$. The sets of $\mathcal{I}$ are called the *independent sets* of the independence system. An independent set B contained in a subset $X \subseteq V$ is a *base* (basis) of X if no other independent set $A \subseteq X$ strictly contains B. By the above terminologies we restate the definition of p-independence system as follows.

Definition 6 *([15]). An independence system $(V, \mathcal{I})$ is a p-independence system if, for every subset $X \subseteq V$ and for any two bases B_1, B_2 of X, we have $|B_1|/|B_2| \leq p$.*

In our model, assume the utility function is characterized by these parameters, and the collection of all feasible subsets constructs a p-independence system. We also assume that there exist a utility function value oracle and an independence oracle; i.e., for any $A \subseteq V$, we can obtain the value of $f(A)$ and know if A in $\mathcal{I}$ or not. The model can be described as follows:

$$OPT \leftarrow \arg \max_{S \subseteq V, S \in \mathcal{I}} f(S), \tag{1}$$

where $(V, \mathcal{I})$ is a p-independence system.

4   Algorithms

In this section, we present some algorithms in dealing with non-submodular maximization. Before providing our main algorithm, we first investigate the efficiency of two sub-algorithms. In Subsect. 4.1, we restate the greedy algorithm for submodular maximization with p-independence system constraint, and show that some good properties are still retained in the non-submodular setting. In Subsect. 4.2, we present a local search for non-monotone non-submodular maximization without any constraint. Finally, we provide the core algorithm in Subsect. 4.3.

4.1   Greedy Algorithm Applied to Non-submodular Optimization

The pseudo codes of the greedy algorithm are presented in Algorithm 1. Let $S^G = \{u_1, ..., u_\ell\}$ be the returned set by Algorithm 1. We start with $S^G = \emptyset$. In each iteration, we choose the element u with maximum gain, and add it to the current solution if it satisfies $S^G + u \in \mathcal{I}$. For clarity, we let OPT be any optimum solution set of maximizing the utility function under p-independence system constraint. Then we can derive a lower bound of $f(S^G)$ by the following theorem.

Theorem 1. *Let S^G be the returned set of Algorithm 1, then we have*

$$f(OPT \cup S^G) \leq \left(\frac{p}{\gamma^2 \check{\gamma}(1 - \check{\alpha})} + 1 \right) f(S^G).$$

Proof. Refer to the full version of this paper.

Algorithm 1. Greedy$(\mathcal{V}, f, \mathcal{I})$

1: $S^G \leftarrow \emptyset, A \leftarrow \emptyset$
2: **repeat**
3:    $A \leftarrow \{e | S^G \cup \{e\} \in \mathcal{I}\}$
4:    **if** $A \neq \emptyset$ **then**
5:       $e \leftarrow \arg\max_{e' \in A} f_{S^G}(e')$
6:       $S^G \leftarrow S^G + e$
7:    **end if**
8: **until** $A = \emptyset$
9: Return S^G

Let $B \in \mathcal{I}$ be any independent set and set $S_i^G = \{u_1, ..., u_i\}$ be the set of the first i elements added by Algorithm 1. We can iteratively construct a partition of B according to S^G. We start with $B_0 = B$ and set $B_i = \{u \in B \setminus S_i^G | S_i^G + u \in \mathcal{I}\}$ for iteration $i \in [\ell] = \{1, ..., \ell\}$, where ℓ denotes the size of S^G in the end. Then the collection of $\{B_{i-1} \setminus B_i\}_{i=1}^{\ell}$ derives a partition of B. Let $C_i = B_{i-1} \setminus B_i$ for any $i \in [\ell]$. The construction can be summarized as Algorithm 2. The properties of the above partition are presented in the following lemma.

Algorithm 2. Construct($\mathcal{V}, f, \mathcal{I}$)

1: $S_0 \leftarrow \emptyset, B_0 \leftarrow B$
2: **for** $i = 1 : \ell$ **do**
3:     $B_i \leftarrow \{u \in B \setminus S_i | S_i + u \in \mathcal{I}\}$
4: **end for**
5: Return $\{B_{i-1} \setminus B_i\}_{i=1}^{\ell}$

Lemma 1. *Let $\{C_i\}_{i=1}^{\ell}$ be the returned partition of Algorithm 2, then*

- *for each $i \in [\ell]$, we have $\sum_{j=1}^{i} p_j \leq p \cdot i$ where $p_j = |C_j|$; and*
- *for each $i \in [\ell]$, we have $p_i \cdot \delta_i \geq \gamma(1 - \check{\alpha})f_{SG}(C_i)$.*

Proof. Refer to the full version of this paper.

4.2 Local Search Applied to Non-submodular Optimization

In this subsection, we present a local search algorithm for the non-monotone non-submodular maximization problem without any constraint. The main pseudo codes are provided by Algorithm 3. Feige et al. [14] introduced the local search approach to deal with the non-monotone submodular optimization problem. We extend their algorithm to the non-submodular setting, and show that the algorithm still keeps a near constant approximation ratio by increasing a factor.

In order to implement our algorithm in polynomial time, we relax the local search approach and find an approximate local solution. Let S^{LS} be the returned set of Algorithm 3 and let OPT^o be any optimum solution without any constraint. We restate the definition of approximate local optimum as follows.

Definition 7 *([14]). Given a set function $f : 2^{\mathcal{V}} \to R$, a set $A \subseteq \mathcal{V}$ is called a $(1 + \lambda)$-approximate local optimum if, $f(A - u) \leq (1 + \lambda) \cdot f(A)$ for any $u \in A$ and $f(A + u) \leq (1 + \lambda) \cdot f(A)$ for any $u \notin A$.*

By the definition of the approximate local optimum solution, we show that there exists a similar performance guarantee in the non-submodular case. The details are summarized in the following theorem.

Theorem 2. *Given $\epsilon > 0, c$ and $\check{\alpha} \in [0, 1)$. Let S^{LS} be the returned set of Algorithm 3 by setting set $\lambda = \frac{c^2 \epsilon}{(1-\check{\alpha})n}$. We have*

$$f(OPT^o) \leq \left(\frac{3}{c^2} + \epsilon\right) \cdot f(S^{LS}).$$

Proof. Refer to the full version of this paper.

Before proving the above theorem, we need the following lemma.

Lemma 2. *If S is a $(1 + \lambda)$-approximate local optimum for a non-submodular function f, then for any set T such that $T \subseteq S$ or $T \supseteq S$, we have*

$$f(T) \leq [1 + \lambda n(1 - \check{\alpha})] \cdot f(S).$$

Algorithm 3. Local Search$(\mathcal{V}, f, \lambda)$

1: $S \leftarrow \arg\max_{u \in \mathcal{V}} f(u), \mathcal{U} \leftarrow \mathcal{V}$
2: **repeat**
3:    **if** there exists $u \in \mathcal{U} \setminus S$ such that $f(S + u) \geq (1 + \lambda)f(S)$ **then**
4:        $S \leftarrow S + u$
5:        $\mathcal{U} \leftarrow \mathcal{U} - u$
6:    **end if**
7: **until**
8: **if** there exists $u \in S$ such that $f(S - u) \geq (1 + \lambda)f(S)$ **then**
9:    $S \leftarrow S - u$, and go back to Repeat loop.
10: **end if**
11: Return $S^{LS} \leftarrow \arg\max\{f(S), f(\mathcal{V} \setminus S)\}$

Proof. Let $S = \{u_1, ..., u_q\}$ be a $(1 + \lambda)$-approximate local optimum solution returned by Algorithm 3. W.l.o.g., we assume $T \subseteq S$, then we construct T_i such that $T = T_1 \subseteq T_2 \subseteq \cdots \subseteq T_r = S$ and $u_i = T_i \setminus T_{i-1}$. For each $i \in \{2, ..., q\}$, we have

$$f(T_i) - f(T_{i-1}) \geq (1 - \check{\alpha})(f(S) - f(S - u_i)) \geq -\lambda(1 - \check{\alpha})f(S),$$

where the first inequality follows by the definition of the inverse generalized curvature and the second inequality follows by the definition of the approximate local optimum. Summing up the above inequalities, we have

$$f(S) - f(T) = \sum_{i=2}^{q} [f(T_i) - f(T_{i-1})] \geq -\lambda q(1 - \check{\alpha})f(S),$$

implying that $f(T) \leq [1 + \lambda q(1 - \check{\alpha})]f(S) \leq [1 + \lambda n(1 - \check{\alpha})]f(S)$, where the second inequality follows from $q \leq n$. Simultaneously, the case of $T \supseteq S$ can be similarly derived by the above process.

4.3 The Core Algorithm

In this subsection, we present the main algorithm, which is an extension of the Repeat-Greedy algorithm introduced in [15]. The pseudo codes are presented as Algorithm 4. We run the main algorithm in r rounds. Let $\mathcal{V}_i$ be the set of candidate elements set at the start of round $i \in [r]$. We first run the greedy step of Algorithm 1. Then, we proceed with the local search step of Algorithm 3 on the set returned from the first step. Simultaneously, we update the candidate ground set as $\mathcal{V}_i = \mathcal{V} \setminus \mathcal{V}_{i-1}$. Finally, we output the best solution among all returned sets. We can directly obtain two estimations of the utility function by Theorems 1 and 2, respectively. The results are summarized as follows.

Lemma 3. *For any iteration $i \in [r]$ of Algorithm 4, we have*

1. $f(S_i \cup (OPT \cap \mathcal{V}_i)) \leq \left(\frac{p}{\gamma^2 \check{\gamma}(1 - \check{\alpha})} + 1\right) f(S_i)$, and
2. $f(S_i \cap OPT) \leq \left(\frac{3}{c^2} + \epsilon\right) f(S_i')$.

Buchbinder et al. [6] derived an interesting property in dealing with non-monotone submodular optimization problems. Now, we extend this property to the non-submodular case, as summarized in the following lemma.

Lemma 4. *Let* $g : 2^{\mathcal{V}} \to R_+$ *be a set function with inverse generalized curvature* $\check{\alpha}_g$, *and* S *be a random subset of* $\mathcal{V}$ *where each element appears with probability at most* pr *(not necessarily independent). Then* $\mathbf{E}[g(S)] \geq [1 - (1 - \check{\alpha}_g)\text{pr}] \cdot g(\emptyset)$.

Proof. Refer to the full version of this paper.

Using this result, we can derive an estimation of $f(OPT)$. Let S be a random set of $\{S_i\}_{i=1}^r$ with probability pr $= \frac{1}{r}$ and set $g(S) = f(OPT \cup S)$ for any $S \subseteq \mathcal{V}$. Then we have $\check{\alpha} = \check{\alpha}_f = \check{\alpha}_g$. By Lemma 4, we yield

$$\frac{1}{r} \sum_{i=1}^r f(S_i \cup OPT) = \mathbf{E}[f(S \cup OPT)] = \mathbf{E}[g(S)] \geq [1 - (1 - \check{\alpha}_g)\text{pr}] \cdot g(\emptyset)$$

$$= \left(1 - \frac{1 - \check{\alpha}}{r}\right) \cdot f(OPT).$$

Multiplying both sides of the last inequality by r, we get

$$\sum_{i=1}^r f(S_i \cup OPT) \geq [r - (1 - \check{\alpha})] \cdot f(OPT). \tag{2}$$

The following lemma presents a property based on zero order approximate submodularity coefficient of the objective function.

Lemma 5 *([15]). For any subsets* $A, B, C \subseteq \mathcal{V}$, *we have*

$$f(A \cup B) \leq \frac{1}{c} \cdot [f(A \cup (B \cap C)) + f(B \setminus C)].$$

Algorithm 4. Repeat Greedy$(\mathcal{V}, f, \mathcal{I}, r)$

1: $i = 1, \mathcal{V}_1 \leftarrow \mathcal{V}$
2: **repeat**
3:   $S_i \leftarrow$ output of Greedy$(\mathcal{V}_i, f, \mathcal{I})$
4:   $S_i' \leftarrow$ output of Local Search(S_i, f, λ_i)
5:   $\mathcal{V}_i \leftarrow \mathcal{V}_i \setminus S_i$
6:   $i \leftarrow i + 1$
7: **until** $i = r$
8: Return $S \leftarrow \arg\max\{f(S_i), f(S_i')\}_{i \in [r]}$

Proof. To prove this lemma, we have the following

$$f(A \cup B) = f(A \cup (B \cap C) \cup (B \setminus C))$$
$$\leq f(A \cup (B \cap C) \cup (B \setminus C)) + f(A \cup (B \cap C) \cap (B \setminus C))$$
$$\leq \frac{1}{c} \cdot [f(A \cup (B \cap C)) + f(B \setminus C)],$$

where the first inequality follows from the nonnegativity of the objective function and the second inequality is derived by the definition of the zero order approximate submodularity coefficient c.

From these lemmas and choosing properly the number of rounds, we conclude that if the parameters of the utility function are fixed, or have a food estimation, then Algorithm 4 yieds a near constant performance guarantee for problem (1). The details are presented in the following theorem.

Theorem 3. *Give an objective function $f : 2^V \to R_+$ with parameters $c, \gamma, \check{\gamma}, \check{\alpha}$, and a real number $\epsilon > 0$, let S be the returned set of Algorithm 4. Set $r = \lceil \Delta \rceil$. Then we have*

$$\frac{f(OPT)}{f(S)} \leq \left[\left(\frac{p}{\gamma^2 \check{\gamma} c (1 - \check{\alpha})} + \frac{3\Delta}{2c^4} + \frac{1}{c} \right) + \frac{\epsilon \Delta}{2c^4} \right] \cdot [1 - (1 - \check{\alpha}) \cdot (\Delta + 1)]^{-1},$$

where

$$\Delta = (1 - \check{\alpha}) + \sqrt{(1 - \check{\alpha})^2 + (1 - \check{\alpha}) \left(\frac{2c^3}{3 + c^2 \epsilon} + 1 \right) + \frac{p}{\gamma^2 \check{\gamma}} \cdot \frac{2c^3}{3 + c^2 \epsilon}}.$$

Proof. Refer to the full version of this paper.

5   Conclusion

We consider the non-submodular and non-monotonic maximization problem with a p-independence system constraint, where the objective utility function is characterized by a set of parameters such as submodularity (supmodularity) ratio, inverse generalized curvature, and zero order approximate submodularity coefficient. We study a greedy algorithm applied to non-submodular optimization with p-independence system constraint, and show the algorithm preserves some good properties even though the objective function is non-submodularity. Then, we investigate the unconstrained non-submodular maximization problem. Utilizing an approximate local search technique, we derive an $O(3/c^2 + \epsilon)$-approximation algorithm, where c is the zero order approximate submodularity coefficient. Finally, combining these two algorithms, we obtain an almost constant approximation algorithm for the non-monotone non-submodular maximization problem with p-independence system constraint.

Acknowledgments. The first two authors are supported by Natural Science Foundation of China (Nos. 11531014, 11871081). The third author is supported by Natural Sciences and Engineering Research Council of Canada (No. 283106). The fourth author is supported by China Postdoctoral Science Foundation funded project (No. 2018M643233) and Natural Science Foundation of China (No. 61433012). The fifth author is supported by Natural Science Foundation of Shanxi province (No. 201801D121022).

References

1. Badanidiyuru, A., Mirzasoleiman, B., Karbasi, A., Krause, A.: Streaming submodular maximization: massive data summarization on the fly. In: 20th ACM SIGKDD International Conference on Knowledge Discovery and Data Mining, pp. 671–680. ACM (2014)
2. Bian, A.-A., Buhmann, J.-M., Krause, A., Tschiatschek, S.: Guarantees for greedy maximization of non-submodular functions with applications. In: 34th International Conference on Machine Learning, pp. 498–507. JMLR (2017)
3. Bogunovic, I., Zhao, J., Cevher, V.: Robust maximization of non-submodular objectives. In: 21st International Conference on Artificial Intelligence and Statistics, pp. 890–899. Playa Blanca, Lanzarote (2018)
4. Buchbinder, N., Feldman, M.: Constrained submodular maximization via a nonsymmetric technique arXiv:1611.03253 (2016)
5. Buchbinder, N., Feldman, M.: Deterministic algorithms for submodular maximization problems. ACM Trans. Algorithms **14**(3), 32 (2018)
6. Buchbinder, N., Feldman, M., Naor, J.-S., Schwartz, R.: Submodular maximization with cardinality constraints. In: 25th Annual ACM-SIAM Symposium on Discrete Algorithms, pp. 1433–1452. Society for Industrial and Applied Mathematics (2014)
7. Buchbinder, N., Feldman, M., Seffi, J., Schwartz, R.: A tight linear time 1/2-approximation for unconstrained submodular maximization. SIAM J. Comput. **44**(5), 1384–1402 (2015)
8. Calinescu, G., Chekuri, C., Pál, M., Vondrák, J.: Maximizing a monotone submodular function subject to a matroid constraint. SIAM J. Comput. **40**(6), 1740–1766 (2011)
9. Chen, W., Zhang, H.: Complete submodularity characterization in the comparative independent cascade model. Theor. Comput. Sci. (2018)
10. Conforti, M., Cornuéjols, G.: Submodular set functions, matroids and the greedy algorithm: tight worst-case bounds and some generalizations of the Rado-Edmonds theorem. Discrete Appl. Math. **7**(3), 251–274 (1984)
11. Das, A., Kempe, D.: Submodular meets spectral: greedy algorithms for subset selection, sparse approximation and dictionary selection. In: 28th International Conference on Machine Learning, pp. 1057–1064. Omnipress (2011)
12. El-Arini, K., Veda, G., Shahaf, D., Guestrin, C.: Turning down the noise in the blogosphere, In: 15th ACM SIGKDD International Conference on Knowledge Discovery and Data Mining, pp. 289–298. ACM (2009)
13. Feige, U.: A threshold of $\ln n$ for approximating set cover. J. ACM **45**(4), 634–652 (1998)
14. Feige, U., Mirrokni, V.-S., Vondrák, J.: Maximizing non-monotone submodular functions. SIAM J. Comput. **40**(4), 1133–1153 (2011)

15. Feldman, M., Harshaw, C., Karbasi, A.: Greed is good: near-optimal submodular maximization via greedy optimization. In: 30th Annual Conference on Learning Theory, pp. 758–784. Springer (2017)
16. Gomes, R., Krause, A.: Budgeted nonparametric learning from data streams. In: 27th International Conference on Machine Learning, pp. 391–398. Omnipress (2010)
17. Gupta, A., Roth, A., Schoenebeck, G., Talwar, K.: Constrained non-monotone submodular maximization: offline and secretary algorithms. In: Saberi, A. (ed.) WINE 2010. LNCS, vol. 6484, pp. 246–257. Springer, Heidelberg (2010). https://doi.org/10.1007/978-3-642-17572-5_20
18. Kempe, D., Kleinberg, J., Tardos, É.: Maximizing the spread of influence through a social network. In: 9th ACM SIGKDD International Conference on Knowledge Discovery and Data Mining, pp. 137–146. ACM (2003)
19. Lee, J., Sviridenko, M., Vondrák, J.: Submodular maximization over multiple matroids via generalized exchange properties. Math. Oper. Res. 35(4), 795–806 (2010)
20. Lin, H., Bilmes, J.: A class of submodular functions for document summarization. In: 49th Annual Meeting of the Association for Computational Linguistics: Human Language Technologies, vol. 1, pp, 510–520. Association for Computational Linguistics (2011)
21. Mirzasoleiman, B., Badanidiyuru, A., Karbasi, A.: Fast constrained submodular maximization: personalized data summarization. In: 33rd International Conference on Machine Learning, pp. 1358–1366. JMLR (2016)
22. Nemhauser, G.-L., Wolsey, L.-A., Fisher, M.-L.: An analysis of approximations for maximizing submodular set functions–I. Math. Program. 14(1), 265–294 (1978)
23. Sviridenko, M.: A note on maximizing a submodular set function subject to a knapsack constraint. Oper. Res. Lett. 32(1), 41–43 (2004)
24. Sviridenko, M., Vondrák, J., Ward, J.: Optimal approximation for submodular and supermodular optimization with bounded curvature. Math. Oper. Res. 42(4), 1197–1218 (2017)

Truthful Mechanism Design of Reversed Auction on Cloud Computing

Deshi Ye$^{(\boxtimes)}$, Feng Xie, and Guochuan Zhang

College of Computer Science, Zhejiang University, Hangzhou 310027, China
{yedeshi,21421114,zgc}@zju.edu.cn

Abstract. In this paper, we study a mechanism design of reversed auction on cloud computing. A cloud computing platform has a set of jobs and would like to rent VM instances to process these jobs from cloud providers. In the auction model, each cloud provider (agent) who owns VM instances will submit a bid on the costs for using such VM instances. The mechanism determines the number of VM instances from each agent, and payments that have to be paid for using the chosen VM instances. The utility of every agent is the payment received minus the true cost. Our proposed mechanism is a deterministic truthful mechanism that the utility of each agent is maximized by revealing the true costs. We first provide the analysis of the approximation ratios and then run experiments using both realistic workload and uniformly random data to show the performance of the proposed mechanisms.

Keywords: Auction and mechanism design · Approximation ratio · Multi-dimensional bin packing

1 Introduction

Cloud computing enables individual users and enterprises to use computing resources on-demand and pay only for the resources and service they use. Cloud computing emerged as a new computing platform such as Amazon EC2 and Microsoft Azure offer fixed-price and auction-based mechanisms to sell virtual machine (VM) instances to users. This allows more and more application service providers (ASP) to deploy their applications in clouds. An application service provider is a company that holds and manages remote software applications such as media services, content, and entertainment that are accessed by users over the Internet using a rental or usage-based transaction-pricing model. Examples include ISP, Yahoo Email, Gmails. If we regard customers' service request as a vector of resource (CPU, Memory, Disk) [21], then a sequence of such vectors will be generated in the ASP system. A practical problem for the ASP is to allocate such demands into VMs rented from cloud providers, and the goal is to minimize the total cost due to the rental of VMs.

Supported in part by NSFC(11671355).

The existing markets focus on the scenario that users or tenants compete for resources in a cloud provider. In this paper, we consider another scenario, namely reversed auction, in which cloud providers will compete for the tasks in an ASP system. In details, we consider such a computing platform that an ASP owns a set of computing jobs that require multi-dimensional computing resources, and there exist a number of cloud providers that can offer computing resources in the form of virtual machine instances for that ASP. The platform is required to determine which type of VM instance and how many such instances will be used such that the total cost incurred is minimized. In the market model, the true cost of each VM instance is private information that is only known to the cloud provider. Agents (cloud providers) will submit biddings of the type of VM instance and the cost to the system. The utility of each cloud provider is the payment received from the ASP platform minus the cost incurred to run the jobs. Usually, the type of instances offered by the cloud provider is fixed. A cloud provider may increase the cost to produce more profit (maximize the utility). However, this is a competitive market, different cloud providers may offer better choices. One goal of the mechanism design is to make sure that no cloud provider can benefit from strategically manipulate the decisions. The optimization of this problem is named as vector bin packing with general costs (VPGC), and the mechanism design version of VPGC is called MD-VPGC.

1.1   Our Contribution

We first design truthful mechanisms for one-dimensional VPGC. To guarantee the truthfulness, we extended the payment function for single parameter [1] to our problem such that any allocation algorithm (bin packing algorithm) satisfies the monotonic property is a truthful mechanism. An algorithm is a monotone if the number of bins used for that type of bin is not increasing if its cost increases. Consequently, we provide a class of algorithms based on the density rank, which is the ratio between the cost and the size of that bin. We show that if the bin packing algorithm is Next Fit (NF), Next Fit Decreasing (NFD), First Fit Decreasing (FFD), then the algorithm is monotone, while the First Fit (FF) is not monotone. Next, we extend the results to general multiple dimensions. The approximation ratios of the proposed mechanisms are also given in this work.

Besides the theoretical results, we implement our mechanisms in real data. We collect items from Google trace. Each item from the Google trace consists of three-dimensional information CPU, Memory, and Disk. The agents are collected from the virtual machine instances from Amazon EC2. We assume that the cost of each instance is a truthful cost. Then, we discuss the approximation ratios according to the experimental results. To the best of our knowledge, this work is the first effort on mechanism design for the vector packing problem.

1.2   Related Work

Recently, mechanism design attracted a great deal in the area of cloud provision problems, which can be regarded as the mechanism design on knapsack problems.

Mu'Alem and Nisan first provided a 2-approximation mechanism [14] for the knapsack problem. Chekuri and Gamzu [4] studied the mechanism design for the multiple knapsack problem via a greedy iterative packing. In general, they gave a truthful $(2+\varepsilon)$-approximate mechanism among single-minded agents, and further improved to $(e/(e-1)+\varepsilon)$ for knapsacks with identical capacity, where ε is an arbitrarily small positive number and e is the natural number. Briest et al. [3] designed a new approach in rounding scheme that leads to a monotone FPTAS for the knapsack problem. Moreover, when the number of knapsacks is a fixed constant, there exists a monotone PTAS for the multiple knapsack problem [3]. Nejad [16] give a greedy mechanism for dynamic virtual machine provision. Mashayekhy et al. [13] provided a PTAS mechanism for single multiple knapsack problem. In [12], a 3-approximated mechanism was proposed for multiple and multi-dimensional knapsacks.

If all agents reveal their truthful costs, it is the classical bin packing with costs problem. For one-dimension bin packing with general costs, Epstein and Levin [8] provided an APTAS. Kang and Park [11] considered this problem with the constraint that $c_i/b_i \leq c_j/b_j$ for $b_i \geq b_j$, and provided an algorithm of asymptotic approximation ratio $3/2$. If the cost of each bin is proportionate to its size, then the problem becomes the variable sized bin packing problem, which was first investigated by Friesen and Langston [9].

For arbitrary d-dimensional vector bin packing, Chekuri and Khanna [5] showed vector bin packing is hard to approximate to within a $d^{1/2-\epsilon}$ factor for all fixed $\epsilon > 0$, and Bansal et al. [2] showed that it is $d^{1-\epsilon}$ in-approximate. A $(d+\epsilon)$-approximate algorithm can be easily extended from the APTAS [7]. Chekuri and Khanna [5] presented an algorithm with approximation of $1+\epsilon d+O(\ln 1/\epsilon)$. For constant d, Bansal et al. [2] provided the current best algorithm with approximation ratio of $0.807+\ln(d+1)+\epsilon$, and it is APX-hard even for $d=2$ [23]. Gabay and Zaourar [10] studied vector bin packing with heterogeneous bins (bins with different size) and the goal is to find a feasible packing of items into the given bins.

2 Preliminaries

We define the studied problem as a vector packing with general costs (VPGC) as follows. The ASP platform has a set of n items (or jobs) $I = (a_1, \ldots, a_n)$. Each item a_i consists of heterogeneous resources such as cores, memory, storage, etc. In this work, we suppose that there are total d types of resources, i.e. $a_i = (a_{i1}, a_{i2}, \ldots, a_{id})$ is a d-dimensional vector.

We consider that there is a set of m agents, $B = (B_1, \ldots, B_m)$, could offer sufficiently large number of bins (virtual machines). Each agent B_i owns a type of bins such that the capacity is $b_i = (b_{i1}, b_{i2}, \ldots, b_{id})$ and the cost is c_i. We denote $B_i = (b_i, c_i)$. In this work, we assume that b_i is publicly known, and c_i is a piece of private information. When agents receive requests of bins from the ASP platform, each agent B_i will submit a bid $R_i = (b_i, \theta_i)$ to the ASP, where θ_i may not equal to c_i. Let $R = (R_1, R_2, \ldots, R_m)$ be the profile of biddings. For simplicity, we let $R_i = \theta_i$ in the following since b_i is a piece of public information.

Once the ASP platform receives the bids R from agents, the ASP platform is required to determine the number of bins from each agent, and the payment to each agent for using those bins. Let $num_i(R)$ be the number of bins bought from the agent B_i and $P_i(R)$ be the payment from the ASP to the agent B_i according to the bid R, respectively. The utility function of agent B_i according to the bidding R is defined as

$$u_i(R) = P_i(R) - c_i \cdot num_i(R).$$

Note that each agent is selfish, and each agent's goal is to maximize his/her utility, which will motivate each agent B_i to manipulate the mechanism by submitting a different bid θ_i from the true cost c_i to increase the utility.

Let B_{iq} be set of items assigned in the qth bin of the agent B_i. A packing is feasible, we require that for each i and q, we have $\sum_{j \in B_{iq}} a_{jk} \leq b_{ik}$ for each coordinate $1 \leq k \leq d$.

We define by $\mathcal{C}(A(I,R))$ the *social cost* of the mechanism $\mathcal{M}$ to pack the item set I with bids profile R, which is the total costs of bins used by all agents, i.e.,

$$\mathcal{C}(\mathcal{M}(I,R)) = \sum_{i=1}^{m} c_i \cdot num_i(R).$$

A mechanism $\mathcal{M}$ is said to be ρ-approximated if $\mathcal{C}(\mathcal{M}(I,R)) \leq \rho \cdot \mathcal{C}(OPT(I))$, where $OPT(I)$ is an optimal solution for the set I. A mechanism is said ρ-approximated in *asymptotic* if

$$\mathcal{C}(\mathcal{M}(I,R)) \leq \rho \cdot \mathcal{C}(OPT(I)) + \alpha, \tag{1}$$

where α is a constant.

Let $R_{-i} = \{R_1, \ldots, R_{i-1}, R_{i+1}, \ldots, R_m\}$ be the bids except agent B_i's bid.

Definition 1. *(Truthfulness): A mechanism $\mathcal{M}$ consisting of a packing function $\mathcal{A}$ and a payment function $\mathcal{P}$ is truthful (or strategy-proof) if for every agent B_i with the true cost c_i cannot increase his/her utility by declaring any other cost θ_i regardless of every bidding of other agents R_{-i}, i.e., it satisfies*

$$u_i(c_i, R_{-i}) \geq u_i(\theta_i, R_{-i}).$$

This definition implies that truthful reporting is a dominant strategy for every agent.

Definition 2. *(Individual rationality): A mechanism $\mathcal{M}$ is said to be individual rationality if every agent always obtains non-negative utility with bidding of the true cost, i.e., $u_i(c_i, R_{-i}) \geq 0$ for any i and any R_{-i}.*

The Vickrey-Clarke-Groves (VCG) mechanism [18,22] is a well-known mechanism that ensures truthful bidding. However, the VCG mechanism requires an optimal solution to pack the items. Note that the vector packing problem is NP-hard, which generalize the classical one-dimensional bin packing problem.

Unfortunately, if an approximation algorithm is applied in the packing, then VCG is not guaranteed to be truthful [17]. As a result, we are going to design a deterministic truthful mechanism, while the property of individual rationality satisfies.

It is worthy to note that if there exists an item that can only be packed into a specific type bin, then that agent of that type bin will claim an arbitrary high cost such that his/her utility is arbitrarily large. To avoid this scenario, we set an upper bound to the cost of any bin in the system. In real applications, the cost of each bin will be in a reasonable interval, and hence we further suppose that for each item, there are at least two types of bins can accommodate it. To design a truthful mechanism, we will follow the framework of Archer and Tardos [1] to find monotone algorithms. The payment of each agent B_i according to the bidding R is given as below.

$$P_i(R) = \theta_i \cdot num_i(R) + \int_{\theta_i}^{\infty} num_i(x, R_{-i})dx. \tag{2}$$

By the assumption that the cost of each bin is upper bounded, we know that $\int_{\theta_i}^{\infty} num_i(x, R_{-i})dx$ is finite.

Definition 3. *(Monotone:) A packing algorithm $\mathcal{A}$ is monotone if $num_i(R)$ is a non-increasing function on the increasing of bid θ_i by fixing R_{-i}.*

Theorem 1. *A mechanism $\mathcal{M}$ with a monotone packing algorithm $\mathcal{A}$, and associated with above payment $\mathcal{P}$ (2) is truthful.*

Proof. This proof can be extended from Theorem 4.2 by Archer and Tardos [1]. □

Corollary 1. *A mechanism $\mathcal{M}$ adopt the payment in Eq. (2) satisfies individual rationality.*

3   Mechanism Design Framework

In this section, we propose a deterministic truthful mechanism design for VPGC. To be convenient, we first address the mechanism design framework for the one-dimensional case, and we extend it to the general multi-dimensional version of the problem in the full version of this work.

3.1   Monotone Algorithms for One-Dimensional VPGC

According to Theorem 1, the key challenge in the design of a truthful mechanism is to propose a packing algorithm in order to satisfy monotonicity. An algorithm in our mechanism consists of a way of methods to select bins and the way to pack items among the selected bins. The core idea of our packing algorithm is to assign an item to a bin with the cheapest unit cost.

Before describing the detailed algorithm, we need to define some notations. For each item j with size a_j, let us define a feasible set of bins $\mathbf{Fs}_j = \{B_i | a_j \le b_i\}$. Let us define the cost density of the ith type of bins (the bins owned by the agent B_i) be $den(B_i) = c_i/b_i$. Our main idea is to select the cheapest feasible bin, i.e., the bin with the smallest density to assign every item. Our algorithm is given in Algorithm 1. The algorithm consists of two phases. In phase one, the algorithm finds a type of bin with the smallest density among the feasible bins. In the second phase, all the items assigned in a specific type of bin are packed according to a classical bin packing algorithm, such as Next Fit (NF), First Fit (FF), First Fit Decreasing (FFD) and so on. We also use $DF+(BinPacking)$ to denote the detailed algorithm, where $BinPacking$ is a bin packing algorithm, for example, DF+NF means we use Next Fit instead of $BinPacking$, and DF+FF means we use First Fit instead of $BinPacking$. For one-dimensional bin packing algorithms (see e.g. [6]), we will describe the detailed algorithms as follows.

- Next-Fit (NF): NF picks an item and packs the item in the currently opened bin if it can fit, otherwise this bin will be closed, and packs the current item in an empty bin.
- Next-Fit Decreasing (NFD): NFD is the same as NF except that items are sorted in non-increasing order, i.e. NFD always picks the item with the largest size among the remaining unpacked items.
- First-Fit (FF): Suppose that bins are ordered in a sequence, and items are sorted in a list. FF picks an item from the list, and pack the current item in the lowest indexed nonempty bin if it can be accommodated. Otherwise, FF packs the current item in an empty bin. The procedure is continuous until there is no item in the list.
- First-Fit Decreasing (FFD): FFD is the same as FF except that the items are sorted in non-increasing order, i.e., FFD always packs the item with the largest size among the remaining unpacked items.

Lemma 1. *The algorithms DF+NF and DF+NFD are monotone.*

Proof. Without loss of generality, we assume that $den(B_1) \le den(B_2) \le \dots \le den(B_m)$. According to the definition of monotone, we suppose agent B_i claims a different bid $B_i' = (b_i, c_i')$, where $c_i' > c_i$. If $den(B_i') \le den(B_{i+1})$, the items assigned to agent B_i do not change, and hence the number of bins keep the same. W.L.O.G. we assume that

$$den(B_i') > den(B_{i+1}).$$

We claim that the items choose agent B_i with bidding $B_i' = (b_i, c_i')$ is in a subset of the items that choose agent B_i with bidding $R_i = (b_i, c_i)$.

It was proved by Murgolo in [15] that NF (or NFD) algorithm does not use more bins to pack any subset of items. As a result, DF+NF and DF+NFD are monotone.   □

Lemma 2. *The algorithm DF+FF is not monotone.*

Algorithm 1. Monotone Algorithm for VPGC: DF+(BinPacking)

Input: A set of items $I = (a_1, a_2, \ldots, a_n)$; m agents $\{B_1, \ldots, B_m\}$, with (b_i, c_i)
for each agent.

1 {First phase}
2 **for** $j \leftarrow 1$ **to** n **do**
3      $\mathbf{Fs}_j \leftarrow \{B_i | a_j \leq b_i\}$ /* For each item j, we calculate the set of feasible bins
    B_i that the item j can be assigned */;
4      $k \leftarrow \min_{B_i \in \mathbf{Fs}_j} den(B_i) = c_i / b_i$, in case of tie, k is the index of the bin with
    the smallest cost;
5      $Assign_k = Assign_k \bigcup \{j\}$ /* $Assign_k$ is the set of items that will be
    assigned in the bins of type B_k*/ ;

6 {Second phase}
7 **for** $i \leftarrow 1$ **to** m **do**
8   /* All the jobs that will be assigned in bin of type B_i*/;
9   Apply a classical bin packing algorithm $BinPacking$ for the set of items in
    $Assign_i$;

Output: Assignment for each item a_j

Proof. Given a list of items with size $I = \{\,0.55, 0.7, 0.1, 0.45, 0.15, 0.3, 0.2, 0.55\,\}$. There are three types of bin $B_1 = (1, 1)$, $B_2 = (0.1, 0.101)$, and $B_3 = (1, 2)$. Note that $den(B_1) = 1$ and $den(B_2) = 1.01$. Then, bin B_1 is the feasible bin for all items and it is the smallest cost density. According to the FF algorithm, we need 3 bins to pack all items in I.

To prove an algorithm is monotone, we need to prove if bin B_1 increases its cost, the number of bins used shall not be increasing. We show it is not correct.

Suppose bin B_1 reports a different bid $B_1' = (1, 1.11)$. Then $den(B_1) = 1.11$. According to the algorithm, the item of size 0.1 shall be assigned to the bin B_2, and all the other items shall be assigned to B_1 with the FF algorithm. Note that the FF algorithm needs 4 bins to pack all items in $I/\{0.1\}$. $\qquad\square$

Lemma 3. *The algorithm DF+FFD is monotone.*

Proof. Similar as Lemma 1, we assume that $den(B_1) \leq den(B_2) \leq \ldots \leq den(B_m)$. The profile of the bidding is $R = \{B_1, \ldots, B_m\}$. Now we suppose agent B_i claims a different bid $B_i' = (b_i, c_i')$, where $c_i' > c_i$. Again we could assume that $den(B_i') > den(B_{i+1})$, otherwise the items assigned to agent i do not change, and hence the number of bins keeps the same.

Let $S_i(R)$ be the set of items that choose the agent of type B_i under the bidding of R. Clearly, $S_i(R_{-i}, B_i') \subseteq S_i(R)$. Namely, items choose agent B_i by bidding B_i' is in a subset of the items that choose the agent with bidding B_i. Let j be the largest number such that $den(B_j) < den(B_i')$, then $j \geq i + 1$. Clearly, any item with size at most of b_j will choose agent j, and the size of any item in $S_i(R_{-i}, B_i')$ is at least b_j. Hence, the size of any items in $S_i(R_{-i}, B_i')$ is at least the size of items in $S_i(R) \backslash S_i(R_{-i}, B_i')$.

According to the algorithm FFD, to pack items in $S_i(R)$, we need first pack all items in $S_i(R_{-i}, B_i')$, and then the remaining items. Thus, the bins used by items in $S_i(R)$ will be at least that of $S_i(R_{-i}, B_i')$. In consequence, DF+FFD is monotone followed by the definition.   □

Lemma 4. *The algorithm DF+BinPacking is at least 2-approximation in asymptotic.*

Proof. For any given $\epsilon > 0$, there are n items, all are with size $1/2+\epsilon$. There are two type of bins $B_1 = (1,1)$ and $B_2 = (1 + 2\epsilon, 1 + 3\epsilon)$. The algorithm uses type bin B_1 to pack all items. The total cost is n. While an optimal algorithm uses type B_2 to pack all items, with total cost approaches to $n/2 + O(\epsilon)$. Therefore no matter which BinPacking algorithm we adopt, the lower bound 2 follows immediately.   □

Theorem 2. *The asymptotic approximation ratio of DF+NF is at most of 2.*

Proof. W.L.O.G. we assume that $den(B_1) \leq den(B_2) \leq \ldots \leq den(B_m)$. Suppose that the item j is packed in bin B_i, which implies that $a_j > b_k$, where $1 \leq k \leq i-1$. It is worth to note that $den(B_i)$ is the cheapest unit cost to pack the item j. Hence, an optimal algorithm to pack the item j need cost at least $a_j \cdot den(B_i)$. Let x_i be the number of bins used from agent i. According to the NF algorithm, the total size of items packed in bins of agent i is at least $x_i b_i/2$ or at least $b_i((x_i - 1)/2) + a_j$, where a_j is the last item in the bin of type i.

Let OPT and ALG be the costs required by an optimal algorithm and the approximation algorithm DF+NF, respectively. From the above analysis, it is clear to have the following inequalities.

$$OPT \geq \sum_{i=1}^{m} \{\frac{1}{2}(x_i - 1)b_i \cdot den(B_i) + b_{i-1}den(B_i)\}$$

$$\geq \sum_{i=1}^{m} \{\frac{1}{2}(x_i - 1)b_i \cdot den(B_i) + b_{i-1}den(B_{i-1})\}$$

$$\geq \sum_{i=1}^{m} \{\frac{1}{2}(x_i - 1)c_i + c_{i-1}\}$$

$$\geq \sum_{i=1}^{m} (\frac{1}{2}x_i c_i) - \frac{1}{2}c_m$$

Note that $ALG = \sum_{i=1}^{m} c_i x_i$, then we have $ALG \leq 2OPT + c_m$. The asymptotic ratio of 2 follows since we suppose that every cost of a bin is bounded by a constant.   □

We can extend the monotone algorithm of one dimension VPGC to the general multiple dimension d-VPGC. The detailed algorithms and proofs will be given the full version of this paper.

Theorem 3. *The asymptotic approximation ratio of the d-VPGC(NF) or (d-VPGC(FFD)) for the d-VPGC problem is at most 2d.*

4   Experimental Results

We perform experiments to investigate the performance of the proposed mechanisms against the performance of the optimal social optimum.

4.1   Data Sets and Simulation Setup

We conduct our simulations based on Google cluster-usage traces [20]. Google trace data has been explored and used extensively in industry and research, such as the characterization analysis [19]. In details, the data is collected from about 12500 machines over 29 days and the chosen data is one of the 500 task-events trace data files. This data includes CPU, memory and disk information. Each record represents a 3-dimensional item, and there are 77409 records in total. Google cluster data provides job requests with a large variety of resource requirements.

The agents' data are collected from publicly available information from Amazon EC2. Currently, there is no auction-based platform available to simulate our proposed mechanism. Instead, we suppose that each instance in the Amazon EC2 is an individual agent. The capacity of each instance and its true cost is the one given on the platform. For example, instance m3.medium is a 3-dimensional vector $(1, 3.75, 4)$ with cost 0.067, where vCPU has 1 core, and memory is 3.75 GiB, and storage is 4 GB, and the unit cost per hour is \$0.067. This data is collected from https://aws.amazon.com/ec2/ on Aug 16,2016, and the actual values may be updated in that web from time to time.

To avoid one agent can dictator by bidding a very large cost, we require in addition that there are at least two types of bins can accommodate an item. In this case, the upper bound cost M in our experiment can be ignored due to the fact that we do not need to use that bin if its cost is high enough. Actually, the payment is a finite number for each agent. The total number of items used in the experiment is ranging from 1000 to 50,000.

4.2   Analysis of Results

In the experiment, we run the designed mechanisms for these two data sets. In the theoretical part, we have proved upper bounds in the asymptotic, there might have some constant α in (1). In this experiment, we only consider absolute competitive ratio, i.e., the constant α is zero. To obtain approximation ratios of the mechanism, we compare the total costs of the approximated mechanisms to the lower bound of an optimal solution in (3). Let us simplify the instance, in which each item is only associated with data in the qth coordinate. To pack the item j, the cost we require is at least $\min_i den(B_{iq}, j)$. Let D_{iq} be the set of items that have the cheapest cost when it is packed in a bin of type i when only considering the qth coordinate. Thus, a lower bound of an optimal algorithm for the general d dimensional instance, is

$$\max_q \sum_i \frac{\sum_{j \in D_{iq}} a_{jq}}{b_{iq}} c_i. \tag{3}$$

We compare the approximation ratios for a different number of items ranging from 1000 to 50,000. We explore several factors that might affect the approximation ratios of the mechanism, such as the number of items, different data sets, different dimensions of data. The payment and the utility of each agent will be calculated as well. Though the truthfulness of mechanism was already proved in theory, we also run the experiment to see how the utility function varies according to different bids.

d-VPGC(NF) vs d-VPGC(FFD). In our mechanisms, we use one-dimensional BinPacking algorithms, in which NF and FFD were shown to be monotone algorithms, while the FF is not monotone. Furthermore, FFD will outperform NFD. Hence, we will compare FFD and NF, and aim to find how far they are different. Figure 1(a) shows the approximations ratios of d-VPGC(NF) and d-VPGC(FFD) in 3-dimensional data with the number of items in 1000 to 50,000.

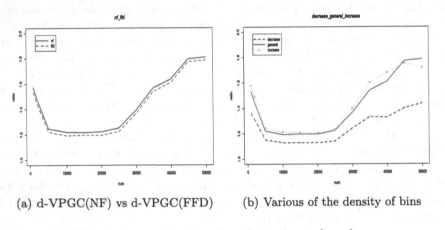

(a) d-VPGC(NF) vs d-VPGC(FFD)   (b) Various of the density of bins

Fig. 1. Approximation ratios of experimental results

The results show that d-VPGC(NF) and d-VPGC(FFD) have similar curves regarding the approximation ratios. The approximation ratios of d-VPGC(FFD) are between [1.19, 1.79], while the approximation ratios of d-VPGC(NF) are between [1.22, 1.81]. The approximation ratios in this data set are much better than the proved theoretical worst-case bounds of 6. In addition, we observe that d-VPGC(FFD) is outperformed d-VPGC(NF). Hence, we only plot the performance of d-VPGC(FFD) in the remaining simulations.

Density of Bins. The density of each type of bins plays a key role in the proposed mechanism. In the experiment, we select agents based on two rules according to the density of bins. One rule is that the density of bins is decreasing, i.e., the larger bin has a smaller density. Another rule is density increasing, i.e., the larger bin has a larger density.

Figure 1(b) shows these two rules compared with a general case that does not have such rules. The decreasing rule means that the larger bin has a smaller density, hence the mechanism will prefer to use a bin with a larger capacity. The results show that decreasing rule will generate the smallest approximation ratios compared with the other two rules. The increasing rule has the largest approximations because using smaller bins might produce more spare spaces, and hence increases the approximation ratios.

5 Conclusions and Remarks

This work presented the first mechanism design for the vector packing with general costs problem. We showed that the proposed mechanisms are individually rational and truthful. Our mechanisms were based on monotonic algorithms that choose bins with the smallest unit cost. The experimental results showed the proposed mechanisms obtain better performance in practice than the theoretical proved results.

There are some interesting problems to investigate which are left open by our paper. In particular, we only consider the static one round bidding, it would be nice to deal with the setting when the agents arrive online.

Acknowledgment. The authors thank anonymous referees for helpful comments and suggestions to improve the presentation of this paper.

References

1. Archer, A., Tardos, E.: Truthful mechanisms for one-parameter agents. In: Proceedings of the 42nd IEEE Symposium on Foundations of Computer Science, pp. 482–491. IEEE Computer Society (2001)
2. Bansal, N., Elias, M., Khan, A.: Improved approximation for vector bin packing. In: Proceedings of the 27th Annual ACM-SIAM Symposium on Discrete Algorithms (SODA), pp. 1561–1579. ACM-SIAM (2016)
3. Briest, P., Krysta, P., Vöcking, B.: Approximation techniques for utilitarian mechanism design. SIAM J. Comput. **40**(6), 1587–1622 (2011)
4. Chekuri, C., Gamzu, I.: Truthful mechanisms via greedy iterative packing. In: Dinur, I., Jansen, K., Naor, J., Rolim, J. (eds.) APPROX/RANDOM -2009. LNCS, vol. 5687, pp. 56–69. Springer, Heidelberg (2009). https://doi.org/10.1007/978-3-642-03685-9_5
5. Chekuri, C., Khanna, S.: On multidimensional packing problems. SIAM J. Comput. **33**(4), 837–851 (2004)
6. Coffman, E.G., Csirik, J., Galambos, G., Martello, S., Vigo, D.: Bin packing approximation algorithms: survey and classification. In: Pardalos, P.M., Du, D.-Z., Graham, R.L. (eds.) Handbook of Combinatorial Optimization, pp. 455–531. Springer, New York (2013). https://doi.org/10.1007/978-1-4419-7997-1_35
7. de La Vega, W.F., Lueker, G.S.: Bin packing can be solved within $1+\varepsilon$ in linear time. Combinatorica **1**(4), 349–355 (1981)
8. Epstein, L., Levin, A.: An APTAS for generalized cost variable-sized bin packing. SIAM J. Comput. **38**(1), 411–428 (2008)

9. Friesen, D.K., Langston, M.A.: Variable sized bin packing. SIAM J. Comput. **15**(1), 222–230 (1986)
10. Gabay, M., Zaourar, S.: Vector bin packing with heterogeneous bins: application to the machine reassignment problem. Ann. Oper. Res. **242**, 161–194 (2015)
11. Kang, J., Park, S.: Algorithms for the variable sized bin packing problem. Eur. J. Oper. Res. **147**(2), 365–372 (2003)
12. Mashayekhy, L., Nejad, M.M., Grosu, D.: Physical machine resource management in clouds: a mechanism design approach. IEEE Trans. Cloud Comput. **3**(3), 247–260 (2015)
13. Mashayekhy, L., Nejad, M.M., Grosu, D.: A PTAS mechanism for provisioning and allocation of heterogeneous cloud resources. IEEE Trans. Parallel Distrib. Syst. **26**(9), 2386–2399 (2015)
14. Mu'Alem, A., Nisan, N.: Truthful approximation mechanisms for restricted combinatorial auctions. Games Econ. Behav. **64**(2), 612–631 (2008)
15. Murgolo, F.D.: Anomalous behavior in bin packing algorithms. Discrete Appl. Math. **21**(3), 229–243 (1988)
16. Nejad, M.M., Mashayekhy, L., Grosu, D.: A family of truthful greedy mechanisms for dynamic virtual machine provisioning and allocation in clouds. In: IEEE CLOUD, pp. 188–195 (2013)
17. Nisan, N., Ronen, A.: Computationally feasible VCG mechanisms. J. Artif. Intell. Res. (JAIR) **29**, 19–47 (2007)
18. Nisan, N., Roughgarden, T., Tardos, E., Vazirani, V.V.: Algorithmic Game Theory, vol. 1. Cambridge University Press, Cambridge (2007)
19. Reiss, C., Katz, R.H., Kozuch, M.A.: Towards understanding heterogeneous clouds at scale: Google trace analysis. ISTC-CC-TR-12-101, Carnegie Mellon University (2012)
20. Reiss, C., Wilkes, J., Hellerstein, J.L.: Google cluster-usage traces: format + schema. Technical report, Google Inc., Mountain View, CA, USA, November 2011. http://code.google.com/p/googleclusterdata/wiki/TraceVersion2. Accessed 20 Mar 2012
21. Stillwell, M., Schanzenbach, D., Vivien, F., Casanova, H.: Resource allocation algorithms for virtualized service hosting platforms. J. Parallel Distrib. Comput. **70**(9), 962–974 (2010)
22. Vickrey, W.: Counterspeculation, auctions, and competitive sealed tenders. J. Finance **16**(1), 8–37 (1961)
23. Woeginger, G.J.: There is no asymptotic PTAS for two-dimensional vector packing. Inf. Process. Lett. **64**(6), 293–297 (1997)

Distance Constrained Vehicle Routing Problem to Minimize the Total Cost

Wei Yu[1($\boxtimes$)], Zhaohui Liu[1], and Xiaoguang Bao[2]

[1] Department of Mathematics, East China University of Science and Technology,
Shanghai 200237, China
{yuwei,zhliu}@ecust.edu.cn
[2] College of Information Technology,
Shanghai Ocean University, Shanghai 201306, China
xgbao@shou.edu.cn

Abstract. Given $\lambda > 0$, an undirected complete graph $G = (V, E)$ with nonnegative edge-weight function obeying the triangle inequality and a depot vertex $r \in V$, a set $\{C_1, \ldots, C_k\}$ of cycles is called a λ-*bounded r-cycle cover* if $V \subseteq \bigcup_{i=1}^{k} V(C_i)$ and each cycle C_i contains r and has a length of at most λ. The Distance Constrained Vehicle Routing Problem with the objective of minimizing the total cost (DVRP-TC) aims to find a λ-bounded r-cycle cover $\{C_1, \ldots, C_k\}$ such that the sum of the total length of the cycles and γk is minimized, where γ is an input indicating the assignment cost of a single cycle.

For DVRP-TC on tree metric, we show a 2-approximation algorithm that is implied by the existing results and give an LP relaxation whose integrality gap has an upper bound of 5/2. In particular, when $\gamma = 0$ we prove that this bound can be improved to 2. For the unrooted version of DVRP-TC, we devise a 5-approximation algorithm and show that a natural set-covering LP relaxation has a constant integrality gap of 25 using the rounding procedure given by Nagarajan and Ravi (2008).

Keywords: Vehicle Routing · Cycle cover · Path cover ·
Approximation algorithm · Integrality gap

1 Introduction

In the Distance Constrained Vehicle Routing Problem (DVRP), we are given a set of n vertices in a metric space, a specified depot r, and a distance bound λ, the aim is to find a set of tours, which start and end at r and have a length of at most λ, for the vehicles to cover all the vertices such that one of the following objectives is minimized: (i) the total distance of the tours; (ii) the number of tours. We denoted by DVRP-D (DVRP-N) the DVRP with the first (second) objective. In unrooted DVRP, the depot r is not specified and the tours are allowed to contain any set of vertices as long as their length do not exceed λ.

The (unrooted) DVRP arises naturally in many practical applications including daily routes scheduling for courier carriers, milkruns from manufacturing facilities, sensor coverage in wireless sensor networks, and so on (see [2,9,11,13]).

© Springer Nature Switzerland AG 2019
D.-Z. Du et al. (Eds.): COCOON 2019, LNCS 11653, pp. 639–650, 2019.
https://doi.org/10.1007/978-3-030-26176-4_53

Both (unrooted) DVRP-D and (unrooted) DVRP-N are NP-hard since they generalize the well-known Traveling Salesman Problem. Moreover, due to the NP-Completeness of the Hamiltonian Cycle Problem [6], (unrooted) DVRP-N cannot be approximated within a ratio less than 2 unless P = NP. The results in [12] imply that (unrooted) DVRP-N defined on a tree metric space still has an inapproximability lower bound of 3/2.

Li et al. [8] showed that DVRP-D and DVRP-N are within a factor of two in terms of approximability. They gave $\left(1 + \frac{\alpha\beta}{\beta-2}\right)$-approximation algorithms for both DVRP-D and DVRP-N, where α is the approximation ratio for the traveling salesman problem and β is the ratio between λ and the distance from the depot r to the farthest vertex. Recently, Liang et al. [9] improved the ratio for DVRP-D to $\frac{\alpha\beta}{\beta-2}$. They also proved that the 2-approximation algorithm for DVRP-N on tree metric in [11,12] is versatile enough to achieve the same approximation factor for DVRP-D on tree metric.

Nagarajan and Ravi [10,11] developed an $O(\min\{\log n, \log \lambda\})$-approximation algorithm for DVRP-N, which was improved to $O\left(\min\left\{\log n, \frac{\log \lambda}{\log\log \lambda}\right\}\right)$ by Frigstad and Swamy [5]. These results also imply the same bounds on the integrality gap of a natural set-covering LP relaxation for DVRP-N. For the problem on tree metric, Nagarajan and Ravi [11] proved that the integrality gap of this LP relaxation is at most 20. Yu et al. [17] gave a new LP relaxation based on the 2-approximation algorithm for DVRP-N on tree metric in [11,12] and derived an upper bound of 5/2 on the integrality gap.

Though it is still open whether DVRP-N admits a constant-factor approximation algorithm, the unrooted DVRP-N, which is also called the Minimum Cycle Cover Problem (MCCP) in the literature, seems easier to approximate. Arkin et al. [1] first presented a 3-approximation algorithm for a related Minimum Tree Cover Problem which implies a 6-approximation algorithm for MCCP by a simple edge-doubling strategy. After a sequence of improvements by Khani and Salavatipour [7], Yu and Liu [15], Yu et al. [16] proposed an algorithm with approximation ratio 32/7. Nagarajan and Ravi [11]'s results imply that an unrooted version of the above-mentioned set-covering LP relaxation has an integrality of 34, as noted by Yu et al. [17], who also obtained an improved LP relaxation for MCCP with integrality gap 6.

In this paper, we introduce the DVRP with the objective of minimizing the total cost (DVRP-TC) that is a natural generalization of both DVRP-D and DVRP-N. We assume w.l.o.g that one unit distance takes one unit cost and the cost of sending one vehicle is γ. Then the objective of DVRP-TC is the sum of the traveling cost, i.e., the total distance of the tours, and the assignment cost of vehicles, i.e., γ times the number of tours.

We mainly obtained the following results. First, we show that the existing results already imply a 2-approximation algorithm for DVRP-TC on tree metric. Secondly, we give an LP relaxation for DVRP-TC on tree metric similar to that in [17] and prove that its integrality gap can be bounded by 5/2. In particular, when $\gamma = 0$ we show an upper bound of 2. Thirdly, we devise a 5-approximation algorithm for unrooted DVRP-TC. Fourthly, we show a natural set-covering LP relaxation for unrooted DVRP-TC has a constant integrality gap of 25 using the rounding procedure in [11]. Lastly, we mention that our results can be applied to the path-version of (unrooted) DVRP-TC.

The rest of the paper is organized as follows. We formally state the problem and give some preliminary results in Sect. 2. In Sect. 3 we deal with DVRP-TC on tree metric, which is followed by the results on unrooted DVRP-TC in Sect. 4. Finally, we give a short discussion on the path-version of DVRP-TC in Sect. 5.

2   Preliminaries

Given an undirected weighted graph $G = (V, E)$ with vertex set V and edge set E, $w(e)$ denotes the weight or length of edge e. If $e = (u, v)$, we also use $w(u, v)$ to denote the weight of e. For a subgraph G' of G, the graph obtained by adding some copies of the edges in G' is called a multi-subgraph of G. For a (multi-)subgraph H (e.g. tree, cycle, path) of G, let $V(H), E(H)$ be the vertex set and edge set of H, respectively. The weight of H is defined as $w(H) = \sum_{e \in E(H)} w(e)$. If H is a multi-subgraph, $E(H)$ is a multi-set of edges and the edges appearing multiple times contribute multiply to $w(H)$.

A cycle C is also called a tour on $V(C)$. A cycle (tree) that contains some special vertex $r \in V$, called the depot, is referred to as an r-cycle (r-tree). An r-path is a path starting from r. A set $\{C_1, \ldots, C_k\}$ of cycles is called a *cycle cover* if $V \subseteq \bigcup_{i=1}^{k} V(C_i)$. If each C_i is an r-cycle, $\{C_1, \ldots, C_k\}$ is called an *r-cycle cover*. For any $\lambda \geq 0$, a cycle (tree, path) is called λ-bounded if its length is at most λ. A cycle cover is called λ-bounded if all the cycles in this cycle cover are λ-bounded. By replacing cycles with paths (trees) we can define *path cover* (*tree cover*) or *r-path cover* (*r-tree cover*) similarly.

We formally state the problems to be studied as follows.

In the Distance Constrained Vehicle Routing Problem with the objective of minimizing the total traveling distance of vehicles (DVRP-D), we are given $\lambda > 0$, an undirected complete graph $G = (V, E)$, a metric weight function $w : E \to \mathbb{N}$ that is nonnegative, symmetric and obeys the triangle inequality, and a depot $r \in V$, the aim is to find a λ-bounded r-cycle cover $\{C_1, \ldots, C_k\}$ such that the total length $\sum_{i=1}^{k} w(C_i)$ of the cycles is minimized.

In the Distance Constrained Vehicle Routing Problem with the objective of minimizing the number of the vehicles used (DVRP-N), we have the same input as in DVRP-D and the objective is to minimize the number k of cycles in the λ-bounded r-cycle cover.

In the Distance Constrained Vehicle Routing Problem with the objective of minimizing the total cost (DVRP-TC), we have an additional input $\gamma \geq 0$, which represents the assignment cost of one cycle, compared to DVRP-D and the objective is to minimize the sum of the total length of the cycles and the total assignment cost, i.e., $\sum_{i=1}^{k} w(C_i) + \gamma k$.

In the unrooted version of DVRP-TC (DVRP-D, DVRP-N), the depot r is removed from the inputs and the r-cycles are replaced by cycles.

The DVRP-TC (DVRP-D, DVRP-N) on tree metric is a special case where G is restricted to the metric closure of a weighted tree T. For an edge e (a vertex u) and a vertex v in T, v is called below e (u) if the unique path from r to v passes e (u). A set $V' \subseteq V$ of vertices is called below e (u) if each vertex in V' is below e (u).

Given an instance of DVRP-TC or its unrooted version, OPT indicates the optimal solution as well as its objective value. We call each cycle in OPT an optimum cycle. By the triangle inequality, we can assume w.l.o.g that any two optimum cycles are vertex-disjoint. By taking $\gamma = 0$ ($\gamma = +\infty$) we derive an instance of DVRP-D (DVRP-N) from an instance of DVRP-TC and define OPT_D (OPT_N) as the corresponding optimal value. Clearly, it holds that $OPT_D + \gamma OPT_N \leq OPT$.

We use n to denote the number of vertices of G. If (IP) is an integer programming for the DVRP-TC (DVRP-D, DVRP-N) or its unrooted version, we denote by OPT_{IP} the optimal value of (IP). OPT_{LP} is defined similarly for an LP relaxation (LP) for the problem.

The following cycle-splitting result on breaking a long cycle into a series of short paths is very useful. The basic idea is to add the edges greedily to a path along the cycle and throw out the last edge once this path has a length more than the target value.

Lemma 1 [1,4,14]. *Given a tour C on V' and $B > 0$, we can split the tour, in $O(|V'|)$ time, into $\max\left\{\left\lceil \frac{w(C)}{B} \right\rceil, 1\right\}$ B-bounded paths such that each vertex is located at exactly one path and the total length of the paths is at most $w(C)$.*

3   Tree Metric

In this section we deal with DVRP-TC on tree metric. We first show that the existing results already imply a 2-approximation algorithm. Then we present an LP relaxation that has an integrality gap of $5/2$.

Given an instance of DVRP-TC consisting of an undirected complete graph $G = (V, E)$ with depot r, $\lambda > 0$ and $\gamma \geq 0$, let $\mathcal{C}_\lambda$ be the set of all λ-bounded r-cycles. We can generalize the set-covering integer programming formulation in [11] for DVRP-N to obtain the following integer programming formulation (IP-TC) for DVRP-TC, where a binary variable x_C is associated with each r-cycle $C \in \mathcal{C}_\lambda$ and the first constraint is to ensure that each non-depot vertex is covered by at least one r-cycle in $\mathcal{C}_\lambda$. The first term of the objective function indicates the total length of the cycles while the second term represents the assignment cost of the cycles.

$$\min \sum_{C \in \mathcal{C}_\lambda} w(C) x_C + \gamma \sum_{C \in \mathcal{C}_\lambda} x_C$$

$$\text{s.t.} \qquad \sum_{C \in \mathcal{C}_\lambda : v \in V(C)} x_C \geq 1, \qquad \forall v \in V \setminus \{r\} \qquad (IP-TC)$$

$$x_C \in \{0, 1\}, \qquad \forall C \in \mathcal{C}_\lambda.$$

The corresponding LP relaxation (LP-TC) is obtained by neglecting the integral constraints on the variables.

$$\min \sum_{C \in \mathcal{C}_\lambda} w(C)x_C + \gamma \sum_{C \in \mathcal{C}_\lambda} x_C$$

$$s.t. \quad \sum_{C \in \mathcal{C}_\lambda : v \in V(C)} x_C \geq 1, \qquad \forall v \in V \setminus \{r\} \qquad (LP - TC)$$

$$x_C \geq 0, \qquad \forall C \in \mathcal{C}_\lambda.$$

Note that the constraints $x_C \leq 1$ for all $C \in \mathcal{C}_\lambda$ is unnecessary due to the first constraint and the minimization objective.

Next we focus on DVRP-TC on tree metric induced by a weighted tree $T = (V, E_T)$. As noted by Nagarajan and Ravi [11,12], we can assume without loss of generality that T is a binary tree rooted at r (otherwise one can add some dummy vertices and zero-weight edges).

Nagarajan and Ravi [11] showed that if $\gamma = +\infty$ (i.e., DVRP-N) the integrality gap of (LP-TC) is at most 20 for tree metric instances. Their proof depends on a crucial concept called *heavy cluster*, which is a set of vertices $F \subseteq V$ such that the induced subgraph of F is connected and all the vertices in F cannot be covered by a single λ-bounded r-cycle. They obtained the following results.

Lemma 2 [11,12]. *(i) There is a polynomial algorithm that finds k disjoint heavy clusters $F_1, \ldots, F_k \subseteq V$ and a λ-bounded r-cycle cover $\mathcal{C} = \{C_1, \ldots, C_p\}$ with $p \leq 2k + 1$; (ii) If there exist k disjoint heavy clusters $F_1, \ldots, F_k \subseteq V$ in the tree T, the minimum number of λ-bounded r-cycles required to cover $\bigcup_{i=1}^k F_i$ is at least $k + 1$.*

On the one hand, the second conclusion in the above lemma implies that $|\mathcal{C}| = p \leq 2OPT_N$. On the other hand, Liang et al. [9] proved that the total length $\sum_{i=1}^p w(C_i)$ of the cycles in $\mathcal{C}$ cannot exceed $2OPT_D$. Therefore, the total cost of $\mathcal{C}$ is at most $2OPT_D + 2\gamma OPT_N \leq 2OPT$.

Theorem 1 [9,11,12]. *There is a 2-approximation algorithm for DVRP-TC on tree metric.*

Note that for DVRP-TC on tree metric, an r-cycle is a multi-subgraph of T that consists of two copies of the edges of some r-tree of T. To effectively exploit the tree structure, Yu et al. [17] observed that covering all the vertices in T is equivalent to covering at least twice all the edges and proposed the following integer programming formulation for DVRP-N on tree metric, where n_e denotes the number of heavy clusters in $F_1, \ldots, F_k$ in Lemma 2(i) that are below e. Since $F_1, \ldots, F_k$ can be computed in polynomial time by Lemma 2(i), n_e can also be determined in polynomial time.

$$\min \sum_{C \in \mathcal{C}_\lambda} x_C$$

$$s.t. \quad \sum_{C \in \mathcal{C}_\lambda : e \in E(C)} x_C \geq 2(n_e + 1), \quad \forall e \in E_T \qquad (IP - T)$$

$$x_C \in \{0, 1\}, \quad \forall C \in \mathcal{C}_\lambda$$

(Note that for $e = (r, v)$ and an r-cycle $C = r - e - v - e - r$, $E(C)$ is a multiset consisting of two copies of e. And x_C contributes twice to the left side of the above inequality.)

After changing the objective function we can obtain the following integer programming and linear programming formulations for DVRP-TC.

$$
\min \sum_{C \in \mathcal{C}_\lambda} w(C) x_C + \gamma \sum_{C \in \mathcal{C}_\lambda} x_C
$$

$$
s.t. \quad \sum_{C \in \mathcal{C}_\lambda : e \in E(C)} x_C \geq 2(n_e + 1), \quad \forall e \in E_T \qquad (IP - T - TC)
$$

$$
x_C \in \{0, 1\}, \qquad \forall C \in \mathcal{C}_\lambda
$$

$$
\min \sum_{C \in \mathcal{C}_\lambda} w(C) x_C + \gamma \sum_{C \in \mathcal{C}_\lambda} x_C
$$

$$
s.t. \quad \sum_{C \in \mathcal{C}_\lambda : e \in E(C)} x_C \geq 2(n_e + 1), \quad \forall e \in E_T \qquad (LP - T - TC)
$$

$$
0 \leq x_C \leq 1, \qquad \forall C \in \mathcal{C}_\lambda
$$

The following theorem gives an upper bound on the integrality gap of (LP-T-TC).

Theorem 2. *The integrality gap of (LP-T-TC) is at most 5/2. In particular, the integrality gap of (LP-T-TC) is at most 2 if $\gamma = 0$.*

Proof. Suppose $(x_C^*)_{C \in \mathcal{C}_\lambda}$ is an optimal solution to (LP-T-TC) and

$$
OPT_{LP-T-TC} = \sum_{C \in \mathcal{C}_\lambda} w(C) x_C^* + \gamma \sum_{C \in \mathcal{C}_\lambda} x_C^*
$$

is the optimal value.

By the first constraint of (LP-T-TC) corresponding to any edge e we have

$$
\sum_{C \in \mathcal{C}_\lambda} x_C^* \geq \frac{1}{2} \sum_{C \in \mathcal{C}_\lambda : e \in E(C)} x_C^* \geq n_e + 1, \tag{1}
$$

where the first inequality follows from the fact that for each $C \in \mathcal{C}_\lambda$ the edge multiset $E(C)$ contains two copies of edges used by C and the second inequality is due to the constraints of (LP-T-TC).

Multiply by $w(e)$ in the first constraint of (LP-T-TC) and take the summation over all $e \in E_T$, we obtain

$$
\sum_{e \in E_T} 2(n_e + 1) w(e) \leq \sum_{e \in E_T} \left(\sum_{C \in \mathcal{C}_\lambda : e \in E(C)} x_C^* \right) w(e)
$$

$$
= \sum_{C \in \mathcal{C}_\lambda} w(C) x_C^*
$$

$$
\leq \lambda \sum_{C \in \mathcal{C}_\lambda} x_C^*, \tag{2}
$$

where the equality holds by exchanging the order of the two summations and the last inequality follows from the definition of $\mathcal{C}_\lambda$.

We distinguish two cases.

Case 1. $n_e = 0$ for any $e \in E_T$. Since T is a binary tree, the depot r has two possible children u_1, u_2. Set $e_i = (r, u_i)(i = 1, 2)$. Let T_{e_i} be the r-tree of T consisting of e_i and the subtree rooted at u_i. Since $n_{e_i} = 0$, we know that $w(T_{e_i}) \leq \frac{\lambda}{2}$ and T can be covered by at most two r-cycles C_1, C_2 in $\mathcal{C}_\lambda$, where $C_i(i = 1, 2)$ is obtained by doubling the edges in T_{e_i}. Therefore we have a feasible integral solution to (IP-T-TC) of objective value at most

$$w(C_1) + w(C_2) + 2\gamma \leq 2 \sum_{e \in E_T} w(e) + 2\gamma$$

$$\leq \sum_{C \in \mathcal{C}_\lambda} w(C)x_C^* + 2\gamma \sum_{C \in \mathcal{C}_\lambda} x_C^*$$

$$\leq 2OPT_{LP-T-TC},$$

where the second inequality follows from (1) and (2).

Case 2. There exists some edge $\tilde{e}$ with $n_{\tilde{e}} \geq 1$, which implies $k \geq 1$. By replacing e with $\tilde{e}$ in (1) we have

$$\sum_{C \in \mathcal{C}_\lambda} x_C^* \geq n_{\tilde{e}} + 1 \geq 2. \tag{3}$$

On the other hand, for $i = 1, \ldots, k$, by definition the induced subgraph of F_i, denoted by $T[F_i]$, is actually a subtree of T. Let $v_i \in F_i$ be the highest vertex in F_i. We obtain an r-cycle $\tilde{C}_i$ by doubling the edges in the r-tree consisting of $T[F_i]$ and the unique path from r to v_i. Let $E' \subseteq E_T$ be the set of edges used by $\tilde{C}_1, \ldots, \tilde{C}_k$. For each $e \in E'$, it is used at most $2(n_e + 1)$ times by the cycles $\tilde{C}_1, \ldots, \tilde{C}_k$, where $2n_e$ is due to the heavy clusters below it and if e happens to be in some $T[F_i]$ it appears two more times in $\tilde{C}_i$. Since F_i is a heavy cluster, we have $w(\tilde{C}_i) \geq \lambda$. Then

$$k\lambda \leq \sum_{i=1}^{k} w(\tilde{C}_i) \leq \sum_{e \in E'} 2(n_e + 1)w(e)$$

$$\leq \sum_{e \in E_T} 2(n_e + 1)w(e)$$

$$\leq \sum_{C \in \mathcal{C}_\lambda} w(C)x_C^*$$

$$\leq \lambda \sum_{C \in \mathcal{C}_\lambda} x_C^*,$$

where the last two inequalities follow from (2). This implies $k \leq \sum_{C \in \mathcal{C}_\lambda} x_C^*$. Since we already have $\sum_{C \in \mathcal{C}_\lambda} x_C^* \geq 2$ by (3), it follows that

$$\sum_{C \in \mathcal{C}_\lambda} x_C^* \geq \max\{2, k\}. \tag{4}$$

By the results in [9], the total length $\sum_{i=1}^{p} w(C_i)$ of the cycles in the λ-bounded r-cycle cover $\mathcal{C} = \{C_1, \ldots, C_p\}$ with $p \leq 2k+1$ in Lemma 2 cannot exceed $\sum_{e \in E} 4(n_e + 1)w(e)$, which implies $\sum_{i=1}^{p} w(C_i) \leq 2\sum_{C \in \mathcal{C}_\lambda} w(C)x_C^*$ by (2). So the total cost of $\mathcal{C}$ is at most

$$2\sum_{C \in \mathcal{C}_\lambda} w(C)x_C^* + (2k+1)\gamma \leq 2\sum_{C \in \mathcal{C}_\lambda} w(C)x_C^* + 2\sum_{C \in \mathcal{C}_\lambda} x_C^* + \gamma$$

$$= 2OPT_{LP-T-TC} + \gamma \tag{5}$$

$$\leq 2OPT_{LP-T-TC} + \frac{\gamma\sum_{C \in \mathcal{C}_\lambda} x_C^*}{\max\{2,k\}}$$

$$\leq \left(2 + \frac{1}{\max\{2,k\}}\right)OPT_{LP-T-TC}$$

$$\leq \frac{5}{2}OPT_{LP-T-TC},$$

where the first two inequalities follows from (4).

By (5), the total cost of $\mathcal{C}$ is at most $2OPT_{LP-T-TC}$ when $\gamma = 0$. This completes the proof.   □

4   General Metric

In this section we deal with unrooted DVRP-TC. We first develop a 5-approximation algorithm by modifying the algorithm in [3] and then show that the unrooted version of (LP-TC) has a constant integrality gap using a similar LP-rounding procedure to that in [11].

4.1   A 5-Approximation Algorithm

We give the following algorithm for the unrooted DVRP-TC. The input is an instance consisting of an undirected weighted complete graph $G = (V, E)$, a distance bound $\lambda > 0$ and an assignment cost γ for a single cycle.

Algorithm $UDVRP - TC$

For each $k = n, n-1, \ldots, 1$, do the following:

Step 1. Find a minimum weight spanning forest F_k of G consisting of exactly k connected components $T_{k,1}, \ldots, T_{k,k}$, which are actually trees. Set $\mathcal{C}_k = \emptyset$.

Step 2. For each $i = 1, \ldots, k$, double all the edges in $T_{k,i}$ to obtain an Eulerian graph and shortcut the repeated vertices of the Eulerian tour of this graph to obtain a cycle $C_{k,i}$. By Lemma 1, we can split $C_{k,i}$ into at most $\max\left\{\left\lceil \frac{w(C_{k,i})}{\frac{\lambda}{2}}\right\rceil, 1\right\}$ $\frac{\lambda}{2}$-bounded paths. After that we connect the two end vertices of each path to obtain the same number of λ-bounded cycles and put all these cycles into $\mathcal{C}_k$.

The algorithm returns one of the λ-bounded cycle cover among $C_1, \ldots, C_n$ with minimum total cost.

Note that F_k can be found by running the well-known Kruskal's Algorithm and stop when there are exactly k connected components (actually each component is a tree). Therefore, all F_k's can be obtained in $O(n^2 \log n)$ time if we compute them in the order $F_n, F_{n-1}, \ldots, F_1$. For each k, Step 2 takes $O(n)$ time by Lemma 1. So the total running time of Step 2 for $k = 1, \ldots, n$ is $O(n^2)$. To sum up, Algorithm $UDVRP - TC$ has a polynomial time complexity of $O(n^2 \log n)$.

By construction it can be seen that C_k is a λ-bounded cycle cover for each $k = 1, \ldots, n$. Assume that OPT consists of $\hat{k}$ optimum cycles $C_1^*, \ldots, C_{\hat{k}}^*$. It follows that $OPT = \sum_{i=1}^{\hat{k}} w(C_i^*) + \hat{k}\gamma$. Since the algorithm returns the cycle cover among $C_1, \ldots, C_n$ with minimum total cost, to show that Algorithm $UDVRP - TC$ is a 5-approximation algorithm it is sufficient to prove that the total cost of $C_{\hat{k}}$ is at most $5OPT$.

Consider the running of Algorithm $UDVRP - TC$ for $k = \hat{k}$. Since the collection of all the optimum cycles $C_1^*, \ldots, C_{\hat{k}}^*$ contains a spanning forest with exactly $\hat{k}$ connected components and all the optimum cycles are λ-bounded, we have

$$w(F_{\hat{k}}) = \sum_{i=1}^{\hat{k}} w(T_{\hat{k},i}) \leq \sum_{i=1}^{\hat{k}} w(C_i^*) \leq \hat{k}\lambda. \tag{6}$$

And the cycles $C_{\hat{k},1}, \ldots, C_{\hat{k},\hat{k}}$ are obtained by doubling the edges in $T_{\hat{k},1}, \ldots, T_{\hat{k},\hat{k}}$, it follows that

$$\sum_{i=1}^{\hat{k}} w(C_{\hat{k},i}) \leq 2 \sum_{i=1}^{\hat{k}} w(T_{\hat{k},i}) = 2w(F_{\hat{k}}), \tag{7}$$

which implies the total length of the $\frac{\lambda}{2}$-bounded paths is at most $2w(F_{\hat{k}})$ by Lemma 1. Consequently, the total length of the cycles in $C_{\hat{k}}$, which are obtained by connecting the two end vertices of these paths, cannot be greater than $4w(F_{\hat{k}})$. Moreover, the total number of cycles in $C_{\hat{k}}$ is identical to the total number of the $\frac{\lambda}{2}$-bounded paths, which is, by Lemma 1, at most

$$\sum_{i=1}^{\hat{k}} \max \left\{ \left\lceil \frac{w(C_{\hat{k},i})}{\frac{\lambda}{2}} \right\rceil, 1 \right\} \leq \sum_{i=1}^{\hat{k}} \left(\frac{w(C_{\hat{k},i})}{\frac{\lambda}{2}} + 1 \right) \leq \frac{4w(F_{\hat{k}})}{\lambda} + \hat{k} \leq 5\hat{k},$$

where the second inequality holds by (7) and the last inequality follows from (6).

Therefore, the total cost of $C_{\hat{k}}$ is no more than

$$4w(F_{\hat{k}}) + 5\hat{k}\gamma \leq 4 \sum_{i=1}^{\hat{k}} w(C_i^*) + 5\hat{k}\gamma \leq 5OPT.$$

Theorem 3. *Algorithm $UDVRP - TC$ is a 5-approximation algorithm for unrooted DVRP-TC.*

Remark 1. Algorithm $UDVRP - TC$ is actually a modification of the algorithm in [3] tailored for the **Distance Constraint Sweep-Coverage with the Minimum Sum of the Number of Mobile Sensors and the Number of Base Stations** (MinDCSC-SB), which is a generalization of unrooted DVRP-N. Compared to DVRP-N, there are two additional inputs, i.e., the velocity v of mobile sensors and the coverage period t, in MinDCSC-SB. The objective is to find a λ-bounded cycle cover $\{C_1, \ldots, C_k\}$ such that $\sum_{i=1}^{k} \left\lceil \frac{w(C_i)}{vt} \right\rceil + k$ is minimized, where $\left\lceil \frac{w(C_i)}{vt} \right\rceil$ represents the number of mobile sensors that are deployed to traveling along C_i periodically and k indicates the number of base stations (each cycle installs one base station). Chen et al. [3] developed a 7-approximation algorithm for MinDCSC-SB. Our result suggests that for a relatively simplified objective function, i.e., $\sum_{i=1}^{k} w(C_i) + \gamma k$, one can have an improved approximation ratio.

4.2   Integrality Gap

Given an instance of unrooted DVRP-TC consisting of $G = (V, E)$ and $\lambda > 0$, let $\mathcal{C}_\lambda$ be the set of all λ-bounded cycles. If we associate a variable x_C with each cycle $C \in \mathcal{C}_\lambda$, the unrooted versions of (IP-TC) and (LP-TC) are given by

$$\min \sum_{C \in \mathcal{C}_\lambda} w(C)x_C + \gamma \sum_{C \in \mathcal{C}_\lambda} x_C$$

$$\text{s.t.} \quad \sum_{C \in \mathcal{C}_\lambda : v \in V(C)} x_C \geq 1, \qquad \forall v \in V \qquad (IP - TC - U)$$

$$x_C \in \{0, 1\}, \qquad \forall C \in \mathcal{C}_\lambda$$

and

$$\min \sum_{C \in \mathcal{C}_\lambda} w(C)x_C + \gamma \sum_{C \in \mathcal{C}_\lambda} x_C$$

$$\text{s.t.} \quad \sum_{C \in \mathcal{C}_\lambda : v \in V(C)} x_C \geq 1, \qquad \forall v \in V \qquad (LP - TC - U)$$

$$x_C \geq 0, \qquad \forall C \in \mathcal{C}_\lambda.$$

We show the following result using a similar LP-rounding procedure in [11] for unrooted DVRP-N.

Theorem 4. *The integrality gap of (LP-TC-U) is at most 25.*

5   Path-Version Problems

In the path-version of (unrooted) DVRP-TC (DVRP-D, DVRP-N), the (r-)cycles are replaced by (r-)paths. Actually, the rounding procedure in [11] was proposed to deal with the path-version of unrooted DVRP-N. We note that our results on DVRP-TC on tree metric or unrooted DVRP-TC can be easily applied to their path-version problems to obtain approximation algorithms (or LP relaxations) with similar bounds on approximation ratios (or integrality gaps).

Acknowledgements. This research is supported by the National Natural Science Foundation of China under grants numbers 11671135, 11701363, the Natural Science Foundation of Shanghai under grant number 19ZR1411800 and the Fundamental Research Fund for the Central Universities under grant number 22220184028.

References

1. Arkin, E.M., Hassin, R., Levin, A.: Approximations for minimum and min-max vehicle routing problems. J. Algorithms **59**, 1–18 (2006)
2. Assad, A.A.: Modeling and implementation issues in vehicle routing. In: Golden, B.L., Assad, A.A. (eds.) Vehicle Routing: Methods and Studies, pp. 7–45. North Holland, Amsterdam (1988)
3. Chen, Q., Huang, X., Ran, Y.: Approximation algorithm for distance constraint sweep coverage without predetermined base stations. Discrete Math. Algorithms Appl. **10**(5), 1850064 (2018)
4. Frederickson, G.N., Hecht, M.S., Kim, C.E.: Approximation algorithms for some routing problems. SIAM J. Comput. **7**(2), 178–193 (1978)
5. Friggstad, Z., Swamy, C.: Approximation algorithms for regret-bounded vehicle routing and applications to distance-constrained vehicle routing. In: The Proceedings of the 46th Annual ACM Symposium on Theory of Computing, pp. 744–753 (2014)
6. Garey, M.R., Johnson, D.S.: Computers and Intractability: A Guide to the Theory of NP-completeness. Freeman, San Francisco (1979)
7. Khani, M.R., Salavatipour, M.R.: Approximation algorithms for min-max tree cover and bounded tree cover problems. Algorithmica **69**, 443–460 (2014)
8. Li, C.-L., Simchi-Levi, D., Desrochers, M.: On the distance constrained vehicle routing problem. Oper. Res. **40**, 790–799 (1992)
9. Liang, J., Huang, X., Zhang, Z.: Approximation algorithms for distance constraint sweep coverage with base stations. J. Comb. Optim. **37**, 1111–1125 (2019)
10. Nagarajan, V., Ravi, R.: Minimum vehicle routing with a common deadline. In: Díaz, J., Jansen, K., Rolim, J.D.P., Zwick, U. (eds.) APPROX/RANDOM -2006. LNCS, vol. 4110, pp. 212–223. Springer, Heidelberg (2006). https://doi.org/10.1007/11830924_21
11. Nagarajan, V., Ravi, R.: Approximation algorithms for distance constrained vehicle routing problems. Tepper School of Business, Carnegie Mellon University, Pittsburgh (2008)
12. Nagarajan, V., Ravi, R.: Approximation algorithms for distance constrained vehicle routing problems. Networks **59**(2), 209–214 (2012)
13. Xu, W., Liang, W., Lin, X.: Approximation algorithms for min-max cycle cover problems. IEEE Trans. Comput. **64**(3), 600–613 (2015)
14. Xu, Z., Xu, L., Zhu, W.: Approximation results for a min-max location-routing problem. Discrete Appl. Math. **160**, 306–320 (2012)
15. Yu, W., Liu, Z.: Improved approximation algorithms for some min-max cycle cover problems. Theoret. Comput. Sci. **654**, 45–58 (2016)

16. Yu, W., Liu, Z., Bao, X.: New approximation algorithms for the minimum cycle cover problem. In: Chen, J., Lu, P. (eds.) FAW 2018. LNCS, vol. 10823, pp. 81–95. Springer, Cham (2018). https://doi.org/10.1007/978-3-319-78455-7_7
17. Yu, W., Liu, Z., Bao, X.: New LP relaxations for minimum cycle/path/tree cover problems. In: Tang, S., Du, D.-Z., Woodruff, D., Butenko, S. (eds.) AAIM 2018. LNCS, vol. 11343, pp. 221–232. Springer, Cham (2018). https://doi.org/10.1007/978-3-030-04618-7_18

Greedy Algorithm for Maximization of Non-submodular Functions Subject to Knapsack Constraint

Zhenning Zhang[1], Bin Liu[2], Yishui Wang[3(✉)], Dachuan Xu[4], and Dongmei Zhang[5]

[1] College of Applied Sciences, Beijing University of Technology, Beijing 100124, People's Republic of China
[2] School of Mathematical Science, Ocean University of China, Qingdao 266100, People's Republic of China
[3] Shenzhen Institutes of Advanced Technology, Chinese Academy of Sciences, Shenzhen 518055, People's Republic of China
ys.wang1@siat.ac.cn
[4] Department of Operations Research and Scientific Computing, Beijing University of Technology, Beijing 100124, People's Republic of China
[5] School of Computer Science and Technology, Shandong Jianzhu University, Jinan 250101, People's Republic of China

Abstract. Although submodular maximization generalizes many fundamental problems in discrete optimization, lots of real-world problems are non-submodular. In this paper, we consider the maximization problem of non-submodular function with a knapsack constraint, and explore the performance of the greedy algorithm. Our guarantee is characterized by the submodularity ratio β and curvature α. In particular, we prove that the greedy algorithm enjoys a tight approximation guarantee of $\frac{1}{\alpha}\left(1 - e^{-\alpha\beta}\right)$ for the above problem. To our knowledge, it is the first tight constant factor for this problem. In addition, we experimentally validate our algorithm by an important application, the Bayesian A-optimality.

Keywords: Non-submodular · Knapsack constraint · Submodularity ratio · Curvature · Diminishing-return ratio

1 Introduction

In recent decades, maximizing submodular function has been well studied [1,13, 14], and has been widely applied into many fields, such as scheduling problem on a single machine [15], maximum entropy sampling problem and column-subset selection problem [20]. Let S be a subset of a finite set $I = \{1, 2, \ldots, n\}$ which usually is considered as a ground set. The set function $f : 2^I \to \mathbb{R}^+$ is (i) submodular if $f(S) + f(T) \geq f(S \cup T) + f(S \cap T)$ for any $S, T \subseteq I$ and (ii) nondecreasing if $f(S) \leq f(T)$ for $S \subseteq T \subseteq I$. The submodular optimization problem

© Springer Nature Switzerland AG 2019
D.-Z. Du et al. (Eds.): COCOON 2019, LNCS 11653, pp. 651–662, 2019.
https://doi.org/10.1007/978-3-030-26176-4_54

is expressed as $\max_{S \subseteq I} f(S)$. For the submodular optimization problem with a K-cardinality constraint, that is $\max_{S \subseteq I, |S| \le K} f(S)$, Nemhauser et al. [12] provide a $(1 - 1/e)$-approximation using the greedy algorithm. Sviridenko [19] gives a $(1 - 1/e)$-approximation algorithm for maximizing submodular set functions with a knapsack constraint, that is $\max_{S \subseteq I} \left\{ f(S) : \sum_{i \in S} c_i \le B \right\}$, where B and c_i ($i \in \{1, 2, \ldots, n\}$) are nonnegative and can be considered as the total budget and each element's budget, respectively. Calinescu et al. [5] obtain a $(1 - 1/e)$-approximation for an arbitrary matroid constraint. The factor of $1 - 1/e$ can be improved by characterizing the curvature of the objective function [7–9,20], which is introduced to measure how close a submodular function to being modular [7].

Since the set function $f : 2^I \to \mathbb{R}^+$ can also be considered as a function defined on a Boolean hypercube $\{0, 1\}^I$, Soma and Yoshida [17] generalize submodularity to functions defined on the integer lattice $\mathbb{Z}_+^I$. They design polynomial-time $(1 - 1/e - \varepsilon)$-approximation algorithms for the submodular optimization problem on the integer lattice with the following three different constraints: cardinality constraint, polymatroid constraint and knapsack constraint, respectively.

However, many objective functions arising from applications are not submodular, such as experimental design, spare Gaussian processes and budget allocation problem [4,11].

For non-submodular function, Bian et al. [4] introduce the submodularity ratio to measure the distance from submodularity of a set function. By characterizing the objective function with submodularity ratio β and curvature α, Bian et al. [4] give a tight ratio of $\frac{1}{\alpha} \left(1 - e^{-\alpha\beta}\right)$ for the greedy algorithm for maximizing a non-submodular function subject to a cardinality constraint. Kuhnle et al. [11] generalize the curvature and the submodularity ratio to the function defined on integer lattice. Two fast modified greedy algorithms (ThresholdGreedy and FastGreedy) are designed for the non-submodular function with a cardinality constraint on the integer lattice.

Compared with the maximization with a cardinality constraint, optimization with a knapsack constraint is more practical [2,13]. It is reasonable that each element has a different budget, and the total budget of the selected subset shouldn't be exceeded. Here, we investigate maximization of objective function with a knapsack constraint, where the set function is non-decreasing and non-submodular. Let $I = \{1, 2, \ldots, n\}$ be the ground set, B, c_i, $i = 1, 2, \ldots, n$ be non-negative integers and the objective function f be non-negative, which means $f(S) \ge 0$, for any $S \subseteq I$. Then, our problem can be formulated as

$$\max_{S \subseteq I, \sum_{i \in S} c_i \le B} f(S). \tag{1}$$

We adopt the greedy method to deal with the above problem. To get the approximation ratio of the algorithm, we introduce the diminishing-return (DR) ratio [11], which can be considered as a generalization of DR-submodular [3,16–18] for non-submodular functions. The DR ratio can be considered as a measure to show how close a set function satisfying the DR property.

Our main contributions in this paper are listed as follows.

- To our knowledge, we give the first tight constant factor approximation guarantees for maximizing non-submodular nondecreasing functions with a knapsack constraint, by characterizing the submodularity ratio and curvature of the objective function.
- We experimentally validate our algorithm on Bayesian A-optimality in experimental design.

This paper is organized as follows. In Sect. 2, we introduce the mathematical definitions of the DR ratio, submodularity ratio and curvature. In Sect. 3, we design a greedy algorithm for any instance. In Sect. 4, we derive a constant factor approximation guarantee for the greedy algorithm. The proofs of Lemmas 2, 3 and Theorem 1 will be given in the journal version. In Sect. 5, we introduce the application of our algorithm on the Bayesian A-optimality. In Sect. 6, we give the numerical experiment for the Bayesian A-optimality in experimental design.

2   Submodularity Ratio and Curvature

In this section, we will express some notations and introduce the mathematical definitions of the DR ratio, submodularity ratio and the generalized curvature of the objective function.

Let $f(\cdot)$ be a set function. For any subset A, $D \subseteq I$, let $\rho_D(A)$ express $\rho_D(A) = f(A \cup D) - f(A)$. This function gives the gain of inserting set D into the set A. Specially, for $i \in I$, $\rho_i(A) = f(A \cup \{i\}) - f(A)$.

Definition 1 (Diminishing-return ratio) [11]. *The diminishing-return (DR) ratio of a non-negative set function f is the largest scalar γ, such that*

$$\gamma \rho_i(T) \leq \rho_i(S), \tag{2}$$

for any subset S, $T \subseteq I$, $S \subseteq T$ and $i \in I - T$.

Definition 2 (Submodularity ratio) [4]. *The submodularity ratio of a non-negative set function $f(\cdot)$ is the largest scalar β, such that*

$$\beta \rho_T(S) \leq \sum_{i \in T \backslash S} \rho_i(S), \tag{3}$$

for any subset $S, T \subseteq I$.

Remark 1. For any non-decreasing set function $f(\cdot)$, it holds that $\beta, \gamma \in [0,1]$, $\gamma \leq \beta$ and $f(\cdot)$ is submodular if and only if $\gamma = 1$ or $\beta = 1$ [4,11].

Definition 3 (Curvature) [4]. *The curvature of a non-negative function $f(\cdot)$ is the smallest scalar α, such that*

$$\rho_i(S \backslash \{i\} \cup T) \geq (1 - \alpha)\rho_i(S \backslash \{i\}), \tag{4}$$

for any T, $S \subseteq I$, $i \in S \backslash T$.

Remark 2. The curvature $\alpha \in [0,1]$, and any non-decreasing non-negative set function $f(\cdot)$ is supermodular if and only if $\alpha = 0$ [4].

3 Greedy Algorithm for Maximizing Non-submodular Function with a Knapsack Constraint

The greedy algorithm (GAMNSK) for the problem (1) is designed as follows.

Phase I. Enumerates all feasible solutions sets with cardinality from one to P, where $P = \lceil \frac{\alpha}{\gamma(\alpha-1+e^{-\alpha\beta})} \rceil$.

Phase II. For all feasible sets Y with cardinality P satisfying $\sum_{i \in Y} c_i \leq B$, carry out the following procedure:

1. Let $S^0 = Y$, $I^0 = I$, $t = 1$;
2. At step t, S^{t-1} is given, find

$$\theta_t = \max_{i \in I^{t-1} \backslash S^{t-1}} \frac{f\left(S^{t-1} \cup \{i\}\right) - f\left(S^{t-1}\right)}{c_i},$$

$$i_t = \arg \max_{i \in I^{t-1} \backslash S^{t-1}} \frac{f\left(S^{t-1} \cup \{i\}\right) - f\left(S^{t-1}\right)}{c_i}.$$

Let c_{i_t} denote the weight associated with i_t.
   - If $\sum_{\tau=1}^{t} c_{i_\tau} \leq B - \sum_{i \in Y} c_i$, let $S^t = S^{t-1} \cup \{i_t\}$, $I^t = I^{t-1}$.
   - If $\sum_{\tau=1}^{t-1} c_{i_\tau} + c_{i_t} > B - \sum_{i \in Y} c_i$, let $S^t = S^{t-1}$, $I^t = I^{t-1} \backslash \{i_t\}$.
3. Stop when $I^t \backslash S^t = \emptyset$.

In Phase I, let S_1 be a feasible set with the cardinality $1, 2 \ldots$, or P, which can make the objective function achieve the maximum value. Let S_2 be the solution obtained in Phase II, which achieve the maximum value of objective function.

Phase III. If $f(S_1) \geq f(S_2)$, output S_1; otherwise, output S_2.

4 Approximation Guarantee

By characterizing the DR ratio, submodularity ratio and curvature of the objective function, we consider the approximation ratio of GAMNSK for problem (1) in this section.

Now, we give some notations. Let $S^0 = Y = \{m_1, m_2, \ldots, m_P\}$. Define $g(S) = f(S) - f(Y)$, where $g(S)$ is nondecreasing and characterized by the submodularity ratio β.

S^t is the tth step solution obtained by GAMNSK. Let S^* denote the optimal solution of problem (1), and $t^* + 1$ be the first step of GAMNSK for which the algorithm does not add element $i_{t^*+1} \in S^*$ to the set S^{t^*}, that is, $S^{t^*+1} = S^{t^*}$ and $I^{t^*+1} = I^{t^*} \backslash \{i_{t^*+1}\}$. Therefore, the intersection between the optimal solution S^* and GAMNSK solution S^{t^*+1} can be expressed as $S^* \cap S^{t^*+1} = \{m_{\vartheta_1}, \ldots, m_{\vartheta_v}, i_{l_1}, \ldots, i_{l_s}\}$, where $l_1, \ldots, l_s$ are consistent with the order of greedy selection and $\{m_{\vartheta_1}, \ldots, m_{\vartheta_v}\} \subseteq \{m_1, m_2 \ldots, m_P\}$.

To get the approximation ratio of GAMNSK, we intend to get the proportion of $g\left(S^{t^*}\right)/g\left(S^*\right)$. In fact, instead of getting $g\left(S^{t^*}\right)/g\left(S^*\right)$ directly, we try to obtain $g\left(S^{t^*} \cup \{i_{t^*+1}\}\right)/g\left(S^*\right)$ firstly. Meanwhile, utilizing the following Lemma 1 and some transformations, we transfer the ratio of $g\left(S^{t^*} \cup \{i_{t^*+1}\}\right)/g\left(S^*\right)$ into a Linear Program (LP) with some constrains.

Lemma 1. *For the iterative step* $t \in \{0, 1, \dots, t^*\}$*, The equation*

$$g(S^*) \leq \alpha \sum_{t:i_t \in S^t \setminus S^*} c_{i_t} \theta_t + \sum_{t:i_t \in (S^t \setminus Y) \cap S^*} c_{i_t} \theta_t + \frac{1}{\beta} \left(B' - \sum_{i \in (S^t \setminus Y) \cap S^*} c_i \right) \theta_{t+1}$$

(5)

holds, where $B' = B - \sum_{i \in Y} c_i$*,* $Y = \{m_1, m_2 \dots, m_P\}$ *and* $(S^t \setminus Y) \cap S^* = \{i_{l_1}, \dots, i_{l_s}\}$*.*

The corresponding weights of the elements $\{i_{l_1}, \dots, i_{l_s}\}$ are $c_{i_{l_1}}, \dots, c_{i_{l_s}}$. Let $B_t = \sum_{j=1}^{t} c_{i_j}$, $B_0 = 0$ and $\xi_r = \theta_t$, $r = B_{t-1} + 1, \dots, B_t$, $t = 1, 2, \dots, t^*$. Equation (5) can be expressed as

$$g(S^*) \leq \alpha \sum_{t:i_t \in S^t \setminus S^*} \sum_{r=B_{t-1}+1}^{B_t} \xi_r + \sum_{t:i_t(S^t \setminus Y) \cap S^*} \sum_{r=B_{t-1}+1}^{B_t} \xi_r$$

$$+ \frac{1}{\beta} \left(B' - \sum_{i \in (S^t \setminus Y) \cap S^*} c_i \right) \xi_{B_t+1}.$$

These equations imply

$$g(S^*) \leq \min_{t=1,2,\dots,t^*} \left\{ \alpha \sum_{t:i_t \in S^t \setminus S^*} \sum_{r=B_{t-1}+1}^{B_t} \xi_r + \sum_{t:i_t(S^t \setminus Y) \cap S^*} \sum_{r=B_{t-1}+1}^{B_t} \xi_r \right.$$

$$\left. + \frac{1}{\beta} \left(B' - \sum_{i \in (S^t \setminus Y) \cap S^*} c_i \right) \xi_{B_t+1} \right\}$$

$$= \min_{s=1,\dots,B_{t^*}+1} \left\{ \sum_{r=1}^{s-1} \zeta_r \xi_r + \frac{(B' - B'')}{\beta} \xi_s \right\},$$

where the coefficient ζ_r is given by

$$\zeta_r = \begin{cases} 1, & r = B_{l_m} - c_{i_{l_m}} + 1, \dots, B_{l_m}, \ m = 1, 2, \dots, s, \\ \alpha, & \text{otherwise}, \end{cases}$$

and B'' is the total amount of the coefficient ζ_r which is equal to 1.

Therefore, Eq. (5) can be written as

$$g(S^*) \leq \sum_{r=1}^{s-1} \zeta_r \xi_r + \frac{(B' - B'')}{\beta} \xi_s,$$

(6)

for $s = 1, \dots, B_{t^*}+1$.

Since

$$g(S^{t^*} \cup \{i_{t^*+1}\}) = \sum_{j=1}^{t^*+1} \left(g\left(S^{j-1} \cup \{i_j\}\right) - g\left(S^{j-1}\right) \right)$$

$$= \sum_{j=1}^{t^*+1} \left(f\left(S^{j-1} \cup \{i_j\}\right) - f\left(S^{j-1}\right) \right)$$

$$= \sum_{j=1}^{t^*+1} c_{i_j} \theta_j$$

and $\xi_r = \theta_j$, $r = B_{j-1} + 1, \ldots, B_j$, $j = 1, 2, \ldots, t^* + 1$, we can get

$$\frac{g\left(S^{t^*} \cup \{i_{t^*+1}\}\right)}{g\left(S^*\right)} = \frac{\sum_{j=1}^{t^*+1} c_{i_j} \theta_j}{g\left(S^*\right)} = \frac{\sum_{r=1}^{B_{t^*+1}} \xi_r}{g\left(S^*\right)}.$$

Denote $x_r := \frac{\xi_r}{g(S^*)}$, $r = 1, \ldots, B_{t^*+1}$. Equation (6) gives B_{t^*+1} constraints over the variables x_r. Therefore, the worst case approximation of $\dfrac{g\left(S^{t^*} \cup \{i_{t^*+1}\}\right)}{g(S^*)}$ can be specified by the following LP,

$$\min \sum_{r=1}^{B_{t^*+1}} x_r \tag{7}$$

$$\text{s. t.} \sum_{r=1}^{h-1} \zeta_r x_r + \frac{(B' - B'')}{\beta} x_h \geq 1,$$

$$x_r \geq 0,$$

$$r, h = 1, 2 \ldots, B_{t^*+1}. \tag{8}$$

In order to get the optimal values of the LP (7), we analyse the key structure of the LP. To illustrate relationships between the optimal solutions of the LP and specific ranks in constraint matrix, we express the objective function as

$$\varphi(\{B_{l_1} - c_{i_{l_1}} + 1, \ldots, B_{l_1}, \ldots, B_{l_m} - c_{i_{l_m}} + 1, \ldots, B_{l_m}, \ldots, B_{l_s} - c_{i_{l_s}} + 1, \ldots, B_{l_s}\})$$

$$= \min \sum_{r=1}^{B_{t^*+1}} x_r.$$

In addition, the corresponding LP can be denote by $LP(\{B_{l_1} - c_{i_{l_1}} + 1, \ldots, B_{l_1}, \ldots, B_{l_m} - c_{i_{l_m}} + 1, \ldots, B_{l_m}, \ldots, B_{l_s} - c_{i_{l_s}} + 1, \ldots, B_{l_s}\})$, which denote that coefficients of following ranks $\{B_{l_1} - c_{i_{l_1}} + 1, \ldots, B_{l_1}, \ldots, B_{l_m} - c_{i_{l_m}} + 1, \ldots, B_{l_m}, \ldots, B_{l_s} - c_{i_{l_s}} + 1, \ldots, B_{l_s}\}$ of the constraint matrix are 1, and others are α.

We achieve a significant property for the optimal solution.

Lemma 2. *Assume that the optimal solution of the constructed LP is $x^* \in \mathbb{R}^{B_{t^*+1}}$, it holds that $x_q^* \leq x_{q+1}^*$, where $q = B_{l_m} - c_{i_{l_m}} + u_m$, $1 \leq u_m \leq c_{i_{l_m}}$, $m = 1, 2, \ldots, s$.*

From this characteristic, we can get another important Lemma as follows.

Lemma 3. *For all* $\{i_{l_1}, i_{l_2}, \ldots, i_{l_s}\} \subseteq S^{t^*+1}$ *with the corresponding columns* $\{B_{l_1} - c_{i_{l_1}} + 1, \ldots, B_{l_1}, \ldots, B_{l_m} - c_{i_{l_m}} + 1, \ldots, B_{l_m}, \ldots, B_{l_s} - c_{i_{l_s}} + 1, \ldots, B_{l_s}\}$ *in constraint matrix, it holds that*

(a). $\varphi(\{B_{l_1} - c_{i_{l_1}} + 1, \ldots, B_{l_1}, \ldots, B_{l_m} - c_{i_{l_m}} + 1, \ldots, B_{l_m}, \ldots, B_{l_s} - c_{i_{l_s}} + 1, \ldots, B_{l_s}\}) \geq \varphi(\emptyset)$,

(b).

$$\varphi(\emptyset) = \frac{1}{\alpha}\left[1 - \left(1 - \frac{\alpha\beta}{B'}\right)^{B_{t^*+1}}\right]. \tag{9}$$

From the first conclusion in Lemma 3, we know the optimal value of the LP is obtained by coefficients of the constraint matrix are all α. In addition, the second conclusion in Lemma 3 gives the optimal value of the LP.

By using the conclusions in Lemma 3, we receive the ratio of GAMNSK.

Theorem 1. *Let* $f(\cdot)$ *be a non-negative nondecreasing set function with DR ratio* γ, *submodularity ratio* β *and curvature* α. *The performance guarantee of GAMNSK for solving problem (1) is equal to* $\frac{1}{\alpha}\left(1 - e^{-\alpha\beta}\right)$.

Since the factor $\frac{1}{\alpha}\left(1 - e^{-\alpha\beta}\right)$ is tight for the non-submodular function with a cardinality constraint, which can be considered as a special case for that with a knapsack constraint, $\frac{1}{\alpha}\left(1 - e^{-\alpha\beta}\right)$ is also tight for the maximization of non-submodular function with a knapsack constraint.

5 Applications: Bayesian A-Optimality in Experimental Design

In classical Bayesian experimental design [6], the purpose is to select a set of experiments such that some statistical criterion is optimized. In [10], the statistical criterion is defined as maximization of the difference of variance between prior distribution and posterior distribution over the parameters.

Let the ground set be constructed by n experimental stimuli $\{\mathbf{x}_1, \ldots, \mathbf{x}_n\}$, where $\mathbf{x}_i \in \mathbb{R}^d$, $i = 1, 2, \ldots, n$, that is, $\mathcal{V} = \{\mathbf{x}_1, \ldots, \mathbf{x}_n\}$. The corresponding index set $I = \{1, 2, \ldots, n\}$ can also be considered as ground set. For any subset $S \subseteq I$, the corresponding stimuli constitutes a matrix $\mathbf{X}_S := [\mathbf{x}_{v_1}, \ldots, \mathbf{x}_{v_s}] \in \mathbb{R}^{d \times |S|}$. The linear combination of $\mathbf{X}_S$ can be expressed as $\mathbf{Y}_S = \mathbf{X}_S^T \theta + \mathbf{w}$, where $\theta \in \mathbb{R}^d$ is the parameter vector, $\mathbf{Y}_S$ is the vector of dependent variables, and $\mathbf{w}$ is the Gaussian noise with zero mean and variance σ^2, that is, $\mathbf{w} \sim \mathcal{N}(0, \sigma^2 \mathbf{I})$. Suppose the prior distribution of the parameter vector θ is $\theta \sim \mathcal{N}(0, \Lambda^{-1})$, $\Lambda = \eta^2 \mathbf{I}$. By calculation, the variance of the posterior distribution of θ is $\Sigma_{\theta|\mathbf{Y}_S} = (\Lambda + \sigma^{-2}\mathbf{X}_S\mathbf{X}_S^T)^{-1}$. The target is selecting subset $S \subseteq I$, such that the following function is maximized,

$$F_A(S) := \text{tr}(\Lambda^{-1}) - \text{tr}((\Lambda + \sigma^{-2}\mathbf{X}_S\mathbf{X}_S^T)^{-1}).$$

However, it costs money to do any experiment, and different experiments cost differently. Therefore, it is reasonable to assume that the objective function has a knapsack constraint. Let $c_i \in \mathbb{R}_+$ be the budget of the experimental stimuli $\mathbf{x}_i$, and the budget of experimental design should satisfy $\sum_{i \in S} c_i \leq B$ for subset $S \subseteq I$, where $B \in \mathbb{R}_+$. Then, the target can be rewritten as

$$\max_{S \subseteq I, \sum_{i \in S} c_i \leq B} F_A(S). \tag{10}$$

The submodularity ratio and curvature are characters for the target function $F_A(S)$, not decided by the constraint. Therefore, the submodularity ratio and curvature of $F_A(S)$ are the same as in Proposition 1 in [4].

Proposition 1. *Let the experimental stimuli be normalized, that is, $\|\mathbf{x}_i\| = 1$, $i = 1, 2, \ldots, n$. Let the spectral norm of $\mathbf{X}$, which is the data matrix constituted by all experimental stimulus, be $\|\mathbf{X}\|$. Thus, we get that the objective function (10) is monotone nondecreasing, the lower bound of the submodularity ratio β is $\frac{\eta^2}{\|\mathbf{X}\|(\eta^2 + \sigma^{-2}\|\mathbf{X}\|)}$ and the upper bound of the curvature α is $1 - \frac{\eta^2}{\|\mathbf{X}\|(\eta^2 + \sigma^{-2}\|\mathbf{X}\|)}$.*

6 Numerical Results: Bayesian A-Optimality in Experimental Design

We implement the greedy algorithm GAMNSK provided in Sect. 3 on some instances of Bayesian A-optimality in experimental design to show the performance. In the algorithm GAMNSK, the number P is calculated by the diminishing-return ratio, submodularity ratio and curvature. Unfortunately, these three measures are not computationally tractable. So we see P as a fixed parameter in our experiment. We set $P = 2, 3$ and 4 respectively for all instances, and compare their performances.

We use the Boston house-price data (http://lib.stat.cmu.edu/datasets/boston) with 506 samples and 14 features. The parameters σ and η are set to 1, $c_i(\forall i = 1, 2, \cdots, n)$ is drawn from an uniform distribution $U[0, 1]$, and $B = C/1, C/2, \cdots, C/10$ where $C := \sum_{i=1}^{n} c_i$. We will compare the objective values returned by GAMNSK, SDP algorithm and exact algorithm (exhausive search). To save time, we only use $n = 10$ and 20 samples (drawn randomly from 506 samples of the data) to do the experiments.

To comparing our GAMNSK algorithm with semidefinite programming (SDP) algorithm, we first give a SDP relaxation of maximizing Bayesian A-optimality function with a knapsack constraint, and then use the SDP relaxation to produce a solution of the problem. Maximizing a Bayesian A-optimality function with a knapsack constraint is equivalent to

$$\min_{S \subseteq I, \sum_{i \in S} c_i \leq B} \mathrm{tr}((\Lambda + \sigma^{-2}\mathbf{X}_S\mathbf{X}_S^T)^{-1}). \tag{11}$$

By introducing 0-1 variables $\xi_i, i = 1, 2, ..., n$, we translate the formulation (11) as

$$\min \ \operatorname{tr}((\Lambda + \sigma^{-2} \sum_{j=1}^{n} \xi_i \mathbf{x}_i \mathbf{x}_i^T)^{-1})$$

$$\text{s. t.} \ \sum_{i=1}^{n} \xi_i c_i \leq B,$$

$$\xi_i \in \{0, 1\}, i = 1, 2, ..., n.$$

Let $\lambda_i = \xi_i/B, i = 1, 2, ..., n$, and relax these variables, we get

$$\min \ \operatorname{tr}((\Lambda + B\sigma^{-2} \sum_{i=1}^{n} \lambda_i \mathbf{x}_i \mathbf{x}_i^T)^{-1})$$

$$\text{s. t.} \ \sum_{i=1}^{n} \lambda_i c_i = 1,$$

$$\lambda_i \in \mathbb{R}_+^n, i = 1, 2, ..., n.$$

According to the Schur complement lemma, we get the SDP relaxation as follows.

$$\min \ \sum_{i=1}^{d} u_j$$

$$\text{s. t.} \ \begin{bmatrix} \Lambda + B\sigma^{-2} \sum_{i=1}^{n} \lambda_i \mathbf{x}_i \mathbf{x}_i^T & e_j \\ e_j & u_j \end{bmatrix} \succeq 0, \ j = 1, 2, ..., d, \tag{12}$$

$$\sum_{i=1}^{n} \lambda_i c_i = 1,$$

$$\lambda_i \in \mathbb{R}_+^n, i = 1, 2, ..., n.$$

The SDP algorithm first solves the SDP relaxation (12) to obtain the relaxed solution $\tilde{\lambda}_i, i = 1, 2, ..., n$, and then sorts them in the decreasing order $\{i_1, i_2, ..., i_n\}$. The final solution is the set of first k items in the order $\{i_1, i_2, ..., i_n\}$ subject to $\sum_{l=1}^{k} c_{i_l} \leq B$ and $\sum_{l=1}^{k+1} c_{i_l} > B$.

We repeat 10 experiments for each size and show the average ratios of objective value and optimal value, and the average running times in Fig. 1. For these instances, the objective values returned by GAMNSK with $P = 2, 3$ and 4 are very close to the optimal values, little better than the SDP algorithm. For the instances with $n = 10$, GAMNSK with $P = 2, 3$ and 4 are faster than the SDP algorithm. For the instances with $n = 20$, GAMNSK with larger P has faster growing of the running time as a function of the increasing in B, while the SDP algorithm has stable running times regardless of how big the number B is.

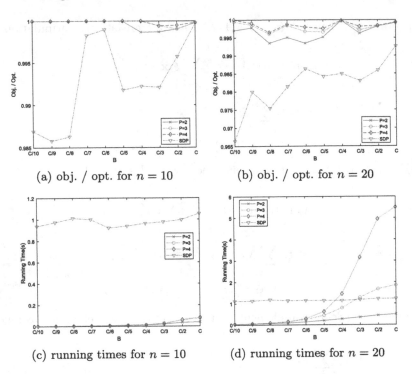

Fig. 1. Results for Bayesian A-optimality in experimental design on Boston house-price data

7 Conclusion

By characterizing the submodularity ratio and curvature of the objective function, we obtain a constant factor for the greedy algorithm for maximization of non-submodular nondecreasing set function with a knapsack constraint. Although in Phase I of GAMNSK, it seems to have to enumerate all feasible solutions with cardinality P, which depends on the submodularity ratio and the curvature. In fact, for practical problems we need not to enumerate all feasible solutions with $P = \lceil \frac{\alpha}{\gamma(\alpha - 1 + e^{-\alpha\beta})} \rceil$. Only enumerating small amount of P, we can get a satisfied solution. In addition, $P = \lceil \frac{\alpha}{\gamma(\alpha - 1 + e^{-\alpha\beta})} \rceil$ is just used for analyse the ratio of GAMNSK.

Acknowledgment. The first author is supported by Science and Technology Program of Beijing Education Commission (No. KM201810005006). The second author is supported by Natural Science Foundation of Shandong Province of China (No. ZR2017QA010). The third author is supported by National Natural Science Foundation of China (No. 61433012), Shenzhen research grant (KQJSCX20180330170311901, JCYJ20180305180840138 and GGFW2017073114031767). The fourth author is supported by National Natural Science Foundation of China (No. 11531014). The fifth author is supported by National Natural Science Foundation of China (No. 11871081).

References

1. Balkanski, E., Singer, Y.: Approximation guarantees for adaptive sampling. In: Proceedings of the Thirty-Fifth International Conference on Machine Learning, pp. 384–393 (2018)
2. Bettinelli, A., Cacchiani, V., Malaguti, E.: A branch-and-bound algorithm for the knapsack problem with conflict graph. INFORMS J. Comput. **29**(3), 457–473 (2017)
3. Bian, A., Levy, K.Y., Krause, A., Buhmann, J.M.: Non-monotone continuous DR-submodular maximization: structure and algorithms. In: Proceedings of the Thirty-First International Conference on Neural Information Processing Systems, pp. 486–496 (2017)
4. Bian, A., Buhmann, J.M., Krause, A., Tschiatschek, S.: Guarantees for greedy maximization of non-submodular functions with applications. In: Proceedings of the Thirty-Fourth International Conference on Machine Learning, pp. 498–507 (2017)
5. Calinescu, G., Chekuri, C., Pál, M., Vondrák, J.: Maximizing a submodular set function subject to a matroid constraint. SIAM J. Comput. **40**(6), 1740–1766 (2011)
6. Chaloner, K., Verdinelli, I.: Bayesian experimental design: a review. Stat. Sci. **10**(3), 273–304 (1995)
7. Conforti, M., Cornuéjols, G.: Submodular set functions, matroids and the greedy algorithm: tight worst-case bounds and some generalizations of the radoedmonds theorem. Discrete Appl. Math. **7**(3), 251–274 (1984)
8. Iyer, R., Bilmes, J.: Submodular optimization with submodular cover and submodular knapsack constraints. In: Proceedings of the Twenty-Sixth International Conference on Neural Information Processing Systems, pp. 2436–2444 (2013)
9. Iyer, R., Jegelka, S., Bilmes, J.: Curvature and optimal algorithms for learning and minimizing submodular functions. In: Proceedings of the Thirtieth International Conference on Machine Learning, pp. 2742–2750 (2013)
10. Krause, A., Singh, A., Guestrin, C.: Nearoptimal sensor placements in Gaussian processes: theory, efficient algorithms and empirical studies. J. Mach. Learn. Res. **9**(Feb), 235–284 (2008)
11. Kuhnle, A., Smith, J.D., Crawford, V.G., Thai, M.T.: Fast maximization of non-submodular, monotonic functions on the integer lattice. In: Proceedings of the Thirty-Fifth International Conference on Machine Learning, pp. 2791–2800 (2018)
12. Nemhauser, G.L., Wolsey, L.A., Fisher, M.L.: An analysis of approximations for maximizing submodular set functions-I. Math. Program. **14**(1), 265–294 (1978)
13. Sakaue, S., Ishihata, M.: Accelerated best-first search with upper-bound computation for submodular function maximization. In: Proceedings of the Thirthy-Second AAAI Conference on Artificial Intelligence, pp. 1413–1421 (2018)
14. Sakaue, S., Nishino, M., Yasuda, N.: Submodular function maximization over graphs via zero-suppressed binary decision diagrams. In: Proceedings of the Thirthy-Second AAAI Conference on Artificial Intelligence, pp. 1422–1430 (2018)
15. Shioura, A., Shakhlevich, N.V., Strusevich, V.A.: Application of submodular optimization to single machine scheduling with controllable processing times subject to release dates and deadlines. INFORMS J. Comput. **28**(1), 148–161 (2016)
16. Soma, T., Yoshida, Y.: A generalization of submodular cover via the diminishing return property on the integer lattice. In: Proceedings of the Twenty-Eighth International Conference on Neural Information Processing Systems, pp. 847–855 (2015)

17. Soma, T., Yoshida, Y.: Maximizing monotone submodular functions over the integer lattice. In: Proceedings of the Nineteenth International Conference on Integer Programming and Combinatorial Optimization, pp. 325–336 (2016)
18. Soma, T., Yoshida, Y.: Non-monotone DR-submodular function maximization. In: Proceedings of the Thirthy-First AAAI Conference on Artificial Intelligence, pp. 898–904 (2017)
19. Sviridenko, M.: A note on maximizing a submodular set function subject to a knapsack constraint. Oper. Res. Lett. **32**, 41–43 (2004)
20. Sviridenko, M., Vondrák, J., Ward, J.: Optimal approximation for submodular and supermodular optimization with bounded curvature. Math. Oper. Res. **42**(4), 1197–1218 (2017)

A Proof System for a Unified Temporal Logic

Liang Zhao[1], Xiaobing Wang[1(✉)], Xinfeng Shu[2(✉)], and Nan Zhang[1]

[1] Institute of Computing Theory and Technology and ISN Laboratory,
Xidian University, P.O. Box 177, Xi'an 710071, People's Republic of China
{lzhao,nanzhang}@xidian.edu.cn, xbwang@mail.xidian.edu.cn
[2] School of Computer Science and Technology, Xi'an University of Posts
and Communications, Xi'an 710061, People's Republic of China
shuxf@xupt.edu.cn

Abstract. Theorem proving is a widely used approach to the verification of computer systems, and its theoretical basis is generally a proof system for formal derivation of logic formulas. In this paper, we propose a proof system for Propositional Projection Temporal Logic (PPTL) with indexed expressions, which is a unified temporal logic that subsumes the well-used Linear Temporal Logic (LTL). First, the syntax, semantics and logic laws of PPTL that allows indexed expressions are introduced, and the representation of LTL constructs by PPTL formulas is shown. Then, the proof system for the logic is presented which consists of axioms and inference rules for the derivation of both basic constructs and indexed expressions of PPTL. To show the capability of the proof system, several examples of formal proofs are provided. Finally, the soundness of the proof system is demonstrated.

Keywords: Theorem proving · Proof system · Temporal Logic · Indexed expression · Soundness

1 Introduction

Projection Temporal Logic (PTL) [3] is an extension of Interval Temporal Logic (ITL) [13] by introducing a new projection construct and supporting both finite and infinite timeline. Within the PTL framework, Propositional PTL (PPTL) is proved to have the full regular expressiveness [18], and its decision problem has been solved [5]. Further, Modeling, Simulation and Verification Language (MSVL) [7], an executable subset of PTL armed with a framing technique, is defined as the language for system modeling. Based on these theoretical work, a unified *model checking* [2] approach with PTL is developed for formal verification of computer systems [4,20].

This research is supported by the NSFC Grant Nos. 61751207, 61732013, 61672403, and 61572386.

D.-Z. Du et al. (Eds.): COCOON 2019, LNCS 11653, pp. 663–676, 2019.
https://doi.org/10.1007/978-3-030-26176-4_55

The advantage of the model checking approach is that verification can be done automatically. However, it suffer from the state explosion problem and thus less suitable to verify data intensive applications. Another approach to system verification widely used in practice is *theorem proving* [10], in which a *proof system* for a specific logic, usually a temporal logic, is constructed in terms of axioms and inference rules. To verify whether a computer system satisfies a desired property, both the system and the property are characterized by formulas S and P of the logic, respectively. Then, the problem is to check whether $S \rightarrow P$ can be proved formally by the axioms and inference rules of the proof system. Generally, such a verification process involves human assistance and can be done semi-automatically. The advantage of theorem proving is that it avoids the state explosion problem and can verify both finite-state and infinite-state systems, including data intensive applications.

In the past three decades, a number of proof systems for Liner Temporal Logic (LTL), Computing Tree Logic (CTL) and other temporal logics have emerged [9,11,12]. However, the expressive power of these logics is weaker than ITL or PPTL. Within the ITL community, several researchers have investigated axiomatic systems with different extensions. Rosner and Pnueli [16] present a proof system for a propositional choppy logic with chop, next and until operators. Bowman and Thompson provide a completeness proof and a tableau-based decision procedure for PITL with projection over finite intervals [1]. Moszkowski proposes an axiom system over finite intervals for PITL, and later extends the work to infinite intervals [14]. Besides, two proof systems are formalized for PPTL [8] and PTL [17], respectively.

A recent study [6] extends PPTL with *indexed expressions* that take the form of $\bigvee_{i \in \mathbb{N}} R[i]$. Although indexed expressions are obtained by applying the syntax rules countably infinitely many times, they have definite semantics and certain good properties. Especially, they can equivalently represent the strong until U and weak until W constructs of LTL. As a result, PPTL with indexed expressions is a unified temporal logic that involves LTL as one of its subsets.

To employ PPTL and the indexed-expression technique to system verification in the theorem proving approach, we develop a proof system Π for the unified logic in this paper. Specifically, Π consists of two sub-systems of axioms and inference rules. The first sub-system is provided for formal proof of basic PPTL constructs, such as next, projection, always and chop plus. The second sub-system is designed especially for formal derivation of formulas with indexed expressions. We provide a few examples to show how the proof system Π works, and demonstrate its soundness that every formula proved by Π is valid.

The paper is organized as follows. The next section introduces PPTL, including indexed expressions and their capability of representing LTL. Then, Sect. 3 presents our proof system in terms of axioms and inference rules and Sect. 4 provides examples of formal proofs with the proof system. In Sect. 5, the soundness of the proof system is demonstrated. Finally, conclusions are drawn in Sect. 6 with a discussion on potential future work.

2   Propositional Projection Temporal Logic

We first introduce basic notions of Propositional Projection Temporal Logic (PPTL) [3]. Let $\mathcal{P}$ be a countable set of atomic propositions and $\mathbb{B} = \{tt, ff\}$ the boolean domain. Usually, we use small letters, possibly with subscripts, like p, q, r_1, to denote atomic propositions in $\mathcal{P}$ and capital letters, possibly with subscripts, like P, Q, R_1, to represent general PPTL formulas. Formally, the formulas of PPTL are inductively defined by the following syntax:

$$P \quad ::= q \mid \neg P \mid P \wedge P \mid \bigcirc P \mid (P^{[+]}) \,\text{prj}\, P$$
$$P^{[+]} ::= P \mid P, P^{[+]}$$

where $P^{[+]}$ represents a finite sequence of PPTL formulas separated by commas. For simplicity, some previous publications of PPTL, such as [3,18], use the syntax defined as follows.

$$P ::= q \mid \neg P \mid P_1 \wedge P_2 \mid \bigcirc P \mid (P_1, \ldots, P_m) \,\text{prj}\, P$$

It is trivial to prove that the two definitions are equivalent. Notice that $\bigcirc$ (next), and prj (projection) are temporal operators, while $\neg$ and $\wedge$ are defined as they are in classical propositional logic. A formula is called a *state formula* if it does not contain any temporal operators.

In the semantics of PTL, formulas are interpreted upon intervals. An *interval* $\sigma = \langle s_0, s_1, \ldots \rangle$ is a non-empty sequence of states, finite or infinite, while a *state* s is a mapping from $\mathcal{P}$ to $\mathbb{B}$. The length of an interval σ, denoted as $|\sigma|$, is the number of states in σ minus one if σ is finite; or the smallest infinite ordinal ω if σ is infinite. Let $\mathbb{N}$ denote the set of natural numbers. To have a uniform notation for both finite and infinite intervals, we use *extended natural numbers* as indices, that is $\mathbb{N}_\omega = \mathbb{N} \cup \{\omega\}$, and extend the comparison operators, $=, <, \leq$, to $\mathbb{N}_\omega$ by considering $\omega = \omega$ and for all $i \in \mathbb{N}$, $i < \omega$. Moreover, we write $\preceq$ as $\leq -\{(\omega, \omega)\}$. To simplify definitions, we denote σ by $\langle s_0, \cdots, s_{|\sigma|} \rangle$, where $s_{|\sigma|}$ is undefined if σ is infinite. With such a notation, $\sigma_{(i..j)}$ $(0 \leq i \preceq j \leq |\sigma|)$ denotes a sub-interval $\langle s_i, \cdots, s_j \rangle$.

To formalize the semantics of the projection construct prj, we define an auxiliary operator $\downarrow$ to get rid of singleton points. For an interval σ and natural numbers $i_0 \leq \ldots \leq i_m$ $(m \in \mathbb{N})$, $\sigma \downarrow (i_0, \ldots, i_m)$ denotes the interval obtained from σ by preserving only states with non-repeated indices from i_l $(0 \leq l \leq m)$. For example, $\langle s_0, s_1, s_2, s_3, s_4 \rangle \downarrow (0, 0, 2, 2, 3) = \langle s_0, s_2, s_3 \rangle$. In addtion, we use $\sigma \cdot \sigma'$ to denote the *concatenation* of two intervals σ and σ', provided σ is finite.

An *interpretation* is a tuple $\mathcal{I} = (\sigma, k, j)$, where $\sigma = \langle s_0, s_1, \ldots \rangle$ is an interval, $k \in \mathbb{N}$ and $j \in \mathbb{N}_\omega$ with $0 \leq k \leq j \leq |\sigma|$. For a PPTL formula P, $\mathcal{I} \models P$ denotes that $\mathcal{I}$ is an interpretation of P, defined inductively as follows. Intuitively, it means that P is interpreted over the subinterval $\sigma_{(k..j)}$ of σ with the current state being s_k.

$$\mathcal{I} \models p \qquad\qquad\qquad \text{iff } s_k(p) = tt$$
$$\mathcal{I} \models \neg P \qquad\qquad\quad \text{iff } \mathcal{I} \not\models P$$
$$\mathcal{I} \models P_1 \wedge P_2 \qquad\quad\; \text{iff } \mathcal{I} \models P_1 \text{ and } \mathcal{I} \models P_2$$
$$\mathcal{I} \models \bigcirc P \qquad\qquad\; \text{iff } k < j \text{ and } (\sigma, k+1, j) \models P$$
$$\mathcal{I} \models (P_1, \dots, P_m)\,\mathsf{prj}\;Q \quad \text{iff there exist (extended) natural numbers } k = i_0 \le \dots \le i_{m-1}$$

$$\preceq i_m \le j \text{ such that } (\sigma, i_{l-1}, i_l) \models P_l \text{ for all } 1 \le l \le m$$
$$\text{and } (\sigma', 0, |\sigma'|) \models Q \text{ for } \sigma' \text{ given by either}$$
$$(1)\ \sigma' = \sigma \downarrow (i_0, \dots, i_m) \cdot \sigma_{(i_m+1, \dots, j)} \quad \text{if } i_m < j, \text{ or}$$
$$(2)\ \sigma' \text{ is a prefix of } \sigma \downarrow (i_0, \dots, i_m) \quad \text{if } i_m = j.$$

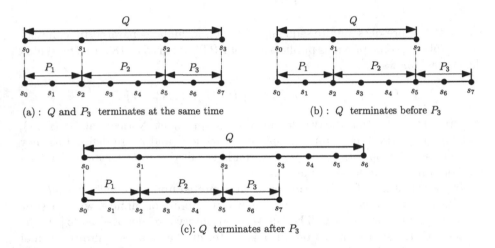

(a) : Q and P_3 terminates at the same time

(b) : Q terminates before P_3

(c): Q terminates after P_3

Fig. 1. Possible interpretations of $(P_1, P_2, P_3)\,\mathsf{prj}\;Q$

The projection construct $(P_1, \dots, P_m)\,\mathsf{prj}\;Q$ is a key operator of PPTL, which has potential applications for compositional reasoning about concurrent systems. The construct allows formulas $P_1, \dots, P_m, Q$ to be autonomous, each interpreted in its own interval. Specifically, two different time scales exist, where $P_1, \dots, P_m$ are interpreted over a series of fine-grained intervals while Q is interpreted over a coarse-grained projected interval. In particular, the sequence of $P_1, \dots, P_m$ and Q may terminate at different time points. The intuition of $(P_1, P_2, P_3)\,\mathsf{prj}\;Q$ is shown in Fig. 1.

A formula P is satisfied by an interval σ, denoted by $\sigma \models P$, if $(\sigma, 0, |\sigma|) \models P$. A formula P is *satisfiable* if $\sigma \models P$ for some σ while a formula P is *valid*, denoted by $\models P$, if $\sigma \models P$ for all intervals σ. The abbreviations such as tt, $f\!f$, $\vee$, $\rightarrow$ and $\leftrightarrow$ are defined as usual. Some derived formulas of PPTL are shown below. They are useful in characterizing various temporal properties.

$$\varepsilon \quad \overset{\text{def}}{=} \quad \neg \bigcirc tt \qquad\qquad P\,;Q \quad \overset{\text{def}}{=} \quad (P,Q)\,\mathsf{prj}\,\varepsilon$$

$$\Diamond P \quad \overset{\text{def}}{=} \quad tt\,;P \qquad\qquad \Box P \quad \overset{\text{def}}{=} \quad \neg\Diamond\neg P$$

$$P^+ \quad \overset{\text{def}}{=} \quad (P^{[+]})\,\mathsf{prj}\,\varepsilon \qquad\qquad P^* \quad \overset{\text{def}}{=} \quad \varepsilon \vee P^+$$

$$\mathsf{halt}(P) \quad \overset{\text{def}}{=} \quad \Box(\varepsilon \leftrightarrow P) \qquad\qquad \mathsf{final}(P) \quad \overset{\text{def}}{=} \quad \Box(\varepsilon \to P)$$

$$\mathsf{keep}(P) \quad \overset{\text{def}}{=} \quad \Box(\bigcirc tt \to P) \qquad\qquad \mathsf{rem}(P) \quad \overset{\text{def}}{=} \quad \Box(\bigcirc tt \to \bigcirc P)$$

$$\mathsf{fin} \quad \overset{\text{def}}{=} \quad \Diamond\varepsilon \qquad\qquad \mathsf{ln}(n) \quad \overset{\text{def}}{=} \quad \begin{cases} \varepsilon & \text{if } n = 0 \\ \bigcirc\mathsf{ln}(n-1) & \text{if } n \geq 1 \end{cases}$$

$$\mathsf{inf} \quad \overset{\text{def}}{=} \quad \Box\bigcirc tt \qquad\qquad P \parallel Q \quad \overset{\text{def}}{=} \quad P \wedge (Q\,;tt) \vee Q \wedge (P\,;tt)$$

The *empty* formula ε means that the current state is the last one of the interval, and the *chop* construct $P;Q$ means the sequential composition of P and Q. Besides, the *sometimes* construct $\Diamond P$ (resp. *always* construct $\Box P$) indicates P holds at some state (resp. every state) from the current state on. The meaning of the *chop plus* construct P^+ (resp. *chop star* construct P^*) is as usual, i.e., P holds repeatedly for one or more (resp. zero or more) times. Then, $\mathsf{halt}(P)$ and $\mathsf{final}(P)$ say that P holds at the last state of the interval, while $\mathsf{halt}(P)$ also requires P holds only at the last state. The formula $\mathsf{keep}(P)$ (resp. $\mathsf{rem}(P)$) means that P holds at every state that has a next (resp. previous) state in the interval. In addition, fin (resp. inf) indicates the interval is finite (resp. infinite), and $\mathsf{ln}(n)$ claims that the length of the remaining interval is exactly n. Finally, the *parallel* construct $P \parallel Q$ means that P and Q are interpreted in parallel.

We denote $\models P \leftrightarrow Q$ by $P \equiv Q$ (*equivalence*), and $\models P \to Q$ by $P \subset Q$ (*strong implication*). Some logic laws of PPTL are provided as follows, whose proofs can be found in [3].

$$\Diamond P \equiv P \vee \bigcirc\Diamond P \qquad\qquad \Box P \equiv P \wedge \varepsilon \vee P \wedge \bigcirc\Box P$$

$$P^+ \equiv P \vee (P\,;P^+) \qquad\qquad Q\,;(P_1 \vee P_2) \equiv (Q\,;P_1) \vee (Q\,;P_2)$$

$$\Box\neg Q \equiv \neg\Diamond Q \qquad\qquad \mathsf{keep}(P) \equiv \varepsilon \vee P \wedge \bigcirc\mathsf{keep}(P)$$

$$\Box P \vee \Box Q \subset \Box(P \vee Q) \qquad\qquad \mathsf{halt}(P) \equiv P \wedge \varepsilon \vee \neg P \wedge \bigcirc\mathsf{halt}(P)$$

$$\Box(P \wedge Q) \equiv \Box P \wedge \Box Q \qquad\qquad \mathsf{final}(P) \equiv P \wedge \varepsilon \vee \bigcirc\mathsf{final}(P)$$

$$tt \equiv \mathsf{fin} \vee \mathsf{inf} \qquad\qquad \mathsf{rem}(P) \equiv \varepsilon \vee \bigcirc P \wedge \bigcirc\mathsf{rem}(P)$$

$$\bigcirc(P \vee Q) \equiv \bigcirc P \vee \bigcirc Q \qquad\qquad \Box(P \wedge \bigcirc tt) \equiv P \wedge \bigcirc\Box(P \wedge \bigcirc tt)$$

$$\bigcirc(P \wedge Q) \equiv \bigcirc P \wedge \bigcirc Q \qquad\qquad \Box(P \to Q) \subset \Box P \to \Box Q$$

$$\bigcirc\neg P \equiv \bigcirc tt \wedge \neg\bigcirc P \qquad\qquad P_1\,;(P_2\,;P_3) \equiv (P_1\,;P_2)\,;P_3$$

2.1 Indexed Expression

Generally, a *well-formed* formula of PPTL is constructed by applying the syntax rules finitely many times. However, some formulas formed by applying the syntax rules countably infinitely many times, such as $\bigvee_{i\in\mathbb{N}} \bigcirc^i P$, have definite semantics and good properties.

A recent work [6] studies a kind of such formulas, namely *indexed expressions*, which are of the form $\bigvee_{i \in \mathbb{N}} R[i]$ where $R[i]$ is a well-formed formula with an index $i \in \mathbb{N}$. Specifically, $R[i]$ may contain sub-formulas such as $\bigcirc^i P$, P^i and $P^{(i)}$. For $i \in \mathbb{N}$, $\bigcirc^i P$ is the application of the next operator to P for i times, P^i means P holds repeatedly for i times, while $P^{(i)}$ means P holds at the consecutive i states from the current state on. Formally, they are defined as follows.

$$\bigcirc^i P \stackrel{def}{=} \begin{cases} P & \text{if } i = 0 \\ \bigcirc\bigcirc^{i-1} P & \text{if } i \geq 1 \end{cases} \qquad P^i \stackrel{def}{=} \begin{cases} \varepsilon & \text{if } i = 0 \\ P^{i-1} ; P & \text{if } i \geq 1 \end{cases}$$

$$P^{(i)} \stackrel{def}{=} \begin{cases} tt & \text{if } i = 0 \\ P & \text{if } i = 1 \\ P \wedge \bigcirc P^{(i-1)} & \text{if } i > 1 \end{cases}$$

The semantics of an indexed expression is clear. Let $\mathcal{I}$ be an interpretation.

$$\mathcal{I} \models \bigvee_{i \in \mathbb{N}} R[i] \qquad \text{iff there exists } i \in \mathbb{N} \text{ such that } \mathcal{I} \models R[i]$$

Intuitively, $\bigvee_{i \in \mathbb{N}} R[i]$ means some $R[i]$ holds for $i \in \mathbb{N}$. From now on, we consider PPTL with indexed expressions in that a formula may contain indexed expression(s) as its sub-formula(s). Nevertheless, it is good that indexed expressions do not occur nested, since $R[i]$ in $\bigvee_{i \in \mathbb{N}} R[i]$ is a well-formed formula.

To avoid excessive use of parentheses, we specify the precedence of operators as: 1. $+$, $*$; 2. $\neg$, $\bigcirc$, $\square$, $\lozenge$, prj; 3. $\wedge$; 4. $\bigvee_{i \in \mathbb{N}}$; 5. $\vee$; 6. $;$; 7. $\|$; 8. $\rightarrow$; 9. $\leftrightarrow$, where 1 = highest and 9 = lowest.

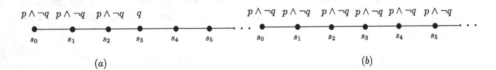

Fig. 2. Intuitive meaning of $p \cup q$ (see (a)) and $p \, W \, q$ (see (a) or (b))

Indexed expressions of the form $\bigvee_{i \in \mathbb{N}} P^{(i)} \wedge \bigcirc^i Q$ are closely related to the recursive equation $X \equiv Q \vee P \wedge \bigcirc X$ [6], shown by the following lemmas.

Lemma 1. *The recursive equation* $X \equiv Q \vee P \wedge \bigcirc X$ *has exactly two solutions:* $\bigvee_{i \in \mathbb{N}} P^{(i)} \wedge \bigcirc^i Q$ *and* $\bigvee_{i \in \mathbb{N}} P^{(i)} \wedge \bigcirc^i Q \vee \square(P \wedge \bigcirc tt)$.

Lemma 2. *Let X be a formula satisfying* $X \equiv Q \vee P \wedge \bigcirc X$, *then*

1. $X \subset \lozenge Q$ *iff* $X \equiv \bigvee_{i \in \mathbb{N}} P^{(i)} \wedge \bigcirc^i Q$, *and*
2. $\square(P \wedge \bigcirc tt) \subset X$ *iff* $X \equiv \bigvee_{i \in \mathbb{N}} P^{(i)} \wedge \bigcirc^i Q \vee \square(P \wedge \bigcirc tt)$.

Many cases of indexed expressions are intrinsically well-formed in that they are equivalent to well-formed formulas. Specifically, we have the following logic laws. Most of the laws are proved using the above lemmas, the others are proved by the fixed-point induction approach [19].

$$\bigvee_{i\in\mathbb{N}} \bigcirc^i Q \equiv \Diamond Q \qquad\qquad \bigvee_{i\in\mathbb{N}} Q^i \equiv Q^*$$
$$\bigvee_{i\in\mathbb{N}} P^{(i)} \wedge \bigcirc^i(P \wedge \varepsilon) \equiv \Box P \wedge \Diamond\varepsilon \qquad \bigvee_{i\in\mathbb{N}} P^{(i)} \wedge \bigcirc^i\varepsilon \equiv \mathsf{keep}(P) \wedge \Diamond\varepsilon$$
$$\bigvee_{i\in\mathbb{N}} (\neg P)^{(i)} \wedge \bigcirc^i(P \wedge \varepsilon) \equiv \mathsf{halt}(P) \wedge \Diamond\varepsilon \qquad \bigvee_{i\in\mathbb{N}} (\bigcirc P)^{(i)} \wedge \bigcirc^i\varepsilon \equiv \mathsf{rem}(P) \wedge \Diamond\varepsilon$$
$$\bigvee_{i\in\mathbb{N}} P^{(i)} \wedge \bigcirc^i(P \wedge \varepsilon) \vee \Box(P \wedge \bigcirc tt) \equiv \Box P$$
$$\bigvee_{i\in\mathbb{N}} P^{(i)} \wedge \bigcirc^i\varepsilon \vee \Box(P \wedge \bigcirc tt) \equiv \mathsf{keep}(P)$$
$$\bigvee_{i\in\mathbb{N}} (\neg P)^{(i)} \wedge \bigcirc^i(P \wedge \varepsilon) \vee \Box(\neg P \wedge \bigcirc tt) \equiv \mathsf{halt}(P)$$
$$\bigvee_{i\in\mathbb{N}} (\bigcirc P)^{(i)} \wedge \bigcirc^i\varepsilon \vee \Box(\bigcirc P \wedge \bigcirc tt) \equiv \mathsf{rem}(P)$$

2.2   Representing Linear Temporal Logic

Linear Temporal Logic (LTL) [15] is a well-known temporal logic, which is based on a linear-time perspective and defined over an infinite interval. Usually, LTL refers to the propositional subset of LTL which has been widely used in practice. In LTL, the most prominent operators are *strong until* U and *weak until* W, which is a weak version of U. Their intuitive meanings are shown in Fig. 2 and more details can be found in [2].

We show that both the strong until and weak until operators can be represented as PPTL formulas with indexed expressions. Suppose $(P \, U \, Q)_\mathsf{P}$ and $(P \, W \, Q)_\mathsf{P}$ are PPTL formulas that have the same meaning as $P \, U \, Q$ and $P \, W \, Q$, respectively. Then, both $(P \, U \, Q)_\mathsf{P}$ and $(P \, W \, Q)_\mathsf{P}$ satisfy the recursive equation $X \equiv Q \wedge \mathsf{inf} \vee P \wedge \neg Q \wedge \bigcirc X$. In addition, they satisfy $(P \, U \, Q)_\mathsf{P} \subset \Diamond(Q \wedge \mathsf{inf})$ and $\Box(P \wedge \neg Q \wedge \bigcirc tt) \subset (P \, W \, Q)_\mathsf{P}$. According to Lemma 2, we have

$$(P \, U \, Q)_\mathsf{P} \equiv \bigvee_{i\in\mathbb{N}} (P \wedge \neg Q)^{(i)} \wedge \bigcirc^i(Q \wedge \mathsf{inf}) \equiv \mathsf{inf} \wedge \bigvee_{i\in\mathbb{N}} (P \wedge \neg Q)^{(i)} \wedge \bigcirc^i Q, \text{ and}$$
$$(P \, W \, Q)_\mathsf{P} \equiv \mathsf{inf} \wedge \bigvee_{i\in\mathbb{N}} (P \wedge \neg Q)^{(i)} \wedge \bigcirc^i Q \vee \Box(P \wedge \neg Q \wedge \bigcirc tt).$$

Therefore, we can simply define the two operators in PPTL in the following way.

$$P \, U \, Q \stackrel{\text{def}}{=} \mathsf{inf} \wedge \bigvee_{i\in\mathbb{N}} (P \wedge \neg Q)^{(i)} \wedge \bigcirc^i Q, \text{ and}$$
$$P \, W \, Q \stackrel{\text{def}}{=} \mathsf{inf} \wedge \bigvee_{i\in\mathbb{N}} (P \wedge \neg Q)^{(i)} \wedge \bigcirc^i Q \vee \Box(P \wedge \neg Q \wedge \bigcirc tt).$$

On the other hand, except U and W, every construct of LTL has a direct counterpart in PPTL. As a result, PPTL with indexed expressions is a unified temporal logic that subsumes LTL.

3   Proof System

This section presents the proof system Π for PPTL with indexed expressions, consisting of a set of *axioms* and *inference rules*. Each axiom defines a formula that can be directly derived by the system, and each inference rule defines a one-step derivation of a conclusion formula from one or more hypothesis formulas.

A *formal proof* (or *formal derivation*) of a formula P is a sequence of formulas $P_0, \ldots, P_n$ $(n \in \mathbb{N})$ such that $P_n = P$ and each P_i $(0 \le i \le n)$ is either an axiom or the conclusion formula of an inference rule, provided every hypothesis formula of the rule has occurred in the preceding formulas $P_0, \ldots, P_{i-1}$. If such a formal proof exists, we say that P is proved by Π or P is a *theorem* of Π, denoted as $\vdash_\Pi P$. When there is no confusion, we omit the subscript and simply write $\vdash P$.

Specifically, the proof system Π is composed of two parts: axioms and inference rules Π_B for basic constructs, such as next and projection, and those Π_I for indexed expressions, i.e. $\Pi = \Pi_B \cup \Pi_I$.

3.1   Axioms and Inference Rules for Basic Constructs

A proof system for basic constructs of PPTL has been proposed in [8], which adopts a relatively complex version of syntax that considers a projection-plus construct $(P_1, \ldots, (P_i, \ldots, P_j)^\oplus, \ldots, P_m)$ prj Q as a basic construct. Here, we provide Π_B, an equivalent but more concise presentation of the proof system, based on the current version of syntax considering the projection construct instead of the projection-plus construct.

Let S denote a state formula and Ω represent a finite sequence of formulas, which is possible the empty sequence τ. For convenience, we define (τ) prj P to be P. The set of axioms of Π_B are presented as follows.

TAU	ψ where ψ is an instance of a propositional tautology
POF	$(\Omega_1, P_1 \lor P_2, \Omega_2)$ prj $Q \leftrightarrow (\Omega_1, P_1, \Omega_2)$ prj $Q \lor (\Omega_1, P_2, \Omega_2)$ prj Q
POB	(Ω) prj $(Q_1 \lor Q_2) \leftrightarrow (\Omega)$ prj $Q_1 \lor (\Omega)$ prj Q_2
PFN	$(\Omega_1, P_1, P_2, \Omega_2)$ prj $Q \leftrightarrow (\Omega_1, P_1 \land \mathsf{fin}, P_2, \Omega_2)$ prj Q
PIN	$(\Omega, P \land \mathsf{inf})$ prj $Q \leftrightarrow (\Omega, P \land \mathsf{inf})$ prj $(Q \land \mathsf{fin})$
PSM	$(\Omega_1, S \land \varepsilon, P, \Omega_2)$ prj $Q \leftrightarrow (\Omega_1, S \land P, \Omega_2)$ prj Q

PSF	$(S \land P, \Omega)$ prj $Q \leftrightarrow S \land (P, \Omega)$ prj Q	PSB	(Ω) prj $(S \land Q) \leftrightarrow S \land (\Omega)$ prj Q
PEF	(Ω, ε) prj $Q \leftrightarrow (\Omega)$ prj Q	PEB	(P) prj $\varepsilon \leftrightarrow P$
PNX	$(\bigcirc P, \Omega)$ prj $\bigcirc Q \leftrightarrow \bigcirc ((P, \Omega)$ prj $Q)$	CPR	$P^+ \leftrightarrow P \lor (P \land \bigcirc tt\,;P^+)$
NXN	$\bigcirc \neg P \leftrightarrow \neg(\varepsilon \lor \bigcirc P)$	NXA	$\bigcirc(P \land Q) \leftrightarrow \bigcirc P \land \bigcirc Q$
CNX	$\bigcirc P\,;Q \leftrightarrow \bigcirc(P\,;Q)$	CAS	$P_1\,;(P_2\,;P_3) \leftrightarrow (P_1\,;P_2)\,;P_3$
STN	$P \land \Diamond \neg P \to \Diamond(P \land \bigcirc \neg P)$	ALR	$\Box P \leftrightarrow P \land (\varepsilon \lor \bigcirc \Box P)$

Intuitively, an axiom is a formula that is supposed to be valid. Especially, the validity of a formula $P \leftrightarrow Q$ means that P and Q are equivalent.

The set of inference rules of Π_B are presented as follows.

MP	$\dfrac{P \quad P \to Q}{Q}$	SUB	$\dfrac{P(Q) \quad Q \leftrightarrow Q'}{P(Q')}$
NXM	$\dfrac{P \to Q}{\bigcirc P \to \bigcirc Q}$	PRM	$\dfrac{P \to P' \quad Q \to Q'}{(\Omega_1, P, \Omega_2)\text{ prj }Q \to (\Omega_1, P', \Omega_2)\text{ prj }Q'}$
ALW	$\dfrac{P}{\Box P}$	CPM	$\dfrac{P \to Q}{P^+ \to Q^+}$
REC	$\dfrac{P \to Q \lor \bigcirc P}{P \to \Diamond Q \lor \Box \bigcirc P}$		

The rule MP is the classic rule of modus ponens for propositional logic. And in the rule SUB, $P(Q)$ denotes a formula P with a sub-formula Q, and $P(Q')$ is the formula obtained from $P(Q)$ by substituting Q with Q'. Intuitively, an inference rule means: if the hypothesis formulas are all valid, the conclusion formula is also valid. More explanations of these axioms and inference rules can be found in [8].

3.2   Axioms and Inference Rules for Indexed Expressions

We propose a proof (sub-)system Π_I to reason about PPTL formulas with indexed expressions. The set of axioms of Π_I are presented as follows. Here, P denotes a formula without any index.

IST $\bigvee_{i\in\mathbb{N}} Q^i \leftrightarrow Q^*$

INS $R[i] \to \bigvee_{i\in\mathbb{N}} R[i]$ INR $\bigvee_{i\in\mathbb{N}} R[i] \leftrightarrow R[0] \vee \bigvee_{i\in\mathbb{N}} R[i+1]$

INA $\bigvee_{i\in\mathbb{N}} P \wedge R[i] \leftrightarrow P \wedge \bigvee_{i\in\mathbb{N}} R[i]$ INO $\bigvee_{i\in\mathbb{N}} (P \vee R[i]) \leftrightarrow P \vee \bigvee_{i\in\mathbb{N}} R[i]$

INN $\bigvee_{i\in\mathbb{N}} \bigcirc R[i] \leftrightarrow \bigcirc \bigvee_{i\in\mathbb{N}} R[i]$ INC $\bigvee_{i\in\mathbb{N}} (P\,;R[i]) \leftrightarrow P\,;\bigvee_{i\in\mathbb{N}} R[i]$

Among the axioms, IST indicates that an indexed expression $\bigvee_{i\in\mathbb{N}} Q^i$ can always be replaced by the star construct Q^*. In fact, both the formulas mean that P holds repeatedly for zero or more times. Then, INS and INR reflect two standard property of the infinite disjunction operator $\bigvee_{i\in\mathbb{N}}$. In addition, INA, INO, INN and INC indicate that the and, or, next and chop operators are distributive over the infinite disjunction operator, respectively.

The set of inference rules of Π_I are presented as follows.

INM $$\dfrac{R[i] \to R'[i]}{\bigvee_{i\in\mathbb{N}} R[i] \to \bigvee_{i\in\mathbb{N}} R'[i]}$$

REF $$\dfrac{R \leftrightarrow Q \vee P \wedge \bigcirc R \qquad R \to \Diamond Q}{\bigvee_{i\in\mathbb{N}} P^{(i)} \wedge \bigcirc^i Q \leftrightarrow R}$$

REI $$\dfrac{R \leftrightarrow Q \vee P \wedge \bigcirc R \qquad \Box(P \wedge \bigcirc tt) \to R}{\bigvee_{i\in\mathbb{N}} P^{(i)} \wedge \bigcirc^i Q \vee \Box(P \wedge \bigcirc tt) \leftrightarrow R}$$

Among the inference rules, INM indicates that the infinite disjunction operator is monotonic. Besides, REF and REI are provided for solving R from the recursive biconditional "equation" $R \leftrightarrow Q \vee P \wedge \bigcirc R$. The solution is in the form of an indexed expression, possibly in disjunction with a specific always formula. Intuitively, REF and REI are in accordance with the two cases of Lemma 2 that the recursion is made for finitely many times and for possibly infinitely many times, respectively.

The approach of theorem proving can be applied to verify properties of systems formally. First, both a system and a desired property are specified by PPTL formulas S and P, respectively. Then, the system satisfies the property if and only if we can find a formal proof of $\vdash S \to P$ by the proof system Π.

4   Examples of Formal Proofs

To show the capability of the proof system Π, we provided a few examples of formal proofs. Notice that once $\vdash P$ is proved, P is a theorem of the system and can be used in the formal proof of other formulas.

Example 1. $\vdash \mathsf{keep}(P) \leftrightarrow \varepsilon \vee P \wedge \bigcirc\mathsf{keep}(P)$. The theorem is denoted as T1, whose formal proof is given as follows. Recall that $\varepsilon = \neg \bigcirc tt$ and $\mathsf{keep}(P) = \Box(\bigcirc tt \to P)$.

(1) $\mathsf{keep}(P) \leftrightarrow (\bigcirc tt \to P) \wedge (\varepsilon \vee \bigcirc\mathsf{keep}(P))$ ALR
(2) $(\bigcirc tt \to P) \wedge (\varepsilon \vee \bigcirc\mathsf{keep}(P)) \leftrightarrow \varepsilon \vee P \wedge \bigcirc\mathsf{keep}(P)$ TAU
(3) $\mathsf{keep}(P) \leftrightarrow \varepsilon \vee P \wedge \bigcirc\mathsf{keep}(P)$ SUB (1) (2)

Example 2. $\vdash \Diamond P \leftrightarrow P \vee \bigcirc\Diamond P$. The theorem is denoted as T2, whose formal proof is given as follows. Recall that $\Box P = \neg\Diamond\neg P$ for any formula P.

(1) $\neg\Diamond\neg\neg P \leftrightarrow \neg P \wedge (\varepsilon \vee \bigcirc\neg\Diamond\neg\neg P)$ ALR
(2) $\neg\neg P \leftrightarrow P$ TAU
(3) $\neg\Diamond P \leftrightarrow \neg P \wedge (\varepsilon \vee \bigcirc\neg\Diamond P)$ SUB (1) (2)
(4) $(\neg\Diamond P \leftrightarrow \neg P \wedge (\varepsilon \vee \bigcirc\neg\Diamond P)) \to (\Diamond P \leftrightarrow P \vee \neg(\varepsilon \vee \bigcirc\neg\Diamond P))$ TAU
(5) $\Diamond P \leftrightarrow P \vee \neg(\varepsilon \vee \bigcirc\neg\Diamond P)$ MP (3) (4)
(6) $\bigcirc\neg\neg\Diamond P \leftrightarrow \neg(\varepsilon \vee \bigcirc\neg\Diamond P)$ NXN
(7) $\neg\neg\Diamond P \leftrightarrow \Diamond P$ TAU
(8) $\bigcirc\Diamond P \leftrightarrow \neg(\varepsilon \vee \bigcirc\neg\Diamond P)$ SUB (6) (7)
(9) $\Diamond P \leftrightarrow P \vee \bigcirc\Diamond P$ SUB (5) (8)

Example 3. $\vdash \bigvee_{i\in\mathbb{N}} \bigcirc^i Q \leftrightarrow \Diamond Q$. This theorem is denoted as T3, whose formal proof is given as follows.

(1) $\Diamond Q \leftrightarrow Q \vee \bigcirc\Diamond Q$ T2
(2) $\Diamond Q \to \Diamond Q$ TAU
(3) $\bigvee_{i\in\mathbb{N}} \bigcirc^i Q \leftrightarrow \Diamond Q$ REF (1) (2)

Intuitively, the indexed expression $\bigvee_{i\in\mathbb{N}} \bigcirc^i Q$ means Q must hold at some state, which is equivalently characterized by the well-formed formula $\Diamond Q$.

Example 4. $\vdash \bigvee_{i\in\mathbb{N}} P^{(i)} \wedge \bigcirc^i Q \to \Diamond Q$. This theorem is denoted as T4, whose formal proof is given as follows.

(1) $P^{(i)} \wedge \bigcirc^i Q \to \bigcirc^i Q$ TAU
(2) $\bigvee_{i\in\mathbb{N}} P^{(i)} \wedge \bigcirc^i Q \to \bigvee_{i\in\mathbb{N}} \bigcirc^i Q$ INM
(3) $\bigvee_{i\in\mathbb{N}} \bigcirc^i Q \leftrightarrow \Diamond Q$ T3
(4) $\bigvee_{i\in\mathbb{N}} P^{(i)} \wedge \bigcirc^i Q \to \Diamond Q$ SUB (2) (3)

The intuition of T4 is that the indexed expression $\bigvee_{i\in\mathbb{N}} P^{(i)} \wedge \bigcirc^i Q$ implies Q must hold at some state.

Example 5. $\vdash \bigvee_{i\in\mathbb{N}} P^{(i)} \wedge \bigcirc^i \varepsilon \vee \square(P \wedge \bigcirc tt) \leftrightarrow \mathsf{keep}(P)$. This theorem is denoted as T5, whose formal proof is given as follows. Recall that $\Diamond P = tt\ ;P = (tt, P)\,\mathsf{prj}\,\varepsilon$ for any formula P.

$$
\begin{array}{llll}
(1) & \mathsf{keep}(P) \leftrightarrow \varepsilon \vee P \wedge \bigcirc\mathsf{keep}(P) & & \text{T1} \\
(2) & \neg(\bigcirc tt \to P) \to \neg(P \wedge \bigcirc tt) & & \text{TAU} \\
(3) & \varepsilon \to \varepsilon & & \text{TAU} \\
(4) & \Diamond\neg(\bigcirc tt \to P) \to \Diamond\neg(P \wedge \bigcirc tt) & & \text{PRM (2) (3)} \\
(5) & (\Diamond\neg(\bigcirc tt \to P) \to \Diamond\neg(P \wedge \bigcirc tt)) \to (\square(P \wedge \bigcirc tt) \to \mathsf{keep}(P)) & & \text{TAU} \\
(6) & \square(P \wedge \bigcirc tt) \to \mathsf{keep}(P) & & \text{MP (4) (5)} \\
(7) & \bigvee_{i\in\mathbb{N}} P^{(i)} \wedge \bigcirc^i \varepsilon \vee \square(P \wedge \bigcirc tt) \leftrightarrow \mathsf{keep}(P) & & \text{REI (1) (6)}
\end{array}
$$

Intuitively, the formula $\bigvee_{i\in\mathbb{N}} P^{(i)} \wedge \bigcirc^i \varepsilon \vee \square(P \wedge \bigcirc tt)$ with an indexed expression means that the current interval is either finite or infinite and P keeps holding at every non-final state, which is equivalently characterized by the well-formed formula $\mathsf{keep}(P)$.

Example 6. $\vdash P\,\mathsf{U}\,Q \leftrightarrow \mathsf{inf} \wedge \bigvee_{i\in\mathbb{N}} P^{(i)} \wedge \bigcirc^i Q$. This theorem is denoted as T6, whose formal proof is given as follows. Recall that $P\,\mathsf{U}\,Q = \mathsf{inf} \wedge \bigvee_{i\in\mathbb{N}}(P \wedge \neg Q)^{(i)} \wedge \bigcirc^i Q$.

$$
\begin{array}{llll}
(1) & \bigvee_{i\in\mathbb{N}} P^{(i)} \wedge \bigcirc^i Q \leftrightarrow Q \vee \bigvee_{i\in\mathbb{N}} P \wedge \bigcirc P^{(i)} \wedge \bigcirc\bigcirc^i Q & & \text{INR} \\
(2) & \bigvee_{i\in\mathbb{N}} P \wedge \bigcirc P^{(i)} \wedge \bigcirc\bigcirc^i Q \leftrightarrow P \wedge \bigvee_{i\in\mathbb{N}} \bigcirc P^{(i)} \wedge \bigcirc\bigcirc^i Q & & \text{INA} \\
(3) & \bigvee_{i\in\mathbb{N}} P^{(i)} \wedge \bigcirc^i Q \leftrightarrow Q \vee P \wedge \bigvee_{i\in\mathbb{N}} \bigcirc P^{(i)} \wedge \bigcirc\bigcirc^i Q & & \text{SUB (1) (2)} \\
(4) & \bigcirc(P^{(i)} \wedge \bigcirc^i Q) \leftrightarrow \bigcirc P^{(i)} \wedge \bigcirc\bigcirc^i Q & & \text{NXA} \\
(5) & \bigvee_{i\in\mathbb{N}} P^{(i)} \wedge \bigcirc^i Q \leftrightarrow Q \vee P \wedge \bigvee_{i\in\mathbb{N}} \bigcirc(P^{(i)} \wedge \bigcirc^i Q) & & \text{SUB (3) (4)} \\
(6) & \bigvee_{i\in\mathbb{N}} \bigcirc(P^{(i)} \wedge \bigcirc^i Q) \leftrightarrow \bigcirc \bigvee_{i\in\mathbb{N}} P^{(i)} \wedge \bigcirc^i Q & & \text{INN} \\
(7) & \bigvee_{i\in\mathbb{N}} P^{(i)} \wedge \bigcirc^i Q \leftrightarrow Q \vee P \wedge \bigcirc \bigvee_{i\in\mathbb{N}} P^{(i)} \wedge \bigcirc^i Q & & \text{SUB (5) (6)} \\
(8) & Q \vee P \wedge \bigcirc \bigvee_{i\in\mathbb{N}} P^{(i)} \wedge \bigcirc^i Q \\
 & \leftrightarrow Q \vee P \wedge \neg Q \wedge \bigcirc \bigvee_{i\in\mathbb{N}} P^{(i)} \wedge \bigcirc^i Q & & \text{TAU} \\
(9) & \bigvee_{i\in\mathbb{N}} P^{(i)} \wedge \bigcirc^i Q \leftrightarrow Q \vee P \wedge \neg Q \wedge \bigcirc \bigvee_{i\in\mathbb{N}} P^{(i)} \wedge \bigcirc^i Q & & \text{SUB (7) (8)} \\
(10) & \bigvee_{i\in\mathbb{N}} P^{(i)} \wedge \bigcirc^i Q \to \Diamond Q & & \text{T4} \\
(11) & \bigvee_{i\in\mathbb{N}}(P \wedge \neg Q)^{(i)} \wedge \bigcirc^i Q \leftrightarrow \bigvee_{i\in\mathbb{N}} P^{(i)} \wedge \bigcirc^i Q & & \text{REF (9) (10)} \\
(12) & (\bigvee_{i\in\mathbb{N}}(P \wedge \neg Q)^{(i)} \wedge \bigcirc^i Q \leftrightarrow \bigvee_{i\in\mathbb{N}} P^{(i)} \wedge \bigcirc^i Q) \\
 & \to (P\,\mathsf{U}\,Q \leftrightarrow \mathsf{inf} \wedge \bigvee_{i\in\mathbb{N}} P^{(i)} \wedge \bigcirc^i Q) & & \text{TAU} \\
(13) & P\,\mathsf{U}\,Q \leftrightarrow \mathsf{inf} \wedge \bigvee_{i\in\mathbb{N}} P^{(i)} \wedge \bigcirc^i Q & & \text{MP (11) (12)}
\end{array}
$$

T6 indicates that the representation of $P\,\mathsf{U}\,Q$ can be simplified by replacing the indexed expression $\bigvee_{i\in\mathbb{N}}(P \wedge \neg Q)^{(i)} \wedge \bigcirc^i Q$ with a relatively concise one $\bigvee_{i\in\mathbb{N}} P^{(i)} \wedge \bigcirc^i Q$. Intuitively, the former indexed expression means that P holds until Q holds for the first time, and the latter indexed expression means that P holds until sometimes Q holds. These meanings are actually equivalent.

5   Soundness

An observation about the examples presented in the previous section is that many theorems of the proof system Π coincide with the logic laws of PPTL.

For instance, $\bigvee_{i \in \mathbb{N}} \bigcirc^i Q \leftrightarrow \Diamond Q$ is a theorem (T3), and there is a logic law $\bigvee_{i \in \mathbb{N}} \bigcirc^i Q \equiv \Diamond Q$ which means the formula $\bigvee_{i \in \mathbb{N}} \bigcirc^i Q \leftrightarrow \Diamond Q$ is valid.

In fact, this is a universal phenomenon. We are going to show that the proof system Π is sound, i.e., each theorem proved by Π is valid.

For this, we first establish the soundness of axioms and inference rules of Π. On the one hand, each axiom is a valid formula. On the other hand, the conclusion formula of each inference rule is valid, provided all the hypothesis formulas are valid.

Theorem 1 (Soundness of Axioms and Inference Rules). *Each axiom or inference rule of Π is sound, in that*

1. $\models P$ *if P is an axiom of Π, and*
2. $\models P$ *if* $\dfrac{P_1 \quad \cdots \quad P_n}{P}$ *is an inference rule of Π $(n \geq 1)$ and $\models P_i$ for each* $1 \leq i \leq n$.

The proof of Theorem 1 is presented in the Appendix of this paper.

As a natural deduction of Theorem 1, every formula proved by Π is valid. That is, the proof system Π is sound.

Theorem 2 (Soundness of Π). *For each PPTL formula P, $\vdash P$ implies $\models P$.*

Proof. $\vdash P$ means there is a formal proof $P_1, \ldots, P_n$ $(n \geq 1)$ with $P_n = P$. According to Theorem 1, $\models P_i$ for each $1 \leq i \leq n$. This involves $\models P$.   □

6   Conclusions

In this paper, we develop a proof system Π for PPTL with indexed expressions, which is a unified temporal logic that subsumes the well-used LTL. Specifically, Π consists of axioms and inference rules for formal derivation of both basic PPTL constructs and indexed expressions. We provide a few examples to show how the proof system works. In addition, we demonstrate Π is sound in that every formula proved by Π is valid.

In the near future, we are going to prove the completeness of Π, i.e., every valid formula can also be proved formally by Π. This may be achieved by studying the normal form of indexed expressions and then making structural induction based on the normal form. We also plan to explore more meaningful styles of indexed expressions other than $\bigvee_{i \in \mathbb{N}} Q^i$ and $\bigvee_{i \in \mathbb{N}} P^{(i)} \wedge \bigcirc^i Q$, including their logic laws and relation with specific recursive equations. Which styles of indexed expressions have equivalent well-formed formulas is still an open question.

Appendix

This appendix presents the proof of Theorem 1.

Proof. We only need to prove the soundness of axioms and inference rules in Π_I. The soundness of axioms and inference rules in Π_B has been proved in [8].

(IST) For any interval σ, we have

$$
\begin{aligned}
&\sigma \models \bigvee_{i \in \mathbb{N}} Q^i \\
\Longleftrightarrow\ &\sigma \models Q^i \text{ for some } i \in \mathbb{N} \\
\Longleftrightarrow\ &\sigma \models \varepsilon \text{ or } \sigma \models Q^i \text{ for some } i \geq 1 \\
\Longleftrightarrow\ &\sigma \models \varepsilon \text{ or } \sigma \models Q^+
\end{aligned}
$$

which indicates $\sigma \models \bigvee_{i \in \mathbb{N}} Q^i \leftrightarrow Q^*$. Recall that $Q^* = \varepsilon \vee Q^+$.

(INS) For any interval σ, we have

$$
\begin{aligned}
&\sigma \models R[i] \\
\Longrightarrow\ &\sigma \models R[i] \text{ for some } i \in \mathbb{N} \\
\Longleftrightarrow\ &\sigma \models \bigvee_{i \in \mathbb{N}} R[i]
\end{aligned}
$$

which indicates $\sigma \models R[i] \rightarrow \bigvee_{i \in \mathbb{N}} R[i]$.

(INR) For any interval σ, we have

$$
\begin{aligned}
&\sigma \models \bigvee_{i \in \mathbb{N}} R[i] \\
\Longleftrightarrow\ &\sigma \models R[i] \text{ for some } i \in \mathbb{N} \\
\Longleftrightarrow\ &\sigma \models R[0] \text{ or } \sigma \models R[i+1] \text{ for some } i \in \mathbb{N} \\
\Longleftrightarrow\ &\sigma \models R[0] \text{ or } \sigma \models \bigvee_{i \in \mathbb{N}} R[i+1]
\end{aligned}
$$

which indicates $\sigma \models \bigvee_{i \in \mathbb{N}} R[i] \leftrightarrow R[0] \vee \bigvee_{i \in \mathbb{N}} R[i+1]$.

(INA) For any interval σ, we have

$$
\begin{aligned}
&\sigma \models \bigvee_{i \in \mathbb{N}} P \wedge R[i] \\
\Longleftrightarrow\ &\sigma \models P \wedge R[i] \text{ for some } i \in \mathbb{N} \\
\Longleftrightarrow\ &\sigma \models P \text{ and } \sigma \models R[i] \text{ for some} i \in \mathbb{N} \\
\Longleftrightarrow\ &\sigma \models P \text{ and } \sigma \models \bigvee_{i \in \mathbb{N}} R[i]
\end{aligned}
$$

which indicates $\sigma \models \bigvee_{i \in \mathbb{N}} P \wedge R[i] \leftrightarrow P \wedge \bigvee_{i \in \mathbb{N}} R[i]$. The proofs of (INO), (INN) and (INC) are similar.

(INM) Suppose $\models R[i] \rightarrow R'[i]$. Then, for any interval σ, $\sigma \models R[i]$ implies $\sigma \models R'[i]$. So, for any interval σ, $\sigma \models R[i]$ for some $i \in \mathbb{N}$ implies $\sigma \models R'[i]$ for some $i \in \mathbb{N}$, which means $\sigma \models \bigvee_{i \in \mathbb{N}} R[i]$ implies $\sigma \models \bigvee_{i \in \mathbb{N}} R'[i]$, or equivalently $\sigma \models \bigvee_{i \in \mathbb{N}} R[i] \rightarrow \bigvee_{i \in \mathbb{N}} R'[i]$.

(REF) Suppose $\models R \leftrightarrow Q \vee P \wedge \bigcirc R$ and $\models R \rightarrow \Diamond Q$. Then, $R \equiv Q \vee P \wedge \bigcirc R$ and $R \subset \Diamond Q$. According to Lemma 2, $\bigvee_{i \in \mathbb{N}} P^{(i)} \wedge \bigcirc^i Q \equiv R$, which means $\models \bigvee_{i \in \mathbb{N}} P^{(i)} \wedge \bigcirc^i Q \leftrightarrow R$. The proof of (REI) is similar. $\qquad\square$

References

1. Bowman, H., Thompson, S.J.: A decision procedure and complete axiomatization of finite interval temporal logic with projection. J. Logic Comput. **13**(2), 195–239 (2003)
2. Clarke, E.M., Grumberg, O., Peled, D.A.: Model Checking. The MIT Press, Cambridge (2000)
3. Duan, Z.: Temporal Logic and Temporal Logic Programming. Science Press, Beijing (2006)
4. Duan, Z., Tian, C.: A unified model checking approach with projection temporal logic. In: Liu, S., Maibaum, T., Araki, K. (eds.) ICFEM 2008. LNCS, vol. 5256, pp. 167–186. Springer, Heidelberg (2008). https://doi.org/10.1007/978-3-540-88194-0_12
5. Duan, Z., Tian, C.: A practical decision procedure for propositional projection temporal logic with infinite models. Theoret. Comput. Sci. **554**, 169–190 (2014)
6. Duan, Z., Tian, C., Zhang, N., Ma, Q., Du, H.: Index set expressions can represent temporal logic formulas. Theoret. Comput. Sci. (2018). https://doi.org/10.1016/j.tcs.2018.11.030
7. Duan, Z., Yang, X., Koutny, M.: Framed temporal logic programming. Sci. Comput. Program. **70**(1), 31–61 (2008)
8. Duan, Z., Zhang, N., Koutny, M.: A complete proof system for propositional projection temporal logic. Theoret. Comput. Sci. **497**, 84–107 (2013)
9. French, T., Reynolds, M.: A sound and complete proof system for QPTL. In: Advances in Modal logic 2002, pp. 127–148 (2002)
10. Kanso, K., Setzer, A.: A light-weight integration of automated and interactive theorem proving. Math. Struct. Comput. Sci. **26**(01), 129–153 (2016)
11. Kesten, Y., Pnueli, A.: Complete proof system for QPTL. J. Logic Comput. **12**(5), 701–745 (2002)
12. Manna, Z., Pnueli, A.: Completing the temporal picture. Theoret. Comput. Sci. **83**(1), 91–130 (1991)
13. Moszkowski, B.C.: Executing Temporal Logic Programs. Cambridge University, Cambridge (1986)
14. Moszkowski, B.C.: A complete axiom system for propositional interval temporal logic with infinite time. Log. Methods Comput. Sci. **8**(3) (2012)
15. Pnueli, A.: The temporal logic of programs. In: Proceedings of the 18th Annual IEEE Symposium on Foundations of Computer Science, pp. 46–57. IEEE Computer Society (1977)
16. Rosner, R., Pnueli, A.: A choppy logic. In: LICS 1986, pp. 306–313 (1986)
17. Shu, X., Duan, Z.: A decision procedure and complete axiomatization for projection temporal logic. Theor. Comput. Sci. (2017). https://doi.org/10.1016/j.tcs.2017.09.026
18. Tian, C., Duan, Z.: Expressiveness of propositional projection temporal logic with star. Theoret. Comput. Sci. **412**, 1729–1744 (2011)
19. Winskel, G.: The Formal Semantics of Programming Languages: An Introduction. The MIT Press, Cambridge (1993)
20. Zhang, N., Duan, Z., Tian, C.: Model checking concurrent systems with MSVL. Sci. China: Inf. Sci. **59**(11), 101–118 (2016)

Author Index

Printed in the United States
By Bookmasters